Science Press   Springer-Verlag

Proceedings of Symposium on Paddy Soil

Proceedings of Symposium on Paddy Soil

Edited by
Institute of Soil Science, Academia Sinica

with 317 figures and 445 tables

Science Press, Beijing
Springer-Verlag
Berlin Heidelberg New York
1981

Institute of Soil Science, Academia Sinica
Beijing
The People's Republic of China

Published by Science Press, Beijing

Distribution rights throughout the world, excluding The People's Republic of China, granted to Springer-Verlag Berlin Heidelberg New York

ISBN-13:978-3-642-68143-1   e-ISBN-13:978-3-642-68141-7
DOI: 10.1007/978-3-642-68141-7

© Science Press, Beijing and Springer-Verlag Berlin Heidelberg 1981
Softcover reprint of the hardcover 1st edition 1981

2131/3140-54321

Science Press Book No. 2452·12

PREFACE

China contributes a large part to rice production, one of the most important crops in the world. It is estimated that in China rice constitutes about half of the total food production, covering an area of about 30% of 10^8 hectares of cultivated land of the whole country.

Owing to the peculiar water regime, paddy soils possess quite different properties physically, chemically and biologically as compared with those of upland soils. Such properties have a conspicuous effect on fertility and management practice of paddy field.

For the purpose of summing up the past work and opening up new prospects, a "Symposium on Paddy Soils" was organized under the auspices of Academia Sinica, held on October 19-24, 1980 in Nanjing, which was followed by a seven-day paddy soil excursion in the lower Changjiang Delta. In addition to 120 Chinese soil scientists, 56 guests coming from America, Asia, Europe and Oceania attended the symposium on invitation. A total of 110 papers were presented either orally or by poster. All these are collected and published in the present proceedings which we hope may be helpful to the scientific exchanges between soil scientists of China and other countries.

Organizing Committee of the
"Symposium on Paddy Soils"
December 1980, Nanjing

CONTENTS

PREFACE .... (i)

PART I. PLENARY SESSION

Soil Factors That Influence Rice Production .... *N.C. Brady* (1)
Some Characteristics of High Fertility Paddy Soils .... *Chen Jia-fang, Li Shi-ye* (20)
Rice Soils for Food Production.......................... *R. Duda 1, J. Hrabovszky, A. Pécrot* (31)
Recent Progress in Studies of Soil Structure, and Its Relation to Properties and
    Management of Paddy Soils.. *D.J.Greenland* (42)
Some Aspects of the Physical Chemistry of Paddy Soils .... *F.N. Ponnamperuma* (59)
Oxidation-Reduction Properties of Paddy Soils...................................... *Yu Tian-ren* (95)
The Role of Inorganic Redox Systems in Controlling Reduction in Paddy Soils ....
    .... *W.H. Patrick, Jr.* (107)
Fertility of Paddy soils in Tropical Asia .... *Kazutake Kyuma* (118)
On the Genetic Classification of Paddy Soils in China............................ *Gong Zi-tong* (129)
The Classification of "Paddy Soils" as Related to Soil Taxonomy ...... *F.R. Moormann* (139)
Principal Types of Low Yield Paddy Soil in China .... *Xiao Ze-hong* (151)
The Fertility and Fertilizer Use of the Important Paddy Soils of China ..... *Lu Ru-kun* (160)
Evaluation of Nitrogen Fertility and Increasing Fertilizer Efficiency in Wetland Rice
    Soils .... *S.K. De Datta, P.J. Stangel, E.T. Craswell* (171)
Cultivation and Application of Green Manure in Paddy Fields of China ....
    .... *Gu Rong-shen, Wen Qi-xiao* (207)
Cropping System in Relation to Fertility of Paddy Soils in China .... *Xu Qi* (220)

PART II. SECTION SESSION

Section 1. Properties

Some Laboratory Observations on the Formation of Black Layer in the Recently Reclaimed
    Saline Soils Under Rice Cultivation...
    .... *Yuan Ke-neng, Huang Chang-yong, Zhu Zu-xiang* (231)
Reducing Organic Substances Responsible for Removal of Fe (III) and Mn (IV) From
    Subsurface Horizon of Lowland Rice Soil...
    .... *Masanori OKAZAKI, Hidenori WADA, Yasuo TAKAI* (235)
Voltammetric Determination of Reducing Substances in Paddy Soils ....
    .... *Ding Chang-pu, Liu Zhi-guang* (251)
Chemical Equilibria of Sulfides as Studied With a Hydrogen Sulfide Sensor
    .... *Pan Shu-zhen* (255)
Determination of Redox Potential (Eh) by Depolarizing Curve Method
    .... *Liu Zhi-guang* (258)

A Preliminary Study on Properties of Organo-Mineral Complex of Paddy Soils in Taihu Lake Region .... *Fu Ji-ping, Zhang Jing-shen* (262)

Determination of Lime Potential of Paddy Soils With Two Ion-Solective Electrodes .... *Wang Jing-hua* (269)

The Availability of Micronutrients in Paddy Soils and Its Assessment by Soil Analysis Including Radioisotopic Techniques .... *K.G. Tiller* (273)

A preliminary Study on Specific Adsorption of Cu Ion by Paddy Soils of Southern Jiangsu Province .... *Wu Mei-ling* (290)

Studies on Accumulation of Strontium-90 in Paddy Soils .... *Kinjiro Kawase, Eizo Yokoyama* (296)

Residual BHC in Paddy Soils and Its Contamination to Rice .... *Zhang Shui-ming, Ma Xing-fa, Li Xun-guang* (302)

Decomposition of Plant Materials in Relation to Their Chemical Composition in Paddy Soil .... *Shi Shu-lian, Lin Xin-xiong, Wen Qi-xiao* (306)

Some Physical Properties of Paddy Profiles in Taihu Lake Region .... *Yu De-fen, Yao Xian-liang* (311)

Hydrologic Status of Paddy Fields During Rice-Growing Period .... *Zhang Bing-gang, Chen Zhi-xiong* (316)

Release of Nitrogen From Decomposing Legume Roots and Nodules .... *John S. Waid, Azizah Chulan* (320)

Investigation on Ecological Distribution of Fungi in Paddy Soils .... *Hao Wen-ying, Yao Hui-qin, Xu Yue-rong* (323)

Mechanisms of Bacterial Iron-Reduction in Flooded Soils .... *J.C.G. Ottow* (330)

Investigation on Nitrogen-Fixing Bacteria in Rice Rhizosphere .... *Qiu Yuan-sheng, Zhou Shu-ping, Mo Xiao-zhen, You Chong-biao, Wang Da-si* (344)

Biological Nitrogen Fixation Associated With Wetland Rice .... *I. Watanabe* (348)

Studies on the Preparation of Dried Inoculum of Blue-Green Algae and Its Application in Paddy Field .... *Huang You-xing, Fang Guang-ru, Yan Yu-zhou, Wang Ting, Su Guo-feng* (356)

Preliminary Studies on Process of Nitrogen Excretion by Azolla .... *Liu Zhong-zhu, Chen bing-huan, Song Wei* (363)

On Ecological Distribution of Bacteria in Paddy Soils ...... *Yin Rui-ling, Hao Wen-ying* (369)

Iron Toxicity in Rice Plants and Nitrogenase Activity in the Rhizosphere as Related to Potassium Application .... *G. Trolldenier* (375)

Section 2. Genesis and Classification

Principles of Classification as Applied to Paddy Soils .... *Klaus W. Flach and Oliver W. Rice, Jr.* (381)

Classifying Rice Soils on the Basis of Their Physiological Characteristics .... *Hou Guang-jiong* (387)

The Recent Trend of Paddy Soil Classification in Japan .... *Takeshi Matsui* (392)

Paddy Soils in Chengdu Plain and Their Classification .... *Zhang Xian-wan* (397)

The Application of Cluster Analysis to the Material Classification of Paddy Soils in Taihu Lake Area .... *Liu Duo-sen, Lu Yan-chun* (405)

Evolution and Development of Alluvial Paddy Soils in Zhujiang Delta .... *Lu Fa-xi, Zhu Shi-qing, Yuan Cai-ting, Shen Dao-ying, Yang Yuan-ying, Luo Lian-xiang* (409)

On the Classification of Basic Categories of Paddy Soils in Taihu Lake Region.............
.... *Liu Yuan-chang* (414)
Characteristic Features of Paddy Soils of Japan............................ *Masanori Mitsuchi* (419)
On the Classification and Use of Some Dark Clayey Paddy Soils In China
.... *Huang Rui-cai, Wu Shan-mei* (428)
On the Genesis, Characteristics and Utilization of the Acid Sulphate Paddy Soil in
China .... *Wei Qi-fan* (439)
Some Chemical Properties of Soils From the Taoyuan Prefecture, China, in Particular
Their Fertility .... *Hideo Okajima, Hiroki Imai* (444)
On the Characteristics of Leaching and Accumulation of Chemical Elements in Two Eutric
Paddy Soils in Taoyuan County, Northern Hunan Province....................................
.... *Chen Zhi-cheng, Jiao Jian-ying, Zhao Wen-jun* (449)
Effects of Neotectonic Movement on Development of Paddy Soils in Changjiang
Delta.. *Lu Jing-gang* (454)
Characteristics of Geographical Distribution of Fertile Paddy Soil in Fujian
Province... *Zhu He-jian,*
Guo Cheng-da, Lin Zhen-sheng, Chen Zhen-gao, Tan Bing-hua, Li Quan-bao (461)
Micromorphological Features as the Indication of the Fertility of Paddy
Soils .... *Cao Sheng-geng, Jin Guang* (466)
Differentiation of Iron Oxide in Paddy Soils in Taihu Lake Region...............................
.... *Xu Zu-yi, Chen Jia-fang* (471)
Changes in Clay Minerals in the Genesis of Paddy Soils .... *Zhang Xiao-nian* (475)
Clay Minerals of Paddy Soils in Taihu Lake Region ....
.... *Xu Ji-quan, Yang De-yong, Jiang Mei-yin* (480)
On the Characteristics of Fe-Organic Coating in Paddy Soils.......................................
.... *Gu Xin-yun, Li Shu-qiu* (486)
The Magnetic Susceptibility Distinction of Paddy Soils in Taihu Lake Region...............
.... *Yu Jin-yan, Zhao Wei-sheng, Zhan Shuo-ren* (493)

Section 3. Management

Effect of Long-Term Rice Cultivation on Some Properties of Soda Solonchaks at Qian
Gorlos Irrigation Region of Jilin Province in China ....
.... *Wang Chun-yu, Wang Ru-yong, Zhang Su-jun, Zhang Xiu-lan, Tian Lin-jie* (498)
On the Tendency of Organic Matter Accumulation in Paddy Soil Under Triple Cropping
System in Suburbs of Shanghai.. *Xi Zhen-bang* (502)
Stagnancy of Water in Paddy Soils Under the Triple Cropping System and Its
Improvement .... *Li Shi-jun, Li Xue-yuan* (509)
Drainage of Paddy Soils in Taihu Lake Region .... *Cheng Yun-sheng* (517)
Improving Rice Seedling Emergence in Flooded Soil by Use of Calcium Peroxide..........
.... *S. Yoshida, F.T. Parao* (524)
Rice Fertilization and Wastewater Nutrients Recycling .... *Paolo Sequi* (531)
The Study of Nitrogen Distribution Around Rice Rhizosphere......................................
.... *Liu Zhi-yu, Qin Sheng-wu* (541)
Fertilizer Management of Paddi Soils With Physical Constraints .... *H. R. von Uexkull* (547)

Rationalization of Fertilizer Application to Rice in a Cool Environment ....
.... *Akira Tanaka and Shinichiro Sekiya* (560)
Transformation and Distribution of Organic and Inorganic Fertilizer Nitrogen in Rice
and Soil System .... *Huang Dong-mai, Gao Jia-hua, Zhu Pei-li* (570)
Transformation of Fertilizer, Straw and Soil Nitrogen in Paddy Soil ....
.... *Hideaki Kai, Sadao Kawaguchi, Wittaya Masayna* (578)
Nitrogen Efficiency Study Under Flooded Paddy Conditions: A Review of Inputs Study
I .... *Yoshio Yamada, Saleem Ahmed, Adelaida Alcantara, and Nurul Huda Khan* (588)
The Fate of Nitrogen Fertilizer in Paddy Soils .... *Chen Rong-ye, Zhu Zhao-liang* (597)
Potassium Status of Paddy Soils in Some Countries of South and East Asia ....
.... *G. Kemmler* (603)
On the Potential of K-Nutrition and the Requirement of K-Fertilizer in Important Paddy
Soils of China .... *Xie Jian-chang, Ma Mao-tong, Du Cheng-lin, Chen Ji-xing* (617)
Biochemical and Nutritional Diagnosis of K-Deficiency of Paddy Rice Plants With Regard
to Soil Fertility .... *Sun Xi, Ma Guo-rui, Lin Rong-xin, Yin Xian-xiang* (621)
Sulphur Content and Distribution in Paddy Soils of South China ....
.... *Liu Chong-qun, Chen Guo-an, Cao Shu-qing* (628)
The Status of Microelements in Relation to Crop Production in Paddy Soils of China: III.
Zinc.. *Zhu Qi-qing, Liu Zheng* (635)
A Report on the Saling-Sodic Soils of Baigezhuang and the Zinc Deficiency of Paddy
Crop............................... *Wang Zhong-lian, Wang Wan-zhang, Qi Ming* (641)
Some Methods to Minimize Zinc Deficiency in Transplanted Wetland Rice ....
.... *R.Y. Reyes, Robert Brinkman* (646)
Development of the Zinc Extension Component for Irrigated Rice in the Philippine
Masagana 99 Production Program .... *Aniceto C. Bautista, Juan C. Bunoan, Jr.,*
.... *Reeshon Feuer* (656)
On Zinc Deficiency of Paddy Soils .... *Xie Zhen-chi, Deng Kai-yu,*
Yang Hai-qing, Gong Yu-xi, Wang Zhen-wen, Wang Qin-sheng (671)
Recovery of Fertilizer-Nitrogen by Rice Grown in A Greenhouse Under Varying Soil-
and Climatic Conditions .... *A.C.B.M. van der Kruijs, J.C.P.M. Jacobs, P.D.J.*
van der Vorm and A. van Diest (678)

PART III. POSTER SESSION

Effect of Water Conditions and Organic Matter on Modulus of Rupture of Paddy Soils
.... *Yuan Jian-fang, Zhou Yue-hua* (689)
The Effect of Water-Stable Aggregates in Various Sizes on Air and Water Regime of
Soils .... *Xu Fu-an* (694)
Influence of Transformation of Iron Oxides on Soil Structure .... *He Qun, Xu Zu-yi* (699)
Determination of Stability Constant of Mn (II)-Complex by Voltammetric
Method .... *Bao Xue-ming, Ding Chang-pu* (704)
Determination of pH of Paddy Soils in situ .... *Cang Dong-qing, Yu Tian-ren* (709)
Mercury Pollution of Some Paddy Soils .... *Yang Guo-zhi, Rong Jie* (716)
Effect of Nitrogen Sources on Some Physiological Characteristics of Azolla ....
.... *You Chong-biao, Li Jing-wei, Song Wei, Wei Wen-xiong* (719)

Studies on Nitrogen Fixation by Association of Bacteria With Rice Root.....................
.... *Lin Cang, Huang Shi-zhen, Tang Long-fei, Liu Zhong-zhu, Li Jing-wei* (726)
Geographical Distribution of Paddy Soils in China............................ *Chen Hong-zhao* (734)
Characteristics of Some Types of Paddy Soil in Northeast China .... *Lan Shi-zhen* (741)
The Formation and Characteristics of Cold Spring Paddy Soil ....
.... *Lin Zeng-quan, Chen Jia-ju* (746)
Genetic and Anthropogenic Characteristics of Paddy Soils Derived From Swampy Land
    in Lixiahe District, Jiangsu Province .... *Lei Wen-jin* (750)
Geochemical Characteristics of Some Transitional and Rare Earth Elements in Paddy Soils
    of Red Earth Region, Guangdong Province....................................... *Yang Xue-yi* (754)
Preliminary Studies on Primary Minerals of Some Paddy Soils in Taihu Lake Region ...
.... *Luo Jia-xian* (759)
Double Cropping of Rice in Triple Cropping System and Soil Fertility.........................
.... *Yang Wen-yuan, Liang Dun-fu, Xie Chun-qing, Wan Zong-yi* (765)
Characteristics of High-Yield Paddy Soils in Suburbs of Shanghai .... *Fu Ming-hua* (769)
On the Forming of Waterlogged Condition in the Cultivated Horizon of Paddy Soils in
    Shanghai During Dry Farming Period and the Approach to Its Elimination
    .... *Yang Jin-lou, Zhu Ji-cheng, Jiang Su-zhen, Shi Nan-chang, Zhu Lian-long* (775)
A Comparative Study on Methods of Tillage of Paddy Soils in Taihu Lake Region.......
.... *Zhao Cheng-zhai, Zhou Zheng-du, Dong Bo-shu* (780)
Effect of Rice Planting on Improvement and Utilization of Saline Soils in Tarim
    Basin.................................. *Li Li-qun, Dong Han-zhang, Wang Zun-qin* (786)
On Rice Nutrient Disorder and Its Diagnosis in China ....
.... *Luo Zhi-chao, Tang Yong-liang, Cui Rong-hao* (789)
Characteristics of Nitrogen Mineralization of Paddy Soils and Their Effect on the
    Efficiency of Nitrogen Fertilizer....... *Cai Gui-xin, Zhang Shao-lin, Zhu Zhao-liang* (793)
Uptake of Nitrogen byRice Plant From Straw Manure, Urea and Soil.........................
.... *Mo Shu-xun, Qian Ju-fang* (800)
Phosphorus Balance on Paddy Soils Reclaimed From the Salinized Waste Land, Northern
    Jiangsu.................................. *Li Qing-kui, Qin Sheng-wu* (805)
A Study on the Responses to Fertilizer Potassium on Paddy Soils in Guangdong
    Province.. *Zhu Wei-he, Wen Ying-chang, Shen Dao-ying* (809)
Iron and Chromium Uptake by Crops on Well Drained Soils and on Poorly Drained
    Wetland Soils....... *Robert Brinkman, R.Y. Reyes, H.W. Scharpenseel, E. Eichwald* (816)
The Status of Microelements in Relation to Crop production in Paddy Soils of China: I.
    Boron.. *Liu Zheng, Zhu Qi-qing* (825)
The Status of Microelements in Relation to Crop Production in Paddy Soils of
    China: II. Molybdenum.. *Tang Li-hua* (832)
Effect of Nitrapyrin on the Inhibition of Nitrification in Some Paddy Soils of
    China .... *Li Liang-mo, Zang Shuang, Zhou Xiu-ru, Pan Ying-hua* (837)
Effect of Soil Fertility on the Growth and Yield of Rice .... *Kaoru Seino* (845)

APPENDICES (Speeches in Closing Session)
China and the International Community of Soil Science................. *Wim G. Sombroek* (856)
Soils and People.. *C. F. Bentley* (858)
Some Remarks on the Development of Soil Science in China................. *James Thorp* (862)

ix

SOIL FACTORS THAT INFLUENCE RICE PRODUCTION

N. C. Brady
(Director General, International Rice Research Institute, Los Baños, Laguna, Philippines)

Perhaps no major food crop rivals rice in the range of hydrologic and climatic conditions under which it is grown. Rice is produced in every continent except Antarctica and thrives in an area ranging in latitude from 53° to 40°(1). It grows as a dryland crop much like maize or wheat, as a rainfed crop under alternately flooded and dry conditions, and as a continuously flooded crop. Farmers grow rice on alluvial plains, flooded valleys, and terraced hillsides. Even though it has less drought tolerance than other cereals, rice grows well in arid areas under irrigation such as in Egypt and Pakistan. Likewise, despite rice's sensitivity to low tempera- ture, yields are high in northern China and Japan and at elevations of more than 3,000 meters in the tropics and subtropics.

The wide range of rice-growing conditions suggests an equally wide variety of soils on which rice is grown. Moormann (1978)(2) identified the major soil taxa in rice-growing areas(Table 1). He points out that globally, the most important suborders are Aquents, Aquepts, Ochrepts, Tropepts, Aqualfs, and Aquults, although locally other suborders such as Uderts are significant. In a subsequent publication(3) rice soils are classed in greater detail using the Soil Taxonomy (USDA)(4) and FAO-UNESCO soil unit systems.

Most paddy soils are located at lower positions in landscapes to permit water easy access from higher irrigation or natural runoff sources. These wetland rice areas may be flat-bottomed valleys or terraced and bunded hillsides of otherwise upland areas. But more often, they are found in alluvial flood plains, deltaic plains, coastal plains, tidal flats, marshes, and major river valleys. Such areas are generally characterized by a natural or induced "aquic" moisture regime that implies high moisture and low oxygen conditions(4). As a consequence of this high moisture-low oxygen require- ment, the soil characteristics for paddy rice culture may vary less than those for upland or dryland rice.

The high moisture content and flooded conditions of many paddy soil areas are not natural but are induced by man. Consequently, a soil that fits into one soil-taxonomy category may in time be modified by paddy rice culture to justify its classification in a different category. Japanese scientists have devoted much attention to this problem(5-7) and suggest a classification scheme specific for paddy soils. This approach should be further considered especially in studies of rice-growing soils of the tropics. Paddy soil classification, especially as it relates to the clas- sification of dryland soils, must receive high priority if results at one location are to be extrapolated to other locations with similar soils.

Table 1   Major rice-growing soils classified according to soil taxonomy
(USDA, 1975)

Order	Suborders in use for rice growing		
	Major importance	Local importance	Minor importance
Alfisols	Aqualfs, Ustults[a]	Udalfs	Xeralfs
Aridisols	-	-	Orthids[b], Argids[b]
Entisols	Aquents	Fluvents[a]	Orthents, Psamments
Histosols	-	-	Hemists, Saprists
Inceptisols	Aquepts, Ochrepts[a] Tropepts[a]	-	Andepts
Mollisols	-	Aquolls	Udolls
Oxisols	-	-	Orthox, Ustox
Spodosols	-	-	Aquods
Ultisols	Aquults, Udults	Humults	Ustults[a]
Vertisols	-	Uderts Usterts	Torrerts, Xererts[b]

a = Aquic subgroup only   b = Exclusively under irrigation

Source: Moorman   1978(2)

PHYSICAL PROPERTIES AND RICE PRODUCTION

The physical properties of paddy soils are generally similar to the
sediments or other materials from which the soils have formed. Because
the prevention of excessive percolation is a necessity for efficient rice
production, paddy soils are usually medium- to fine-textured; clays, clay
loams, silt loams, and silty clay loams are common. Kawaguchi and Kyuma
(1977)(8) found that 40 % of the paddy soils studied in South and Southeast Asia contain at least 45 % clay. They suggest from experience in
Japan that physical conditions that permit percolation of 10-20 mm water/
day are necessary for high yields. The coarser-textured soils can be used
for paddy production if some means (such as a high water table) are available to reduce soil percolation.

Although it is difficult to clearly separate the effects of physical
and chemical properties of paddy soils on rice production, the former
are thought to be of lesser significance(9). But a puddled or compact
soil is generally preferred for paddy rice production(10), both for

efficiency of water utilization and for yield.

The growing tendency of farmers in the tropics to grow a sequence of one or more rice crops followed by an upland crop has implications in relation to physical properties of paddy soils. The puddled condition conducive to good rice production on heavy-textured soils is not conducive for upland crops. A challenge of the future is to develop techniques that permit farmers with limited access to mechanical power to successfully make transitions between upland and wetland cropping conditions.

Proper management and control of water is essential to attain the high yields common in Japan and China. This requires good soil physical properties. But for rice yields of 4-5 tons per hectare common in the tropics less attention needs to be paid to physical properties(9).

CHEMICAL PROPERTIES AND RICE PRODUCTION

Electrochemical changes are dramatic when soils are submerged in water. These changes largely control both the beneficial and detrimental effects of chemicals on rice production. Ponnamperuma (1978)(11) identified the most significant electrochemical changes that affect the rice plant as:

1) Soil reduction or decrease in redox potential (Eh);
2) Increase in pH of acid soils and decrease in pH of alkaline soils;
3) Increase in specific conductance and ionic strength;
4) Ionic equilibria; and
5) Sorption and desorption.

Table 2 Shows the low redox potential (Eh) commonly found in water-

Table 2   Range in redox potential (Eh) commonly found in well-drained and in waterlogged soils

Kind of soils	Redox potential (Eh) mV
Aerated (well drained)	+700 to +500
Moderately reduced	+400 to +200
Reduced	+100 to -100
Highly reduced	-100 to -300

Source: Patrick and Mahapatra 1968(55)

logged soils. For most of the root zones of paddy soils, the redox potential varies from +200 to −300 mV. But the potential of the upper few millimeters of soil is higher -- +300 to +500 mV-- indicating an oxidized layer[12]. This difference in potential is significant, especially in relation to nitrogen transformations in the soil.

In the absence of molecule oxygen (low Eh) other compounds or elements act as electron acceptors; in doing so they are changed from oxidized to reduced forms (Table 3). Lowering the Eh increases the availability of

Table 3   Examples of oxidized and reduced forms in redox systems operative in paddy soils and the approximate redox potential (Eh) at which the oxidized forms become unstable

Oxidized form	Redox potential (Eh) for instability	Reduced form
O_2	+380 to 320	H_2O
NO_3^-	+280 to 220	N_2O, N_2
$SO_4^=$	−120 to −180	S, $S^=$
Fe^{+++}	+180 to +150	Fe^{++}
Mn^{++++}	+280 to +220	Mn^{++}
CO_2	−200 to −280	CH_4
H^+		H_2

Source: Patrick and Mahapatra 1968(55)

N, P, Si, Fe, Mn, and Mo and reduces the availability of S, Cu, and Zn[12].

The equilibrium pH of paddy soils usually ranges from 6.5 to 7.5[13]. The pH of more acid soils is increased to about 6.5, which results in an increase in availability of P, Si, and Mo; a reduction in possible toxicities from Al, Mn, Fe, CO_2, and organic acids; and an increase in microbial release of nutrients[11]. The decrease in the pH of alkaline soils increases the availability of P, Ca, Fe, Mn, Cu, and Zn.

The combined effects of lowered redox potential and soil buffering (narrowing the pH range) are generally beneficial to rice production. Submergence and concomitant reduction tend to minimize soil differences, to reduce chemical toxicities, and to increase nutrient availability. Therefore, chemical constraints on rice production are generally less severe under paddy conditions than under upland crop culture.

Two other factors encourage good productivity of paddy soils. Because those soils are generally found in lower levels of the landscape, some soil enrichment commonly occurs from erosion of the surrounding uplands. Likewise, inputs of nutrients from irrigation water are significant in meeting the nutritional requirement of the rice plant. For these reasons, paddy soils are generally more fertile than their upland counterparts. Together with higher biological fixation of nitrogen in wetlands, these factors help account for the fact that rice has been grown continuously in some areas of Asia for centuries without serious soil deterioration and with maintenance of reasonable yield levels.

SOIL DEFICIENCIES

Because of the wide variations in characteristics of soils on which wetland rice is grown, the same general nutrient deficiencies and toxicities prevail as in upland-areas. The most significant of these deficiencies and toxicities will be briefly considered.

NITROGEN

Deficiency of nitrogen probably limits rice production over wide areas more than deficiency of any other element. Nitrogen deficiencies are especially critical in the tropics where paddy soils are generally lower in N than those in temperate regions (Table 4)(8). The level of total soil

Table 4   Major nutrient contents in 410 soils in Tropical Asia

Total N (%)		Available P[1] (mg P_2O_5/100g)		Exchangeable K (me/100g)	
Range	% of samples	Range	% of samples	Range	% of samples
$<$ .05	13	$<$ 1.5	53	$<$ .15	31
.05 - .10	35	1.5 - 3.0	17	.15 - .30	24
.10 - .15	30	3.0 - 4.5	9	.30 - .45	16
.15 - .20	9	4.5 - 6.0	5	.45 - .60	10
$>$ .20	13	$>$ 6.0	16	$>$ .60	19

1) Using Bray and Kurtz no.2 extraction

Source: Kawaguchi and Kyuma 1977(8)

nitrogen is a critical factor in rice production because even if fertilizer application is high, about 2/3 of the N taken up by the rice plant comes from the soil(14,15).

Two other factors influence nitrogen availability to the rice plant; biological nitrogen fixation and biological transformation of nitrogen under submerged conditions. Blue-green algae in the paddy water fix significant quantities of N; Venkataraman (1975)(16) and Singh (1978)(17) observed 20-30 kg N/ha additions from inoculation of paddies with these organisms. Nitrogen fixation by heterotrophic organisms in the rice paddy may account for an even larger nitrogen addition(18-20), although rates of 5 to 10 kg/ha are assumed to be more reasonable(21,22). Also N-fixing organisms associated with other aquatic plants such as Azolla are of considerable importance locally. Their use may be extended as research identifies adapted species and improved cultural practices. Biological fixation undoubtedly accounts for much of the N available for unfertilized rice crops and could be a reason for the wide-spread production of rice in Asia.

Biological transformations account both for increases and losses in available N, not only from soil organic matter but also from applied fertilizer. The ammonium forms released by breakdown of soil organic matter or added in fertilizers is a major source of N for the rice plant. In contrast, nitrates formed during periods of nonsubmergence or in the upper 2-mm oxidized layer of submerged soils are subject to denitrification, providing a net loss of this element.

This loss is of great significance and largely accounts for the fact that efficiency of utilization of applied chemical fertilizers is generally lower in paddy than in upland soils(23). Also, losses of gaseous ammonia may be more important than previously thought to be the case(24). These factors account for the relatively low efficiency of utilization of N fertilizers (30-40 %) commonly measured.

Root-zone placement of applied chemical fertilizer minimizes both ammonia volatilization and denitrification losses by keeping the applied N surrounded by reduced soil(25,26). Also, localized root-zone placement of N minimizes the adverse effects of high mineral nitrogen on biological N fixation.

Murayama (1970)(27) reported that 120-150 kg N/ha are required for high rice yields (5-6 t/ha) in Japan. Comparable N levels are required for modern varieties to yield similarly in the tropics. Yet farmers in developing countries of the tropics apply only a fraction of the N fertilizers needed for such high yields. Stangel (1979)(28) classifies three groups of N-consuming countries. Group I (Japan, South Korea, and Egypt) currently uses nitrogen rates (149 kg N/ha)as high as can be justified (potential rate). Group II (China, Iran, Indonesia, and West Malaysia) uses only 30 kg N/ha, about 40 % of the assumed potential (81 kg N/ha). Group III (the other rice-growing countries of Asia) uses an average of only 13 kg/N/ha but their assumed potential is 50 to 66 kg N/ha. Obviously, marked increases in nitrogen fertilizer use will be required if these countries are to increase their rice production.

The recent availability of lodging-resistant semidwarf rices has greatly increased the economic benefits from the application of N fertilizers. Such benefits have been at least partly achieved in the temperate zones, but are yet to reach much of the tropics where the potential for achievement has been well demonstrated. Inadequate N continues to be among the most serious

factors constraining rice yields in the tropics (IRRI 1979)(29). As stated
by Patnaik and Rao (1978) "Fertilizer nitrogen may be considered the king-
pin in rice farming"(21).

PHOSPHOROUS

The net effects of submergence on phosphorous availability are favorable
to rice production(30,31). Increases in the pH of acid soils induced by
submergence reduce the levels of iron, aluminum, and manganese, elements that
hold phosphorous in unavailable forms. Furthermore, the combined effects
of changes in pH and Eh caused by submergence result in reduced sorption and
occlusion of phosphorous in compounds of iron, manganese, and aluminum(12).

But the problem of phosphate availability is aggravated by the generally
low level of total phosphorous in many Asian rice soils. Kawaguchi and Kyuma
(1977)(8) found that the total phosphorous content of almost a fourth of the
410 tropical soil samples they studied were less than 400 ppm, generally considered a critically low level for rice soils (Table 4).

At rice yield levels common in the tropics (1-3 t/ha) responses to
phosphorous application have not been great, except where total P levels of
the soils are low or where fixation is high. But at higher yield levels,
responses to phosphorous have been significant, especially on Uitisols,
Oxisols, Sulfaquepts, Andosols, and Vertisols(31). The spread of high yielding
semidwarf varieties in the tropics and increased use of nitrogen fertilizers
have provided yields comparable to those obtained in Japan, China, and other
temperate-zone countries. Significant responses to phosphorous additions are
common at these higher yield levels.

Phosphorous also plays a critical role in influencing N fixation in paddy
fields. Apparently the phosphorous requirement of the N-fixing microbes is
higher than that of the rice plant(32). Practical experience in China and
Vietnam suggests the need for regular application of phosphate fertilizers
for optimal Azolla yields(33,34). Similarly, phosphorous is important for
optimum growth of blue-green algae(35).

POTASSIUM

Rice response to applied potassium fertilizer has generally been much
lower than to nitrogen or even phosphorous fertilizers. This is probably
because of the higher potassium levels in paddy soils associated with their
high clay content, of the young parent materials of these soils, and of the
additions of sediments and nutrients in irrigation water. Although significant responses to potassium have been reported in the tropics (e.g.
Mahapatra and Prasad 1970)(36), the element has usually not been limiting.

Experience in Japan and China suggests that as yields increase in the
tropics, needs for added potassium will increase. Rice removes much more
potassiumthan phosphorous. A 5-6 t crop removes about 135 kg K_2O/ha(37).
Modern rice varieties remove more than 4 times as much potassium as do the
lower-yielding traditional rices(38). Unless the rice straw is returned to
the soil, either directly or through compost, depletion of available soil
potassium is to be expected, especially in areas where 2 or 3 crops are grown
annually. Increased cropping intensities coupled with high potassium removal
will probably force greater use of potassium fertilizer to achieve high yields

of rice in the tropics(39).

ZINC

Deficiencies of zinc are more widespread than of any other micronutrient in wetland rice production. Ponnamperuma (1974)(40) reported that about 2 million hectares of Asian riceland are believed to be deficient in zinc. Soils from calcareous materials are more likely to be zinc-deficient than those from acid materials(41). Saline and saline-alkaline soils in India have responded dramatically to zinc application.

Zinc deficiency can be overcome in several ways(42). Dipping rice roots in a 1-4 % suspension of zinc oxide alleviates minor zinc deficiencies. Similarly seed treatment has sometimes proved effective. But zinc-sulfate applications are generally most effective.

OTHER MICRONUTRIENTS

Under specific soil conditions deficiencies of Cu, Fe, B, and Mo have been noted(42,43). Copper deficiencies may occur on some high-pH soils and on organic soils. Iron and manganese deficiencies have been found on alkaline or sodic soils low in organic matter. Boron and molybdenum deficiencies have been noted, but only under restricted conditions.

SULFUR

Although widespread and serious sulfur deficiencies have not been noted in rice, response to sulfur has been found in paddy soils. Several factors suggest that sulfur may become more limiting in the future. Total sulfur contents of paddy soils in the tropics are low compared to comparable soils in the temperate zone(44). Furthermore, submergence generally decreases sulfur availability. Under low Eh conditions, sulfates can be reduced to sulfides which in turn can be tied up as insoluble iron or manganese compounds. Under extreme conditions H_2S may build up to toxic levels.

The quantity of sulfur removed by rice plants approximate that of potassium. Because rice straw is not commonly returned to the soil but is burned in the tropics, much of the sulfur is lost. As rice yields increase, the quantity of sulfur removed will also increase. The supply of sulfur from the burning of fossil fuels and its return to the soil through rain water in the tropics is minimal compared with that near large cities and industrial plants in the more industrialized countries. Furthermore, the fertilizers being used to supply nitrogen, phosphorous, and potassium are lower in sulfur today than a decade earlier. Thus, sulfur deficiency will probably be more widespread in the future than has been the case up to now.

Sulfur deficiency in rice has been noted in India and in East Java, Indonesia(44,45). The sulfur deficiency can be overcome by adding elemental sulfur or gypsum,or by the choice of a fertilizer containing sulfur to supply nitrogen or phosphate or both (e.g. ammonium sulfate and ordinary superphosphate.

SILICON

Although silicon is not normally classified as an element essential for plant growth, it appears necessary for good rice yields(46). In Japan and Korea significant quantities of silicate-containing fertilizers are used(47,48). The silicon enhances resistance to lodging, to pests such as stem borers, and to diseases such as rice blast. Slag that contains calcium silicate is used as a silicon source; rates of 1.5 to 2.0 per ha are applied.

The extent to which silicon deficiencies occur in tropical rice is still not known, although responses to silicate applications have been noted in the Philippines. This research area should be explored.

TOXICITIES (PROBLEM SOILS)

Several adverse soil conditions are encountered in areas where rice is being or, if the adverse conditions could be modified, could be grown. These include iron-toxic soils, saline and sodic soils, acid sulfate soils, and peat soils.

1. Iron-toxic Soils

Iron toxicity is a serious constraint on certain acid paddy soils. Van Breemen and Moormann (1978)(49) suggest that iron toxicity may be a problem on soils with pH below 5.8 when aerobic and below 6.5 when anaerobic. They suggest that these conditions prevail in young acid-sulfate soils (sulfaquepts) in parts of Indonesia, India, Vietnam, Malaysia, and Sierra Leone, as well as poorly drained, light-textured soils in valleys receiving interflow water from adjacent highlands in Sri Lanka, parts of India, and Sierra Leone. Certain "lateritic" soils are also known to exhibit iron toxicities. Table 5 shows characteristics of the three groups of soils where iron deficiency has been observed.

Iron toxicity can be ameliorated by liming, soil drainage, and by adding green manure or compost(49). Unfortunately, the quantities of liming materials required are often prohibitive as is drainage to remove excess acids or to prevent upwelling of water from side slopes. The breeding of high yielding rice varieties which will tolerate high iron levels is an approach which is being pursued.

2. Acid Sulfate Soils

Millions of hectares of potential rice lands in tidal swamp areas of most rice-growing countries are found that are not developed because of acid-sulfate soils (Table 6). These soils have pH values below 3.5 (if Entisols) or 4 (if Inceptisols) in the upper 50 cm. Except for their low pH, acid-sulfate soils have other characteristics that suit them well for rice production(50). The acid-sulfate areas are usually well supplied with plant nutrients and water, and are otherwise topographically and hydrologically favorable for rice production. There are about 5 million hectares of these soils in South and Southeast Asia, about 3.7 million hectares in Africa, and about 2 million hectares in South America(51)

Table 5   Summary of soil data and nutrient deficiencies (inferred from
          lower-than-critical levels in plants) for sites where iron
          toxicity was observed in the field

Description	Soil				Plants deficient in
	pH	Active Fe(%)	Organic matter (%)	CEC (me/100g)	
Acid sulfate soils	3.4-4.2	0.6-1.0	3.5-5.7	12-18	none
Sandy soils	5.2-5.8	0.035-0.5	0.5-4.0	3-9	Ca and Mg (Thailand) P (India) K (Sri Lanka)
So-called "Lateritic" soils	4.5-5.8	0.2-0.8	1.5-5.3	8.1-16	P (India) none (Malaysia)

Source: Tanaka and Yoshida 1970(41)

Table 6   Approximate area of acid sulfate soils in certain South &
          Southeast countries from estimates made by different authors

Country	Approximate area (thousands of ha)
Indonesia	2,000
Vietnam	1,000
Bangladesh	700
Thailand	670
India	390
Khmer (Kampuchea)	200
Burma	180
Malaysia	150

Source : Van Breemen and Pons 1978(50)

The high degree of acidity (or potential acidity) of acid-sulfate soils makes reclamation through liming economically impractical in most cases. The only practical way to manage these soils is by proper drainage and water management. Maintaining high water table can control the oxidation of pyrite (the sulfur source in these (soils). Where feasible, drainage and removal of acidity by leaching can be used. Proper water management and controlled drainage have made rice production possible in large acid-sulfate areas of Thailand and Vietnam. Great care must be exercised, however, in attempting to repeat these success in areas with greater potential for acidity development.

3. Saline and Sodic Soils

Rice could be grown on about 65 million hectares of saline and alkali soils if the salt and pH levels could be lowered, or if tolerant rice varieties could be developed (Table 7).

Table 7 **Current and potential problem rice soils of Asia cover more than 100 million hectares**

Kind of soil	Extent
	(million ha)
Saline	62.5
Alkali	2.2
Acid sulfate	9.8
Organic	29.0

Source : IRRI 1977(56)

Of the 48 million hectares of saline soils in South and Southeast Asia, 27 million are coastal saline soil areas with continuous water supply(52). Coastal areas that are subject to high daily tides may not be suitable for crop production without high inputs, but areas at river mouths where water is brackish and areas not influenced by tides but only by upward seepage of saline water can be used for rice production.

Inland saline and sodic soils are found in arid and semi-arid regions. When irrigated and properly drained, they make excellent paddy soils, with rice yields among the world's highest if the salt and alkalinity levels are not too high. High solar radiation and low disease and insect incidence in these arid and semi-arid areas are conducive to high yields.

Egypt is an excellent example of appropriate use of saline and potentially saline soils. Egypt's climate is arid and crop growth is possible only with irrigation. But salt would build up in these soils if only upland crops were grown and if the soils were not properly drained. The Egyptians have developed a unique system of alternate cropping; an upland crop such as cotton is grown for 2-3 years, followed by two or more rice crops. Irrigation of the cotton results in some salt accumulation in the upper soil zones. Paddy rice culture leaches these salts downward while simultaneously producing 5 to 6 t/ha of rice per year. This crop rotation, supplemented by the installation of extensive tile drainage system, provides reasonable yield stability, permits the reclamation of some naturally salty soil areas, and assures rational cropping systems without danger of soil deterioration. Those techniques should be usable in similar situations in other countries.

Amelioration of sodic soils usually requires drainage along with some chemical treatments and leaching to reduce the level of exchangeable sodium. Gypsum is the chemical most commonly used to treat sodic soils, although elemental sulfur is alos used. Care must be taken in the incorporation of nitrogen fertilizers into the soil to reduce the possibility of ammonia volatilization during the cropping season.

4. Peat Soils

There are about 32 million hectares of tropical peat soils; twothirds are in Asia(53). Peats have organic matter contents of at least 65 % in the upper 50 cm, are characterized by high water tables, and are found mostly along low-lying coastal areas. Wide differences in the physical and chemical properties of peats accounts for the great variability in the success of attempts to crop them.

Potential for rice production on peats may be considerable but equally impressive problems constrain production on these soils. Chemical deficiencies and toxicities are common; so are toxicities due to organic constituents such as phenols. Although the percentage content of essential elements such as nitrogen appears to be quite satisfactory, the absolute amounts in the crop rooting volume are low because of their low bulk densities. Sterility of wetland rice also presents a serious problem on some peats. No explanation for this sterility is available although copper deficiency or nonavailability may play a role(53).

To effectively reclaim peat soils, land must be cleared of trees and other plants and the area must be drained. Excessive drainage should be avoided to prevent the organic materials from oxidizing too rapidly. Specific cultural and management practices depend on the problems encountered in each area. Thus although the size of the peat areas and their physiographic and hydrologic conditions suggest considerable potential as rice soils, much research is yet to be done to determine if this potential is practically achievable.

SOIL FERTILITY EVALUATION

The soil-related constraints to rice production are important as single factors but their interaction and mutual reinforcement are of even greater significance in determining rice yields. The work of Kawaguchi and Kyuma (1977)(8) is a good illustration of an attempt to analyze the various soil

factors that affect rice yields. They collected 410 surface soil samples from
9 rice-growing countries in South and Southeast Asia and characterized them
by their chemical and mineralogic properties as well as the geologic materials
from which the soils were formed. They then made correlation analyses of the
soil data and came to the following conclusions:

1) Characters relating to base status, texture, and clay
   mineralogy are mutually highly correlated.
2) Characters related to organic matter status are mutually
   highly correlated but are correlated only slightly to
   insignificantly with other characters.
3) The same is true for the characters related to phosphorous
   status.

Accordingly, Kawaguchi and Kyuma identified three factors or fertility
components that could be used to make rational comparisons among the 410
soil samples. These three components have the following characteristics(9):

1) Inherent potentiality (IP) -- the soil character determined
   by the nature and amount of clay and base status.
2) Organic matter and nitrogen status (OM) -- the soil character
   related to the organic matter and nitrogen reserve.
3) Available phosphorous status (AP) -- the soil character related
   to available phosphorous supplying power.

These scientists developed a numerical rating for each of the above
three fertility components so that the overall mean for the 410 samples was
zero and the variance was one. Then they grouped the samples by countries
and by geographic and geological characteristics of the sites from which the
samples came. Comparisons among these groupings indicate how the group
means compare with the mean of all 410 samples. Samples with scores greater
than zero had above-average ratings; samples with below-average ratings had
scores of less than zero. Likewise, variances greater than one showed less
variability than average and those higher than one greater than average.

Table 8 summarizes the ratings by country. The 16 samples from Cambodia
(now Kampuchea), and the 80 from Thailand showed lower-than-average ratings
for each of the three soil fertility components and about average variability
among data from a given country. In contrast, the 54 Philippine samples
showed higher-than-average ratings for two of the three components and about
an average rating for the third (phosphorous availability).

But more significant than country averages are those related to samples
from areas with similar geological materials. For example, samples from
regions of active volcanism showed high inherent potentiality (IP) ratings
while those from sandy old alluvial and acidic rocks showed low IP ratings.
Similarly the organic matter (OM) ratings were related primarily to climatic
differences and available phosphorous (AP) component ratings to parent material.
These ratings are very helpful in rationally evaluating soil-related constraints
to paddy rice production.

Table 8   Means and Standard Deviations of three factor scores used to evaluate fertility of 410 soil samples collected in different countries of South and Southeast Asia

Country	No. of samples	IP		OM		AP	
		Mean	S.D.	Mean	S.D.	Mean	S.D.
Bangladesh	53	-0.438	0.708	0.176	0.704	0.459	0.800
Burma	16	0.118	0.710	-0.128	0.646	0.308	1.121
Cambodia	16	-0.231	0.990	-0.155	0.954	-1.277	0.841
India	73	0.449	0.837	-0.780	0.619	0.581	0.906
Indonesia	44	0.618	0.766	-0.014	0.746	0.031	0.734
Malaysia	41	-0.545	0.719	1.398	0.927	0.026	0.726
Philippines	54	0.618	0.703	0.337	0.673	0.022	1.041
Sri Lanka	33	-0.510	0.853	0.150	1.064	-0.110	0.650
Thailand	80	-0.364	1.195	-0.347	0.939	-0.641	0.769
Total	410	0.00	1.00	0.00	1.00	0.00	1.00

IP = Inherent Potentiality;   OM = Organic Matter-Nitrogen;

AP = Available Phosphorous.

Source: Kawaguchi and Kyuma 1977(8)

INTERNATIONAL COOPERATION

    Research performed by Drs. Kawaguchi and Kyuma and by their cooperators in Asian rice-growing countries are excellent exmaples of the tupe of research that must be done if we are to gain a better understanding of the role of soils in relation to paddy rice production. Wetland soils are generally under-investigated in relation to their importance because they are not the ones on which most of the world's agricultural and forestry crops are produced. No highly industrialized country except Japan places much emphasis on wetland soils. Furthermore, much of the research being accomplished on these soils in the different rice-producing countries is less well-coordinated than is desired. Also, communications among soil scientists and crop production specialists in a given country are often lacking, and are far from adequate among the different rice growing countries.

Steps are being taken to seek support for a simple but sharply focused international mechanism to encourage coordination of soils research in relation to food crop production. An International Board with a small planning coordinating staff and data bank capabilities is envisaged (Fig. 1)(54).

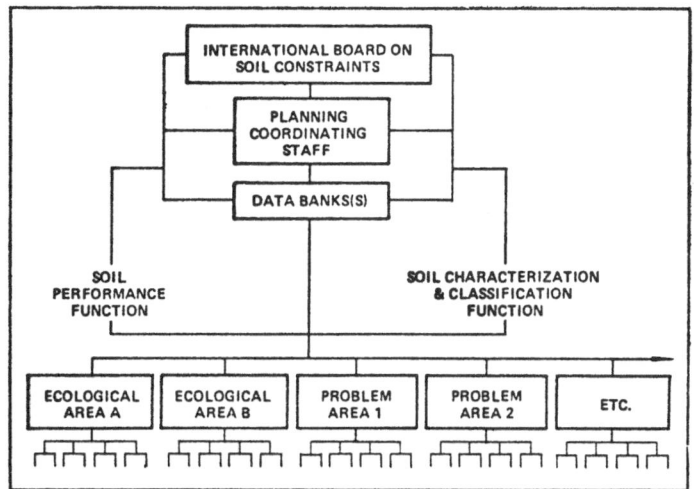

Fig. 1 General concept for the organization and function on an international board on soil-related constraints to world food production (Brady 1980)(54)

More detailed planning and actual implementation would be accomplished through ecological- or problem-area work groups or task forces. The task forces would probably require little new financial support, since their primary function would be to refocus and to improve coordination and planning of research that is already underway.

The status of our knowledge base for wetland soils is entirely adequate in relation to the need for such knowledge. Unlike other fields of endeavor, soil scientists in rice-growing countries cannot place heavy reliance on research conducted in the more developed countries. Nor can scientists in the developing countries afford the luxury of overlapping and uncoordinated activities.

Examples of research areas on which international coordinated planning and implementation would be appropriate for wetland rice soils are:

1) Classification of wetland soils in correlation with classification systems pertinent for dryland cropped areas.

2) Nitrogen balance studies with emphasis on fixation, transformations, and losses.

3) Characterization of adverse soils areas having some implied potential for paddy rice production with the objective of removal or bypassing of the crop production constraints.

4) Development of a fertility and productivity evaluating mechanism for wetland soils to enhance estimates of crop response to chemical inputs.

5) Organic matter cycling and management in rice-growing areas of the tropics.

Some international cooperation is already underway on evaluation of soil fertility and fertilizers. An International Network on Soil Fertility and Fertilzier Evaluation for Rice (INSFFER) has been established which permits cooperation among scientists from Asian rice-producing countries and two international centers, the International Rice Research Institute (IRRI) and the International Fertilizer Development Center (IFDC). Under the leadership of IFDC, ^{15}N balance studies are being conducted and annual planning, reporting, and training sessions are held.

The possibility of the development of a well-coordinated international approach to research on rice-producing soils seems obvious, with some minor modifications of existing programs. First, a concerted international effort is needed to classify paddy soils in relation to uplant soils. Second, soil management aspects must receive attention. One approach would be to broaden the focus of INSFFER and to make this a starting point of a specific soils research effort relating to rice production. This broadening would include not only the research areas of focus but would provide mechanisms to attract the interest of scientists from China, Japan, and other more developed nations to work on paddy soil problems of the developing countries.

I would suggest this as one of the future research strategies considered at this symposium. The solutions to soil problems constraining rice production are important to all of us. But those solutions are of even greater importance to the world's rice farmers and to the billions of people who consume this important crop. We should not let them down .

REFERENCES

(1) Lu, J.J., and T.T. Chang, Rice in its temporal and spatial perspectives. In B.S. Luh, ed. Rice: production and utilization. AVI Publishing Co. Inc. Westport (1980).

(2) Moormann, F.R., Morphology and classification of soils on which rice is grown. Pages 255-272 in International Rice Research Institute. Soils and rice. Los Baños, Philippines (1978).

(3) Moormann, F.R., and N. van Breemen, Rice: soil, water, land. International Rice Research Institute. Los Baños, Philippines (1978).

(4) USDA (United States Department of Agriculture) Soil Conservation Service. Soil classification, a comprehensive system. Washington, D.C.(1975).

(5) Kanno, I.,A new classification system of rice soils in Japan. Pedologist 6:2-10 (1962).

(6) Kyuma, K., and K. Kawaguchi, Major soils of Southeast Asia and the classification of soils under rice cultivation. Southeast Asian Stud. 4:290-312 (1966).

(7) Mitsuchi, M., Pedogenetic characteristics of paddy soils and their significance in soil classification. Bull. Natl. Inst. Agric. Sci, B, 25:29-115 (1974).

(8) Kawaguchi, K., and K. Kyuma, Paddy soils in tropical Asia, their material nature and fertility. The University Press of Hawaii, Honolulu, Hawaii (1977).

(9) Kyuma, K., Productivity of lowland soils. International Rice Research Institute, Los Baños, Philippines (1980), (in press)

(10) Ghildyal, B.P., Effects of compaction and puddling on soil physical properties and rice growth. Pages 317-336 in International Rice Research Institute. Soils and rice Los Baños, Philippines (1978).

(11) Ponnamperuma, F.N., Electrochemical changes in submerged soils and the growth of rice. Pages 421-441 in International Rice Research Institute. Soils and rice. Los Baños, Philippines (1978).

(12) Ponnamperuma, F.N., The chemistry of submerged soils. Adv. Agron. 24:29-96 (1972).

(13) Patrick, W.H., Jr., and C.N. Reddy, Chemical changes in rice soils. Pages 361-379 in International Rice Research Institute. Soils and rice. Los Baños, Philippines (1978).

(14) Koyama, T., C. Chamek, and N. Niamsrichand, Nitrogen application technology for tropical rice as determined by field experiments using ^{15}N tracer technique. Trop. Agric. Res. Cent., Tech. Bull. 3 (1973).

(15) Broadbent, F. E., Nitrogen transformation in flooded soils. Pages 543-559 in International Rice Research Institute. Soils and Rice. Los Baños, Philippines (1978).

(16) Venkataraman, G.S., The role of blue-green algae in tropical rice cultivation. In W.D.P. Stewart ed. Nitrogen fixation by free-living microorganism Cambridge University Press, Cambridge (1975).

(17) Singh, P.K., Nitrogen economy of rice soils in relation to nitrogen fixation by blue-green algae. In Indian Council of Agricultural Research. National symposium on increasing rice yield in kharif (February 8-11, 1978). Central Rice Research Institute, Cuttack, India (1978).

(18) Yoshida T., and R.R. Ancajas, Nitrogen-fixing activity in upland and flooded rice fields. Soil Sci. Soc. Am. Proc. 37:42-46 (1973).

(19) Dommerguez, Y.R., J. Balandreau, G. Rinaudo, and P. Weinhard, Non-symbiotic nitrogen fixation in the rhizosphere of rice, maize and different tropical grasses. Soil Biol. Biochem. 5:83-89 (1972).

(20) Koyama, T., and A. App, Nitrogen balance in flooded rice soils. Pages 95-104 in International Rice Research Institute. Nitrogen and rice. Los Baños, Philippines (1979).

(21) Patnaik, S., and M.V. Rao, Sources of nitrogen for rice production. Pages 5-23 in International Rice Research Institute. Nitrogen and rice. Los Baños, Philippines (1979).

(22) Watanable, I., and W. Cholitkul, Field studies on nitrogen fixation in paddy soils. Pages 223-239 in International Rice Research Institute. Nitrogen and rice. Los Baños, Philippines (1978).

(23) Mitsui, S., Inorganic nutrition, fertilization and soil amelioration for lowland rice. Yokendo Ltd., Tokyo (1954).

(24) Bouldin, D. R. and B.V. Alimagno, NH_3 volatilization from IRRI paddies following broadcast applications of fertilizer nitrogen. International 'Rice Research Institute, Los Baños, Philippines, 51 P (1976).

(25) Mitsui, S., Recognition of the importance of denitrification and its impact on various improved and mechanized applications of nitrogen to rice plant. In Society of the Science of Soil and Manure, Japan. Proceedings of the international seminar on soil environment and fertility management in intensive agriculture (SEFMIA) Tokyo-Japan, 1977. Tokyo (1977).

(26) De Datta, S.K., Fertilizer management for efficient use in wetland rice soils. Pages 671-701 in International Rice Research Institute. Soils and rice. Los Baños, Laguna, Philippines (1978).

(27) Murayama, N. Development of fertilization for rice culture in Japan. In Forestry & Fisheries Research Council, Ministry of Agriculture and Forestry. Japan. Proceedings of a symposium on tropical agriculture research (1970).

(28) Stangel, P.J. Nitrogen requirement and adequacy of supply for rice production. Pages 45-69 in International Rice Research Institute. Nitrogen and rice. Los Banos, Philippines (1979).

(29) International Rice Research Institute, Research Highlights for 1978. Los Baños, Philippines, 118 p (1979).

(30) Mitsui, S.,Inorganic nutrition, fertilization and soil amelioration for lowland rice. Yokendo Ltd., Tokyo (1960).

(31) Goswami, N.N., and N.K. Banerjee, Phosphorus, potassium, and other macroelements. Pages 561-580 in International Rice Research Institute. Soils and rice. Los Baños, Philippines (1978).

(32) Watanabe, I., C.R. Espinas, N.S. Berja, and B.V. Alimagno, Utilization of the azolla-anabaena complex as a nitrogen fertilizer for rice. IRRI Res. Pap. Ser. 11 (1977).

(33) Liu, C.C.,Use of azolla in rice production in China. Pages 375-394 in International Rice Research Institute. Nitrogen and rice. Los Baños, Philippines (1979).

(34) Tuan D.T., and T.Q. Thuyet, Use of azolla in rice production in Vietnam. Pages 395-405 in International Rice Research Institute. Nitrogen and rice. Los Baños, Philippines (1979).

(35) Fogg, G.E., W.D.P. Stewart, P. Fay, and A.E. Walsby, The blue-green algae. Academic Press, London (1973).

(36) Mahapatra, I.C., and R. Prasad, Response of rice to potassium in relation to its transformation and availability under waterlogged condition. Fert. News 15(2): 34-41 (1970).

(37) Kemmler, G.,Potash fertilizer of rice in Japan. Fert. News 15(2): 57-63 (1970).

(38) Von Uexkull, H.R.,Role of fertilizer in the intensification of rice cultivation. In Role of fertilization in the intensification of agricultural production. Proc. 9th Cong. Int. Potash Inst., Antibes (1970).

(39) De Datta, S.K., and K.A. Gomez, Changes in soil fertility under intensive rice cropping with improved varieties. Soil Sci. 120:361-366 (1975).

(40) Ponnamperuma, F.N.,Micronutrient limitations in acid tropical rice soils. In E. Bornemisza and A. Alvarado, eds. Soil Management in Tropical America. North Carolina State University, Raleigh, North Carolina (1974).

(41) Tanaka, A., and S. Yoshida, Nutritional disorders of the rice plant in Asia. IRRI Tech. Bull. 10 (1970).

(42) Randhawa, N.S., M.K. Sinha, and P.N. Takkar, Micronutrients. Pages 581-603 in International Rice Research Institute. Soils and Rice. Los Baños, Philippines (1978).

(43) Ponnamperuma, F.N.,Behavior of minor elements in paddy soils. IRRI Res.Pap. Ser. 8. 15 p (1977).

(44) Blair, G.J., C.P. Mamaril, and M. Ismunadji,Sulfur deficiency in soils in the tropics as a constraint to food production. Pages 233-251 in International Rice Research Institute and New York State College of Agriculture and Life Sciences, Cornell University. Priorities for alleviating soil-related constraints to food production in the tropics, Los Baños, Laguna, Philippines (1980).

(45) Ismunadji, M., I. Zulkarnaini, and M. Miyake, Sulfur deficiency in lowland rice in Java. Contrib. Cent. Res. Inst. Agric. Bogor 14:1-17 (1975).

(46) Yoshida, S., The physiology of silicon in rice. Food Fert. Technol. Cent. for Asia and the Pacific Council, Taipei, Tech. Bull. 25 (1975).

(47) Takahashi, N., Silica as a nutrient to the rice plant, JARQ 3(3):1-4 (1968).

(48) Park, C.S., Practice of fertilizer application for rice under the shortage of chemical fertilizer in Korea. Paper given at the seminar on maintaining rice production in the face of a shortage of chemical fertilizers. Food and Fertilizer Technology Center, Taipei, Taiwan (1975). (Unpubl. mimeo.)

(49) Van Breemen, N., and F.R. Moormann., Iron-toxic soils. Pages 781-800 in International Rice Research Institute. Soils and rice. Los Baños, Philippines (1978).

(50) Van Breemen, N., and L.J. Pons., Acid sulfate soils and rice. Pages 739-761 in International Rice Research Institute. Soil and rice. Los Baños, Philippines (1978).

(51) Van Breemen, N., Acidity of wetland soils, including Histosols, as a constraint to food production. Pages 189-202 in International Rice Research Institute and New York State College of Agriculture and Life Sciences, Cornell University. Priorities for alleviating soil-related constraints to food production in the tropics. Los Banos, Laguna, Philippines (1980).

(52) Ponnamperuma, F.N., and A.K. Bandyopadhya, Soil Salinity as a constraint on food production in the humid tropics. Pages 203-216 in International Rice Research Institute and New York State College of Agriculture and Life Sciences, Cornell University. Priorities for alleviating soil-related constraints to food production in the tropics. Los Baños, Laguna, Philippines (1980).

(53) Driessen P.M., Peat soils. Pages 763-779 in International Rice Research Institute. Soils and rice. Los Baños, Laguna, Philippines (1978).

(54) Brady, N.C., Concluding remarks. Pages 463-465 in International Rice Research Institute and New York State College of Agriculture and Life Sciences, Cornell University. Priorities for alleviating soil-related constraints to food production in the tropics. Los Baños, Laguna, Philippines (1980).

(55) Patrick, W.H., Jr., and I.C. Mahapatra, Transformation and availability to rice of nitrogen and phosphorus in waterlogged soils. Adv. Agron. 20:323-359 (1968).

(56) International Rice Research Institute, Research Highlights for 1976. Los Banos, Philippines, 108 p (1977).

SOME CHARACTERISTICS OF HIGH FERTILITY PADDY SOILS

Chen Jia-fang
(Institute of Soil Science, Academia Sinica, Nanjing)

Li Shi-ye
(Institute of Soils and Fertilizers, Agricultural
Academy of Zhejiang, Hangzhou)

THE NATURE OF THE DEVELOPMENT OF FERTILITY OF PADDY SOILS

In the course of the development of fertility of paddy soils, it is frequently observed that the influence of cultivation practices (including rotation) on soil properties is gradually strengthened and that of natural factors weakened, and as a consequence the properties of various soils become more and more common but less and less different though their original properties were quite different. Taking paddy soils derived from swamp soils of the Lixia River region and from meadow soils of southern Jiangsu Province as the examples, their contents of organic matter and total nitrogen and cation exchange capacity gradually approach each other following the development of fertility(Table 1). The coefficients of variation of some chemical and physical properties for the high fertility and low fertility paddy soils in Tables 2 and 3 also show the same tendency. The coefficients of variation in contents of organic matter and total nitrogen and volume weight for low fertility paddy soils developed on alluvial materials of the Zhujiang River and the Hanjiang River Deltas are 42%, 43% and 15% respectively, whereas for fertile soils the coefficients reduced to 17%, 22% and 9% respectively due to the intensification of common properties(1). In Zhejiang Province, though the organic matter content for high-fertility paddy soils can vary by a factor of 6.5, 60% of the samples are in the range of 2-4%. Likewise, 2/3 of the samples have total nitrogen content of 0.15-0.3% and 3/4, a P content of 0.1-0.2%, although the corresponding ranges are 0.14-0.45% and 0.05-0.23% respectively(2). This phenomenon of "wide range with proper concentration" also indicates the strengthening of common properties and the weakening of individual properties with the development of fertility of paddy soils. As a matter of fact, some similar morphological features could be also discovered on the profile of high fertility paddy soils with various genesis(3,4). Therefore, we consider that the basic characteristics of the development of fertility of paddy soils are the progressive intensification of the effect of man-made factors on properties of the soil and the increase of common properties.

Nevertheless, there still exist some zonal or regional characteristics in high fertility paddy soils(5,6). For example, it is seen from Table 4 that there is a tendency of gradual decrease in CEC of the clay fraction from north to south for paddy soils derived from similar parent materials, presumably reflecting the zonal influence in the composition of clay minerals. Regional characteristics include influences of parent material on soil texture and topography on water condition(5,6). Table 5 shows the effect of micro-topography on water condition and the content and composition of soil organic matter.

Table 1 Changes of properties during the development of paddy
soils derived from different preceding soils

Preceding soil	Fertility	O. M.(%)	N (%)	CEC (m.e./100g)
Swamp soil	Low	4.32	0.257	31.2
	Medium	3.53	0.190	27.3
	High	2.67	0.126	19.2
Meadow soil	High	2.22	0.127	20.1
	Medium	2.11	0.112	17.9
	Low	1.34	0.087	-

Table 2 Variation of chemical properties of low fertility
and high fertility paddy soils

Fertility	Item	pH	O.M.(%)	N (%)	C:N	CEC (m.e./100g)
Low	Range	4.6-7.3	1.0-4.7	0.05-0.26	5.7-11.6	8.3-31.2
	Coefficient of variation (%)	14.2	62.4	47.2	20.1	59.9
High	Range	5.8-7.6	2.2-3.2	0.13-0.18	9.4-13.4	8.5-19.2
	Coefficient of variation (%)	9.3	14.4	14.0	11.3	29.1

Table 3 Variation of physical properties of low fertility and
high fertility paddy soils

Fertility	Item	Volume wt. in water(g/ml)	Coefficient of structure	Clay (%)	Clay/silt
Low	Range	0.35-0.81	0.05-0.83	4.2-38.6	0.40-9.00
	Coefficient of variation (%)	30.2	43.0	61.4	105
High	Range	0.50-0.69	0.40-0.89	5.0-39.3	0.40-7.10
	Coefficient of variation (%)	10.5	23.2	61.1	93.8

Table 4 Comparison of CEC of the clay fraction of paddy
soils derived from the same parent material

Parent material	Location	CEC (m.e./100g)
Lacustrine deposit and alluvium	Jiangsu	42.2 ± 2.6
	Zhejiang	31.5 ± 1.7
	Fujian	25.4 ± 3.0
Quaoternary red clay	Zhejiang	25.8 ± 1.1
	Hunan	21.8 ± 1.2
	Guangxi	17.7 ± 2.1
Weathered granite	Jiangsu	21.8 ± 1.0
	Fujian	16.4 ± 1.8
	Guangdong	12.2 ± 1.0

Table 5   Effect of topography and water condition
on organic matter of paddy soils

Location	Topography	No. of samples	Water condition	O.M.(%)	N (%)	C:N
Jiangsu	Plain	11	Ground water	3.47	0.165	12.3
		12	good drainage	2.20	0.126	10.0
Jiangxi	Mountain	12	Ground water	3.23	0.154	12.1
		19	good drainage	3.08	0.158	11.3
Guangdong	Hill	6	Ground water	3.12	0.150	12.1
		4	good drainage	2.21	0.130	9.8

SOME CHARACTERISTICS OF HIGH FERTILITY PADDY SOILS

In general, high fertility paddy soils are suitable either for rice
or for upland crops, and are easily to be managed and regualted. The
following are their common characteristics.

1. Adequate and Harmonized Nutrients

The important point for high fertility paddy soils with respect to
organic matter and nutrients is not "the more the better", but adequate
in amounts and harmonized one another(3,4). Some studies in China indicate that under the present agricultural conditions the proper organic
matter content of high fertility paddy soils is generally in the range
of 2-4%. If the content is below 2% the soil may be insufficient in organic
matter, whereas a content of over 4% is frequently caused by a strong anaerobic condition of the soil. The proper total nitrogen content may be taken
as 0.13% to 0.23%, and the contents of total phosphorus and total potassium
should be over 0.1% and 1.5% respectively(2,4,6-10). It was observed that
in the yield range of 2,250 to 6,000 kg/ha, there was a significant positive
correlation between rice yield and organic matter or nitrogen content of the
soil, but this correlation disappeared when rice yield was over 6,000 kg/ha(2)
This seems to imply that in addition to adequate quantities of organic matter
and nitrogen, good supplying power of nutrients is important for high ferti-
litypaddy soils. For instance, although the contents of organic matter and
nitrogen of high fertility paddy soils in the suburbs of Shanghai are not
more than that of the infertile slightly gleyed paddy soils, yet its nitrogen supplying intensity is doubled than the latter, and is characterized by
steadiness with time in supplying(9). Generally, high fertility paddy soils
have a high nitrogen supplying intensity (Table 6) and an early priming effect
(Table 7), and as a consequence rice grown on these soils is early tillering
(4,11,12).

23

Table 6 The nitrogen content and supplying intensity
of paddy soils(21)

Soil type	Character	N (%)	Supplying intensity (mg NH$_4$-N/100g/day)			
			6/8–6/13	6/13–6/23	6/27–7/8	7/8–7/23
Permeable	Fertile	0.169	1.33	0.54	0.25	0.10
Slightly gleyed	Wet & clayey	0.119	0.59	0.17	0.08	0.05
Bleached	Sedimenting	0.086	0.37	0.03	–	–

Table 7 Priming effect of soil nitrogen by urea for
paddy soils with different fertility

Soil character	Method of application	Liberated N (mg/cylinder)				Nitrogen supply of soil
		May 24	May 29	June 8	June 28	
Fertile	Surface applied	78	52	–	39	Early
	granule	91	76	82	68	primed
Stiffy	Surface applied	64	17	–	54	Late
	granule	22	47	80	97	primed

    With respect to the supplying and reserving of nutrients, high ferti-
lity paddy soils are characterized by large buffering capacity. That is to
say, the rice neither suffers from starvation caused by inadequate application of fertilizer, nor shows luxurious growth when heavily fertilized(3).
Some studies showed that this feature was related to the high cation exchange
capacity of the soil. As shown in Table 8, for high-fertility paddy soils
two thirds of the samples have a CEC of 10–20 m.e./100g, or slightly high-
er(6,13). On the other hand, for low-fertility paddy soils half of the
samples have a CEC of lower than 10 and one fifth are higher than 20. The
former case may occur in sandy soils with low retaining power for nutrients,
while in the latter case the heavy texture of the soil may retard the supplying of nutrients for plants(4,12).

                        2. Proper Permeability

    In practice farmers frequently take the proper percolation rate of
the surface water over paddy fields as an important criterion in evalua-
tingthe fertility of paddy soils. For, it is the over-all reflection

of a good pedon structure, an ideal soil structure of the plowed layer and an unobstructed drainage.

Table 8 Distribution of CEC in the plowed
layer of paddy soils

Fertil-ity	No. of samples	Percentage of the total			
		<10.0 m.e.	10.1–15.0 m.e.	15.1–20.0 m.e.	>20.0 m.e.
High	39	25.6	46.2	20.5	7.7
Low	34	47.0	23.5	8.8	20.6

The water conditions of the soil has an important effect on nutrients, soil air and soil temperature. Consequently, the farmers pay adequate attention to regulating soil air and soil temperature by controlling soil water, and thus to controlling nutrient supply for plants(3,14–16). The important point in this respect is that the soil should have a good structure and a proper ratio between non-capillary and capillary porosity. According to some measurements in the Taihu Lake region and Shanghai, the non-capillary porosities for high fertility and low fertility paddy soils during dry-farming period are 8–10% and less than 4% respectively, and the latter soils have a capillary porosity higher by 5% than the former(3,8). In Zhujiang Delta the non-capillary porosity of high fertility paddy soils may be as high as 14%, while for low fertility paddy soils it is only 3.6%(3). Table 9 shows that high fertility paddy soils have a ratio of air-filled porosity to total porosity of o.22 $\pm$ o.05 àt pF 2, while in low fertility soils the ratio is only 0.13 $\pm$ 0.02, indicating the retention of too much water.

Table 9 Soil porosity in the plowed layer of high
fertility and low-fertility paddy soils

Fertility	Porosity (%)		Air porosity /total porosity	Samples of total porosity		
	Water-filled	Air-filled		<49%	49–51%	>51%
Low	41.1 $\pm$ 2.1	6.4 $\pm$ 0.9	0.13 $\pm$ 0.02	7	2	
High	40.2 $\pm$ 1.9	11.2 $\pm$ 2.8	0.22 $\pm$ 0.05	6		5

In addition to the effect on air-filled porosity and air exchange, excessive capillary porosity can cause the soil to have a high water-retaining capacity. The results are the difficulties in drainage of superfluous water and plowing of the field. An experiment showed that fall-plowing under excessive water conditions caused the soil to have a compact structure and a high volume weight, and a high amount of large clods in the plowed layer(Table 10). The results of another fall-plowing experiment showed that no clods greater than 6 cm would be discovered under favourable plowing, but there were more than 51% of large clods over 6 cm and 11% of smaller clods less than 2 cm if plowed under excessive water condition, and the residual effects might be remained next summer(8).

Table 10   Effect of preceding crop on distribution of clods
of various sizes in paddy fields(22)

Preceding crop	No. of fields	Soil water (%)	Distribution of clods (%)				
			< 0.5 cm	0.5-1.0 cm	1-3 cm	3-8 cm	> 8 cm
Late rice	6	29-39	0.8	6.2	19.3	46.6	27.1
Medium-matured rice	6	22-26	13.3	11.8	34.9	38.3	1.7
Cotton	2	22-24	30.0	19.6	31.0	19.4	0

Paddy soils with excessive capillary pores often show strong shrinkage after drying, especially for the plowed layer. As a result there appear large crackings and hard clods in the field. For instance it was observed that there was a positive correlation between shrinkage after drying and the ratio of water-filled porosity to total porosity for a series of soils with similar clay content (24.0 $\pm$ 1.9%) (Fig. 1). This ratio also has a positive relationship with the modulus of rupture of the soil (r = 0.585** , n = 19) (17).

The percolation water carries some dissolved oxygen to the active root zone. It promotes the movement of nutrients within the soil and thus aids in their supplying to rice roots. It has the favorable effect of diluting or eliminating the toxic substances and renewing the soil environment(18). As a result, the assimilation of nitrogen by rice plant is enhanced (Table 11) Of course, it is also possible that the percolating water may carry away some nutrients. For this reason, a proper permeability is advantageous for paddy fields. According to some investigations in China there is a proper percolation rate for high fertility paddy soils, the actual rate being dependent on local conditions. For instance, according to the measurements in some irrigation experimental stations of Jiangsu Province, the percolation rate for high fertility paddy soils is in the range of 9-15 mm per day(18). The figure in Shanghai is similar(8). It delta of the Zhujiang River it is 15-20 mm or 7-15 mm(3,7). In Zhejiang Province, the daily percolation rate of paddy soils with annual yields of 15 ton per hectare is about 10-20 mm(3).

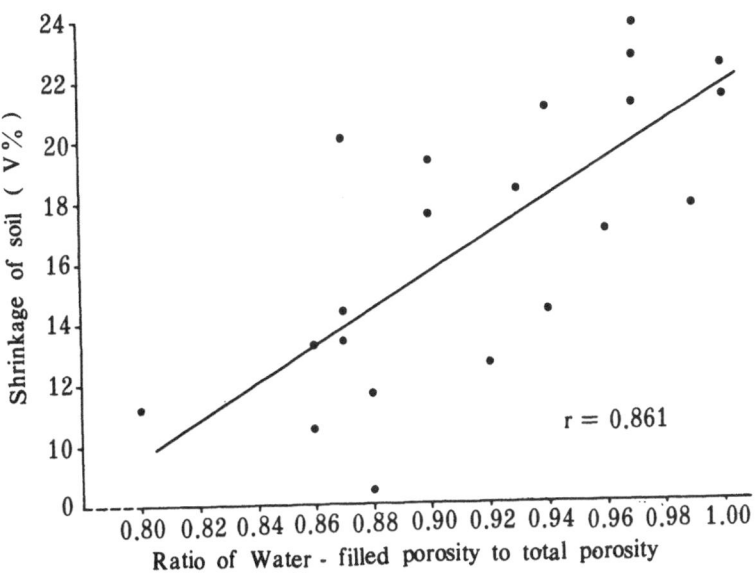

Fig. 1 Effect of the ratio of water-filled porosity to
total porosity on shrinkage of soils

Table 11 Effect of permeability of the soil on assimilation of
nitrogen by rice plant(23)(highest tillering stage)

Sample no.	Plant part	Assimilation (mg N/pot)	
		Permeable	Non-permeable
19	Above-ground	358	246
	roots	80	66
17	Above-ground	270	257
	roots	61	55

3. Good Pedon Structure

The morphological features of a soil profile reflects comprehensively
the environment of soil formation and many properties of the soil. Some
investigations on high fertility paddy soils showed that their pedon structure generally consisted of four horizons, namely, a thick plowed layer, a
well-developed plowed sole, a mottled horizon with apparent vertical cleavages and a substratum with good retaining capability for water. These
four horizons constitute a harmonic pedon, and represent the morphological

features of a high fertility paddy soil(3,6-8,10,19).

   Plowed layer: It usually has a thickness of 18-22 cm, dark grayish
brown in color, with good structure, permeable to water and air. Generally
80% of the rice roots distribute in this layer, although the actual depth
of root-concentrating zone differs with the species. Consequently, the
thickness of this plowed layer is closely related to the amount of roots
and the depth of the distribution of them(20) (Fig. 2). The plowed layer
of high fertility paddy soils may be divided into three sublayers, as contrast sharply with that of low fertility ones. For example, the differentiation in the plowed layer of gleyed paddy soils is not distinct, while for
the sedimenting bleached paddy soils the plowed layer can be distinguished
as two sublayers(7,19).

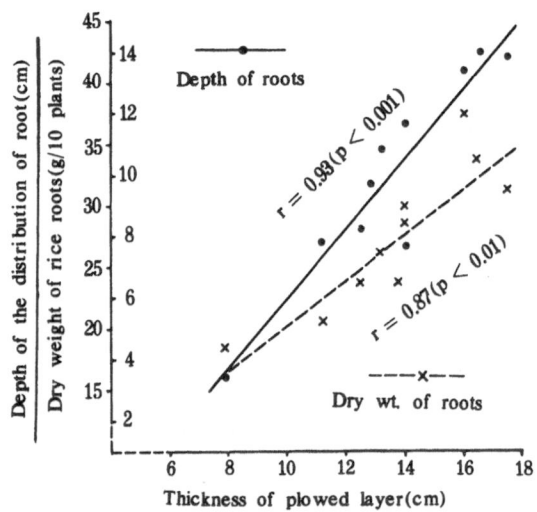

Fig. 2 Relationship between the growth
of rice roots and thickness of plowed layer

   Plowed pan: 5-10 cm in thickness, dark brown in color. The content
of organic matter corresponds to 40-80% of the plowed layer. The compact
blocky structure can form fine crackings after drying and close together
when wetted, thus is beneficial for the retaining of water and nutrients.
However, a too compact plowed pan is not suitable for foot penetration.
In such cases it is advisable to make deep-plowing in autumn so as to avoid
the formation of a compact plowed pan(4). This practice is beneficial for
both the dry-farming crops and the succeeding rice the next year(3,4).

   Mottled horizon: This horizon is characterized by an apparent vertical
cleavage and by many rusty mottlings. The presence of such a horizon is a
criterion of good permeability to air and water.

   Substratum: The morphological features of this horizon are influenced
by the topography and parent material. In Taihu Plain it is gray in color
and with blocky structure and gray coatings. The substratum of paddy soils

derived from red earth in the low hilly regions often have a red color inherited from the parent material.

The morphological features of high fertility paddy soils may be influenced to a certain extent by the rotation system. For instance, in regions with double-cropped rice or triple-cropped rice where the soil is submerged for a relatively long time throughout the year, the paddy soils are darker in color and the pedon structure is not so distinct, whereas in regions with rice-wheat rotation the soil is lighter in color and has a well-developed pedon structure due to the intermittent alternation of wetting and drying.

CONCLUDING REMARKS

A high fertility paddy soil provides the potentiality for obtaining high yield of rice as well as upland crops. The farmers in various districts of China know how to create such a soil and to get high yields conforming to local conditions. It is generally the economic restrictions which prevent the enlarging in area of such soils. It can be anticipated, that the total area of such soils will increase considerably with the development of irrigation and drainage facilities and the increase in application of fertilizers and manures.

REFERENCES

(1) Investigation Group of Guangdong, Institute of Soil Science, Academia Sinica, Characteristics of aggregates in arable horizon of paddy soils as related to the tilth and soil fertility in Guangdong Province. Turang Tongbao, 3, 24-30(1961). (in Chinese)
(2) Faculty of Soils, Zhejiang Agricultural University, Rediscussion about some agricultural properties of fertile paddy soil. Zhejiang Agricultural Sciences, 1, 10-15(1976).(in Chinese)
(3) Institute of Soil Science, Academia Sinica, Nanjing(ed.), Soils of China. Science Press, Beijing, 23-35(1978). (in Chinese)
(4) Institute of Soil Science, Academia Sinica, The Environmental Conditions of Soils for High Yield of Rice. Science Press, Beijing(1961). (in Chinese)
(5) Research Group of High Yield Crops, Institute of Soil Science, Academia Sinica, Soil condition of high yield rice and its regulation. Acta Pedologica Sinica, 8, 79-100(1960). (in Chinese with Russian summary)
(6) Zhu He-jian, Guo Cheng-da, Lin Zhen-sheng, Chen Zhen-gao, Tan Bing-hua, Li Quan-bao, Studies on the characteristics of high-yielding paddy soil in Fujian Province(1979, ms). (in Chinese)
(7) Institute of Soils and Fertilizers, Agricultural Academy of Guangdong, Soil conditions and cultivation practices of high-yielding in Guangdong Province. Guangdong Agricultural Sciences, 6, 7-12(1975). (in Chinese)
(8) Soil Section of the Institute of Soil, Fertilizer and Plant Protection, Shanghai Academy of Agricultural Sciences, Characteristics of the soil fertility and its improvement of the high-yielding paddy soil in the Shanghai suburbs. Scientia Agricultura Sinica, 2, 66-72 (1978). (in Chinese)
(9) Institute of Soil, Fertilizer and Plant Protection, Shanghai Academy of Agricultural Sciences, Soil fertility and its management and improvement in suburbs of Shanghai. Turang, 5, 181-184(1974). (in Chinese)

(10) Institute of Soils and Fertilizers, Agricultural Academy of Hunan. Preliminary investigation on the pedological and agrochemical criterion of high-yield paddy soils. Turang Tongbao, 2, 11-13(1979). (in Chinese)

(11) Zhu Zhao-liang, Chen Rong-ye, Xu Yong-fu, Xu Yin-hua, Zhang Shao-lin, The effect of forms and methods placement of nitrogen fertilizer on the characteristics of the nitrogen supply in paddy soils. Acta Pedologica Sinica, 16(3), 218-233(1979). (in Chinese with English summary)

(12) Liu Zhi-yu, Peng Qian-tao, Yin Chu-liang, Chen Jia-fang, Wu Shi-zhong, Zhu Zong-yu, Studies on the growth of late rice as affected by the nutrient supply of the soil in "Lianhu Lake" farm, Jiangsu Province. Acta Pedologica Sinica, 13, 387-394(1965). (in Chinese with Russian summary)

(13) Lu Fa-xi, Zhu Shi-qing, The level of fertility and cultivation practices of main types of paddy soils in Guangdong Province. Guangdong Agricultural Sciences, 6, 1-6(1979). (in Chinese)

(14) Chen Jia-fang, Preliminary study on the high yield experience of Chen Yong-kang for nutrients regulation of rice by water management. Turang Tongbao, 4, 1-6(1962). (in Chinese)

(15) Chen Jia-fang, Management and promotion of fertility of paddy soil in the plain of Suzhou, Jiangsu. Turang, 1, 13-20(1976). (in Chinese)

(16) Jiang Bo-fan, Lu Ru-kun, Gu Yi-chu, Li A-rong, The content of iron phosphates in the paddy soils of southern China and their significance to the phosphorus nutrition of rice plant. Acta Pedologica Sinica, 11(4), 361-369(1963). (in Chinese with English summary)

(17) Chen Jia-fang, Zhao Cheng-zhai, Zhou Zheng-du, Characteristics of pore space in arable horizon of compact paddy field in Suzhou district, Jiangsu. Turang, 3, 81-84(1978). (in Chinese)

(18) Yang Guo-zhi, Chen Jia-fang, On the significance of constant renewal of soil condition as affected by the permeability of paddy soil. Acta Pedologica Sinica, 9(1), 65-71 (1961). (in Chinese with English summary)

(19) Xu Qi, Morphological designation of fertility of paddy soil in the Taihu Lake region. Turang, 4, 162-165(1975). (in Chinese)

(20) Lin Jing-liang, Characteristics of plowpan in paddy soils and their relations to the yield of rice. Turang Tongbao, 4, 49-51(1965). (in Chinese)

(21) Zhu Zhao-liang, Wang Zu-qiang, Xu Yin-hua, Investigation of nitrogen supplying regime of soils. II. The transformation of ammonium sulfate in rice soil and its influence on the soil nitrogen supplying status. Acta Pedologica Sinica, 11(2), 185-195(1962). (in Chinese with English summary)

(22) Yang Wen-yuan, Soil fertility and double-cropping of rice in triple crop system. Sichuan Turang Tongxun, 2, 44-54(1974). (in Chinese)

(23) Wan Chuan-bin, Cheng Yun-sheng, Analysis of Chen Yong-kang's experience of cultivation practice, fertilization and water-control based on the soil conditions. Jiangsu Nongxuebao, 3, 36-41(1962). (in Chinese)

RICE SOILS FOR FOOD PRODUCTION

R. Dudal, J. Hrabovszky, A. Pécrot
(Food and Agriculture Organization of the United Nations, Rome)

Projections reveal that agricultural production will have to increase by 60 percent in the next 20 years to sustain the world's population in the year 2000. In a study dealing with agriculture in 90 developing countries towards the end of this century, FAO estimates that 72 percent of this increase in production will need to come from intensification of existing agriculture and 28 percent from a further expansion of the cultivated area (FAO, 1979). These figures are averages for all regions and all major food crops.

For rice the increase in production required is 92 percent as a result of high population increases in the rice consuming regions. It is estimated that this production increase could be obtained for 39 percent from an expansion of rice area harvested and for 61 percent from yield increases. In absolute figures this means an additional 25 million hectares of rice harvested, above the 101 million hectares presently cultivated in the countries under study in 1975. The intensification effort required is reflected by the quantities of fertilizers estimated to be needed for rice production: 19 million tons NPK nutrients in the year 2000 as compared to the 3 million tons applied in 1975. Main support for yield increases should come from irrigation on 53 percent of the rice lands as compared to 37 percent of rice land irrigated in 1975. There should be a considerable reduction in "partially irrigated" in favour of a large increase of "fully irrigated" land.

The overriding question in these projections is whether presently cultivated land will lend itself to more intensive production, and if sufficient soil and water resources are available for a horizontal expansion of rice cropping.

Appraisals of the global extent of potentially arable land vary from 3 to 7 billion hectares out of which only 1.5 billion are being used at present. Although these estimates indicate that large tracts of potentially arable land are still available, they are too global to permit practical conclusions with regard to planning future land use.

Concrete answers are needed to the following questions:

- where are potential arable areas located,

- for which crops are they suitable,

- are they arable on a sustained basis,

- which level of technology is required,

- what are the investments needed,

- where can maximum returns from increased inputs be obtained,

- how much land is being lost yearly because of various degradation processes,

- are there critical areas where land resources cannot support present
  or future populations.

While sufficient arable land is available worldwide, its distribution
and quality varies greatly between and within countries. Land resources
are fixed and cannot be moved. Some areas are better endowed than others
and there is an urgent need for inventories and resources by country and
region.

LAND EVALUATION

In order to obtain a more precise assessment of the production potential
of the world's lands and so provide the physical data base necessary for
planning future agricultural development, FAO has recently initiated a
study of potential land use by major agro-ecological zones (FAO, 1978).

The methodology to assess the agricultural potential of the world's
land resources uses six basic principles (FAO, 1976):

i.   land suitability is only meaningful in relation to a specific use,
     e.g. land suited to the cultivation of cassava is not necessarily
     suited to the cultivation of pearl millet;

ii. the evaluation of production potential needs to be made in respect
     of specified input levels, e.g. whether fertilizers are being applied
     or not, if pest control is effected, if machinery or hand tools are
     being used;

iii. suitability must refer to use on a sustained basis, that is the
     envisaged use of land must not result in its depletion, e.g. through
     wind erosion, water erosion, salinization or other degradation
     processes;

iv. evaluation involves comparison of more than one alternative type
     of land use, e.g. suitability for millet or sorghum or maize, and
     not just for a single crop;

v.   different kinds of land use are compared at least on a simple economic
     basis; that is, suitability for each use is assessed by comparing the
     values of the produce to the cost of production;

vi. a multidisciplinary approach is adopted, the evaluation being based
     on inputs from crop ecologists, agronomists, climatologists and
     economists, in addition to those from soil scientists.

At first sight the principles proposed do not seem unprecedented but
in practice they call for considerable changes in traditional resource
interpretation. First of all, the concept of land evaluation is wider
than the appraisal of soil qualities to which previous suitability classifications were sometimes limited. Land is defined geographically as a
specific area, the attributes of which relate to soils, geology, hydrology,
plant and animal population, climatic conditions and the results of past
and present human activities to the extent that these attributes exert a
significant influence on present and future uses. Furthermore, the
multidisciplinary approach requires the matching of the physical resource
base with prevailing social and economic conditions which may considerably

influence production perspectives.

The appraisal is made in terms of current suitability that is for a specific use of land areas in their present condition, or with modest modifications which are within the reach of the farmer. Two levels of inputs - low and high - are considered. The former is a low technological level involving hand cultivation, no or insufficient fertilizer application, local cultivars, no chemical pest, disease and weed control, fallow periods, small farm holdings and some untimely operations because of labour bottle-necks.

The high input level involves mechanical cultivation, sufficient fertilizer, high yielding cultivars, chemical pest, disease and weed control, timely operations and a generally high standard of management with simple conservation practices.

The information needs therefore refer to the various elements of the landscape as well as physical and socio-economic.

INFORMATION NEEDS

For large parts of the world precise information on land resources is still missing. A special effort will be necessary to obtain the basic data which are needed for the further development of agriculture in the decades to come.

The information to be collected should provide the parameters which are relevant for planning purposes at an appropriate level of aggregation, taking into account the time and financial resources available. The information needed may be divided into three types: the basic parameters, the dynamic properties and the interrelationships between the different variables (Hrabovszky, 1980). The basic parameters refer mainly to geo-based information, geology, physiography, hydrology and soils. However, the interpretation of this information calls for bringing in dynamic properties such as soil degradation, droughts and floods, modifications of land use, that is changes which take place over time and which follow a certain path. Information on interrelationships involves an analysis of the interaction between physical, economic, technological and human factors, which result in a production function of land use. The needs for additional information in the years to come include inventories of the different components of the land: soils, water, relief, geology, land form, vegetation, land use, soil degradation and climate.

SOIL RESOURCES

On a global scale the FAO/Unesco Soil Map of the World (FAO, 1971-1980) provides an overall basis for the inventory of land resources. The legend comprises 106 soil units which have been clustered into 26 major groupings (FAO, 1974).

The map units are associations of soils which occur within the limits of physiographic entities. Each association is composed of a dominant soil and of associated soils, the latter covering at least 20 percent of the area of the mapping unit. Important soils which cover less than 20 percent of the areas are added as inclusions. The textural class of the dominant soil and the general slope class are given for each association. Important land characteristics, not reflected by the soil associations themselves, are shown as phases, such as the occurrence of salinity or of hard layers at shallow depth. Areas of dunes, shifting sands, salt flats and rock debris are shown separately.

In comparison to other global soil maps, the FAO/Unesco inventory is unique in that it is the result of wide international cooperation and that it is based - to the extent possible - on actual survey information and correlation. The reliability of the map differs from one area to another depending on the accuracy and detail of the material used for its compilation. Table 1 shows the sources of information available and provides a picture of soil survey coverage in different parts of the world. Class I areas are systematic soil surveys in which the boundaries are based on field studies; for Class II, boundaries are derived from physiographic data while Class III information is derived from the interpretation of general data on land forms, geology, climate and vegetation.

It appears that only about a fifth of the world's soils have actually been surveyed. The highest percentage of survey coverage is found in Europe; it is the lowest in Africa. When percentages are calculated after deduction of arid regions and permafrost areas, which are surveyed only in few instances, Class I surveys cover respectively 10.8, 23.3, 15.4, 80.2, 46.1, 15.0 percent of the regions listed in Table 1 and 28.2 percent of the world.

Table 1   Soil survey coverage (percent)

	Class I	Class II	Class III
Africa	7.5	38.0	54.5
Asia	19.0	49.0	32.0
Australasia	11.0	61.0	28.0
Europe	76.3	23.7	–
North and Central America	28.0	16.0	56.0
South America	14.6	45.9	39.5
World	21.0	40.0	39.0

These figures clearly point to the need of an intensive programme of soil resources appraisals especially in areas where an expansion of arable land is envisaged.

A minimum programme is the preparation of soil maps at a scale of 1:1 000 000. These exploratory types of surveys provide an overall view of the soil resources of a country, allow for the delineation of areas with development potential which qualify for investigation at the reconnaissance scale, in the order of 1:250 000. In turn these reconnaissance surveys provide the necessary information to select specific project areas which deserve a detailed study. It should be stressed that soil surveys should no longer be confined to classifying soils and drawing soil boundaries, but need to consider all the ancillary studies which assist in the interpretation of soil maps. The task of the soil surveyor is not only to map soils but more and more to interpret the data in terms of agricultural potential of the land taking into account climate, land use, farming systems and accessibility.

CLIMATIC RESOURCES

The climatic inventory used in assessing land suitability takes into account the crop's climatic requirements such as rainfall, soil moisture, temperature and radiation. The inventory, therefore, refers to the growing period – that is the number of days when water availability and temperature regime permit crop growth – and to major climatic divisions which cater for photosynthetic and temperature requirements of the crops used in the assessment (FAO, 1978). The growing period is the continuous period during the year, from the time when rainfall exceeds half potential evapotranspiration (calculated by the Penman method) until the time when rainfall falls below full potential evapotranspiration, plus a number of days required to evaporate an assumed 100 mm of soil moisture reserve. Consequently, a normal growing period must exhibit a humid phase, i.e. a period in which rainfall is greater than potential evapotranspiration. Additionally, it excludes any period when crop growth is not possible in the growing period because of low temperatures.

Length-of-growing-period data are calculated and zones with similar lengths of growing period are delineated by constructing isolines at intervals of 30 days (e.g. 90–119 days, 120–149 days, 150–179 days, etc.). Zones with a humid phase are designated as normal. Zones without a humid phase and consequently unable to meet full crop water requirements from rainfall, are designated as intermediate. An additional isoline for a growing period of 75 days was also included to cover possible interpretations for pearl millet in drier areas.

For each zone thus delineated by the length of growing period isolines, average values of major climatic elements (radiation, day and night-time temperature, etc.) characterizing the growing period, are calculated for subsequent potential biomass and yield calculations.

SUITABILITY ASSESSMENT

The suitability assessment is made by matching the soil and climatic inventory with the soil and climatic requirements of alternative types of land use. Such proved to be difficult because of the lack of precise information on soil requirements of various crops in specific climatic conditions, and the lack of detailed information on climatic requirements, particularly moisture, under specific soil conditions. The definition of such parameters is an integral part of land evaluation. The climatic matching exercise is, in essence, by consecutively comparing the photosynthesis temperature requirements of each crop group with the main temperature regimes during the growing period in the major climates; ascertaining if the prevailing length of growing periods will permit any yield of the crop and, if so, calculating potential net biomass and yield of crops in the various adequate lengths of growing periods inventoried, according to the climatic factors therein. The assessment also takes into account yield reductions due to rainfall variability, moisture stress, waterlogging and losses due to pests, diseases and weeds.

The climatic suitability assessment for each crop was defined in terms of a percentage range of the maximum attainable yield without constraints. Growing period zones, capable of yielding 80 percent or more of the maximum yield attainable, were classified as very suitable; zones yielding less than 80 percent to 40 percent as suitable; zones yielding less than 40 percent to 20 percent as marginally suitable and zones yielding less than 20 percent as not suitable. This activity results in a climatic crop suitability assessment of each main climatic division and length of growing period. The yields so calculated are for high input/ideal soil conditions. Agronomically attainable low input yields are calculated in a similar manner, the climatically potential yield under low inputs being 25 percent of the climatically potential yield under high inputs.

The actual land suitability is obtained by superimposing the soil assessment on the climatic appraisal. If the soil unit largely meets the crop requirements no change is made in the climatic suitability assessment. If the soil unit only partly meets the crop's requirements, i.e. if the soil does not allow the full climatic yield potential to be attained, the climatic suitability assessment is downgraded by one class. Areas of soils which fail to meet the crop's minimum soil requirements are classified as not suitable since severe soil limitations override the climatic attributes.

SOIL REQUIREMENTS FOR RICE

The soil characteristics which are significant for rice growing are briefly reviewed below: drainage and permeability, texture, fertility and chemical composition, organic matter content, reaction and salinity (Moormann and Dudal, 1968).

Drainage and Permeability

The natural drainage status of the soils is as important for irrigated lands as for those which are cultivated under dry conditions. The drainage conditions largely determine the reduction stage in the soil and the availability of both nitrogen and phosphate. When the soil is too oxidized, soil nitrate and phosphate are less available. On the other hand, when the soil is highly reduced, free hydrogen sulphide develops which is harmful to root development, and the concentration of ferrous ions may become toxic.

Excessively and well drained soils, with a rapid permeability are usually inferior, because too much water is lost by percolation. Under rainfed conditions yields may fail on such soils. On imperfectly and poorly drained soils temporary water shortage in the growing season is not nearly as serious. However, very poorly drained soils, as characterized by permanent high groundwater and reductive conditions throughout the year are lower yielding or may be unsuitable for rice production.

Texture

Optimum yields are generally obtained on soils with a medium texture. Significant differences in yield levels between the sandy and clayey soil series were noted in the Mekong Delta, the clayey soils being superior.

Not only is the texture of the surface layers important but also the layers below should be taken into account. If the subsurface horizons are clayey, the influence of sandy surface horizons diminishes. However a medium textured surface horizon may be more favourable because the soil can be more easily worked. A sandy subsoil, appearing at a depth of 40 cm or more, has little bearing on the production level if the surface layers are clayey.

Fertility and Chemical Composition

The inherent fertility of the soil, conditioned by its mineralogical and chemical composition, is an important factor in assessing the suitability of land for rice. Although fertilizers can to a high degree supplement a low fertility level, the fact is that in many countries where rice is grown, fertilizers are not widely applied owing to economic constraints.

The availability of plant nutrients on non-fertilized rice soils depends largely on the nature of the parent materials and on the degree of weathering of the soil. Fluvisols, on which large areas of paddy lands occur are formed on recent deposits, and show little or no profile differentiation. In such soils, texture often reflects their inherent fertility, the sandy profiles being the least favourable. Not all alluvial deposits, even when clayey, are fertile, especially in the tropics and subtropics. In these areas, the origin of the sedimentary material is important. The fertility level of Regosols is also to a large degree related to the nature of the parent material. Regosols on medium textured volcanic ash give excellent rice yields; sandy ashes are less suitable.

Vertisols, Luvisols, Cambisols and their hydromorphic associates are relatively fertile and so are Xerosols and other soils of the dryer range of conditions under which rice is grown. Acrisols, Ferralsols, and the hydromorphic associates related to these groups, are relatively poor in plant nutrients and show various deficiencies. Andosols are frequently deficient in magnesium. However, more research is needed on the natural fertility of the soils on which rice is grown, in order to make precise appraisals based on soil classification units.

Organic Matter

Rice is grown on soils which very widely in their organic matter content, ranging from 30 percent or more in peaty soils to 0.8 percent and less in certain Gleysols. In Africa, the organic matter content is used as an indication of natural fertility for irrigated rice, in connexion with the availability of nitrogen. The presence of too much organic matter may become a limiting factor. Rice on strongly peaty soils

37

has the tendency to lodge and transplanting in such soils under water-logged conditions is difficult.

Organic matter in the form of manure or green manure is not often applied on rice lands. Recycling of organic wastes for use as manure in partial substitution of mineral fertilizers is receiving increased interest in the present economical circumstances.

Soil Reaction

Rice grows under a wide variation of pH, e.g. from 3.5 to 8.5. Large extents of rice land have a pH varying from 4.5 to 6.5 which seems to be the most suitable range for rice production.

Negative effects related with high pH are often more the result of a high concentration of soluble salts, such as Na and Mg. Negative effects associated with low pH's in strongly leached soils are more the result of low fertility than of the pH itself.

In acid sulphate soils the development of acid from iron polysulphides may result in aluminium and iron toxicity. Although pH as such does not seem to influence rice growth to a great extent, it is an indicator of fertility status, of harmful levels of certain elements, e.g. iron and aluminium in the acid range, or of soluble salts in the high pH range.

Soil Salinity

Rice is generally reported to have a medium tolerance to salt. With high salinity yield depressions up to 50 percent may be expected. Varietal differences as to salt tolerance are well noted. Also the effect of salinity on the growth of rice depends upon the stage of development of the plant. Rice is most tolerant during the germination stage and after flowering, while it is most sensitive during the young seedling stage.

RICE LANDS IN DEVELOPING COUNTRIES

The soil characteristics described above are included or implied in the definitions of the 106 soil units of the legend of the Soil Map of the World. For bunded rice land was evaluated in terms of three suitability classes (suitable, marginally suitable, not suitable) at two levels of inputs. Slope, which is indicated on the map, was also taken into consideration: slopes of 0 to 4 % were evaluated as suitable; 4 to 8 % as marginally suitable and more than 8 % as not suitable. This assessment refers to land under natural conditions. With terracing, sloping land can successfully be used. The importance of physiography and position in the landscape is well known and reflects on soil characteristics such as depth, salinity, drainage, waterlogging and nutrient deficiencies (Moormann and Van Breemen, 1979). Furthermore land evaluation for irrigated rice is closely related to availability and quality of water resources. However these aspects could not be considered at the level of generalization of the study.

Under rainfed conditions, it was considered that lengths of growing periods shorter than 180 days, corresponding roughly to rainfall regimes of 1,000 mm or less, are not suitable for bunded rice. The 180 day growing period may be assumed to provide three consecutive months with more than 200 mm precipitation per month, which is the minimum acceptable distribution for cultivation of paddy rice under bunded conditions. In this broad regional study it was not possible to consider irrigation because information at country or subcontinent level on surface and groundwater availability is not available. In a second step of the study more detailed investigations, including irrigation potential, are now underway.

Rice yields in climatically suitable areas are, to a large degree, dependent on complete water control. This is considered not possible under purely rainfed conditions and therefore no "very suitable" (VS) land was recognized for this crop from the climatic viewpoint.

Difficulties of land preparation and harvesting may be expected to exclude cultivation of the crop in all year round humid-growing periods, i.e. 365 days length-of-growing-period areas.

Based on the above assumptions and the combination of soil and climate assessment, the following results were obtained for Africa and Southeast Asia (Table 2).

Table 2   Extents ($X10^3$ ha) of land variously suited
to the production of rainfed rice

| | Africa | | | | Southeast Asia | | | |
| | High inputs | | Low inputs | | High inputs | | Low inputs | |
	S	Ms	S	MS	S	MS	S	MS
Warm tropical lowlands	132 208	147 923	61 018	164 160	109 025	28 707	51 341	52 082
Warm subtropics (summer rainfall)	1 239	540	402	404	13 910	2 475	10 283	3 500
Total	133 447	148 463	61 420	164 564	122 935	31 182	61 624	55 582

S = Suitable
MS = Marginally suitable

According to the FAO Production Yearbook, the area currently under paddy cultivation in Africa is estimated at 4,350,000 ha. This continent has therefore very large reserves of suitable and marginally suitable land for extensions of rice cultivation, particularly if a high level of inputs is applied. Land reserves for rice cultivation are also, though to a lesser extent, available in Southeast Asia.

It is estimated (FAO, 1979) that the following distribution of rice lands should be aimed at in the 90 developing countries under study, spread in the different agro-ecological zones of the world (Table. 3).

Table 3 Land use structure for rice (million ha)

	Uplands	Lowlands (not irrigated)	Fully irrigated	Partially irrigated	Marginal areas
1975	4.3	50.0	19.8	17.4	9.6
2000	1.7	46.6	53.1	12.4	12.6

The figures in Table 3 reflect a strategy to achieve a large share of additional rice production through yield increases. This has been suggested for two reasons. The first and foremost is that in many of the major rice producing countries the opportunities for large scale increases for the area of rice are not feasible because there are no reserves of suitable lands to be brought under rice cultivation. The second consideration stems from the fact that additional production coming from yield increases in most situations is cheaper to achieve than bringing new land under cultivation. As a result emphasis is given to irrigated areas, and within the non-irrigated areas to an expansion of rice area under high rainfall conditions. These irrigation development proposals advocate major improvements on existing schemes which would take them from the partially irrigated into the fully irrigated class. It can also be seen that between 1975 and 2000 it is proposed to expand the irrigated area by 28.3 million ha. Part of this would come from conversion of already cultivated upland areas, but for a certain share new areas would have to be made arable.

The results of the extension in area and improvements in the production capability of rice lands together with higher input use and improved management could bring about a paddy output in the year 2000 of 389 million tons as compared to 193 million tons in 1975. Underlying this would be an increase in the harvested area of rice from 98 million ha in 1975 to 126 million ha and an increase in yields from 2 tons to 3.1 tons of paddy per hectare.

CONCLUSIONS

High yielding varieties have been developed with a genetic production potential of 6 tons per hectare and more. However, such yields are obtained only in a few countries and the world average is still only 2 tons. Major constraints for increasing yield are related to soil conditions, water management and efficient use of agricultural inputs. Considerable progress remains to be done in developing techniques and to launch action programmes to overcome these constraints at the farm level.

The experience gained in China over many centuries in labour intensive rice production, recycling organic wastes, biological nitrogen fixation will be a source of new information offering an original and well adapted approach to intensification of rice production which could be promoted in a number of densely populated countries in Southeast Asia.

This Symposium is therefore most timely and will contribute to an increased exchange of experience, to the identification of gaps in research and the promotion of cooperation between countries and research institutions.

REFERENCES

(1) Food and Agriculture Organization of the United Nations. Soil Map of the World, Ten volumes, eighteen map sheets, Unesco, Paris (1971-1980).
(2) Food and Agriculture Organization of the United Nations. Soil Map of the World, Vol, I, Legend, Unesco, Paris (1974).
(3) Food and Agriculture Organization of the United Nations. A Framework for Lnad Evaluation, FAO Soils Bulletin No 32, Rome (1976).
(4) Food and Agriculture Organization of the United Nations. Report on the Agro-ecological Zones Project, World Soil Resources Report No 48, Rome (1978).
(5) Food and Agriculture Organization of the United Nations. Agriculture towards 2000, Rome (1979).
(6) Hrabovszky, J. Information and Planning: Striving for the Best while Living with Feasible Compromises. Proceedings of Workshop on Information Requirements for Development Planning in Developing Countries, ITC, Enschede (1980).
(7) Moormann, F. R., Dudal, R. Characteristics of Soils on which Paddy is Grown in relation to their Capability Classification. 11th Session International Rice Commission: Proceedings of the Working Party on Rice Soils, Water and Fertilizer Practices, Kandy (1968).
(8) Moormann, F. R., Van Breemen, N. Rice: Soil, Water, Land. International Rice Research Institute. Los Baños (1979).

RECENT PROGRESS IN STUDIES OF SOIL STRUCTURE, AND ITS RELATION TO PROPERTIES AND MANAGEMENT OF PADDY SOILS

D.J. Greenland

(The International Rice Research Institute, Los Baños, Philippines)

INTRODUCTION

Soil structure has been defined as "the arrangement of the soil particles and of the pore spaces between them (1). It includes the size, shape and arrangement of the aggregates formed when primary particles are clustered together into larger, separable units." The structural organization of the soil particles determines such important properties as water movement and storage in soils, gas exchange between roots and atmosphere, the ease of tillage, the extent to which the soil is explored by plant roots, and erodibility. In relation to paddy rice production, the aspects which are important are those factors related to water movement and root growth. Where upland crops are grown in rotation with flooded rice, then the condition of the soil after draining from the flooded state and the ease with which seeds can be drilled, and can emerge and establish their root systems, become of critical importance.

The arrangements which soil particles adopt in the field are a resultant of their particle size, shapes and flexibility, the forces which act between them, and the various forces to which they are subjected. Particle size distributions of soils always influence the packing arrangements which are possible, and for large, inert particles of relatively regular shape packing theory (e.g. 2) can give an approximation to the structures developed. In most soils, however, a range of particle sizes occurs, and between the smaller clay sized particles there are very strong interactions, so that 'structure' becomes dependent on these forces, and the various stresses to which the soil is, or has been, subjected. For subsoils, the principal stresses are those associated with the shrinkage and swelling, which occurs when the soil dries and rewets, or, in cold areas, thaws and freezes. For surface soils, in addition to these processes the effects of biological action, such as those due to worms and various soil animals, and of traffic and tillage implements, as well as raindrop impact, become important. These various structure-forming processes lead to the development of peds or aggregates, usually characteristic of the soil in which they occur.

In paddy soils, the subsoil may seldom become dry, and never freeze; few roots or soil animals penetrate it, and thus its characteristic appearance, unless allowed to dry prior to examination, is of apedality. However within a generally uniform appearance pores and channels may exist, produced by pedogenic processes, such as the dissolution and precipitation of different mineral constituents, or the infilling of soil pores or cracks by the process of lessivage, possibly operating during an earlier climatic regime. Many studies have been made relating to the pedogenic processes operating in paddy soils. These were recently summarized by Kanno (3) and Moormann (4). They give little attention to specific structural characteristics, although the common development of a densely packed soil layer or plough pan, at the lower limit of cultivation in the profile, is mentioned as a common characteristic feature.

While the various moulding processes contribute to structural development in soils, it is the physico-chemical forces acting between the particles
which determine the stability of the particle associations. The smallest associations are commonly referred to as domains (5, 6). They are clusters of
clay crystals, not normally larger than about 5 µm equivalent spherical diameter (e.s.d.). The forces between the particles are electrochemical. The
domains are usually clustered to form microaggregates (5 µm to 1 mm e.s.d.)
which in turn may be clustered with silt and sand grains to form aggregates
(Fig. 1). The packing of aggregates and microaggregates determines the porosity of the soil. When there are well formed and stable aggregates, a relatively high proportion of large pores will be present in the soil. If the
aggregates are unstable and fall apart into microaggregates, these may pack
closer so that fewer large pores are present. If the microaggregates are unstable, then domains and individual silt and soil particles will separate
causing a densely packed, or massive structure to develop. If the clay particles disperse from the domains, they tend to flow rather readily, forming
cutans, and blocking partially or completely the pores which they enter. Soils
with dispersed clays are those which give rise to the most difficult to manage
structural conditions.

Thus it is not only the structure of a soil which is important, but its
stability. Many early attempts to measure the favourable effects of organic
matter additions to soils were based on determination of aggregate stability.
The value of such studies was limited because the stresses to which the aggregates were subjected were not related to those to which they are exposed in
the field. Similarly strength varies considerably with water content of the
aggregates, and the water content at the time the stress is applied was often
inadequately defined. Further, the significance of aggregate breakdown depends
on how it is related to porosity changes in the soil. The necessary proportion of large pores for air and water transmission and easy root growth, has
generally been inadequately defined. A better understanding of the needed
porosities for unrestricted root growth is still needed. For upland crops,
adequate 'storage pores' which retain available water (0.5-50 µm equivalent
cylindrical diameter) as well as adequate transmission pores (>50 µm e.c.d.)
are necessary. In addition roots grow easily only into relatively very large
pores, wider than about 50 µm, and must be able to enlarge these, as root diameters of most arable crops are of the order of 250 µm and upwards. The low
strength of saturated soils means that there is usually little problem of pore
expansion in flooded paddies, provided opportunity exists for initial root
penetration.

A major reason for characterizing soil structure in terms of aggregate
size and shape is the relative ease with which this can be done. Pore size
distributions on the other hand are relatively difficult to determine, and
the continuity and stability of pores difficult to define and measure.

The present paper discusses recent work on the surface properties of clays
which relates to particle interactions, and the stability of structural units.
The influence of organic materials on those interactions is also discussed
briefly. Problems of measurement of pore size distribution are described.
Structural attributes are then considered in relation to the physical properties of paddy soils, and various processes related to production of rice and
other crops grown in succession to paddy rice.

STRUCTURAL PROPERTIES AND CLAY CHARACTERISTICS

The forces which act between adjacent clay particles have been extensively studied (7, 8). These forces can be described in terms of the "diffuse double layer" theory, adjusted to take account of changes which occur when clay plates are in very close proximity. These latter forces are not as yet fully understood (9). For the normal soil condition of calcium-ion saturation, attractive forces are dominant, so that domains, or "quasi-crystals" persist as structural entities, unless a suitable stress is applied which can enable the energy barrier to separation of the clay particles to be overcome (5). This may happen for low charge density 2:1 type of clay minerals even when calcium saturated if a mechanical stress is applied to the wet soil. In the presence of calcium carbonate or calcium sulphate it does not occur, presumably because these salts maintain a sufficient electrolyte concentration in the soil solution to prevent dispersion. Also in soils where the organic carbon content exceeds about 2 per cent, little dispersion occurs (9).

For soils dominated by kaolinite and hydrous oxides of iron and aluminium, and perhaps also those where chloritic minerals are present, domain and microaggregate structures appear to be particularly stable (11), probably due to interparticle cementation by the hydrous oxides which acts in addition to the electrostatic forces (12). The major differences between the constant charge surfaces of the mica-type minerals, and the constant-potential hydrous oxide type surfaces (8) undoubtedly account for much of the difference in the structural behaviour of soils dominated by the two types of clay.

Although microaggregates in many red soils are particularly stable, dispersion of the clay fraction of some of the less acid red soils (Alfisols) occurs rather readily. In such soils clay translocation is frequently observed (13). The reason for this appears to be that specific adsorption of anions by the hydrous oxides (14) can lead to a substantial lowering of the isoelectric point. Their dispersion behaviour then becomes similar to that of the 2:1 minerals.

The effects of changes in the redox potential of soils after flooding on their dispersion-flocculation behaviour has been inadequately studied. Koenigs (15, 16) examined the behaviour of several Indonesian soils in this respect, and discusses the important differences between soils whose clay fractions are dominated by 2:1 type clay minerals (Vertisols, or margalitic clays in Koenig's terminology) and those dominated by 1:1 type clays and hydrous oxides of iron and aluminium (Oxisols, and Oxic Ultisols and Alfisols, or Latosols). Kita and Kawaguchi (17) and Ahmed (18) mention that flooding tends to reduce the structural stability of soils. In these and more recent studies (13, 19), the important differences between the strongly flocculated clay fractions of Oxisols, where active aluminium exerts a dominant role, and Alfisols and Ultisols where clay particles may be rather easily dispersed and translocated, have been described. Deshpande et al. (20) found that the iron oxides in several well oxidized red soils were structurally inactive. A much higher proportion of active iron was found in soils subjected to repeated oxidation and reduction (21). Kyuma (22) has reviewed information relating to the clay mineralogy of soils used for rice production in Asia, but the changes in their physico-chemical properties associated with alternation of oxidizing and reducing conditions has been inadequately studied. The opportunity to improve our understanding of paddy soil behaviour through such studies is considerable. A specific example is the improvement to our appreciation of processes operating in Planosols provided by Brinkman's description of Ferrolysis (23).

STRUCTURAL PROPERTIES AND SOIL ORGANIC MATTER

There have been many studies to show that organic matter in soil tends to
stabilize the soil aggregates (24). In soils of small organic contents, the
polysaccharide fraction of the organic matter often exerts a dominant role
(25) but where larger organic matter contents occur (>4 per cent) other components also contribute to aggregate stability (26, 27). The organic materials
act by forming interparticle bonds at the periphery of domains and microaggregates (Fig. 1). The mechanisms of bonding of the organic materials to the inorganic particles have been extensively studied (28, 29, 30). Polysaccharides
adsorb through both ionic associations, and by virtue of van der Waals and
entropy interactions (31). The importance of the non-ionic interactions has
been fully demonstrated by studies of polyvinyl alcohol (32, 33) and polyethylene glycol adsorption (34). Humic materials are adsorbed primarily through
association of their carboxyl groups with polyvalent metal ions at the clay
surface (35, 36). Although organic polymers are strongly adsorbed by clay
minerals, they do not eliminate swelling behaviour, which indicates that the
polymers are adsorbed at the e ternal edges of domains and microaggregates,
helping them to retain their identity, but allowing osmotic forces to operate
between the particles surrounded by the polymer (37, 38.)

Turchenek and Oades (39) have shown that organic matter in several soils
is concentrated in microaggregates which are difficult to disperse. The organic matter and clay particles are associated by a range of bonding mechanisms, and so are exceptionally stable. Much remains to be learnt of the
details of microaggregate stabilization in soils, and particularly of the
changes in that stability which may arise as organic matter levels change
under the influence of cultivation. The effects of strongly reducing conditions, and alternating oxidation and reduction conditions on the clay-
organic association also require further study.

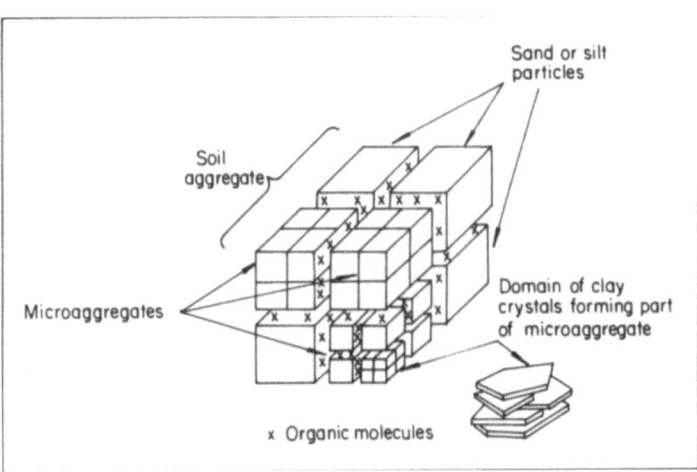

Fig. 1   Schematic illustration of structural organization of soils (85)

MEASUREMENTS OF POROSITY

Total porosity of soil is related to bulk density by the equation

$$\text{Total porosity } (f_t) = 1.0 - \frac{\text{Bulk density}}{\text{Particle density}}$$

Bulk density and particle density are relatively easy to determine, and so total porosity can be readily assessed (40). Changes in total porosity are accompanied by changes in bulk density, and increases of bulk density are often considered to be indications of structural deterioration. It is usually assumed that such changes reflect a loss of the largest, and therefore the weakest, pores in the soil. In most instances this is correct. Compaction of soils usually results in the loss of transmission pores and occasionally of some storage pores. However, shear forces applied to wet soils, in which interparticle separations are greatest and so stability least, can lead to dispersion of clay particles, and more drastic structural changes than are produced by compaction forces. Depending on the nature of the electrostatic forces operating at the clay surfaces, and the way in which the shearing force is applied, either a gel structure may be created, or a particularly well oriented structure (Fig. 2). If a gel structure is formed from an initially partially oriented structure, then the bulk density is likely to decrease, whereas if the structure changes from partially oriented to well oriented, then a very high bulk density will be produced. A gel structure will tend to be formed where interparticle forces are weak, and a well oriented structure when the attractive forces are strong.

Thus mechanical manipulation applied to a wet clay soil may be expected to produce a decrease in bulk density, and an increase in total porosity, if a gel-type structure is created. On the other hand, if the degree of parallel orientation of platy particles is increased, an increase in bulk density and a decrease in porosity can be expected. Different results can therefore arise from wet soil tillage, depending on the nature of the soil colloids present and the ionic concentration of the soil solution, which determine the particle interactions, and the precise method of tillage employed.

The information on clay behaviour has been inferred from studies of clay swelling (41, 42, 43) and the nature of clay fabrics before and after mechanical shear (44). Direct studies of pore size distributions in swollen clay soils have been few. The reason for this has been the lack of appropriate techniques for conducting such studies. For a rigid porous system, the pore size distribution can be derived from the volume of a liquid desorbed from the system when different suctions are applied, as the force required to remove the liquid is related to the size of the pore in which it is held by the capillary rise equation:

$$r = \frac{2\,\gamma\cos\theta}{\rho\,g\,h}$$

where

r = equivalent pore radius
γ = surface tension of liquid
θ = contact angle of liquid with soil particles
ρ = density of the liquid
g = acceleration due to gravity
h = suction, expressed as height of a column of liquid

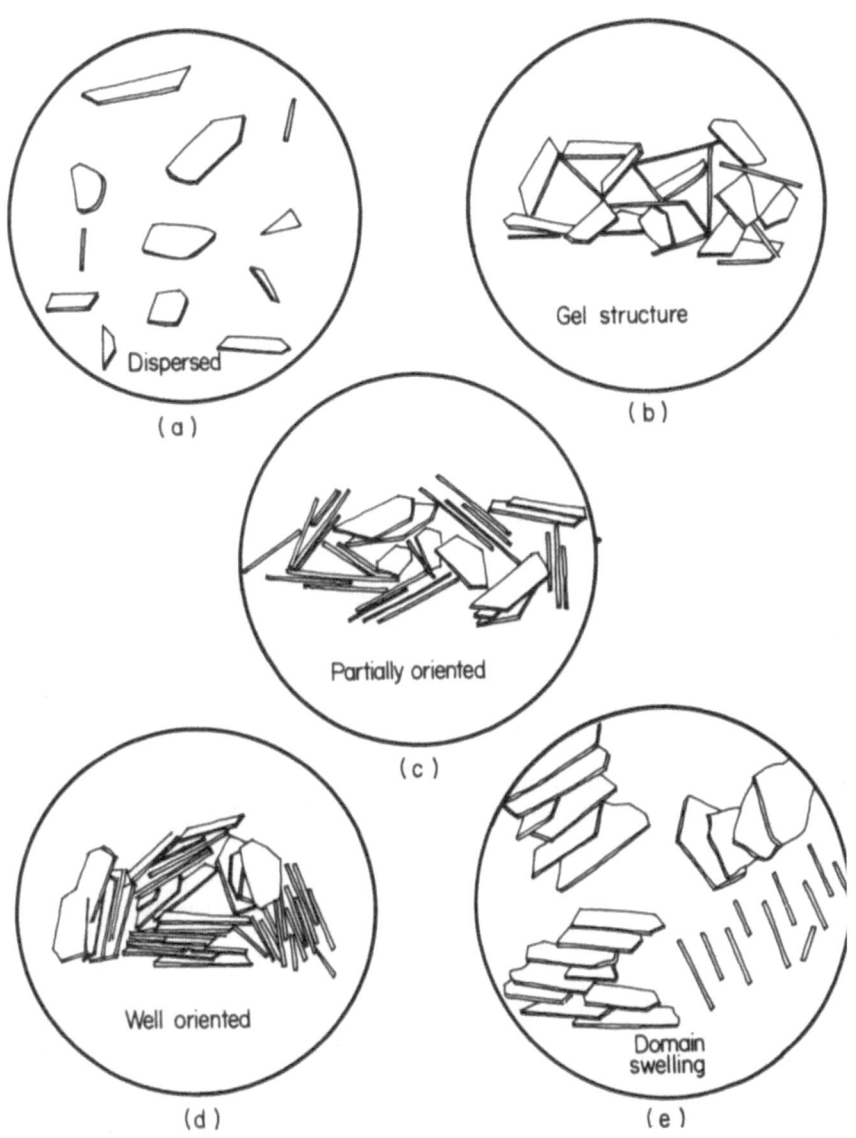

Fig. 2   Clay particle arrangements in disordered, gel-like, structures, and in well oriented domains (43)

The commonly determined moisture characteristic curve of a soil is often used to derive a pore size distribution, since the volume of water extracted between successive suctions should represent the volume contained in pores with radii between limits corresponding to those suctions. Departures from this relationship arise from lack of pore continuity, and hysteresis associated with pore constrictions. For a non-rigid system, a much greater problem arises because the structure changes as water is withdrawn, i.e. the clay shrinks.

To determine the pore size distribution of a swollen clay soil is therefore difficult. It requires the replacement of water by a technique that avoids shrinkage. Greene-Kelly (45) suggested that minimal shrinkage was produced when large soil samples were impregnated with polyethylene glycol solutions of increasing strength and molecular weight. Optical examination of the thin sections of impregnated material enabled the pores larger than about 50 μm to be studied. Automated optical scanning using instruments such as the "Quantimet optical analyser" enables rapid measurement of the number of such pores to be made (46). However even using automated scanning techniques it is difficult to develop statistically valid assessments of pore sizes as they relate to field properties, because of the variability of porosity in the field, and the need to examine very large numbers of thin sections to obtain statistically valid results.

An alternative method of water removal is by critical point drying (47), in which water is successively replaced by liquids of lower dielectric constant (usually methanol-water mixtures) and then methanol by liquid carbon dioxide, and finally the carbon dioxide allowed to evaporate above its critical point. Compared with direct drying which leads to shrinkages of the order of 50 per cent or more, this method causes shrinkages of less than 15 per cent (48). Freeze drying has been suggested as an alternative (6) but has been shown to lead to serious artifacts (48, 49). After water has been removed injection of a non-polar liquid enables a more satisfactory pore size distribution to be obtained. Of the techniques available, mercury injection is probably the most satisfactory (50, 51).

The potential of these techniques has still to be fully explored. At present mercury injection methods have only been applied to studies of aggregate porosity, because it is difficult to develop equipment to handle samples larger than about 5 cm^3. Interaggregate porosities have therefore to be determined by optical methods. Where both optical and water extraction methods have been applied to the same sample, the agreement obtained has not been particularly satisfactory (46). As yet, no comparative studies have been made of porosities determined by both optical methods and critical point drying/mercury injection techniques.

Mercury injection methods can be used to measure pores down to an equivalent cylindrical diameter of about 10 nm. Beyond this point the pressures required for mercury injection become extremely large. Hence porosities corresponding to particle separations of the order of a few molecular layers have been studied primarily by determining the adsorption and capillary condensation of inert gases, usually nitrogen or krypton, at low temperatures (52, 53).

The majority of the pores in soils which are smaller than 20 nm e.c.d. are those between clay crystals within domains. Almost certainly some losses of internal pores occur during the pre-drying that is necessary prior to inert gas adsorption, and this must again cause some structural reorganization (48, 54).

In general water desorption curves indicate that there is a higher proportion of storage pores present than exists in reality. The difference is large for fine textured soils, but small or negligible in sandy soils (55).

STRUCTURAL CHARACTERISTICS OF PADDY SOILS

Puddling of some heavy textured soils leads to an increase in total soil volume, and a decrease in bulk density (56, 57). This is due to the destruction of aggregates and the corresponding loss of inter-aggregate or transmission pores, apparently accompanied by an increase in the inter-microaggregate and interdomain pores. A small increase has also been observed in residual porosity. While there is no doubt about the increase in total porosity, the allocation of the total porosity between different size classes has to be treated with caution, as it was derived from water extraction, which as we have seen is likely to give erroneous results because of clay shrinkage.

The structure developed after cultivating the wet soil depends on the soil texture and the stability of the structural components of the soil. As discussed above, these are related to the surface characteristics of the clay fraction, and the effects of organic matter and hydrous oxides which may form interparticle glues or cements, and the electrolyte concentration of the soil solution. Those soils in which the clay fraction disperses when the soil is worked wet, probably form a random gel-like structure of the individual soil particles. Where domains and microaggregates are more stable, considerably greater mechanical work will be required to develop a well-puddled condition. The dispersion of clay by wet manipulation will usually be followed by reflocculation, and a gradual consolidation or 'ripening' of the structure.

Where iron oxides and hydroxides are involved in the stabilization of the structural units, either alone or through their association with organic matter, changes in the structure of the puddled soil may occur over a period of weeks, as the ferric iron is reduced. As the domains and microaggregates are 'destabilized' their components are able to adopt closer alignments with respect to each other. Thus the soft, puddled soil will tend to "harden," especially if further cultivations producing shear forces in the puddled soil are conducted.

A hypothetical example of the changes in pore size distribution occurring when a clay soil is puddled is shown in Fig. 3. Bulk densities may either increase or decrease on puddling depending on the proportions of transmission pores lost and residual and storage pores gained.

Puddling is unlikely to cause complete dispersion of the clay, unless there is a very high proportion of sodium present on the exchange complex. Nevertheless partial dispersion, which releases sufficient clay to fill many of the pores between the structural units, will be sufficient to create a characteristically 'puddled' structure.

Many of the earlier attempts to characterize the structure of puddled soils in fact described the stability of aggregates formed when the puddled soil was dried (58). The aggregates were in fact usually made by breaking clods, and were weak. Very little critical attention appears to have been given to the important 'restructuring' process in paddy soils which are to be used for upland crops in rotation with rice. Proper management of organic matter level, and of the formation of structurally active ferric hydroxides when the reduced soil is reoxidized, should give opportunities for creating a favourable structural condition after drying from the puddled condition. Many "self-mulching" clay soils, which develop a friable structure on drying, are known. The factors which determine whether a cloddy or friable condition is formed on drying have still to be determined.

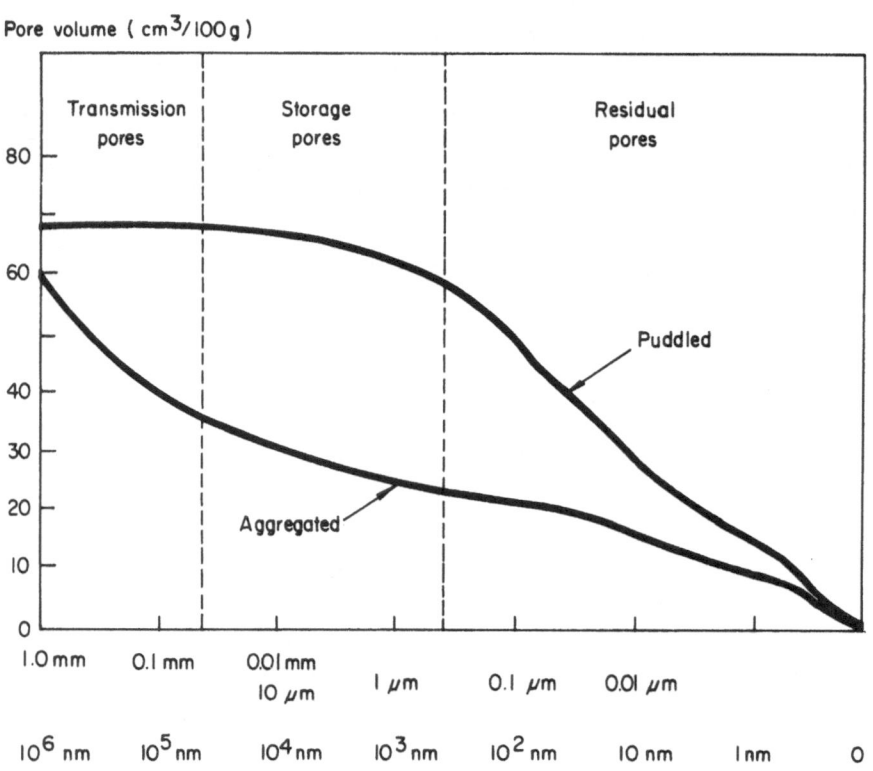

Fig. 3 Hypothetical example of the change in pore size distribution due to puddling of a clay soil. In the aggregated soil particles are closely aligned. After puddling, the clay particles disperse temporarily, and assume a gel structure, in which the particles are not aligned. Thus there is a significant increase in the residual pore volume. Aggregates are destroyed by puddling, and so the interaggregate or 'transmission pore' volume diminishes substantially. The change in storage pore volume will depend on the stability of microaggregates and domains

PORE SIZE DISTRIBUTION IN PADDY SOILS

Few detailed studies of pore size distributions in paddy soils appear
to have been published. Yao Hsiang-Liang et al. (59) and Xiung Yi et al. (60)
have presented data for porosity of several Chinese paddy soils, including
data for aeration (transmission) and total porosity. The aeration porosity
is low in all of the soils. Many papers report bulk densities, and the
effects of puddling or compaction on these values, but as noted above the
important processes related to crop production relate to certain pore sizes,
not to total porosity. The effect of tillage by various methods, and in
wet and dry conditions, on bulk density has been reviewed by several authors.
None of the work they describe includes a study of pore size distribution or
soil strength, or the relation of these properties to clay mineralogy or
organic matter content of the soil (58, 61, 62, 63).

POROSITY AND ROOT GROWTH IN PADDY SOILS

Using artificial media Kar and Ghildyal (64) showed that penetration
of rice roots depended on both the pore diameter of the media and their
rigidity, as Wiersum (65) found for other crop species. They suggested that
a pore size of about 75 µm would allow penetration by the rice root tip. Subsequently by mixing clay, silt and sand sized quartz powders, Kar, Varade and
Ghildyal (66) prepared synthetic "soils" in which the pore size distribution
had a sharp maximum. In the sand and sandy loam soils where there was a
greater frequency of pores with radii larger than 75 µm, root penetration was
most extensive. It was severely restricted in the clay and clay loam. Considerable root penetration into the silt and silty clay loam soils where pore
sizes were also mostly less than 75 µm was interpreted in terms of the greater
total porosity of these soils, and so greater particle mobility and lesser
rigidity of the structure.

Root growth of rice has also been examined by Obermueller and Mikkelsen
(67) using pots filled with sieved aggregates 2-5 mm, or 5-15 mm e.s.d. of the
"Stockton clay adobe," and flooded. The aggregates were stabilized with HPAN
soil conditioner. The soil in one set of pots was not stabilized but puddled.
No data is given for the porosity or density of the aggregates or the puddled
soil, but on the reasonable assumption that an 'adobe' derived clay is highly
porous, it is not surprising that little effect of differences in aggregate
size were observed, and that puddling tended to improve rice growth.

Kar et al. (68) also examined root penetration into a clay, a loam and a
sandy loam compacted to different densities. They found that penetration was
reduced as the bulk density increased, but root volume and number showed a
more complex relationship to bulk density, influenced also by temperature.
Without more detailed information on the relationship between bulk density
and porosity and consistence of the soil, it is difficult to interpret such
studies.

POROSITY AND WATER USE EFFICIENCY IN PADDY SOILS

A major reason for puddling many rice soils is to reduce water losses
by percolation to a low level (69, 70). How low it should be is still uncertain. A minimum rate of 10 mm per day is often considered to be necessary if
high yields are to be obtained. The reason for this percolation rate is
believed to be to ensure that toxic materials are removed from the root zone,
and oxygen rich water allowed to move in. Recirculation of drainage water

has been shown by Ghildyal and Patel (71) to have an adverse effect on yields, presumably due to toxic materials it contains, or which develop if poorly oxygenated water is recirculated. No toxic materials could be identified in the drainage water. Certainly if the redox potential falls too low (below about -100 mV) toxins are likely to be present in the saturated soil, but their formation, and the decrease in E_h, is dependent on the presence of decomposable organic materials in the soil, the oxygen concentration in the water, the rate of respiration of the rice roots, temperature and other factors. Hence it is not surprising that the need for drainage of paddy soils, and the optimum rate of drainage, remain undefined for most paddy soils.

For lighter textured soils which are well aggregated, and which overlie a permeable subsoil, it is obviously important that tillage of the wet soil should be used to remove, or at least destroy the continuity of, the majority of transmission pores in the soil. For heavier soils, where percolation rates are often well below 10 mm per day whether or not they are puddled, the reasons for puddling are less clear, although most studies described in the literature indicate that improved yields are obtained after puddling (62, 63, 69, 70, 72, 73). Possibly the softening of the soil allows greater root proliferation, but this has still to be proved, and is likely to differ between soils. The change in porosity will also influence the rate at which nutrient ions are able to diffuse to the plant roots, a factor which may be of major importance in relation to phosphorus nutrition. Where percolation is significantly reduced, leaching losses of plant nutrients will also be reduced.

The importance of making percolation studies in well characterized field sites needs to be stressed. Water movement in the soil profile is strongly influenced by landscape position, and the position of the natural water table. Transfer of samples to tubes or large drums removes influences of this sort, as well as modifying considerably the existing structure. There is a need for a major study of the influence of drainage and aeration conditions on rice in properly simulated field conditions where water table position is controlled, as well as the oxygen content of water entering and leaving the soil. Results of studies using drums often greatly exceed values obtained by field measurements (74). Differences between saturated hydraulic conductivities measured in the field and the laboratory also differ substantially. Lal found for an Alfisol that the laboratory methods gave results 22 to 40 times greater than determinations made in situ (75).

SOIL CONSISTENCE AND SOIL TRAFFICABILITY

Repeated puddling of the soil, in situations where continuously high groundwater prevents drying of the subsoil, or where continuous year round irrigation is practiced and drainage poor, can lead to the soil becoming so soft that it is difficult to cultivate, because people, animals and implements sink too deeply into the mud. In Japan, the generally low bulk densities (below 0.9 g cm^{-3}) developed in less well drained paddy soil regions, distinguish such soils from those which are better drained (76). In these poorer situations a period free of irrigation to allow the soil to dry to a sufficient depth is usually recommended (77, 78). Alternatively attempts may be made to create a 'harder' consistence by manipulation of the soil at the appropriate water content (79). A simpler process which is used at the International Rice Research Institute is to alternate cultivation by a buffalo-pulled plough, which over a period of several years tends to soften the soil until the animal sinks to its belly, with cultivation using light power tillers, which creates a hardened plough pan at relatively shallow depth.

At the present time this type of soil manipulation is an art, and not a science. In none of the reviews of tillage methods and soil physical properties published in recent years has any attempt been made to relate the information obtained to specific soil characteristics, although Sanchez attempted to assess the earlier literature in relation to soil type (58). A considerable volume of empirical data is accumulating, but because the essential details of soil and land characteristics which might enable it to be understood, and used to develop a proper scientific understanding of paddy soil behaviour are not determined and recorded, it is of limited value.

NEEDED RESEARCH RELATED TO THE MANAGEMENT OF THE PHYSICAL PROPERTIES OF PADDY SOILS

The objective of research on the management of the physical properties of paddy soils should be to develop a rationale which will enable optimum management procedures to be related to soil characteristics. Tillage methods and water management, including drainage before or during the growth of the rice crop, as well as the desirability of the use of organic amendments, are largely based on empirical knowledge. At present excessive tillage and much waste of water occurs in many areas because the objectives of tillage in terms of producing changes in soil structure and soil physical conditions have not been defined. The response of different soils to different tillage procedures differs greatly. Emerson (80) described a simple system for the classification of soils according to their physical behaviour when wet, and subsequently showed (81) how this classification could be applied to predict the response of soils to wet compaction.

In many parts of Asia there is an acute need to increase food crop production. Because of land shortages this requires more intensive use of land on which rice is grown. This frequently involves production of an upland crop on land which has previously been puddled and a rice crop grown in flooded conditions. As the upland crop is grown at the end of the monsoon, it may be largely dependent on residual moisture in the soil. Unless the structural condition of the soil is such that roots can develop adequately and moisture be released to them, the crop is likely to be a failure. Systematic studies of the suitability of paddy soils for production of upland crops after rice are few. Morris and Zandstra (82) have discussed some physical factors of landscape and climate related to this, but physical characteristics of the soil have received limited attention.

A recent conference at the International Institute of Tropical Agriculture in Ibadan, Nigeria, reviewed much of the information available regarding soil physical conditions and production of upland crops (83). Although a range of studies of topics such as aggregate stability and the influence on it of organic matter have been reported, there is a great need for more information regarding topics such as the significance of short term water logging and drought incidence on production of crops. Such problems are likely to occur frequently in soils which have previously been puddled for rice production. Wien, Lal and Pulver (84) found that the yield of cowpeas flooded twice during their growth period was reduced by 91 per cent. The cloddy condition typical of many paddy soils after drying is unsuited to root proliferation, so that crops are more liable to suffer from drought than in soils where conditions favour development of an extensive root system.

Undoubtedly tillage, artificial drainage and modified irrigation management methods can alter soil physical properties, not only in relation to conditions during the production of the rice crop, but also in relation to the condition of the soil after the rice crop has been harvested (77).

Soil properties, such as proportion and nature of the clay fraction and presence of salts or calcium carbonate, which can influence the 'restructuring' process have been little investigated. A systematic study of soil characteristics in relation to the redevelopment of a structure favourable to upland crop production after paddy rice is badly needed.

Techniques are now available which enable us to determine the porosity of clays in the wet state, and the influence of tillage on that structure. The response of the soil differs considerably according to its mineralogy. Critical studies of the changes in the structural condition of different paddy soils, before and after puddling, and before and after drainage, should enable a better basis to be established for recommending appropriate management techniques for rice production, and also the establishment and production of upland crops grown in succession to paddy rice. Undoubtedly Chinese experience in soil management for paddy rice production can provide a wealth of information not presently available to observers from other parts of the world. Hopefully exchange of experiences can be mutually advantageous to all concerned with improvement of the productivity of paddy soils.

ACKNOWLEDGEMENT

I am grateful to Dr. P.A. Sanchez for helpful comments in the preparation of this paper.

REFERENCES

(1) Marshall, T.J., The nature, development and significance of soil structure. Trans. Comm. IV and V, Int. Soc. Soil Sci., New Zealand, 243-257 (1962).
(2) Dexter, A.R., Hewitt, J.S., The structure of beds of spherical particles. J. Soil Sci., 29, 146-155 (1978).
(3) Kanno, I., Genesis of rice soils with special reference to profile development. Soils and Rice, International Rice Research Institute, Philippines. 237-253 (1978).
(4) Moormann, F.R., Morphology and classification of soils on which rice is grown. Soils and Rice, International Rice Research Institute, Philippines, 255-272 (1978).
(5) Aylmore, L.A.G., Quirk, J.P., Domains and quasicrystalline regions in clay systems. Soil Sci. Soc. Amer. Proc., 35, 652-654 (1971).
(6) Quirk, J.P., Some physico-chemical aspects of soil structural stability -- a review. Modification of Soil Structure. Wileys, Chichester, 3-16 (1978).
(7) Israelachivilli, J.N., Ninham, B.W., Intermolecular forces -- the long and short of it. J. Colloid and Interface Sci. 58 (1), 14-25 (1977).
(8) Arnold, P.W., Surface-electrolyte interactions. The Chemistry of Soil Constituents. Wileys, Chichester, 355-404 (1978).
(9) Adams, G.E., Israelachivilli, J.N., Measurement of forces between two mica surfaces in aqueous potassium nitrate solutions. Modification of Soil Structure. Wileys, Chichester, 27-34 (1978).
(10) Greenland, D.J., Rimmer, D., Payne, D., Determination of the structural stability class of English and Welsh soils, using a water coherence test. J. Soil Sci., 26, 294-303 (1975).
(11) Ahn, P.M., Microaggregation in tropical soils: its measurement and effects on the maintenance of soil productivity. Soil Physical Conditions and Crop Production in the Tropics. Wileys, Chichester, 75-86 (1979).
(12) Tama, K., El-Swaify, S.A., Charge, colloidal and structural stability interrelationships for oxidic soils. Modification of Soil Structure. Wileys, Chichester, 41-49 (1978).

(13) Moormann, F.R., Representative toposequences of soils in southern Nigeria and their pedology. Characterisation of Soils in Relation to their Classification and Management for Crop Production: Examples from Some Areas of the Humid Tropics. Oxford Univ. Press, Oxford, 10-29 (1980).

(14) Hingston, F.J., Posner, A.M., Quirk, J.P., Anion adsorption by goethite and gibbsite. I. The role of the proton in determining adsorption envelopes. J. Soil Sci. 23, 177-192 (1972).

(15) Koenigs, F.F.R., The mechanical stability of clay soils as influenced by moisture conditions and some other factors. Versl. landb. oriderz. 67.7. (1961).

(16) Koenigs, F.F.R., The puddling of clay soils. Neth. J. Agric. Sci. 11, 145-156 (1963).

(17) Kita, D., Kawaguchi, K., The effects of both the reduction of the soil under waterlogged condition and the dehydration of the reduced soil upon soil structure. J. Sci. Soil Manure, Japan, 31, 375-379 and 495-498 (1960).

(18) Ahmed, N., The effect of evolution of gases and reducing conditions in a submerged soil on its subsequent physical status. Trop. Agric. (Trin.), 40,205-209 (1963).

(19) Hughes, J.C., Mineralogy. Characterisation of Soils in Relation to their Classification and Management for Crop Production: Examples from Some Areas of the Humid Tropics. Oxford Univ. Press, Oxford (1980).

(20) Deshpande, T.L., Greenland, D.J., Quirk, J.P., Changes in soil properties associated with the removal of iron and aluminium oxides. J. Soil Sci., 19, 108-122 (1968).

(21) Habibullah, A.K.M., Greenland, D.J., Brammer, H.M., Clay mineralogy of some seasonally flooded soils of East Pakistan. J. Soil Sci., 22, 179-190 (1971).

(22) Kyuma, K., Mineral composition of rice soils. Soils and Rice, International Rice Research Institute, Philippines. 219-236 (1978).

(23) Brinkman, R., Ferrolysis, a soil forming process in hydromorphic conditions. Pudoc, Wageningen (1979).

(24) Allison, F.E., Soil organic matter and its role in crop production. Elsevier, Amsterdam, 637 (1973).

(25) Greenland, D.J., Lindstrom, G.R., Quirk, J.P., Organic materials which stabilise natural soil aggregates. Soil Sci. Soc. Amer. Proc. 26, 366-371 (1962).

(26) Stefanson, R., Effect of periodate and pyrophosphate on the seasonal changes in aggregate stabilization. Aust. J. Soil Research 9, 33-42 (1971).

(27) Hamblin, A.P., Greenland, D.J., Effect of organic constituents and complexed metal ions on aggregate stability of some East Anglian soils. J. Soil Sci. 28, 410-416 (1977).

(28) Greenland, D.J., Interaction between clays and organic compounds in soils. Part I. Mechanisms of interaction between clays and defined organic compounds. Soils and Fert. 28, 415-425 (1965).

(29) Theng, B.K.G., The Chemistry of Clay-Organic Reactions. London, Hilger. (1974).

(30) Theng, B.K.G., Formation and properties of clay-polymer complexes. Elsevier Scientific Pub. Co., Amsterdam and New York. Developments in Soil Science 9, 362 (1979).

(31) Parfitt, R.L., Greenland, D.J., Adsorption of polysaccharides by montmorillonite. Soil Sci. Soc. Amer. Proc. 34, 862-866 (1970).

(32) Greenland, D.J., Adsorption of polyvinyl alcohols by montmorillonite. J. Colloid Sci. 18, 647-664 (1963).

(33) Emerson, W.W., Raupach, M., The reaction of polyvinyl alcohol with montmorillonite. Aust. J. Soil Res. 2, 46-55 (1964).
(34) Parfitt, R.L., Greenland, D.J., Adsorption of poly(ethylene glycols) by
clay minerals. Clay Minerals 8, 305-315 (1970).
(35) Edwards, A.P., Bremner, J.M., Dispersion of soil particle by sonic vibration; Microaggregates in soils. J. Soil Sci. 18, 45-63; 64-73 (1967).
(36) Greenland, D.J., Interactions between humic and fulvic acids and clays.
Soil Sci. 111, 34-41 (1971).
(37) Emerson, W.W., The effect of polymers on the swelling of montmorillonite.
J. Soil Sci. 14, 52-63 (1963).
(38) Theng, B.K.G., Greenland, D.J., Quirk, J.P., Swelling in water of complexes
of montmorillonite with polyvinyl alcohol. Aust. J. Soil Res. 5, 69-76
(1967).
(39) Turchenek, L.W., Oades, J.M., Organo-mineral particles in soils. Modification of Soil Structure. Wileys, Chichester, 137-144 (1978).
(40) McIntyre, D.S., Pore space and aeration determinations. Methods for
Analysis of Irrigated Soils. Commonwealth Agric. Bureau, Farnham Royal,
Bucks, 67-74 (1974).
(41) Croney, D., Coleman, J.D., Soil structure in relation to soil suction
(pF). J. Soil Sci. 5, 75-84 (1954).
(42) Holmes, J.W., Water sorption and the swelling of clay blocks. J. Soil
Sci. 6, 200-208 (1955).
(43) Aylmore, L.A.G., Quirk, J.P., The structural status of clay systems. Clay
and Clay Miner. 9, 104-130 (1962).
(44) Smart, P., Electron microscope methods in soil micromorphology. Proc. 4th
Int. Working Meeting, Soil Micromorphology, Kingston, Ontario (1973).
(45) Greene-Kelly, R., The shrinkage of clay soils during impregnation by polyethylene glycols. J. Soil Sci. 22, 191-202 (1971).
(46) Bullock, P., Thomasson, A.J., Rothamsted studies of soil structure.
II. Measurement and characterisation of macroporosity by image analysis
and comparison with data from water retention measurements. J. Soil Sci.
30 (3), 391-413 (1979).
(47) Greene-Kelly, R., The preparation of clay soils for the determination of
structure. J. Soil Sci. 24, 277-283 (1973).
(48) Lawrence, G.P., Payne, D., Greenland, D.J., Pore size distribution in critical point and freeze dried aggregates from clay subsoils. J. Soil Sci.
30, 499-516 (1979).
(49) Murray, R.S., Quirk, J.P., Clay-water interactions and the mechanism of
soil swelling. Colloids and Surfaces 1, 17-32 (1980).
(50) Lawrence, G.P., Measurement of pore sizes in fine-textured soils: a
review of existing techniques. J. Soil Sci. 28, 527-540 (1977).
(51) Newman, A.C.D., Thomasson, A.J., Rothamsted studies of soil structure.
III. Pore size distributions and shrinkage processes. J. Soil Sci. 30
(3), 415-439 (1979).
(52) Gregg, S.J., Sing, K.S.W., Adsorption, surface area and porosity. Academic
Press, London, (1967).
(53) Greenland, D.J., Mott, C.J., Surfaces of soil particles. The Chemistry of
Soil Constituents. Wileys, Chichester, 321-355 (1978).
(54) Aylmore, L.A.G., Quirk, J.P., The micropore size distribution of clay
mineral systems. J. Soil Sci. 18, 1-17 (1967).
(55) Nagpal, N.K., Boersma, L., DeBacker, L.W., Pore size distributions of soils
from mercury-intrusion porosimeter data. Soil Sci. Soc. Amer. Proc. 36,
264-267 (1972).
(56) Campbell, R.B., Freezing point of water in puddled and unpuddled soils at
different moisture tension values. Soil Sci. 73, 221-229 (1952).

(57) Jamison, V.C., Changes in air-water relationships due to structural improvement of soils. Soil Sci. 76, 143-151 (1953).

(58) Sanchez, P.A., Rice performance under puddled and granulated soil cropping systems in southeast Asia. Ph.D. thesis, Cornell Univ., New York, (1968).

(59) Yao Hsiang-liang, Chao Wei Ching, Yu Teh-fen, Hsu Hsiu-yun, Preliminary investigation of structural characteristics of fertile paddy soil. Acta Pedologica Sinica 15, 12-22 (1978).

(60) Xiung Yi, Xu Qi, Yao Xian-liang, Zhu Zhao-liang, Effect of cropping system on the fertility of paddy soils. Acta Pedologica Sinica 17, 116-119 (1980).

(61) Curfs, H.P.F., System development in agricultural mechanization with special reference to soil tillage and weed control -- a case study for West Africa. H. Veenman and Zonen B.V. Wageningen, 179 (1976).

(62) Ghildyal, B.P., Soil water flux and evapotranspiration in the presence of a shallow water table in a Mollisol. Soil Physical Conditions and Crop Production in the Tropics. Wileys, Chichester, 159-172 (1979).

(63) Wickham, T.H., Singh, V.P., Water movement through wet soils. Soils and Rice, International Rice Research Institute, Philippines. 337-360 (1978).

(64) Kar, S., Ghildyal, B.P., Rice root growth in relation to size, quantity and rigidity of pores. Plant and Soil 43, 627-637 (1975).

(65) Wiersum, L.K., The relationship of the size and structural rigidity of pores to their penetration by roots. Plant and Soil 9, 75-85 (1957).

(66) Kar, S., Varade, S.B., Ghildyal, B.P., Pore size distribution and root growth relations of rice in artificially synthesized soils. Soil Sci. 128 (6), 364-368 (1979).

(67) Obermueller, A.J., Mikkelsen, D.S., Effects of water management and soil aggregation on the growth and nutrient uptake of rice. Agron. J. 66, 627-632 (1974).

(68) Kar, S., Varade, S.B., Subramanyam, T.K., Ghildyal, B.P., Soil physical conditions affecting rice root growth: bulk density and submerged soil temperature regime. Agron. J. 68 (1), 23-26 (1976).

(69) Sanchez, P.A., Puddling tropical rice soils. Soil Sci. 115, 149-158 and 303-308 (1973).

(70) De Datta, S.K., Kerim, M.S.A.A.A., Water and nitrogen economy of rainfed rice as affected by soil puddling. Soil Sci. Soc. Amer. Proc. 38 (3), 515-518 (1974).

(71) Ghildyal, B.P., Patel, C.L., Effect of varying drainage conditions on water use and growth of rice. Soil Physical Conditions and Crop Production in the Tropics. Wileys, Chichester, 199-204 (1979).

(72) Savant, N.K., De Datta, S.K., Movement and distribution of ammonium-N following deep placement of urea in a wetland rice soil. Soil Sci. Soc. Amer. 44 (3), 559-565 (1980).

(73) De Datta, S.K., Barker, R., Land preparation of rice soils. Soils and Rice, International Rice Research Institute, Philippines. 623-648 (1978).

(74) Sanchez, P.A., Properties and management of soils in the tropics. Wileys, New York. 420 (1976).

(75) Lal, R., Physical characteristics of soils of the tropics; determination and management. Soil Physical Properties and Crop Production in the Tropics. Wileys, Chichester, 7-44 (1979).

(76) Terasawa, S., Physical properties of paddy soil in Japan. JARQ 9 (1), 18-23 (1975).

(77) Maeda, K., Minami, M., Studies on physical and chemical properties and improvement of soil productivity in heavy clayey paddy fields. II. Drainage acceleration technique in ill-drained paddy fields. Bull. Hokkaido Prefect. Agric. Expt. Stn. No. 37, (English summary), 34 (1977).

(78) Anyoji, H., Improvement of soil hardness in paddy fields for the mechanization of harvesting -- study in the Muda irrigation project area, Malaysia. Bull. Nat. Res. Inst. of Agric. Engg. No. 17, (English summary), 21-22 (1978).

(79) Fujio, F., Rheological properties of paddy soil on the sedimentation and hardening by kneading of paddy soil in the paste. Soil Physical Conditions and Plant Growth, Japan. (1975).

(80) Emerson, W.W., A classification of soil aggregates based on their coherence in water. Aust. J. Soil Res. 5, 47-57 (1967).

(81) Emerson, W.W., Aggregate classification and the hydraulic conductivity of compacted subsoils. Modification of Soil Structure. Wileys, Chichester, 239-248 (1979).

(82) Morris, R.A., Zandstra, H.G., Soil and climatic determinants in relation to cropping patterns. Proc. Int. Rice Research Conf., International Rice Research Institute, Philippines (1978).

(83) Lal, R., Greenland, D.J., Soil physical properties and crop production in the tropics. Wileys, Chichester (1979).

(84) Wien, C., Lal, R., Pulver, E.L., Effects of transient flooding on growth and yield of some tropical crops. Soil Physical Properties and Crop Production in the Tropics. Wileys, Chichester, 235-248 (1979).

(85) Williams, B.G., Greenland, D.J., Quirk, J.P., Adsorption of polyvinyl alcohol by natural soil aggregates. Aust. J. Soil Res. 4, 131-143 (1966).

SOME ASPECTS OF THE PHYSICAL CHEMISTRY OF PADDY SOILS

F. N. Ponnamperuma
(The International Rice Research Institute
Los Baños, Laguna, Philippines)

INTRODUCTION

Paddy soils are soils that are managed in a special way for the wet cultivation of rice. The management practices include flooding, puddling, maintaining a layer of standing water while the crop is on the land, draining and drying the fields, and reflooding for the next rice crop. Paddy soils are usually Entisols and Inceptisols and occur in landscapes where both surface and internal drainage are poor. Thus most paddy soils are found in deltas and adjacent flood plains, in valleys, and coastal plains. The chemical composition of paddy soils is variable but their hydrology and management practices confer on them many distinct properties (1, 2).

During the period of flooding or soil submergence, the oxygen supply to the soil is virtually cut off. Within a day or two of flooding, aerobic microbes use up the trapped oxygen and render the soil (except a thin surface layer) virtually free of molecular oxygen. Then facultative and anaerobic organisms use oxidized soil components as electron acceptors in their respiration reducing the soil in the sequence predicted by thermodynamics (Table 1). Draining and drying reverse those changes.

Soil reduction is accompanied by important chemical and physicochemical processes:

1. decrease in redox potential (Eh) or pE
2. changes in pH
3. changes in specific conductance
4. denitrification
5. accumulation of ammonium and fixation of nitrogen gas
6. reduction of Mn(IV)
7. reduction of Fe(III)
8. reduction of sulfate
9. production of organic acids
10. accumulation of carbon dioxide
11. changes in concentration of water-soluble iron, manganese, phosphorus, silicon, boron, copper, molybdenum, and zinc (3, 4).

These processes in paddy soils will be discussed under the following headings:

1. Electrochemical changes
2. Redox equilibria
3. Mineral equilibria
4. Ion exchange
5. Sorption and desorption
6. Chemical kinetics
7. Formation of organo-metallic complexes.

Table 1   Some redox systems in surface media

System	E_o (V)	pE_o	pE_o^7
$1/4 \ O_{2g} + H^+_{aq} + e = 1/2 \ H_2Ol$	1.229	20.80	13.80
$1/5 \ NO^-_{3aq} + 6/5 \ H^+_{aq} + e = 1/10 \ N_{2g} + 3/5 \ H_2Ol$	1.245	21.06	12.66
$1/2 \ NO^-_{3aq} + H^+_{aq} + e = 1/2 \ NO^-_{2aq} + 1/2 \ H_2Ol$	0.834	14.11	7.11
$1/2 \ MnO_{2c} + 2 \ H^+_{aq} + e = 1/2 \ Mn^{2+}_{aq} + H_2Ol$	1.229	20.80	6.80
$1/2 \ CH_3COCOOHaq + H^+_{aq} + e = 1/2 \ CH_3CHOHCOOHaq$	0.256	4.33	-2.67
$Fe(OH)_{3s} + 3 \ H^+_{aq} + e = Fe^{2+}_{aq} + 3 \ H_2Ol$	1.057	17.87	-3.13
$1/2 \ CH_3CHOaq + H^+_{aq} + e = 1/2 \ CH_3CH_2OHaq$	0.221	3.74	-3.26
$1/8 \ SO^{2-}_{4 \ aq} + 5/4 \ H^+_{aq} + e = 1/8 \ H_2Saq + 1/2 \ H_2Ol$	0.303	5.12	-3.63
$1/8 \ CO_{2g} + H^+_{aq} + e = 1/8 \ CH_{4g} + 1/4 \ H_2Ol$	0.169	2.86	-4.14
$1/6 \ N_{2g} + 4/3 \ H^+_{aq} + e = 1/3 \ NH^+_{4aq}$	0.274	4.64	-4.69
$1/8 \ HPO^{2-}_{4aq} + 5/4 \ H^+_{aq} + e = 1/8 \ PH_{3g} + 1/2 \ H_2Ol$	0.212	3.59	-5.16
$1/2 \ NADP^+_{aq} + 1/2 \ H^+_{aq} + e = 1/2 \ NADPHaq$	-0.106	-1.79	-5.29
$1/2 \ NAD^+_{aq} + 1/2 \ H^+_{aq} + e = 1/2 \ NADHaq$	-0.123	-2.08	-5.58
$H^+_{aq} + e = 1/2 \ H_{2g}$	0.000	0.00	-7.00
$Ferredoxin(ox)_{aq} + e = Ferredoxin(red)_{aq}$	-0.432	-7.31	-7.31

Source: Ponnamperuma (1972)

ELECTROCHEMICAL CHANGES IN PADDY SOILS

Changes in Eh or pE

Soil Eh.   A decrease in redox potential is the most striking electrochemical change caused by flooding a soil. Eh falls sharply upon flooding,
reaches a minimum within a few days, rises rapidly to a maximum, and then
decreases asymptotically with time (Fig. 1). Presence of organic matter, a
low content of nitrate and manganese dioxide, and a temperature of 35°C
favor the decrease in Eh, and a value as low as -0.25 V may be attained
within 2 weeks of flooding. Influence of soil properties on Eh decrease is
shown in Figure 2 and of temperature in Figure 3. Soils low in active iron
and manganese and high in organic matter showed the quickest Eh decrease.
Adding nitrate or manganese dioxide to reduced soils raised Eh (5).

At first sight Eh measurement appears to be the simplest, quickest, and
most meaningful method of characterizing soil reduction. But in practice
intrinsic and extrinsic errors deprive soil Eh measurements of theoretical or
practical significance. The errors include electrode malfunctioning, soil
heterogeneity, lack of reproducibility, liquid junction potential error, and
the difference between soil Eh and solution Eh. For these and other reasons,
Ponnamperuma stated that soil redox potentials were of little diagnostic
value in rice culture. But Yu and Liu found that strongly reducing conditions
depressed the growth of rice; Aomine reported that a mosaic of high and low
redox spots in rice fields benefited rice; and Ponnamperuma proposed that an
Eh of +0.2 to -0.2 V is good for rice (6-11).

Soil solution Eh or pE.   In contrast to soil, the interstitial solution
in a paddy soil is a homogeneous phase in dynamic equilibrium with the solid
and gas phases of the soil, gives reproducible potentials, and reflects the
changes in activities of the ionic species that affect Eh or pE. So soil
solution measurements are thermodynamically meaningful and can be used to
study redox equilibria in paddy soils.

The   solutions of submerged soils have Eh values of -60 to 170 mV or pE
values of -1 to +3 at pH 7. The optimum soil solution Eh for rice is 10 to
120 mV or pE 0.2 to 2.0 at pH 7.0 (11).

Theoretical relationships

For the equilibrium,

$$Ox + ne + mH^+ \rightleftharpoons Red \tag{1}$$

$$Eh = Eo + \frac{RT}{nF} \ln \frac{(Ox)}{(Red)} + m \frac{RT}{nF} \ln(H^+) \tag{2}$$

or $$Eh = Eo + 2.303 \frac{RT}{nF} \log \frac{(Ox)}{(Red)} - \frac{2.303\ RTm}{nF} pH \tag{3}$$

or $$pE = pEo - \frac{1}{n} p(Ox) + \frac{1}{n} p(Red) - \frac{m}{n} pH \tag{4}$$

where (Ox) and (Red) are activities of the oxidized and reduced phases,
$pE = Eh/2.303\ RTF^{-1} = Eh/0.0591$ at 25°C, and the other symbols have their
usual significance.

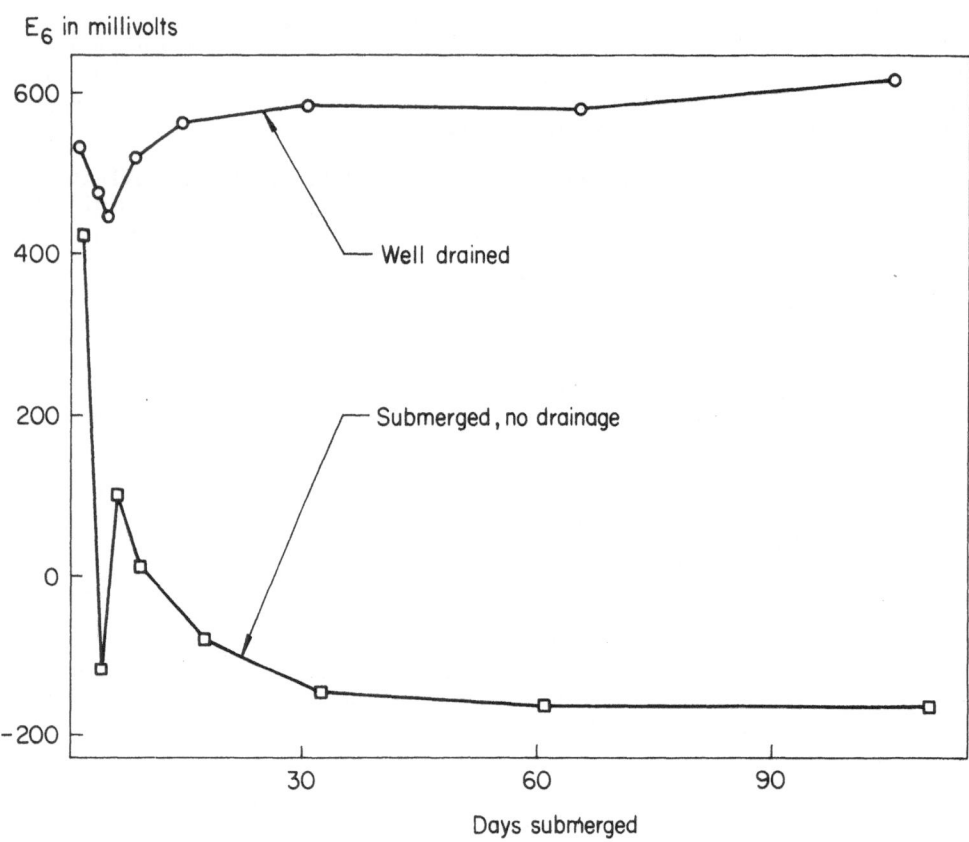

Fig. 1   Changes in redox potential of a well-drained and a submerged soil with time

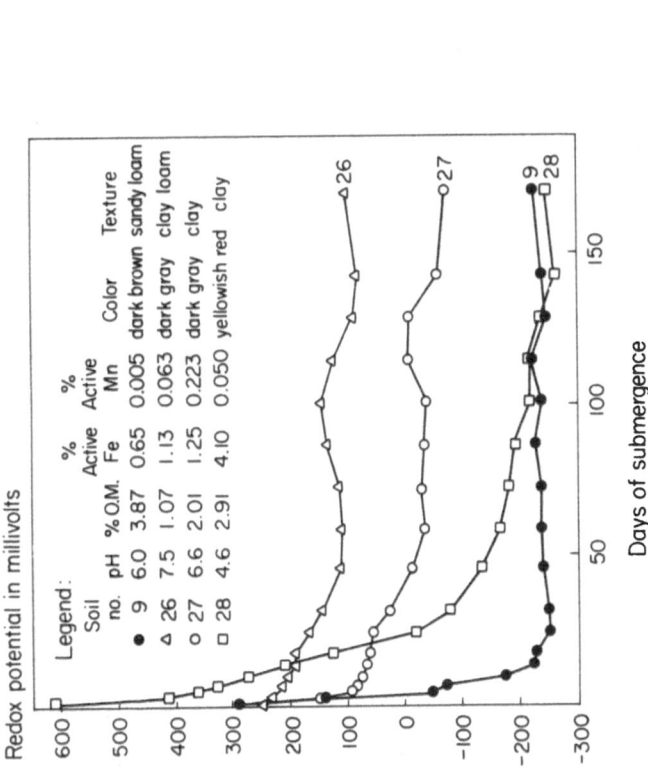

Redox potential in millivolts
600
500
400
300
200
100
0
-100
-200
-300
Legend:
Soil no. pH % O.M. % Active Fe % Active Mn Color Texture
● 9 6.0 3.87 0.65 0.005 dark brown sandy loam
△ 26 7.5 1.07 1.13 0.063 dark gray clay loam
○ 27 6.6 2.01 1.25 0.223 dark gray clay
□ 28 4.6 2.91 4.10 0.050 yellowish red clay
26
27
9
28
50
100
150
Days of submergence

Fig. 2 Changes in the redox potential of four soils kept submerged

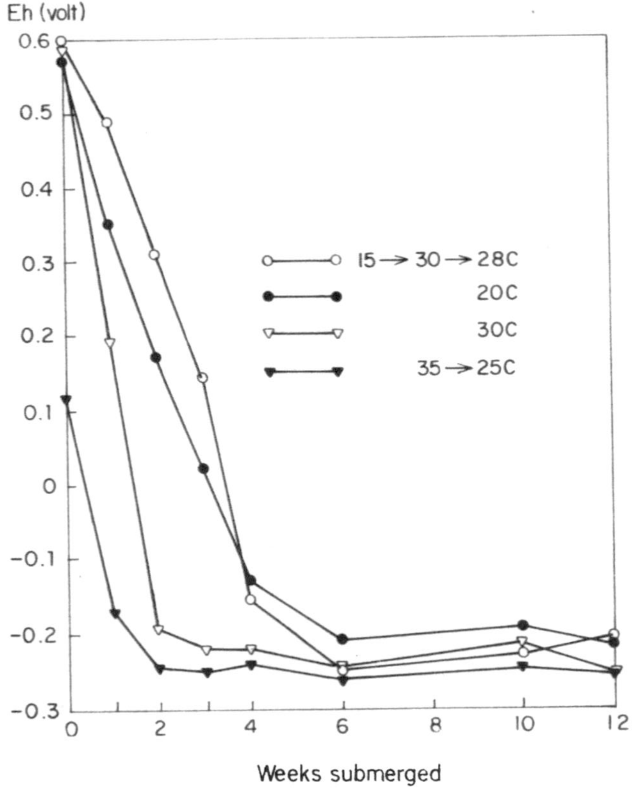

Eh (volt)
0.6
0.5
0.4
0.3
0.2
0.1
0
-0.1
-0.2
-0.3
15 → 30 → 28C
20C
30C
35 → 25C
0
2
4
6
8
10
12
Weeks submerged

Fig. 3 Influence of temperature on the kinetics of Eh in Luisiana clay

Because Eh depends on pH in reactions involving H^+ ions, the term "rH" was proposed as a convenient way of expressing oxidation-reduction intensity without having to specify both Eh and pH.

$$rH = - \log P_{H_2}$$

The misuse of rH prompted Clark, who introduced the concept, to say "rH has become an unmitigated nuisance" (12).

The decrease in Eh on flooding is due to a decrease in activity of the oxidized phase, an increase in activity of the reduced phase, or an increase in pH. If the oxidized and reduced phases in equilibrium are solids, then the middle term in Eqs. 3 and 4 vanishes and Eh or pE varies inversely as pH.

The decrease in Eh or pE of most acid soils not high in organic matter can be explained quantitatively in terms of the potential of the $Fe(OH)_3$-Fe^{2+} system (5, 13).

$$Fe(OH)_3 + 3 H^+ + e = Fe^{2+} + 3 H_2O \qquad (5)$$

$$\text{with } Eh = 1.06 - 0.059 \log Fe^{2+} - 0.177 \text{ pH} \qquad (6)$$

$$\text{or } \quad pE = 17.87 + pFe^{2+} - 3 \text{ pH} \qquad (7)$$

Changes in pH

The pH of acid soils increases on flooding whereas the opposite occurs in calcareous and sodic soils (Fig. 4). For most soils the fairly stable pH value attained after several weeks of submergence is between 6.7 and 7.2 in the soil or 6.5 and 7.0 in the soil solution. Yu and Liu reported that the final pH of waterlogged soils with initial pH values of 4.5 to 8.0 was 6.3-7.5. Peat soils and some acid sulfate soils may have pH values of 5 even in the flooded state. The pH increase in iron-deficient soils is small (3, 5, 8).

Soils high in organic matter and in reducible iron attain a pH of 6.5 within a few weeks of flooding at temperatures above 30°C. Low temperature retards the pH increase in acid soils (Fig. 5).

The increase in pH of most acid soils is largely due to the reduction of Fe(III) to Fe(II) and can be described quantitatively by the equations (13):

$$Eh = 1.06 - 0.059 \log Fe^{2+} - 0.177 \text{ pH} \qquad (6)$$

$$\text{or } \quad pE = 17.87 + pFe^{2+} - 3 \text{ pH} \qquad (7)$$

Although the increase in pH of acid soils is brought about by soil reduction, the fairly stable pH attained after a few weeks of flooding is regulated by the partial pressure of carbon dioxide (P_{CO_2}). For ferruginous soils the empirical relationship is

$$pH = 6.1 - 0.58 \log P_{CO_2} \qquad (8)$$

The pH values of flooded calcareous and sodic soils are lower than those of the well drained soils because of carbon dioxide accumulation. The pH values of calcareous and alkali soils are highly sensitive to P_{CO_2} as shown below.

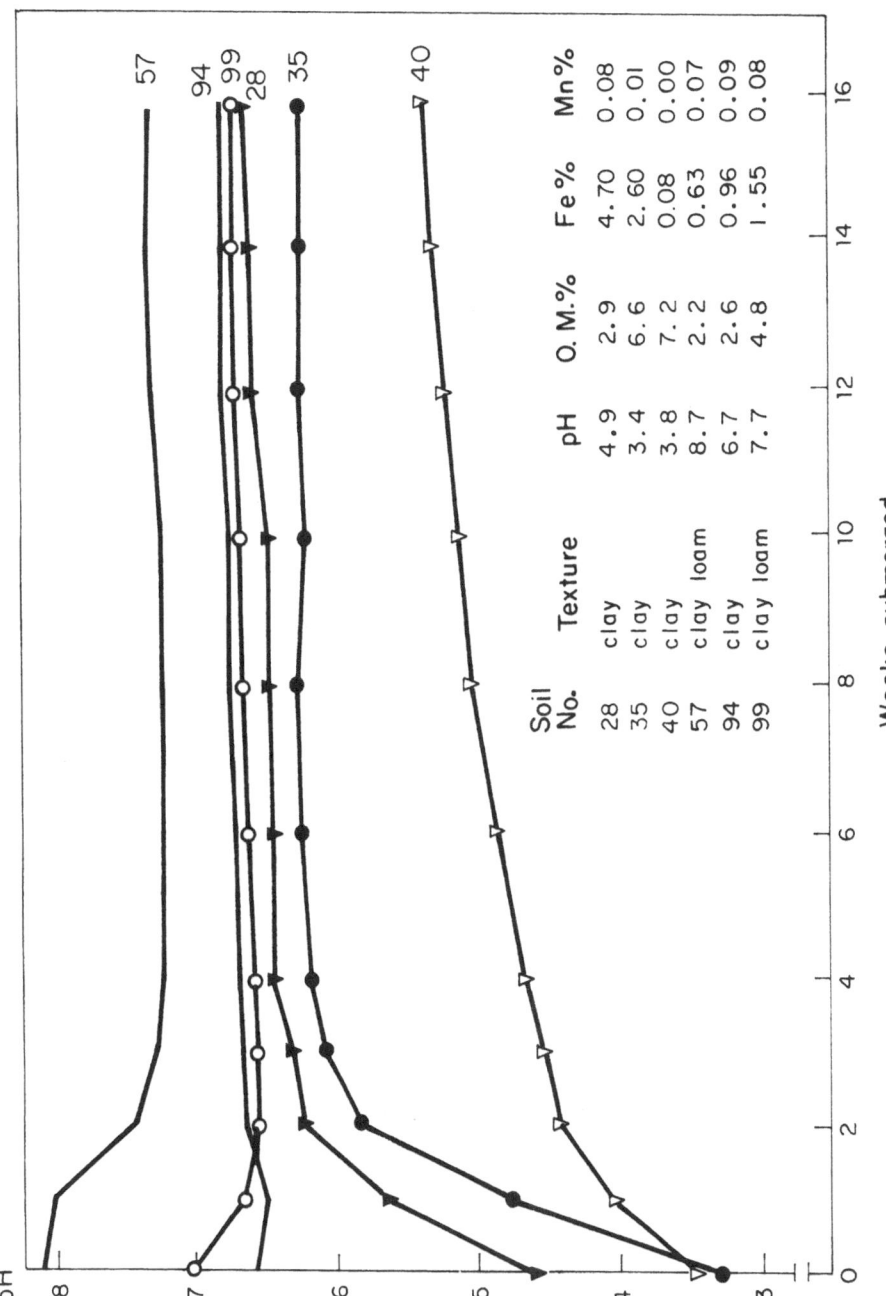

Fig. 4 Kinetics of the pH values of some submerged soils

Soil No.	Texture	pH	O. M.%	Fe%	Mn%
28	clay	4.9	2.9	4.70	0.08
35	clay	3.4	6.6	2.60	0.01
40	clay	3.8	7.2	0.08	0.00
57	clay loam	8.7	2.2	0.63	0.07
94	clay	6.7	2.6	0.96	0.09
99	clay loam	7.7	4.8	1.55	0.08

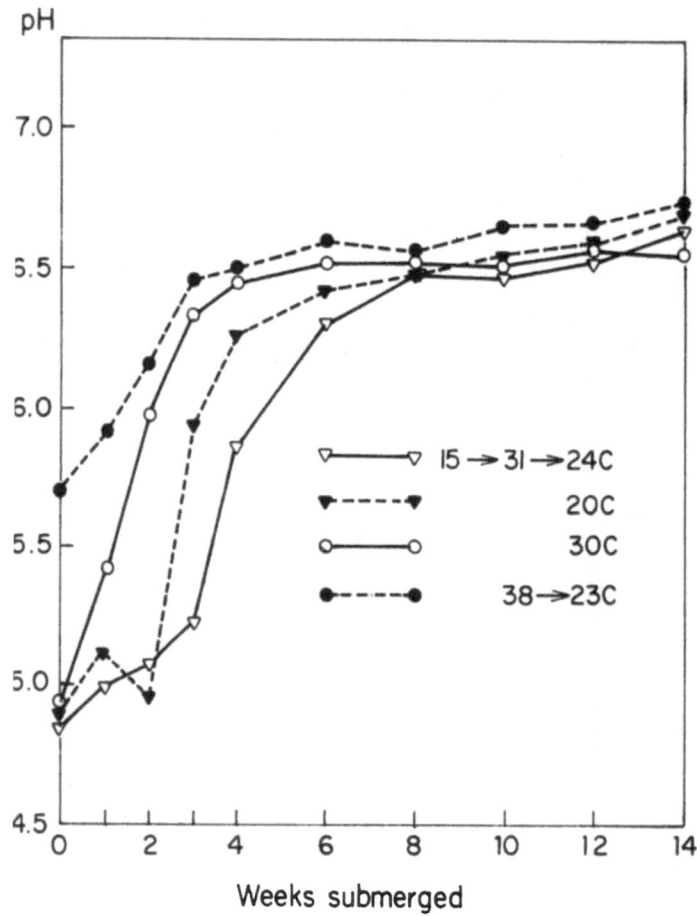

Fig. 5 Influence of temperature on the kinetics of pH in Luisiana clay

Calcareous soils: pH = 6.0 - 2/3 log P_{CO_2} (9)

Sodic soils   :   pH = 7.85 + log (alkalinity) - log P_{CO_2} (10)

The pH values of most reduced soils equilibrated with carbon dioxide at 1 atm is 6.1 (14).

The pH profoundly influences hydroxide, carbonate, sulfide, phosphate, and silicate equilibria in paddy soils. Those equilibria regulate the precipitation and dissolution of solids, the sorption and desorption of ions, and the concentration of such nutritionally significant elements or substances such as phosphorus, iron, aluminum, copper, zinc, hydrogen sulfide, carbonic acid, and organic acids.

I consider a pH of 6.6 in the soil solution most suited to rice because at this pH the availability of most macro- and micro-nutrients is high, and injurious concentrations of aluminum, manganese, iron, carbon dioxide and organic acids are absent (15).

Changes in Specific Conductance

The specific conductance of the solution of most soils increases after submergence, attains a maximum, and declines to a fairly stable value which varies with the soil. The changes reflect the balance between reactions that produce ions and those that inactivate them. There is a close similarity between the kinetics of specific conductance and that of other cations (Fig. 6). In normal paddy soils the peak specific conductances are of the order of 2-4 mmho/cm at 25°C. But sandy soils high in organic matter and acid sulfate soils may attain specific conductance values exceeding 4 mmho/cm, the harmful limit for rice.

As specific conductance increases so does ionic strength. Ionic strength is important because it governs the activity coefficients of ecologically important ions. The ionic strength of a natural solution in mole/l is numerically equal to 16 times the specific conductance in mmho/cm at 25°C up to ionic strengths of 0.05 (16).

REDOX EQUILIBRIA IN PADDY SOILS

In recent years, the application of equilibrium thermodynamics to the quantitative study of chemical transformations in nature has assumed increasing importance. Despite many theoretical and practical difficulties, equilibrium thermodynamics has been useful in understanding mineral associations in nature, the composition of sea water, other natural waters, and chemical changes in submerged soils. The difficulties are minimal in submerged paddy soils because submerged soils are more or less closed, isobaric, isothermal systems in which many reactions are catalyzed by enzymes. So equilibrium thermodynamics can be applied to chemical changes in paddy soils. Of the chemical changes reduction is the most important. The reduction process can be described quantitatively in terms of several redox systems (13, 14, 17-20).

The O_2-H_2O System

This system dominates the redox picture in all environments exposed to atmospheric oxygen. Oxygen at a P_{O_2} of 0.21 atm is theoretically such a powerful oxidant (Equations 11 and 12) that CO_2, H_2O, NO_3^-, SO_4^-, Fe_2O_3 and

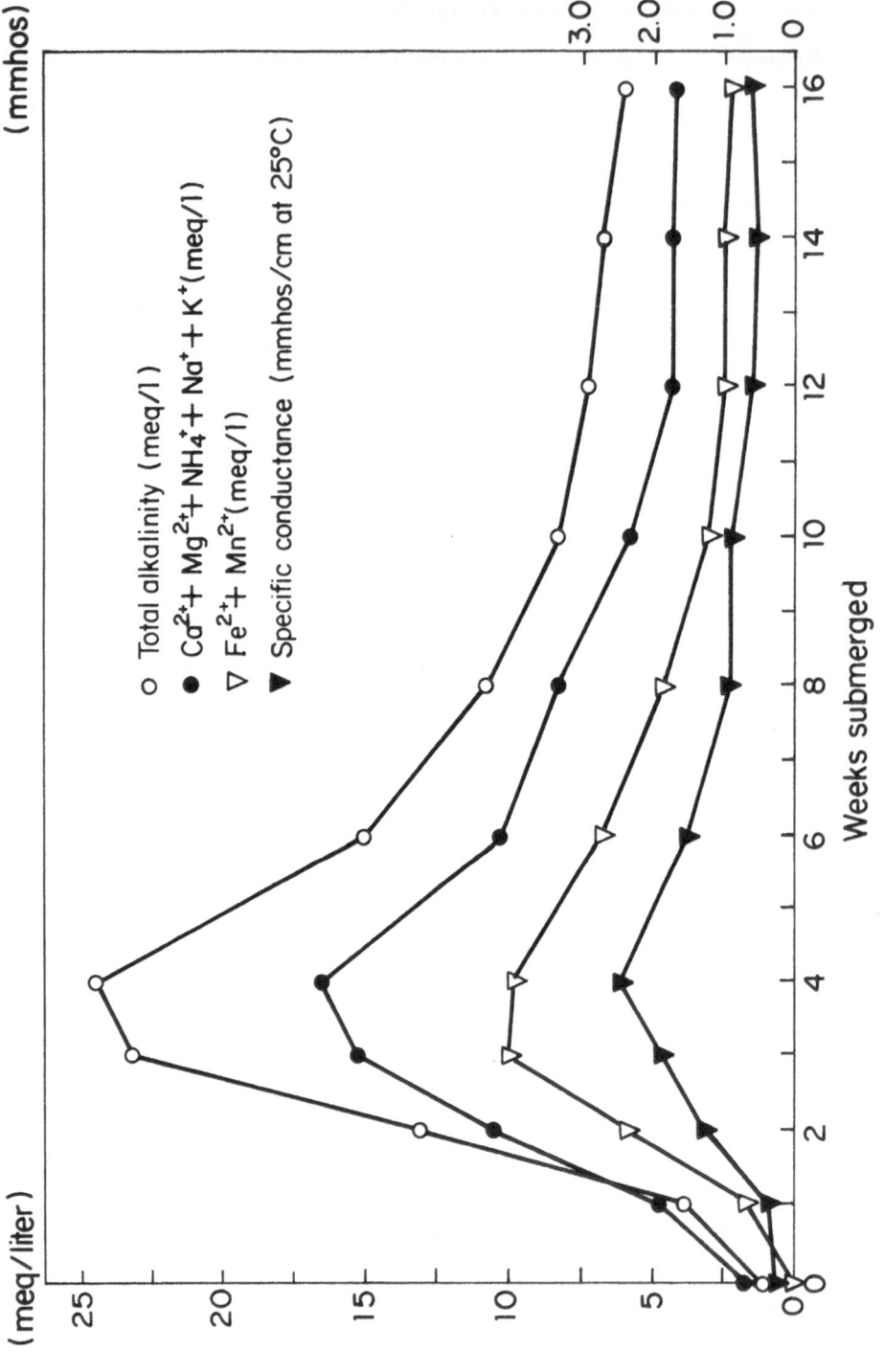

Fig. 6 Kinetics of specific conductance and cation concentrations in a submerged ferrallitic soil

MnO_2 are the stable forms of the elements. The equations for the O_2–H_2O systems are

$$1/4\ O_2 + H^+ + e = 1/2\ H_2O \qquad (11)$$

$$\text{and} \quad pE = 20.80 - 1/4\ pO_2 - pH \qquad (12)$$

when pO_2 is the negative logarithm to base 10 of the partial pressure of oxygen in atm.

According to Turner and Patrick the potential at which oxygen becomes undetectable is 0.33 V or a pE of 5.63. Thus very small concentrations are sufficient to keep a soil oxidized. But within a day of flooding, there is little free oxygen left in paddy soils (21).

The Nitrogen Systems

The NO_3^-–N_2 system. The conversion of nitrate to oxides of nitrogen or nitrogen gas is denitrification and is a result of nitrate respiration induced by oxygen deficiency. In paddy soils, nitrate disappears within a few days of flooding.

The equations for the NO_3^-–N_2 system are

$$1/5\ NO_3^- + 6/5\ H^+ + e = 1/10\ N_2 + 3/5\ H_2O \qquad (13)$$

$$\text{with } pE = 21.06 - 1/5\ pNO_3^- + 1/10\ pN_2 - 6/5\ pH \qquad (14)$$

where pNO_3^- is the negative logarithm to base 10 of the NO_3^- activity and pN_2 is the negative logarithm of the partial pressure of nitrogen in atm. This equation shows that at a pE of 5.63 at pH 7.0 (the pE at which oxygen becomes undetectable) P_{N_2} in equilibrium with 10^{-2} $\underline{M}$ NO_3^- is 10^{66}. This mean that nitrate is highly unstable in anaerobic media.

For 280 Philippine paddy soils, the loss of nitrate on flooding ranged from 10–78 kg N/ha per season with a mean of 26 (22).

Denitrification is one main path of nitrate loss in both fertilized and unfertilized paddy fields.

The NO_3^-–NH_4^+ system. In aerobic soils the end product of the mineralization of organic nitrogen is nitrate; in paddy soils it is ammonium. The equations

$$1/8\ NO_3^- + 5/4\ H^+ + e = 1/8\ NH_4^+ + 3/8\ H_2O \qquad (15)$$

$$pE = 14.91 - 1/8\ pNO_3^- + 1/8\ pNH_4^+ - 5/4\ pH \qquad (16)$$

show that at the low potentials of paddy soils (pE = -1 to 3) the concentration of nitrate that can be in equilibrium with ammonium is infinitesimal. Thus ammonium accumulates in paddy soils, and concentrations as high as 630 ppm have been reported (1).

The N_2–NH_4^+ system. The low pE of flooded paddy soils helps maintain nitrogen fertility by encouraging nitrogen fixation. The reduction of N_2 to NH_4^+ needs a high electron activity as shown by:

$$1/6 \ N_2 + 4/3 \ H^+ + e = 1/3 \ NH_4^+ \qquad (17)$$

$$pE = 4.64 - 1/6 \ pN_2 + 1/3 \ pNH_4^+ - 4/3 \ pH \qquad (18)$$

The sources of electrons at high activity are NADH, NADPH, and ferredoxin produced by anaerobic bacteria in the bulk of the soil and by phototrophic organisms at the soil surface (3).

The Mn(IV)-Mn(II)

The MnO_2-Mn^{2+} system. One of the earliest detectable chemical changes after a soil is flooded is the increase in concentration of water-soluble manganese. The reduction is both chemical and biological, and precedes that of iron. The complexity and variability of the composition of manganese oxides makes quantitative studies of hte manganese systems difficult. For pyrolusite (β-MnO_2) the electrochemical equations are

$$1/2 \ MnO_2 + 2 \ H^+ + e = 1/2 \ Mn^{2+} + H_2O \qquad (19)$$

$$pE = 20.80 + 1/2 \ pMn^{2+} - 2 \ pH \qquad (20)$$

At a pE of 5.63 at pH 7.0 (at which molecular oxygen becomes undetectable) the concentration of Mn^{2+} in the solution phase should be about $10^{2.34}$ mole/l. The actual concentration in paddy soils seldom exceeds $10^{-2.7}$. This means that the complex manganese oxides found in paddy soils are less reactive than pyrolusite (20).

The Mn_3O_4-$MnCO_3$ system. In almost all paddy soils the concentration of water-soluble manganese increases, reaches a peak and decreases. Ponnamperuma attributed this to the precipitation of manganous carbonate. The equations for the Mn_3O_4-$MnCO_3$ system are

$$1/2 \ Mn_3O_4 + 3/2 \ CO_2 + H^+ + e = 1/2 \ MnCO_3 + 1/2 \ H_2O \qquad (21)$$

$$\text{with } pE = 18.57 - 3/2 \ pCO_2 - pH \qquad (22)$$

The observed relationship was

$$pE = 10.86 - 1.41 \ pCO_2 - pH \qquad (23)$$

The Fe(III)-Fe(II) Systems

The $Fe(OH)_3$-Fe^{2+} system. The most important chemical change that takes place when a soil is flooded is the reduction of iron and an increase in its solubility in water. The concentration of water-soluble iron increases, reaches peaks as high as 600 ppm within 1-3 weeks of flooding and show a steep roughly exponential decrease to levels of 50-100 ppm. In neutral and calcareous soils the concentration of water-soluble iron rarely exceeds 20 ppm.

The increase in concentration of water-soluble iron is related to pE and pH of the $Fe(OH)_3$-Fe^{2+} system.

$$Fe(OH)_3 + 3 \ H^+ + e = Fe^{2+} + 3 \ H_2O \qquad (5)$$

$$pE = 17.87 + pFe^{2+} - 3 \ pH \qquad (7)$$

The $Fe_3(OH)_8$-Fe^{2+} and $Fe(OH)_3$-$Fe_3(OH)_8$ systems. The decrease in concentration of Fe^{2+} is probably due to the precipitation of $Fe_3(OH)_8$ or or $Fe_3O_4 \cdot nH_2O$. After that two more systems come into operation: $Fe_3(OH)_8$-Fe^{2+} and $Fe(OH)_3$-$Fe_3(OH)_8$ for which the equations are

$$1/2 \; Fe_3(OH)_8 + 4 \; H^+ + e = 3/2 \; Fe^{2+} + 4 \; H_2O \qquad (24)$$

$$pE = 23.27 + 3/2 \; pFe^{2+} - 4 \; pH \qquad (25)$$

$$3 \; Fe(OH)_3 + H^+ + e = Fe_3(OH)_8 + H_2O \qquad (26)$$

and $\quad pE = 7.26 - pH \qquad (27)$

The SO_4^{2-}-H_2S System

In aerated soils and waters elemental sulfur, sulfides, and organic sulfur compounds are oxidized by bacteria to sulfate. In paddy soils, sulfate is reduced to hydrogen sulfide which reacts with iron, manganese, copper, and zinc forming insoluble sulfides.

The equations for the SO_4^{2-}-H_2S system are

$$1/8 \; SO_4^{2-} + 5/4 \; H^+ + e = 1/8 \; H_2S + 1/2 \; H_2O \qquad (28)$$

$$pE = 5.12 - 1/8 \; pSO_4^{2-} + 1/8 \; pH_2S - 5/4 \; pH \qquad (29)$$

It shows that reduction of sulfate requires a low pE. Fermentation reactions in paddy soils provide the high electron activity needed for sulfate reduction.

In neutral and alkali soils concentrations as high as 1500 ppm SO_4^{2-} may be reduced to zero within 6 weeks of submergence.

MINERAL EQUILIBRIA IN PADDY SOILS

The main kinds of mineral equilibria in flooded paddy soils involve

1. Carbonates
2. Hydroxides
3. Sulfides
4. Silicate clays

Carbonate Equilibria

Carbon dioxide accumulates in flooded soils because diffusion of gases in submerged and saturated soils is extremely slow. A highly water-soluble and chemically reactive substance like carbon dioxide that accumulates in large amounts must profoundly influence chemical equilibria in paddy soils with their high content of divalent ions such as Ca^{2+}, Mg^{2+}, Fe^{2+}, and Mn^{2+}. Other divalent ions of nutritional interest are Cu^{2+} and Zn^{2+}. The interactions among these cations, carbon dioxide, and water in reduced soils, and with sodium bicarbonate and carbonate in sodic soils produce several carbonate systems.

The Na_2CO_3-H_2O-CO_2 system. The P_{CO_2} and the concentration of Na^+ associated with HCO_3^-, CO_3^- and OH^- (alkalinity) determine the pH of this system according to the equation (23).

71

$$pH = 7.85 + \log \{Na^+\} - \log P_{CO_2} - 0.51 \ I^{1/2} \qquad (30)$$

The pH values of soils containing $NaHCO_3$ equilibrated with different P_{CO_2} values conformed closely to the above equation (13).

The $CaCO_3$-H_2O-CO_2 system. The theoretical equations for this system can be simplified and written as follows:

$$pH = 6.03 - 2/3 \log P_{CO_2} \qquad (31)$$

$$\text{and} \quad pH + 1/2 \log Ca^{2+} + 1/2 \ P_{CO_2} = 4.92 \qquad (32)$$

Calcareous soils, both aerobic and anaerobic conformed closely to the above equations. But the slightly high value (6.1) obtained for the constant indicated that calcium carbonate in these soils was slightly more soluble than calcite (13, 14, 24).

The $MnCO_3$-H_2O-CO_2 system. When a manganiferous soil is submerged, the concentration of water-soluble manganese increases, reaches a peak, and decreases. This decrease has been attributed to the precipitation of $MnCO_3$. The simplified equations for the system are

$$pH = 5.4 - 2/3 \log P_{CO_2} \qquad (33)$$

$$\text{and} \quad pH + 1/2 \log Mn^{2+} + 1/2 \log P_{CO_2} = 4.1 \qquad (34)$$

Equilibration studies with reduced manganiferous soils gave slopes of 0.64 to 0.67 for the first equation but values of 6.1 for the constant, showing that $MnCO_3$ was present in a form more soluble than rhodocrosite. The observed coefficients of $\log Mn^{2+}$ in equation 34 were 0.46 to 0.53 and the constant was 4.37 to 4.56 (20, 24, 25).

The $FeCO_3$-H_2O-CO_2 system. Reduced ferruginous soils do not satisfy the two equations for carbonate equilibria (20).

$$pH = 5.2 - 2/3 \log P_{CO_2} \qquad (35)$$

$$pH + 1/2 \log Fe^{2+} + 1/2 \log P_{CO_2} = 3.75 \qquad (36)$$

Hydroxide Equilibria

Because the pH of reduced paddy soils is about 7, the concentration of OH^- ions is sufficient to precipitate the hydroxides of Al^{3+}, Fe^{2+}, and perhaps the hydroxy carbonates of Cu^{2+} and Zn^{2+}.

Aluminum hydroxide. The equation for the precipitation of aluminum hydroxide in the pH range of paddy soils may be written as (26):

$$Al(OH)^{2+} + 2 \ OH^- = Al(OH)_3 \qquad (37)$$

$$\text{with} \quad pH - 1/2 \ pAl = 2.2 \qquad (38)$$

At pH 3.5, $\{Al\} = 2.6 \times 10^{-1.6}$

pH 5.4, $\{Al\} = 2.6 \times 10^{-3.6}$

Ferrosoferric hydroxide. The concentration of water-soluble iron in paddy soils increases, reaches a peak, and then declines, or reaches a plateau, some weeks after flooding. The fairly stable concentrations reached appear to be governed by the solubility of ferrosoferric hydroxide.

$$Fe_3(OH)_8 = Fe^{2+} + 2\ OH^- + 2\ Fe(OH)_3 \tag{39}$$

with $\qquad pH - 1/2\ pFe^{2+} = 5.4 \tag{40}$

The iron hydroxide potential ($pH - 1/2\ pFe^{2+}$) appears to be characteristic for a given soil and remains constant over long periods of submergence. If its value is determined for a soil, it can be used to calculate the activity of Fe^{2+}. The only additional parameter required is the specific conductance from which the ionic strength is derived (27).

Table 2 shows that soils at the same pH and ionic strength can have widely differing concentrations of water-soluble iron. It also shows that at pH 6.5 soil no. 21 has a toxic concentration of water-soluble Fe^{2+} whereas at pH 7.0 soil no. 26 may be iron-deficient for wetland rice. Bao et al found a unit change in pH resulted in a change of 0.7-0.9 in pFe^{2+} (28).

Sulfide Equilibria

Hydrogen sulfide formed by sulfate reduction or the anaerobic decomposition of organic matter reacts with Fe^{2+}, Mn^{2+}, Cu^{2+}, Cu^+, and Zn^{2+} to form insoluble sulfides as shown below:

Reaction	pKs	
$Fe^{2+} + S^{2-} = FeS$	18.4	(41)
$Mn^{2+} + S^{2-} = MnS$	15.2	(42)
$Zn^{2+} + S^{2-} = ZnS$ (sphalerite)	25.2	(43)
$Zn^{2+} + S^{2-} = ZnS$ (wurtsite)	22.8	(44)
$Cu^{2+} + S^{2-} = CuS$	36.1	(45)
$2\ Cu^+ + S^{2-} = Cu_2S$	48.2	(46)

Ayotade obtained the following pKs values for 25 soils from the activity of the cations and the activity of S^{2-} measured with a sulfide electrode, 8 weeks after submergence (29).

$$pFe^{2+} + pS^{2-} = 17.1 - 19.5$$

$$pMn^{2+} + pS^{2-} = 17.0 - 20.2$$

$$pZn^{2+} + pS^{2-} = 22.3 - 25.8$$

$$pCu^{2+} + pS^{2-} = 20.2 - 22.7$$

$$2\ pCu^+ + pS^{2-} = 26.6 - 29.5$$

These figures suggest that the sulfides of iron, manganese, and zinc, but not those of copper were present in those soils. Gilmour and Kittrick confirmed the presence of ZnS in a flooded soil using solubility criteria. Bao et al found that the coordinates of the experimental pH-pFe^{2+} curves for paddy soils fell between the theoretical lines for FeS and $Fe(OH)_2$ (28, 30).

Sulfide equilibria have important nutritional implications for rice:

1. hydrogen sulfide toxicity is averted by its removal as insoluble sulfides
2. the availability of zinc decreases
3. rice roots need to oxidize the sulfides to sulfate to absorb sulfur
4. loss of sulfur by leaching is prevented

Equilibria Involving Silicate Clays

According to Shoji the most fertile paddy soils in Japan have a high montmorillonite content. Recent studies with Philippine paddy soils suggest that soils dominant in montmorillonite do not appear to have nitrogen, potassium or phosphorus fixation problems (31, 32, 33).

A study of the clay fractions of 214 rice soils in the Philippines revealed the presence of vermiculite, beidellite, montmorillonite, kaolinite, and x-ray amorphous material (33).

All soils with vermiculite and beidellite as dominant clays had an exchangeable K^+ content of <2 mmol/kg. In virtually all other soils the exchangeable K^+ content was >2 mmol/kg.
All soils with dominant halloysite or x-ray amorphous material had Olsen phosphorus values which were <10 mg/kg. The mean for samples with dominant halloysite was 2 mg/kg. For soils with x-ray amorphous material or halloysite as secondary component, the mean was 5 mg/kg. The values for all other soils ranged from 10-50 mg/kg.

In laboratory equilibration studies of clays of wetland rice soils the following observations were made:

1. K^+ fixation was highest in beidellitic clays, moderately high in vermiculitic clays, and lowest in clays consisting of montmorillonite, chlorite, hydrous mica, halloysite, kaolinite and x-ray amorphous material.
2. NH_4^+ fixation was highest in beidellitic and vermiculitic clays, moderately high in montmorillonitic clays and lowest in clays that did not contain those three minerals.
3. Phosphate fixation was highest in clays dominant in kaolinite or halloysite, moderately high in clays with either of these two minerals and x-ray amorphous material, and lowest in vermiculitic clays (33).

ION EXCHANGE

In aerobic soils ion exchange plays an important role in the replenishment of the soil solution with anions and cations absorbed by plant roots. In submerged soils the large amounts of Fe^{2+} and Mn^{2+} are brought into solution by soil reduction and displace cations from the clay complex, increasing the concentrations of Na^+, K^+, NH_4^+, Ca^{2+}, Mg^{2+} in the soil solution. The close parallelism between the kinetics of water-soluble $Fe^{2+} + Mn^{2+}$ and other cations (Fig. 6) illustrates the role of Fe^{2+} and Mn^{2+} in cation exchange.

Bao et al reported that the amount of exchangeable Fe^{2+} was influenced by the cation exchange capacity of the soil, the amount of sulfide present and pH (35).

Sims and Patrick found that greater amounts of iron, manganese, zinc, and copper were present in exchangeable form at low Eh and pH than at higher Eh and pH (34).

The displacement and loss of bases by Fe^{2+} may cause acidification or ferrolysis (36).

$$Ca^{2+}\text{---clay} + Fe^{2+} \rightarrow Fe^{2+}\text{---clay} + Ca^{2+}$$

$$Fe^{2+}\text{---clay} + O_2 + H_2O \rightarrow H^+\text{---clay} + Fe_2O_3$$

A study of the Q/I relationship for NH_4^+ revealed that a part of the NH_4^+ in flooded soils is adsorbed specifically and associated with exchange sites from which it cannot easily be replaced by Ca^{2+} and Mg^{2+} (37).

Because salt and alkali profoundly affect cation exchange reactions and the availability of potassium, Pasricha and Ponnamperuma studied the influence of varying concentrations of sodium chloride and sodium bicarbonate on the activity ratio of K^+, $a_K/(a_{Ca} + a_{Mg})^{1/2}$, at varying periods of submergence. The activity ratio decreased with increase in salinity in spite of the increase in K^+ concentration, but increased with alkalinity. The relationship between labile K^+ and the activity ratio (Q/I relationship) was rectilinear. That indicates that the exchange reactions obeyed Schofield's ratio law (38).

SORPTION AND DESORPTION

Sorption and desorption play an important part in the fixation and release of plant nutrients. The sorbing agents are clay, organic matter, silicic acid, and the hydrous oxides of iron, aluminum, and manganese. Sorption may be due to electrostatic attraction, covalent bonding, or isomorphous replacement in the crystal lattice. Because of the changes in surface properties brought about by changes in reduction of Fe(III) and Mn(IV) hydrous oxides and pH alterations, sorbed ions may be released into the soil solution in flooded soils. Among the plant nutrients involved are phosphorus, sulfur, copper, zinc, molybdenum, and boron.

Phosphate is strongly adsorbed by clay, hydrous oxides of iron and aluminum, and calcium carbonate. When a soil undergoes reduction some of the adsorbed phosphate is released into the soil solution because of reduction of iron oxide and the increase in pH. Khalid et al observed that phosphate was released under reduced conditions but at high levels of added phosphate, more phosphate was adsorbed under reduced conditions than under oxidized conditions (4, 10, 39, 40).

Sulfate is adsorbed by both crystalline and amorphous iron oxides. The initial increase in concentration of water-soluble iron observed when acid soils are flooded has been ascribed to desorption of SO_4^{2-} following reduction of the oxides (3, 39).

Silica is sorbed by hydrous oxides of iron and aluminum, and adsorption increases with pH of up to 9. When a soil undergoes reduction, the concentration of water-soluble silica increases apparently because of the reduction of Fe(III) to Fe(II)(39).

Cu^{2+} and Zn^{2+} are adsorbed by silicic acid and by hydrous oxides. Boron and molybdenum are adsorbed by clays and hydrous oxides of iron and aluminum (41, 42).

Generally the concentrations of water-soluble phosphorus, silicon and molybdenum increase, whereas those of copper, zinc, and boron decrease when a soil is submerged. Copper and zinc are probably released on soil reduction but later resorbed or precipitated.

CHEMICAL KINETICS OF PADDY SOILS

Flooding a soil sets in motion a series of chemical and biochemical changes whose courses and rates are determined by soil properties temperature, and duration of submergence.

Denitrification

The kinetics of denitrification varies markedly with soil and temperature. Figure 7 shows that denitrification at 35°C was fastest in Pila clay (pH 7.1, O.M. 3.6%), moderately fast in Luisiana clay (pH 4.7, O.M. 3.4%) and slowest in Keelung silt loam (pH 7.7, O.M. 6.8%) despite its high pH and organic matter content. In all soils denitrification was hardly detectable at 5°C and increased rapidly with increase in temperature. Figure 8 illustrates the influence of temperature on the kinetics in Maahas clay (pH 6.6, O.M. 2.0%)(43).

The kinetics of nitrate showed roughly exponential trends in all four soils suggesting that denitrification followed first order kinetics ($-dc/dt = kc$). The rectilinear regressions of log NO_3 on time (Fig. 9) for each soil at four temperatures reveal first order kinetics. The variation of the rate constant with temperature obeyed the Arrhenius law ($k_T = Ae^{-E/RT}$) and the temperature coefficient of denitrification (Q_{10}) varied with the soil and the temperature range but was of the order for biochemical reactions (43, 44).

In neutral, tropical paddy soils nitrate disappears within a few days of flooding and surface-applied nitrogen fertilizers are partly lost by nitrification-denitrification.

Ammonification

The kinetics of ammonification varied with the soil and temperature (Fig. 10). The rate was lowest at 15°C and highest at 45°C in all soils except Luisiana clay in which the rate at 45°C was less than at 35°C (43).

Figure 11 shows the kinetics of ammonification in 7 flooded soils over 16 weeks. Soils high in total nitrogen released more NH_4^+ than others, and more than half the NH_4^+ mineralized in 16 weeks was released in the first 2 weeks. The kinetics can be described by

$$y = a + b \log x$$

where y = total NH_4^+-N and x is weeks submerged, "a" is the NH_4^+-N content of the soil 1 day after flooding, and "b" is a measure of the velocity of NH_4^+-N

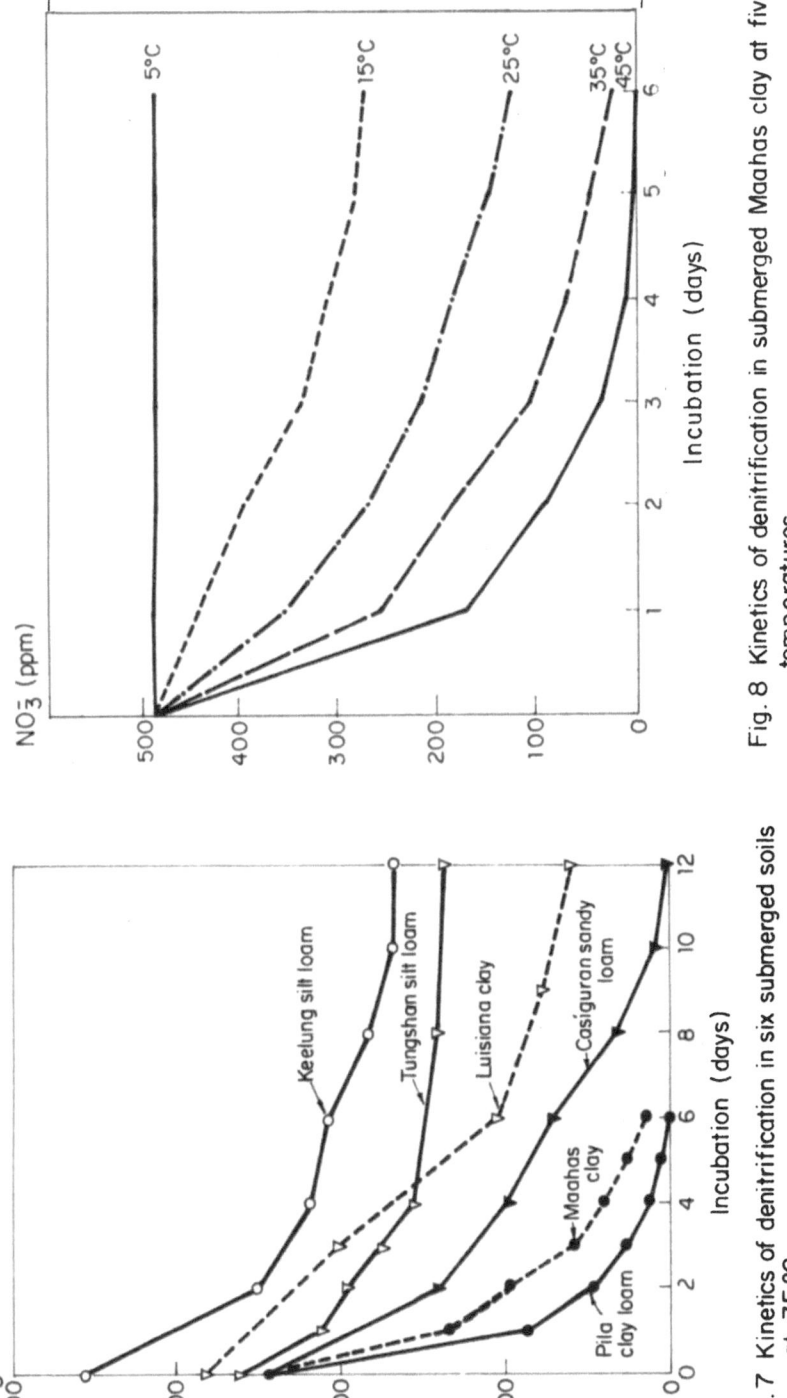

NO$_3^-$ (ppm)

5°C

15°C

25°C

35°C
45°C

Incubation (days)

Fig. 8 Kinetics of denitrification in submerged Maahas clay at five temperatures

NO$_3^-$-N (ppm)

Keelung silt loam

Tungshan silt loam

Luisiana clay

Casiguran sandy loam

Maahas clay

Pila clay loam

Incubation (days)

Fig. 7 Kinetics of denitrification in six submerged soils at 35°C

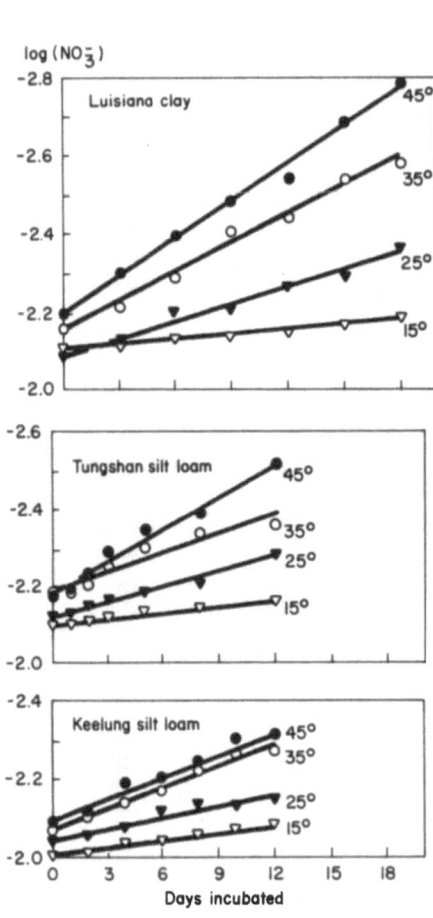

Fig.9 Relationship between log (NO₃⁻) and time of anaerobic incubation at four temperatures in Luisiana clay, Tungshan silt loam, and Keelung silt loam

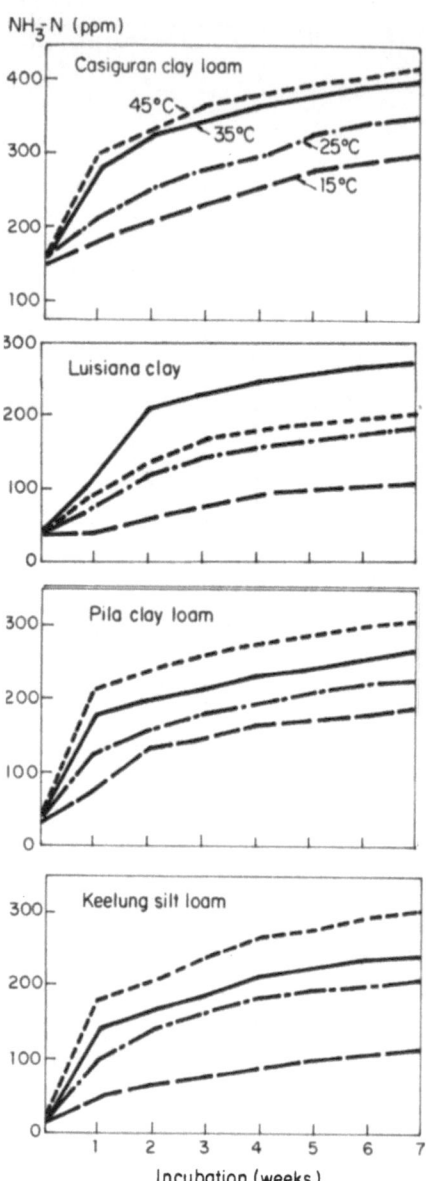

Fig. 10 Influence of temperature on ammonification in four submerged soils

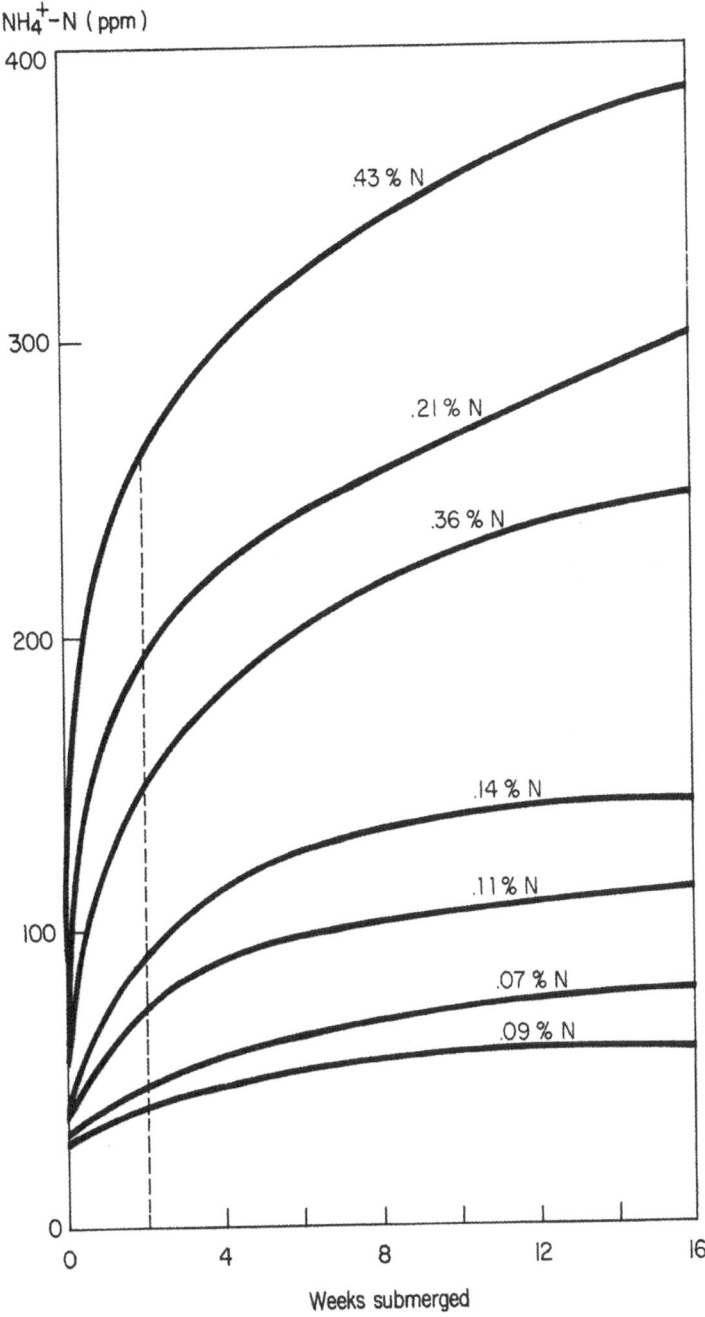

NH$_4^+$-N (ppm)

.43 % N

.21 % N

.36 % N

.14 % N

.11% N

.07 % N

.09 % N

Weeks submerged

Fig. 11   Kinetics of NH$_4^+$ release in 7 submerged soils

release in a flooded soil.

The implications for rice fertilization are that soils low in nitrogen need basal and top dressings, soils moderate in total nitrogen may need only a topdressing, and soils high in nitrogen may need no nitrogen fertilizer for yields of 5 to 6 t/ha (45).

Kinetics of Water-soluble Manganese

When a soil is flooded the concentration of water-soluble manganese increases, reaches a peak, and declines. The rate of increase in concentration as well as the peak and final stable concentrations are determined mainly by the manganese and organic matter content of the soil (Fig. 12).

Kinetics of Carbon Dioxide

One to three tons of carbon dioxide are produced in the plowed layer of 1 ha of a paddy soil during the first few weeks of submergence. The partial pressure of carbon dioxide (P_{CO_2}) is a good measure of carbon dioxide accumulation (25).

P_{CO2} in a soil increases after submergence reaches a peak of 0.2-0.8 atm 1-3 weeks later, and declines to a fairly stable value of 0.05-0.2 atm. Acid soils high in organic matter but low in iron and manganese give higher P_{CO_2} values than neutral soils throughout the season. Temperature markedly affects the kinetics of P_{CO2} (Fig. 13)(25).

Practical implications for rice culture are that carbon dioxide injury may occur on acid soils low in iron and manganese and that low temperature aggravates it.

Kinetics of Organic Acids

The main organic acids in flooded paddy soils are formic, acetic, propionic, and butyric acids. Of these, acetic acid is the most abundant.

When a soil is flooded the concentration of water-soluble organic acids increases, reaches a peak value of 10-40 mmole/liter in 1-2 weeks, and then declines to less than 1 mmole/liter a few weeks later. Organic acids persist longer in cold soils than in warm soils (Fig. 14)(46).

Organic acid injury to rice may occur on cold acid soils and soils heavily manured with organic materials.

Kinetics of Water-soluble Iron

The increase in concentration of water-soluble iron is one important change caused by flooding a soil. Soil properties and temperature markedly influence the kinetics of water-soluble iron (Fig. 15).

Acid soils at 30°C build up concentrations as high as 600 ppm within 2-4 weeks of flooding and then decline to levels of 50-100 ppm. Soils high in organic matter but low in active iron give high concentrations that persist for months. Neutral soils reach a plateau of about 20 ppm. Low temperature retards the peaks and broaden the area under them (44, 46).

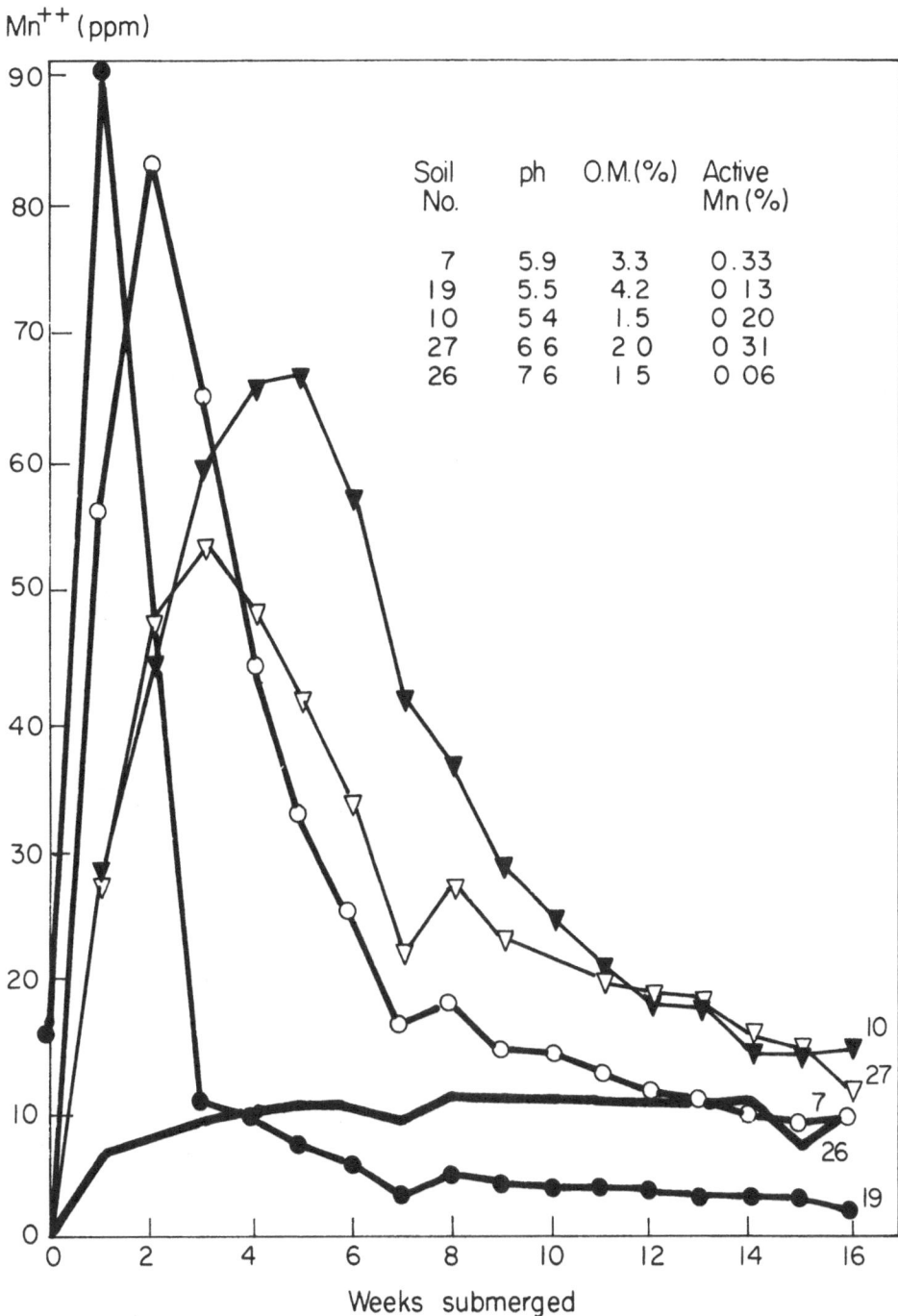

Fig. 12. Kinetics of water-soluble Mn++

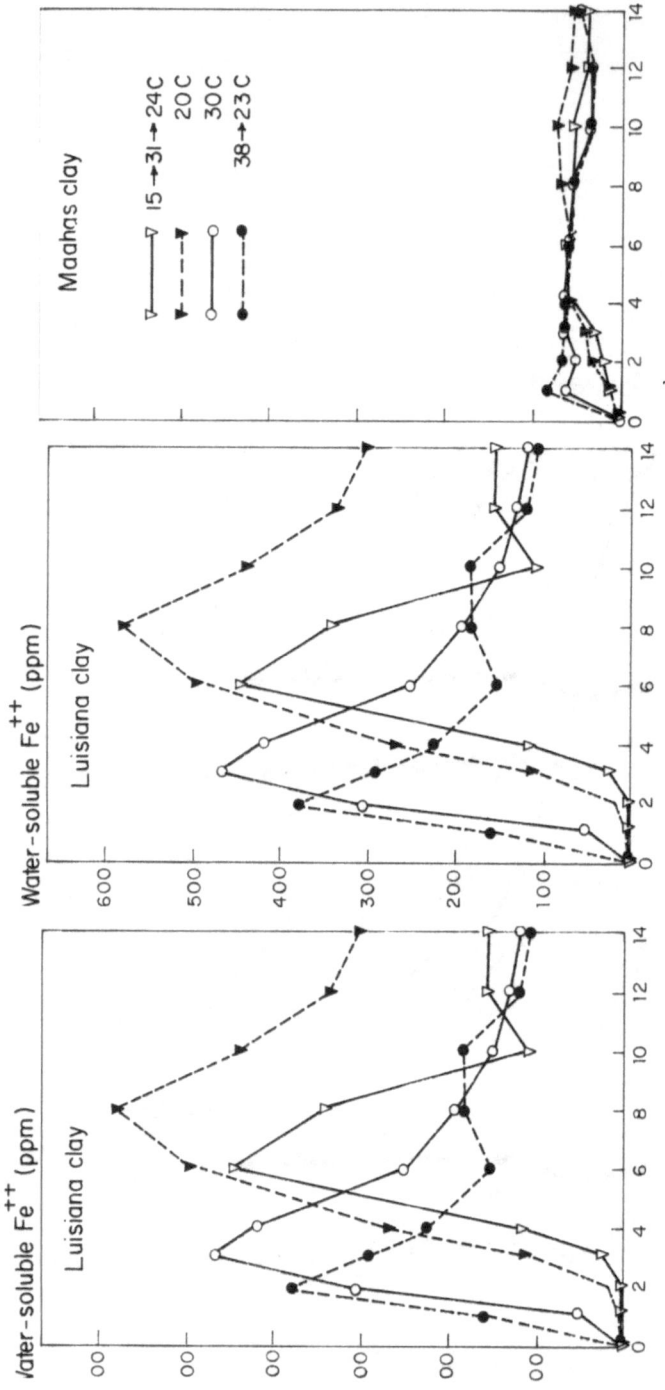

Fig. 13 Influence of temperature on the kinetics of Fe^{++} in three soils

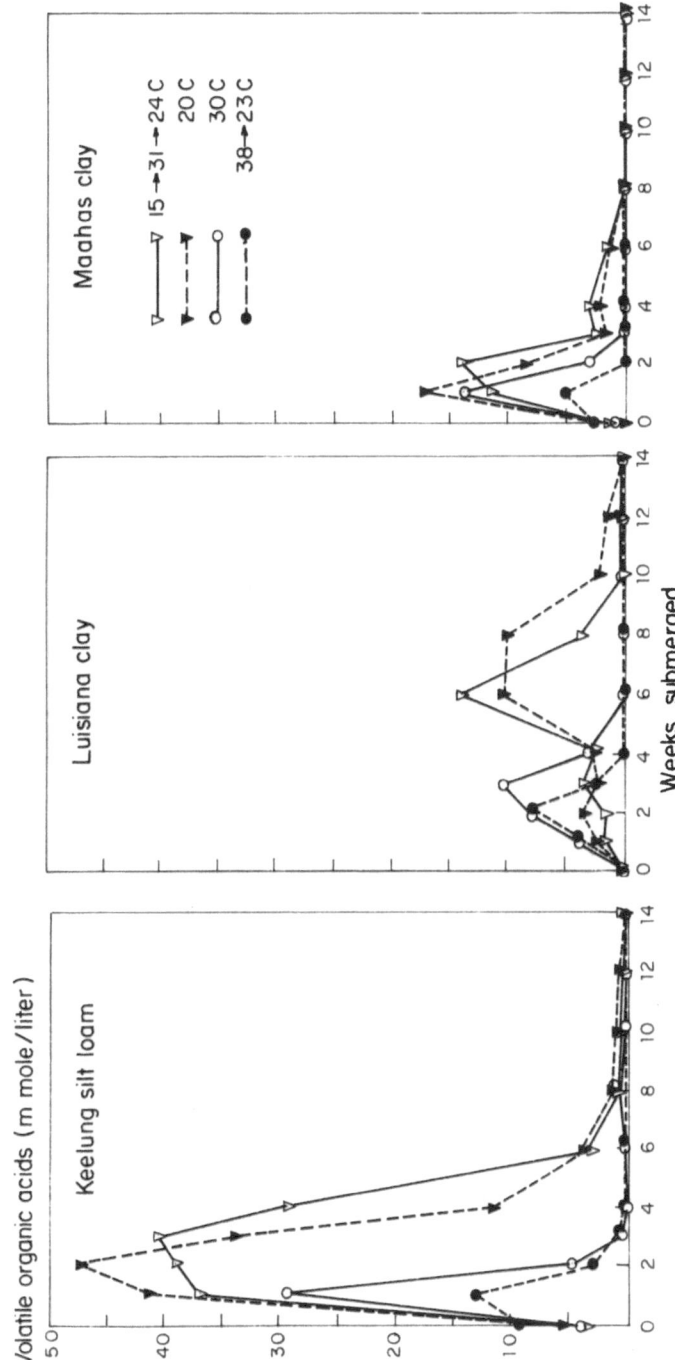

Fig. 14 Influence of temperature on the kinetics of volatile organic acids in three soils

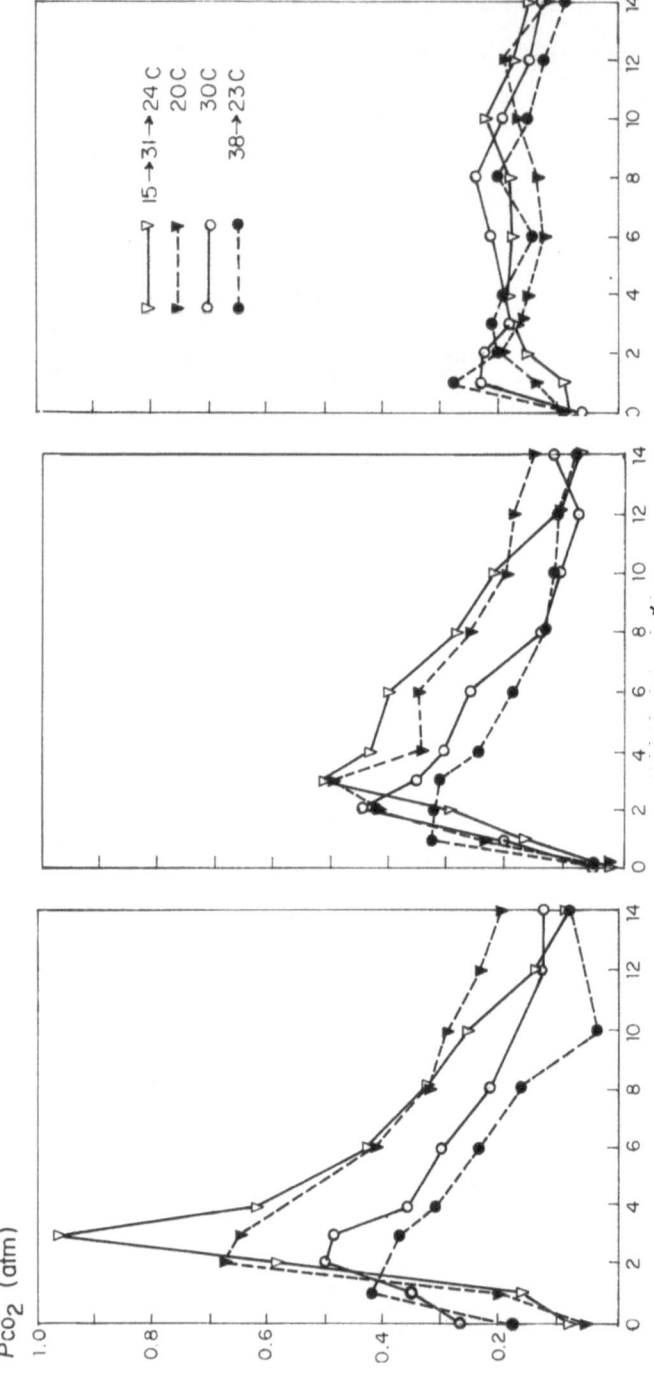

P_{CO_2} (atm)

15 → 31 → 24 C
20 C
30 C
38 → 23 C

Fig. 15 Influence of temperature on the kinetics of P_{CO_2} in three soils

Table 2   Influence of soil properties on the Fe(II) hydroxide potential $pFe(OH)_2$ and the concentration of water-soluble Fe^{2+} at pH 6.5 and ionic strength of 0.03 mol/liter

Soil no.	pH	O.M. (%)	Fe (%)	Mn (%)	$pFe(OH)_2$	Fe^{2+} (ppm)
21	4.6	4.1	2.8	0.02	5.41	629
7	5.9	3.3	1.7	0.33	5.09	195
27	6.6	2.0	1.6	0.31	4.90	64
26	7.6	1.5	0.3	0.06	4.83	43

Table 3   Equations for kinetics of water-soluble SO_4^{2-} in five soils

Soil No.	pH	Equation
1	7.6	$\log (SO_4^{2-}) = -2.48 - 0.48\,t$
39	8.1	$\log (SO_4^{2-}) = -2.14 - 0.44\,t$
26	7.6	$\log (SO_4^{2-}) = -3.34 - 0.36\,t$
23	5.7	$\log (SO_4^{2-}) = -2.57 - 0.17\,t$
35	2.8	$\log (SO_4^{2-}) = -0.91 - 0.018\,t$

The increase in concentration of water-soluble iron benefits rice but an excess is harmful. Iron toxicity occurs on strongly acid soils (47).

Kinetics of Water-soluble Sulfate

The kinetics of water-soluble sulfate varies widely with soil properties (Fig. 16). The neutral and alkaline soils, regardless of initial SO_4^{2-} content lost SO_4^{2-} rapidly and contained less than 20 ppm, 4 weeks after flooding. But the acid soils showed first an appreciable increase of water-soluble SO_4^{2-}, followed by a slow decline to a final concentration of 1 ppm, 16 weeks after flooding. In the acid sulfate soil, SO_4^{2-} declined steadily from 12,000 ppm at flooding to 5000 ppm, 16 weeks later.

The rapid disappearance of sulfate in the slightly alkaline soils confirms earlier findings that a high pH favors sulfate reduction. The slight increase in SO_4^{2-} concentration in the acid soils in the early stages of flooding may be ascribed to release of SO_4^{2-} from anion exchange sites as pH increased, and the subsequent decrease, to SO_4^{2-} reduction.

Sulfate reduction followed first order kinetics with velocity constants that varied with the soil as shown in Table 3. The velocity constants were high in the alkaline soils and low in the acid soils (24).

The influence of temperature on sulfate reduction is shown in Figure 17. Sulfate reduction may cause:

1. sulfur deficiency in neutral and alkaline soils low in SO_4^{2-};
2. zinc deficiency due to precipitation of ZnS; and
3. H_2S toxicity in soils low in active iron.

Kinetics of Water-soluble Phosphate

When a soil is flooded the concentration of water-soluble phosphorus increases. But soil properties markedly affect the kinetics (Fig. 18). The increases were rapid and considerable in the sandy calcareous soil low in iron and least in the acid clay, high in iron.

Kinetics of water-soluble Boron, Copper, Molybdenum, and Zinc

Flooding a soil causes a decrease in the concentration of boron, copper, and zinc, but an increase in that of molybdenum (Table 4). The ions $B(OH)_4^-$, Cu^{2+}, MoO_4^{2-} and Zn^{2+} are present sorbed on hydrous oxides and clays. The changes in concentration on soil reduction may be due to sorption and desorption depending on changes in point of zero charge of the absorbing substances brought about by reduction of Fe(III) and Mn(IV) oxides.

FORMATION OF ORGANO-METALLIC COMPLEXES

To ascertain the role of organic manures in soil fertility, organo-mineral complexes in paddy soils were studied.

An appreciable proportion of water-soluble Fe^{2+} in paddy soils manured with organic materials is present as organic complexes. Bao et al found a positive correlation between complexed Fe^{2+} and the organic matter content of the soil (35).

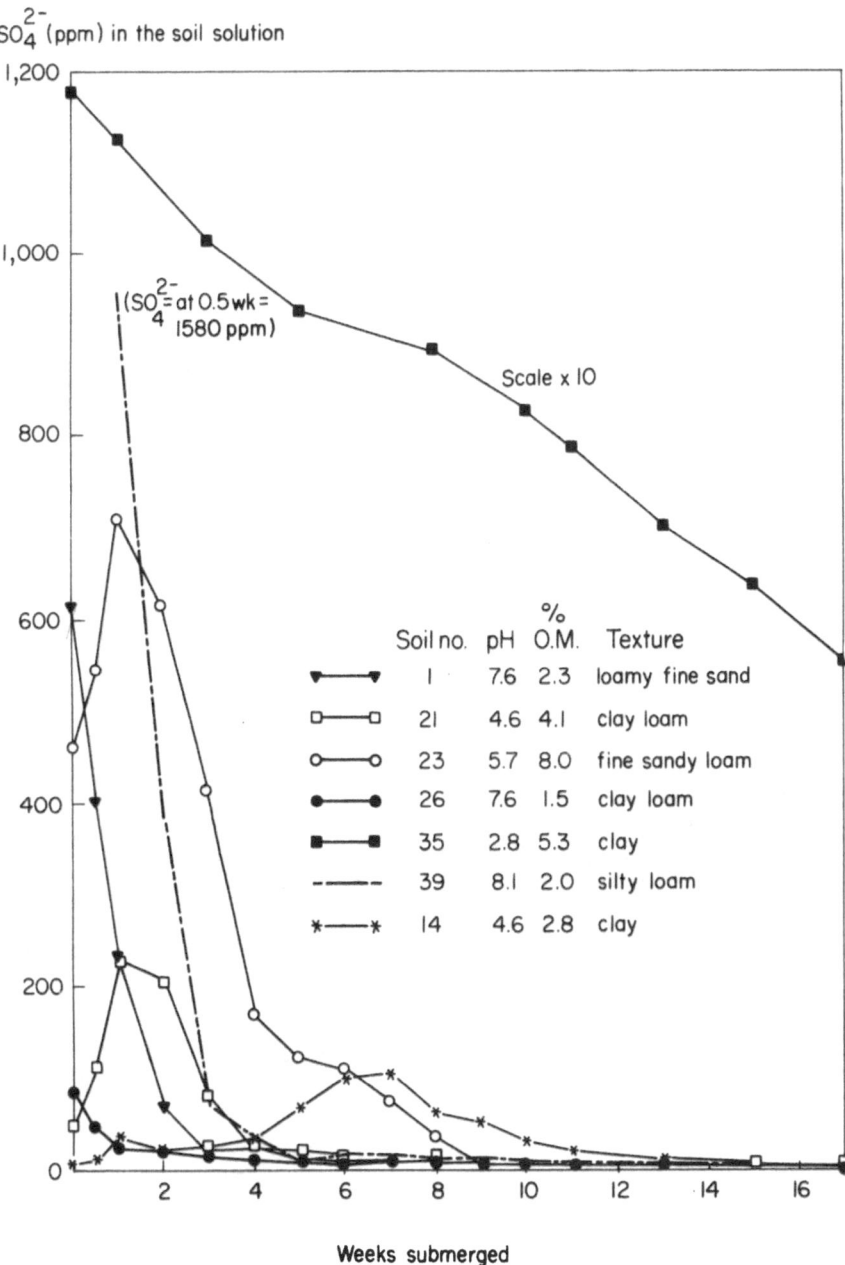

SO_4^{2-} (ppm) in the soil solution

$I (SO_4^{2-}$ at 0.5 wk = 1580 ppm)

Scale x 10

	Soil no.	pH	% O.M.	Texture
▼——▼	1	7.6	2.3	loamy fine sand
□——□	21	4.6	4.1	clay loam
○——○	23	5.7	8.0	fine sandy loam
●——●	26	7.6	1.5	clay loam
■——■	35	2.8	5.3	clay
————	39	8.1	2.0	silty loam
——	14	4.6	2.8	clay

Weeks submerged

Fig. 16 Influence of soil properties on sulfate reduction in flooded soils

Table 4   Kinetics of water-soluble boron, copper, molybdenum, and
           zinc in two submerged soils

Element	Concentration (ppm)				
	1 wk	2 wk	3 wk	4 wk	6 wk

Luisiana clay (pH: 4.8; O.M.: 2.8%)

Element	1 wk	2 wk	3 wk	4 wk	6 wk
B	0.55	0.49	0.19	0.46	0.30
Cu	0.11	0.11	0.03	0.04	0.02
Mo	0.18	0.06	0.28	0.15	0.24
Zn	0.30	0.09	0.05	0.08	0.03

Maahas clay (pH: 6.6; O.M.: 2.0%)

Element	1 wk	2 wk	3 wk	4 wk	6 wk
B	1.80	1.01	1.15	1.15	1.18
Cu	0.06	0.05	0.03	0.03	0.02
Mo	0.04	0.09	0.08	0.17	0.12
Zn	0.18	0.08	0.04	0.06	0.03

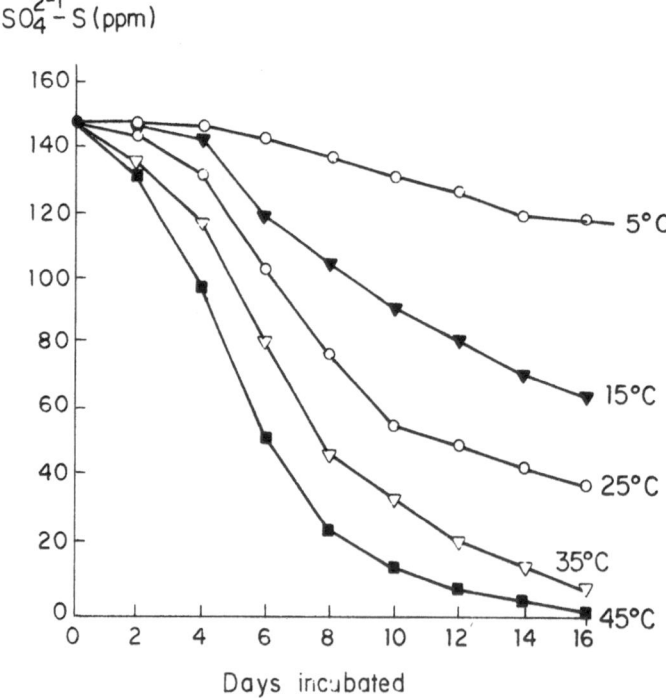

Fig. 17   Kinetics of sulfate reduction in Maahas clay
           incubated at different temperatures

P (ppm)

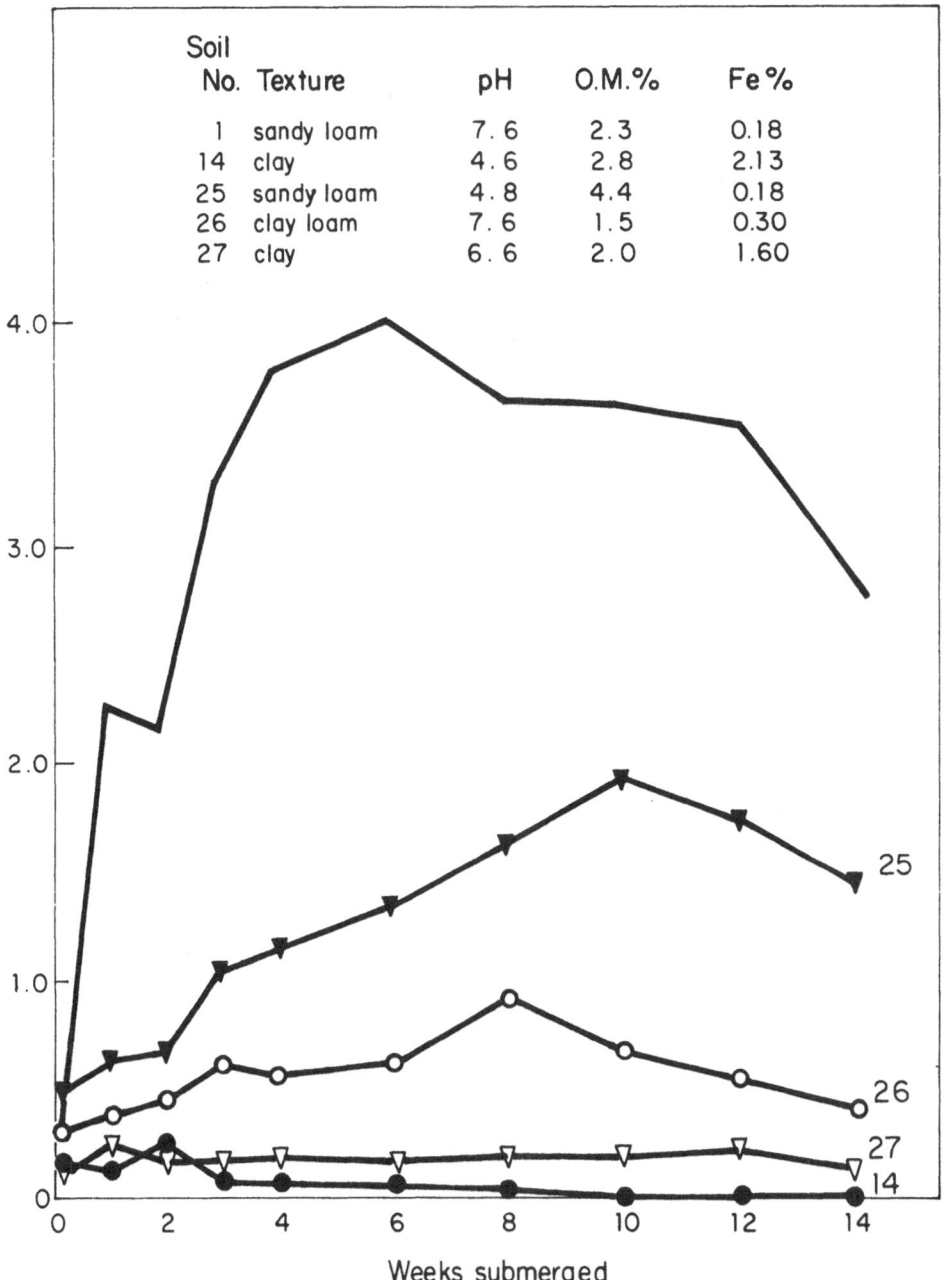

Soil No.	Texture	pH	O.M.%	Fe%
1	sandy loam	7.6	2.3	0.18
14	clay	4.6	2.8	2.13
25	sandy loam	4.8	4.4	0.18
26	clay loam	7.6	1.5	0.30
27	clay	6.6	2.0	1.60

Weeks submerged

Fig.18 Kinetics of water-soluble phosphate in 5 submerged soils

Pao et al reported that in paddy soils

1. ten to thirty percent of water-soluble Fe^{2+} was present as organic complexes;
2. the apparent stability constants of the complexes ranged from 2.4 to 4.0;
3. the concentration and binding power of the organic ligands were highest at the stage of intensive organic matter decomposition;
4. pH strongly affected the equilibrium between Fe^{2+} ions and the complexes;
5. Mn^{2+} ions competed with Fe^{2+} ions for the organic ligands.

Pao and Yu observed that 87% of the water-soluble organic complexes of Fe^{2+} was present as charged moieties (48, 49).

PRACTICAL SIGNIFICANCE IN RICE CULTURE

The decrease in Eh or pE caused by flooding: destroys nitrate; increases the availability of iron, phosphorus, and silicon; decreases the availability of sulfur and zinc; and promotes nitrogen fixation.

The increase in pH of acid soils following soil reduction: increases the availability of phosphorus and molybdenum; decreases the availability of copper and zinc; and depresses the toxicity of manganese, aluminum, iron, organic acids, and carbon dioxide. The decrease in pH of sodic and calcareous soils increases the availability of phosphorus, iron, and manganese in them.

The displacement of basic cations from the colloidal complex by Fe^{2+} and Mn^{2+} may increase the concentration of nutrients temporarily, but may lead to soil acidification if the basic cations are lost from the system.

Nutrient elements, such as phosphorus, molybdenum, and silicon, sorbed by soil materials may be released into the soil solution following soil reduction or pH changes.

Temperatures over 25°C favor the loss of nitrate, the release of ammonium and sulfate reduction. Low temperatures favor the production and persistence of high concentrations of organic acids, carbon dioxide, and water-soluble iron in acid soils.

An Eh of 70 to 120 mV or a pE of about 0.2 to 2.0, a pH of about 6.6, a specific conductance of about 2 mmho/cm at 25°C, and a temperature of 30 to 35°C (all in the soil solution) favor nutrient uptake by rice.

REFERENCES

(1) Kawaguchi, K., Kyuma, K., Paddy soils in tropical Asia. Univ. of Hawaii Press (1977).

(2) Moormann, F. R., van Breemen, N., Rice: soil, water, land. International Rice Research Institute, Los Baños, Philippines (1978).

(3) Ponnamperuma, F. N., The chemistry of submerged soils. Adv. Agron., 24, 29-96 (1972).

(4) Patrick, W. H. Jr., Reddy, C. N., Chemical changes in rice soils. Soils and rice. International Rice Research Institute. Los Baños, Laguna, Philippines, pages 361-379 (1978).

(5) Yamane, I., Electrochemical changes in rice soils. Soils and rice. International Rice Research Institute, Los Baños, Laguna, Philippines, pages 381-398 (1978).

(6) Hsuan Cha-Hsiang, Yu Tien-jen, Studies of electrochemical properties of soils. IV. Causes of the suspension effect. Acta Pedologica Sinica, 12, 307-320 (1964).

(7) Ponnamperuma, F. N., Dynamic aspects of flooded soils. The mineral nutrition of the rice plant. John Hopkins Press, pages 295-328 (1965).

(8) Yu Tien-jen, Lieu, W. L., Studies on oxidation-reduction processes in paddy soil. III. Influence of oxidation-reduction conditions of the soil on the growth of rice. Acta Pedol. Sinica, 5, 292-304 (1957).

(9) Aomine, S., A review of research on redox potentials of paddy soils in Japan. Soil Sci., 94 (1), 6-13 (1962).

(10) Ponnamperuma, F. N., Soil chemical and fertility characteristics important to land evaluation for wetland rice. Paper presented at the IRRI-CSEAS Workshop on Land Evaluation for Rice-Based Cropping Systems, International Rice Research Institute, Los Baños, Philippines (1979).

(11) Ponnamperuma, F. N., Electrochemical changes in submerged soils and the growth of rice. Soils and rice. International Rice Research Institute, Los Baños, Laguna, Philippines, pages 421-444 (1978).

(12) Clark, W. M., Oxidation-reduction potentials of organic systems. The Williams & Wilkins Company (1960).

(13) Ponnamperuma, F. N., Martinez, E., Loy, T., Influence of redox potential and partial pressure of carbon dioxide on pH values and the suspension effect of flooded soils. Soil Sci., 101 (6), 421-431 (1966a).

(14) Ponnamperuma, F. N., Castro, R. U., Valencia, C. M. Experimental study of the influence of the partial pressure of carbon dioxide on the pH values of aqueous carbonate systems. Soil Sci. Soc. Amer. Proc., 33, 239-241 (1969a).

(15) Ponnamperuma, F. N., Physicochemical properties of submerged soils in relation to fertility. IRRI Research Paper Series No. 5 (1977).

(16) Ponnamperuma, F. N., Tianco, E. M., Loy, T. A., Ionic strengths of the solutions of flooded soils and other natural aqueous solutions from specific conductance. Soil Sci., 102 (6), 408–413 (1966b).

(17) Garrels, R. M., Christ, C. L., Solutions, minerals, and equilibria. Harper and Row (1965).

(18) Sillen, L. G., The physical chemistry of sea water. Oceanography, Publ. No. 67, Amer. Ass. Advance. Sci., Washington, D. C., page 549–582 (1961).

(19) Stumm, W., Morgan, J. J., Aquatic chemistry. Wiley Interscience (1970).

(20) Ponnamperuma, F. N., Loy, T. A., Tianco, E. M., Redox equilibria in flooded soils: II. The manganese oxide systems. Soil Sci., 108, 48–57 (1969b).

(21) Turner, F. T., Patrick, W. H. Jr., Chemical changes in waterlogged soils as a result of oxygen depletion. 9th International Congress of Soil Science, IV, 53–65 (1968).

(22) IRRI (International Rice Research Institute), Annual report for 1972. Los Baños, Philippines (1973).

(23) Ponnamperuma, F. N., A theoretical study of aqueous carbonate equilibria. Soil Sci. 103 (2), 90–100 (1967).

(24) IRRI (International Rice Research Institute), Annual report for 1965. Los Baños, Philippines (1966).

(25) IRRI (International Rice Research Institute), Annual report for 1964. Los Baños, Philippines (1965).

(26) Raupach, M., Solubility of simple aluminum compounds expected in soils. III. Aluminum ions in soil solutions and aluminum phosphate in soils. Aust. J. Soil Research, 1 (1), 46–54 (1963).

(27) Arden, T. V., The solubility products of ferrous and ferrosic hydroxides. J. Chem. Soc., pages 882–885 (1950).

(28) Bao, Hsuo-ming, Liu Chi-Kuang, Wu Chun, Yu Tien-Jen, Studies on oxidation-reduction processes in paddy soil. VII. Forms of the ferrous iron. Acta Pedologica Sinica 12, 297–306 (1964).

(29) Ayotade, K. A., Studies on hydrogen sulfide measurement levels and equilibria in solutions of submerged soils. Unpublished terminal report submitted to the International Rice Research Institute, Los Baños, Philippines (1972).

(30) Gilmour, J. T., Kittrick, J. A., Solubility and equilibria of zinc in a flooded soil. Soil Sci. Soc. Am. J., 43, 890–892 (1979).

(31) Shoji, S., Some notes on clay minerals in relation to soil fertility and rice production in Japan. The fertility of paddy soils and fertilizer applications for rice. Food and Fertilizer Technology Center for the Asian and Pacific Region. Taipei, Taiwan, China (1976).

(32) Bajwa, I., Ponnamperuma, F. N., Clay mineralogies of some Philippine rice soils and their relationship to available phosphorus and exchangeable potassium status. Paper presented at the Crop Science Society of the Philippines Meetings, Baybay, Leyte, Philippines (1980).

(33) Bajwa, I., Unpublished terminal report submitted to the International Rice Research Institute, Los Baños, Philippines (1980).

(34) Sims, J. L., Patrick, W. H. Jr., The distribution of micronutrient cations in soil under conditions of varying redox potential and pH. Soil Sci. Soc. Am. J., 42 (2), 258-262 (1978).

(35) Bao Hsuo-ming, Liu Chi-Kuang, Yu Tien-Jen, Studies on oxidation-reduction processes in paddy soils. VI. Determination of ferrous iron chelated by organic matter. Acta Pedologica Sinica 12, 216-221 (1964).

(36) Brinkman, R., Ferrolysis, a soil-forming process in hydromorphic conditions. Centre for Agricultural Publishing and Documentation, Wageningen (1979).

(37) Pasricha, N. S., Exchange equilibria of ammonium in some paddy soils. Soil Sci. 121 (5), 267-271 (1976).

(38) Pasricha, N. S., Ponnamperuma, F. N., Ionic equilibria in flooded saline, alkali soils: the K^+-$(Ca^{2+} + Mg^{2+})$ exchange equilibria. Soil Sci., 122 (6), 315-320 (1976).

(39) Parfitt, R. L., Anion adsorption by soils and soil materials. Adv. Agron. 30, 1-50 (1978).

(40) Khalid, R. A., Patrick, W. H. Jr., Delaune, R. D., Phosphorus sorption characteristics of flooded soils. Soil Sci. Soc. Am. J., 41 (2), 305-310 (1977).

(41) Iler, R. K., The chemistry of silica. John Wiley & Sons (1979).

(42) Jenne, E. A., Controls on Mn, Fe, Co, Ni, Cu, and Zn concentrations in soils and water: the significant role of hydrous Mn and Fe oxides. Trace inorganics in water. Adv. in Chem. Ser. 73, Am. Chem. Soc., Washington, D. C., pages 337-396 (1968).

(43) Gupta, G. P., The influence of temperature on the chemical kinetics of submerged soils. Unpublished Ph. D. thesis. Indian Council of Agricultural Research, New Delhi (1974).

(44) IRRI (International Rice Research Institute), Annual report for 1973. Los Baños, Philippines (1974).

(45) Ponnamperuma, F. N., The nitrogen supply in tropical wetland rice soils. Paper presented at a special workshop on nitrogen fixation and utilization in rice fields, International Rice Research Institute, Los Baños, Philippines (1980).

(46) Cho, D. Y., Ponnamperuma, F. N., Influence of soil temperature on the chemical kinetics of flooded soils and the growth of rice. Soil Sci., 112, 184-194 (1971).

(47) van Breemen, N., Moormann, F. R., Iron toxic soils. Soils and rice.

International Rice Research Institute, Los Baños, Laguna, Philippines, pages 781-800 (1978).

(48) Pao Hsuen-ming, Liu Chih-kuang, Yu Tien-jen, Studies on oxidation-reduction processes in paddy soils. IX. The forms of water-soluble ferrous iron. Acta Pedoligica Sinica, 15, 174-181(1978).

(49) Pao Hsuo-ming, Yu Tien-jen, Studies on oxidation-reduction processes in paddy soils. VIII. Characterization of the water-soluble ferrous iron. Acta Pedologica Sinica, 15, 13-22 (1978).

OXIDATION-REDUCTION PROPERTIES OF PADDY SOILS

Yu Tian-ren
(Institute of Soil Science, Academia Sinica, Nanjing)

In this paper, a brief review is made about studies on oxidation-reduction properties of paddy soils carried out in the Department of Soil Electrochemistry, Institute of Soil Science.

PHYSICO-CHEMICAL EQUILIBRIA OF REDOX SYSTEMS

1. Ferrous Iron

We distinguish the ferrous iron as: (1) water-soluble; (2) exchangeable; (3) complexed with the organic matter of the solid phase; (4) precipitated (1-5). It is seen from Fig. 1 that pH of the medium plays a dominant role in determining the relative proportions among the various forms of ferrous iron. As expected, the percentage of water-soluble iron increases steadily with the decrease of pH, at the expense of precipitated iron. It is to be noted that although the percentages of exchangeable and complexed iron also increase with the fall of pH, they decrease again at pH values below about 4. We think that this is due to the competition of hydrogen and especially aluminum ions for the exchange sites and ligands with the ferrous iron.

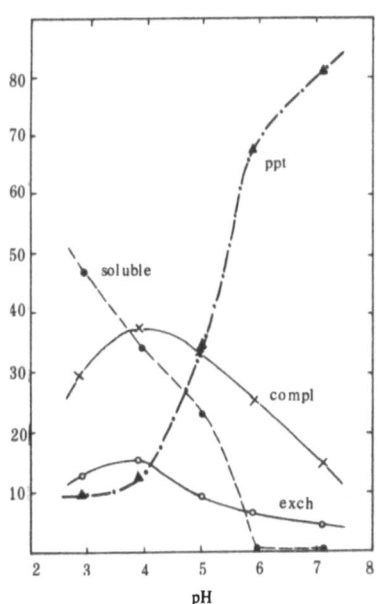

Fig. 1 Effect of pH on the form
of ferrous iron (acid
paddy soil)

The water-soluble ferrous iron may be distinguished as ionic and chelated, and the latter may be further characterized as those carrying positive charge and those carrying negative charge (Table 1). The ionic iron generally accounts for more than 70% of the total water-soluble iron. But, for strongly reduced paddy soils rich in organic matter, the chelated ferrous iron may be as high as 40% of the total.

Table 1 Forms of water-soluble ferrous iron

Soil type	Treatment	Ferrous iron (ppm)				
		chelated			Free ionic	Total
		Positive charge	Negative charge	Sum		
Str. r.		72.7	9.7	82.4	103.0	185.4
Med. r.	Original	5.5	6.0	11.5	26.5	38.0
Weak r.		2.3	2.2	4.5	15.0	19.5
Acid		8.3	5.4	13.7	82.9	96.6
Glei	5% O.M.	15.3	0.4	15.7	78.0	93.7
Sandy		1.1	0.2	1.3	8.4	9.7

It is seen from Table 1 that the relative proportions of iron-chelates carrying positive charge and those carrying negative charge vary with the soil.

The equilibrium of chelation-dissociation of water-soluble iron obeys the mass-action law. It is evidenced from Fig. 2 that for a given amount of ferrous iron the larger the amount of chelating agent, in this case the decomposition products of green manure, the higher the percentage of chelated iron. If sufficient chelating agent is present, as high as 90% of the ferrous iron may be chelated. And, as can be seen from Fig. 3, the presence of manganese ions may cause a part of the chelated iron to dissociate into free ions. The effect of hydrogen ions in competing for the ligands with ferrous iron is even more striking. As can be seen from Fig. 4, for the three soils there is a linear relationship between the logarithm of chelated iron and the pH of the solution, with a proportional increase of ionic iron as the hydrogen ions are increased.

With respect to the stability of the chelates of ferrous iron, it was found that the stability constant varies with the source of the organic substances. From Fig. 5 it can be seen that the constants for the decomposition products of Astragalus and vetch are higher than that of radish. For a given kind of green manure the constants are highest at the period of intensive decomposition of the organic matter.

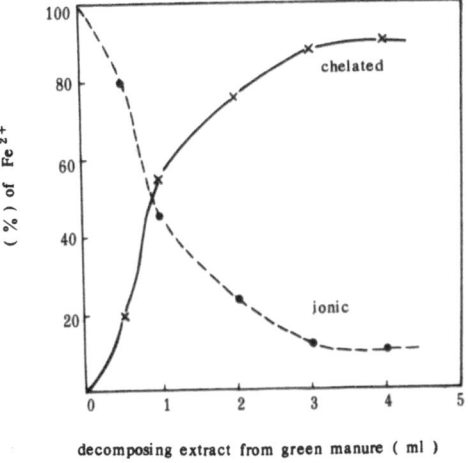

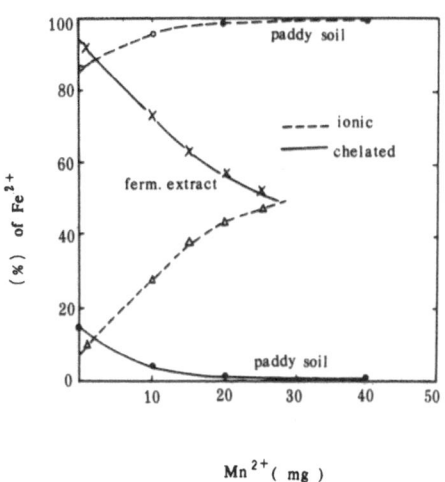

Fig. 2 Relationship between
amount of chelated ferrous
iron and amount of chela-
ting agent

Fig. 3 Effect of Mn^{2+} on the
equilibrium of chelated
ferrous iron

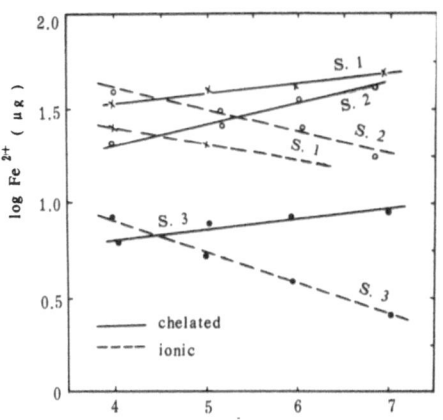

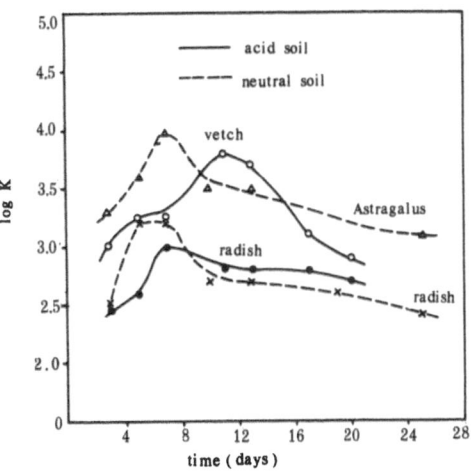

Fig. 4 Effect of pH on the
equilibrium of chelated
ferrous iron

Fig. 5 Stability constant of
Fe^{2+}-chelate with
decomposition products
of different green manures

The amount of exchangeable ferrous iron increases with the cation-exchange capacity of the soil (Table 2). And, owing to the precipitating effect, it decreases as the amount of sulfide ions increases.

Table 2 Relationship between exchangeable ferrous iron
and cation-exchange capacity of the soil

Soil	Fe^{2+} added (m.e./100g)	Original C E C (m.e./100g)	C E C (m.e./100g)	Exch. Fe^{2+} (m.e./100g)	% of total Fe^{2+}
Acid sandy paddy soil	5.37	7.36	1.84	0.30	5.6
			3.68	0.56	11.6
			7.36	0.83	19.4
Neutral paddy soil	7.16	21.0	5.3	0.86	20.1
			10.5	1.92	32.4
			21.0	2.42	38.6

We think that the presence of large amount of sulfide ions and the relatively high pH in reduced soils are the reasons why the amount of exchangeable ferrous iron is not so high as generally expected.

We distinguish the complexed ferrous iron from the above-mentioned water-soluble chelated iron in that it is associated with the solid phase of the soil organic matter. This fraction accounts for 16-36% of the total ferrous iron. When the organic matter was removed from the soil, no complexed ferrous iron could be found. From Fig. 6 it is seen that there is a linear correlation between the amount of this iron and the organic matter content of the soil, and it can be calculated from the slope of the straight line that one gram of soil organic matter can complex 6-8 mg of ferrous iron.

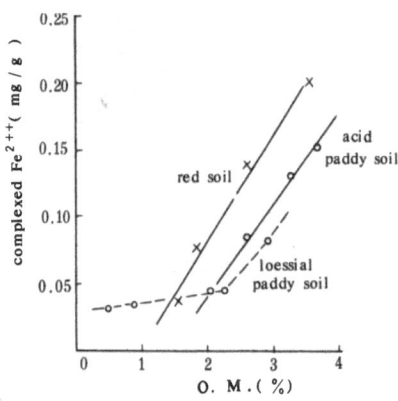

Fig. 6 Relationship between
complexed ferrous iron
and organic matter content
of the soil

98

2. Manganese

The behavior of manganese is very similar to that of ferrous iron
(6,7). The stability constants of water-soluble manganese-chelates are
of the same order of magnitude as that of ferrous-chelates, namely from
2.5 to 5.4, and there is also a peak value of log K for organic substances
produced at the stage of intensive decomposition of plant materials
(Table 3).

Table 3 Log K of Mn-complex with decomposition products of plant
materials different stages of decomposition

Treatment	Plant material	log K			
		2-5 days	5-6 days	12-14 days	20 days
no	Vetch	5.0	5.4	3.0	3.0
	Milk vetch	4.7	4.8	3.0	2.9
	Rice straw	4.0	4.0	2.7	2.9
Kaolin	Vetch	3.8	4.1	3.8	2.6
	Milk vetch	3.5	4.1	3.3	2.5
	Rice straw	3.0	3.3	2.9	2.5

3. Sulfides[8]

Among the factors which control the equilibria:
$$FeS \rightleftharpoons Fe^{2+} + S^{2-}$$
$$Fe^{2+} + 2OH^- \rightleftharpoons Fe(OH)_2$$
$$S^{2-} + H^+ \rightleftharpoons HS^-$$
$$HS^- + H^+ \rightleftharpoons H_2S$$
the pH of the medium is determinative with respect to both free sulfide
ions and molecular hydrogen sulfide. Fig. 7 shows this clearly. Since
the pH_2S value as measured with an electrochemical sensor and that
calculated from the pS^{2-} as determined with an ion-selective electrode
coincide fairly well, irrespective of the fact that the two methods are
based on different principles, it should be considered that there do exist
quantitative dependences of sulfide ion and free hydrogen sulfide on pH in
the soil.

Fig. 7 pH_2S and pS^{2-} as
function of pH for
an acid paddy soil

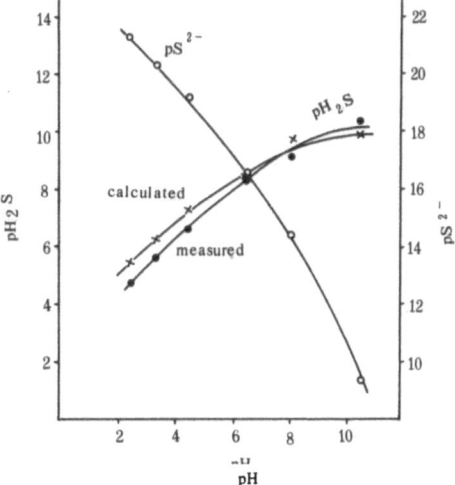

As expected, the addition of ferrous iron makes the concentration of hydrogen sulfide decreased.

4. Organic Reducing Substances

We don't know very much about the nature and behavior of this category of redox system. A limited number of work (9,10) shows that:

(1) Contrary to ferrous iron and sulfides, the extracted amount of organic reducing substances is practically independent of the pH of the equilibrium solution.

(2) The active groups with respect to chelating capacity for those carrying negative charge are generally larger in amount than those groups carrying positive charge (Table 4). The concentration of ligands as determined by potentiometric titration is nearly equal to the chelating capacity. On the other hand, the concentration of reducing groups of the decomposition products of organic matter is rather high.

Table 4 Concentration of active groups of decomposition products of various plant materials

Plant material	Concentration (N x 10^-)			
	Negative charge			Positive charge
	Reducing group*	Ligands	Chelating capacity	Chelating capacity
Vetch	—	3.5	3.8	2.2
Crotalaria	79	2.7	2.9	2.7
Astragalus	47	2.4	2.2	2.6
Alfalfa	30	1.1	1.8	0.8
Rice straw	91	2.1	1.8	2.2
Hairy vetch	43	2.7	1.6	1.6

* Titration with $KMnO_4$

(3) Owing to the reducing power metioned above, they can be character ized by voltammetric methods. It is seen from Fig. 8 that the composition of the water-soluble organic matter is rather complicated, especially at the stage of active decomposition of the green manure.

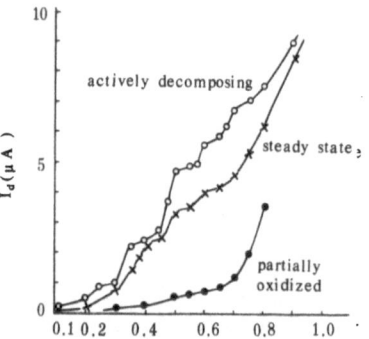

Fig. 8 I–V curves of decomposition products of green manure

In summarizing the physico-chemical equilibria of redox systems, we are deeply impressed with the mass-action law operating in many cases, even for so complicated a system as the soil.

ELECTROCHEMICAL CHARACTERISTICS

In this respect, discussion is only made of the relationship between the intensity factor (redox potential) and the capacity factor (amount of reducing substances) of redox status of the soil.

In addition to chemical analyses, several electrochemical means have been employed.

1. Potentiometric Titration

From the titration curves of soil extracts (11) it was found that for soils containing 7.6 and 4.6 m.e. of reducing substances per 100 grams of soil, the corresponding Eh were 30 and 140 mV respectively.

2. Depolarization Curve

An attempt has been made to utilize the principle of depolarization to characterize the capacity factor of the redox systems of soils (12, 13). For, if the platinum electrode is polarized for a certain time and then allowed to be depolarized by the redox substances of the soil, it may be assumed that the larger the amount of depolarizer near the electrode surface, the shorter the time required for the electrode to attain its equilibrium potential. Comparison of typical depolarization curves for a paddy profile (Fig. 9) shows that for the surface, plowpan and gley horizons with Eh of 480, 460 and -170 mV the required times are about 7, 5 and 1 minute respectively.

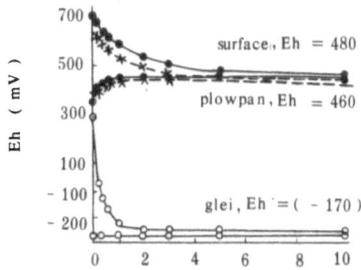

Fig. 9 Depolarization
curves for soils
of various horizons
(Nanjing)

3. Voltammetric Measurements

The diffusion current at a carbon electrode under an applied positive potential was used as an index of the amount of reducing substances of the soil (8,9). In a series of experiments with a carbon paste electrode it was found that there was a linear relationship between the Eh and the log_arithm of the amount of reducing substances expressed as micro l. In another series of experiments with a graphite electrode the relationship was even more pronounced, with a correlation coefficient of 0.905 (Fig. 10).

From what have been said above it may be concluded that the intensity factor and the capacity factor of redox status are closely interrelated, even for so complicated a system as the soil with various redox systems.

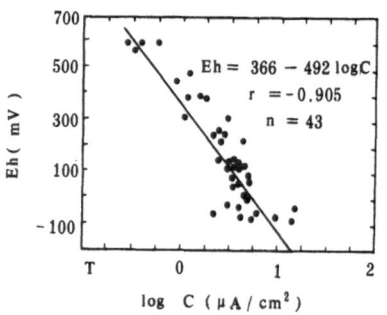

This gives rise to one theoretical question. In considering the Nernst equation it is generally assumed that in addition to the E^O term it is the ratio of the amount of oxidized form to that of reduced form that determines the Eh of the system(8,9). In paddy soils with various systems(14,15), if the relative proportions of the oxidized forms are so low that they contribute very little to the electron exchange current between the platinum electrode and the solution, it may be that it is mainly the amounts of the reduced forms that are more important with respect to the measured mixed potential at the platinum electrode.

Fig. 10 Relationship between Eh and amount of reducing substances(C)

SOME PRACTICAL IMPLICATIONS

In the following, three questions are to be discussed.

1. Soil Genesis

In China, some soil scientists suppose that between the plowpan and the illuvial horizon there is a horizon called percogenic horizon, namely percolation-developed horizon. But from the redox potential during the whole rice-growing season (16) (Fig. 11), it should be considered that even in the plowpan horizon the processes are predominantly illuvial with respect to oxidation-reduction, especially for its lower part.

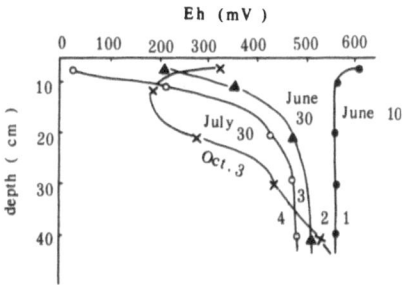

Fig. 11 Dynamics of redox potential
of a paddy profile
(bleached paddy soil)

2. Leaching of Nutrients

Table 5 indicates that the adsorption of ferrous iron causes an
equivalent exchange of calcium from the soil. Thus, the presence of
large amount of ferrous iron will enhance the leaching of nutrient
cations (17). In the field it has been observed that the leaching of
calcium from the surface layer after the application of lime is very intense. The leaching of potassium is even more striking. It is believed
that these are partly caused by the presence of ferrous and manganese
ions under reduced conditions.

Table 5 Exchange of calcium by ferrous iron

Soil	C E C (m.e./100g)	Fe^{2+}added (m.e./100g)	Fe^{2+}addsorbed (m.e./100g)	Ca^{2+}replaced (m.e./100g)	$\dfrac{Fe^{2+}\text{adsorbed}}{Ca^{2+}\text{replaced}}$
Acid		0.25	2.13	2.19	0.97
mountain		0.50	2.41	3.20	0.75
paddy	8.82	1.00	3.10	3.52	0.88
soil		2.00	3.44	3.67	0.94
		4.00	5.17	5.43	0.95
Acid		0.25	1.85	1.32	1.20
red		0.50	2.24	2.22	1.01
paddy	9.32	1.00	3.10	2.85	1.09
Soil		2.00	3.47	3.23	1.07
		4.00	4.83	4.61	1.05
Mean					0.99

3. Soil and Plant

It is shown is Table 6 that contrary to the case of wheat in which the Eh of root-zone region is lower than that of the bulk soil, the Eh of the root-zone region of rice is much higher than that of the bulk soil (12, 13,18). This is apparently due to the relative abundance of oxygen caused by the secretion from the rice roots.

Table 6 Effect of plant roots on Eh of the soil (mV)

Depth (cm)	Wheat Root zone	10cm*	20cm	Rice Root zone	5cm	10cm	20cm
0-3	425	485	605	250	80	15	-30
3-8	545	535	615	155	20	-10	-70
8-15	615	610	610	70	-15	-55	-100
15-20	620	620	610	55	-30	-65	-95
20-30	-	625	605	30	-45	-80	-100

*Distance from root zone

On the other hand, the redox condition of the soil affects the growth of rice in several ways. The redox potential of the soil is closely correlated with that of leaf sap of rice (11) and the percentage of black roots of rice plants (10). The tillering of rice is retarded in soils rich in strongly reducing substances (Table 7). The toxic effect of ferrous iron on rice growth deserves special attention. Based on field data, we tentatively take 50-100 ppm of water-soluble ferrous iron as the critical limit for rice growth (10). As for the toxicity of hydrogen sulfide, we have observed with an electrochemical sensor that for some acid paddy soils rich in easily decomposable organic matter the pH_2S may be 4.7, which in terms of molecular hydrogen sulfide corresponds to an amount of more than ten times higher than the critical level for the normal growth of rice. Thus, although rice plant possesses a peculiar physiological function, its growth can be retarded by strongly reducing conditions of the soil.

Table 7 Growth of rice in relation to amount of reducing substances (I_d) of the soil

Rice growth	I_d (μA)	
	Strongly reducing	Weakly reducing
Moist soil	0.07	1.23
Normal growth	0.42	2.21
Late tillering	0.67	1.74
Non-tillering	1.38	2.85
Black mud	2.5	6.2

In summarizing this section, we are of the opinion that from the practical point of view the control of redox conditions of soils is very important (19, 20).

CONCLUDING REMARKS

Although we have done a little on oxidation-reduction properties of paddy soils, we are aware that many problems remain to be elucidated. For instance, the quantitative relationship between redox potential and various systems is not known yet. We know very little about the nature and behavior of organic systems. And, the mechanism of the harmful effects under various reduced conditions is unclear. We believe that further studies on these and other subjects may aid in our understanding pertaining to redox properties of paddy soils, and of upland soils as well.

REFERENCES

(1) Yu Tian-ren, Ling Yun-xiao, Ding Chang-pu, Mou Run-sheng, Liu Wan-lan, Oxidation-reduction status of soils in subtropical regions. Kexue Tongbao, 11, 338-339 (1957). (in Chinese)

(2) Bao Xuo-ming, Liu Zhi-kuang, Yu Tian-ren, Studies on oxidation-reduction processes in paddy soils. VI. Determination of complexed ferrous iron. Acta Pedologica Sinica, 12, 216-221 (1964). (in Chinese with English summary)

(3) Bao Xuo-ming, Liu Zhi-kuang, Wu Jun, Studies on oxidation-reduction processes in paddy soils. VII. Forms of the ferrous iron. Acta Pedologica Sinica, 12, 297-306 (1964). (in Chinese with English summary)

(4) Bao Xuo-ming, Yu Tian-ren, Studies on oxidation-reduction processes in paddy soils. VIII. Characterization of the water-soluble ferrous iron. Acta Pedologica Sinica, 15, 13-22 (1978). (in Chinese with English summary)

(5) Bao Xuo-ming, Liu Zhi-kuang, Yu Tian-ren, Studies on oxidation-reduction processes in paddy soils. IX. Forms of the water-soluble ferrous iron. Acta Pedologica Sinica, 15, 174-181 (1978). (in Chinese with English summary)

(6) Ding Chang-pu, Yu Tian-ren, Studies on oxidation-reduction processes in paddy soils. IV. Activities of iron and manganese in paddy soils derived from red earth. Acta Pedologica Sinica, 6, 99-107 (1958). (in Chinese with English summary)

(7) Yu Tian-ren, Ling Yun-xiao, Mou Run-sheng, Liu Wan-lan, Effect of soil reaction on the activity of manganese. Turang Zhuanbao, 33, 16-30 (1958). (in Chinese with English summary)

(8) Yu Tian-ren, Zhang Xiao-nian, Electrochemical Methods and Their Applications in Soil Research. Science Press, Beijing, 408-432 (1980). (in Chinese)

(9) Yu Tian-ren (ed.). Electrochemical Properties of Soils and Their Research Methods. Science Press, Beijing, 399-464 (1976). (in Chinese)

(10) Institute of Soil Science, Academia Sinica, Nanjing (ed.), Soils of China. Science Press, Beijing, 345-359 (1978). (in Chinese)

(11) Yu Tian-ren, Liu Wan-lan, Studies on oxidation-reduction processes in paddy soils. III. Effect of oxidation-reduction on the growth of rice. Acta Pedologica Sinica, 5, 292-304 (1957). (in Chinese with English summary)

(12) Yu Tian-ren, Li Sung-hua, Studies on oxidation-reduction processes in paddy soils. I. Conditions affecting redox potential. Acta Pedologica Sinica, 5, 97-110 (1957). (in Chinese with Russian summary)

(13) Yu Tian-ren, Li Sung-hua, Studies on oxidation-reduction processes in paddy soils. II. Mutual influences between soil and plant. Acta Pedologica Sinica, 5, 166-174 (1957). (in Chinese with Russian summary)

(14) Yu Tian-ren, Xie Jian-chang, Yang Guo-zhi, On system determining redox potential in paddy soils. Kexue Tongbao, 6, 205-206 (1959). (In Chinese)

(15) Liu Zhi-kuang, Yu Tian-ren, Studies on oxidation-reduction processes in paddy soils. V. Determination of the reducing compounds. Acta Pedologica Sinica, 10, 13-28 (1962). (in Chinese with English summary)

(16) Yu Tian-ren, Xie Jian-chang, Yang Guo-zhi, Gao Zi-qin, Chen Jia-fang Shen Ren-shui, Ding Chang-pu, Zhou Qi-kun, Formation and reclamation of low-yield "white paddy soil" in Tai Lake region. Acta Pedologica Sinica, 7, 42-58 (1959). (in Chinese with English summary)

(17) Yu Tian-ren, Ding Chang-pu, On the status of exchangeable bases and its relation to the genesis of paddy soils derived from red earth. Turang Zhuanbao, 33, 31-43 (1958). (in Chinese with English summary)

(18) Liu Zhi-kuang, Yu Tian-ren, Studies on electrochemical properties of soils. II. Application of micro-electrodes in soil research. Acta Pedologica Sinica, 11, 160-170 (1963). (in Chinese with English summary)

(19) Yu Tian-ren, Liu Zhi-kuang, Oxidation-reduction processes in paddy soils and their relations to the growth of rice plant. Acta Pedologica Sinica, 12, 380-389 (1964). (in Chinese with English summary)

(20) Yu Tian-ren, Development of soil physical chemistry in China, a critical review. Acta Pedologica Sinica, 16, 203-210 (1979). (in Chinese with English summary)

THE ROLE OF INORGANIC REDOX SYSTEMS IN CONTROLLING REDUCTION IN PADDY SOILS

W. H. Patrick, Jr.
Boyd Professor
(Laboratory for Wetland Soils and Sediments, Center for Wetland Resources,
Louisiana State University, Baton Rouge, Louisiana 70803)

The unique water regime of paddy soils affects the reactivity of inorganic redox systems which usually remains inactive in drained soils. Oxygen, the inorganic nitrogen compounds, and sometimes manganese compounds, can undergo oxidation-reduction reactions in typically well drained soils, but other redox systems that are more difficult to reduce than these ordinarily are not involved in redox reactions. Thus inorganic ferric oxyhydroxide compounds, and especially sulfate and carbon dioxide do not normally undergo microbial reduction in well drained soils. Under excess water conditions in paddy soils, however, the supply of oxygen into the soil is curtailed and facultative anaerobic microorganisms and strict anaerobic microorganisms use these oxidized systems as electron acceptors and convert them to reduced forms. These reductions are prevented as long as the soil is porous enough to allow entry of oxygen from the atmosphere since the reduction of ferric iron, sulfate and carbon dioxide cannot take place in the presence of oxygen. The major effect of flooding on reduction processes is to saturate the pore space of the soil with water, thereby stopping gaseous diffusion of oxygen into the soil.

The reduction of the inorganic redox systems in the soil following flooding can be described in both intensity and capacity terms. The intensity factor determines the relative ease of the reduction, whereas the capacity factor denotes the amount of the redox system undergoing reduction. The capacity factor of a redox system probably can be best described in terms of its oxygen equivalent. The intensity factor can be represented by the free energy of the reduction, or more commonly by the equivalent electromotive force (EMF) of the reactions. In natural systems where there is biological activity and where several redox systems function, such as soils, the oxidation-reduction or redox potential is ordinarily used to denote the intensity of reduction.

The capacity factor of the various redox systems will vary from one soil to another. In general, the amount of oxygen in the soil at the time of flooding of a well drained soil is very low, consisting of the oxygen in the trapped air spaces plus that dissolved in the water occupying the pore space The amount of nitrate present at flooding is likely to be more variable than oxygen, but is usually only a few parts per million. Reducible manganese oxides are present in much higher concentrations in most soils than is oxygen or nitrate, but the manganese oxide concentration is variable with some soils having less than 100 ppm reducible manganese and others having over ten times as much.

The objective of the present paper is to describe in a quantitative way the intensity and capacity components of the major soil inorganic redox systems and show their relative importance in controlling the redox status of the soil. For rice culture the redox condition of the soil is an important detriminant of plant growth and grain yield. Moderate reducing conditions are known to enhance growth through a number of mechanisms, while intense reducing conditions produce substances that are toxic to the plant or which require a significant amount of the plant's energy to overcome.

Table 1 shows the redox system that will be dealt with in this paper. These systems are ranked on the basis of ease of reduction from the oxygen-water system to the carbon dioxide-methane system. Oxygen readily accepts electrons from decomposing plant material while the reduction of carbon dioxide to methane occurs only under very reducing conditions. The oxidized and reduced components of these various inorganic redox systems are shown as well as an indication of their condition in the soil solution.

The author and his associates have devoted considerable attention to studying the intensity aspects of these inorganic redox systems. The approach taken has been to determine the redox potential at which the oxidized component becomes unstable and accepts electrons from respiring microorganisms. The redox potential was chosen as an index of intensity for these studies because it covers the entire range over which these various inorganic redox systems function. As shown in Figure 1, the redox potential range encountered in waterlogged soils extends from approximately -300 millivolts on the reducing end to approximately +700 millivolts on the oxidizing end or about the entire range encountered in all biological systems. In well drained soils that are permeated with oxygen from the atmosphere the normal range of redox potential encountered is much less and occurs in a narrow range at the oxidizing end of the redox scale.

Our experimental approach to determining the redox intensity at which each of these inorganic systems function has been to set up a system in which the redox potential is closely controlled and the reduction of the oxidized component of the various inorganic redox systems studied. This technique involves the use of stirred soil suspensions that have their redox potential closely controlled at any point in the range indicated in Figure 1. A suspension of soil ranging from a 2/1 to 4/1 soil-water mixture (w/w) is incubated in a sealed chamber and the desired redox potential obtained by automatically adding very small amounts of oxygen to the system when a monitoring platinum electrode indicates that the redox potential has decreased below the set value. A detailed description of this method is given in (1, 2). The pH as well as the redox potential can be closely controlled in this system which allows the two major parameters involved in Nernst type redox reactions to be utilized in the study of redox systems. Several papers (3, 4, 5) report results in which the effects of both redox potential and pH on various soil systems were studied.

A summary of the results of a number of studies of the critical redox potential at which the various inorganic redox systems become unstable is shown in Figure 2. Going from the most easily reduced to the most difficultly reduced systems shows that oxygen is reduced first, followed by nitrate and oxidized manganese compounds, and then followed by ferric iron compounds. After the reduction of ferric iron the next system to become unstable is sulfate followed by the reduction of carbon dioxide to methane. The order of these reduction reactions is the same as that indicated by thermodynamic considerations as the data in Table 2 indicates. The reduction of several of these oxidized redox systems is completely sequential, i.e., all of one system is reduced before the next system begins to undergo reduction. As examples of this sequential reduction or lack of overlap in the reductions of several

Table 1   Oxidized and reduced forms of inorganic redox
components of soils that undergo reduction following flooding

OXIDIZED	REDUCED
O_2 (volatile)	H_2O (soluble)
NO_3^- (soluble)	N_2 (volatile)
Mn^{4+} (insoluble)	Mn^{2+} (soluble)
Fe^{3+} (insoluble)	Fe^{2+} (soluble)
SO_4^{2-} (soluble)	S^{2-} (insoluble)
CO_2 (soluble)	CH_4 (volatile)

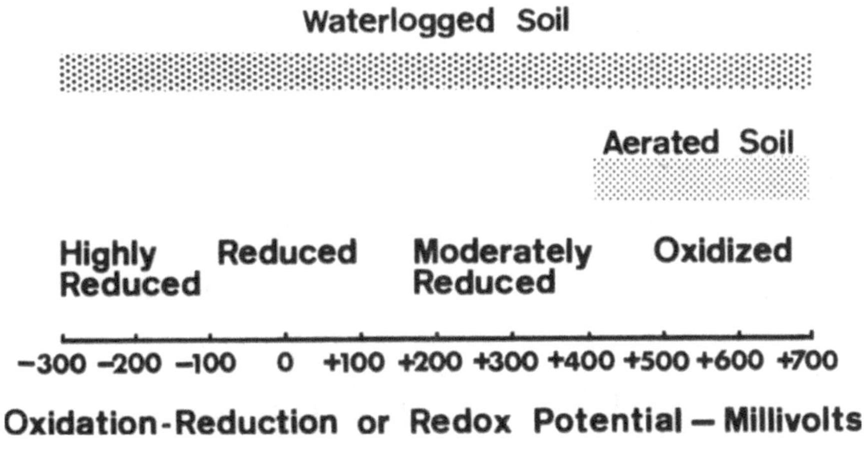

Fig. 1   The range of redox potentials encountered in waterlogged soils and aerated soils

Table 2   Thermodynamic sequence of several reduction
reactions involving inorganic oxygen, nitrogen,
manganese, iron and sulfur systems

Electrochemical Reaction	E_0^7
$O_2 + 4H^+ + 4e \rightleftarrows 2H_2O$	0.83
$NO_3^- + H_2O + 2e \rightleftarrows NO_2^- + 2OH^-$	0.43
$MnO_2 + 4H^+ + 2e \rightleftarrows Mn^{+2} + 2H_2O$	0.41
$Fe(OH)_3 + e \rightleftarrows Fe^{+2} + 3OH^-$	-0.13
$SO_3^{2-} + 3H_2O + 6e \rightleftarrows S^{2-} + 6OH^-$	-0.20

From Ponnamperuma (6).

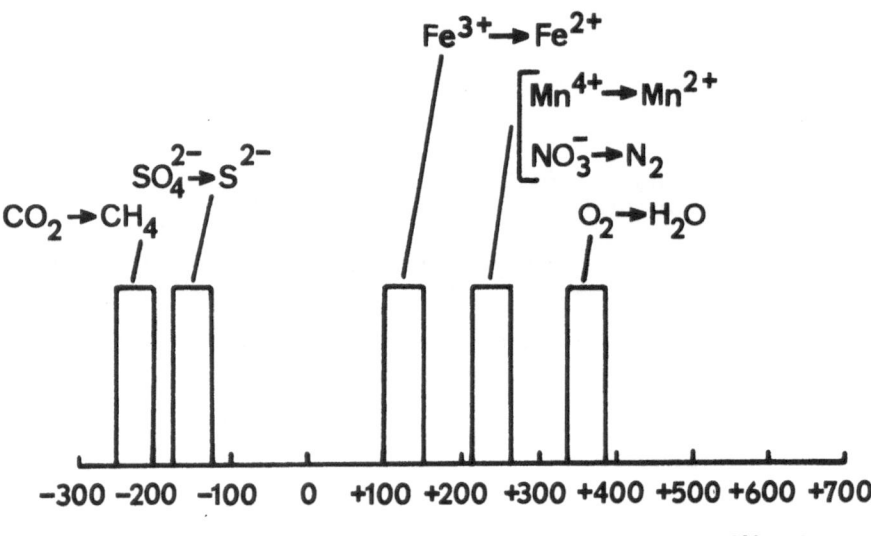

$$Fe^{3+} \rightarrow Fe^{2+}$$

$$\left[Mn^{4+} \rightarrow Mn^{2+} \right.$$

$$\left. NO_3^- \rightarrow N_2 \right]$$

$$SO_4^{2-} \rightarrow S^{2-}$$

$$CO_2 \rightarrow CH_4$$

$$O_2 \rightarrow H_2O$$

-300 -200 -100   0   +100 +200 +300 +400 +500 +600 +700

Oxidation-Reduction or Redox Potential, Millivolts
(Corrected to pH 7)

Fig. 2   The critical redox potential at which oxidized inorganic redox
systems begin to undergo reduction in flooded soils

of the systems no ferric iron is reduced to the ferrous form as long as any
oxygen or nitrate is present in the soil. Likewise, sulfate and carbon
dioxide will not be reduced if oxygen or nitrate are present. At the more
reducing end of the scale, almost all of the sulfate must be reduced to
sulfide before any methane appears.

The above results show that the soil inorganic redox systems differ
considerably in their reduction intensities with the systems at the oxidized
end of the redox scale accepting electrons from respiring microorganisms much
more readily than the oxidants shown on the reducing end of the redox scale.
On the basis of the energy made available for microbial respiration more energy
is released per electron transferred where an easily reduced redox system is
involved (oxygen, for example) as compared to reduction of the oxidized iron
system. There are many uses of such information on reduction intensity, one
example is that the knowledge of the redox of a soil will give an indication
of the oxidation-reduction status of these various components. For example,
a redox potential of zero millivolts indicates that oxygen and nitrate are
not likely to be present and that the bioreducible iron and manganese compounds
are in a reduced state. At this same potential, however, sulfate is stable
in the soil with no sulfide being formed and there also will be no methane
produced at this potential.

Reduction of the various inorganic redox systems are carried out by
different types of microorganisms. Reduction of oxygen to water is carried
out by true aerobic microorganisms. Reduction of nitrate to nitrogen,
reduction of manganic manganese to the manganous form, and reduction of
ferric iron to the ferrous form are carried out by facultative anaerobes.
These are microorganisms that normally function with oxygen as the terminal
electron acceptor but which have the capacity to switch over to other electron
acceptors when oxygen is limiting or absent. Common examples of this type of
microorganism are the denitrifiers and the iron reducers. In general, the
more difficult the reduction, the fewer the species that will carry out the
reduction reaction. Consequently, there are many more species of aerobes than
of facultative anaerobes in a soil, and among the facultative anaerobes many
more species can reduce nitrate than are capable of reducing ferric iron to
the ferrous form. The reduction of sulfate to sulfide and carbon dioxide
to methane are carried out by true anaerobes that cannot function in the
presence of oxygen or nitrate.

Although a knowledge of the critical redox potential at which the inorganic
redox systems become unstable and are reduced provides valuable information, it
provides no indication of the total capacity of the system to accept electrons
and thereby support respiration. For this reason, it is essential that an
understanding of the capacity factor in redox reactions in soils be obtained.
This capacity factor is equivalent to the total amount of electrons accepted
by the oxidants in support of microbial respiratory activity. As would be
anticipated, those redox systems present in lowest amount generally have the
smallest capacity for supporting microbiological respiration.

Two experiments are reported here that were designed to evaluate the
capacity of the inorganic redox systems to support respiration. In the first
experiment, samples of a Crowley silt loam soil from the coastal prairie rice
growing area of Louisiana and Sharkey clay from the alluvial rice growing
area of Louisiana were waterlogged and the reduction of the various inorganic

redox systems measured with time. The amount of oxidants available for
bacterial reduction in the two soils was for the Crowley: O_2, 10 ppm (on
dry soil basis); NO_3^--N, 43 ppm; Mn^{4+}, 250 ppm; and Fe^{3+}, 2500 ppm. For the
Sharkey clay the values were O_2, 10 ppm; NO_3^--N, 34 ppm; Mn^{4+}, 330 ppm; and
Fe^{3+}, 2400 ppm. In order that these inorganic oxidants be expressed on an
equivalent basis they were converted to the equivalent amount of oxygen that
could support glucose oxidation. Equations showing the reaction of O_2, NO_3^-,
MnO_2 and Fe_2O_3 with glucose are shown in Table 3. These equations were used
as the basis for calculating the oxygen equivalent of each of the inorganic
redox components. These oxygen equivalents are shown in Table 4 and are given
on both a molar basis in column 3 and on a microgram per gram of soil or parts
per million (ppm) basis in column 4. For example, the figures in column 4
show that .35 ppm NO_3^--N is equivalent to 1 ppm O_2 while 3.49 ppm Fe^{3+} iron is
equivalent 1 ppm O_2.

Using these calculated values, the oxygen equivalent of each of the redox
systems that support the respiration of microorganisms decomposing soil organic
matter are shown in Figures 3 and 4 for the Crowley and Sharkey soils. The
total oxygen equivalent of oxygen itself was very low which, along with its
ease of reduction, was responsible for the rapid disappearance of oxygen
following flooding. The oxygen equivalent of nitrate in these two particular
soils with their fairly high initial nitrate concentration was appreciably
higher than that of oxygen. Nitrate is a very effective electron acceptor
since five electrons are required to reduce a nitrate ion to its elemental
form. The oxygen equivalent of manganic manganese was about the same as that
of nitrate in these two particular soils. In some cases where soil nitrate
is lower manganese will be relatively more important in supporting respiration.
The iron system is by far the most important inorganic redox system in soils
insofar as its capacity to support microbial respiration is concerned. For
both soils the oxygen equivalent of oxidized iron was more than twice that
of the other redox systems combined.

The relative contributions of the manganese and iron systems to supporting
facultative anaerobic respiration was determined in a second experiment. In
this experiment suspensions of a Crowley soil were maintained at various redox
potentials over most of the possible redox range encountered in soils using
the same technique mentioned above. The redox potential was initially set on
the control apparatus at +500 millivolts and then decreased by 50 millivolt
increments to very reducing potentials. At each of the selected redox
potential values a two week incubation period was allowed for equilibration
to take place. After equilibrium samples of the soil suspension were withdrawn and the exchangeable plus soluble Mn^{2+} and Fe^{2+} determined. The
experiment was carried out at pH 7 and extractions of iron and manganese made
with a one normal sodium acetate solution of the same pH. The results are
expressed in the oxygen equivalent amounts of manganese and iron that were
reduced per unit change in redox potential. This provided curves for manganese
and iron redox buffering in the soil at the redox potential at which these
systems function.

The results of this study are shown in Figure 5 and illustrate two
important conditions. The first is the difference in the redox ranges at
which the two systems buffer the redox potential. The second is the difference
in capacity of the two systems to serve as electron acceptors. The major
effect of manganese is seen in the range +200 to +300 millivolts while the

Table 3   Reaction of inorganic oxidants with glucose

Oxidant	Reaction
O_2:	$C_6H_{12}O_6 + 6\ O_2 \longrightarrow 6\ CO_2 + 6\ H_2O$
NO_3^-:	$5\ C_6H_{12}O_6 + 24\ NO_3^- + 24\ H^+ \longrightarrow 30\ CO_2 + 12\ N_2 + 42\ H_2O$
MnO_2:	$C_6H_{12}O_6 + 12\ MnO_2 + 24\ H^+ \longrightarrow 6\ CO_2 + 12\ Mn^{2+} + 18\ H_2O$
Fe_2O_3:	$C_6H_{12}O_6 + 12\ Fe_2O_3 + 48\ H^+ \longrightarrow 6\ CO_2 + 24\ Fe^{2+} + 30\ H_2O$

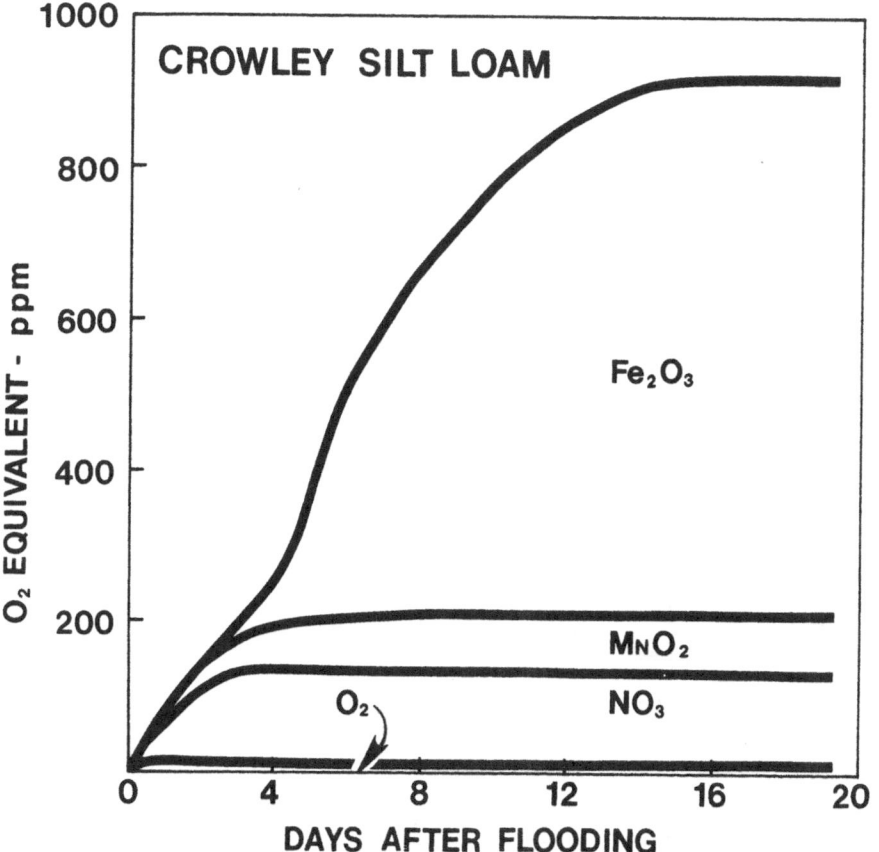

Fig. 3   The capacity of different inorganic redox systems to accept electrons from decomposing organic matter on basis of oxygen equivalents for a Crowley silt loam soil

Table 4   Oxygen equivalent of inorganic oxidants

Oxidant	Moles oxidant/ moles glucose	O_2 equivalents on mole basis	Equivalents on µg/g or ppm basis
O_2	6	1	1 ppm O_2/ppm O_2
NO_3^--N	24/5	0.8	0.35 ppm N/ppm O_2
MnO_2	12	2	3.43 ppm Mn/ppm O_2
Fe_2O_3	12	2	3.49 ppm Fe/ppm O_2

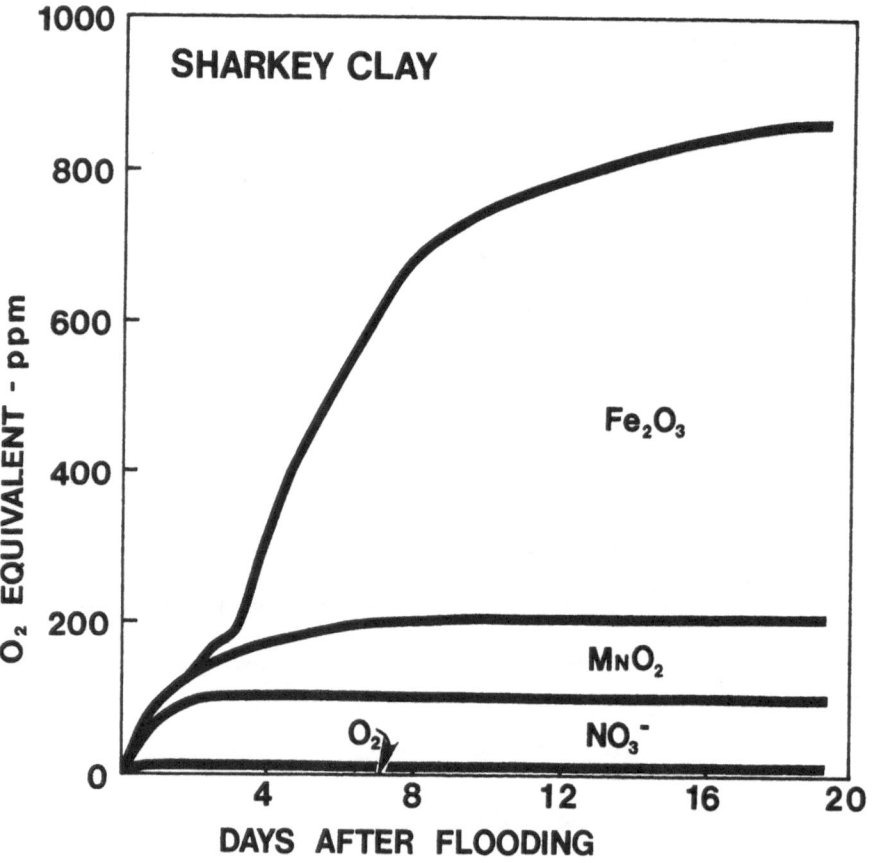

Fig. 4   The capacity of different inorganic redox systems to accept electrons from decomposing organic matter on basis of oxygen equivalents for a Sharkey clay soil

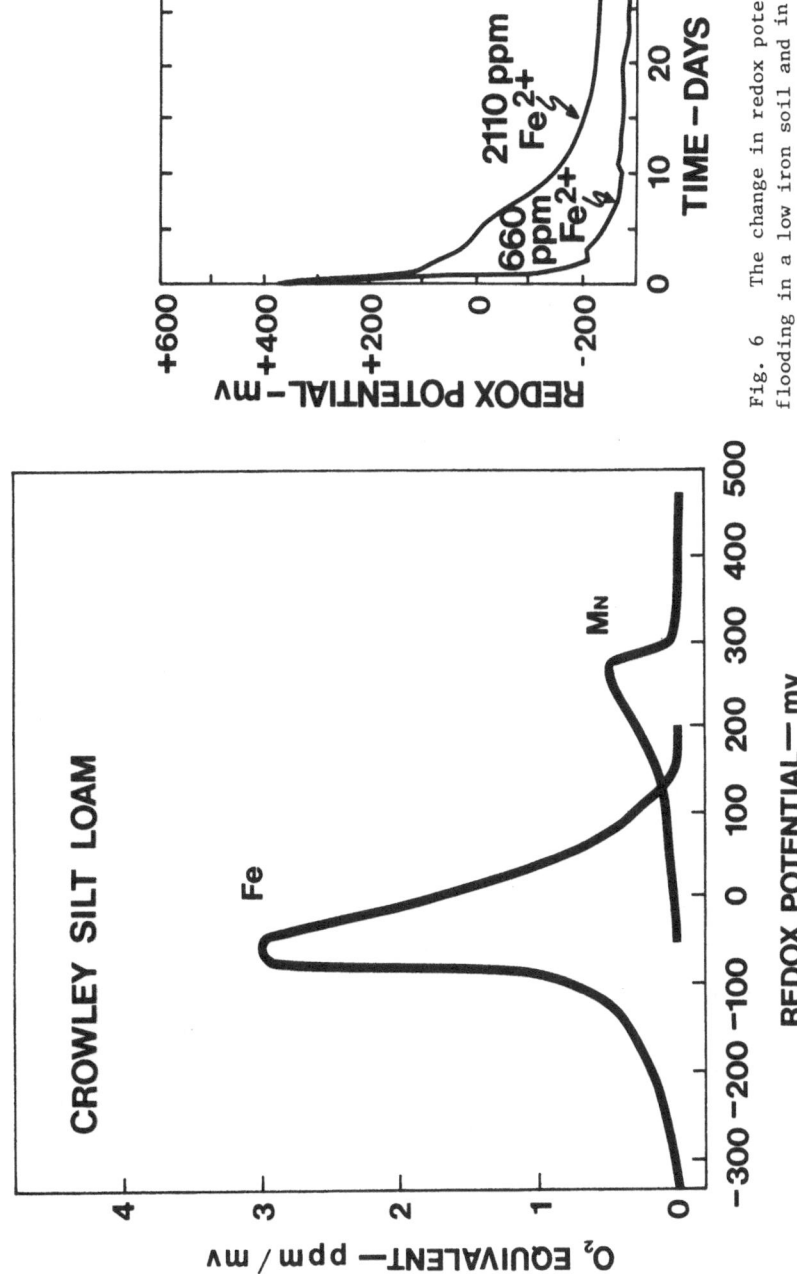

Fig. 6　The change in redox potential after flooding in a low iron soil and in a high iron soil

Fig. 5　The redox buffering capacity of the manganese and iron systems in a Crowley silt loam soil

major effect of the iron system is seen in the range -100 to +100 millivolts. When a soil is flooded and the redox potential decreases from a highly oxidized value of about +500 millivolts to about +100 millivolts almost all of the manganese is in the reduced form while little of the ferric iron has been affected. The oxidized manganese serves to maintain the redox potential in the range +300 millivolts to +200 millivolts by accepting electrons from the decomposing organic matter and preventing the redox potential from decreasing to very negative values until all of the bioreducible manganese has been converted to the manganous form. Likewise, oxidized iron maintains the redox potential in the range +100 to -100 millivolts. The buffering effect of iron is essentially depleted at about -150 millivolts and it is in this range that the facultative anaerobes cease to function and true anaerobes such as those that reduce sulfate and carbon dioxide take over.

The total capacities of the two systems to accept electrons is denoted by the area under the curves. It is apparent that, as in the previously described experiment, the capacity of the iron system is much greater than the manganese system. These results show the relatively high redox buffering capacity of the iron system and explain why soils with a large amount of bioreducible iron do not undergo a rapid decrease in redox potential. The effect of soil iron content on redox buffering as measured by the decrease in redox potential is illustrated in Figure 6 which shows the change in redox potential following flooding of two soils that have different iron contents. The soil with only 660 ppm of bioreducible iron showed a rapid drop in redox potential with the potential reaching -200 millivolts in less than three days. The high iron soil underwent a much slower decrease in redox potential and required approximately two weeks for the redox potential to reach -200 millivolts. This redox buffering by the iron system can help prevent the onset of toxic reducing conditions in flooded soils.

Another role of the inorganic redox systems in flooded soils is to support organic matter decomposition. The decomposition of organic matter supported by the nitrate, manganese and iron systems is similar to the decomposition supported by oxygen since carbon dioxide and the reduced oxidant are the major products of this type of decomposition. True anaerobes, on the other hand, produce organic acids, aldehydes and organic sulfur compounds that under certain conditions are toxic to rice plants. The presence of these substances is one of the major reasons why intense reducing conditions, such as those characterized by redox potentials below that at which all of the reducible iron has been converted to the ferrous form, are considered to be undesirable for optimum rice growth.

Another beneficial function of the large amount of the inorganic oxidants in the soil is the nutritional effect of the nitrogen, sulfur and phosphorus released from the decomposing organic matter. For the Sharkey clay, for example, an O_2 equivalent of 860 ppm will provide the electron accepters for the oxidation of organic matter equivalent to 806 ppm glucose. If it is assumed that 1 mole of nitrogen is released per 20 moles of carbon dioxide produced, facultative anaerobic respiration in this soil should mineralize approximately 16 ppm of N. This is probably more decomposition than can be supported by the amount of oxygen than can diffuse into a flooded soil during the growing season. It is likely that much of the decomposition of organic matter that takes place in flooded soils is carried out by the facultative anaerobes that reduce nitrate, manganic manganese compounds and especially ferric iron compounds.

In coastal areas that receive sea water containing large amounts of sulfate the reduction of sulfate to sulfide provides a few species of true anaerobes with a respiratory system that can support considerable oxidative activity. In non-cultivated coastal marshes where most of the active iron is in the reduced form it has been estimated that much of the respiratory activity is due to sulfate. In these areas the soil often stays wet most of the year preventing the iron from oxidizing to the ferric form, while sulfate is continuously supplied from the sea. In inland areas where rice growing is more important the more limited amount of sulfate in the soil does not support significant anaerobic respiration.

CONCLUSIONS

Both the intensity and the capacity factors must be considered when evaluating the role of inorganic reductants in supporting the microbial decomposition of organic matter in flooded soils. While oxygen and nitrate are extremely active as electron accepters and function at a high redox potential (thereby preventing the redox potential from falling to a low level as long as they are present) their low concentrations in flooded soils limit their activity to a few hours or a day or so at most. Most of the anaerobic organic matter decomposition taking place in soils is supported by oxidized bioreducible ferric compounds. These oxidized iron compounds provide a large reservoir for accepting electrons during the microbial decomposition of organic matter by facultative anaerobes that function after oxygen has been depleted in the soil. Although occupying a relatively negative position on the redox potential scale the high capacity for accepting electrons of the oxidized iron compounds make iron the key redox element in flooded soils.

LITERATURE CITED

(1) Patrick, W. H., Jr. 1966. Apparatus for controlling the oxidation-reduction potential of waterlogged soils. Nature 212:1278-1279.

(2) Patrick, W. H., Jr., B. G. Williams, and J. T. Moraghan. 1973. A simple system for controlling redox potential and pH in soil suspensions. Soil Science Society of America Proceedings 37:331-332.

(3) Gotoh, S., and W. H. Patrick, Jr. 1974. Transformation of iron in a waterlogged soil as influenced by redox potential and pH. Soil Science Society of America Proceedings 38:66-71.

(4) Patrick, W. H., Jr., S. Gotoh, and B. G. Williams. 1973. Strengite dissolution in flooded soils and sediments. Science 179:564-565.

(5) Moraghan, J. T., and W. H. Patrick, Jr. 1974. Selected metabolic processes in a submerged soil at controlled pH values. Transactions of 10th International Congress of Soil Science, II:264-269.

(6) Ponnamperuma, F. N. 1965. Dynamic aspects of flooded soils and the nutrition of the rice plant. In the Mineral Nutrition of Rice, The Johns Hopkins Press, 1965.

FERTILITY OF PADDY SOILS IN TROPICAL ASIA

Kazutake Kyuma
(Faculty of Agriculture, Kyoto University, Japan)

INTRODUCTION

More than ninety percent of the world rice lands occur in monsoon Asia, of which nearly two-thirds are distributed in monsoon tropical Asia that extends to the east of the Indus and to the south of the Chinese border, including insular southeast Asia.

We have been studying paddy soils in tropical Asia with the aim to clarify their material and fertility characteristics finally to contribute to an increase in rice production in the region, which still remains at present at a low level of 1.5-2 tons of paddy per hectare. So far some 600 paddy soil samples from this region were studied both in the field and in the laboratory.

Although the number is still too small to cover the entire region of tropical Asia in some detail, it is at the same time large enough to allow us to draw some generalizations. Thus, we attempted to set up a scheme of fertility evaluation for tropical Asian paddy soils. In this paper I will explain how we approached to this problem and what the results of fertility evaluation were.

CORRELATION ANALYSIS

When we study many soils both in the field and in the laboratory the number of data obtained is so large that it is almost impossible to draw useful information either on soil fertility or on soil genesis by a manual manipulation of the data. Thus, it is imperative to use a computer for the data processing. For handling such a big data matrix there are many kinds of multivariate statistical methods. The basic prerequisite for these various methods is the correlation between variables that are obtained for the object of study, i.e. soil in our case. Therefore, the first step of our study is to look into the correlation of the data we obtained.

In Table 1 correlation coefficients between all pairs of 29 variables for 410 tropical Asian paddy soils are given in a matrix form. The variables may be grouped into six, that is, those related to base status, mechanical composition, clay mineralogical composition, organic matter status, phosphorus status, and total chemical composition. To facilitate distinction of the different degrees of correlation, the table is transformed into a figure by dividing the range of correlation coefficients into five grades, each designated by a specific pattern as shown in the legend of Fig. 1. A careful examination of the correlation matrix has revealed a few important points as summarized below.

i) Base status characters are highly correlated not only among themselves but also with textural composition, clay mineralogy, and part of the total chemical composition. Silt, 10 A minerals, total titanium and potassium contents are rather exceptional, showing only low to insignificant correlations with base status characters.

ii) Of the mutually correlated characters referred to in i), sand content in the soil, 7 A mineral content in the clay fraction, and total silica content

	pH	Ex-Ca	Ex-Mg	Ex-(Ca+Mg)	Ex-Na	Ex-K	CEC	Avail. SiO₂	Sand	Silt	Clay	7 A Min.	10 A Min.	14 A Min.	TC	TN	NH₄-N	TP	Bray-P	HCl-P	TSIO	TFEO	TALO	TCAO	TMGO	TMNO	TTIO	TKAO
pH																												
Ex-Ca	622																											
Ex-Mg	289	534																										
Ex-(Ca + Mg)	569	943	786																									
Ex-Na	298	275	579	430																								
Ex-K	324	448	633	577	609																							
CEC	399	855	781	934	371	595																						
Avail. SiO₂	625	701	527	721	293	514	718																					
Sand	-016	-463	-524	-545	-243	-436	-662	-351																				
Silt	-041	-027	016	-014	-032	015	014	-074	-564																			
Clay	045	577	623	668	313	542	791	470	-854	052																		
7 A Min.	-537	-564	-474	-600	-236	-378	556	-412	283	-209	-211																	
10 A Min.	093	-125	-164	-156	-023	064	-221	-302	-066	280	-096	-230																
14 A Min.	462	639	585	699	253	330	688	527	-282	053	308	-759	-325															
TC	-254	002	152	062	044	139	233	070	-316	110	313	-008	-159	040														
TN	-264	-057	085	-008	022	122	154	026	-297	166	255	-018	-105	-012	958													
NH₄-N	-117	009	084	040	021	115	143	222	-190	108	162	008	-305	100	514	586												
TP	245	194	155	203	148	320	244	415	-198	135	154	-201	-023	080	292	333	323											
Bray-P	268	079	026	068	145	350	039	154	078	-059	-057	-144	166	038	010	009	-047	333										
HCl-P	379	154	057	136	257	356	047	159	094	-001	-113	-290	253	114	-007	002	-059	436	806									
TSIO	-322	-512	-502	-573	-218	-456	-631	-603	655	-223	-651	365	002	-326	-212	-209	-243	-509	-020	-119								
TFEO	389	557	504	607	202	384	638	685	-512	100	555	-304	-133	373	045	013	114	481	-008	066	-841							
TALO	066	351	436	429	162	404	544	457	-716	244	712	-198	-030	183	330	330	321	432	-032	010	-919	657						
TCAO	553	418	157	368	109	174	278	418	-010	018	000	-449	-079	402	-063	-050	092	243	075	210	-406	289	159					
TMGO	494	434	415	482	328	379	434	414	-324	226	250	-501	110	377	038	065	089	342	102	272	-627	502	411	600				
TMNO	404	466	431	511	345	331	485	637	-262	039	292	-249	-239	376	-069	-107	099	393	049	144	-525	686	351	300	350			
TTIO	102	195	241	238	052	095	274	231	-166	-066	242	-065	-128	158	024	-007	014	201	-024	007	-350	499	217	-006	150	385		
TKAO	136	-014	-017	-017	052	158	-066	-175	-133	227	018	-234	666	-151	004	049	-153	036	183	285	-224	-042	192	-015	260	-186	-113	
TPHO	173	193	144	198	085	269	239	380	-118	054	109	-216	-082	073	344	376	387	734	249	357	-471	392	409	278	275	333	224	043

NOTE: Abbreviations for variable names are as in the preceding; TSIO to TPHO are for total elemental oxides in the order of SiO₂, Fe₂O₃, Al₂O₃, CaO. MgO. MnO₂. TiO₂. K₂O and P₂O₅.

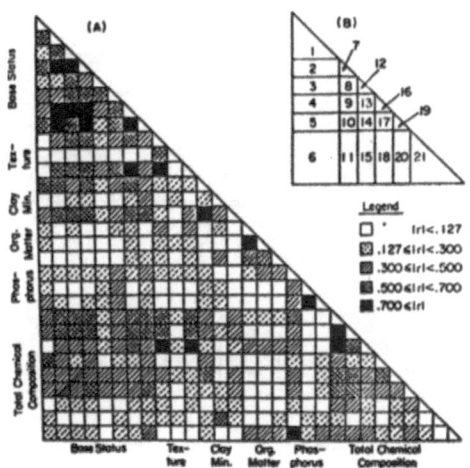

Fig. 1 Patternized expression of the correlation matrix given Table 1

in the soil are negatively correlated with most of the other characters, though their mutual correlations are positive.
iii) Characters representing organic matter are not highly correlated with any of the character groups, though their mutual correlations are high.
 iv) The same can be said of the characters related to available phosphorus status.

Such correlations between soil data are basically held for any soil groups, although a slight hindrance may be expected due to human interference, such as liming and fertilizer application, in the case of intensively managed cultivated soils.

NUMERICAL TAXONOMY

In an attempt to summarize the results of studies on tropical paddy soils, numerical taxonomy was first applied to our own data, because the concept and statistical method of numerical taxonomy was easy to comprehend.

Numerical taxonomy may be defined as "the numerical evaluation of the affinity or similarity between taxonomic units and the ordering of these units into taxa on the basis of their affinities" (1). It was originally proposed for general or natural classification of such objects as plants, insects, microbes, etc. In this study, however, the method is used to create groups on the basis of the similarity of a limited number of characters which are relevant to the fertility of the soil.

The actual procedure of numerical taxonomy consists of the following steps; standardization of the data to make them dimensionless, computation of between-sample similarity coefficients, sorting or clustering, and formulation of a dendrogram. Usually, between-sample correlation coefficient and taxonomic distance (or Euclidean distance) are used as similarity coefficients to represent similarities in the pattern and in the magnitude, respectively.

$$r_{jk} = \frac{\sum_{i=1}^{n} (X_{ij} - \bar{X}_j) (X_{ik} - \bar{X}_k)}{\sqrt{\sum_{i=1}^{n}(X_{ij} - \bar{X}_j)^2 \ \sum_{i=1}^{n} (X_{ik} - \bar{X}_k)^2}}$$

$$d_{jk} = \left\{ \frac{\sum_{i=1}^{n} (X_{ij} - X_{ik})^2}{n} \right\}^{\frac{1}{2}}$$

As a representative sorting or clustering method we used the weighted pair-group method, which allows the two mutually nearest operational taxonomic units(a soil or a soil group) to join in one clustering cycle.

An example of the dendrogram is shown in Fig. 2. The results of numerical taxonomy were statisfactory in that soils apparently similar in our field observation were put together in a cluster with a high similarity coefficient. Useful as the result was for our purpose of summarizing the soil data, it could not show the fertility relations among the established clusters, that is, it did not show which one is more fertile or less fertile relative to others. Thus search for a method to look into the fertility relations among the soil samples was the next step of our effort.

In the course of manipulating the data for refining the numerical

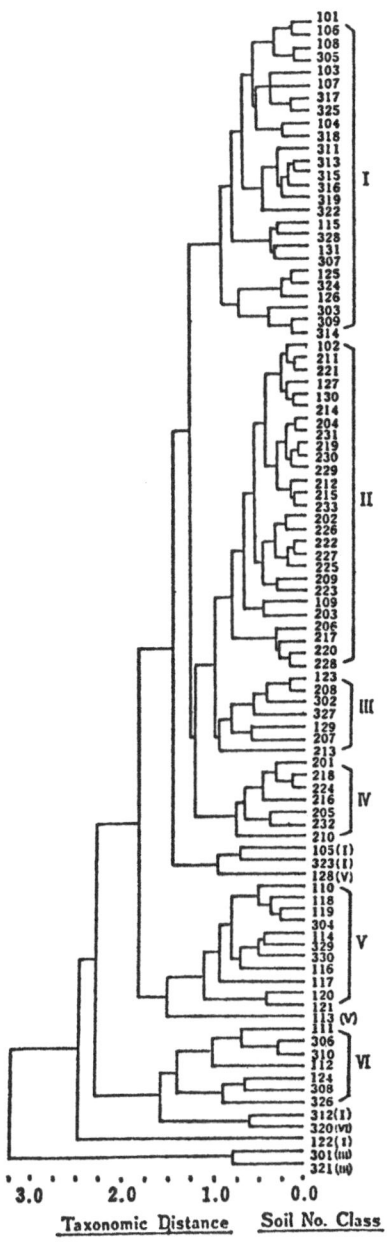

Fig. 2 Dendrogram showing relationship
       among the 94 surface soils with
       respect to fertility based on
       taxonomic distance

taxonomy, we found that principal component analysis used to extract mutually independent compound characters could give a sort of chemical potentiality rating of the sample soils. Thus, we furthered our effort along this line and finally came to application of factor analysis, as a refinement of principal component analysis, for attaining fertility evaluation for tropical Asian paddy soils.

PRINCIPAL COMPONENT ANALYSIS AND FACTOR ANALYSIS

The method of principal component analysis is used for attaining a "parsimonious summarization of a mass of observation" (2). In other words, it is used to extract the hidden essence of a thing or material that is not directly measurable.

Given $\underline{n}$ samples, each of which is defined by $\underline{p}$ characters, they can be expressed as $\underline{n}$ points scattered in a $\underline{p}$-dimensional space. The principal component analysis aims at reducing the $\underline{p}$-axes to orthogonal $\underline{m}$-axes, where $\underline{m} < \underline{p}$, with a minimum of loss in information. Mathematically this produces a set of new $\underline{m}$ variables from the original $\underline{p}$ variables by an orthogonal transformation.

To illustrate the above mentioned principle of principal component analysis, the schematic diagram of the two variables case is shown in Fig. 3.

When the two variables X_1 and X_2 are highly correlated, the axes can be rotated to the position of Y_1 and Y_2, so as that the variance along the Y_1 axis becomes maximum and that along the Y_2 axis minimum. If the latter is sufficiently small, we can neglect the Y_2 and regard the Y_1 alone as a compound character of X_1 and X_2. Thus the number of axes is reduced from 2 to 1 with a minimum loss of information.

The new compound variables Y_1 and Y_2 can be expressed linearly as:

$$Y_1 = a_{11}x_1 + a_{12}x_2$$

$$Y_2 = a_{21}x_1 + a_{22}x_2$$

in terms of x_1 and x_2 multiplied by coefficients, a_{11}, a_{12} and a_{21}, a_{22}, which are called factor loadings.

Factor analysis has aims and procedures similar to those for principal component analysis. There are, however, certain differences between the two. In the case of principal component analysis, it is not necessary to have any previous assumptions concerning the number and character of the principal components or factors to be extracted, whereas in the case of factor analysis the following must be assumed (3):

1) The number of common factors to be considered,
2) The extent of contribution of each variable to the common factors.

In other words, the fundamental model of factor analysis is expressed as follows:

$$x_i = a_{i1}f_1 + a_{i2}f_2 + \cdots\cdots + a_{ik}f_k + \cdots + a_{im}f_m + e_i$$

Where f_k (k=1, 2, $\cdots$, m) is the score of $\underline{m}$ factors for each sample, $\{a_{ik}\}$ (i=1,2, $\cdots$, p; k=1,2,$\cdots$, m) is the factor loading for the ith variable and k^{th} common factor, and e_i (i=1,2,$\cdots$, p) is error or specific character of each variable that is not explained by the $\underline{m}$ factors. Given $\underline{n}$ samples having $\underline{p}$ variables, factor analysis aims at attaining the best estimates of factor loading matrix and error variance simultaneously upon certain assumptions.

Another feature of factor analysis is that the estimated factor axes can be rotated freely so as to make interpretation of the factor easier. This is possible because of what is called indeterminacy of factor axes.

122

If the factors, after rotation, are interpretable, computation of factor scores (f_k in the model) follows.

There are several methods available for computing factor scores, but the principle underlying these is the least square estimation.

FERTILITY RATING

In our study of fertility evaluation for 410 tropical Asian paddy soils we used 11 characters listed in Table 2. As the result of principal component analysis and successively adopted factor analysis, we obtained the factor loadings as shown in Table 3.

The first factor is highly correlated only with the characters related to base status and parent material, such as CEC, exchangeable cations, available silica, total phosphorus and sand, with the latter being opposite in sign to the rest. Thus, the first factor may be termed inherent potentiality (IP), and is determined primarily by the nature and amount of clay and base status.

The second factor is related to TC, TN, and NH_3-N. Moderately high loadings on TP and Sand are interpretable in terms of organic phosphorus and textural control on organic matter accumulation, respectively. Therefore, the second factor may be termed organic matter and nitrogen status (OM).

The third factor can clearly be interpreted as available phosphorus status (AP). Factor loading on TP is much less than on Bray-P and HCl-P. Contribution of other variables to this factor is minor.

It was observed that these three mutually orthogonal factors are in accordance with the result of correlation analysis referred to earlier. This leads to the interesting and important inferences that soil fertility of tropical Asian paddy soils is composed of at least three major components, and that both organic matter status and available phosphorus status of these soils are independent of what we call inherent potentiality.

The factor scores were computed for individual soil samples so that quantitative evaluation of the three components of soil fertility may be made. The scores thus computed for the samples are standardized with a mean of zero and a variance of unity. Therefore, positive score values indicate above-average status with reference to the overall mean for the 410 sample soils, and negative values indicate below-average status.

Although some reservation regarding the sampling procedure adopted in this study is necessary, a rough estimation of fertility status can be made for each country by calculating a mean score for the samples concerned. Table 4 shows the mean values of the three factors for each country.

Inherent potentiality is highest for the soils of Indonesia and the Philippines, followed by those of India. The first two countries are situated in a region influenced by volcanic activity and the parent material of the soil is continuously rejuvenated by fresh volcanic ejecta. India is located in semiarid to subhumid climatic regions, and the weathering and leaching process of the soil material has not been very intensive, especially in the basaltic rock area of the Deccan Plateau that constitutes the catchment of the Godavari-Krishna rivers.

The soils of Malaysia and Sri Lanka, which are situated in permanently humid to monsoonal climatic regions of the low latitudes, are among the poorest with respect to inherent potentiality. The soils of Bangladesh, Cambodia, and Thailand are mostly on the poorer side of the overall mean.

Organic matter-nitrogen status is by far the highest for Malaysian soils, with a mean as high as 1.40. The second highest is for the Philippines, with a mean of 0.34. Conversely Indian soils, with the lowest mean of -0.73, are the poorest. The soils of the other countries are more or less similar, with mean scores clustering around the overall mean. It

Table 2 List of characters used for analysis

Character No.	Name	Brief Description
1	TC (Total Carbon)	as percent of air-dried soil, Tyurin's wet combustion method.
2	TN (Total Nitrogen)	as percent of air-dried soil, Kjeldahl digestion and steam distillation.
3	NH_3-N	in mg N/100g of air-dried soil, after incubation for 2 weeks at 40°C.
4	Bray-P	in mg P_2O_5/100g of air-dried soil, Bray-Kurtz No. 2 method.
5	Ex-K	in me/100g of air-dried soil, N NH_4-acetate extraction, flame photometry.
6	CEC	in me/100g of air-dried soil, buffered neutral N $CaCl_2$ medium.
7	Av-Si (Available Silica)	in mg SiO_2/100g of air-dried soil, pH 4 Acetic acid extraction at 40°C.
8	TP (Total Phosphorus)	in mg P_2O_5/100g of air-dried soil, either HF-H_2SO_4 or HNO_3-H_2SO_4 digestion.
9	HCl-P	in mg P_2O_5/100g of air-dried soil, 0.2N HCl extraction at 40°C for 5 hrs.
10	Sand	as percent of organic matter-free dried soil, sum of coarse and fine sands.
11	Ex-Ca+Mg	in me/100g of air-dried soil, N NaCl extraction, EDTA titration.

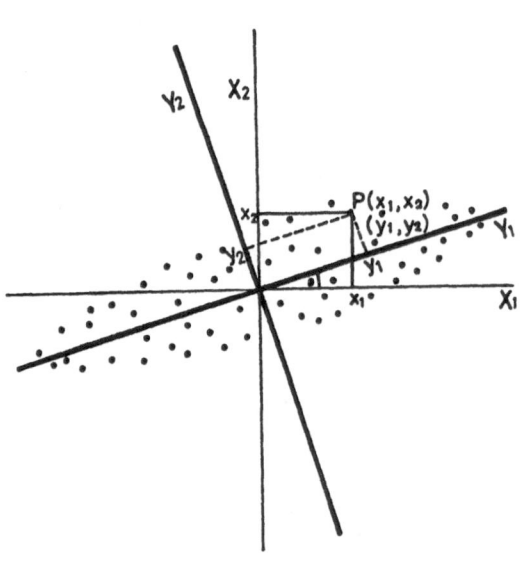

Fig. 3 Schematic diagram showing the principle of principal component analysis

Table 3 Terminal factor loading matrix for three factors after varimax rotation

	Factor 1	Factor 2	Factor 3
(TC)	0.288	0.913	0.118
(TN)	0.173	0.965	0.151
(NH$_3$-N)	0.113	0.668	0.046
(TP)	0.478	0.407	0.531
(Bray-P)	0.089	0.104	0.943
(HCl-P)	0.182	0.042	0.881
(Ex-Ca + Mg)	0.936	0.276	0.108
(Ex-K)	0.944	0.053	0.147
(CEC)	0.777	0.246	0.303
(Av-Si)	0.796	0.118	0.218
(Sand)	−0.486	−0.323	0.086

Table 4 Means and standard deviations of the three factor scores for the respective countries

Country	No. of Samples	IP		OM		AP	
		Mean	S.D.	Mean	S.D.	Mean	S.D.
Bangladesh	53	−0.438	0.708	0.176	0.704	0.459	0.800
Burma	16	0.118	0.710	−0.128	0.646	0.308	1.121
Cambodia	16	−0.231	0.990	−0.155	0.954	−1.277	0.841
India	73	0.449	0.837	−0.780	0.619	0.581	0.906
Indonesia	44	0.618	0.766	−0.014	0.746	0.031	0.734
Malaysia	41	−0.545	0.719	1.398	0.927	0.026	0.726
Philippines	54	0.618	0.703	0.337	0.673	0.022	1.041
Sri Lanka	33	−0.510	0.853	0.150	1.064	−0.110	0.650
Thailand	80	−0.364	1.195	−0.347	0.939	−0.641	0.769

Table 5 Means and standard deviations of the three factor scores for selected regions

REGION	NO. OF SAMPLES	IP		OM		AP	
		Mean	S.D.	Mean	S.D.	Mean	S.D.
Sri Lanka Wet & Interm. Zone	14	−1.07	0.64	0.84	1.10	−0.10	0.64
Sri Lanka Dry Zone	19	−0.10	0.76	−0.36	0.70	−0.12	0.67
Bangladesh Ganges	15	0.33	0.35	0.04	0.82	0.90	0.50
Bangladesh Madhupur-Barind	9	−0.75	0.60	0.35	0.64	−0.14	0.40
Bangladesh Marginal	16	−0.87	0.62	0.36	0.66	−0.06	0.77
Bangladesh Brahmaputra	13	−0.57	0.48	−0.01	0.64	1.01	0.66
W. Malaysia Kedah-Perlis	10	0.25	0.44	1.21	0.56	0.02	0.54
W. Malaysia East Coast	10	−1.23	0.20	0.78	0.59	−0.54	0.57
India Godavari-Krishna	10	1.38	0.29	−0.73	0.47	1.11	0.95
Thailand NE Plateau	32	−1.18	1.27	−1.14	0.77	−0.94	0.70
Thailand Intermontane Basin	4	−0.27	0.52	0.06	0.69	−0.72	0.80
Thailand Upper Central Plain	14	−0.04	0.76	−0.05	0.49	−0.31	0.69
Thailand Bangkok Plain	24	0.53	0.64	0.24	0.66	−0.51	0.86
Thailand South	6	−0.40	0.74	0.58	0.48	−0.29	0.39

Table 6 Mean contents of clay mineral species and selected total elemental oxides for the samples in each inherent potentiality class

IP CLASS	1	2	3	4	5	F-VALUE
No. of Samples	(88)	(92)	(77)	(70)	(83)	
7 Å	27.84	40.27	44.03	52.71	67.83	51.34
10 Å	8.13	15.00	17.60	18.00	11.15	7.95
14 Å	64.03	44.73	38.38	25.50	19.82	78.09
SiO_2	63.86	66.80	71.61	74.94	84.89	73.46
Fe_2O_3	9.24	7.42	5.75	4.52	2.26	74.89
Al_2O_3	20.31	19.31	16.22	15.60	9.78	41.84
CaO	2.13	1.72	1.69	1.09	0.37	11.25
MgO	1.25	1.20	1.09	0.68	0.31	28.14
TiO_2	1.36	1.25	1.13	1.07	0.87	8.91
K_2O	1.50	2.00	2.28	1.90	1.53	5.90

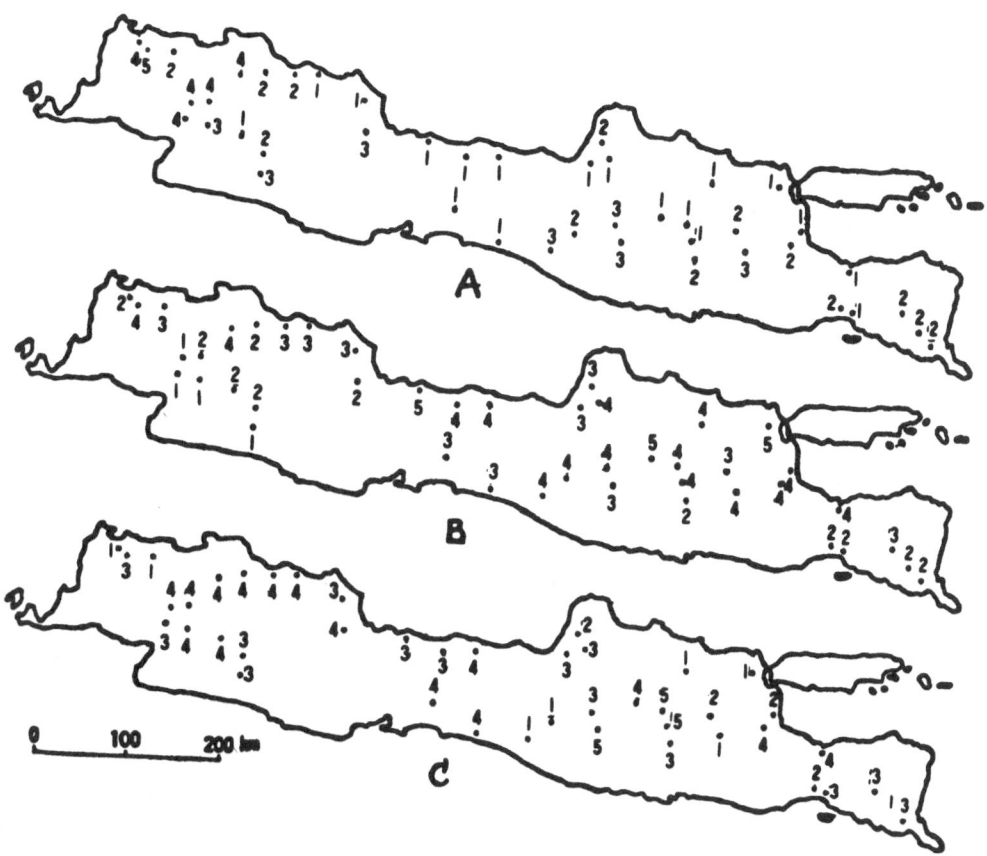

Fig. 4 Map of Java, Indonesia, showing distribution of samples
in terms of: A, inherent potentiality grade; B, organic
matter and N grade; C, available P grade

seems that high OM scores are associated with humid climate and fine soil texture in addition to a low terrain condition.

Available phosphorus status is high for the soils of India and Bangladesh, while Cambodian as well as Thai soils are the poorest. The regionality observed in this property is difficult to explain, but may be ascribed to the nature and degree of weathering of the parent rocks.

Similar calculations were done for regions that can be defined more or less discretely with respect to climate, parent material, and areal extension. The result is given in Table 5. The highest scores for IP and AP are for the soils of Godavari-Krishna region, which, however, have the second lowest OM score. The soils of the Northeast Plateau region of Thailand are characterized by very low scores for all the three fertility components.

FERTILITY CLASSIFICARION AND ITS MAPPING

To effect a fertility classification, the whole range of computed scores was divided into classes, with arbitrary class limits of $\pm$ 0.25 and $\pm$ 0.84. The assumption underlying the selection of the limits is that, if the distribution of the scores is normal, five classes of almost equal size should occur. The potentiality of each fertility component at different class levels could be designated as follows:

Class No.	Class limits	Potentiality
1	> 0.84	very high
2	0.84 ~ 0.25	high
3	0.25 ~ -0.25	intermediate
4	-0.25 ~ -0.84	low
5	-0.84 >	very low

Although clay mineral composition and total chemical composition were not directly used in the computation of fertility component scores, they are well represented by inherent potentiality. This is clear from Table 6, which shows the mean contents of clay mineral species and selected elemental oxides for each of the five inherent potentiality classes. The difference in the means is statistically highly significant. Of the variables listed in the table, 10 Å mineral and total potash content have a peculiar pattern with their maximum in the intermediate classes.

The fertility class number of each sample was plotted in a map at the corresponding sampling site. A set of three maps for IP, OM, and AP, respectively, was prepared for each country. Figure 4 shows the maps of Java, Indonesia. From such maps we can locate problem areas for each of the three fertility components.

REFERENCES

(1) Sokal, R. R. and Sneath, P.H.A. Principles of Numerical Taxonomy, Freeman & Co., San Francisco (1963)
(2) Seal, H. L. Multivariate Statistical Analysis for Biologists, Methuen & Co., London (1964)
(3) Asano, C. 因子分析法通論 (An Introduction to Internal and External Factor Analyses), 共立出版, 東京 (1971)

ON THE GENETIC CLASSIFICATION OF PADDY SOILS IN CHINA

Gong Zi-tong
(Institute of Soil Science, Academia Sinica, Nanjing)

Rice has been cultivated in China for about seven thousand years. In
the third century A. D. rice cultivation spread in the valleys of the Changjiang and the Huanghe Rivers. At present, paddy soils are distributed
almost all over the country, although they mainly concentrate in the plains
and hilly and mountainous regions to the south of the Qinling mountains and
the Huahe River (Fig. 1).

SOME REMARKS ON CLASSIFICATION OF PADDY SOILS IN CHINA

1. Great Diversity of Paddy Soils

Rice can be planted on various soils so long as the climate is suitable
and water supply assured. Paddy soils are widely dispersed from the tropical
to the frigid zones, from the humid monsoon region to the continental arid
region, and from the coastal plains to the high plateau over 2,400 meters.
The hydrothermal regime of the soil is usually changed by land leveling or
terracing. The alternation of oxidation and reduction and the translocation
of corresponding materials in the soil always take place due to the periodical and seasonal irrigation. The land surface is often raised up to 0.25
to 1 cm annually owing to the application of muds as manures in some polder
areas (Table 1). The application of liming materials and wood ashes on acid
soils may lead to the increase in base-saturation (Fig. 2). Because of the
complexity of natural conditions and the profoundness of the influence of
human activities, paddy soils in China are characterized by their diversity
of types.

Table 1 Effect of warping and application
of mud on depth of soil

Location	Method	Raising of land surface(cm/year)
Tianjin	Warping	1.0
Lixiahe, Jiangsu	Application of mud	0.5
Zhujiang Delta	Application of mud	0.25 - 0.5

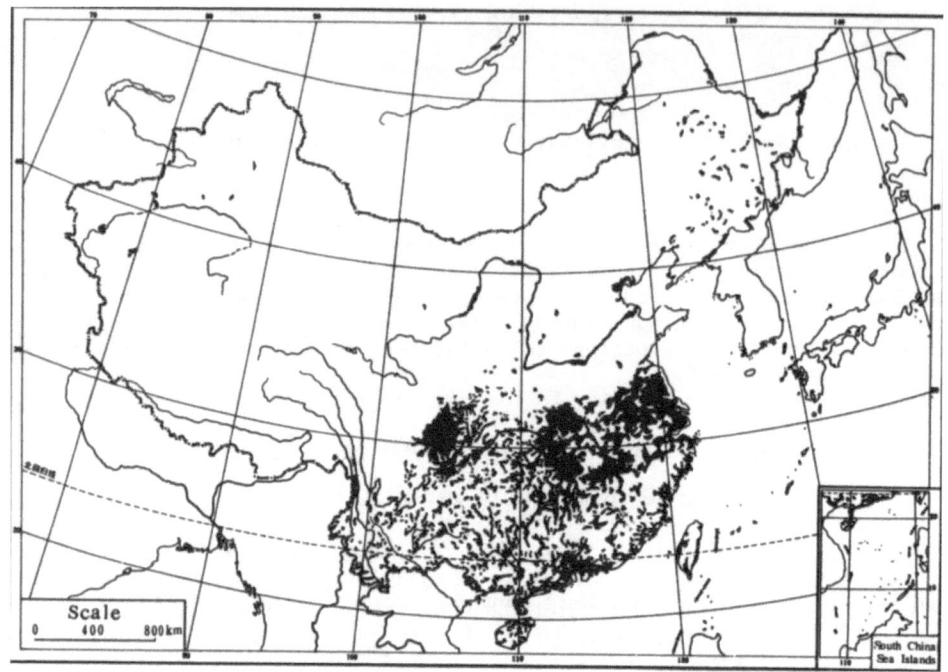

Fig. 1 Distribution of paddy soil in China
(From Zhang Jun-min and Wang He-lin)

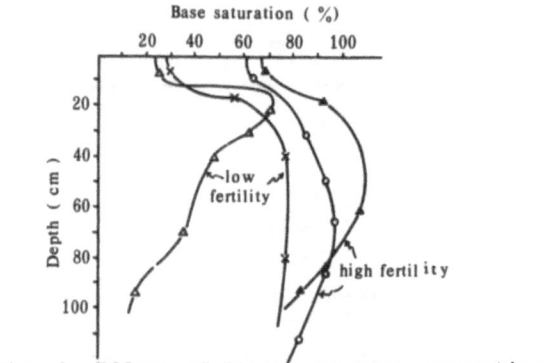

Fig. 2 Effect of liming on base saturation
of lateritic paddy soil

2. One of the Earlist Countries in Studying the Classification of Paddy Soils

The study of genetic classification of paddy soils in China started in the 1930's. While, in the literature of soil science, some authors still obscure the conception between process of podsolisation and paddy soil formation, the Chinese soil scientists Chu, Ma, Sung and Hou (1938(1) have already raised the conception for the delimitation of horizons of paddy soils, and defined a W horizon characterizing the paddy soil as the distinction from the B horizon of podzolic soil. Hseung Yi (1941)(2) noted the essential distinction between the formation of paddy soil and the process of podzolization chemically, for he showed that there is only the translocation of iron and manganese in paddy soil, whereas aluminum also undergoes translocation in podzolic soil. Zhu Lian-qing et al. (1940)(3) demonstrated the uniqueness of paddy soils based on their formation, characteristics and morphology.

3. Recognition of the Important Place of Paddy Soil in Soil Classification

Due to the two characteristics mentioned above, most of the soil scientists in China recognize the important place of paddy soil in soil classification. The commonly held reasonings are as follows: From the standpoint of factors of soil formation, human activities have a profound influence on the paddy soil. The processes occuring in the paddy soil are quite different from those in other great groups. Well-developed paddy soils have their characteristic profiles. The area of paddy soils in China is so large that it is impossible to work out a complete soil classification system without the special consideration on paddy soil. Therefore, from the very beginning of soil classification in China paddy soil has attracted the attention of most soil scientists.

SOME VIEWPOINTS ON THE CLASSIFICATION OF PADDY SOILS IN CHINA

The classification of paddy soils reflects the understanding of them by soil scientists. Therefore, classification systems vary with the difference in viewpoints on which the classification is adopted. In China, there are three main viewpoints in the classification of paddy soils, i.e. the classification based on geographical distribution, the classification based on genetic factors and the classification based on genetic processes.

1. Systems Based on Geographical Distribution

Paddy soils spread widely throughout various climatic zones. In accordance with the diversities in composition of humic substances, clay minerals and cropping systems, soil scientists with this viewpoint made an attempt to divide the paddy soils into two great groups. Lately, they were further classified into three great groups: southern, northern and that with Fe-organic coating (Table 2). The advantage of the system is that the soils are classified according to their distinctive conditions of soil formation, while its shortage is the somewhat obscureness between soil regionization and soil classification. In this system, it only emphasizes the geographical conditions, but overlooks the genetic processes and the properties of the soil.

Table 2 Classification systems based on geographical distribution

Zones	Soil conditions of high yield rice(1961)(4)	Cited from "a draft classification of the soils of China" (1978) (5)	
Humid temperate	- - - - - - - - - -	Northern	Northern
North subtropical	Shanxue paddy soils(with Fe-organic coating)	With Fe-organic coating	Southern
South to middle subtropical	Nirou paddy soils (Ferrallitic)	Southern	

2. Systems Based on Genetic Factors

The factors considered include original soil, water regime and soil acidity.

Classification based on original soils on which paddy soils are formed is one of the systems on this viewpoint. In the 1950's, the paddy soils were classified into three types: paddy soils derived from zonal soils, from meadow soils and from swampy soils. In the 1970's they were further subdivided. For example, those from meadow soils were divided into three subgroups, i.e., that of acid meadow, of neutral meadow and of calcareous meadow. In another system, soils were denominated as rice alluvial soil, rice bleached soil and rice peaty soil etc. (Table 3). This viewpoint emphasizes the in-

Table 3 Classification systems based on original soil

Original soil	"Genetic classification of paddy soils in southern China" (1959)(6)	"A draft of classification of the soils of China" (1978)	Wang Ru-yong yong (1959)(7)
Automorphous	From zonal soils	Lateritic Yellow-brown Purple	- - - - - -
Meadow	From meadow soil	Acid meadow Neutral meadow Calcareous meadow	Rice alluvial Rice bleached Rice black clayey
Swampy	From swampy soils	Gleyed Swampy	Rice gleyed Rice peaty
Saline	- - - - - - -	Salty	

fluence of the original soils on which paddy soils are formed, Undoub-tedly, original soil is of importance, especially in the initial stage of soil formation. However, with the continuation of long term cultiva-tion under submerging conditions, the characteristics of the paddy soil will become more and more obvious, while those of the original soil will gradually disappear. Systems on this viewpoint only pay attention to the "origin" of the soil, but not to its "development". That is to say, they only emphasize the residual properties of the original soil, but neglect the processes being proceeding.

As to the system based on water regime, in the 1950's, some subgroups were classfied, such as rainfed paddy soil, water leaking paddy soil, drought-resistant paddy soil, water logged paddy soil and cold spring paddy soil.

In the book "The Soils of China", published in 1978, they are divided into the following subgroups: surface submergic, ground water, well-drained and cold spring water paddy soils etc. Recently, some soil scientists pro-posed the side-bleached, permeable, percolating, water-logged, stagnating paddy soils as the great groups (Table 4). This is the most influential system of paddy soil classification based on the viewpoint of genetic factors.

Table 4 Classification systems based on water regime

Bureau of Agriculture, Jiangxi Province (1959)(8)	"The soils of China" (1978)(9)	Xu Qi et al (1980)(10)
Rainfed	Surface water	Side bleached
Drought-resistant	Well-drained	Permeable
Water-logged	Ground water	Water-logged
Cold spring	Cold spring	———
Water leaking	———	Percorlating
———	———	Stagnating

In the system based on soil acidity (Table 5) the paddy soils are divided into acid, neutral and calcareous paddy soils. On the similar viewpoint, Yu Tiar-ren has classified them as strongly eluvial, moderately eluvial and weakly eluvial paddy soils(11). This system virtually re-flects the important property of the soil, but it is not connected with the stages of soil development.

Table 5 Classification systems based on acidity

Cited from "Genetic classification of paddy soils in south China"	Yu Tian-ren (1976)(11)
Acid	Strongly eluvial
Neutral	Moderately eluvial Weakly eluvial
Calcareous	

3. Systems Based on Genetic Processes

The above-mentioned two viewpoints about paddy soil classification have a similar defect of neglecting the genetic processes of the soil. On the basis of morphological studies. Chu, Ma, Sung and Hou (1938) raised the conception of subhydrogenic, hydromorphic percogenic and submergenic processes in paddy soil formation, which were used as the basis of nomenclature for subgroups. Lately, some revisions have been made. For example, the term subhydrogenic has been changed to the gleyed, and the hydrogenic has been further divided. Besides, the bleached paddy soils has been added in the system(Table 6). No doubt, this system of classification is an important achievement in the classification of paddy soils. However, it

Table 6 Classification systems based on genesis

Chu, Ma Sung and Hou (1938) (subgroup)	Symposium on soil classification (1963) (great group)	Gong Zi-tong et al. (1980)(12) (subgroup)
Submergenic	Surface gleyed	Oxidizing
Percogenic Hydrornorphic	Bleached Hydromorphic	Redoxing
Subhydrogenic	Gleyed	Reducing

seems that some of the conceptions lack experimental verification. For example, the hydrogenic horizon denoted originally the horizon in which the whole horizon is occupied by surface water or temporarily stagnating ground water. However, field observations have shown that the so-called "hydrogenic horizon" is often unsaturated with water in the irrigation period. Considering the effect of oxidation-reduction process on the genetic features of paddy soils, we have proposed a classification system in accordance with the oxidation-reduction process in soils. For, following the eluviation and illuviation of materials resulting from the oxidation and reduction in the soil, there form the unique features of paddy profile. As shown in Fig. 3, for the submergic paddy soil (3a), except the plowed horizon which is in reduction stage in rice-growing season, all other horizons are in

oxidation state. There is a distinct oxidative B horizon in the profile.
For the gleyed paddy soil (3b), except the plowed horizon which is in oxidation state in the growing season of dry-farming crops, all other horizons
are in reduction state. There is a predominantly reductive Bg horizon
in the profile. In hydromorphic paddy soils, though the ground water
table varies (Fig. 3c, 4) and so is the intensity of oxidation or reduction,
the common characteristic is that there is an alternation of oxidation and
reduction in WB horizon of the soil. On the basis of the unique properties
and processes of paddy soils, we suggest that paddy soils with an oxidative
B horizon should be termed oxidizing paddy soils, those with a reductive
Bg horizon reducing paddy soils, and those with an alternatively oxidative
and reductive WB horizon redoxing paddy soils.

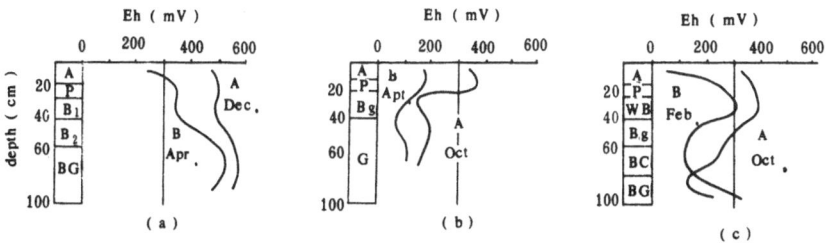

Fig. 3 Soil profiles of different oxidation reduction
status A Season of dry-farming crop B season of irrigation

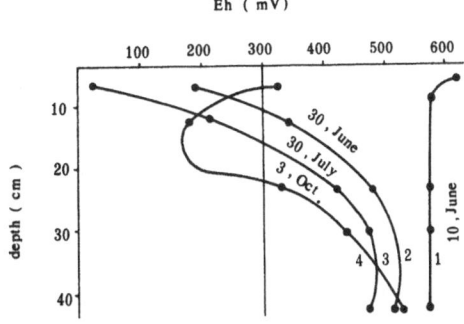

Fig. 4 Change of oxidation-reduction
potential in a profile of bleached paddy
soil

DISCUSSION ON SOME PROBLEMS OF CLASSIFICA-
TION OF PADDY SOILS

What is paddy soil? Can any soil be termed paddy soil just when rice is cultivated, or only soils with unique diagnostic horizons, for instance A-P-W or A-P-B horizon, can be regarded as paddy soils? We consider that the identification of a soil should inevitably be done according to the integration of natural conditions, processes and properties of the soil. Of course, soils with A-P-W(8) horizons may be considered as paddy soil. On the other hand, if a soil is under the condition of alternation of oxidation and reduction and has some features of material translocation resulting by the reductive eluviation and oxidative illuviation, but not well-developed WB, B or Bg horizon below A and P horizons, it may still be considered as the transitional type of a paddy soil. As to the soil under submergence all over the year, without alternation of oxidation and reduction and differentiation in the profile, it should be classified into the subgroup of its original soil.

In the following, in accordance with researches of some predecessors and our own work, I shall talk about some opinions on classification of paddy soils for discussion.

Paddy soil, due to the alternation of processes of oxidation and reduction in it, is different from the automorphous and aquatic soils. In paddy soil, there is not only the downward translocation of Fe and Mn resulting from the influence of irrigation water, but also the upward translocation of these elements due to the rising of ground water. Thus, these characteristics give the distinction between the paddy soil and the meadow soil. The translocation of mineral elements also significantly differs from those occuring in podzolic soils. And what is more important, the formation of paddy soil is closely related to the human activities of cultivation. Therefore, we consider that the paddy soil is an independent soil order in the classification.

Soil great group reflects the genetic stage of soil formation. As mentioned above, both the viewpoint of classification based on geographical distribution and the viewpoint of classification based on soil factors neglect the integration of the delimitation of soil great group with the genetic stages of soil development. We have attempted to integrate the environmental conditions, process of soil formation and properties of the soil and classify the paddy soils into three great groups—the dystric (acid). eutric (neutral) and the calcaric (calcareous). Their distinction is shown in Table 7

Table 7   Some characteristics of different great groups

Great group	pH*	Base saturation(%)*	$CaCO_3$	Clay mineral
Dystric	<6.0	<70	——	1:1 type
Eutric	6.0-7.0	70-100	——	Intermediate
Calcaric	7.5	100	+	2:1 type

*Average in 1 m solum

Subgroup should reflect the genetic phase and additional processes in soil formation. It is seen from Fig. 5 that all the paddy soils are interrelated with one another, and that different subgroups actually represent

136

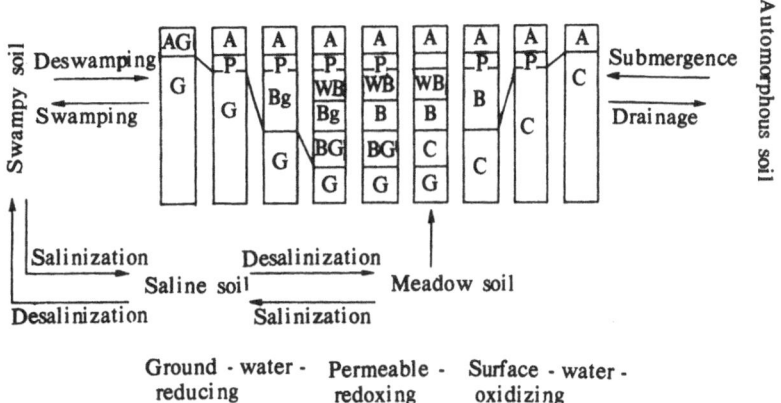

Fig. 5 Patterns of paddy soil

the different phases of soil development. Some of them have typical features, and some are accompanied with additional processes. For instance, redoxing paddy soils may be regarded as the typical subgroup, while soils with simultaneous glization as an additional process are classified as reducing paddy soils, and those accompanied with the process of automorphous soil formation are considered as oxidizing paddy soils. Besides, those with process of severe reducing eluviation and with albic horizon (WL) are called bleached paddy soils (Table 8). The four subgroups may also

Table 8 Genetic features of different subgroups

Subgroup	Genetic features	Typical profile	Water regime
Redoxing	Typical oxidation-redoction process	A–P–WB–B–G	Permeable
Reducing	With simultaneous gleization	A–P–Bg–G	Water-logged
Oxidizing	Accompanied with automorphous soil formation	A–P–B–BC	Surface submerged
Bleached	With severe reducing eluviation	A–P–LW–Bg (B)	Stagnating

be termed hydromorphic, gleyed, submergic and bleached, but they should be endowed with new implication.

The comparison of different classification systems of paddy soils is shown in Table 9.

137

Table. 9 Comparison of systems of paddy soil classification

Present author		Chu, Ma,Sung and Hou	Symposium on soil classification
A	B(with new implication)		
Reducing dystric	Acid gleyed	Subhydrogenic	Gleyed
Reducing eutric	Neutral gleyed		
Reducing calcaric	Calcareous gleyed		
(Reduced) bleached	Bleached	– – – – – –	Bleached
Redoxing dystric	Acid hydromorphic	Permeable	Hydromorphic
Redoxing eutric	Neutral hydromorphic	hydromorphic	
Redoxing calcaric	Calcareous hydromorphic		
Oxidizing dystric	Acid submergic	Submergic	Surface gleyed
Oxidizing eutric	Neutral submergic		
Oxidizing calcaric	Calcareous submergic		

REFERENCES

(1) Chu L.T., Ma Y.T., Sung T.C., Hou K.C., The nomenclature of the various horizon of paddy soils. Special Soil Publication, China, series B 4, 73-78(1938).
(2) Hseung Y., Some chemical properties of paddy soils. Special Soil Publication, China, series A 4, 1-22(1941).
(3) Zhu Liang-qing, Is the paddy soil as a main soil type. Soils Quarterly, China, 1(3), 48-58(1940). (in Chinese)
(4) Institute of Soil Science, Academia Sinica, The Environmental Conditions of Soils for High Yield of Rice. Science Press, Beijing, 8-51(1961). (in Chinese)
(5) Gong Zi-tong, Zhao Qi-guo, Zen Zhao-shun, Lin Pei, Wang Ren-chao, A draft classification of the soils of China. Turang, 5, 68-69(1978). (in Chinese)
(6) Soil Survey Group, Institute of Soil Science, Academia Sinica, Genetic classification of paddy soils in southern China. Acta Pedologica Sinica, 7, 28-41(1959). (in Chinese)
(7) Wang Ru-yong, Agricultural characteristics and their improvment of rice soil in Jilin Province(1959, ms.). (in Chinese)
(8) Bureau of Agriculture, Jiangxi Province, A tentative classification of paddy soils of Jiangxi Province. Turang Tongbao, 5, 46-48(1959). (in Chinese)
(9) Institute of Soil Science, Academia Sinica, Nanjing(ed.), Soils of China. Science Press, Beijing, 441-448(1978). (in Chinese)
(10) Xu Qi, Lu Yan-chun, Zhu Hong-guan, The genesis and classification of the paddy soils, Tai-lake basin, Jiangsu Province, China. Acta Pedologica Sinica, 17, 131-142(1980). (in Chinese with English summary)
(11) Yu Tian-ren(ed.), Electrochemical Properties of Soils and Their Research Methods. Science Press, Beijing, 280, 422(1976). (in Chinese)
(12) Gong Zi-tong, Wei Qi-fan, Tang Xin-nong, Chen Zhi-cheng, The Soils and Their Reasonable Utilization in Taoyuan County. Hunan Science & Technology Press, Changsha(1980). (in Chinese)

THE CLASSIFICATION OF "PADDY SOILS" AS RELATED TO
SOIL TAXONOMY

F.R. Moormann
(Soil Science Dept., State University of Utrecht,
3508 TA Netherlands)

Paddy soils connotes a class of soils, occurring in land that is used
for the growing of rice and of which the surface is submerged during all
or part of the growing season. As has been pointed out by various authors,
the term "paddy soil" is a generic name used in connection with a specific
land utilization type. The common pedological denominator among the various
paddy soils resides in their soil moisture regime: paddy soil are hydromorphic either naturally, or artificially or both.

Classification of paddy soils has historically followed two different
approaches. A first approach, termed "utilitarian" (1) classifies soils in
accordance with characteristics which are a direct result of a specific
landuse and/or which have management implications related to that specific
landuse - in this case the cultivation of paddy rice. This approach was,
since the 1930's, developed primarily in Japan. Best known internationally
are the classification proposals made by Kanno (2,3) based on hydrological
conditions of distinguished soil taxa in the class of artificial hydromor-
phous soils, or mineral paddy soils. A summary description of the most important of the Japanese systems is given by Kawaguchi and Kyuma (4). Elsewhere in Asia, the utilitarian approach has been used on a national or
regional scale (5). In more recent work in Japan (e.g. 6,7) and elsewhere
in Asia (8) there has been a trend to make a linkage between various taxa
of paddy soils and established taxa of international soil classification
systems.

A second approach to the classification of paddy soils can be termed
the "pedological" approach, whereby the periodically submerged soils on
which rice is grown are classified as taxa or members of an overall soil
classification system, irrespective of the specific landuse or vegetation.
This approach is relatively young; few - if any - comprehensive and internationally valid proposals for the incorporation of paddy soils in general
soil classification systems are available. Possibly the firmest statements
on classification of paddy soils within the framework of an overall system
have been made by proponents of genetic soil classifications.

Cultivation of soils is considered as a process: a natural genetic
soil type therefore changes it's taxon when the process of soil formation
is qualitatively changed due to cultivation (9,10). This is mainly so in
well drained soils, brought under rice cultivation. Soils where the effects
of flooding for rice only result in minor, quantitative changes of the soil
forming processes, are distinguished at a lower categorical level but not
at that of the genetic soil type. This is commonly the case in soils which
were originally hydromorphic: e.g. dark meadow soils (10).

In Japan, Matsui (11) recently developed a system in which he proposes
the paddy soils as part of comprehensive soil system, based mainly on the
ideas of the Central-European genetic school (12).

In a morphometric system based on soil genesis, soils are classified
not on the basis of processes or factors of soil formation, but on measurable characteristics, resulting thereof. In the currently internationally most utilized morphometric systems, i.e. Soil Taxonomy (13) and the

139

Legend of the World Soil Map (14) no provision has been made as yet to introduce diagnostic characteristics, which could be used to distinguish some or all paddy soils as distinct and specific taxa.

A review of the literature however indicates that attention has been given to the problem of incorporating soils that have been altered by flooding in a morphometric system, and more specifically in Soil Taxonomy.

For well drained soils, used for wet rice cultivation, the existence of "inverted gley" was noted (15-17). It was proposed that antraquic subgroups of various taxa be introduced for soils, showing distinct and measurable inverted gley characteristics.

For paddy soils, in which under flooded conditions, illuvial Fe or Fe-Mn horizons are formed below the Ap horizon, the term "Aquorizem" has been proposed by Kyuma and Kawaguchi (8). In a later publication by these authors (4) it is stated that, because the Aquorizem morphology is "a mere superposition on the more fundamental soil nature", it seems inappropriate to set up an independent class above the subgroup level.

The general trend in recent decennia has been to single out those paddy soils on which intermittend flooding has changed the characteristics of the original soil profiles, and to establish separate taxa for such soils in existing taxonomic systems. Various names for such taxa, mostly proposed at the subgroup level, have been suggested. Already mentioned were the names anthraquic and aquorizem (ic). Another is "hydragric" (6,18). There appears to exist a fairly wide-spread trend to classify soils that are used for growing rice under conditions of flooding in one of the two main taxonomic classification systems i.e. Soil Taxonomy (13) or the FAO Legend of the Soil map of the world (14). What is clearly lacking in these efforts is the quantitative characterization of those characteristics, which can and should be used as differentiae to establish relevant taxa for those paddy soils which require a separate place in a taxonomic system.

THE TAXONOMIC APPROACH TO THE CLASSIFICATION OF PADDY SOILS

In Soil Taxonomy polypedons are classified by their inherent properties. The grouping of soils into distinct and unique classification units or taxa is based on the determination of the measurable morphological and analytical soil properties. For the higher categories of the classification system, the differentiae used preferentially are either those that result from soil genesis or those that affect soil genesis, or both. Contrary to genetic classification systems, the processes and factors of soil formation themselves cannot be used as differentiae, but the measurable and lasting changes brought about by these processes are used, provided such changes meet with defined quantitative requirements.

The specific hydrologic regime to which soils, flooded for the growing of rice are submitted cannot, by itself, be used as a taxonomic differentia. However if, due to this specific hydrological regime, the original soil acquires new, quantitatively measurable properties, such properties can be introduced as taxonomic differentiae. In a large proportion of soils submitted to flooding for the growing of rice, no or little changes occur, which could be used as new taxonomic differentiae. For instance, flooding for rice growing of naturally hydromorphic, clayey Aquepts of the wide deltas of South and East Asia (19) will generally not impose any measurable changes on the original pedons. Mostly therefore, such soils can be classified in one of the existing taxa of Soil Taxonomy. Also, changes on naturally well drained soils, used for irrigated rice cultivation are not necessarily well expressed: their taxonomic classification does not change.

On the other hand, many of the pedogenetic effects of the specific hydrological regime in paddy lands are not exclusive. Quite obviously for instance in lands where crops other than rice are grown under flooded conditions, the pedogenetic processes may be identical to those operative in flooded rice lands. Processes such as ferrolysis will, under certain conditions be enhanced by the use of the land for paddy cultivation, but they are by no means exclusive for such conditions. In such cases, the specific landuse and its concurrent hydrologic regime, are irrelevant for the taxonomic classification of the polypedons involved.

In summary, the taxonomic approach to the classification of paddy soils at the four highest categories of the system requires consideration of following points:
- do paddy soils acquire new pedogenetic properties due to the specific landuse?
- are one or more of the new properties useable as taxonomic differentiae?
- which new taxa (if any) are required to accommodate soils, which have been altered by flooding for rice in the framework of Soil Taxonomy.

PEDOGENETIC PROCESSES AND THEIR TAXONOMIC CONSEQUENCES

Pedogenetic changes, superimposed on the original or natural pedons which are submitted to flooding for growing of rice have been the subject of many studies. The results of these studies are linguistically not all readily accesible, but for a review we refer to Kanno (20) and Moormann (21). In the following, we briefly enumerate the various processes, their degree of specificity for paddy land conditions and the measure to which they influence the morphogenetic characteristics cf "paddy soils".

Moormann and Van Breemen (19) distinguished temporary changes and (semi) permanent changes in soils growing flooded rice. Temporary changes are limited to surface soils and are associated with puddling as practized in many paddy fields, and also with alternating chemical oxidation and reduction. Such temporary surficial edaphic changes, though important from the point of view of crop management and production, do not by themselves change the taxonomic classification of soils. The cumulative effect of these recurrent temporary changes however, may lead to more permanent changes of the pedon and possibly, therefore, to a change in the taxonomic classification. We can group the changes, which may lead to the alteration of the taxonomic classification of soils under flooding for rice, in three categories, i.e.:
- changes due to agricultural engineering
- changes in the soil moisture regime
- changes due to alteration or neoformation of genetic horizons.

Changes due to Agricultural Engineering

Soils, associated with wet rice cultivation, are commonly changed by the effects of man's activities to create the specific environment of leveled, bunded and, frequently, irrigated fields. Many soils in rice fields are "anthropogenic" (9). One should, however, bear in mind that such changes are by no means general and that large areas of rice fields have not undergone more change than the average cultivated fields under dryland crops.

The *formation of an irrigation cover* is a common occurrence in paddy fields which are irrigated or naturally flooded by water, rich in mud. This is the case in certain river plains flooded by water, high in silt and clay, but also in certain terraced rice fields on side slopes where the irrigation water is muddy. In the latter case, important irrigation covers may be formed in the fields close to the irrigation canals (22).

Often the formation of an irrigation cover does not incur a change in the taxonomic classification of the soil at a high categorial level. A taxonomic change is however incurred, if the soil so covered by irrigation sediments belonged to taxa characterized by a distinct profile development. By definition, a pedon with a recent irrigation cover of 50 cm or more would become an Entisol, irrespective of the original soil.

In many southern and eastern Asian countries paddy fields on sloping land are created by *leveling and terracing*. Back-slope digging and foot-slope filling, followed by leveling, may disturb the original horizon sequences of the soils (16,19,20). The taxonomic effect of changes in the soil by cut and fill depends on the slopes. Although changes are strongly locale-specific, it is estimated that on slopes up to 5%, changes in the profile remain mainly limited to the epipedon over the larger part of the newly created terrace. If the diagnostic sub-surface horizons are thick, no aberrant taxa will normally be formed. Deeper disturbance on steeper slopes may have a distinct taxonomic effect, if the diagnostic sub-surface horizons originally were thin, or were occurring at shallow depth, or both. More fundamental changes occur when the soil material for terracing is brought from elsewhere, as in the Banaue terraces in Northern Luzon, Philippines (23). Here, new soils belonging to the order of Entisols are created, irrespective of the original soil. The same effect can be observed in paddy fields on landslides of areas with terraces on steep slopes. Commonly the soils on such restored terraces are also artificial Entisols.

Reclamation and drainage of marine and fresh water swamplands for the creation of new paddy fields frequently will incur changes in the taxonomic classification of the soils so affected, especially if the original pedons belonged to the great group of Hydraquents. Upon reclamation of such soils, ripening will start (24,25), and the soils will evolve from Hydraquents towards other great groups of the Aquent suborder or, eventually, of the Aquept suborder.

Considering changes brought about by agricultural engineering, it should be emphasized that these changes are by no means specific for the paddy soil condition. All changes discussed here, which have taxonomic implications are found under other types of agricultural landuse. Moreover, all these changes can be accommodated in the present Soil Taxonomy; and according to our present knowledge, no specific new taxa are required at the levels of generalization which we are discussing in this paper.

Changes in the Soil Moisture Regime

In Soil Taxonomy, the soil moisture regime is used as an important soil property and is considered as determinant of processes that affect the morphology, and hence the taxonomic classification of soils. To be recognized in the higher categories of this classification the effects of an aquic moisture regime must be measurable both in terms of the presence during some part of the year of reducing conditions and of the specific hydromorphic morphology of the soil profile, i.e. the presence of grayish or bluish colors with a low chroma. For differentiation in the highest categories that have an aquic moisture regime the whole soil must be saturated. In the subgroups only the lower horizons are saturated. It should be pointed out that in Soil Taxonomy no specific taxa are foreseen for soils which fall under the heading "Pseudogley" of the European soil literature (for a review, see Schlichting and Schwertmann, 18). The soil moisture regime of pseudogleys, whereby a surficial periodic wetness is not matched by high groundwater, is of importance for the taxonomic classification of certain paddy soils, as we will discuss below.

The effect of impounding water in rice fields is generally an increase in soil wetness. In terms of Soil Taxonomy, soils of paddy fields or of other fields on which water is impounded by man, will have a tendency to acquire supplementary characteristics of an aquic soil moisture regime. The degree of change of the soil moisture regime and the degree to which this change results in the modification of the taxonomic classification, depends on the nature of the original soil, as well as on the soil- and water management. In soils belonging in their natural state to Histosols or to suborders with an aquic or peraquic moisture regime, changes in this regime are slight to negligeable. Concurrently, other morphogenetic changes - if at all present - will be small and will not normally cause a shift in the taxonomic classification at the higher categorical levels. It is a generally established fact that poorly drained lowland soils do not change much, when used as paddy land (see e.g. 10,26,27). Thus, a large proportion of the riceland-soils in major and minor alluvial plains of South and East Asia, belonging to various great groups of the Aquent and Aquept suborders will retain their "original" taxonomic classification unchanged.

Alteration of diagnostic characteristics may, however, take place locally in the pedons of wet ricelands which remain submerged for long periods. Here, the sub-surface horizons may develop neutral grey colors due to continuous waterlogging and absence of oxygen (19). This is seen in double and triple cropped irrigated riceland on fine clayey Inceptisols. Such soils may lose the characteristics required to classify them as Inceptisols, particularly the diagnostic Cambic horizon, and they may become Aquents. This phenomenon is not wide-spread.

More profound changes in the classification according to Soil Taxonomy occur in soils which in their natural status belong to an aquic subgroup of one of several great groups. Such pedons are characterized by periodic water saturation in the lower horizons with concomitant gley characteristics. When submitted to ponding of water, such soils usually develop aquic characteristics in the surface and sub-surface horizons. Where these induced aquic morphological characteristics form a continuum with the deeper horizons having the properties of a natural aquic moisture regime, there is a transfer in the classification from an aquic subgroup to an aquic suborder. This is a common change in soils under sustained cultivation belonging to the orders of Inceptisols, Alfisols and Ultisols. Such landuse-induced Aquepts, Aqualfs, Aquults and others belong, with few exceptions, to clayey families with a low hydraulic conductivity and a restricted downward movement of the surface-water (26,28).

In Vertisols, and more specifically in Usterts that are characterized by cracks that are open > 90 cumulative days, the prevalent diagnostic characteristic may be lost under semi-continuous cultivation for wet paddy. Such Usterts would, by definition, become Uderts. Similarly, vertic subgroups of several great groups may loose the diagnostic differentiae based on cracking.

The change in soil moisture regime under paddy cultivation is most fundamental in moderately well to well drained soils where a surface and sub-surface gley develops which is not connected continuously with a subsoil gley.

The development of this inverted gley has been described by various authors (e.g. 2,8,15,16). The specific soil moisture regime involved can be considered as an anthropic modification of the natural moisture regime; it has been named "anthraquic" by Moormann and Van Breemen (19). In taxonomic terms it can be defined as periodic man-induced water saturation and reduction of the solum to a depth of at least 40 cm without corresponding periodic water saturation and reduction in the horizon(s) below. The soil material in the superficial horizons, submitted to an anthraquic moisture regime shows the same colors of low chroma in the matrix or in mottles as defined

143

for the aquic suborders in Soil Taxonomy. The lower horizons, however, do not show these low chroma colors, indicating the absence of longer periods of water saturation and reduction.

The development of inverted gley due to an anthraquic moisture regime is quite common in bunded and leveled paddy fields on lands where the groundwater table is either deep in the solum or absent and which are situated distinctly above the level of valley bottoms and plains. Inverted gley is strongest developed in pedons with a low permeability (26,28). Frequently such pedons have medium to fine textured soil materials, while in most cases a traffic plan (ploughplan) is present.

In Soil Taxonomy, no specific provision has been made for the classification of soils, showing superficial gley characteristics due to the presence of poorly permeable horizons or layers (pseudogley). Tentative proposals have been made to introduce the anthraquic soil moisture regime in Soil Taxonomy (21), and to recognize at the subgroup level, taxa which show the measurable effects of this moisture regime. A revision of Soil Taxonomy in this sense should, however, pay attention also to the general problem of the classification of soils in which the pseudogley phenomenon is a dominant characteristic. Other changes in soil moisture regime, more specifically those in which an existing aquic regime becomes reinforced, can be accommodated in the present version of Soil Taxonomy and hence do not require a revision of the current system.

Changes due to Alteration or Neoformation of Genetic Horizons

Alteration of soil properties under the specific waterregime in flooded ricelands has been the subject of many studies (for a review see e.g. 19,20). What interests us in the context of the present paper is in how far such changes may affect or determine the classification according to Soil Taxonomy of the soils in which such changes have taken place. Only such changes therefore, which alter the diagnostic horizons and other diagnostic properties of the pedon involved should have our special attention here.

As in the case of changes in soil moisture regime, alterations due to wet rice cultivation in naturally hydromorphic soils are mostly slight and quantitatively insufficient to bring about a change in the taxonomic classification. Thus, most of the soils belonging to the various aquic suborders will not undergo any quantifyable changes of their diagnostic properties other than those directly related to their soil moisture regime.

Permanent changes in soils of paddylands can roughly be divided in soil physical changes resulting from the cultural practice of rice growing, and in soil chemical and mineralogic changes, which are commonly considered as part of soil formation or soil genesis.

Ripening. Changes occurring in newly reclaimed and drained Sulfaquents and Hydraquents were already mentioned above (agricultural engineering changes). Due to irreversible water loss upon drainage the n-value of the horizons between 20 and 50 cm decreases below the critical diagnostic value of 0.7 (24). Some time after drainage and oxidation, sulfaquents may become sulfaquepts due to the formation of a diagnostic sulfuric horizon that has its boundary at less than 50 cm (29).

Changes in the taxonomic classification due to ripening are not exclusive for the paddy soil conditions. In fact, this process and the resulting change in diagnostic characteristics is known to be relatively slow in seasonally flooded soils such as in paddy fields.

Traffic pan. The formation of a so called plough pan, for which a more general term traffic pan was proposed (19) is common to puddled ricelands.

Compared to the surface soil, the traffic pan, occurring normally between 10 and 40 cm, has a higher (dry) bulk density and less medium-to-large seized pores (30).

Although the presence of a traffic pan will strongly influence the surficial soil moisture regime, it does in itself not have a diagnostic significance in Soil Taxonomy. Indeed, a pan of this nature is considered as a transient edaphological feature. It should be pointed out, moreover, that the presence of a traffic pan is by no means limited to the wet riceland condition.

Changes in the organic matter fraction. Depth, color, organic matter content and saturation of the A horizon determine the kind of epipedon, the nature of which is used as a differentia in various taxa at several categorical levels. Changes in the amount of organic matter - both increase and decrease - and of its composition have been reported under wet rice cultivation (see e.g. 19,31,32). Changes are most marked in soils which in their natural status are more or less freely drained, but are minor or even absent in the majority of poorly drained lowland soils under rice (see e.g. 10).

Increase in organic matter may lead to the formation of a mollic or umbric epipedon. Decrease in organic matter may cause mollic or umbric epipedon to become an ochric epipedon, with concurrent changes in classification. However, from available data it is not clear if taxonomically diagnostic changes of the epipedon occur in paddy soils. At most therefore, changes in the organic matter fraction, though genetically important will have very limited taxonomic implications. Such changes are moreover not specific for wet rice lands.

Changes in base status. Under seasonal flooding as in paddy fields, the base status of soils is frequently influenced. Both decrease and increase of base saturation may occur, depending on the balance between the influx of bases from irrigation, floodwater and interflow water, and the outflux of bases by leaching and surface water runoff.

Both biological fixation of Calcium Bicarbonate and decalcification have been observed under paddy conditions (10,19,33).

Loss of bases in soils, seasonally flooded with water low in bases, may ultimately lead to paddy soils with a lower clay content, a low pH and a low base saturation near the surface. Concurrently the surface soil may also be depleted in iron and manganese oxydes and in organic matter.

Most commonly, the change in base status of the superficial horizons will not lead to a change in the taxonomic classification of the pedons involved. Strong weathering however may occur under seasonal flooding; the specific hydromorphic soilforming process involved was named "ferrolysis" by Brinkman (34), but is known under other names in the literature (for a review, see 35).

Ferrolysis involves displacement of exchangeable bases by dissolved ferrous iron produced during (surface) soil reduction, removal of displaced bases by leaching or runoff, and oxidation of adsorbed ferrous iron to ferric oxide and exchangeable hydrogen during the following dry season. The hydrogen-saturated clay, thus formed, will partly decompose to give aluminium interlayered clay (soil chlorite) and silica (34-36). In paddy lands this process is mainly active in soils that are acid in their natural status and that are flooded by water, poor in dissolved bases (e.g. rainwater). Ferrolysis is not specific for paddy soils, and the effects of an advanced degree of ferrolysis have been recognized in various taxa of Soil Taxonomy (37).

A diagnostic horizon, formed by ferrolysis, is the albic horizon, although it should be noted that by no means all albic horizons are a consequence of ferrolysis. When such an albic horizon is formed in paddy soils, the taxonomic designation of the original, non flooded soil, changes either at the subgroup or the great group level.

145

It was pointed out by Brinkman (34) that in certain cases no satisfactory classification according to Soil Taxonomy is possible of pedons in which an albic horizon due to ferrolysis is present. For these cases new taxa will have to be introduced; e.g. Albaquepts.

In general, it seems probable that most of the so called "degraded" paddy soils of Chinese and Russian authors may be soils which show the effect of ferrolysis. It is difficult however, to quantify the term "degraded" morphometrically since no precise definition is known (4).

Changes due to illuviation of soil material. In paddy soils on parent materials of a certain age, mainly situated outside of present flood plaines, a diagnostic illuvial argillic horizon is often present. The argillic horizon is as far as can be ascertained from the literature and from field observations, inherited from the original pedons and there is no clear evidence that illuviation of clay is caused or enchanced by wet rice cultivation. Even in cases, where a decrease of clay in the surface horizon has taken place in rice soils on recent alluvium (31,38), the removal of clay from the surface was not associated with a clear clay increase at some depth. Removal of clay of the surface horizon, either by lateral transport in puddled rice lands or by breakdown of clay as in ferrolysed soils is not in itself diagnostic for the classification according to the present version of Soil Taxonomy.

In certain instances, the fact of flooding of lands irrespective whether they are paddy fields or not, may promote the downward movement of clay, silt and organic matter through cracks and pores. This is seen in the so called flood coatings of the subsoil, which are composed of surface soil material (26,33,39). No provision has been made in Soil Taxonomy to recognize the presence of flood coatings as a differentia.

Changes due to migration of iron and manganese. A process collateral to the seasonal reduction of the surface horizon in superficially gleyed soils is the mobilization of soluble ferrous iron and of manganese (if present) followed by a downward migration and accumulation in the deeper, less reduced or even oxidized subsoil horizons. Although this phenomenon is known for pseudogley soils not in use for rice growing (40,41) it has been more extensively studied under paddy soil conditions. Accumulation of iron or iron and manganese in the subsoil is either direct by oxidation and immobilization in the oxidized subsoil or indirect, via the adsorption of the bivalent ions in the reduced subsoil, and oxidation when the profile dries out (27).

Both degree and nature of the accumulation are strongly variable in paddy soils. Some measurable increase in total Fe in the subsoil may occur in as little as 40 years, but a considerable increase, combined with the formation of distinct Fe or Fe and Mn accumulation horizons can be found only in older paddy soils with an anthraquic moisture regime (2,3,15,26-28, 42). In its most advanced stage, a thin iron (or iron and manganese) pan can be formed which morphologically appears identical to the placic horizon of Soil Taxonomy (19). It should be pointed out that this type of placic horizon is not exclusive for paddy soil conditions, it has been found in mudflats among the Dutch coast below a 30-50 cm layer of reduced clay, resting on sandy, oxygenated subsoil (personal communication J.J. Reynders).

Soils with a distinct Fe or Fe/Mn accumulation horizon have been termed "Aquorizem" by Kyuma and Kawaguchi (8) while the process leading to this specific morphology is called aquorization.

Moormann and Van Breemen (19) proposed the term "hydroferric" for distinct horizons of iron or iron and manganese accumulation that are not indurated. For the indurated form of accumulation the same authors proposed the term "placic", on condition that the definition of this horizon in Soil

Taxonomy is adapted to accommodate such horizons in paddy soils and in other soils with a similar water regime.

The "aquorizem" morphology, with or without a pan, is not recognized explicitly in Soil Taxonomy. In a more general context, it should be stated that no specific provisions have been made for morphological, chemical and mineralogical characteristics of soils with "pseudogley". Most authors that have discussed the place of soils with an aquorizem- or aquorizem-like morphology in Soil Taxonomy are in favour of the recognition of separate taxa for such soils. Names such as "anthraquic" (17), "hydragric" (6,18) and "hydroferric" (19) have been proposed to be used at the subgroup level for soils that show a distinct accumulation of iron or iron and manganese. However, as stated by Kawaguchi and Kyuma (4) more careful examination and quantitative characterization of the differentiating characteristics of soils with an "aquorizem" morphology are necessary before firm recommendations for the introduction of new taxa in Soil Taxonomy can be made.

CONCLUDING REMARKS

Modern soil surveys in rice-growing countries have considerably increased our knowledge of soils, used for the growing of rice under a flooding regime. Mostly, such studies have been accompanied by the more or less detailed examination of the pedons occurring in such rice lands. Moreover, an increasing number of publications is available on the genetic and morpho-genetic implications on pedons, of a regime of periodic superficial water saturation and accompanying reductive processes. Although more data of this nature are essential for the organization of our knowledge on paddy soils in a complete taxonomic system of soil classification, some generalizing conclusions on the subject matter of this paper are warranted:

1. A major proportion of pedons in lands, used for rice under periodic submergence (paddy soils) can be readily accommodated in the classification system of Soil Taxonomy as it now stands. This is true for the four highest categories of the system: order, suborder, great group and subgroup. We can conclude that mostly no changes in diagnostic horizons or properties occur or that such changes as may occur are provided for in Soil Taxonomy.

2. A minor proportion of paddy soil pedons have acquired characteristics which as yet have not been recognized as diagnostic in Soil Taxonomy. It is however possible to propose and adapt new taxa pertaining to the class of "altered" paddy soils. Following the concepts and the mechanism of the system, new taxa required would, in our opinion, pertain to following changes which may occur under the hydrological regime in paddy soils:
   - The development of an anthraquic moisture regime in pedons which in their natural hydrological status were freely drained throughout most of the solum.
   - The formation of an albic horizon in Aquepts.
   - The formation of an accumulation horizon of Fe, or Fe and Mn through a process which is intrinsically different from the cheluviation process leading to the formation of a spodic horizon (35).
   - The formation of a placic horizon.

3. Most, if not all, changes in morphological and analytical characteristics due to the existence of an (imposed) surface gley, are not exclusive for the "paddy soil condition". Under different vegetation/land-use patterns similar processes may be operational, and the soils may acquire comparable characteristics. More specifically, most conditions in which the upper parts of the profiles are periodically water saturated and reduced, without a concommittant reduction in the subsoil

(pseudogley), will in principle have the same effects, though usually not as well expressed as in paddy soils.
4. For the introduction of new taxa in Soil Taxonomy, based on new or aberrant diagnostic horizons or properties, such differentiae need to be quantified with the aid of field- and laboratory studies. This appears to be the task ahead if Soil Taxonomy is to be used to classify all soils which have been altered by flooding for rice.

ACKNOWLEDGEMENT

Thanks are due to Prof.Dr. Kazutake Kyuma for his information on relevant recent Japanese publications.

REFERENCES

(1) Segalen, P., Les classifications des sols. ORSTOM, Paris (1977).
(2) Kanno, I., A scheme for soil classification of paddy fields with special reference to mineral soils. Bull. Kyushu Agr. Exper. Station 4: 261-273 (1956).
(3) Kanno, I., A new classification system of rice soils in Japan. Pedologist 6: 2-10 (1962).
(4) Kawaguchi, K., K. Kyuma, Paddy soils in Tropical Asia, their material nature and fertility. Monogr. Centre SE Asian Studies, no. 10, Kyoto University (1977).
(5) Xu Qi, Lu Yan-chun, Zhu Hong-guan, The genesis and classification of the paddy soils, Tai-lake basin, Jiangsu province, China. Acta Pedologica Sinica 17-2: 120-132 (1980).
(6) Wada, H., A possible classification of paddy soils according to the 7th Approx. Pedologist 10 (2): 141-146 (1966).
(7) Matsuzaka, Y., Study on the classification of paddy soils in Japan. Bull. Nat. Inst. Agr. Sci. Ser. B, 20: 155-349 (1969).
(8) Kyuma, K., Kawaguchi, K., Major soils of southeast Asia and the classification of soils under rice cultivation (paddy soils). Southeast Asian Studies 4: 290-312 (1966).
(9) Grigoriyev, G.I., V.M. Fridland, Classification of soils according to their degree of cultivation. Soviet Soil Science: 445-454 (1964).
(10) Karmanov, I.I., Changes in tropical soils under agricultural use. Soviet Soil Science 1: 31-42 (1966).
(11) Matsui, T., A tentative scientific classification and systematics of Japanese soils (in Japanese). Pedologist, 22-1: 56-70 (1978).
(12) Kubiena, W.L., Bestimmungsbuch der Böden Europas. Stuttgart, Enke (1950).
(13) Soil Survey Staff, Soil Taxonomy, a basic system of soil classification for making and interpreting soil surveys. Agriculture Handbook no. 436, SCS-US Dep. of Agriculture, Washington DC (1975).
(14) FAO - Unesco, Soil map of the world, vol. I, Legend. Unesco, Paris (1974).
(15) Koenigs, F.F.R., A "sawah" profile near Bogor. Contrib. Gen. Agr. Res. Stn. no. 105 (1950).
(16) Dudal, R., Paddy soils. I.R.C. Newsletter, Bangkok 7 (2): 19-27 (1958).
(17) Dudal, R., F.R. Moormann, Major soils of southeast Asia. J. Trop. Geogr. 18: 54-80 (1964).
(18) Otowa, M., Morphological changes of soils by paddy rice cultivation, pages 383-391 in: Schlichting, E., U. Schwertmann (eds.): Pseudogley and gley. Trans Comm. V and VII, Int. Soil Sc. Soc. (1973).

(19) Moormann, F.R., N. van Breemen, Rice, soil, water, land. Int. Rice Res. Institute. Los Baños, Philippines (1978).
(20) Kanno, I., Genesis of rice soils with special reference to profile development. Pages 237-253 in: Soils and Rice. International Rice Research Institute, Los Baños, Philippines (1978).
(21) Moormann, F.R., Morphology and classification of soils on which rice is grown. Pages 256-272 in: Soils and Rice. Internat. Rice Research Institute. Los Baños, Philippines (1978).
(22) Hauser, G.F., R. Sadikin, The productivity of the soils of East Central Java, bases on the yields of Jawa rice. Contrib. Gen. Agr. Res. no. 144 (1956).
(23) Breemen, N. van, L.R. Oldeman, W.G. Wielmaker, The Ifuago rice terraces - pages 39-74 in: Aspects of rice growing in Asia and the Americas. Misc. Paper 7 Landbouw Hogeschool Wageningen (1970).
(24) Pons, L.J., I.S. Zonneveld, Soil ripening and soil classification. Initial soil formation in alluvial deposits and a classification of the resulting soils. ILRI bull. No. 13, Wageningen, The Netherlands (1965).
(25) Motomura, S., F.M. Lapid, H. Yokoi, Soil structure development in asiatic polder soils in relation to iron forms. Soil Sci. Plant Nutr. 16: 47-54 (1970).
(26) Mitsuchi, M., Pedogenic characteristics of paddy soils and their significance in soil classification. Bull. Nat. Inst. Agr. Sc., ser. B, no. 25: 29-115 (Japan., English summary) (1974).
(27) Wada, H., S. Matsumoto, Pedogenic processes in paddy soils. Pedologist 17: 2-15 (1973).
(28) Mitsuchi, M., Permeability series of lowland paddy soils in Japan. Jap. Agr. Res. Quarterly 9: 28-33 (1975).
(29) Kevie, W. van der, Physiography, classification and mapping of acid sulphate soils. Proc. Int. Symp. on Acid Sulphate Soils I, ILRI Publ. 18, vol. I, Wageningen: 204-222 (1972).
(30) Leung, K.W., C.Y. Lai, The characteristics and genesis of paddy soils in northern part of Taiwan. Agr. Res. 22: 77-97 (1973).
(31) Kostenov, H., Genetic and chemical characteristics of meadow gley soils in the rice fields of Primoyre. Soviet Soil Science 7: 291-299 (1975).
(32) Mitsuchi, M., Characters of humus formed under rice cultivation. Soil Sci. Plant Nutr. 20 (3): 249-259 (1974).
(33) Brammer, H., Soil survey project Bangladesh. Technical Reports 2 and 3, FAO Rome (1971).
(34) Brinkman, R., Ferrolysis, a hydromorphic soil forming process. Geoderma 3: 199-206 (1970).
(35) Brinkman, R., Ferrolysis, a soil forming process in hydromorphic conditions. Agr. Res. Rep. 887. Centre for Agr. Publ. and Docum., Wageningen (1979).
(36) Mitsuchi, M., Chloritization in lowland paddy soils. Soil Sci. Plant Nutr. 20: 108-116 (1974).
(37) Brinkman, R., Surface-water gley soils in Bangladesh: genesis. Geoderma 17: 111-144.
(38) Kanno, I., Y. Konyo, S. Arimura, S. Tokudome, Genesis and characteristics of rice soils developed on polder lands of Shiroishi area, Kyushu. Soil Sci. Plant Nutr. 10: 1-20 (1964).
(39) Breemen, N. van, Genesis and solution chemistry of acid sulfate soils in Thailand. Centre for Agr. Publ. and Document, Wageningen (1976).
(40) Blume, H.P., U. Schwertmann, Genetic evaluation of profile distribution of Aluminium, Iron and Manganese Oxides. Soil Sc. Soc. Amer. Proc. 33: 438-444 (1969).

149

(41) Zonn, S.V., A.F. Kostenkova, G.P. Musorok, N.V. Khavkina, Pseudo-podzo lization and its identification from the composition of free forms of iron. Sov. Soil Sci. 7: 531-546 (1975).
(42) Kawaguchi, K., Y. Matsuo, Re-investigation on the distribution of active and inactive oxides along soil profiles in time series of dry rice fields in polder lands of Kojima basin. Soil Plant Food 3: 29-34 (1957).

Sponsored by the Netherlands Organization for the Advancement of Pure Research (Z.W.O.)

PRINCIPAL TYPES OF LOW YIELD PADDY SOIL IN CHINA

Xiao Ze-hong
(Institute of Soils and Fertilizers, Hunan Academy of
Agricultural Sciences, Changsha)

Paddy soil is widely distributed in China, with a total area of 26 million hectares. Due to different natural and cultivative conditions, the rice yield varies in different regions, ranging from 10 tons per ha annually in highly productive regions to only 2 - 3 tons per ha in regions of low productivity. Besides the agricultural techniques, unfavorable climatic and soil conditions are the causes of its low productivity. Among these factors, the soil conditions are frequently especially important in affecting the yield.

On the basis of the formation, fertility and measures of improvement, the low yield paddy soils are roughly distinguished into three categories (table 1)(1-3).

The 1st category or the highest category: According to the dominant factors limiting soil fertility and the possible direction of soil improvement, the low yield paddy soils are divided into 4 units, namely, cold muddy paddy soils with excessive water; settling compact paddy soils with large proportion of sands including coarse silts; heavy paddy soils with large proportion of clay, and toxic paddy soils. Owing to the different dominant factors limiting the fertility of the soil, the improvement and utilization of the four types of soil vary with each other.

The 2nd category: At this level, the soils with similar dominant factor of restriction are subdivided according to the differential in degree of action of the restriction factor. For example, cold muddy paddy soils are subdivided into permanent submergic paddy soils and cold water paddy soils on the basis of the differential in the cause of low temperature and puddling of soils.

The 3rd category: Soils with the similarities of 2nd category are subdivided in accordance with the differential in pratical measures of improvement suitable for the soils. For example, according to the peculiarities in soil improvement, the cold water paddy soils are subdivided into cold marshy paddy soil, cold ground water paddy soil, gleyed paddy soil and rusty water paddy soil.

Because of the diversity of low yield paddy soils, the present article mainly deals with the four types of low yield paddy soil of the 1st category.

As shown in Table 2, it is estimated that the area of the low yield paddy soils in China is about 26% of the total area of paddy soils. The area of settling compact paddy soils is about 40% of the area of low yield paddy soils, 30% for that of cold muddy paddy soils, 25% for that of heavy clayey paddy soils and 5% for the others including toxic paddy soils. The total increase of food grain of the country as a whole would amount at least to 10 million tons annually should the rice yield be increased by 1500 kg per ha through the improvement of the low yield paddy soils.

Table 1 Classification of low yield paddy soils [1]

1st category	2nd category	3rd category
Cold muddy	Permanent submergic	Permanent submergic
	Cold water	Cold marshy Cold ground water Rusty water Gleyed
Heavy clayey	Clayey Sticky clayey	Clayey Sticky clayey
	Calcareous compact	With lime concretion With lime hardpan
Settling compact	Settling muddy Settling sandy Leaking sandy	Settling muddy Settling sandy Leaking sandy
Toxic	Saline Acid sulphate	Saline Acid sulphate
	Mineral toxic	Manganese toxic Sulphur toxic Coal mine water

Table 2 Area of low yield paddy soil in China

Soil type	Area (1000 ha)	% in low yield paddy soil
Cold muddy	1,960	30
Heavy clayey	1,630	25
Settling compact	2,610	40
Toxic and others	330	5

COLD MUDDY PADDY SOILS(1,4-6)

Cold muddy paddy soils are widely distributed in the valley land of mountainous area, depressions of hilly area and adjacent area around lakes in southern China. As a result of ill-drainage, such soils developed under a strongly reducing condition and, therefore, belong to the type of swampy paddy soil. In accordance with the conditions of water and temperature, the soils are subdivided into 5 types as mentioned above and their features of main types are shown in Table 3.

In addition to the deep muddy and puddling layer of the soils on which is difficult for man or cattle or implements to do normal cultivation, the

Table 3 Major types of cold muddy soils

Soil type	Distribution	State of soil water	Thickness of muddy layer	Soil profile
Permanent submergic	Swampy land	Submergence all the year around	thick	A_g-G
Cold marshy	Valley of mountainous region, especially in the region of granite	Spring all over the field	thick	A_g-G or A-G
Cold ground water	Valley in hilly region, especially in the region of limestone	Spring of spot-like distribution	thin	A-G
Gleyed	Entrance to hilly valley or low land	Ground water table found in the depth of 30-40 cm	thin	A-P-G

main causes of low yield of these soils may be summarized as follows.

1. Low Temperature of Soil and Water

The temperature of the soil and water is lower than that of normal paddy soils, bearing a close relation to such conditions as the local climate, topography, vegetation, ground-water and distance from the spring. In gerenal, the variation of temperature in a same rice field of cold ground water paddy soil is about 4-5°C (Table 4), but in extreme cases, the temperature of some of the soils may be 8-9°C lower than that of the normal paddy soil. Owing to the low temperature of the soil and water, microbiological activities are inhibited, which in turn induce the decrease of the activity of root system and the nutrients uptake by the root system of rice.

Table 4 Variation of temperature in a same rice field of cold ground water paddy soil (Zhejiang)

Location	Soil temperature (°C)		Growth of rice seedlings
	2-3 cm	10-12 cm	
Inner side of the field, with spring	20.0	11.5	Inhibited
Center of field	21.5	19.5	No harm
Outer side of the field, without spring	24.5	20.0	Normal

2. Deficiency of Available Nutrients

Owing to the slow decomposition, the organic matter in cold swampy paddy soil may be up to more than 5%, and C:N is higher than 12. The contents of available nutrients, especially those of phosphorus and potassium, are extremely low. In cold swampy paddy soil, because of the intensive leaching of clay and the destruction of soil colloids, the cation exchange capacity is lowered (Table 5). At the same time, a large amount of ferrous iron in the soil induce a portion of nutrients to transfer into soil solution, which intensifies the deficiency of nutrient in soils.

Table 5 Variation of cation exchange capacity in cold
water paddy soil (Jiangxi)

Depth (cm)	No spring		Near spring	
	Exch. bases (m.e./100g)	CEC (m.e./100g)	Exch. bases (m.e./100g)	CEC (m.e./100g)
0-20	4.78	9.67	2.71	7.72
20-35	5.45	9.00	2.83	7.80
35-70	5.90	7.79	1.99	6.74

3. Toxicity of Reducing Substances

A large quantity of reducing substances usually appear in soils with strong reduction. In cold swampy paddy soils, the pH usually ranges between 5-6, which is conducive to the formation of free sulfides or hydrogen sulphide. At the same time, the toxicity of ferrous iron should not be neglected. In cold muddy paddy soil, soluble ferrous iron generally ranges from 20-70 ppm; and 40 -260 ppm in cold water paddy soil. Therefore, rice suffers in different degrees from iron toxicity in these soils, and there is a low yield as a result of stunted seedling, few tellering and low fruiting rate.

HEAVY CLAYEY PADDY SOILS(7,8)

This kind of soil is dispersed in Guangdong, Guangxi, Yunnan, Guizhou, Hunan provinces and Qamdo Prefecture of Sichuan Province. The genesis of the soils are closely related totheir parent materials of which the shale, purple shale, weathering products of limestone and fluvial and lacustrine clayey deposites are predominant; some of them are also developed on the weathering products of basic rocks such as basalt.

In addition to calcareous compact paddy soil as a result of overliming and heavy clayey texture, heavy clayey paddy soils are subdivided, according to their physical properties and clay mineral composition, into two types, namely, muddy clayey paddy soil and non-muddy clayey soil (Table 6).

Heavy texture, unfavorable tilth and low contents of available nutrients are the major disadvantages of these soils (Table 7). In their mechanical composition, clay amounts to more than 30%,and physical clay about 80%. Because of the heavy texture and the low content of organic matter, the soil is very compact, ill-structured and difficult for tillage. Being shrunk

154

Table 6 Main types of heavy clayey soil

Soil type	Coef. of flocculation	Type of clay mineral
Muddy clayey	26.0	2:1
Non-muddy clayey	71.3	1:1

and cleaved seriously when dry, the soil can not be recovered when submerged again. Therefore, the water and nutrients are liable to leak downward from the cleavage of the soil.

Low content of available nutrients is another feature of the soil. Phosphorus is liable to be fixed in the soil, especially so after the soil is dried. For example, in clayey muddy paddy soils, the content of available phosphorus is 2.67 mg per 100g soil when the soil is wet, while it is only 0.65 mg when the soil is dry.

Since the soils have a high coherence and high plastic index, they are difficult for tillage, and always induce a stunted seedling of rice because of low available nutrient content.

SETTLING COMPACT PADDY SOILS (1,2)

These soils are very sandy in texture or excessive in contents of silt, hence low in productivity. The soils are widely distributed in the southern part of China, especially at the middle and lower reaches of the Changjiang River. In addition to the effect of parent material, the formation of these soils is the result of unreasonable "crossed irrigation" and reducing eluviation and leaching of large amount of clay. And, owing to the lateral seepage of water in soil, a whitish layer is formed in various depth. If the whitish layer appears in plough horizon, the rice growth will be inhibited seriously. In accordance with the different characteristics of the soils, they are divided into 3 types, i.e. settling muddy paddy soil, settling sandy paddy soil and leaking sandy paddy soil (Table 8).

The sandy texture and compact structure, deficiency of nutrients and low retention ability for water and nutrients are the causes of low yield of these soils.

Very high content of SiO_2 is found in settling sandy paddy soils, usually amounting to more than 70%, of which 90% are quartz. In settling

Table 7 Mechanical composition of clayey soils (Yunnan)

Soil type	Fertility	O. M. (%)	Clay (%)	Phys. clay (%)
Red sticky clayey	Low	1.10	34.6	79.0
Gray sticky clayey	Low	1.19	58.5	83.5
Fertile sticky	High	2.89	22.4	61.1

Table 8 Types of settling compact paddy soil

Soil type	Distribution	Texture	O. M. (%)
Settling muddy	Hilly terrace or around Lakes	Coarse silt 40-60%	1.5
Settling sandy	Hilly regions of granite, gneiss, sandstone	Sandy parti-cles about 60%	1.4
Leaking sandy	Sandy alluvium and allu-vium on river-bed	Sandy, someti-mes with gra-vels in power layer	1.4

muddy paddy soil, the ratio of coarse silt to clay is above 2; while in settling sandy paddy soil and leaking sandy paddy soil, the ratio of sand to clay is above 4 even more than 5. Micromorphological studies of the settling muddy paddy soil showed that the soil has a compact fabric, and a compact arrangement of coarse silts with few voids. Settling sandy paddy soil has a compact arrangement of sands as its skeleton, with few matrix.

The cation exchange capacity of the soil ranges from 5-10 m.e./100g, thus the nutrients in the soil are liable to be leached.

Owing to the conditions mentioned above, settling compact paddy soils have very low contents of total and available nutrients. The organic matter content in plough horizon is below 1.5% and the N-content is lowest. The content of available phosphorus and potassium is also very low (Table 9).

Rapid settling of the soil particles after tillage under submergic conditions usually induces the compactness of the soil, which is unfavorable for transplanting of rice seedlings.

Owing to the poor retention ability of water and nutrients, and the resulting low content of available nutrients, there is a deficiency in nutrient in the later growing stage of rice in these soils.

Table 9 Organic matter and nutrient contents of settling
compact paddy soil (Hunan)

Depth (cm)	O. M. (%)	N (%)	P_2O_5 (%)	K_2O (%)	Avail. K (ppm)	CEC (m.e./100g)
0-14	1.19	0.072	0.16	2.15	67	7.14
14-31	0.50	0.030	0.15	2.14	47	
31-40	1.26	0.072	0.17	2.56	67	
40-90	1.15	0.079	0.18	2.87	100	

TOXIC PADDY SOILS (9)

In low yield paddy soils containing toxic substances for rice plant, there are coastal saline paddy soil, mineral toxic paddy soil and acid sulphate paddy soil. Since the coastal saline paddy soil belongs to saline soil, it will not be discussed in this paper. In mineral toxic paddy soils, there is toxic paddy soil as a result of coal mine water, heavy metal mine or arsenic ore, etc, scattering within a small area around the mining district. In this paper is dealt with only the acid sulphate paddy soil.

Acid sulphate paddy soil is distributed in the regions of river outlet of Guangdong, Guangxi, Fujian and Zhejiang provinces covering an area of about 1% of the low yield paddy soils in China.

Acid sulphate paddy soils are developed on the seashore of former mangrove swamp where the soil is covered alternatively with saline and fresh water. Owing to the stretehing out to the sea and the raising of alluvium, the mangroves gradually disappeared due to the change of environmental conditions, and buried by alluvial deposit. As a result of slow anaerobic decomposition of the buried mangrove residues under submergence of ground water, there is among others large amount of sulphides and hydrogen sulphide in the subsoil. After drainage, the soil is in oxidizing state, the sulphides in the soil are oxidized into sulphuric acid, thus inducing the strongly acid reaction of the soil.

The injury to rice is the result of the "acid reaction" (Table 10). Because of the strongly acid reaction of the soil, microbiological activities are greatly inhibited. The presence of large amount of aluminum, ferrous iron and manganese ions are harmful to rice growth. Strongly acid reaction may directly destroy the tissues of rice plant, inhibit the physiological activity, and even cause the death of rice plant.

Owing to the location of S-layer in the profile, acid sulphate paddy soil may be subdivided into 3 types as shown in Table 11. In the newly reclaimed strongly acid sulphate soil, the buried depth of mangrove residues is not more than 25 cm, and the whole profile is affected by the buried layer of the mangrove. The pH of the soil is below 2.5. The contents of active acid and soluble Fe and Mn are very high. Ground water table ranges from 30-40 cm. Under these conditions, rice growth is severely inhibited and no harvest could be expected.

In moderately acid sulphate paddy soil, due to the thickening of the covered soil layer, the buried layer of mangrove residues is lowered to 25 cm or below. The amount of active acid and soluble Fe and Mn is decreased

157

Table 10 Chemical properties of an acid sulphate paddy
soil (Guangdong)(9)

Depth (cm)	pH	O.M. (%)	S (%)	Water extractable component (m.e./100g)								
				Ca^{2+}	Mg^{2+}	Fe^{2+}	Mn^{2+}	Al^{3+}	H^+	K^++Na^+	SO_4^{2-}	Cl^-
0–14	3.1	3.33	1.19	0.52	0.14	0.61	0.02	0.37	0.07	2.63	3.21	2.15
14–26	2.8	3.73	0.41	1.25	2.59	0.82	0.08	1.56	0.08	4.28	7.16	3.50
26–50	2.3	7.23	1.12	2.18	5.91	0.86	0.13	10.71	0.71	3.07	19.81	3.76
50–70	2.2	6.60	2.36	2.70	9.02	2.44	0.33	23.71	1.48	0.25	34.75	4.84
70–100	3.0	1.90	0.93	2.91	11.20	1.00	0.60	2.42	1.50	1.38	16.65	3.36

Table 11 Location of S-layer in the profile of acid sulphate
soil in relation to rice growth (Guangdong)

Soil type	S-layer		Rice growth
	Depth (cm)	Water insoluble sulphur (%)	
Strongly acid	18–42	1.15	Severely inhibited
Moderately acid	50–70	1.99	Harmful when improperly managed
Slightly acid	80–95	0.46	No harm

with ground water table around 50 cm. The rice can be cultivated under
proper water management and strict control of the ground water.

In slightly acid sulphate paddy soil, the contents of active acid,
soluble Fe and Mn are even lower. Generally, it can be used for rice
cultivation. When the horizon containing sulphides is lowered to 75 cm or
below, there is not any unfavorable effect on rice growth, and the soil may
be developed into highly fertile paddy soil.

CONCLUSIVE REMARKS

In terms of the improvement of various low yield paddy soils, there
is not only the similarities, but also the peculiarities. The main points
are summarized as follows:

(1) Reasonable irrigation and drainage are the most direct and effective measures for the improvement of cold muddy paddy soil and acid sulphate
paddy soils, with a good effect in a shorter period.

(2) Remarkable effect of rational manuring has been found on heavy
clayey paddy soils and settling sandy paddy soils. Phosphate fertilizers

158

are usually efficient for the low yield paddy soils. It has been shown through farming practice that applying powdered rock phosphate to acid sulphate soil brings about good results.

(3) Rational cultivation or crop rotation system is the most important measures for the improvement of low yield paddy soils except the acid sulphate paddy soil.

(4) Regulation of soil texture is an effective measure for the improvement of both settling compact paddy soils and heavy clayey paddy soils. In agricultural practice, an addition of clay to sandy soil or sand to clayey soil has met with good results in the improvement of poor soils, though it costs much labor in a large area.

REFERENCES

(1) Institute of Soil Science, Academia Sinica, Nanjing (ed.), Soils of China. Science Press, Beijing, 36-48, 490-495 (1978). (in Chinese)
(2) Soil Survey Committee, Ministry of Agriculture, China, Agricultural Soils of China (1964, ms.). (in Chinese)
(3) Institute of Soils and Fertilizers, Hunan Academy of Agricultural Sciences, A preliminary analysis of the improvement measures of low yield paddy soils. In "Selective Proceedings of Improvement Measures of Low Yield Paddy Soil in Hunan", Hunan People's Publishing House, Changsha,54-62(1973). (in Chinese)
(4) Yu Tian-ren, Ding Chang-pu, On some problems of cold spring paddy soil. Turang Tongbao, 4, 22-26 (1958). (in Chinese)
(5) Yu Tian-ren, Liu Zhi-guang, Oxidation-reduction processes in paddy soils and their relation to the growth of rice. Acta Pedologica Sinica, 12, 380-389 (1964). (in Chinese with English summary)
(6) Institute of Soils and Fertilizers, Hunan Academy of Agricultural Sciences, The soil in Dongting Lake region and its improvement and utilization (1980, ms.). (in Chinese)
(7) Department of Soils and Fertilizers, Yunnan Institute of Agricultural Sciences, Studies on the utilization of sticky clayey soil in Yunnan and improvement of its fertility (1964, ms.). (in Chinese)
(8) Zhao Qi-guo, Zou Guo-chu, Clayey paddy soils and their amelioration in Yunnan Province. Acta Pedologica Sinica, 7, 59-67 (1959). (in Chinese)
(9) Gong Zi-tong, Zhou Rui-rong, Genesis of strongly acid saline paddy soils of southern Guangdong. Acta Pedologica Sinica, 12, 183-191 (1964). (in Chinese with English summary)

THE FERTILITY AND FERTILIZER USE OF THE
IMPORTANT PADDY SOILS OF CHINA

Lu Ru-kun
(Institute of Soil Science, Academia Sinica, Nanjing)

INTRODUCTION

1. The Area of Paddy Soils and the Yield of Rice
in China and Their Situation in Rice
Production Throughout the World

The area of paddy soils in China approximates to 25.3 million hectares and the sown area is about 34.60 million hectares that is about 1/4
of the total harvested area of rice field all over the world. The yield
of rice grain, however, accounted for over 1/3 of the total yield throughout the world in 1978 (Table 1), and ranked first of all countries thereby[1].

Table 1 World rice production and China share[*]
(1978)

Country	Area of paddy soils	Total rice production	Yield
	Million ha	Million t	t/ha
World	144.9	384.5	2.65
China	25.3 [**] (34.6)	136.9	3.98
India	40.2	80.7	2.01
Japan	2.5	16.4	6.42
Developed Countries	———	———	5.56
Developing Countries	———	———	2.50

[*] All figures after FAO except China

[**] Harvested area

In 1978, the mean yield of rice grain was estimated as 3.98 t/ha in
China which was lower than that of 5.56 t/ha in developed countries but
higher than that of 2.5 t/ha in developing countries. The yield of rice
grain in different locality of China varied greatly (Table 2). For
example, the mean yield of rice grain in Shanghai amounted to 5.6 t/ha
in 1978, whereas it attained only 3.2 t/ha in Zhuang Autonomous Region
of Guangxi. In Suzhou Prefecture of Jiangsu Province the sum of total
grain yield of a double cropping rice and a single upland crop could
amount to 10-12 t/ha in one year rotation.

Table 2 Illustrations on the variation in rice yield of China
(1978)

Region	Grain
	t/ha
Average of rice yield for the whole country	3.98
Mean rice yield for a single crop of rice, Shanghai	5.6
Mean rice yield for a single crop of rice, Guangxi	3.2
Average yield for 3 cropping (rice-rice-barley or wheat), (Suzhou, Jiangsu	10-12

2. The Fertilizing Level on Rice Field in the Region of Double Cropping Rice in the South of China

What with high complexity of organic manure in kind and quality and what with great difference between various locality, there is not accurate statistic for fertilizing level of rice field in China (Table 3). Table 3 shows an example of fertilizing level in some production brigades of middle standard in Zhejiang province which might serve as a certain representative in this area. It is seen from Table 3 that the total amount of nutrients (N P K) applied is about 133 kg/ha for a single crop rice, of which about half from organic manure, and other from chemical fertilizers. The average yield of rice grain in this area is about 3.75 t/ha per cropping season.

Table 3 Rate of fertilizer Application for
double cropping rice, Zhejiang Province

Average rate	Nutrients applied		
	N	P	K
	kg/ha		
Annual, for 2-crop rice			
Organic manures:			
Milk vetch, 7.5t/ha	24.8	2.64	14.5
Pig manure, 7.5t/ha	33.8	6.20	37.4
Inorganic fertilizers:			
NH_4HCO_3, 750 kg/ha	127.5	——	——
Phosphate, 300 kg/ha	——	19.8	——
Rate for a single crop rice			
From organic	29.3	4.4	26.0
From Inorganic	63.8	9.9	——
Total (N P K)	133.4		

3. The General Contents of Nutrients in
Some Paddy Soils of China

The rice field spreads throughout the whole country, even so, more than 90% of its area is situated on the south of the Changjiang River. These southern paddy soils have been mainly deriving from laterite, red earth and alluvial deposit (Table 4). Table 4 presents the ranges of nutrient content in important type of Chinese paddy soils, by which we can see the trend of the phosphorous and potassium variation that their contents are higher in the north whereas lower in the south owing mainly to the influence of their parent materials on them.

Table 4 Nutrient ontents in some important paddy soils of China

Type of paddy soil	O.M.	N	P	K
		%		
Paddy soil developed on:				
Black soil, Northeast China	3.2–6.9	0.15–0.35	0.06–0.13	1.4–2.1
Meadow soil, North China	0.9–1.4	0.04–0.09	0.04–0.1	1.6–2.0
Secondary loess, Northwest China	0.7–1.4	0.06–0.12	0.05–0.09	0.8–1.4
Alluvia and lacustrine deposits, middle and lower Changjiang River valley	1.2–3.5	0.08–0.19	0.04–0.07	1.2–2.3
Red soil, Central China	1.0–2.9	0.07–0.15	0.02–0.05	0.7–1.0
Lateritic soil, South China	1.5–3.0	0.08–0.12	0.01–0.04	0.2–0.7

As a result of the anaerobic conditions and the higher level of fertilization in waterlogged soils, the contents of various nutrients, in particular of organic matter and nitrogen, are often higher than those in corresponding soils of upland[2] (Table 5); The soil which has been utilizing as rice field is of advantage to the accumulation of organic matter and nitrogen in it thereby.

Table 5 Variation of soil-N as affected by rice cultivation

Soil type	Cultivation	No. of samples	O.M.	N
			%	
Yellow brown soil	Upland	49	0.8–1.76	0.05–0.12
	Paddy	524	1.20–3.48	0.08–0.19
Red soil	Upland	115	0.83–1.98	0.06–0.12
	Paddy	271	1.04–2.97	0.07–0.19

THE NITROGEN SUPPLY ABILITY OF SOIL AND
THE NITROGEN FERTILIZING

1. The nitrogen Supply Ability of Soils

The nitrogen supply ability of soil includes both the amount of nitrogen that may be provided to each cropping and the characteristic of nitrogen supply. Investigation carried out in the provinces or autonomous regions of Jiangsu, Zhejiang, Guangdong and Guangxi shows that in case we regard roughly the amount of nitrogen accumulated in aboveground biomass of rice from no nitrogen fertilizer plot as the nitrogen that may be provided by soil itself under local condition, the amount of nitrogen supplied one cropping of rice in paddy soil is around 52.5-107.5 kg N/ha corresponding to 1.4-3.2% of the total nitrogen content in soil[*].

The ability of nitrogen supply in soil, as mentioned before, not only depend on the amount of nitrogen supply, but also on the characteristic of its supply. According to farmer's experience, the characteristic of nitrogen supply (N release process in soil) when there is no nitrogen fertilizer applied, might be divided into four types as following:

1) The soil can release nitrogen adequately and steadily all over the growing season —— The most soils of this type contains more organic matter and have better physical properties, thus causing the good growth of rice through the whole growing season and being the soils with high yielding ability.

2) The nitrogen release is only a little in the early stage of rice growth while it will greatly increase in later period —— The soils belonged to this type have also a higher content of organic matter in general, but most of them are clayey and of poor draining. Owing to the insufficiency of nitrogen supply in early stage, the rice take less tillering; and because of a great deal of nitrogen release in later period the mature time of rice may be delayed. Thus causing a decrease in grain yield.

3) Nitrogen releases in large quantities in early stage but very little in later period, thus often the rice grows vigorously in the early stage but decline quickly soon after —— These soils usually contain not much organic matter, but they are of light texture and of better physical properties.

4) The soil release some few nitrogen through the whole growing season of rice —— In general the soils of this type are of low productive field with lower natural fertility and thinner cultivated horizon.

It is obvious that the various characteristic of nitrogen release (i.e. supply) relate to the mineralization rate of nitrogen in varying stages of rice growth and its resulting amount in soil, whereas these factors are affected by total nitrogen content, C:N, physical properties of soil and climatic conditions. This knowledge is useful for an expert farmer or a adviser to make his nitrogen fertilizing plan.

---

[*] After Zhu Zhao-liang.

163

2. Response of Rice to Nitrogen Fertilizer and the Plant Recovery of Fertilizer Nitrogen

In view of nitrogen deficiency in nearly all paddy soils of China, reat importance has been attached to the manufacture of nitrogen fertilizer. The output of nitrogen fertilizers was about 8.8 million tons of N in 1979. At the present time urea, ammonium sulphate and ammonium bicarbonate are the common used nitrogen fertilizers (Table 6). Table 6 presents response of rice plant to fertilizers of nitrogen. These data were obtained at the rate of about 60 kg N/ha with certain amount of organic manure as basal.

Table 6 Response of rice plant to nitrogen fertilizers[*]

Locality	Increment in yield
	(kg grain/kg N)
Taihu Lake area, Jiangsu	11.9-16.7
Guangdong	14.3-19.0
Guizhou	15.7-20.8
Jiangxi	16
Whole country[**]	14.0-23.8

[*]Data of NH_4HCO_3 not including

[**]Data collected from experiments conducted in different parts of this country

The efficiency of different kinds of nitrogen fertilizer varies greatly. The % recovery of fertilizer N by rice plant are in the following order: Ammonium sulphate >urea >ammonium bicarbonate. The efficiency of ammonium sulphate or urea applied is about 30-50% whereas it is only 20-30 % for ammonium bicarbonate in general.

The important organic manures for rice crop are pig manure, compost and green manure. Table 7 is the example concerning the efficiency of nitrogen of organic manure[3]. Owing to the great variation in the quality of those organic manures, the efficiency may be markedly different.

Table 7 Recovery of manure-N by rice plant

Manure	No. of trial	N recovery
		%
Leguminous green manure	10	25.3 ± 10.2
Pig manure	9	16.7 ± 9.0
Compost and water-logged compost	10	16.6 ± 5.6

3. Improvement on the Efficiency of
Nitrogen Fertilizers

A lot of investigation has been carried out in order to improve the efficiency of nitrogen fertilizers especially of ammonium bicarbonate in late years. The effective practices are reforming the fertilizer into granulated or ball-like form and deep application (Table 8). Table 8 is the data of trials conducted by our Institute[4]. In general, the effect of deep application with granulated fertilizer is significant, but there is still a problem of lacking in fertilizing machinery or implement for its application.

Table 8 The recovery of $^{15}NH_4HCO_3$ by rice

Treatment	Recovery range	Average
		%
Powder, top dressed	10.8–31.2	22.3±6.0 (n = 16)
Powder, incorporated with soil of surface layer	17.6–50.3	37.4±11.1 (n = 5)
Pill, deep placement at 6 cm below soil surface	40.5–78.8	64.8±16.6 (n = 5)

The problems relative to the fate of nitrogen fertilizer after their application shall be discussed in other essay at this meeting[5].

THE PHOSPHOROUS STATUS OF SOIL AND THE
PHOSPHOROUS FERTILIZING

1. The Forms of Phosphorous and its Distribution
in Some Paddy Soils

A study[2] revealed that the content of occluded phosphorus (O-P) in soils of China increases from north to south with the increasing degree of soil weathering, whereas that of calcium phosphate decreases. In acidic paddy soil of southern China, the occluded phosphorus account for about 40-70% of the total inorganic phosphorus in soil. The calcium phosphate are dominate in soils of northern China and the scattered alluvial paddy soils. The content of iron phosphate is high in acidic paddy soils, usually constituting more than 50% of the non-occluded inorganic phosphorous (Table 9).

165

Table 9 Phosphorus constituents in some
important paddy soils of China

Type of paddy soil	Total P	Inorganic P			
		Al–P	Fe–P	Ca–P	Occluded P
		ppm P			
Paddy soil developed on:					
Loessial soil	856	48.2	229	125	335
Red soil	521	25.7	147	27.5	298
Lateritic soil	391	51.2	43.5	21.0	249
Alluvial soil (Changjiang River valley)	926	10.8	50.9	41.4	111
Alluvial soil (Zhujiang Delta)	568	41.1	19.1	101	197

2. The Significance of Iron Phosphate on Rice Nutrition

Table 10 showed that the iron phosphate content was highly correlated
to the "A" value (Table 10), thus it was thought that iron phosphate is the
principal source of phosphorous for rice in paddy soils[6].

Table 10 Correlation between the form
of soil–P and "A" value

Form of Soil–P	Correlation coefficient to "A" value
Occluded	0.23
Non–occluded	0.98
Fe–P	0.95
Al–P	0.28
Ca–P	0.65

3. The Response of Rice Plant to Phosphorous Fertilizer

In China the utilization of phosphorous fertilizers in large scale
did not take place until the sixties. The main kinds of phosphorous
fertilizer are superphosphate (mainly for calcareous soils in the North)
and calcium magnesium phosphate (for acidic soil in the South). The
output of phosphorous was about 1.80 million tons of P_2O_5 in 1979. Table
11 shows the increment in yield of rice caused by phosphorous applied in
some provinces of China.

Because the efficiency of phosphorous fertilizer applied is so low
that our scientists and peasant often pay special attention to the
economical ways of phosphorous fertilizing. Two examples in this respect
are as follow.

Table 11 Response of rice plant to
        P-fertilizer on paddy soils
        deficient in phosphorus

Locality	Increment in yield
	kg grain/kg p
Hunan	14.2-35.5
Guizhou	18.6
Hebei	43-80.2
Hubei	28.4-51.1
Whole country	17.0-40.9

4. "Increase Nitrogen by Phosphorous Applied"

So called "Increase nitrogen by phosphorous applied" is a colloquial
speech concerninga method of phosphorous fertizing. The methods is that
in a "upland crop-rice" rotation, the P-fertilizer should be applied mostly
to the up-land crop. In case of leguminous green manure crop, it means
that the phosphate fertilizer should be used first to increase the N fixed
by legumes. Thus when the green manure is ploughed in more yield of rice
grain might be obtained from the same amount of phosphate fertilizer
applied as compared to direct use of P fertilizer to the rice crop.
Numerous experiments showed that each kg of P applied by this method could
earn a surplus of N fixed about 3.5-4.1 kg N[2]. And about additional 42-57
kg of rice grain per kg of P applied might be obtained. Another example
is to make fertilizer adhere to seedling roots. These kinds of method
were used long ago for the more expensive organic manure such as soybean
cake and others. Since 1960 it began to be use with phosphorous fertilizer
and got best results. The method used is that mix phosphorous fertilizer
with 3-4 times its weight of rotten organic manure or fertile soil, and
adequate water to make paste, and then to dip the rice seedling roots
before transplanting. For calcium magnesium phosphate it might be adhere
to seedling roots directly without soil added. Table 12 is an example of
the result due to this method used to a paddy soil derived from red earth
with very high phosphorus fixation capacity[7].

Table 12 Comparison of methods of P application on the efficiency
        to rice plant* (red soil, Jiangxi)

Treatment	Average of 3 years	
	Yield, t/ha	increment kg grain/kg fert.
CK	1.56	———
Broadcasting 75 kg/ha	2.45	11.2
Broadcasting 300 kg/ha	3.2	5.5
Root dipping 75 kg/ha	2.93	18.2

* superphosphate

5. The Determination of Available Phosphorous
   in Paddy Soils

A lot of research concerning the determination of available phosphorous in paddy soils has been conducted. The investigation of late

years demonstrated that Olsen's method is applicable to acidic paddy soil in southern China[8]. For reasons of its wide adaptability this method is being put to more and more use. Table 13 presents a preliminary criterion of Olson-P suggested by our Institute[9].

Table 13 Level of soil available-P and crop Response

Level of available P(Olsen's method)	Response to p-fertilizer
<3 P ppm	Crop fail without P-fertilizer
4-5	Good response for all crops
6-10	No response for rice plant
11-15	No response for all cereal crops
16-20	P-fertilizer needed only by legumes and certain crucifer crops
>20	P-nutrition sufficient

THE STATUS OF SOIL POTASSIUM AND POTASSIUM FERTILIZING

In the early sixties a series of research on potassium in red earth has been carried out in China[10], nevertheless, the potassium fertilization at large scale is just carried on since the seventies. In late years potassium is becoming increasingly important in agriculture, and, in several provinces of China a lot of investigation has been carrying on.

At this meeting there is another essay[11] to introduce the problem of potassium of Chinese soils, so it is only a concise illustration in this paper.

1. The Response of Rice to Potassium Fertilizer

Table 14 lists some representative values of response of rice to potassium fertilizer in several provinces of our country.

Table 14 Response of rice plant to K-fertilizer on paddy soils deficient in K

Locality	No. of experiment	Increment of yield
		kg grain/kg K
Mean value of red soil region	1361	9.3
Mean value of important paddy soils in the country	716	6.6

Some investigation revealed that the effectiveness of potassium fertilizer is related to amount of nitrogen applied. The yield increased is higher while rice crop was provided with sufficient nitrogen. Besides, the response to potassium fertilizer on later rice is higher than on early rice.

2. The Evaluation of Available Patassium in Soil

The method of 1 N NH_4Ac extraction for available potassium determination and the 1 N HNO_3 boiling method for slow available potassium are in common use in China. So far as concerns the nation-wide extent, it might be proper to take both available and slow available potassium into account so as to describe the potassium supplying ability in various soils whereas for some provinces or regions with similar kind of soil it is enough to take the available potassium as an index alone. Table 15 is an example for describing the potassium supply ability of soil and the response of crops[12].

Table 15 Soil-K supply capacity and crop response

Index	Available K	Slowly available K	K supply capacity	Response to K-fertilizer
	mg/100g	mg/100g		
1	3.3-6.6	16.6-33.2	very low	Good response for all crops
2	3.3-6.6 6.6-10	33.2-66.4 16.6-33.2	low	Good response for crops of high K-requirement
3	6.6-10	33.2-66.4	medium	Moderat response for crops of high K-requirement
4	10-13.3	33.2-66.4	medium -high	No definite response for crops
5	13.3-16.6	66.4-100	high	Usually no response for crops
6	16.6	100	very high	No response

Indeed, in order to reduce the production costs in agriculture of China there is still much work to be done for efficient and economic use of fertilizer. To some extent this is also a problem facing the other countries all over the world.

REFERENCES

(1) FAO monthy bulletin of statistics, 3, March, 26 (1980).
(2) Institute of Soil Science, Academia Sinica, Nanjing (ed.), Soils of China, Science Press, Beijing, 360-391 (1978). (in Chinese)
(3) Zhu Zhao-liang, Nitrogen nutrition in rice production in China (1980, ms.). (in Chinese)
(4) Li Ching-kwei, Chen Rong-ye, Ammonium bicarbonate used as a nitrogen fertilizer in China. Fert. Res. 1(3), 125-136 (1980).

(5) Chen Rong-ye, Zhu Zhao-liang, The fate of nitrogen fertilizer in
     paddy soils. In this Proceedings.
(6) Jiang Bo-fan, Lu Ru-kun, Gu Yi-chu, Li A-rong, The content of
     iron phosphate in paddy soils of southern China and their sign-
     ificance in rice plant nutrition. Acta Pedologica Sinica, 11,
     361-369 (1963). (in Chinese with English summary)
(7) Institute of Soil Science, Academia Sinica, The Environmental
     Conditions of Soils for High Yield of Rice. Science Pre
     Beijing, 325 (1961). (in Chinese)
(8) Shi Tao-jun, Zhu Yin-mei, Lu Ru-kun, Studies on methods for
     determination of available phosphorus in acid paddy soils. Acta
     Pedologica Sinica, 16, 409-413 (1979). (in Chinese with English
     summary)
(9) Lu Ru-kun, Soil phosphorus. Turang Tongbao, 2, 47-49 (1980).
     (in Chinese)
(10) Li Ching-kwei, Wang Mei-zhu, Zhang Xiao-nian, On the status of
     soil potassium and the sequence of transformation of potassium
     bearing minerals in some important soil types in the red earch
     region of China. Acta Pedologica Sinica, 9, 22-35 (1961). (in
     Chinese with English summary)
(11) Xie Jian-chang, Ma Mao-tong, Du Cheng-lin, Chen Ji-xing, On the
     potential of K-nutrition and the requirement of K-fertilizer in
     important paddy soils of China. In this Proceedings.
(12) Zhang Xiao-pu, Du Cheng-lin, Ma Mao-tong, Chen Ji-xing, Jia Yi, Xie
     Jian-chang, The supply of soil potassium and the effect of
     potassium fertilizer on crop response in Kiangsu Province. Acta
     Pedologica Sinica, 15, 61-74 (1978). (in Chinese with English
     summary)

EVALUATION OF NITROGEN FERTILITY AND INCREASING FERTILIZER EFFICIENCY IN WETLAND RICE SOILS

S. K. De Datta,
(International Rice Research Institute,
P.O. Box 933, Manila, Philippines)

P. J. Stangel and E. T. Craswell
(International Fertilizer Development Center,
P.O. Box 2040, Muscle Shoals, AL 35660, USA)

Reactions of fertilizer nitrogen in wetland rice soils are influenced by the presence in the atmosphere above the floodwater of oxygen that moves slowly by diffusion and convection currents to the soil surface. With time, two distinct soil layers are developed: (1) an oxidized surface layer; and (2) an underlying reduced layer. Their thickness depends upon the properties of the soil and the type of water management. Their development under wetland rice culture triggers several physicochemical and microbiological processes. The resultant soil-water-atmosphere system is highly complex and heterogeneous.

For example, the ammonium nitrogen present in the surface aerobic layer can be readily oxidized to nitrate nitrogen. The nitrate formed during nitrification but not absorbed by the crop moves down to the anaerobic layer and is lost by denitrification. Furthermore, flooding the soil for wetland rice culture causes ammonium to accumulate and lowers the nitrogen requirement for organic matter decomposition. Flooding is also associated with inefficient utilization of applied nitrogen. The efficiency of applied nitrogen is only 30%-50%, and, in many cases, even less (1-6).

The efficiency of fertilizer nitrogen is low because of:

* loss by ammonia volatilization,
* nitrification followed by denitrification,
* biological immobilization, especially by algae,
* fixation of ammonium nitrogen by clays,
* leaching,
* runoff, and
* seepage.

A number of reviewers have summarized nitrogen transformation processes in flooded soils (7-11). A better understanding of such processes would provide a sound basis for developing ways of regulating nitrogen losses from wetland soils and increasing the availability of soil nitrogen to rice.

With regards to management of nitrogen fertilizer, the low fertility of many rice soils and the limited supply of inorganic fertilizer are serious constraints to higher rice yields in South and Southeast Asia.

Rice yields are particularly low on farms where modern varieties are grown with no fertilizer nitrogen but with adequate nitrogen (in the absence of deficiency or toxicity of other nutrient elements and of any limitation of production factors), the new variety yields much more than the old one[5,12].

The rising importance of inorganic fertilizers to increased rice production has prompted a rapid expansion in the capacity for nitrogen fertilizer production in many Asian countries, particularly China, India, and a number of countries in Southeast Asia.

China's initial efforts (1949-70) to increase rice production focused primarily on the effective and intensive use of organic manure, improved irrigation, use of early-maturing modern varieties, and intensive use of existing land. Chemical fertilizers supplemented organic and biological N sources. In recent years additional increases in food production have been boosted through increased use of chemical fertilizers, made possible through increased supplies. Faced with a world food and fertilizer shortage in 1973-74, China stepped up fertilizer imports and increased local production by purchasing thirteen 1,000-mtpd modern ammonia plants with a collective capacity to produce nearly 3 million metric tons of nitrogen per year[13]. These plants, which began coming in on stream in 1976, were largely responsible for the 42% increase in nitrogen consumption in the 2-year period from 1975-76 to 1977-80 [14]. Similar emphasis on nitrogen production and use has taken place in India, Indonesia, and other Asian countries with large populations and large food requirements.

Most of the nitrogen imported or produced in Asia is and will continue to be produced in the form of urea. Japan, a major producer of all nitrogen fertilizers, uses large quantities of nitrogen as NPK compound fertilizers for domestic purposes. Similarly, farmers of Korea and Taiwan of China use considerable nitrogen as compounds. As a result, production of that fertilizer in continental Asia has been relatively high and stable over the years. On the other hand, China, Vietnam, and Democratic Republic of Korea supply crops with nitrogen separately from phosphorus and potassium. The result is a steady increase in the use of urea since 1976 (Fig. 1), which also occurred in tropical Asia. The rapid and large expansion in nitrogen fertilizer production and the increased use of nitrogen for rice, are an important development in the region. However, unless high losses and low efficiency of nitrogen fertilizers are corrected, most of the potential benefits of increased use of fertilizer nitrogen may not be realized.

Furthermore, national planners now realize that a continuously adequate supply of fertilizer at the farm level, when properly focused on rice production, promises stable economic development for many countries.

Fertilizer prices, particularly those of urea, were relatively stable and changed little over the period 1965-73 (Fig. 2). Prices rose sharply because of the food and fertilizer shortage in 1973-74 but quickly dropped back to very near the precrisis level. Many of the factors that affected supply in 1973-74 are also contributing to the current price rise, which may be the beginning of a steady climb in price, particularly for nitrogen, because projected production costs and future capacity are likely to be at or above current international prices for urea.

Furthermore, the price rise might accelerate unless new capacity above that now under construction is scheduled to come on stream before 1985. International prices for urea, currently at $190/mt f.o.b. U.S. Gulf[13], are likely to exceed the U.S. $215 forecast by UNICO[15] for 1982 and match the U.S. $272/mt forecast by The World Bank for 1985.

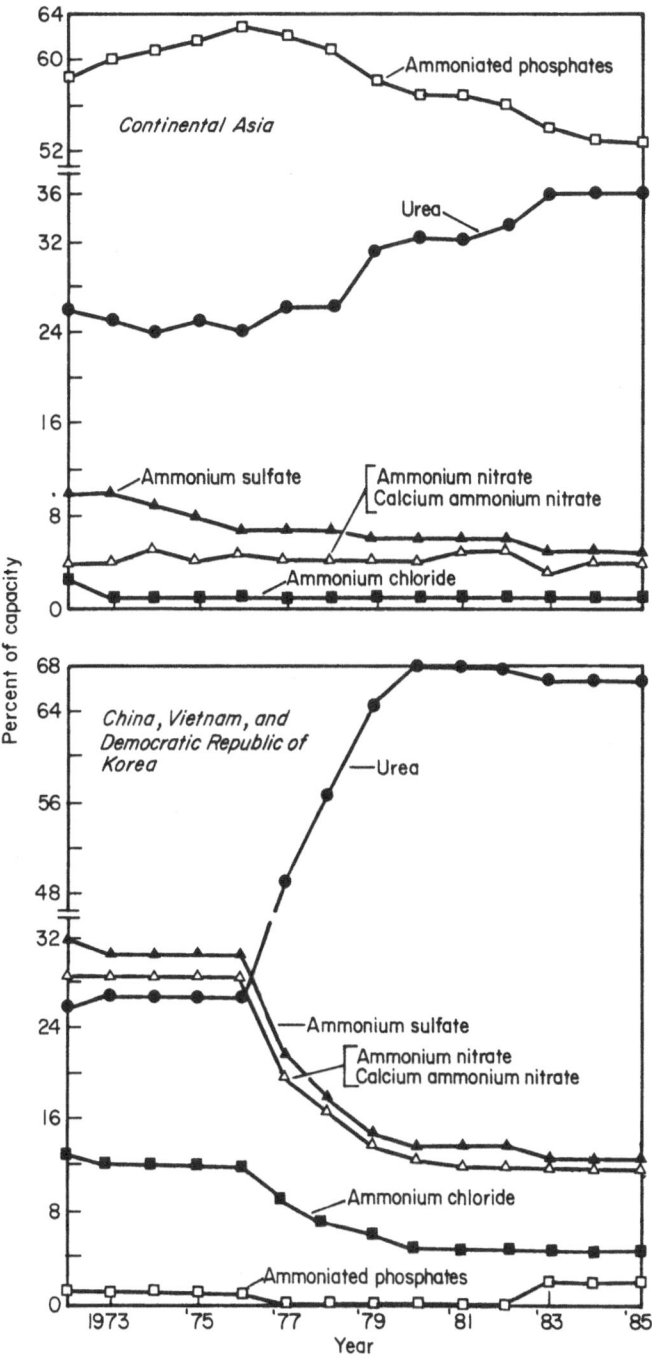

Figure 1   Changes in capacity to produce major nitrogen products (1972-85)
in China, Vietnam, and Democratic Republic of Korea. (From
International Fertilizer Development Center and Tennessee Valley
Authority, unpublished data)

173

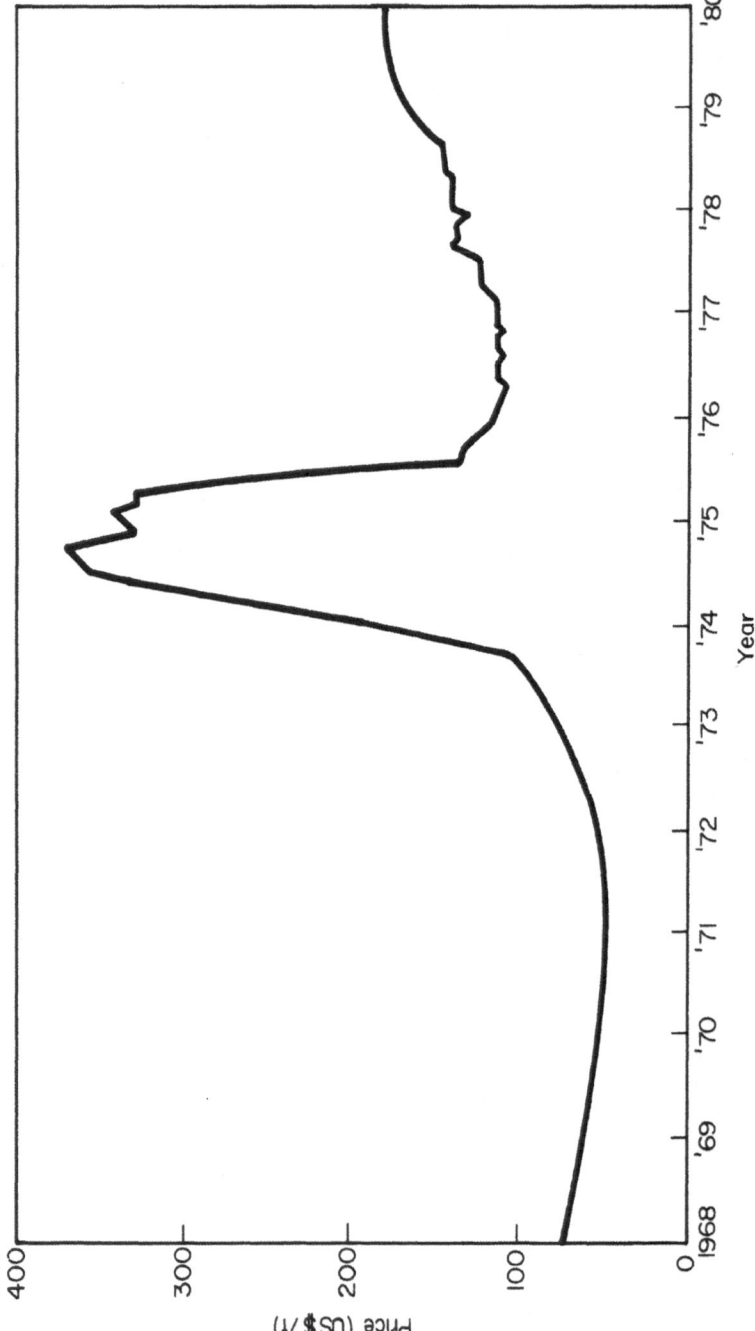

Figure 2   Export prices of urea fertilizers (1968–80).   (Adapted from Stangel[13])

ROLE OF SOIL IN SUPPLYING NITROGEN TO RICE

The higher yields of rice now needed and obtainable with the use of modern varieties require a great quantity of plant nutrients, particularly nitrogen. In traditional rice production systems, nutrients removed by the crop were replenished from biological nitrogen fixation or returned to the soil in the form of compost and manure. In modern agriculture, the nutrient needs of high-yielding crops are supplied through application of chemical fertilizers. However, rice still depends heavily on soil nitrogen for growth and yield. One study in Thailand[16] showed that more than 60% of the nitrogen taken up by the rice crop are derived from mineralization of soil organic matter. By using ^{15}N-labelled fertilizer, the same authors estimated the amount of mineralized soil organic nitrogen during the cropping season to be 131 kg N/ha in a brackish water alluvial soil near Bangkok (1.4% organic carbon and 0.13% total nitrogen). The amount was slightly lower than that in Japanese paddy soil with low fertility. In Thailand, the mineralized soil nitrogen absorbed by the crop was distributed as follows:

* 60% from transplanting to end of tillering,
* 36% from the end of tillering to panicle initiation, and
* 4% during ripening.

The corresponding figures for Japanese soils were 45, 27, and 28%, respectively. For example, it was reported that in Japan[17], the soil nitrogen absorbed by the rice plants averaged 75 kg/ha, which constituted 77% of the total nitrogen in the plant at harvest. Further, it was concluded that differences were attributable to the differences in temperature regimes in the tropics and temperate regions[16]. Another factor to consider is that drying the soil before flooding increased the amount of nitrogen initially released from soil organic matter. This effect may be very important in the tropics, particularly where a distinct dry season occurs.

Some of the nitrogen removed by a rice crop is partially returned to the soil, directly as stubbles, roots, and straw, and indirectly as manure. This soil nitrogen is highly important in producing low to moderate, but stable, yields in tropical Asia. Its importance in wetland rice culture is emphasized in Japan and is reflected in the statement, "Grow paddy with soil fertility, grow barley with fertilizers"[18]. Even in the fertilized fields, up to two-thirds of the nitrogen absorbed by rice is derived from the soil[9,19]. However, nitrogen fertility of lowland rice soils must be maintained if not increased. Using data from Thailand[16], it was calculated that if the total nitrogen content in soil is 0.13%, which is equivalent to 1,960 kg N/ha in a 10-cm surface soil with a bulk density of 1.3 g/cm^3, that amount of nitrogen can be used up in 13 years if fertilizer nitrogen or biological nitrogen fixation does not compensate for the lost fertility[20]. Of course, that does not happen in the actual farm situation. It can then be concluded that nitrogen fertility is maintained by some nitrogen fixers that operate in wetland rice soils. Similar data were obtained in nitrogen balance studies at International Rice Research Institute (IRRI)*. The possible existence of fertility-maintaining mechanisms specific to wetland rice soils has led to a great deal of research on the role of nitrogen-fixing organisms[21] which is beyond the scope of this review.

Basic studies of nitrogen transformation processes in wetland soils have been made primarily by laboratory incubation in the absence of plants. Furthermore, most studies have been limited to one specific transformation process.

---

*App et al 1980, unpublished.

However, some laboratory and field studies with ^{15}N-labelled fertilizers have provided, besides knowledge on plant uptake and efficiency of fertilizer nitrogen in a wetland rice soil [for reviews see (22) and (11)], useful information regarding the transformation of soil nitrogen.

NITROGEN-SUPPLYING CAPACITY OF WETLAND RICE SOILS

The predominant form of inorganic nitrogen in the wetland soil is NH_4^+. The concentration (or activity) of NH_4^+ in soil solution (intensity parameters) is less than that in the sorbed phase in soils with high cation exchange capacity (CEC). However, the soil organic nitrogen pool provides the main reservoir of nitrogen, which must be mineralized to ammonium before it can be absorbed by the rice plant. Thus, a measurement of the exchangeable ammonium content of a soil at any time will not reflect the nitrogen-supplying capacity of the soil. According to (23), after the first flush of mineralization after flooding, NH_4^+ is gradually and continuously released throughout the growing season to meet the needs of the developing plant (Fig. 3). Nitrogen so released is not subject to appreciable losses if a rice crop is present and actively taking up nitrogen.

Methods to Determine Available Soil Nitrogen

The status of available nitrogen in wetland rice soils is influenced by many chemical and biological processes. Mineralization of organic nitrogen in submerged soils depends upon the chemical environment and the microbial population in the soil.

Both chemical and biological methods are generally used to measure the nitrogen-supplying capacity of wetland rice soils. Biological methods have the advantage of using the same process that is responsible for the release of nitrogen in the soil during the growing season (mineralization) to estimate available nitrogen. However, they require a relatively long incubation period. On the other hand, a chemical method is not likely to simulate the activities of microorganisms or to selectively release the fraction of soil nitrogen that is made available for plant growth by these organisms[24]

It has been suggested that the determination of NH_4^+-N mineralized under anaerobic conditions may give a good estimate of the amount of nitrogen available to the wetland rice crop[7,25,26].

Recently reviews are available[9,26] on the methods of measuring the nitrogen-supplying capacity of wetland soils.

1. Anaerobic incubation method. Earlier papers[25,27] confirmed the suitability of the anaerobic incubation technique for determining the availability of soil nitrogen to plants in wetland rice culture. Further, NH_4^+ production in the soil during the 6-day incubation under waterlogged conditions best predicted rice grain yield. They obtained little improvement in the correlation by including the initial NO_3^- or NH_4^+ in the soil. Table 1 shows the wide variation in available nitrogen of some Philippine wetland rice soils, measured using a similar incubation technique.

An earlier report[28] suggests that soils vary widely in the capacity to supply nitrogen. On the basis of Philippine data, the following schedule of nitrogen fertilizer application for a 5 t/ha grain crop is suggested[29]:

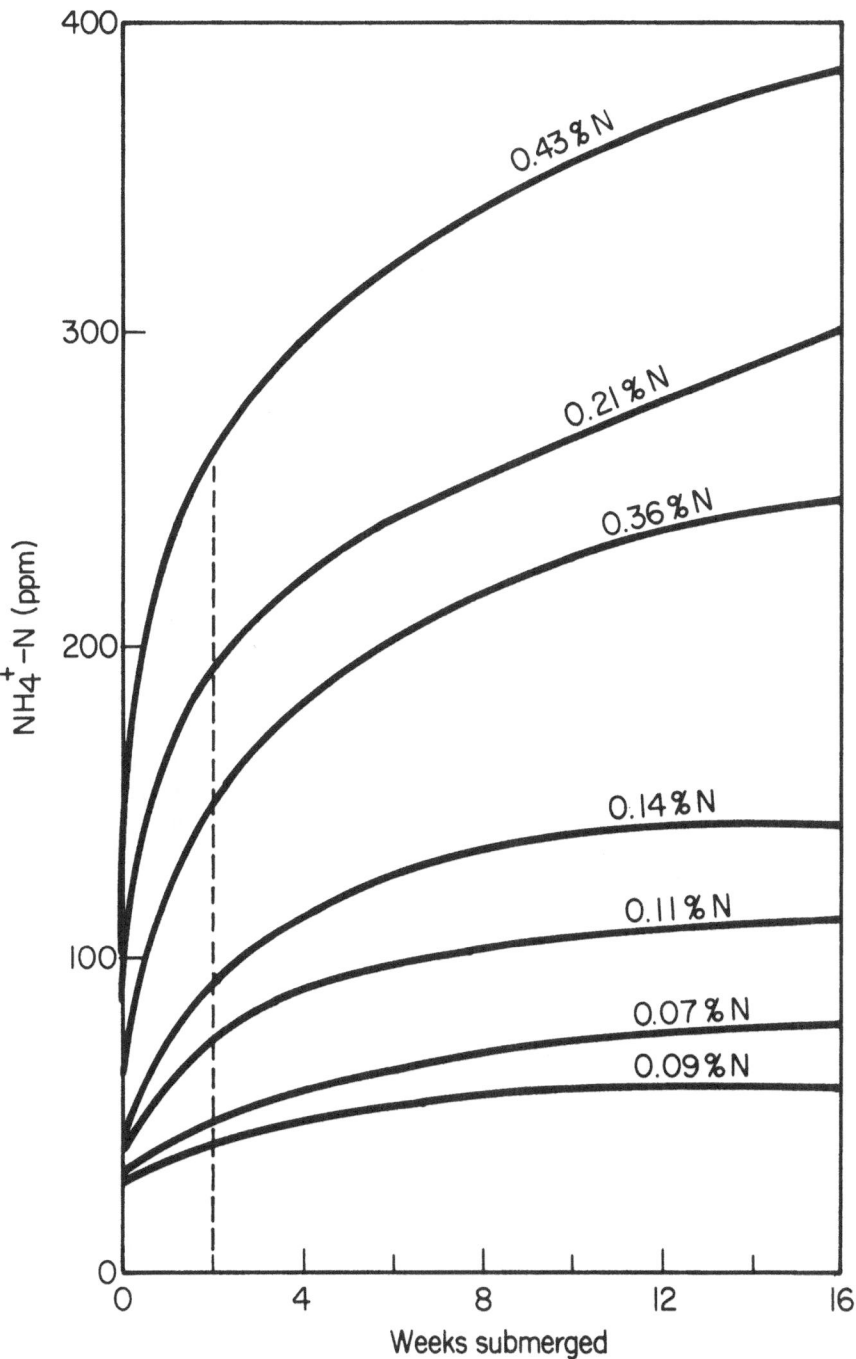

Figure 3   Kinetics of NH_4^+ release in 7 submerged soils
(From Ponnamperuma[23]).

177

Table 1   Distribution of Available N Values of Some Philippine
Wetland Rice Soils and Group Means
(From Ponnamperuma[23])

Province	Samples (no.)	Available N[a] (ppm dry soil)			
		<100	100–200	200–300	>300
Agusan del Norte	30	1 (94)	10 (160)	10 (257)	9 (363)
Albay	20	5 (72)	7 (147)	6 (246)	2 (477)
Bulacan	50	48 (54)	2 (113)	0	0
Cavite	4	2 (72)	2 (159)	0	0
Camarines Sur	16	0	11 (167)	5 (237)	0
Laguna	129	42 (80)	61 (149)	17 (241)	9 (404)
Misamis Oriental	11	0	3 (176)	8 (235)	0
Nueva Vizcaya	34	34 (49)	0	0	0
Pampanga	13	4 (61)	7 (158)	2 (217)	0
Pangasinan	13	13 (55)	0	0	0
Rizal	23	15 (70)	0	8 (215)	0
Sorsogon	20	0	4 (180)	10 (257)	6 (331)
Tarlac	4	4 (61)	0	0	0

[a]Figures in parentheses denote means for each group

1. Soils that need no fertilizer nitrogen (available N > 150 ppm).
2. Soils that need about 50 kg N/ha only at panicle primordia initiation (available N, 100-150 ppm).
3. Soils that need about 50 kg N/ha at planting and 50 kg N/ha at the panicle initiation stage (available N, 50-100 ppm).

From these and subsequent studies, it was further concluded that the process of NH_4^+ release resembles closely that in flooded tropical soils[23]. Furthermore, variations in available nitrogen values can be minimized by standardizing sampling and handling procedures.

In most tropical wetland soils supporting an actively growing crop, almost all the NH_4^+ produced is retained without being leached or denitrified.[20] Furthermore, available nitrogen content of tropical wetland rice soils varies from 3-630 mg/kg or from 6 to 1,260 kg/ha (Table 2).

2. Available nitrogen and field response. Data on available nitrogen in wetland rice soils are meaningful only if they correlate well with field response. Variations due to cropping history, sampling techniques, moisture regime, temperature, and handling procedures greatly affect the final useful-ness of the data. Recently using various techniques, the relationship between available soil N and grain yield of rice in Louisiana was determined[24]. Although the methods differed in the amount of N extracted, almost all appeared highly significantly related.

The available soil N, determined by the anaerobic incubation method correlated highly with the total soil N (by Kjeldahl method), with an r value of 0.814, followed by aerobic incubation methods ($r = 0.764$).

A reasonable estimate of increase in rice yield from the application of 112 kg N/ha at various levels of soil nitrogen can be obtained on the basis of the relationship between yield increase from nitrogen application and available soil nitrogen, determined by the anaerobic incubation method (Fig. 4). Furthermore, the soils with available-nitrogen values above 100 ppm would be unlikely to respond to fertilizer nitrogen[24].

3. Electro-Ultrafiltration technique. Ammonium nitrogen is stable under reduced soil conditions and represents the main nitrogen form for the nutrition of wetland rice. That suggests that nitrogen availability in wetland soils may be governed by the ammonium dynamics in soils. To determine the ammonium dynamics in soils, the electro-ultrafiltration (EUF) technique, which has successfully characterized the K dynamics in soils, was used in a recent study using three Philippine soils[30]. Field-moist soil samples were used for all extractants. Laboratory investigation showed almost quantitative recovery of added ammonium (2-H incubation time) in the EUF extract after 45 minutes of EUF, indicating that no significant ammonium losses occur during the EUF process.

The $EUF-NH_4$ desorption curves depicting the ammonium dynamics in three wetland rice soils in the Philippines are in Fig. 5. A distinctive peak was obtained - after 10 minutes of EUF. It presumably represents the very loosely bound soil ammonium. A much higher peak was recorded for the Pili clay (from Camarines Sur, Bicol) than for the Maahas clay (from IRRI) and Maligaya silty clay loam (from Nueva Ecija), indicating that ammonium mobility decreased in the following order: Pili clay >> Maahas clay >> Maligaya silty clay loam. The higher values of total $EUF-NH_4$ fractions, compared with those of exchangeable ammonium, suggest that some of the ammonium extracted at 400 V

Table 2   Available N Content of Wetland Rice Soils in Tropical Asia
(Adapted from Kawaguchi and Kyuma[20])

Country	Available N (mg/kg)		
	Range	$\overline{X}$	S.E.
Bangladesh	16–320	61	6.6
Burma	4–60	20	4.0
India	5–73	27	1.6
Indonesia	36–366	141	13.0
Philippines	33–630	172	21.0
Thailand	3–164	52	3.6

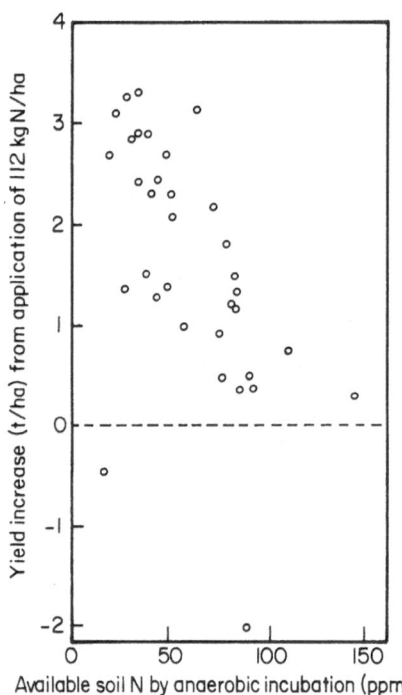

Figure 4   Yield increase from application of 112 kg N/ha
as a function of the available soil N by
anaerobic incubation. The two cases where
yield decreased are the result of adding
nitrogen to plots severely infested
with blast disease   (Adapted from Dolmat
et al.[24])

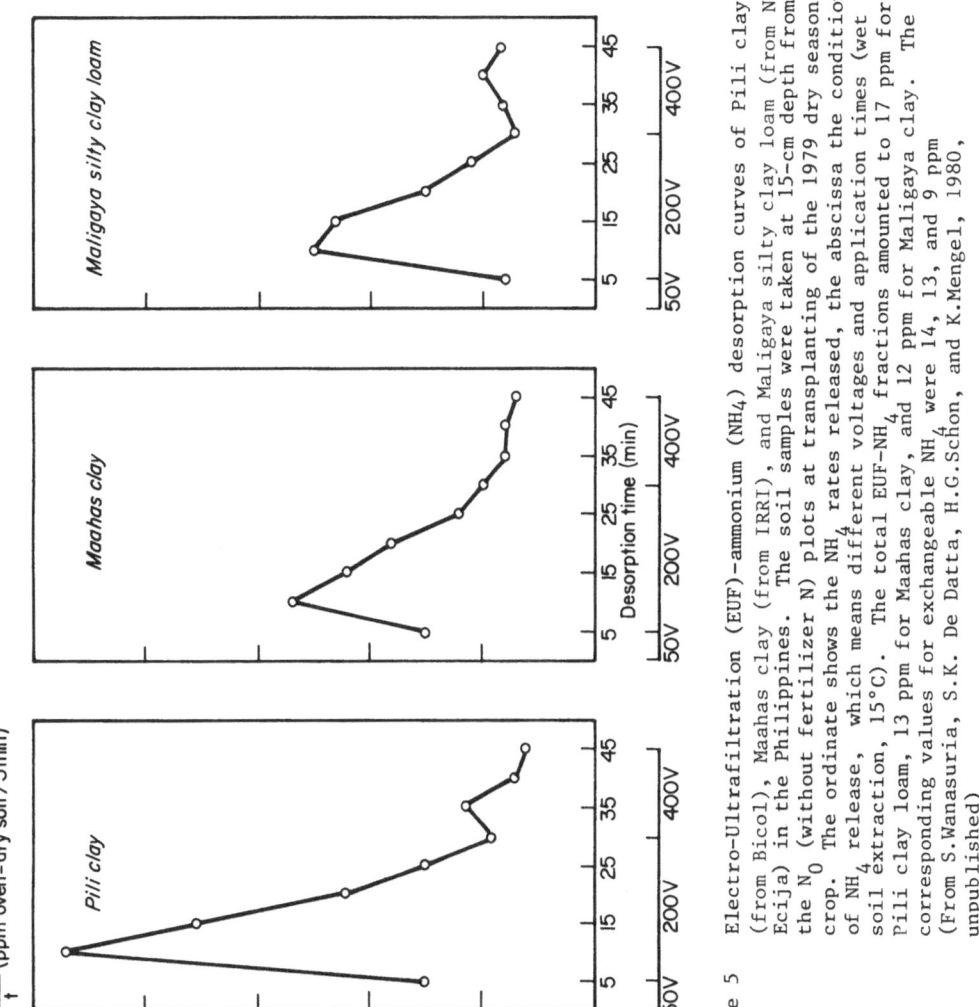

Figure 5   Electro-Ultrafiltration (EUF)-ammonium (NH_4) desorption curves of Pili clay
(from Bicol), Maahas clay (from IRRI), and Maligaya silty clay loam (from Nueva
Ecija) in the Philippines. The soil samples were taken at 15-cm depth from
the N_0 (without fertilizer N) plots at transplanting of the 1979 dry season
crop. The ordinate shows the NH_4 rates released, the abscissa the conditions
of NH_4 release, which means different voltages and application times (wet
soil extraction, 15°C). The total EUF-NH_4 fractions amounted to 17 ppm for
Pili clay loam, 13 ppm for Maahas clay, and 12 ppm for Maligaya clay. The
corresponding values for exchangeable NH_4 were 14, 13, and 9 ppm
(From S.Wanasuria, S.K. De Datta, H.G.Schon, and K.Mengel, 1980,
unpublished)

181

might have originated from the nonexchangeable sites. At 400 V, the three soils showed distinctly different ammonium desorption patterns. That suggests that ammonium selective binding sites were present in the Maligaya silty clay loam (2d peak after 40 minutes), and in Pili clay (2d peak after 35 minutes) but absent in the Maahas clay (no 2d peak).

The EUF-NH_4 in the soil and nitrogen in the plant was monitored during the 1978 wet season on Pili clay. The EUF-NH_4 at 10 days after transplanting (DT) increased considerably after the basal nitrogen fertilization, then declined sharply, reaching minimum level at 40 DT, and remained practically the same until maturity (Fig. 6). The decline in EUF-NH_4 coincided with the rapid nitrogen uptake by the plant, indicating that the depletion of EUF-NH_4 mainly resulted from the nitrogen uptake by the crop.

The results strongly suggest that the measurement of EUF-NH_4 at the beginning of the growing season may yield useful information about the ability of the soil to satisfy the nitrogen requirement of rice.

BASIC STUDIES ON NITROGEN FERTILITY OF WETLAND RICE SOILS

The unique transformations of nitrogen in wetland rice soils are brought about by the influence of water on soil aeration and by the dissolution of broadcast fertilizer nitrogen in the floodwater. Among the various nitrogen transformation processes mentioned earlier, ammonia volatilization from the floodwater surface is unique because other agricultural soils are not covered by free water when fertilizers are applied.

Magnitude of Nitrogen Losses

Research in this field is difficult to conduct because techniques for directly measuring most of the losses under natural conditions in the field are difficult or simply not available. Nitrogen losses can, however, be indirectly measured by the stable isotope ^{15}N.

^{15}N-labelled fertilizer is applied to the soil-plant system and, after a period, the soil + roots and the plant are analyzed for ^{15}N. Any difference in the ^{15}N balance from 100% is termed a loss. Techniques for conducting such research must be extremely accurate and precise to avoid <u>losses</u> that may actually be due to inappropriate techniques.

During the past 5 years IRRI and International Fertilizer Development Center (IFDC) have used the ^{15}N balance technique in the field. Initial studies involved ^{15}N-labelled ammonium sulfate with two soils[31]. In subsequent studies, the fertilizers used were the same as those used in the INSFFER network: urea, SCU, and urea for deep placement as urea supergranules*.

Complete data from both these series of field experiments are not yet available. However, a plan to extend the use of the ^{15}N technique in some Asian countries as part of a cooperative program between IFDC, IRRI, and the national research agencies concerned has been developed. Through ^{15}N balance studies with rice using fertilizer materials such as those mentioned, we hope to determine how the soil characteristics and management practices at a few well-chosen sites affect nitrogen losses.

---

*Craswell et al 1980, <u>unpublished.</u>

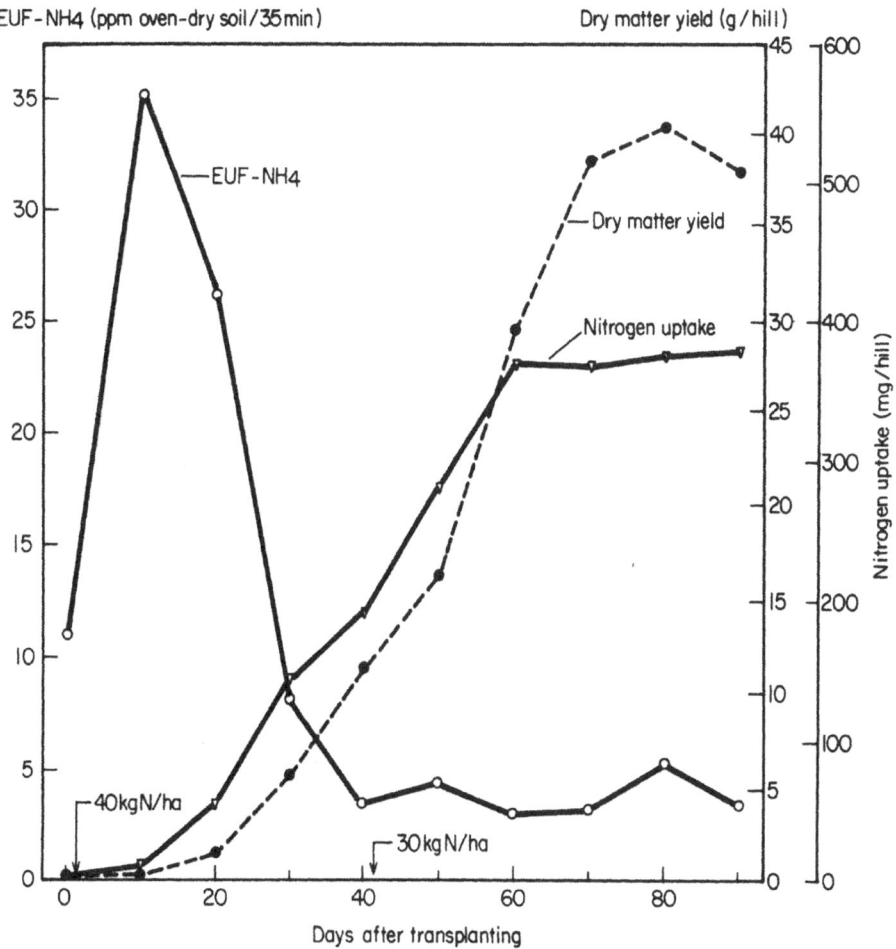

Figure 6   Electro–Ultrafiltration (EUF)-ammonium (NH_4), dry matter
yield and nitrogen uptake (above ground parts of IR36 during
the 1978 wet season in an experiment at the Bicol Rice and
Corn Experiment Station in the Philippines (av. of 9 plots).
All plots received N at 70 kg N/ha applied in 2 doses:
40 kg N/ha basal (shortly after the 1st soil sampling), and
30 kg N/ha topdressed at panicle initiation (shortly after
the 5th soil sampling)   (From S.Wanasuria, S.K. De Datta,
H.G.Schon, and K.Mengel, 1980, unpublished)

Mechanisms of Nitrogen Loss

To help interpret high nitrogen losses and poor fertilizer nitrogen efficiency and to provide a rational basis for the development of improved fertilizers, research to determine the causes of poor nitrogen fertilizer efficiency and the fate of modified urea produced in wetland rice soils has been initiated.

1. <u>Ammonia volatilization</u>. Ammonia volatilization is mainly influenced by the floodwater's aqueous ammonia concentration, fertilizer source, management, soil cation exchange capacity, and algal growth in floodwater[32].

Ammonia volatilization loss is very difficult to measure directly in the field under natural conditions. However, a recently developed micrometeorological technique has been used to measure ammonia losses from ammonium sulfate broadcast on a wetland soil[33]. Volatilization began immediately after the fertilizer was applied and incorporated during the last harrowing before transplanting (Fig. 7). The rate of volatilization reached a maximum in midday, when the temperature and wind speed were maximal. Of the 80 kg N/ha applied, 5% was totally lost. In the same study, 11% of 40 kg N/ha broadcast directly into the floodwater at panicle initiation stage was lost. More such field studies are needed to measure ammonia volatilization from urea, which may be lost more extensively than ammonium sulfate[34].

The micrometeorological technique requires sophisticated instrumentation However, the potential for ammonia loss can be estimated by studying the concentration of urea- and ammonium-nitrogen in the floodwater after fertiliz application, which is particularly influenced by the cation exchange capacity (CEC) of the soil[34].

The effect of soil texture and CEC on the nitrogen content of the floodwater and on ammonia volatilization from five soils in drums was recently studied using inverted bottles through which air was circulated to acid traps to collect the ammonia*. Five treatments were compared: prilled urea SCU broadcast and incorporated, surface-applied prilled urea, urea supergranule placement (8-10 cm), and unfertilized control.

The different soils had similar trends in floodwater pH after fertilizer application (Fig. 8). The values were higher with surface-applied urea than with broadcast and incorporated SCU. The average nitrogen content of floodwater was high (180-200 µg N/ml) on the second day after urea application, but dropped to 40 µg N/ml or less on the fourth day. The nitrogen content of floodwater after incorporation of SCU was extremely low (10 µg N/ml).

High nitrogen content and high pH of the floodwater aggravated nitrogen loss through ammonia volatilization. Such loss was always higher with surface-applied urea than with SCU broadcast and incorporated (Fig. 9). The loss was highest (13.2%) in San Manuel sandy loam with low CEC and lowest (6.2%) in Maahas clay with the highest CEC (Fig. 10). This indicates that soil texture and CEC greatly affect loss of nitrogen as ammonia. Nitrogen loss from SCU was negligible in all the soils probably because of the low nitrogen content and low pH of the floodwater.

Placement of urea supergranule in the coarse-textured San Manuel sandy loam soil with low CEC caused a nitrogen loss of 6.5%, which increased

---

*De Datta and Obcemea 1979, <u>unpublished</u>.

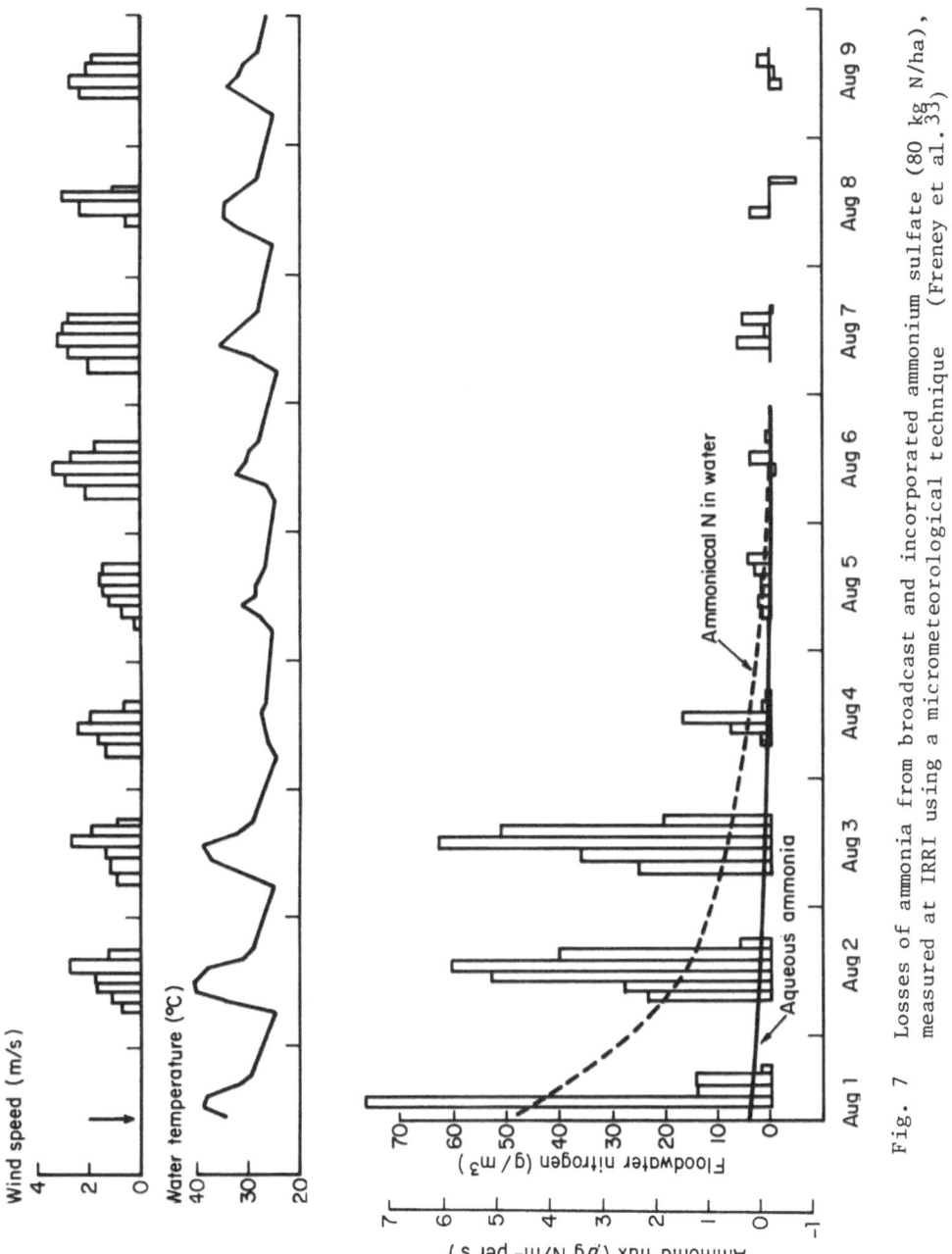

Fig. 7   Losses of ammonia from broadcast and incorporated ammonium sulfate (80 kg N/ha), measured at IRRI using a micrometeorological technique   (Freney et al.[33])

185

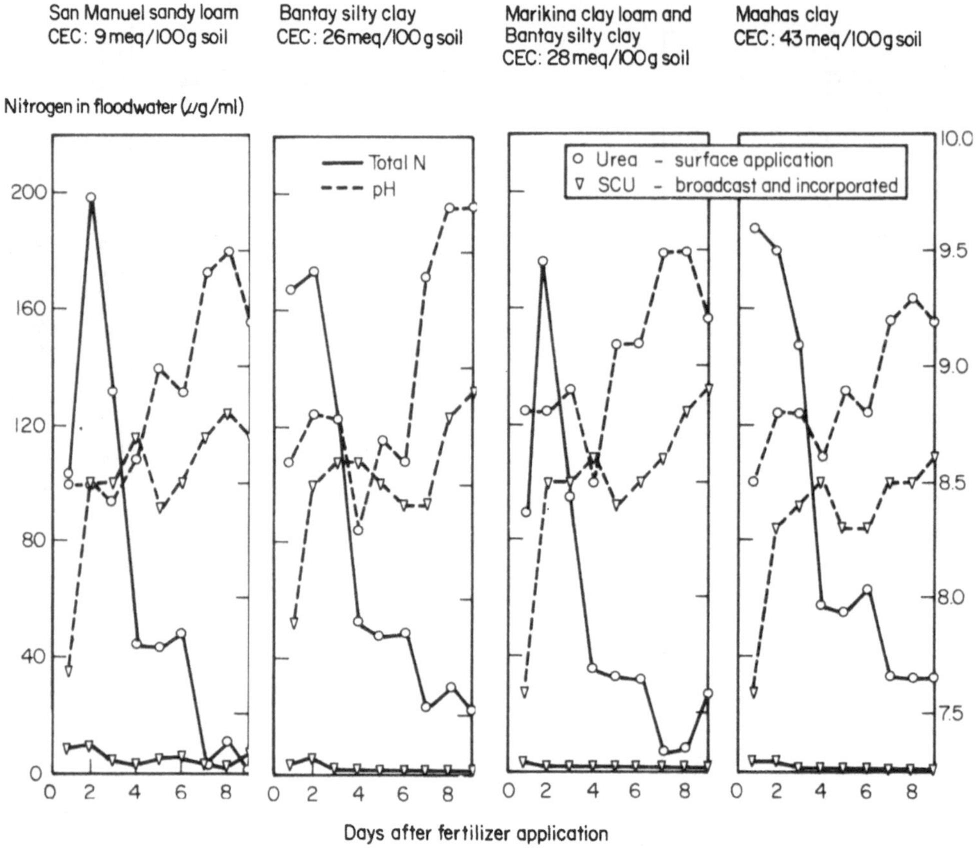

Figure 8 Changes in urea–nitrogen + ammonium nitrogen concentration and pH of floodwater, as affected by cation exchange capacity and texture of the soil, in a large-drum experiment. IRRI, 1979 late dry season (From S.K. De Datta and W.N. Obcemea, 1979, unpublished)

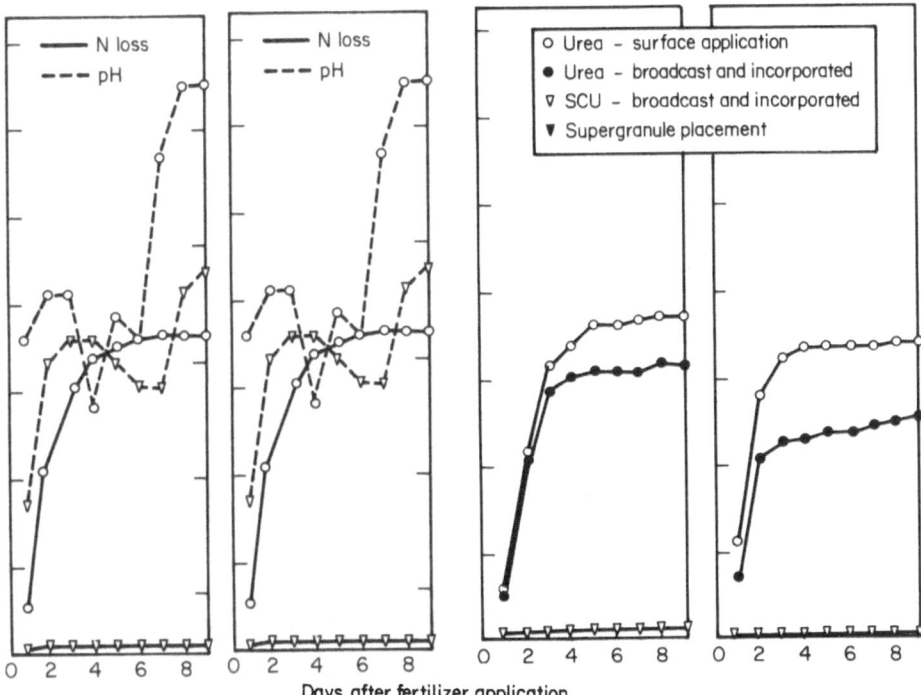

Figure 9   Ammonia volatilization loss and pH of floodwater, as affected by cation exchange capacity and texture of the soil, in a large-drum experiment. IRRI, 1979 late dry season   (From S.K. De Datta and W.N. Obcemea, 1979, unpublished)

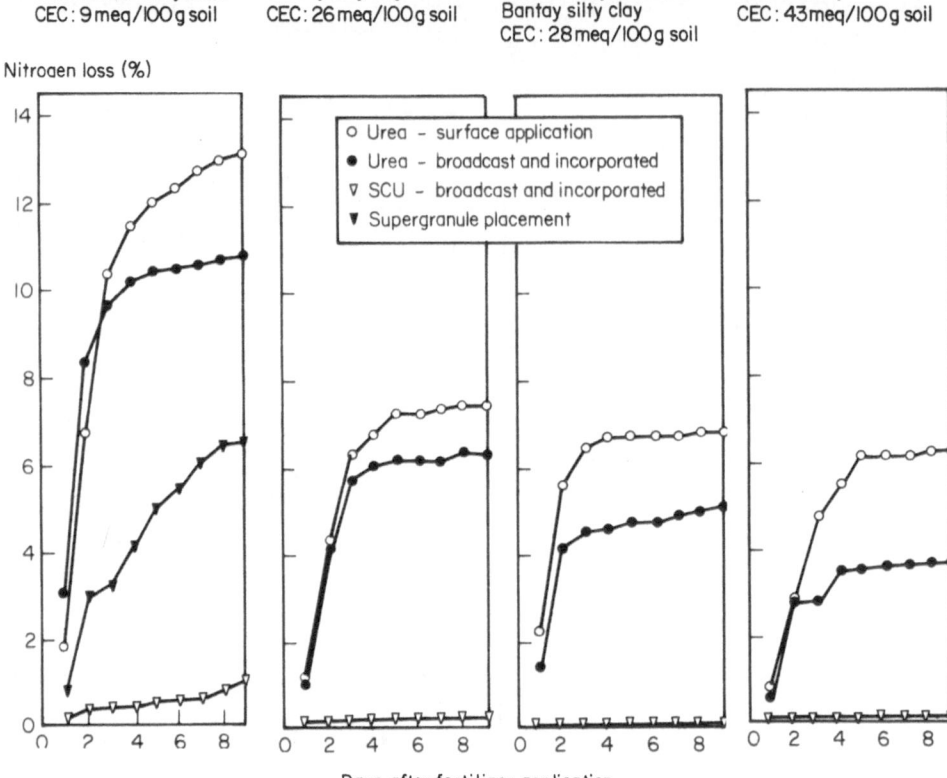

Figure 10   Ammonia volatilization loss from different forms of urea, as affected by cation exchange capacity and texture of the soil in a large-drum experiment. IRRI, 1979 late dry season. Nitrogen loss as ammonia was negligible with supergranule placement in silty clay, clay loam, and clay soils   (From S.K. De Datta and W.N. Obcemea, 1979, unpublished)

gradually from the day of application. The data indicate that modified urea products -- SCU and supergranule -- minimize ammonia volatilization loss in clayey soil but that, in coarse-textured soil, a slow-release nitrogen fertilizer is better for minimizing ammonia loss.

There is evidence of the importance of accumulated ammonia volatilization as a pathway of fertilizer nitrogen loss from flooded soils under certain conditions. Inhibition of urease activity in soils may delay the formation in the floodwater of conditions favorable to ammonia loss, giving the soil and the plant a better chance to compete with the atmosphere as a sink for nitrogen[35]. The addition of phenylphosphorodiamidate (PPD) at the rate of 2% (w/w of urea) delayed the appearance of aqueous ammonia in the floodwater after urea application under controlled conditions (Fig. 11). An algicide also reduced aqueous ammonia concentrations in the floodwater, confirming the significant role of algae in encouraging ammonia loss, shown earlier[31,32]. The research under controlled conditions should be repeated in the field; PPD may be particularly useful, for example, in reducing ammonia losses from broadcast applications of urea at panicle initiation or maximum tillering stage.

2. Nitrification-denitrification. Nitrification-denitrification was traditionally thought to be the major mechanism of nitrogen loss from submerged soils. This conclusion is not, however, based on direct measurements of the gaseous products of denitrification. The elemental nitrogen gas produced from denitrification is extremely difficult to measure against a background of 78% nitrogen in the atmosphere. However, the other product of denitrification, nitrous oxide, is normally only 300 ppb in the atmosphere and therefore losses as nitrous oxide can be measured relatively easily. Recently the nitrous oxide flux in Maahas clay soil at the IRRI farm was determined*. Fluxes were generally less than 80 ng $N/m^2/sec$ and highest with broadcast applications of urea. However, since nitrous oxide is only a minor product of denitrification, the data do not provide an estimate of the importance of denitrification.

3. Movement and distribution of applied nitrogen. Some research on the spatial distribution of nitrogen applied deep in the soil has also been done at IRRI[36]

The computer-plotted isoconcentration lines showed the extremely high concentration of ammonium at the actual placement site. These concentrations dropped off rapidly in the soil planted to rice, whereas in the fallow soil, where nitrogen is dispersed through diffusion, the concentrations were reduced slowly.

4. Compatibility of fertilizer nitrogen application and biological nitrogen fixation. The consequences of expanded nitrogen fertilizer use on the nitrogen balance of wetland rice soils is not clearly understood[37]. Fertilizer placement can strongly influence algal nitrogen fixation. Recently it was shown[38] that the deep placement of urea supergranules in soils does not affect the capability of natural nitrogen-fixing algal flora (Table 3). Nitrogen produced by the blue-green algae is a bonus associated with deep placement of nitrogen; that broadcast applications of urea, in contrast, increase the growth of green algae. These deleterious algae compete with the plant for broadcast fertilizer nitrogen and promote ammonia volatilization by

---

*Craswell et al 1979, unpublished data.

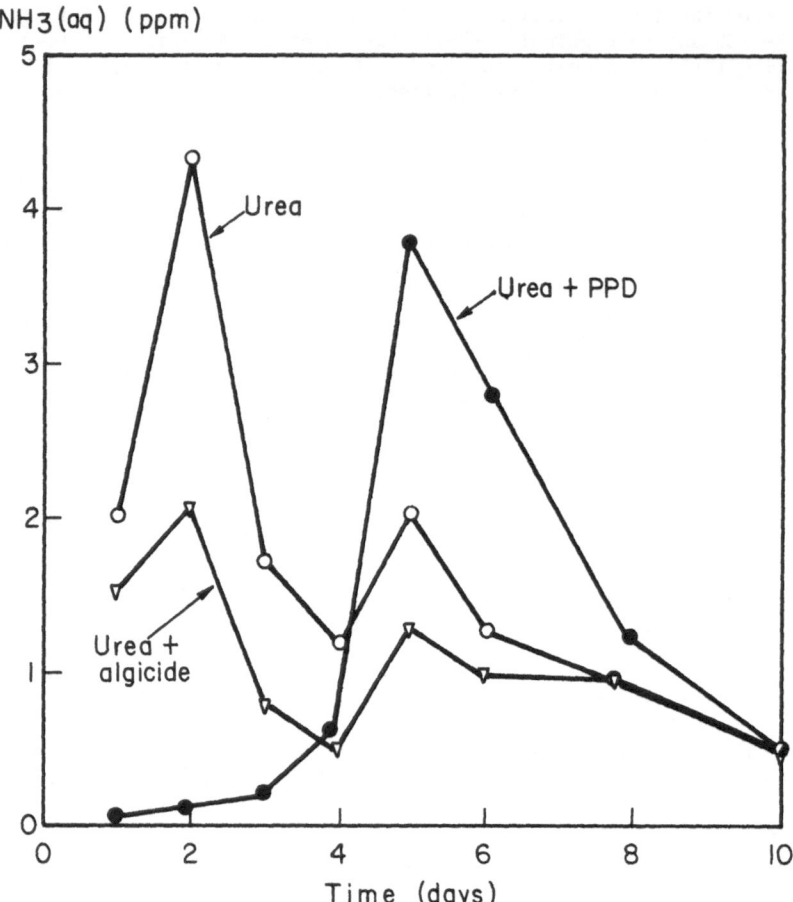

Fig. 11 Effect of an urease inhibitor (PPD) and an algicide on the production of aqueous ammonia in floodwater after broadcast and incorporation of urea -- a greenhouse study by Vlek et al [35]

190

Table 3 Effects of Fertilizer Placement on the Algal Flora and N_2-Fixation in a Field Experiment at IRRI (Roger et al.[38])

Treatment	Control	Urea supergranule (deep placement)	Urea (broadcast)	Ammonium sulfate (broadcast)	SCU (broadcast)
Acetylene reduction activity[a] μmol C_2H_4/m^2 per hour (% of the control)	70 (100)	48 (69)	0 (0)	17 (24)	27 (39)
Chlorophyll[a] ($\mu g/cm^2$)	12.4	12.3	21	15.2	11.8
N_2-fixing blue-green algae (no./cm^2 of soil)	2.0×10^5	1.7×10^5	0.7×10^5	1.4×10^5	1.6×10^5
Green algae (no./cm^2 of soil)	$<10^4$	5.0×10^5	1.0×10^7	1.2×10^7	1.2×10^6

[a]Measured to provide an index of N_2-fixing activity

raising the pH of the floodwater during the day. There is need for further research on systems for the integrated management of nitrogen for rice in which the efficiency of biological nitrogen fixation and of the applied nitrogen fertilizer is maximized.

These results and those reported recently[39] summarize available information on the mechanisms of some nitrogen transformation processes in wetland rice. However, careful studies involving ^{15}N-labelled urea and modified urea fertilizers will be needed to interpret data collected from INSFFER sites differing in soils, water regime, and climatic environment. Furthermore, nitrogen transformation processes should be evaluated on carefully monitored rainfed sites where the problem of nitrogen losses, the unused potential for increased rice production, and poor fertilizer nitrogen efficiency are greater than in irrigated rice culture.

NITROGEN FERTILIZER MANAGEMENT

With the need for higher rice yields and production becoming more urgent, and with the prices of energy-related materials such as fertilizers constantly increasing, efforts to minimize nitrogen losses and increase fertilizer nitrogen efficiency in wetland rice should include detailed studies on fertilizer nitrogen use and crop management practices.

Recent reviews[5,39,40] summarize information on the subject. The following are relevant and recent research findings on the improvement of nitrogen fertilizer efficiency in wetland rice.

Nutrient Removal by a Rice Crop

The analysis of various nutrient elements in rice grown in wetland culture would give a fairly good indication of the removal of essential elements by the crop. About 135 kg N/ha (grain + straw) in the dry season (Table 4) and 133 kg N/ha in the wet season were removed by an early maturing crop of IR36. The amount of nitrogen removed was the highest considering all the other nutrients except silicon. Therefore, the nitrogen fertility of soils should be maintained if high yields are to be maintained every cropping season. The management of nitrogen fertility has become more critical as the newer short-season varieties are introduced and greater cropping intensity is followed in both irrigated and rainfed rice. One of the quickest and simplest way to maintain high soil fertility for high grain yields is to apply a heavy dose of nitrogen fertilizers with other nutrient elements such as phosphorus, potassium, zinc, and sulfur depending on the inherent nutrient status of the soil and nutrient-supplying capacity under wetland rice culture. This is the practice widely followed in most temperate rice-growing countries. In tropical Asia, it is essential to maximize the limited resources of the farmers by increasing the efficiency of modest levels of applied nitrogen fertilizers.

Increasing Fertilizer Nitrogen Efficiency

Insufficient amount and improper timing of nitrogen application are serious constraints to increased rice yields with modern varieties.

Table 4   Nutrient Removal of a Rice Crop (Variety IR36) Yielding
7.2 t/ha Rough Rice[a] at IRRI, Philippines, 1979 Dry Season

| Nutrient element | Mineral content in | | Amount of mineral removed | | | |
| | Straw (%) | Grain (%) | by the crop at harvest (kg/ha) | | per ton of rice production (kg) | |
			Straw	Grain	Straw	Grain
N	0.64	1.43	32	104	6.4	14.3
P	0.05	0.18	2.5	13.0	0.5	1.8
K	2.03	0.21	100	15	20.3	2.1
Mg	0.14	0.09	6.9	6.5	1.4	0.9
Ca	0.29	0.05	14.3	3.6	2.9	0.5
Fe	0.039	0.006	1.9	0.42	0.39	0.06
Mn	0.018	0.003	0.90	0.20	0.18	0.028
Zn	0.002	0.001	0.07	0.09	0.02	0.012
Cu	0.0004	0.0003	0.018	0.019	0.004	0.003
B	0.004	0.003	0.19	0.24	0.04	0.03
Si	8.8	1.1	433	79	88	11
Cl	0.12	0.11	6.0	8.2	1.2	1.1
S	0.063	0.099	3.13	7.16	0.63	0.99

[a]Straw yield = 4.9

Research aimed at increasing the efficiency of nitrogen fertilizer continued to receive major attention in soil fertility research at IRRI. In collaborative studies with the IFDC, new fertilizer materials developed by IFDC and those by commercial sources were evaluated with various cultural practices.

1. Forms of urea and application methods. The system of fertilizer application most commonly recommended to wetland rice farmers is the so-called best-split application. Two-thirds of the fertilizer is broadcast and harrowed into the soil before transplanting and one-third is broadcast at around panicle initiation stage of crop growth. Incorporation of fertilizer reduces the concentration of ammonia in the floodwater and thus reduces losses. Two other concepts can be used to increase fertilizer nitrogen efficiency in wetland rice:

1. Use of controlled-release nitrogen fertilizers.
2. Deep placement of nitrogen fertilizer.

Of the coated fertilizers, SCU has been most widely tested in rice. The slow-release characteristic of SCU limits the concentration of nitrogen in the soil and floodwater at any one time and thus reduces nitrogen losses.

Recently IFDC has developed silica-polymer (sodium silicate and acrylic latex)-coated urea to obtain the same controlled-release effects in wetland rice culture. Both SCU and silica-polymer-coated urea were tested with placement of urea supergranules in an experiment on Maahas clay at IRRI during the 1980 dry season. The so-called best-split application of urea served as a control. Additional treatments were placement of SCU super-granules, a fertilizer material combining both controlled-release and deep-placement concepts. The highest grain yield (7.8 t/ha, with IR44 rice) was obtained with the two controlled-release fertilizers (SCU and silica-polymer-coated urea) broadcast and incorporated. When the grain yields were averaged for the three rates of nitrogen application (27, 54, and 87 kg/ha) there was no significant difference between controlled-release and deep-placement of urea (Fig. 12). Both controlled-release and deep-placed nitrogen fertilizers gave significantly higher yield than the so-called best-split application.

In another experiment, also at the IRRI farm, during the 1980 dry season, fertilizer nitrogen efficiency was evaluated at two seedling ages (20 and 40 days) of two varieties (early maturing IR36 and intermediate-growth-duration IR42). The experiment evaluated two timings of split application and con-trolled-release fertilizer SCU broadcast and incorporated during land prepa-ration. Results suggest that significant yield reductions were more pro-nounced in older seedlings of early maturing IR36 than in those of IR42 (Table 5). Results further suggest that slightly longer growth duration may be partially compensated for by using older seedlings without sacrificing grain yield.

2. Equipment for deep placement. An IRRI-designed plow-sole applicator that attaches to an ordinary plow combines land preparation with fertilizer nitrogen application and increases fertilizer nitrogen efficiency. A side benefit is the reduction in labor needed to apply a basal dose of fertilizer. An experiment conducted during the 1980 dry season at the IRRI farm showed that when 54 and 108 kg N/ha were applied in the plow-sole as urea, rice yield was significantly higher than when urea nitrogen was applied in split-dose systems suggested for high nitrogen efficiency (Table 6). Furthermore, at a low rate of application (54 kg N/ha) deep placement of urea with a plow-sole applicator gave a yield similar to those obtained with hand place-

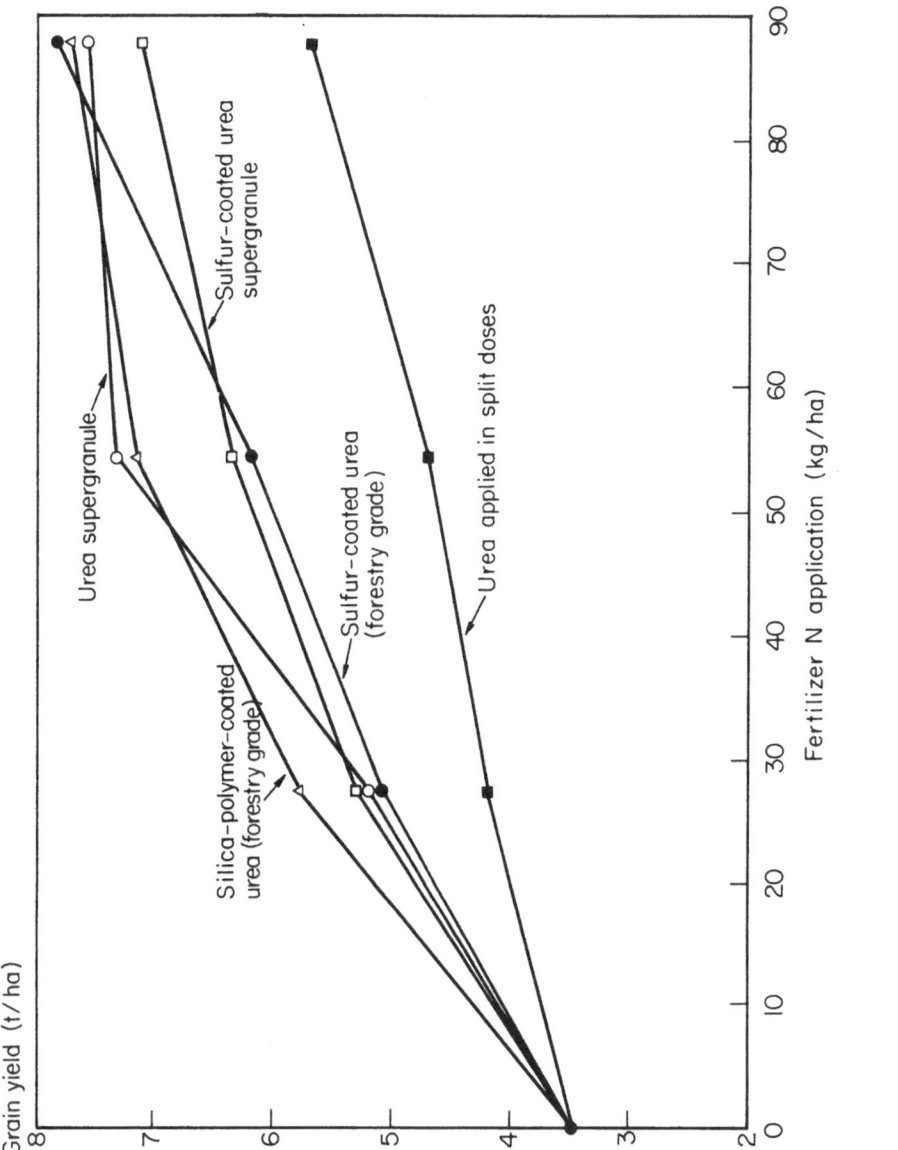

Fig. 12   Yield response to different modified urea fertilizers, IRRI 1980 dry season (De Datta et al 1980, unpublished)

195

Table 5 Effect of Various Fertilizer Management Practices on the Grain Yield of IR36 and IR42 Transplanted as 20- and 40-Day-Old Seedlings, IRRI, 1980 Dry Season (From S.K. De Datta and J. Calabio, unpublished)

Treatment[a]	Grain yield[b] (t/ha)		Difference[c]
	20-day old	40-day old	
IR36			
No nitrogen	3.9 b	3.6 c	0.3^{ns}
87 kg N/ha			
2/3 BI + 1/3 5-7 DBPI	6.4 a	4.9 b	1.5**
1/2 15 DT + 1/2 40 DT	6.4 a	4.8 b	1.6**
Ordinary SCU, broadcast & incorporated	6.7 a	5.0 b	1.7**
150 kg N/ha			
2/3 BI + 1/3 5-7 DBPI	6.7 a	5.6 a	1.1**
IR42			
No nitrogen	4.4 c	4.0 b	0.4*
87 kg N/ha			
2/3 BI + 1/3 5-7 DBPI	6.4 b	6.5 a	-0.1^{ns}
1/2 15 DT + 1/2 40 DT	6.6 b	6.6 a	0.0^{ns}
Ordinary SCU, broadcast & incorporated	7.0 a	6.6 a	0.4*
150 kg N/ha			
2/3 BI + 1/3 5-7 DBPI	7.0 a	6.8 a	0.2^{ns}

[a] BI = basal and incorporated, DBPI = days before panicle initiation, DT = days after transplanting

[b] In a column under each variety, means followed by a common letter are not significantly different at 5% level by Duncan's multiple range test

[c] ns = not significantly different; * = significantly different at 5% level; ** = significantly different at 1% level

Table 6   Grain Yield of IR50 as Affected by Method of Application
of Different Forms of Urea, IRRI, 1980 Dry Season
(From S.K. De Datta and B. Cia, unpublished)

Treatments	Yield[a] (t/ha)
No fertilizer N	3.0 e
54 kg N/ha	
Plowsole	6.3 a
Researchers' split	5.3 c
Farmers' split	4.6 d
Supergranule placement	5.9 abc
Sulfur-coated urea	6.2 ab
108 kg N/ha	
Plowsole	6.0 abc
Researchers' split	6.0 abc
Farmers' split	5.4 bc
Supergranule placement	5.9 abc
Sulfur-coated urea	6.3 a
150 kg N/ha	
Researchers' split	6.3 a

[a] In a column, any 2 means followed by a common letter are not
significantly different from each other at the 5% level

ment of urea supergranules, and broadcast and incorporation of controlled-release SCU. At 108 kg N/ha, most fertilizer nitrogen-treated IR50 plots had severe lodging problems.

      3. <u>Collaborative research program</u>. Concern over fertilizer shortages and high prices in 1974 caused IRRI and leaders of national programs to focus studies on increased fertilizer efficiency in rice. During the 1975 International Rice Research Conference at IRRI, an informal collaborative program on fertilizer nitrogen efficiency in rice in 10 countries in Asia was initiated. After making several changes in the name of the network, the collaborating scientists agreed to call it the International Network on Soil Fertility and Fertilizer Evaluation for Rice (INSFFER). INSFFER is now a truly international network with three-way collaboration between national programs, IRRI, and IFDC.

      The main objectives of the international trials on fertilizer nitrogen efficiency in rice under the INSFFER program are:

1. To obtain an insight into the relationships between nitrogen source, management, and efficiency under a variety of environmental conditions.
2. To monitor nitrogen availability and uptake patterns throughout the growing season to help explain observed differences in nitrogen efficiency under different soil and climatic conditions.

      The data for the third international trials on nitrogen fertilizer efficiency in rice are summarized in a recent report[41]. When the data were averaged over the two nitrogen rates, in three out of nine trials in the dry season, the grain yield of the treatment in which SCU was broadcast and incorporated was significantly higher, and, in one trial, significantly lower than the best-split treatment with urea (Fig. 13). Placement of supergranule urea at 10- to 12-cm soil depth, in general, gave promising results similar to those of SCU, compared with the best-split treatment with urea (Fig. 14).

      INSFFER data clearly show that, on the average, both deep-placement of urea briquets or supergranules and the use of sulfur-coated urea (SCU), compared with the traditional techniques of split application of prilled urea, effectively increase rice yields. However, the relative response to fertilizers at individual sites varies markedly. This variability prevents confident recommendation of a particular fertilizer for a particular soil type or season. For example, urea supergranules appear to be unsuitable for light-textured soils.

      Further research is needed to confirm and clarify the site-fertilizer interactions found by INSFFER, but interpretation of the results is difficult without a better understanding of the fate of the modified urea products in wetland rice soils. The expanded agronomic research by INSFFER to improve nitrogen fertilizer efficiency has therefore been complemented by more basic research.

      The fate of fertilizer in the soil-plant system is being studied to determine the causes of poor efficiency. In a recent paper, strategies on nitrogen fertility and fertilizer management research in wetland rice soils are outlined[39].

Yield difference (t/ha) from best split urea

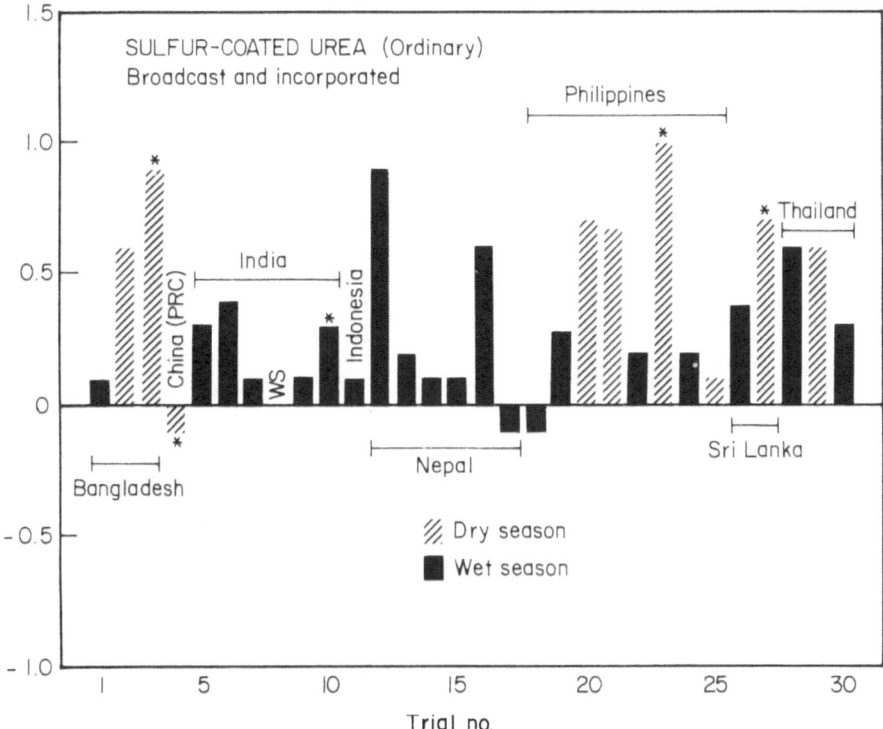

Figure 13   Significant increase or decrease in grain yield in
30 trials with N (av. of 2 rates) as SCU (ordinary)
broadcast and incorporated, compared with urea applied
in split doses at the same rate.   The Third International Trial on Nitrogen Fertilizer Efficiency in
Rice, 1978-1979 (INSFFER)   (From IRRI [41])

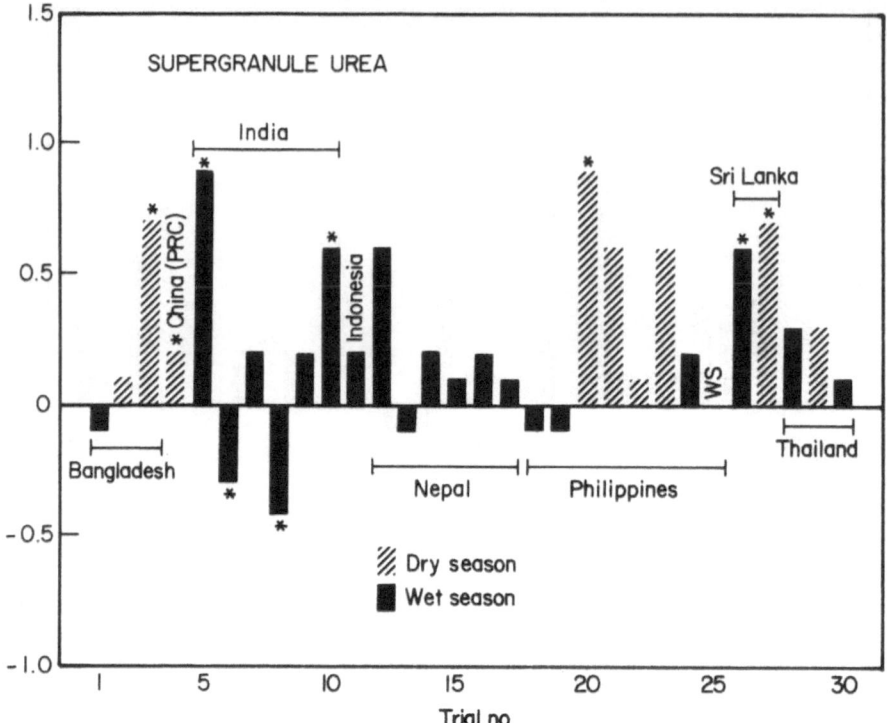

Yield difference (t/ha) from best split urea

Figure 14   Significant increase or decrease in grain yield in
30 trials with N (avg. of 2 rates) as urea supergranule
placement, compared with urea applied in split doses at
the same rate.   The Third International Trial on Nitrogen
Fertilizer Efficiency in Rice, 1978-1979 (INSFFER)
(From IRRI[41])

MAINTENANCE OF SOIL FERTILITY WITH STRAW AND COMPOST

The basic principle in the maintenance of soil fertility is to return to the soil plant nutrients removed from it by crops. With the introduction of modern varieties, complete dependence on plant residues and various forms of organic manures to maintain soil fertility is no longer considered essential or even desirable. However, with increased cost of fertilizers, an integrated management of nitrogen that involves application of inorganic and organic sources of plant nutrients should be considered an essential component of management practices for rice cultivation provided agronomic data and socioeconomic conditions of a given country justify it. In this regard, China has clearly been the leader in using organic residues, compost, azolla, and chemical fertilizers to maintain soil fertility in rice fields. Following are some examples of recent research results in the tropics.

Effect of Straw on Status of Nitrogen and Other Nutrients

Experiments by the IRRI Soil Chemistry Department showed that leaving the straw in the field increased the total nitrogen and other major nutrient elements. Table 7 shows the mean increase after the 12th season, averaged for three water regime treatments (dry fallow, dry fallow with midseason soil drying, and flood fallow).

Returning the straw to the field either as long straw or as compost increased nitrogen content (Table 8). In two separate long-term experiments, it led to a nitrogen gain of 0.023% or 460 kg N/ha in 6 years. Data averaged for 7 years show that nitrogen content of Maahas clay soil generally increased progressively over the years, more so when rice straw was incorporated (Fig. 15). Increases in a straw-management field experiment over 14 seasons (7 years) averaged 57 kg N/ha per season[23].

Use of Compost

Evidence from Japan suggests that farmers who obtain high rice yields use a substantial quantity of compost. However, there is no basis for concluding that the high yields resulted from the complementary effects of compost over the inorganic fertilizers. In fact, because of high labor costs in East Asia (Japan and Korea), the use of compost has steadily declined. Fig. 16 shows the relationship between the amounts of compost and inorganic fertilizers used and rice yields in Japan. Obviously, use of compost played a minor role in steadily increasing rice yields in Japan. China still makes compost from recycled plant, animal, and human wastes and uses it in rice fields. In the Asian tropics, the use of compost could play an important role where alternatives to use of compost are limited and labor is still relatively inexpensive.

At IRRI, a long-term experiment for the past 15 years using compost at the rate of 10 t/ha per season showed no improvement in grain yield when the results were compared on equal-nitrogen basis[42].

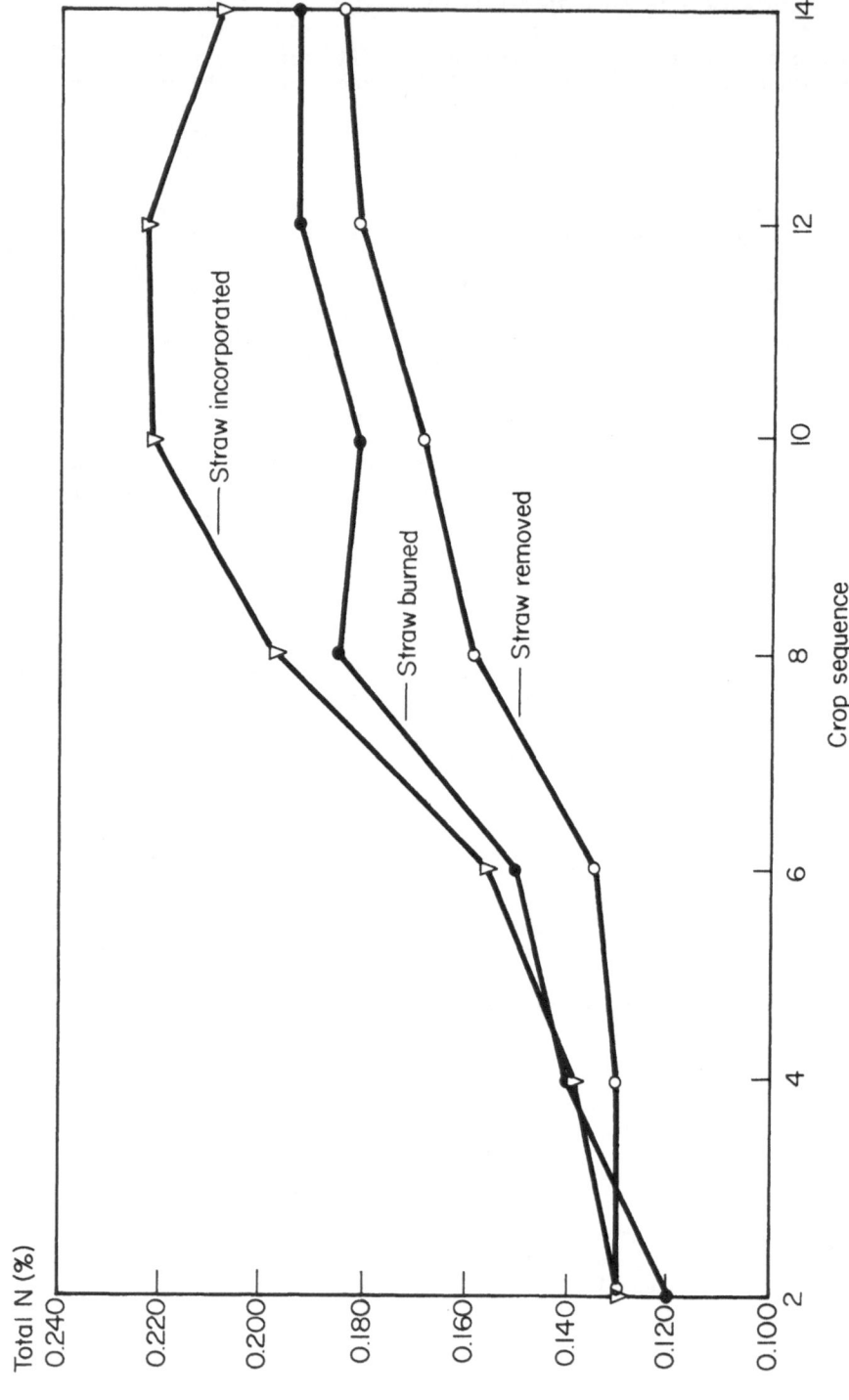

Figure 15   Effect of straw management on the nitrogen content of Maahas clay during 7 years, averaged for 5 varieties, in the field   (From Ponnamperuma[23])

202

Table 7   Effect of Straw on the Nutrient Status of Maahas Clay
After the 12th Season, IRRI, 1978
(IRRI[42])

Treatment	Organic matter (%)	Nitrogen (%)	Exchangeable potassium (meq/100 g)	Exchangeable phosphorus[a] (ppm)
Straw removed	3.00	0.195	1.24	8
Straw incorporated	3.50	0.218	1.41	10
Difference	0.50**	0.023**	0.17**	2**

[a] By Olsen method.

Table 8   Effects of 4 Straw Treatments on the Nutrient Status of
Maahas Clay after the 12th Season, IRRI, 1978 Wet Season
(IRRI[42])

Treatment	Organic matter (%)	Nitrogen (%)	Exchangeable potassium (meq/100 g)	Exchangeable phosphorus[a] (ppm)
Straw removed	3.26 b	0.187 b	1.19 c	12 c
Straw burned	3.17 b	0.187 b	1.35 ab	13 bc
Straw incorporated	3.73 a	0.210 a	1.45 a	15 b
Straw composted	4.01 a	0.211 a	1.27 c	28 a

[a] By Olsen method.

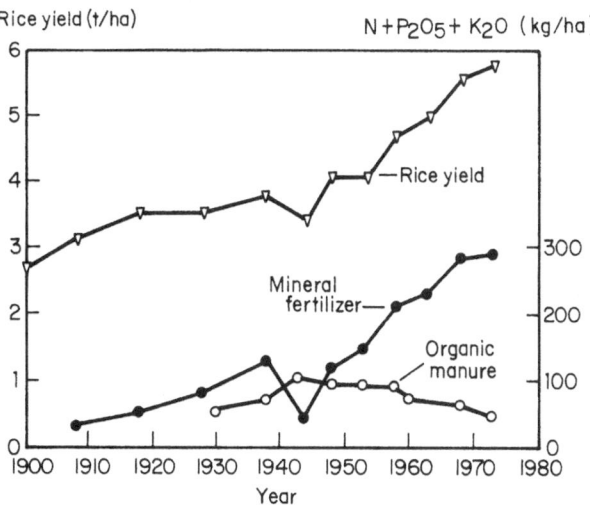

Figure 16   Yield and use of fertilizer and compost
in rice in Japan   (From Kemmler[43])

REFERENCES

(1) Pande, H. K., Bhan, V.M., Effect of row spacings and levels of fertilization on growth, yield and nutrient uptake of upland paddy and on
associated weeds. Riso, 15, 47-67(1966).

(2) De Datta, S. K., Magnaye, C.P., Moomaw, J. C., Efficiency of fertilizer
nitrogen (^{15}N-labelled) for flooded rice. Trans. 9th Int. Soil Sci.
Congr., Adelaide, Australia, 4, 67-76(1968).

(3) Prasad, R., Rajale, G. B., Lakhdive, B. A., Effect of time and method
of application of urea and its treatment with nitrification inhibitors on
the yield and uptake by irrigated upland rice. Indian J. Agric. Sci.,
40, 1118-1127(1970).

(4) De Datta, S. K., Saladaga, F. A., Obcemea, W. N., Yoshida, T., Increasing
efficiency of fertilizer nitrogen in flooded tropical rice. Proc. FAI-
FAO Seminar on Optimising Agricultural Production Under Limited Availability of Fertilizers, New Delhi, 265-288(1974).

(5) Prasad, R., De Datta, S. K., Increasing fertilizer nitrogen efficiency
in wetland rice. In Nitrogen and Rice, International Rice Research
Institute, Los Baños, Philippines, 465-484(1979).

(6) Craswell, E. T., De Datta, S. K., Recent developments in research on
nitrogen fertilizers for rice. IRRI Res. Pap. Ser. 49, 1-11(1980).

(7) Patrick, W. H., Jr., Mahapatra, I. C., Transformation and availability
to rice of nitrogen and phosphorus in waterlogged soils. Adv. Agron.
20, 323-359(1968).

(8) Tusneem, M. E., Patrick, W. H., Jr., Nitrogen transformations in waterlogged soil. La. State Univ. Bull, 657, 1-75(1971).

(9) Broadbent, F. E. Transformations of soil nitrogen. In Soils and Rice,
International Rice Research Institute, Los Baños, Philippines, 543-
559(1978).

(10) Hauck, R. D., Methods for studying N transformations in paddy soils:
review and comments. In Nitrogen and Rice, International Rice Research
Institute, Los Baños, Philippines, 73-93(1979).

(11) Craswell, E. T., Vlek, P. L. G., Fate of fertilizer nitrogen applied to
wetland rice. In Nitrogen and Rice, International Rice Research
Institute, Los Baños, Philippines, 175-192(1979).

(12) Shastry, S. V. S., Freeman, W. H., Pillai, K. G., New rice varieties need
rational management for optimum yields. Yojana, 18(2), 33-36(1974).

(13) Stangel, P. J., The world fertilizer sector -- at a crossroads. Paper
presented at the Symposium on Food Situation in Asia and the Pacific
Region held April 24-29, 1980, at the Asian and Pacific Council, Food
and Fertilizer Technology Center, Taipei, Taiwan.

(14) FAO (Food and Agriculture Organization), Current world fertilizer situation
and outlook -- 1977/78-1983/84. Food and Agriculture Organization of the
United Nations, Rome, Italy, (1980).

(15) UNICO (Universal Corporation), Long-term forecast for world nitrogen
fertilizer demand and supply. Gonowan, Japan, (1977).

(16) Koyama, T., Chammek, C., Niamsrichand, N., Nitrogen application technology
for tropical rice as determined by field experiments using ^{15}N tracer

technique. Tropical Agricultural Research Center Tokyo Tech. Bull., 3, 1-79(1973).

(17) Suzuki, S., Sakai, H., Dei, Y., Proceedings on soil fertility investigations [in Japanese]. J. Sci. Soil Manure, Japan, 39(1), (1968).

(18) Takahashi, J., Natural supply of nutrients in relation to plant requirements. In The Mineral Nutrition of the Rice Plant, The Johns Hopkins Press, 271-293(1965).

(19) Tanaka, A., Role of organic matter. In Soils and Rice, International Rice Research Institute, Los Baños, Philippines, 605-618(1978).

(20) Kawaguchi, K., Kyuma, K., Paddy soils in tropical Asia: their material nature and fertility, The University Press of Hawaii, Honolulu, 1-258(1977).

(21) Kulasooriya, S. A., Roger, P. A., Barraquio, W. L., Watanabe, I., Biological nitrogen fixation by epiphytic microorganisms in rice fields. IRRI Res. Pap. Ser. 47, 1-47(1980).

(22) IAEA (International Atomic Energy Agency), Isotope studies on rice fertilization. Tech. Rep. Series 181, 1-131(1978).

(23) Ponnamperuma, F. N., The nitrogen supply in tropical wetland rice soils. Paper presented at a Special Workshop on Nitrogen Fixation and Utilization in Rice Fields during the 20th anniversary celebration of the International Rice Research Institute, April 28-30, 1980, Los Baños, Philippines.

(24) Dolmat, M. T., Patrick, W. H., Jr., Peterson, F. J., Relation of available soil nitrogen to rice yield. Soil Sci, 129(4), 229-237(1980).

(25) Bremner, J. M., Nitrogen availability indexes. In Methods of soil analysis, 2. C.A. Black, ed., Amer. Soc. Agron., Madison, Wisconsin, 1324-1345(1965).

(26) Chang, S. C., Evaluation of the fertility of rice soils. In Soils and Rice, International Rice Research Institute, Los Baños, Philippines, 523-541(1978).

(27) Sims, J. L., Wells, J. P., Tackett, D. L., Predicting nitrogen availability to rice: 1. Comparison of methods for determining available nitrogen to rice from field and reservoir soils. Soil Sci. Soc. Am. Proc. 31, 672-675(1967).

(28) Kawaguchi, K., Kyuma, K., Lowland rice soils in Thailand. Center for Southeast Asia Studies, 1-270(1969).

(29) Ponnamperuma, F. N., Nitrogen supplying capacity of wetland rice soils. Paper presented at the International Rice Conference on Rainfed Lowland Rice, April 17-21, 1978, International Rice Research Institute, Los Baños, Philippines.

(30) Wanasuria, S., Schon, H. G., De Datta, S. K., Mengel, K., Determining ammonium dynamics of wetland rice soils using electroultrafiltration technique. American Society of Agronomy Meetings, Nov. 30- Dec. 5, 1980, Detroit, Michigan. (Abstr.)

(31) Mikkelsen, D. S., De Datta, S. K., Nitrogen balance (N^{15}-labelled) in wetland rice soils. American Society of Agronomy Meetings, Nov. 30-Dec. 5, 1980, Detroit, Michigan. (Abstr.)

(32) Mikkelsen, D. S., De Datta, S. K., Ammonia volatilization from wetland rice soils. In Nitrogen and Rice, International Rice Research Institute Los Baños, Philippines, 135–156(1979).

(33) Freney, J. R., Denmead, O. T., Watanabe, I., Craswell, E. T., Ammonia and nitrous oxide losses following applications of ammonium sulfate to flooded rice. Aust. J. Agric. Res., (1980). (in press)

(34) Vlek, P. L. G., Craswell, E. T., Effect of nitrogen source and management on ammonia volatilization losses from flooded rice-soil systems. Soil Sci. Soc. Am. Proc. 43, 352–358(1979).

(35) Vlek, P. L. G., Stumpe, J. M., Byrnes, B. H., Urease activity and inhibition in flooded soil systems. Fertilizer Research 1(31), (1980). (in press)

(36) Savant, N. K., De Datta, S. K., Movement and distribution of ammonium-N following deep placement of urea in a wetland rice soil. Soil Sci. Soc. Am. J. 44(3), (1980).

(37) Watanabe, I., Craswell, E. T., App, A. A., Nitrogen cycling in wetland rice soils in East and Southeast Asia. In Nitrogen cycling in Southeast Asian ecosystems, Chieng Mai, Thailand, (1979). (in press)

(38) Roger, P. A., Kulasooriya, S. A., Tirol, A., Craswell, E. T., Deep placement: a method of nitrogen fertilizer application compatible with algal nitrogen fixation in wetland rice soils. Plant and Soil, (1980). (in press)

(39) De Datta, S. K., Craswell, E. T., Nitrogen fertility and fertilizer management in wetland rice soils. Paper presented at the Special Symposium on Rice Research Strategies for the Future during the 20th Anniversary Celebration of the International Rice Research Institute, April 21–25, 1980, Los Baños, Philippines, (1980). (in press)

(40) De Datta, S. K., Fertilizer management for efficient use in wetland rice soils. In Soils and Rice, International Rice Research Institute, Los Baños, Philippines, 671–701(1978).

(41) IRRI (International Rice Research Institute), Preliminary report on the third international trial on nitrogen fertilizer efficiency in rice, 1978–1979. International Network on Soil Fertility and Fertilizer Evaluation for Rice (INSFFER), Los Baños, Philippines, (1980).

(42) IRRI (International Rice Research Institute), Annual report for 1978. Los Baños, Philippines, 1–478(1979).

(43) Kemmler, G., Advances in fertiliser use to rice in South and East Asia 1958–1978. Fertiliser News, February 1980.

CULTIVATION AND APPLICATION OF GREEN MANURE IN PADDY FIELDS OF CHINA

Gu Rong-shen
(Jiangsu Academy of Agricultural Sciences)

Wen Qi-xiao
(Institute of Soil Science, Academia Sinica, Nanjing)

China has a very long history of cultivation of green manure crops in paddy fields. Early in the third century, there were records of the rotation of rice and milk vetch in the local chronicles*. In the 1940s-1950s, a system of rice-green manure crop rotation was employed as an effective measure to increase nitrogen and maintain soil fertility in the region between the Five Ridges (Mt. Wuling) in the south and the Changjiang River in the north. Since 1960, thanks to the improvement of fertilization, inoculation of rhizobium and improved cultivation techniques, the milk vetch, vetch, etc. have been introduced successfully into Guangdong and Guangxi provinces to the south of the Five Ridges, and the Huaihe River valley to the north of the Changjiang River. At the same time, following the solution of such problem as survival through winter and summer for Azolla and the selection of more adaptable varieties of green manure crop in North, Northeast and Northwest China respectively, the total acreage of green manure crop of paddy fields has reached 8 million hectares.

This article deals mainly with a general aspect of the research work of green manure in paddy fields in China.

THE MAIN VARIETIES AND THEIR REGIONAL DISTRIBUTION

The winter green manure crops are the principal part of green manure crops in China, with an area of up to 83% of the total area of green manure crop of paddy field. The area for Azolla cultivation makes up 8.5% of the total area. The ares of summer green manure crops is not more than 500,000 hectares, though distributed widely.

Milk vetch is dominant in the winter green manure crops, covering an area of approximately 74.6% of the total area of winter green manure crops. It is characterized by its higher adaptability to wetness and shade in the seedling stage and is conducive to under-crop sowing before the rice is harvested. Its tender and soft stems and leaves are liable to decompose, therefore, the nutrients it contains are more readily available to the rice plant. According to experimental results from Jiangyin County, Jiangsu Province(1), through a submergence for 24 hours, no dead seedlings of milk vetch were found; and in 120 hours, thereafter, the rate of dead seedlings began to increase with the longer time of submergence ($r = 0.9753**$). In accordance with the data of Zhejiang Province(2), the seedlings of milk vetch could be survived only under a condition that the light intensity at the height of 20 cm from the soil surface is more than 3.5% (2500-3000

---

* "Annals of Guangzhi" by Guo Grong-yi.

Lux, equal to that under a cluster of rice with a yield < 5250 kg/ha);
otherwise, the seedlings would either stop in growth or be liable to die.
In Jiangsu, a yield of fresh weight of 30 t/ha, has been still obtained
from the late rice field with the yield of 4.5-5.0 ton/ha, when the milk
vetch is interplanted under the rice plant 40-50 days before harvesting.

In the past, azolla was cultured with local varieties belonging to
A. imbricata which could not survive over winter without protection in the
isothermal area at less than 5°C in January and could not propagate very
quickly in spring; therefore, it could not be extensively used for the
early rice in the area of double cropping of rice. Recently, A. filiculoides
capable of surviving the winter on the natural water surface in the area to
the south of the isotherm of 3°C in January has been introduced, with results
that the azotase activity of symbiotic blue algae associated with A. filiculoides is 2.7 times higher than that of the local varieties at 15°C(3).
However, a further extension of cultivation of azolla in paddy field is
still restricted by such problems as the method applied, summer surviving
and insect injuries, especially Polypedium illinense.

Sesbania is characterized by its strong tolerance to wetness. It was
reported by Shanghai Academy of Agricultural Sciences that the root system
of sesbania developed better under submerged conditions than under upland
conditions; and under submerged conditions the weight of nodules per plant,
the azotase activity and the dry weight were increased by 137%, 67-80% and
95% respectively as compared with those under upland conditions. Sesbania,
if planted between wheat and late rice for about 40 days, will produce fresh
plant 7.5 t/ha in weight(4). It is planted either in the interval between
early rice and late rice in Guangdong and Fujian or intercropped in the rice
seedling beds in Shanghai and may often give a yield of fresh weight of
15 t/ha.

Besides milk vetch, there are other important winter green manure
crops such as vetches, medic, broad bean. They very in geographical distribution due to their different adaptabilities (Fig. 1). The distribution
of azolla is generally similar to that of milk vetch. Sesbania is usually
dispersed in the coastal region to the south of latitude 40°N.

SOME CULTIVATION TECHNIQUES OF WINTER GREEN MANURE CROPS

In dealing with the raising of the yield of winter green manure crops,
we use milk vetch as an example, and here is a discussion about it .

1. Suitable Time and Rate of Seeding

To postpone the seeding time will lead to the decrease of the yield of
milk vetch significantly (Fig. 2). Bur-clover, smooth vetch and vicia cracca
are similar to milk vetch in this respect(5).

Likewise, excessively early seeding will make the growth of the winter
green manure inhibited by its preceding crop, or the germination damaged
by high temperature or arid climate; too luxuriance in the growth of green
manure in the early stage will aggravate frozen damage in winter. For
instance, the safety growth rate of milk vetch in winter in Jiangsu is as
follows: 5-12 cm in plant height, with approximately 2-4 branches per plant,
covering an area of about 80%. In order to reach this growth rate, an
accumulated temperature of 550-900°C above 5°C is needed: therefore, the

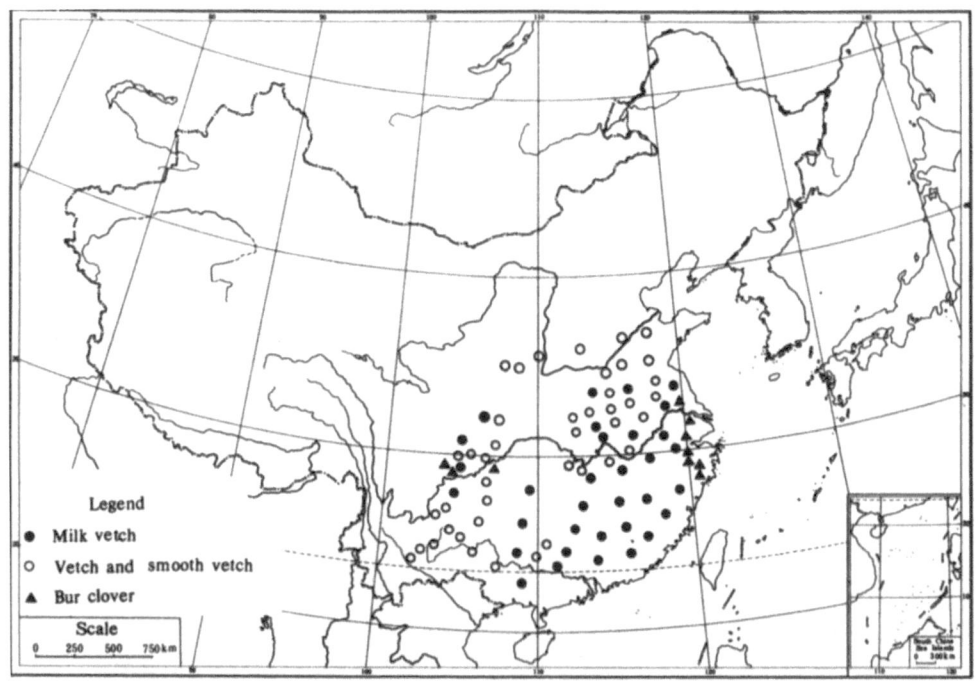

Fig. 1 Geographical distribution of some winter
green manure crops in China

suitable seeding period should be
arranged on September 5-25, and
should be adequately postponed
southward, for example, in Guangxi
the seeding time is on October 7-
24.

Either the increase of plant
weight or the number of stems and
branches is both conducive to the
yield of milk vetch. Plant weight
is related to the seeding period
and the climate in winter and spring. Therefore, the seeding rate
of milk vetch is mainly determined
by the local climate and the variety
of preceding rice. In Jiangsu, the
seeding rate is generally 60 kg/ha
and the seedling numbers are prefera-
bly about 8 million per hectare(6).
And in Jiangxi the suitable seeding
rate should be 30-35 kg/ha. In
southern Guangxi, because the plant

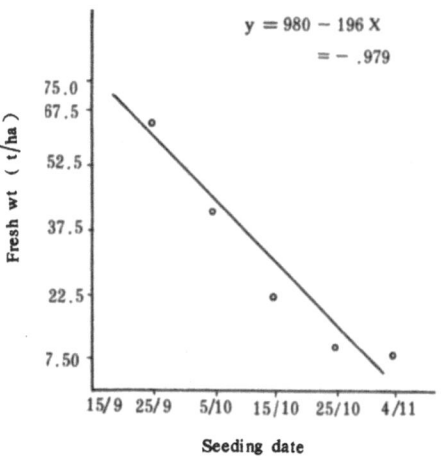

Fig. 2 Yield of milk vetch
as a function of seeding date

may grow up to 100-130 cm. in height, the seeding rate should reduced to 22.5 kg/ha.

2. Inoculation of Rhizobium

The rhizobium of milk vetch has a high specificity, therefore the milk vetch planted in a new area must be inoculated. Only by means of inoculation could a good yield be obtained(7), no matter how the soil fertility is or whether the soil is fertilized or not (Table 1).

Table 1   Effect of inoculation of rhizobia on the yield of milk vetch

Location of experiment	Soil fertility	Year of planting milk vetch	Yield (g/ha) increase		
			CK	Inoculation	(%)
Yancheng	High	1st	2900	39800	1270
Yancheng	Low, dressed with $(NH_4)_2SO_4$	1st	6900	32800	375
Yangzhou	Medium	2nd	16700	36800	120
Wujiang	High	Many	40200	50100	24.6

* Fresh wt.

It is shown in Table 1 that in the area where milk vetch has been cultivated for a very long time, inoculation of good rhizobium strain also has a favorable effect on the increase of yield. Experimental results showed that a good strain is liable to degenerate after a long time of cultivation; therefore, selection and regeneration of good strains from time to time are very important(8).

3. Application of Phosphorous and Nitrogen Fertilizer

In case the available phosphorous content is below 15 ppm in acid and neutral paddy soils or below 10 ppm in calcareous soil, the phosphorous fertilizer would be markedly efficient. A good response of vetch and broad bean to phosphorous fertilizer is still found on the soils with a phosphorous content even more than that mentioned above(9-11).

Table 2 illustrates the effect of different phosphatic fertilizer on the increment of the yields of green manure crops. In the table, 1 kg of P_2O_5 gives an increase of 1.2 kg of nitrogen in green manure. The application of phosphorous fertilizer in such a way is so-called "Increase nitrogen with phosphorous". Trials on bleached paddy soil showed that applying 1 kg of superphosphate to vetch to be plowed into soil as a basal manure of rice gave an increase of 5 kg of grain, while 1 kg of superphosphate applied directly to rice would only give an increase of 1.5 kg of grain(12).

Rational application of nitrogen fertilizer can increase the nitrogen fixation rate of green manure crops. It was reported by Jiangsu Academy of Agricultural Science that 1 kg of fertilizer nitrogen could increase

1.69 $\pm$ 0.91 kg of nitrogen in green manure. The effect of nitrogen fertilizer varies greatly with the following two factors: the soil fertility
and the time of fertilization. On the low-yielding soils with a shallow
plowed horizon and deficient in phosphorous and organic matter, nitrogen
fertilizer should be applied in combination with phosphorous or organic
manure; otherwise, its effect will be restricted. As illustrated in
Fig. 3, the most favorable time of application is at the stage of initial
elongation of stems of milk vetch(13).

Table 2   Effect of various phosphorous sources on
the yield of milk vetch(9,10,12)

P fertilizer	No. of ex-periments	Dosage (P_2O_5, kg/ha)	Increment of yield* per kg P fertilizer	Increment of N per kg P_2O_5
Superphosphate	311	25.5–38.3	347	1.21
Ca–Mg–phosphate	84	36–48	250–438	1–1.75
Rock phosphate	56	112–270	74	0.30

* Fresh wt.

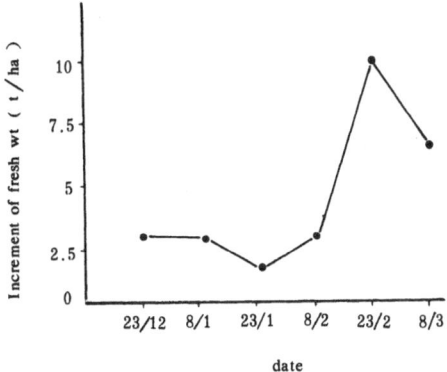

Fig. 3   Increment of milk vetch yield as a
function of time of N-fertilizer application
(25.5 kg N/ha) (yield of CK: 25.6 t/ha)

Experiment with ^{15}N labelled $(NH_4)_2SO_4$ also showed that the utilization rate fo fertilizer-N varied with the time of application. The plant
recovery of the fertilizer-N applied in the seedling stage was 32.1%; that
applied after regreening stage up to 78%(14).

In addition, according to the experiments in Zhejiang and Jiangsu
provinces, application of molybdenum fertilizer on milk vetch may give an
increase of 4.6–4.9 t/ha in fresh weight. Good effect of potassium fertilizer on the yield of green manure has been found on some soils to the

south of the Changjiang River (Table 3).

Table 3 Response of milk vetch to K fertilizer(15,16)

Location of experiment	No. of experiment	K fertilizer, dosage (kg/ha)		Increase (%)	Increment of yield** per kg K_2O (kg)
Zhejiang	10	K_2SO_4	150	18.7	56.3
Jiangsu	17	KCl	114	21.5	131
Hunan	70	Cement* dust	750	33.4	172

  * K_2O - 6 - 8 %
 ** Fresh wt.

4. Control of Soil Moisture Content

    In 9 days after seeding, the water-saturated surface soil is most
favorable for the germination of milk vetch. After germination, the moisture content of the surface soil (0-10 cm) should be maintained till
winter at about 70% W.H.C., in order to speed up the growth of root system
and nodule formation (Table 4). In the period from winter to flowering
stage, the moisture content should be maintained at 60-70% W.H.C. Table 5
shows the relationship between the soil moisture content over the period from
regreening stage to flowering stage and the yield of milk vetch(18).

Table 4 Effect of soil moisture content on the growth
of seedlings of milk vetch(17)

Soil moisture content*(%)	Plant height (cm)	No. of leaves per plant	No. of nodules per plant
20	3.16	0.68	0.07
30	4.73	1.64	0.83
40	4.50	1.51	1.16
50	3.62	0.74	0.60

  * Saturated moisture content-40%, wilting moisture content-
11.9%

Table 5 Effect of soil moisture content on the yield
of milk vetch (Sand loam)

Soil moisture content (%)	Yield* (g/pot)
10-12	1.48
16-20	12.56
24-28	17.61
32-36	4.07

  * Dry wt.   W.H.C. = 40%

EFFECTS OF GREEN MANURE

Green manuring plays an important role in the maintainence of soil fertility. Cultivation of green manure crops is one of the most important measures to improve low-yielding paddy soils and expand the area of rice cultivation.

1. Effect on Crop Yield and Utilization Rate of Nitrogen

The effect of green manure on rice yield varies with the level of soil fertility and variety of green manure crops. In general, 1000 kg of green manure (fr. wt.) can give a yield increase of rice grain ranging from 35-85 kg (Table 6). The effect of milk vetch is highest in the green manure crops.

Table 6 Response of rice to green manure(19-21)

Location of experiments	Soil fertility	No. of experiments	Dosage (t/ha)	Grain yield (kg/ha)	Increase kg	Increase %	Increment of grain per 1000 kg green manure
Guangxi	High	26	15.9	5120	910	21.6	57.3
	Medium	32	16.5	3320	1110	50.0	67.0
Zhejiang	Low	16	14.4	2640	1160	78.1	80.5
Zhejiang	---	422	18.7	---	718	18.6	38.3
Jangsu	---	92	26.0	---	927	17.8	45.0

The residual effect of green manures on the late rice in double cropping of rice is insignificant while that on the wheat is often remarkable (Table 7).

Table 7 Residual effect of green manure on the yield of succeeding crop(5,19)

Crop	Treament	Grain yield (kg/ha)	Increase (%)
Late rice	CK	2990	—
	Vetch*	3150	5.4
	Milk vetch*	3080	3.1
	Horse bean*	3080	3.2
	CK	3930	—
	Azolla	3990	1.5
Wheat	CK	1080	—
	Milk vetch*	1360	25.7

* 75 kg N per hectare

The utilization rate of nitrogen of leguminous winter green manure crops under field conditions is 25.3$\pm$10.2%. Under pot culture conditions, the utilization rate of N of milk vetch, vetch, bur-clover, etc. is much higher than that of Azolla, the former being 42.2$\pm$15.2%, while the latter 27.3$\pm$15.8%. The lower recovery rate of nitrogen of Azolla is closely related with its high lignin content (Table 8).

Table 8   Recovery of nitrogen from green
manure by rice crop

Green manure	C/N	Lignin (%)	recovery (%)	
			Field exp.	Pot exp.
Winter green manure	10-15	8.5-13	25.3$\pm$10.2 (n=10)	42.2$\pm$15.2 (n=10)
Azolla	8.5-12	18-24	——	27.3$\pm$15.8 (n=6)

A portion of the nitrogen of green manure, except that either absorbed by the first crop or lost, may remain in the soil. The quantity and availability of the residual nitrogen varies according to the chemical composition of green manures. It is illustrated in Table 9 that the milk vetch has not only the highest nitrogen utilization rate, but also the highest positive priming effect, and hence the least amount of net residual nitrogen. Azolla has a lower nitrogen utilization rate, a less positive priming effect, and a higher amount of residual nitrogen which is 2 times as much as that of milk vetch. Common water hyacinth has a lower nitrogen utilization rate, a higher negative priming effect and a highest net residual nitrogen. The availability of residual nitrogen of Azolla is lowest, only slightly higher than that of the native soil organic nitrogen, while that of milk vetch is highest, and is still much higher than that of the native soil organic nitrogen even after 3 yields of crops (Table 10).

Table 9   Role of green manure in the maintenance
of soil N

Green manure	Soil N mineralized due to priming effect (mg/pot)	Residual ^{15}N from green manure (mg/pot)	Net residual N(mg/pot)
Milk vetch*	52.8	141	+ 88
Azolla*	11.4	189	+178
Sun-hemp**	17.9	137	+119
Common water hyacinth*	-30.5	196	+227

*300 mg^{15}N/3 kg white bleached soil was added

**250 mg^{15}N/3 kg white bleached soil was added

Table 10   Nitrogen availability ratios(22)

Green manure	2rd crop (buckwheat)	3rd crop (barley)	Closed incubation*
Milk vetch	2.62±0.001	3.00±0.15	1.67
Azolla	1.39+0.00	1.04±0.12	0.81
Common water hyacinth	1.71±0.01	1.63±0.09	1.10

* Soil taken after harvest of barley

2. Effect on the soil Organic Matter Content

So far we have no data from long-term experiment concerning the effect of green manure on soil organic matter content. It has been shown from the decomposition experiment in the field that the humification coefficients (defined as fraction of added plant carbon retained in soil after one year's decomposition) of various green manures are different because of their different composition (Table 11).

Table 11   Humification coefficient of green manures (23)
(Wuxi County, Jiangsu Province)

Material	Humification coefficient
Milk vetch	0.18
Azolla	0.43
Sesbania	0.37
Vetch	0.30
Common vetch	0.27

The replenishment of soil organic matter decomposed is partly determined by the quantity and composition of plant residue. Mixed cultivation of leguminous and gramineous green manure crops often gives a higher biomass of green manure, and provides the plant materials with a higher humification coefficient (Table 12), being thus advantageous to the accumulation of organic matter in soil. This may be a promising way of cultivating green manure crop under the condition that production of chemical fertilizer is considerably on the increase but there is not enough quantity of straw to be returned to the soil.

It is well known that the newly formed organic matter helps improve soil physical properties, and that mixed cultivation of leguminous and gramineous green manure crops has an even better effect in this respect. However, it was found that green manuring was of no help to the improvement of the permeability of paddy soils under ill-drained conditions (Table 13).

215

Table 12 Effect of legume-grass mixed culture on
the yield of green manures (24)

| Treatment | Fresh wt. (kg/ha) | | | Total N |
	Legume	Grass	Total	(kg/ha)
Vetch	58800	——	58800	276
Vetch+ryegrass	55100	27600	82700	302
Milk vetch	54000	——	54000	183
Milk vetch+rye grass	49500	12400	61900	190

Table 13 Effect of application of green manure on physical
properties of a paddy soil*(25)

Treatment	Pressure resistance (kg/cm^3)	Volume wt. (g/cm^3)	Total porosity	Capillary porosity (%)	Water holding capacity (%)
CK	31.4	1.23	53.5	4.4	37.6
Azolla	27.9	1.11	58.0	4.5	46.3
CK	30.7	1.25	52.8	7.0	34.6
Common water hyacinth	26.0	1.20	54.9	6.8	39.7
CK	31.6	1.18	55.6	8.8	39.2
Rice straw	26.0	1.14	56.8	6.1	41.6

* After 3 annual applications

3. "Concentration" and "Activation" of Nutrients in Soil

The absorbing area and cation exchange capacity of the root system
of green manure crops are greater than those of cereal crops (Table 14).
Therefore, they have a stronger ability to absorb the difficultly available nutrients such as phosphorous and potassium in soil (Table 15). The
ability of the green manure crops which can concentrate and activate the
nutrients in soil may be interpreted mainly as why they are used as the
pioneer crops for the improvement of low-yielding paddy soils.

4. Use for Livestocks' Feeding Stuff

Leguminous green manure crops are also used as feeding stuff. Results
from experiment showed that when air-dried and then fed to the pig, 1250 kg
of fresh milk vetch could give to the pig a weight increase of 26.5 kg, and
about 76% of N, 88% of P_2O_5 and 78% of K_2O in the green manure could be recovered in the excreta. As seen from Table 16, the application of this excreta
to the rice gave an grain increase of 27 kg(27). It seems that the utilization
of green manure in such a way should be widely carried out.

216

Table 14 Characteristics of roots of some green
manure crops(26)

Crop	CEC (m.e./100g dry root)	Absorbing area m²/1g dry root	Respiratory intensity (CO_2 mg/1g dry root/hr)
Radish	65.3	0.61	9.53
Milk vetch	47.0	0.41	5.01
Vetch	41.9	0.36	3.91
Wheat	20.5	0.54	1.86

Table 15 Utilization of difficultly available nutrients by
green manure crops (26)

Crop	Increment of dry matter (%)		
	Apatite	Orthoclase	Serpentine
Radish	200	15	54
Milk vetch	312	40	30
Vetch	183	15	53
Wheat	102	12	21

Table 16 Response of rice to method of
utilization of milk vetch

Treatment *	Grain yield** (kg/ha)	Increment (kg/ha)
Control	224	——
Fresh. Digging in	262	38
Air-dried Digging in	259	34
Fresh, Composted with mud	271	47
Air-dried, Fed pig, then excreta used	252	28

* 1250 kg/ha
** $L.S.D_{5\%}$ = 11.1 kg/ha $L.S.D_{1\%}$ = 15.3 kg/ha

REFERENCES

(1) Jiangyin Agricultural Experimental Station, On the early sowing practice of milk vetch(1977, ms.). (in Chinese)
(2) Institute of Soils and Fertilizers, Agricultural Academy of Zhejiang, Effect of Light intensity on the growth of Astragalus sinicus and M. denticulata seedlings(1980, ms.). (in Chinese)

(3) Institute of Soils and Fertilizers, Agricultural Academy of Guangdong, Preliminary investigation on biological characteristics and vegetative propagation of _Azolla filiculoides_(1979, ms.). (in Chinese)

(4) Ding Qian-fa, Wang Zu-an, Investigation on the nitrogenase activity of sesbania nodules under waterlogged conditions(1980, ms.). (in Chinese)

(5) Gu Rong-shen, Peng Kun, Investigation on winter green manure crops in East China. In "Soil and Fertilizer Studies", Shanghai Science & Technology Press, Shanghai, 193-244(1958). (in Chinese)

(6) Gu Rong-shen, Relationship between population and individual of milk vetch, its proper density of planting. In "Studies on Crop Population", Jiangsu People's Publishing House, Nanjing, 137-148(1962). (in Chinese)

(7) Huaiyin Agricultural Experimental Station, Main techniques of milk vetch introduced northward to the Changjiang River(1977, ms.). (in Chinese).

(8) Yangzhou Institute of Agricultural Sciences, Notes on the genetics and selection of rhizobium of milk vetch(1977, ms.). (in Chinese)

(9) Zhao Ren-san, Tang Jian-min, On the application of phosphorous fertilizer and its effect on the yield of milk vetch. Turang Tongbao, 5, 1-8(1964). (in Chinese)

(10) Yuan Cong-hui, Gu Rong-shen, You De-min, Bai Gang-yi, Qiu Jia-zhang, On the amelioration of the whitish bleached paddy soils. Scientia Agricultura Sinica, 8, 44-48(1963). (in Chinese)

(11) Cai Da-tong, Qin Sheng-wu, Dong Bai-shu, Effect of phosphatic fertilizer and green manure on the yield of rice and on the available phosphorous status of a calcareous soil. Jiangsu Agricultural Sciences, 5, 42-47(1979). (in Chinese)

(12) Zhejiang Research Group on Techniques of Phosphorous Application, Report on techniques of phosphrous application to winter crops in Zhejiang Province (1962-1963). Zhejiang Agricultural Sciences, 10, 421-426(1963). (in Chinese)

(13) Agricultural Bureau of Wujin County, A summary account on cultivation technique of green manure crops(1979, ms.). (in Chinese)

(14) Cai Da-tong, Fu Tang-gui, Effect of nitrogen fertilizer on the yield of milk vetch(1980, ms.) (in Chinese)

(15) Soil and Fertilizer Research Institute, Jiangsu Academy of Agricultural Sciences, Experiment on the application of potassium fertilizer. Jiangsu Agricultural Sciences, 4, 53-59(1978). (in Chinese)

(16) Institute of Soils and Fertilizers, Agricultural Academy of Hunan(ed.), Cultivation of Green Manure. Hunan People's Publishing House, Changsha (1975). (in Chinese)

(17) Yulin Agricultural Experimental Station, Effect of soil water regime on the growth of green manure(1963, ms.) (9n Chinese)

(18) Liuzhou Agricultural Experimental Station, Effect of different water regime and fertilization on the yield of vetch and milk vetch(1964, ms.). (in Chinese)

(19) Yulin Agricultural Experimental Station, On the response of early rice to green manure(1963, ms.). (in Chinese)

(20) Institute of Soils and Fertilizers, Agricultural Academy of Zhejiang (ed.), Cultivation and Utilization of Azolla. Agriculture Press, Beijing(1974). (in Chinese)

(21) Agricultural Bureau of Jiangsu Province(ed.), Cultivation Technique of Azolla. Jiangsu People's Publishing House, Nanjing(1973). (in Chinese)

(22) Shi Shu-lian, Wen Qi-xiao, Liao Hai-qiu, The availability of nitrogen of green manures in relation to their chemical composition. Acta Pedologica Sinica, 17, 241-246(1980). (in Chinese with English summary)

(23) Lin Xin-xiong, Cheng Li-li, Shi Shu-lian, Wen Qi-xiao, Characteristics of decomposition of plant residues in soils of southern Jiangsu. Acta Pedologica Sinica, 17, 319-328(1980). (in Chinese with English summary)
(24) Lu Bing-zhang, Investigation on the mixed cultivation of green manure crops(1980, ms.). (in Chinese)
(25) Shi Shu-lian, Cheng Li-li, Lin Hsin-hsiong, Shu Chong-li, Wen Chi-hsiao, Effect of Azolla on the fertility of paddy soils. Acta Pedologica Sinica, 15, 54-60(1978). (in Chinese with English Summary)
(26) Institute of Soil Science, Academia Sinica, The Environmental Conditions of Soils for High Yield of Rice. Science Press, Beijing, 458 (1961). (in Chinese)
(27) Gu Rong-shen, The profits of green manure crop (Astragalus sinicus) under different ways of utilization. Nongye Keji Tongxun, 12, 3 (1977). (in Chinese)

CROPPING SYSTEM IN RELATION TO FERTILITY
OF PADDY SOILS IN CHINA

Xu Qi
(Institute of Soil Science, Academia Sinica, Nanjing)

China has a long history of rice cultivation. It is demonstrated that rice has been cultivated for more than five thousand years in the country. Following the social development and scientific and technological progress, the area of rice cultivation is expanded gradually. Now, it is found almost all over the country. It is estimated that the total area of rice fields approximates to 25,400,000 ha in China, about 25.8% of the cultivated area; and the total output of rice grain accounts for 45% of the total output of food crops of the country.

FORMATION OF CROPPING SYSTEM IN RELATION TO NATURAL CONDITIONS

The distribution of paddy soil in China is similar to that of other places of the world. It is widely dispersed from Heilongjiang of cool temperate zone in the north, to Hainan Island of the tropical zone in the south; and from Taiwan Province and the coastal plains in the east to the oasis of the desert in the west, and it is also distributed even in the canyon of Qinghai-Xizang (Tibet) Plateau up to 2000 meters above sea level.

Most part of China is of monsoon climate, its hydrothermal conditions are changed markedly not only from south to north, but also from east to west. As s result of the remarkable difference in frost-free period, the crops suitable for cultivation and the cropping indices of land are varied in different regions. Thus, various patterns of cropping systems are formed. There are three main patterns of cropping systems, namely single cropping system of rice, double cropping system of rice and wheat, and triple cropping system of rice-rice-wheat (or three yields of rice annually). The required climatic conditions for the three cropping systems are showed in Table 1.

As to the geographical distribution of the cropping systems, the single cropping system of rice is mainly scattered to the north of the Qinling mountains and the Huaihe River and on the Qinghai-Xizang (Tibet) Plateau; the double cropping system is mainly distributed from the Qinling mountains and the Huaihe River southward to the Changjiang River and the Qiantang River, and the triple cropping system is widely spread in the region north to the middle subtropical zone (Fog. 1). The crops and cropping indices included in the cropping systems are changed greatly from the north to the south. For example, there are only two rotation patterns in the region of single cropping of rice. while, in the region of double cropping system of rice and wheat, there are three patterns of three-year rotation. Main rotation systems in the region of simple cropping of rice are successive cropping of rice and 2-year rotations of rice-upland crop (wheat, cotton or corn).

Table 1 Cropping system in relation to natural conditions

Cropping system	Annual precipitation (mm)	Annual temp. (°C)	Accumulated temp. (≥ 10°C)	Frost-free period (day)
Single cropping system of rice	100–750	6–15	1,700–4,500	100–200
Double cropping system of rice-wheat	750–1,000	15–16	4,500–5,000	200–240
Triple cropping system of rice-rice-wheat or rice-rice-rice	>1,000	16–24	5,000–10,000	>240

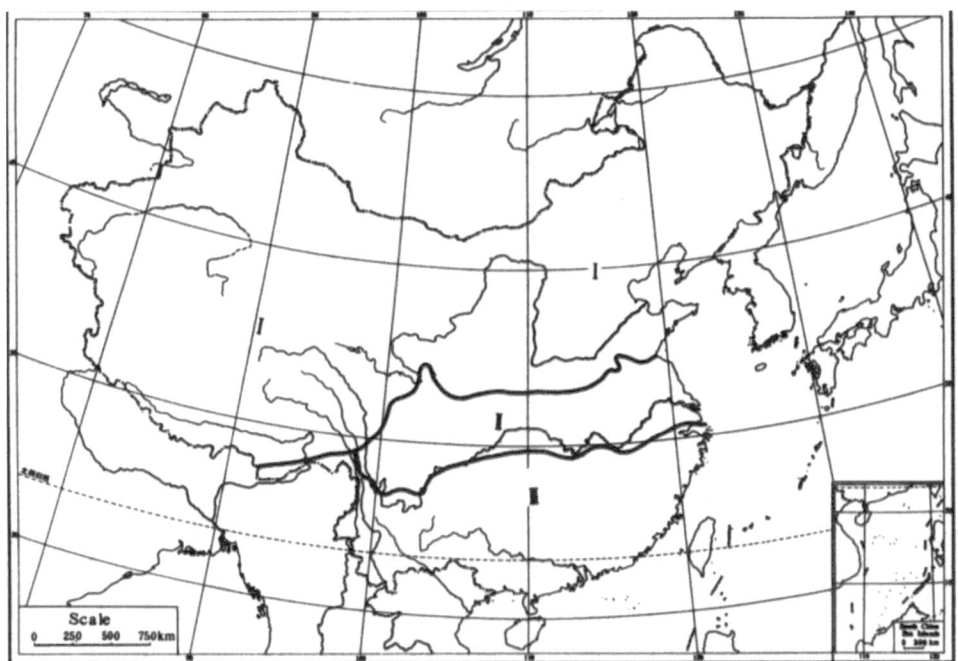

Fig. 1 Regions of cropping systems of rice in China
   I. Region of single cropping system of rice;
   II. Region of double cropping system of rice and wheat;
   III. Region of triple cropping system of rice-rice-upland crop (or rice-rice-rice)

Rotation systems in the region of double cropping system are rice-wheat-rice-wheat-rice-wheat, rice-wheat-rice-rape-rice-wheat and rice-wheat-rice-green manure-rice-wheat.

In the region of triple cropping system there are seven or even more patterns of 3-year rotation (Table 2).

Table 2 Rotation systems in the region
of triple cropping system

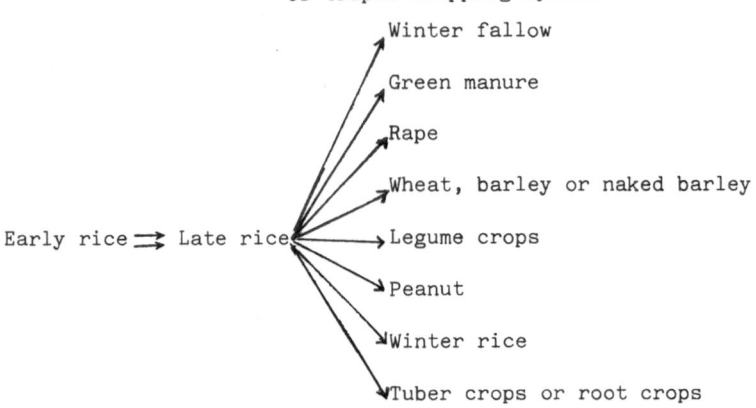

Early rice ⇒ Late rice
- Winter fallow
- Green manure
- Rape
- Wheat, barley or naked barley
- Legume crops
- Peanut
- Winter rice
- Tuber crops or root crops

Rice is a crop with good adaptability. It can be planted on any soils so long as the irrigation is provided. In the north, due to the limitation of water sources, the rice fields are often distributed in the depression or lower parts of plains. In the area southward from the Huaihe River with sufficient rainfall and abundant of water sources, rice can be planted not only on the plains and depressions, but also on the terraced field in hilly area. Therefore, most of the rice fields in China distributed to the south of the Huaihe River (Table 3).

Table 3 Distribution of rice field in China

Cropping system	Area of rice field (ha)	Percentage in total rice field area (%)	Percentage in total arable land area (%)
Single cropping of rice	2.13×10^6	8.2	45.0
Double cropping of rice-wheat	10.0×10^6	39.5	5.2
Triple cropping	13.0×10^6	52.3	49.8

Owing to different conditions of relief, hydrology and soil, there is difference of cropping system even in a limited area. There are also double cropping system of rice-wheat or single cropping of rice in the region of triple cropping system. However, generally speaking, the rotation system becomes more and more complicated from the north to the south, and is more closely related to the variation of soil fertility from the north to the south.

RELATIONSHIP BETWEEN CROPPING SYSTEM AND SOIL FERTILITY

The cropping system is chiefly determined by natural conditions,

especially the hydrothermal condition. In an area with similar hydrothermal condition, the rotation system, cropping index and crop yield
are closely related to soil condition. The evolution of soil fertility
is determined by the material cycling, which is the result of rotation
system. Therefore, farmers always tend to develop the production
potentiality of cropping system by means of reasonable rotation of crops
and corresponding agricultural practices.

1. General Characteristics of the Fertility of Paddy Soils
in Different Regions of Cropping System

The fertility of paddy soil is determined not only by the original
soil, but also by agricultural practices under submergic cultivation.
As to the nutrient content in the plowed horizon, the highest contents of
organic matter and total nitrogen are found in soils of northeastern
region of the country; and the contents tend to be increasing from northern
China to the south. Owing to abundant cold spring paddy fields in the
mountain areas in southern China, the contents of organic matter and total
nitrogen in soils of mountain area are higher than those in the plain area,
phosphorus and potassium contents of soils are closely related to the
parent materials. The content of phosphorus of soils in hilly area is
lower than that in plain area; while the content of potassium of soils in
mountainous area is higher than that in plain area. In the region of triple
cropping system, though the nitrogen content is high (Table 4), the deficiency of nitrogen generally occurs due to the greater need by three yields
of crops, and deficiency of phosphorus is also often found in the region.

Table 4 Nutrient contents in plowed horizon of
paddy soils in different areas of
cropping systems (mean value)[1,2]

Cropping system	area	Content (%)			
		O.M.	N	P_2O_5	K_2O
Single cropping region	Northeast China Plain	3.20	0.15	0.24	2.2
	North China Plain	1.79	0.10	0.14	2.0
Double cropping region	Plains along middle & lower Changjiang River	2.20	0.13	0.11	1.2
	Hills along middle & lower Changjiang River	1.59	0.10	0.08	1.4
Triple cropping region	Plains	2.40	0.16	0.12	1.9
	Mountains	2.70	0.14	0.09	2.3

2. Water Requirement of Different Cropping Systems

Water requirement of rice is related to cropping system and soil
texture. Experiment in various areas showed that the total water
requirement of rice tended decreasing from the north to the south, and
so was the evaporation transpiration and percolation. The paddy fields
in northern China are mostly newly used for rice plantation, and the
soils are mostly sandy. In addition, the climate is dry, and therefore
the evaporation, transpiration and percolation are greater (Table 5).

Table 5 Water requirement of rice under
different cropping systems[3]

Cropping system	Region	Irrigation (day)	Water requirement m^3/ha/year			
			Trans.	Evap.	Perc.	Total
Single cropping	North China	90–130	3750	2850	4800	11400
Single cropping of late rice	Middle & lower	120–150	3600	2400	1350	7350
Succession cropping of early rice	South to Changjiang River	75–95	2100	1650	600	4350
Succession cropping of late rice	"	75–110	2550	1950	900	5400

3. Nutrient Balance under Different Cropping Systems

The cropping system may influence nutrient cycling and nutrient
balance of the soil. The nutrient balance is affected by many factors,
such as the output brought away by the crop harvest, input brought in
by fertilization and the materials brought in or brought out by water.

The chemical composition of the grain and straw of crops varies with
the kind and variety of the crop. From the estimation on the basis of
ordinary or higher level of crop yield, the amounts of nitrogen, phosphorus and potassium brought away by crop harvests per year are rather
high (Table 6).

Table 6 Nutrients brought away by crop harvests
under different cropping systems[2]

Cropping system	Level of yield t/ha	Nutrient (kg/ha)		
		N	P_2O_5	K_2O
Simple cropping of rice	6.0	114.8	56.3	165.8
	7.5	143.3	63.8	207.0
Rice-wheat	7.5	150.8	79.5	222.8
	11.25	231.0	121.5	337.5
Rice-rice-wheat	11.25	225.8	112.5	372.8
	15.0	297.8	158.3	496.5
Triple cropping of rice	11.25	188.3	114.0	375.0
	15.00	255.0	142.5	450.0

There are some nutrient in irrigation water and percolation water. Generally, the amounts of surface run-off and percolation water account for ten percent of the irrigation water respectively. It is seen from the estimation in Table 7 that the amount of nutrients brought in soil by the water can not complement that brought away by crop harvests, especially in the region of triple cropping system.

Table 7 Nutrients brought in soil by irrigation water[2]

Cropping system	Type of water	Nutrient (kg/ha)		
		N	P_2O_5	K_2O
Double cropping of rice & wheat	Irrigation water	5.1	0.33	20.3
	Percolating water	0.68	0.06	3.0
Triple cropping of rice-rice-wheat	Irrigation water	14.3	0.23	13.5
	Percolating water	1.5	0.06	3.3

In China, applications of farm manure, straw and green manure are the important ways to complement the organic matter of soil. The content of nitrogen in green manure ranges from 2.5-4%, of which one third may be used by the crop in the first season. It is seen from the estimation in Table 8 that the effect of green manure on the nitrogen balance of soils is significant. This is the reason why the triple cropping system of rice, rice and green manure crops is predominant in southern China at present.

Table 8 Nitrogen balance in different cropping systems*[4]

Rotation pattern	Yield (kg/ha/year)				Nitrogen balance		
	Rice	Wheat	Rape	Green manure	Outlay	Input	Increment
Rice-rice-milk vetch	9675	—	—	36750	165.5	147.0	-16.5
Rice-rice-rape	9600	—	2123	—	162.8	81.0	-81.8
Rape, wheat and green manure-rice-rice	10200	1035	818	10800	204.0	71.3	-132.8
Rice-rice-wheat (barley or naked barley)	9075	4275	—	—	266.3	—	-266.3

*With the exception of rapeseed, all harvests returned to the field

4. Problems Occured under new Triple Cropping System

In recent years changes of cropping system has brought about marked effect on the increase of yield. These changes include using japonica variety instead of indica variety, planting early rice instead of late rice, and transforming single cropping into double cropping. Transforming double cropping into triple cropping has also produced good effect on increase of yield. However, in some areas where drainage facilities cannot be furnished simultaneously, the double cropping of rice and rice or triple cropping of rice-rice-wheat may often induce the deterioration of soil environmental and nutritional condition which is unfavorable for yield increase of rice and upland crops. The affected soil properties are as follows:

1) Soil environmental factor: Under the condition of successive cropping of rice, the tilth gets badly, plowpan becomes puddling and compact, soil permeability is reduced, and modulus of rupture is increased (Table 9). These changes influence not only the tillering of rice, but also the yield of upland crops.

Table 9 Effect of surface gleyization on properties of plowed horizon of paddy soil[5]

Soil	Eh (mV)		Modulus of rupture (kg/cm)	Permeability (mm/day)
	0–15cm	15cm–		
Normal	292	229	23.5	7.8
With Ag horizon	214	133	35.1	1.5

2) Soil nutritional conditions: The nutrients brought away by crops from the soil are increased with the increase of crop yield and the raising of cropping index. If the applied fertilizers cannot complement the continuous increasing amount of nutrients brought away by crops, the nutrient defficiency will be induced. For example, the phosphorus and potassium fertilizers were once less effective on some soils in the Taihu Lake region, but they become more effective at present.

UTILIZATION OF LAND AND MAINTENANCE OF SOIL FERTILITY

A good cropping system is closely related with rational arrangement of crops suitable for the local soil condition, rational fertilization and water management.

1. Arrangement of Crops Suitable for Soil Conditions

Any variety of crop has its own definite adaptability to environmental conditions. Therefore, it is important that a rational cropping system should be arranged in accordance with the local soil, water and fertilizer supplying conditions. The patterns of rotation systems and the corresponding soil conditions for a triple cropping area in Fujian Province are listed in Table 10.

Table 10 Soil fertility and rotation systems
(Fujian Province)[7]

Soil type	Patterns of rotation system	Relative proportion				Rice: Upland crop
		Food	Beans	Rape	Green manure	
Ill-drained	Rice-rice-winter fallow	6 :	0 :	0 :	0	2:0
Fertile	Triple cropping system of multiple cropping rice and one cropping upland crop	7 :	1 :	1 :	0	3:1
Medium fertility	Triple cropping system of rice-rice-upland crop	6 :	1 :	1 :	1	2:1
Infertile	Triple cropping system of two cropping rice or two cropping upland crops	5 :	2 :	1 :	1	1:1

2. Rational Rotation and Fertilization

The fertility of most paddy soils in China cannot meet the requirement of high yield of rice. Therefore, fertilization is of most significance.

Low organic matter and nitrogen contents are usually found in most soils in the area of single cropping system of rice. The paddy soils in Northeast China are mostly developed from black soil, white bleached soil or swampy soil. Their organic matter content generally ranged from 6–10% in the initial stage for rice cultivation and rapidly decreased to 2–3% after several decades. The organic matter content of paddy soil in other areas got still less. The organic matter of paddy soils developed from alluvial soils has a tendency of accumulation, but its content is only about 2% for rice cultivation of more than one hundred years. Therefore, it is better to grow sweet clover as green manure in rotation and to increase the amount of straw returned to the field.

In the area of double cropping system of rice and wheat, and the area of triplc cropping system of rice-rice-upland crop, green manure crop is of importance in the rotation. Through the rotation of rice and upland crops, soil structure of the plowed horizon may be improved, the contradiction between air and water in soil can be solved, and the requirements of rice and upland crops for soil environments are thus satisfied. (Table 11)

227

Table 11 Effect of rotation of rice and upland
crop on porosity of the soil[8]

Rotation system	Porosity (v%)		
	Capillary	Non–capillary	Total
Successive cropping of rice	46.1	5.1	51.4
Rice–cotton (rice)	39.5	19.2	58.7
Rice–cotton (3 years)	35.9	18.3	54.2

In the area of double cropping rice, good effect of legume crops in the rotation system is found both on the improvement of soil environmental condition and soil nutritional condition (Table 12)

Table 12 Effect of rotation of rice and beans
on properties of the soil[6]

Rotation	Volume weight (g/cm^3)	Porosity (%)	Aggregates 0.25–0.5mm (%)	O.M. (%)
Rice–rice–winter fallow	1.31	41.0	6.1	1.25
Roce–rice–broad bean (5 year rotation)	1.23	49.8	16.2	2.71

3. Improvement of Drainage and Regulation of Air and Water in Soil

Drainage is an important problem for rice cultivation, especially in the double cropping system of rice and wheat and triple cropping ststem including two successive croppings of rice. A proper percolation rate of the soil is one of the most necessary conditions for high yield of rice. Recently, in some area, under–ground pipe drainage or mole channel drainage has been simultanously adopted for rice field with the improvement of water management. It shows that underground pipe drainage is effective in lowering the ground water level and regulating the air and water condition in soil (Fig. 2), and is favorable for the growth of root system of rice (Table 13), as well as effective in preventing the upland crops from waterlogging injury. Therefore, its effect on yield increase is remarkable (Fig. 3).

In a word, cropping system is closely related to soil fertility, At the same time, cropping system, rate of photosythesis, and soil fertiliry constitute an inseparable entirity. Adoptation of rational cropping system and continuous promotion of soil fertility should be the important measures for the increase of crop yield.

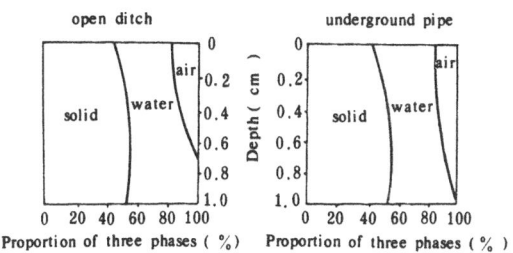

open ditch   underground pipe

Proportion of three phases (%)   Proportion of three phases (%)

Fig. 2 Effect of open ditch and
underground pipe drainage
on air and water regime
in soil

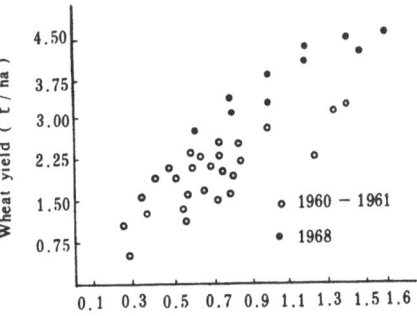

Ground- water table (m)

Fig. 3 Relationship between
wheat yield and ground
water table

Table 13 Comparison of the colors of rice roots (9, 10)

Locality	Treatment	Root (%)		
		White	Yellow	Black
Guangdong	Mole channel drainage	37.3	58.7	4.1
	Surface drainage	15.2	65.6	19.2
Jiangsu	Underground pipe drainage	13.3	86.7	0
	Surface drainage	3.5	51.4	45.1

REFERENCES

(1) Institute of Soil Science, Academia Sinica, Nanjing (ed.), Soils of
China. Science Press, Beijing, 361 (1978). (in Chinese)
(2) Institute of Soil Science, Academia Sinica, The Environmental
Conditions of Soils for High Yield of Rice. Science Press, Beijing,
426-432 (1961). (in Cginese)
(3) Ding Ying (ed.), Cultivation of Rice in China. Agriculture Press,
Beijing, 407-416 (1961). (in Chinese)
(4) Institute of Soils and Fertilizers, Agricultural Academy of Sichuan,
Effect of triple cropping system on soil fertility (1978, ms.).
(in Chinese)
(5) Hseung Yi, Xu Qi, Yao Xian-liang, Zhu Zhao-liang, Effect of cropping system on the fertility of paddy soil. Acta Pedologica Sinica,
17, 101-119 (1980). (in Chinese with English summary)
(6) Lin Jing-liang, Effect of rice-bean rotation system on fertility of
paddy soil and increase of crop yield. Fujian Agricultural Sciences,
4, 22-24 (1964). (in Chinese)

(7) Chen De-shun, Yu Fu-yun, Lin Yu-da, A experience about effect of rice-upland crops rotation on the yield of crops. Fujian Agricultural Sciences, 4, 2-11 (1964). (in Chiense)

(8) Chen Jia-fang, Water-air regime of paddy soil and its influence of the factor in the Taihu Lake region (1979, ms.). (in Chinese)

(9) Institute of Soils and Fertilizers, Agricultural Academy of Guangdong, Effect of channel drainage on improvement of soil and increase of rice yield. Guangdong Agricultural Sciences, 2, 23-29 (1978). (in Chinese)

(10) Nanjing Hydraulic Research Institute, A study of aeration in rice field (1978, ms.). (in Chinese)

(11) Jiangsu Hydraulic Research Institute, A study of stagnancy control by underground drainage (1978), ms.). (in Chinese)

SOME LABORATORY OBSERVATIONS ON THE FORMATION
OF BLACK LAYER IN THE RECENTLY   RECLAIMED
SALINE SOILS UNDER RICE   CULTIVATION

Yuan Ke-neng, Huang Chang-yong,
Zhu Zu-xiang
(Zhejiang Agricultural University, Hangzhou)

In the coastal plain region of Zhejiang Province, rice is the main crop grown on the recently reclaimed saline soils. During the rice growing season, upon submergence of the soil under irrigation water, the subsurface layer of the most areas may soon turn dark black in color. The roots growing in this layer are also mostly black in appearance, having no rust-colored coatings of iron oxides on them. Quite often, death of the seedlings due to necrosis of root tissues may devastate the whole field. The soils liable to the black layer formation are usually characterized by relatively high organic matter content, low percolation rate,   rich in soluble sulfates and low in redox potentials (Table 1).

Table 1   Properties of some representative samples of
recently reclaimed saline soils in
Xiaoshan Region

Location	Depth of black layers (cm)	Texture	O.M. (%)	Red. iron* (ppm)	Perco-lation rate (mm/day)	pH**	Eh**	Soluble salts	
								Total (%)	SO_4^{2-} (m.e./100g)
Hongth commune	0.3–10	fsl	2.20	7160	2.1	7.98	−104	0.155	70.8
Quanjin commune	0.4–6	fsl	1.88	5910	2.9	7.99	− 87	0.090	23.3
Reclaim-ed Farm	None	fsl	1.18	5710	6.0	8.20	—	0.076	9.1

  * Reducible iron is determined by the H_2S reduction method
 ** Average value of the 1-5 cm layer

In order to elucidate the conditions for the black layer formation, a series of laboratory experiments has been carried out using soil columns by packing the soil samples in glass tubes being about 4 cm in diameter and 20 cm in length. The soil samples varying in texture from sandy loam to clay and with variable content of soluble salts ranging from 0.172% to 0.498%, were mixed up with 1% rice straw powder which was used to afford enough energy source. The soil column in the tube was percolated by continuous flooding over the surface with either 0.2% or 0.05% Na_2SO_4 solution. The percolation rate was controlled with a clamp on the outlet rubber tubing underneath the soil column. The experiments were conducted under incubation at different temperatures. The results demonstrated that the black layer was much easier to form at higher temperatures as long as there were sufficient energy materials in the soil for biological activities. The presence of Na_2SO_4 in the irrigation water even as low as 0.05% in concentration hastened up the process markedly. The black layer first appeared as a very thin layer right below the surface oxidized layer (usually at a depth of 3-10 mm below the surface), and thence extended downward as percolation continued to proceed. By the end of the experiments (2-3 months), it eventually reached a thickness of about 5-10 cm. Some of the physico-chemical properties and the reduced constituents of the differentiated layers in soil columns relevant to the black layer formation are summarized in the Table 2.

Table 2 Some physico-chemical properties and reduced
constituents of the differentiated layers in
the soil column experiments (Averages of 17
soil columns)

Soil layers	pH	Eh (mV)	S^{2-}(ppm)*	Fe^{2+}(ppm)**
Oxidized layer	8.11	361	17	190
Black layer	7.34	-156	1456	2216
Gray layer	7.81	0.5	225	907

* Extracted with 4 N HCl
** Extracted with 0.1 M $Al_2(SO_4)_3$, pH 2.5

From the results in the Table 2, it is obvious to note that the black layer is characterized by its strong reducing conditions as indicated by its extreme low average of Eh values. In fact, all individual values of Eh measured in the black layer of the soil column experiments as well as from the fields revealed that they all lay between -100 to -200 mV. This was about the range reported favorable for sulfate reduction(3,4) and for iron reduction at pH 8 (5). The concurrent presence of ferrous iron and sulfides should naturally lead to the precipitation of iron sulfides (FeS. nH_2O), which gave the characteristic black color to the soil layer.

Numerous analytical data showed that most of the soil samples in the black layer contained large amount of sulfides. The contents were all at the levels above 700 ppm, and some even reached as high as 1300-1800 ppm, whereas those in the underlying layer decreased sharply, being all below the 500 ppm level. With regard to the ferrous iron contents, there also existed striking contrasts among different layers (Table 2). The average molar ratio of Fe^{2+} to S^{2-} in the black layer was found to be in approximate correspondance to the FeS formula.

The dynamics of the formation of black layer can be illustrated by Fig. 1. The decline of Eh in the black layer was accompanied by simultaneous increase of reduced iron from the very beginning of the experiment. But on the other hand, the reduction of sulfate was hardly noticeable as long as the Eh value was above zero. And it was not until the Eh dropped to a

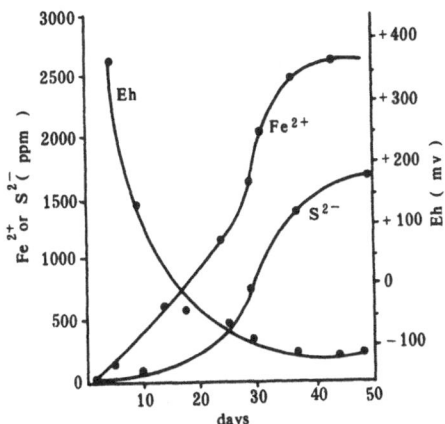

Fig. 1 The trend of changes
of Eh and S^{2-}, Fe^{2+} contents
in the black layer

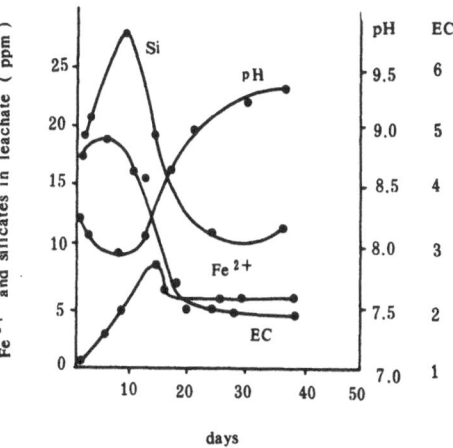

Fig. 2 Changes of pH, EC, concentrations of Fe^{2+} and silicates
in the leachate with time

level below -100 mV, the reduction of sulfates then proceeded with an
increasing rate. It was observed from the soil column experiments that
the thickness of the oxidized layer never exceeded about 1.0 cm, and the
transitions of the color change in time and between soil layers were
usually quite sharp.

The percolates coming out from the lower end of the soil columns
were analyzed for their ferrous iron and silicate concentrations. From
the results shown in Fig. 2, it can be seen that the two curves were more
or less similar in pattern. But whether the initial concurrent uprise of
the Si-curve with the Fe^{2+} curve should be attributed to the breakdown of
the silicate minerals or to their dissolution in the alkali solutions still
remain to be proven. Nor is there enough evidence to suspect the probable
existence of ferrolysis during leaching in the soil column experiments.

In the process of percolating, the
pH values of the oxidized and the reduced
layers were both decreasing but in different rates (Fig. 3). At the end of the
experiments, the average pH attained by
the black layer was 7.34, which was the
lowest among the three layers (Table 2).
The trend of the change of pH of the
percolates was quite different from
that of the soil. As the Na_2SO_4 solution
percolated through the column, the sulfate
was reduced and precipitated as hydrated
ferrous sulfides in the black layer, Thus,
as the biological reduction of sulfates
was getting more and more intense in the
later stages of the experiments, the pH
of the percolates became higher and higher
due to their increment of excess of the

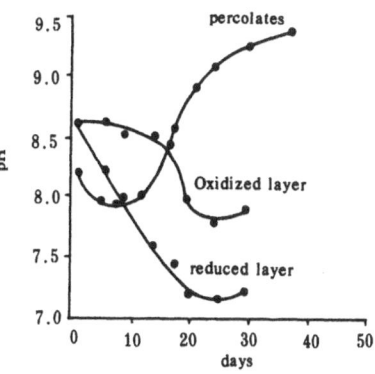

Fig. 3 The change of pH in
soil percolation experiments

sodium over sulfate ions. If the soil columns used in this experiments were not limited to only 20 cm in length, there might have deep-lying soil layers with pH values over 9. In fact, numerous field measurements have proved the existence of such alkali layers with high sodium adsorption ratios deep below the balck layer.

REFERENCES

(1) Yamane, I., Reduction of nitrate and sulfate in submerged soils with special reference to redox potential and water-soluble sugar content of soil. Soil Sci. Plant Nutr., 15,139-148(1969).
(2) Connell, W.E., Patrick, W.H.Jr., Sulphate reduction in soil: effects of redox potential and pH. Science, 159, 86-87(1968).
(3) Engler, R.M., Patrick, W.H.Jr., Sulfate reduction and sulfide oxidation in flooded soil as affected by chemical oxidants. Soil Sci. Soc. Amer. Proc., 37, 685-688(1973).
(4) Gotoh, S., Patrick, W.H.Jr., Transformation of iron in a waterlogged soil as influenced by redox potential and pH. Soil Sci. Soc. Amer. Proc., 38, 66-71(1974)
(5) Bloomfield, C., Sulphate reduction in waterlogged soils. J. Soil Sci., 20, 207-221(1969).
(6) Patrick, W.H.Jr., Delaune, R.D., Characterization of the oxidized and reduced zones in flooded soil. Soil Sci. Soc. Amer. Proc., 36, 573-576 (1972).
(7) Starkey, R. L., Oxidation and reduction of sulfur compounds in soils. Soil Sci., 101, 297-306(1966).
(8) Connell, W.E., Patrick, W.H.Jr., Reduction of sulfate in waterlogged soil. Soil Sci. Soc. Amer. Proc., 33, 711-715(1969).
(9) Ogata, G., Bower, C.A., Significance of biological sulfate reduction in soil salinity. Soil Sci. Soc. Amer. Proc., 29, 23-25(1965).

REDUCING ORGANIC SUBSTANCES RESPONSIBLE FOR REMOVAL OF Fe(III)
AND Mn(IV) FROM SUBSURFACE HORIZON OF LOWLAND RICE SOIL

Masanori OKAZAKI
(Faculty of Agriculture, Tokyo University of Agriculture and
Technology, Tokyo, Japan)

Hidenori WADA,   Yasuo TAKAI
(Faculty of Agriculture, The University of Tokyo, Tokyo, Japan)

Lowland rice soil is a kind of hydromorphic soil which is subjected to
human activities to a various degree.   Especially lowland rice soil with low
water table is frequently regarded as an anthropogenic soil due to its characteristic profile caused by artificial control of irrigation and drainage.
Its pedogenic process has been studied and considered to be closely related
with fertility of lowland rice soil.

During the period of rice cultivation in summer, several soil components
are made soluble in the submerged surface horizon and moved down to subsurface
horizons with percolating water.   For instance, ferrous iron which is soluble
in the reduced surface horizon is eluted from surface horizon and retained in
subsurface horizon by immediate oxidation of ferrous iron to ferric iron oxide.

Recently some investigators have pointed out from detailed field observations and experiments in the laboratory that the classic theory of forming
process of lowland rice soil can hardly be interpreted without consideration
of the presence of reducing organic substances in percolating water (1,2).
They proposed that reducing organic substances in percolating water were responsible for the formation of gray-colored channel or ped cutans on the surface of macro voids in subsurface horizons which were poor in iron and manganese content and rich in easily decomposable organic substances content.

With respect to the reduction of ferric iron in the submerged soil, two
different mechanisms have been proposed: (1) direct biological reduction of
ferric iron acting as a final electron acceptor of a respiratory chain, and
(2) indirect reduction of ferric iron by certain substances released from
heterotrophic bacteria.

HAMMANN and OTTOW (3) insisted on direct biological reduction of ferric
iron by iron-reducing bacteria containing two electron donating systems: a
nitrate reductase system and a "ferrireductase system" in cytoplasmic membrane.   However, STARKEY and HALVORSON (4) and ROBERTS (6) regarded the reduction of ferric iron as a result of indirect microbial reduction, although
they found that almost all facultative anaerobic bacteria was capable of reducing iron compounds.   It has been reported that reducing agents were produced in lowland rice soil incubated under submerged condition.   KUMADA (6)
and IRI (7) found that large amount of reducing agents was contained in the
flooded water of excessively eluviated lowland rice soil than in gray brown
lowland rice soil using indigocarmine as a redox indicator.

By elucidating nature and quantity of water-soluble reducing organic
substances we can understand more precisely the possibility of the indirect
reduction of ferric iron and the removal of Fe and Mn from subsurface horizons by this substances, resulting in the formation of A2g horizon or gray-
colored channel and ped cutans in subsurface horizons.

The purpose of this study is to measure the iron-reducing capacity of
reducing organic substances and to identify reducing organic substances in
leachate from the submerged surface soil of lowland rice soil.

MATERIALS AND METHODS

1. Soil Samples

Soil samples used were taken from surface horizon in two contrasting lowland rice fields with low water table; a excessively eluviated sandy lowland rice soil (degraded lowland rice soil) at Nyuzen-cho in Toyama Prefecture, Nyuzen soil, and a gray brown lowland rice soil at Nagano Agricultural Experiment Station in Nagano City, Nagano soil. Soil samples were air-dried and passed through a 2 mm sieve. Some chemical properties of these soil samples are shown in Table 1.

2. Cation Exchange Resin, Ferric Hydroxide and Aluminum Hydroxide

Cation exchange resin, Amberlite CG 400, was converted into Ca type. Ferric hydroxide was freshly precipitated on celite by rapid neutralization of $2/10$ N $Fe(NO_3)_3$ solution suspending celite with NH_4OH solution to pH 7. Aluminum hydroxide was also precipitated on celite from 10 % $AlCl_3$ solution with 4N NH_4OH solution. Ferric hydroxide and aluminum hydroxide were X-ray amorphous. Two to 3 g of each of these adsorbents was suspended in water and packed in a glass column.

3. Incubation Method

One hundred g of air-dried soil sample was packed in a glass column and kept under submerged condition for incubation at $30°$ C (Fig. 1). One hundred ml of leachate was collected every other day, while keeping soil samples under submerged condition by supplying distilled water to the top of the soil column. In the case of B, C, D, E and F plot, leachate was allowed to pass through the column of adsorbent. After sampling, leachate was acidified to pH 3-4 with sulfuric acid, stored in a cold room and analyzed as soon as possible.

4. Ultrafiltration

Diaflo membranes, UM 05, UM 10, and XM 100 were used for ultrafiltration of organic substances present in leachate. Fifty ml of acidified leachate was placed in the ultrafiltration cell and fractionated according to molecular weight by passing it through each of these membranes following to the method of OGURA (8).

5. Determination of Organic Carbon in Leachate

Organic carbon in leachate and in the ultrafiltered fractions were determined by the wet combustion method of MENZEL and VACCARO (9).

6. Estimation of Iron-Reducing Capacity

The iron-reducing capacity of leachate could be estimated by comparing the amount of ferrous iron in leachate between plots. For instance, the iron-reducing capacity of A plot was considered to be equal to the difference in the amount of ferrous iron in leachate between A and B plot. The effect of cation exchange resin or aluminum hydroxide on the iron-reducing capacity of leachate was calculated by substracting the amount of ferrous iron in leachate of C or F plot from that of D plot. The iron-reducing capacity of organic substances adsorbed with aluminum hydroxide was estimated in the following way. To 2 ml of 10 ppm ferric

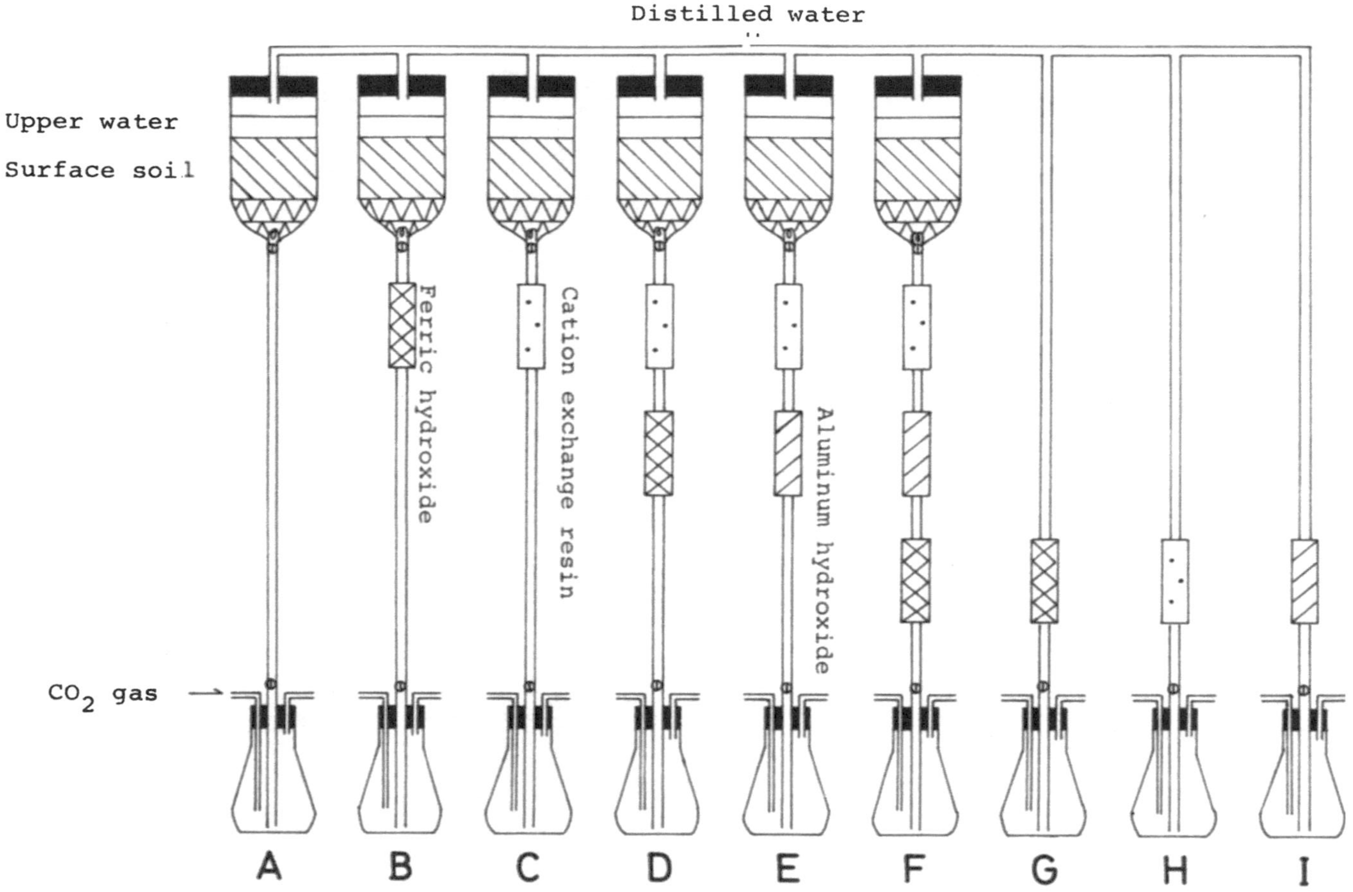

Distilled water

Upper water

Surface soil

CO_2 gas

Aluminum hydroxide

Cation exchange resin

Ferric hydroxide

A   B   C   D   E   F   G   H   I

Fig. 1   Experiment apparatus for percolation experiment.

chloride solution was added 1 ml of 2.5 % o-phenanthroline solution and pH 4.7 sodium acetate - acetic acid solution followed by an aliquot of 0.1 N hydrochloric acid extract from aluminum hydroxide. The amount of ferric iron reduction was regarded as an index of the iron-reducing capacity of the organic substances.

7. Isolation and Identification of Aldehydes

Into 50 ml of saturated solution of 2,4-dinitrophenylhydrazine in 2 N HCl solution, 300 ml of distillate obtained by steam-distillating an appropriate sample was directly dropped and the mixture was allowed to stand at a room temperature for 24 hr. Then precipitate formed collected and vacuum-evapolated to dryness. The precipitate was dissolved in $CH_3COOC_2H_5$ or CH_3OH and analyzed by a gas chromatograph. Shimadzu GC 6A gas chromatograph equipped with a flame ionization detector, packed with OV-17 (1.5 %) on Chromosorb W 80 to 100 mesh, was used. Mass spectrometric analysis was performed on LKB 9000 connected with a gas chromatograph.

RESULTS AND DISCUSSION

1. The Iron-Reducing Capacity of Leachate

Changes in ferrous iron and organic carbon in leachate from the submerged soil with time is shown in Fig. 2. The amount of ferrous iron in leachate of the two soils was high in the early stage of incubation and gradually decreased afterward. Content of ferrous iron in leachate of Nyuzen soil was much lower than that of Nagano soil.

The amount of organic carbon in leachate of these soils was also high in the early stage of incubation and remarkably decreased with increase in the period of incubation.

The iron-reducing capacity of leachate of A plot was much higher for Nyuzen soil than for Nagano soil, as shown in Fig. 3.

The addition of ferric hydroxide to iron-poor Nyuzen soil was drastically decreased the iron-reducing capacity of leachate. On the contrary, the addition of easily decomposable organic substances; rotted manure, rice straw, starch and casin, remarkably increased the iron-reducing capacity. These effects of amendments on the iron-reducing capacity were much more evident in Nyuzen soil than in Nagano soil. From these results it seems reasonable to presume that the iron-reducing capacity of leachate principally depends on content of easily decomposable organic substances and easily reducible iron in the soil, and that ferric iron is reacted with leachate and reduced non-biologically. Accordingly, a soil poor in easily reducible iron is liable to accumulation of reducing organic substances under submerged condition.

The dissolution of ferric iron from ferric hydroxide column was remarkable at the upper part, weak at the middle part and slight at the lower part (10). This result also indicated that non-biological reduction of ferric hydroxide could be brought about by leachate containing reducing organic substances and that most of reducing organic substances was exhausted at the upper part of ferric hydroxide column.

Fig. 3 also shows that the iron-reducing capacity of leachate could not be altered with a Ca type of cation exchange resin. On the contrary,

238

it was almost completely eliminated with aluminum hydroxide. Accordingly, they indicated that reducing organic substances was not cationic substances, but had functional groups combined easily with aluminum hydroxide.

2. Fractionation of Reducing Organic Substances in Leachate

Organic substances (8.8-66.0 mgC/1) in leachate from Nyuzen soil under submerged condition was fractionationed according to molecular weight: molecular weight smaller than 500 (2.1-38.3 mgC/1), 500-10.000 (1.5-7.9 mgC/1), 10.000-100.000 (3.2-31.5 mgC/1), and larger than 100.000 (0.1-16.4 mgC/1), as shown in Fig. 4. Total amount of organic carbon in each fraction was 32.8, 11.2, 33.0 and 23.0 % of sum of total organic carbon for 21 days of incubation. All molecular weight fractions, however, could reduce ferric iron. This indicated that reducing organic substances in leachate was quite heterogenous with respect to molecular weight.

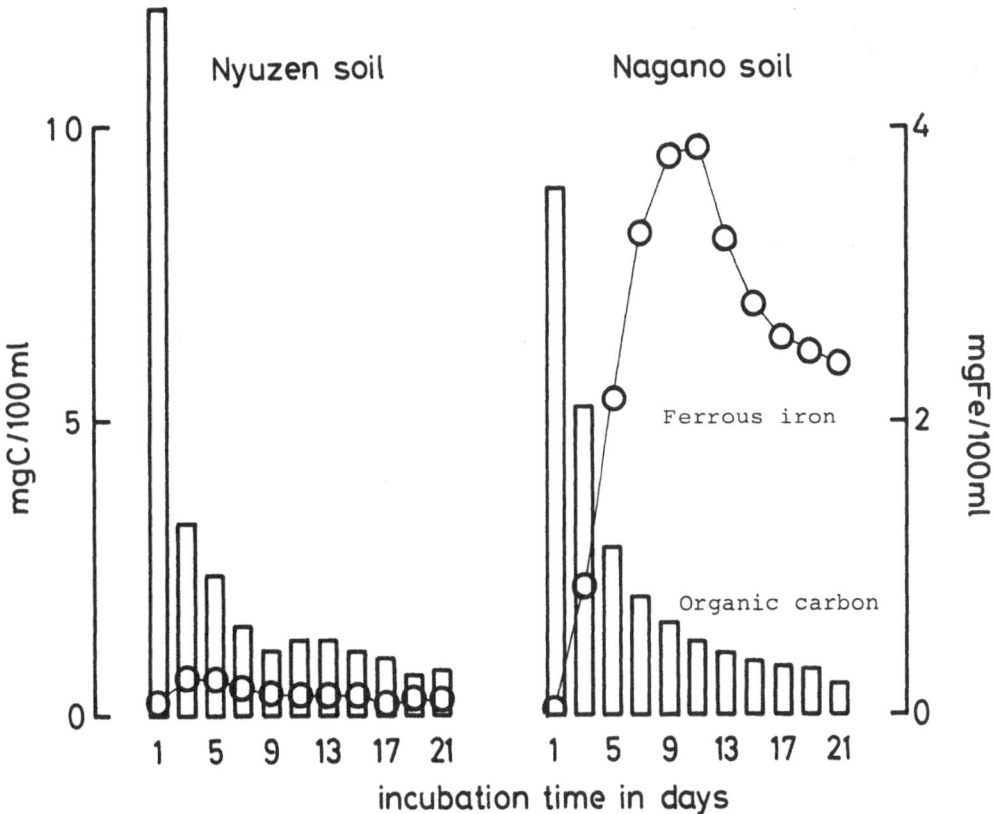

Fig. 2   The amount of ferrous iron and organic carbon in leachate

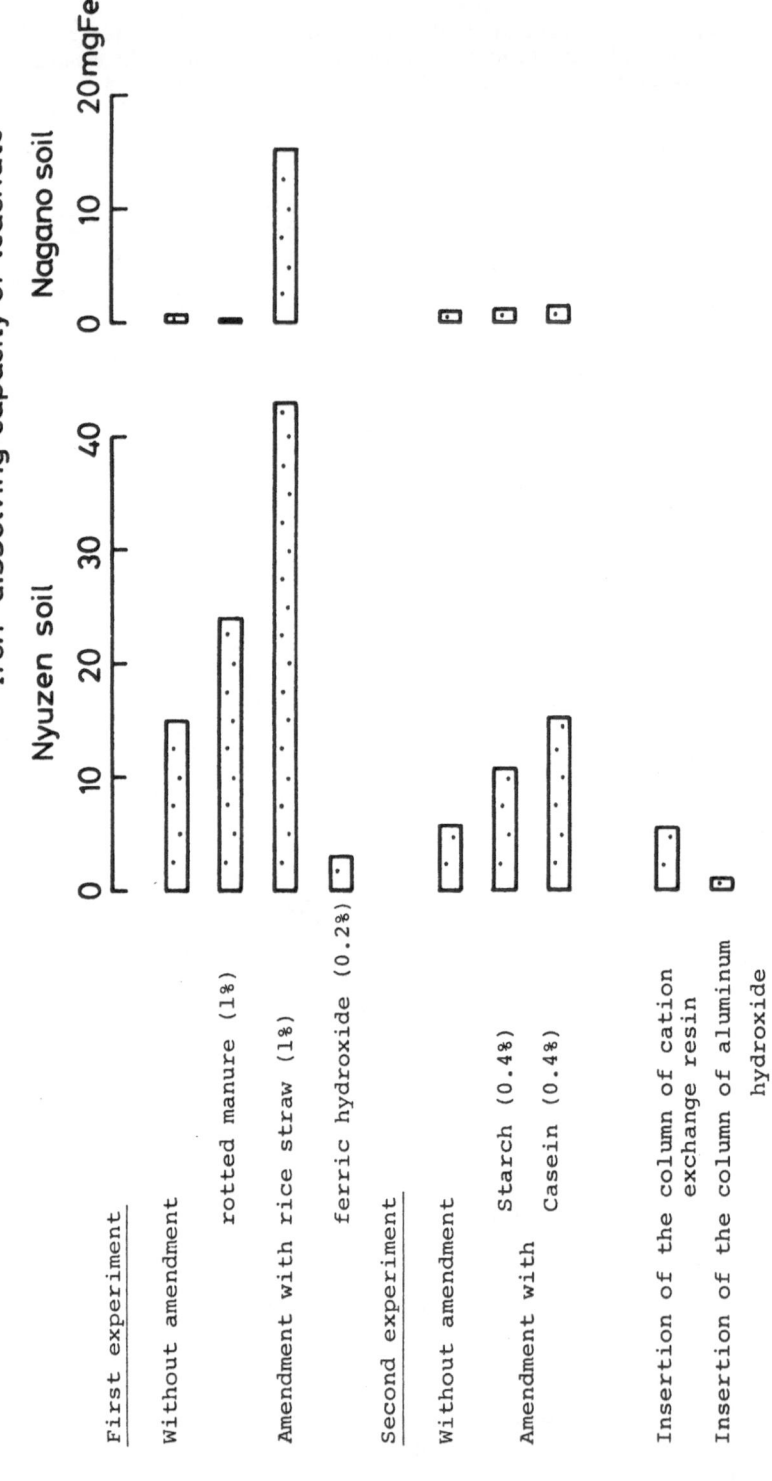

Fig. 3   Iron reducing capacity of leachate

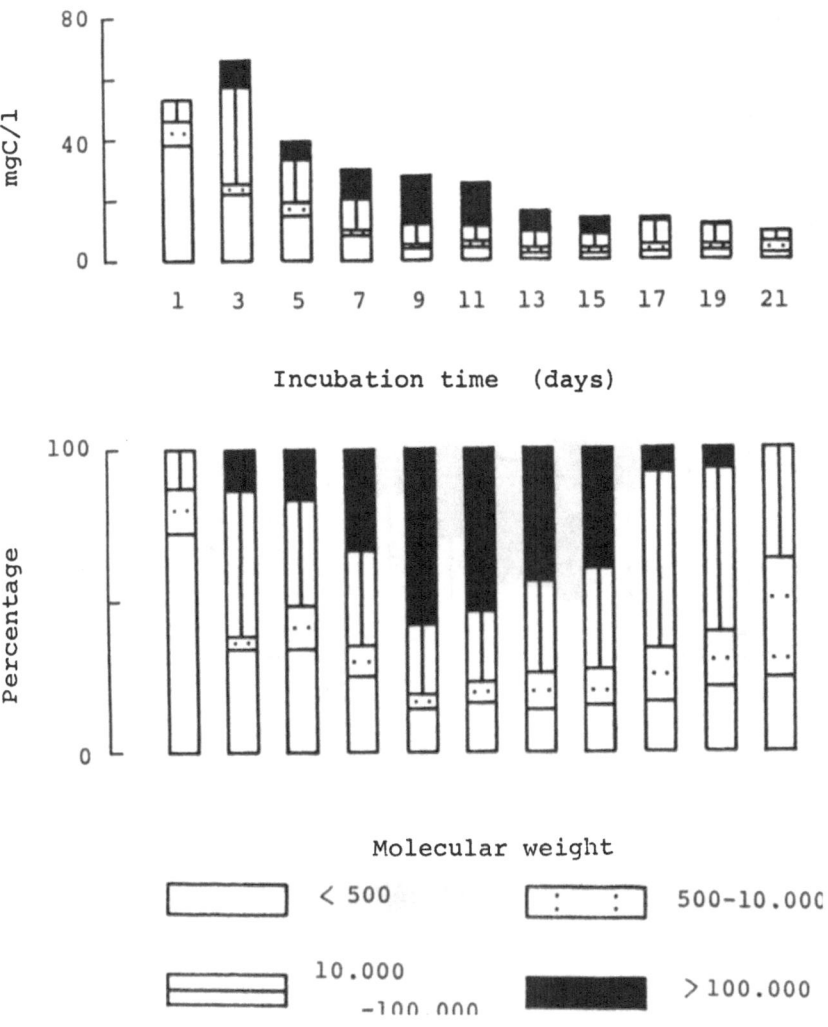

Fig. 4   Changes in molecular weight distributions of

organic carbon in leachate   from   Nyuzen soil

In addition, organic substances adsorbed with aluminum hydroxide, which was responsible for iron-reduction, was still quite heterogenous in regard to molecular weight (Fig. 5).   These results suggested that reducing organic substances was not a chemical compounds with a definite molecular weight but a group of compounds of similar nature with wide range of molecular weight.

As reducing organic substances is efficiently adsorbed with aluminum hydroxide and some of reducing organic substances are low molecular weight compounds, isolation and identification of reducing organic substances with low molecular weight which is extracted with 0.1 N hydrochloric acid from aluminum hydroxide column was conducted.

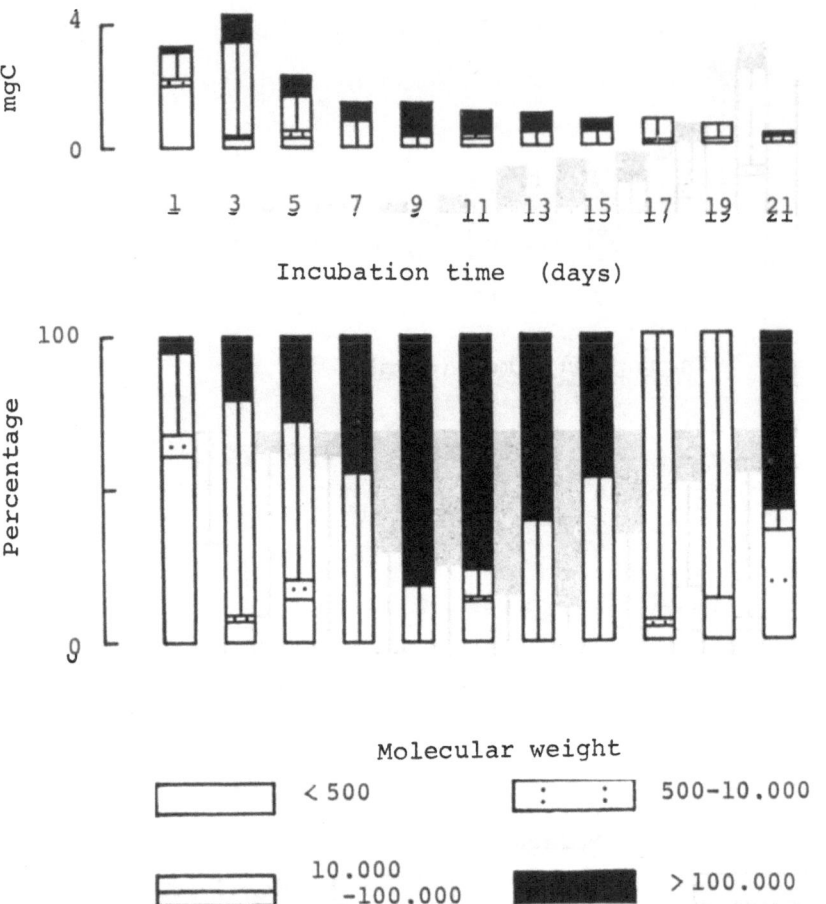

Fig. 5   Changes in molecular weight distribution of organic
substances adsorbed with aluminum hydroxide (Nyuzen soil)

3. Isolation and Identification of Reducing Organic Substances

Organic substances extracted with 0.1 N HCl solution from aluminum hydroxide column represented 33.5 % (Nyuzen soil) and 22.8 % (Nagano soil) of total amount of organic substances adsorbed with aluminum hydroxide (Table 2). The iron-reducing capacities of both acid extract and steam-distillate were much higher in Nyuzen soil than Nagano soil.

Acid extract and steam-distillate readily and quantitatively reacted with 2,4-dinitrophenylhydrazine and 3-methyl-2-benzothiazolonehydrazone (11). Fig. 6 shows the absorption spectra of MBTH complexes of reducing organic substances and formaldehyde. They were characterized by a doublet at approximately 635 and 670 nm. This clearly indicated that acid extract contained aldehydes.

2,4-Dinitrophenylhydrazone derivatives formed were analyzed by using a gas chromatograph combined with a gas chromatograph. Fig. 7 shows the gas chromatograms of 2,4-DNPH derivatives of steam-distillate. All peaks except for peak 1 were composed of two peaks of respective isomers. Identical peaks were recognized in both Nyuzen and Nagano soil.

Mass spectra of 2,4-DNPH derivatives as shown in Fig. 8 were compared with that of authentic 2,4-DNPH derivatives. Five aliphatic aldehydes; formaldehyde, acetaldehyde, propanal, n-butanal and n-pentanal, were identified by these mass spectra. These aldehydes had the ability to reduce ferric iron. These results suggested that percolating water from the submerged soil contained large amount of aliphatic aldehydes and that reducing organic substances in percolating water might be aldehydes with varying molecular weight.

Absolute and relative amount of aliphatic aldehydes identified in leachate are summarized in Table 3. Both of them were different between Nyuzen and Nagano soil. However, it is not yet clearly understood whether such differences in the amount and composition of aliphatic aldehydes with low molecular weight between Nyuzen and Nagano soil have any implication for the fact that the iron-reducing capacity of percolating water is higher for Nyuzen soil than for Nagano soil.

At any events, it may be concluded that percolating water from the submerged soil contained aldehydes which are responsible for the dissolution and removal of Fe and Mn from subsurface horizons of excessively eluviated lowland rice soil as well as gray brown lowland rice soil to smaller extent.

A pedogenic process in lowland rice soil, which the authors proposed beforehand, is schematically shown in Fig. 9. On the bases of the above-mentioned experiments, it can be said that indirect reduction of iron oxide is more important in pedogenic process in paddy field with low water table, because reducing substances can be transferred into subsurface horizons where ferric iron is reduced and removed non-biologically, result-ing in the formation of A2g horizon (excessively eluviated lowland rice soil) and gray colored channel or ped cutans (gray brown lowland rice soil).

ACKNOWLEDGEMENTS

The authors are grateful to Dr. Ryunosuke Hamada, Tokyo University of Agriculture and Technology, for his correction of this manuscript.

Table 1 Chamical properties of soil samples

Soil	Texture	pH H_2O	Org-C %	T-N %	C/N	CEC	Free iron %	Easily reducible Mn mg/100g
Nyuzen (Degraded lowland rice soil)	SL	5.32	1.34	0.12	10.7	6.8	0.29	0.005
Nagano (Gray brown lowland rice soil)	CL	5.26	1.87	0.18	10.1	19.0	1.13	0.008

Table 2 The amount of organic carbon and the iron-reducing capacity of organic substances adsorbed with aluminum hydroxide

Soil	Organic carbon mgC/100 g			Iron-reducing capacity mgFe/100 g	
	(1)	(2)	(3)	(2)	(4)
Nyuzen	22.64	7.58	1.35	1.93	0.22
Nagano	40.25	9.18	0.99	0.98	0.10

(1) Organic carbon adsorbed with aluminum hydroxide
(2) Acid extract
(3) Total aldehydes (MBTH method)
(4) Steam distillate

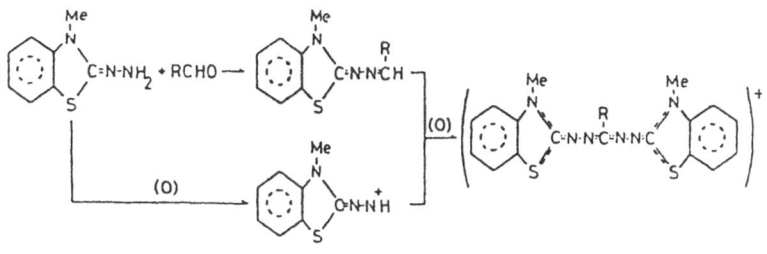

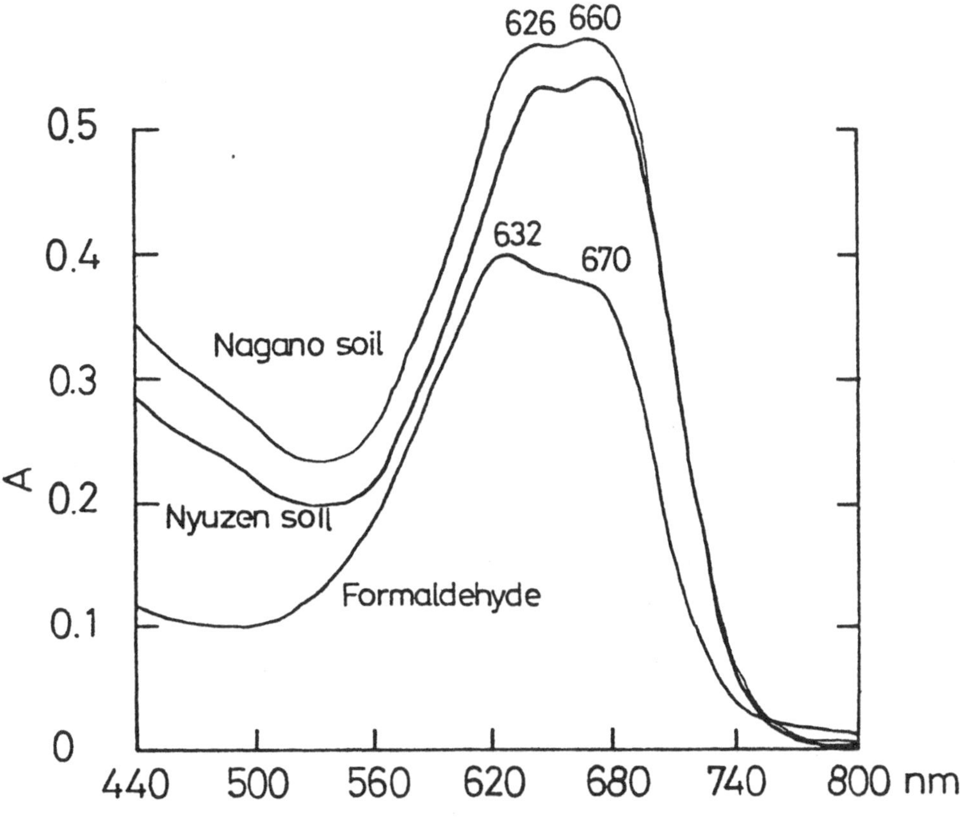

Fig. 6   Absorption spectra of MBTH complexes

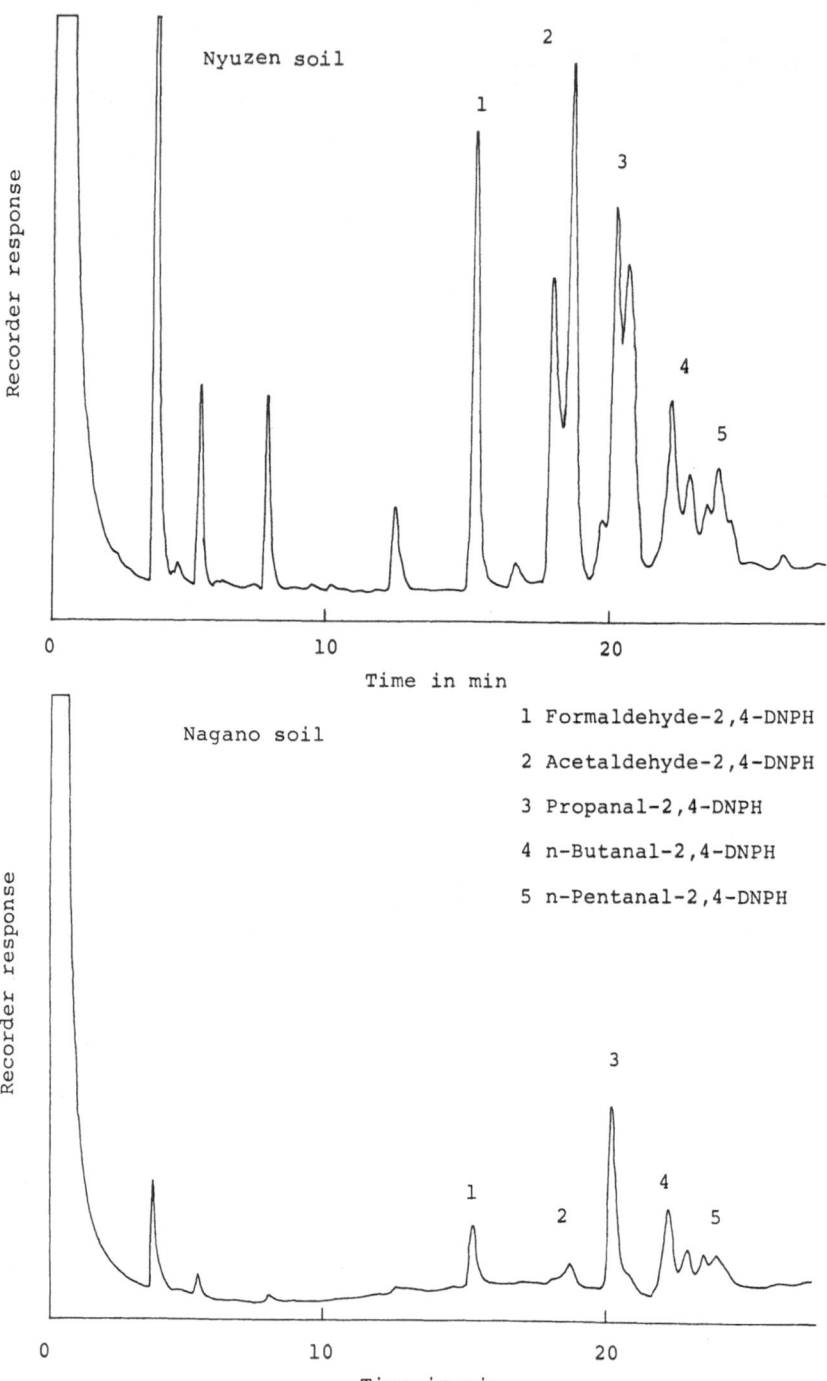

Fig. 7   Gas chromatograms of 2,4-DNPH derivatives

246

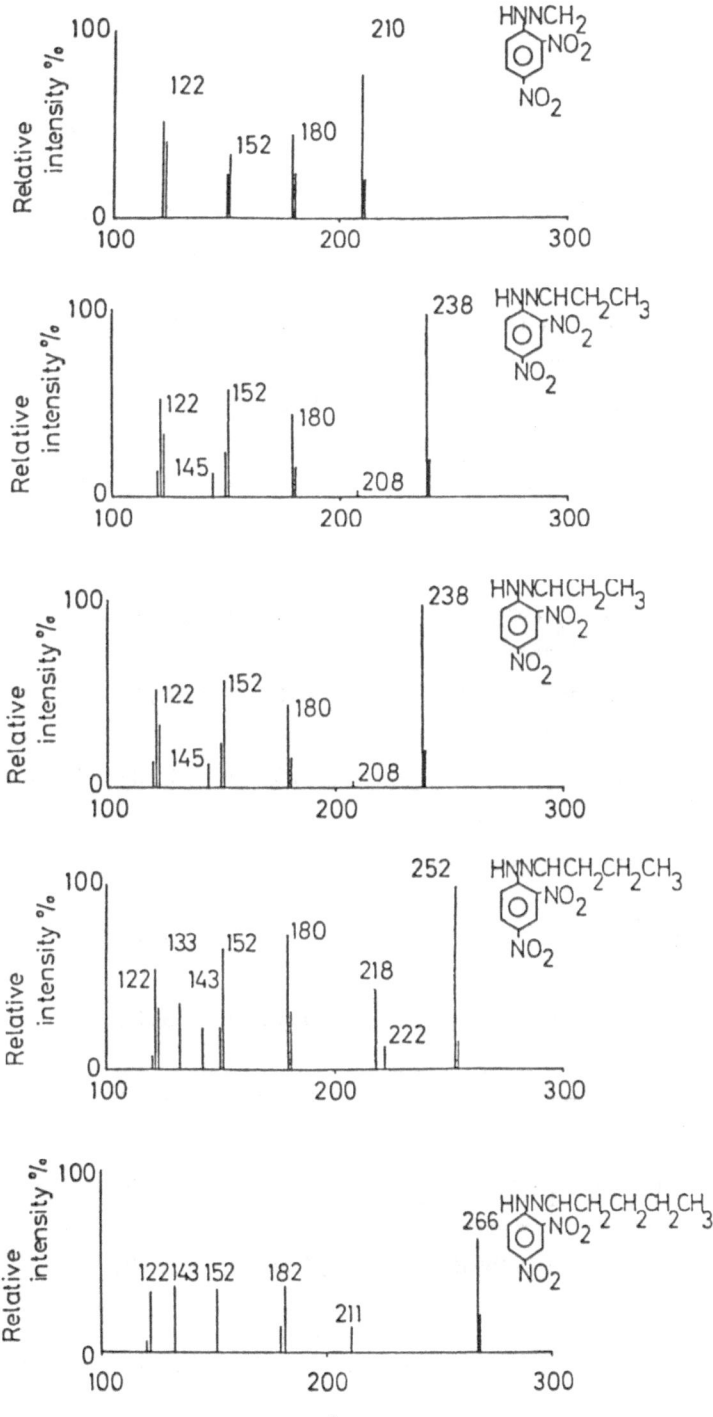

Fig. 8　Mass spectra of 2,4-DNPH derivatives

247

Table 3 Composition of aliphatic aldehydes identified

Soil	2,4-DNPH Derivatives		mgC/100g oven-dry soil
	C_1	Formaldehyde	0.09 (7.0)*
	C_2	Acetaldehyde	0.25 (19.5)
Nyuzen	C_3	Propanal	0.25 (19.5)
	C_4	n-Butanal	0.09 (7.0)
	C_5	n-Pentanal	0.07 (5.5)
	$C_1 + C_2 + C_3 + C_4 + C_5$		0.75 (58.6)
	Total aldehyde (MBTH)		1.28 (100.0)
	C_1	Formaldehyde	0.02 (2.3)
	C_2	Acetaldehyde	0.02 (2.3)
Nagano	C_3	Propanal	0.07 (8.0)
	C_4	n-Butanal	0.05 (5.7)
	C_5	n-Pentanal	0.02 (2.3)
	$C_1 + C_2 + C_3 + C_4 + C_5$		0.18 (20.7)
	Total aldehyde (MBTH)		0.87 (100.0)

* Percentages of aldehyde in total aldehydes (MBTH method)

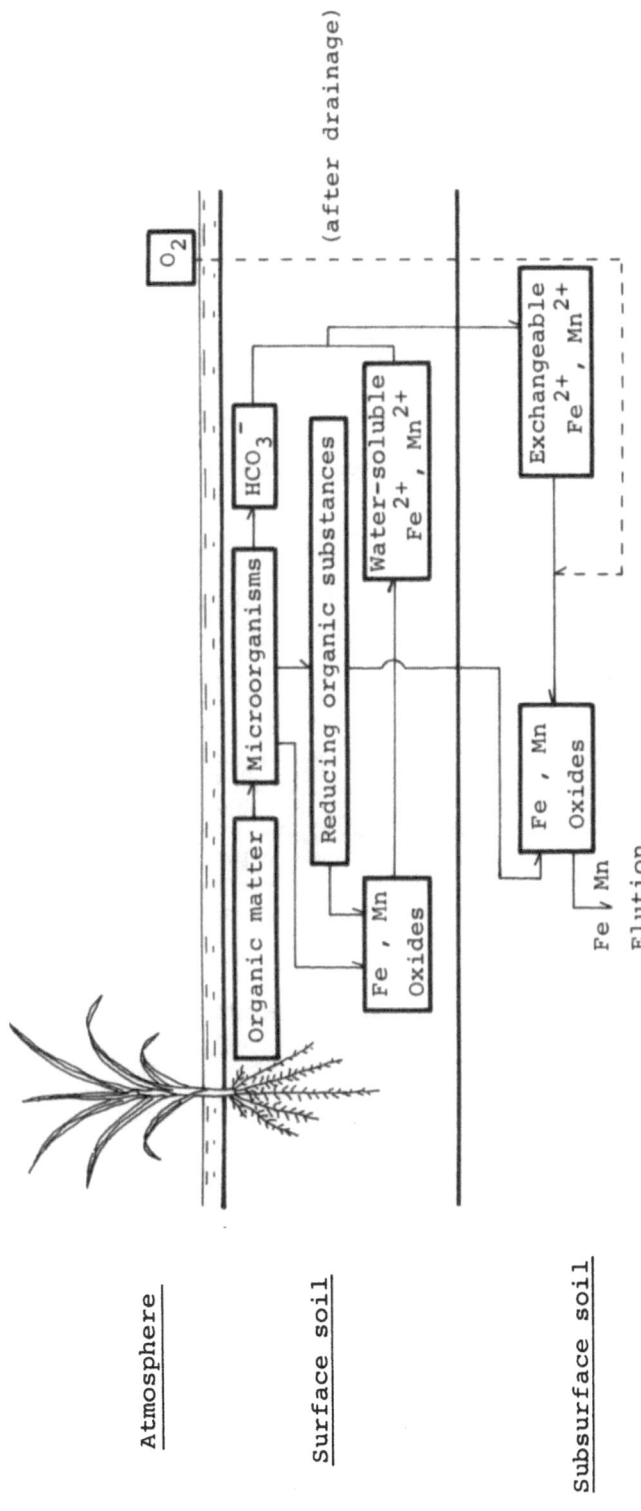

Fig. 9   Pedogenic process in lowland rice soil

249

REFERENCES

(1) WADA, H., YOSHIDA, H. and TAKAI, Y. (1971) Cutans developed in subsurface horizons of clayey paddy soils (Part 1), J. Sci. Soil Manure,
Jpn., 42, 12-17 (in Japanese)

(2) WADA, H., YOSHIDA, H. and MATSUMOTO, S. (1971) Cutans developed in
subsurface horizons of clayey paddy soils (Part 2), J. Sci. Soil
Manure, Jpn., 42, 65-68 (in Japanese)

(3) HAMMANN, R. and OTTOW, J. C. G. (1974) Reductive dissolution of
Fe_2O_3 by Saccharolytic Clostridia and Bacillus Polymyxa under
anaerobic conditions, Z. Pfl. Bodenkd., 137, 108-115

(4) STARKEY, R. L. and HALVORSON, H.O. (1927) Studies on the transformation of iron in nature, II, Concerning the importance of
microorganisms in the solution and precipitation of iron, Soil Sci.,
24, 381-402

(5) ROBERTS, J. L. (1947) Reduction of ferric hydroxide by strain of
bacillus polymyxa, Soil Sci., 63, 135-140

(6) KUMADA, K. (1948) Studies on the rhizosphere of rice seedling (Part
2), On the oxygen consuming power of paddy soils, J. Sci. Soil
Manure, Jpn., 19, 124-128 (in Japanese)

(7) IRI, M. (1961) Studies on edaphological properties of wet paddy soils,
J. Niigata Prefectural Agr. Sta., No 12, 1-51 (in Japanese)

(8) OGURA, N. (1974) Molecular weight fractionation of dissolved organic
matter in coastal seawater by ultrafiltration, Marine Biology, 24,
305-312

(9) MENZEL, D. W. and VACCARO, R. E. (1964) The measurement of dissolved
organic and particulate carbon in sea water, Limnol. Oceanogr., 9,
138-142

(10) WADA, H., OKAZAKI, M. and TAKAI, Y. (1975) Dissolution of ferric
hydroxide with the leachate from the water-logged soil (Part 1),
Roles of water-soluble substances in dynamics of paddy soil, J.
Sci. Soil Manure, Jpn., 46, 201-209 (in Japanese)

(11) SAWICKI, E., HAUSER, T. R., STANLEY, T. W. and ELBERT, W. (1961)
The 3-methyl-2-benzothiazolonehydrazone test, sensitive new methods
for the detection, rapid estimation, and determination of aliphatic
aldehydes, Analy. Chem., 33, 93-96

VOLTAMMETRIC DETERMINATION OF REDUCING SUBSTANCES IN PADDY SOILS

Ding Chang-pu, Liu Zhi-guang
(Institute of Soil Science, Academia Sinica, Nanjing)

For the quantitative determination of reducing substances, potentiometric titration and chemical methods have been employed(1-3). However, these methods are tedious in operation. And, what is more important, the reducing substances may undergo some changes during the determination, with the result that the obtained data do not represent the actual situation in the field. Therefore, we tried to apply the voltammetric method, using a graphite rod as the polarized electrode(4,5), to determine the reducing substances directly in situ.

CHOICE OF APPLIED VOLTAGE

The reducing substances can produce an anodic current at the graphite electrode. The half-wave potentials are different for various reducing substances. Thus, it is possible to distinguish them by controlling the applied voltage.

It is seen from the current-voltage curves of Fig. 1 that there are two plateau regions at 0.3-0.4 V and 0.6-0.7 V vs. N Ag-AgCl electrode respectively. A supplementary experiment shown in Fig. 2 indicates that these correspond to the voltages at which all of the ferrous and manganous ions are oxidized respectively in a solution of pH 6-7. Therefore, we tentatively choose 0.35 and 0.7 V as the applied voltages. The diffusion current at 0.35 V represents ferrous ions and some easily oxidizable organic substances, and that at 0.7 V includes manganous ions and some difficultly oxidizable organic substances. In practical applications the current density at 0.7 V is used as a relative measure for estimating the amount of reducing substances in soils.

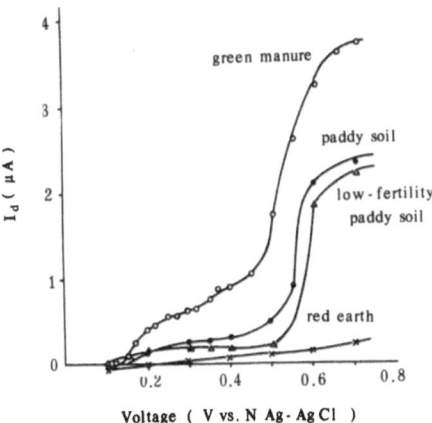

Fig. 1 I-V curves for different soils

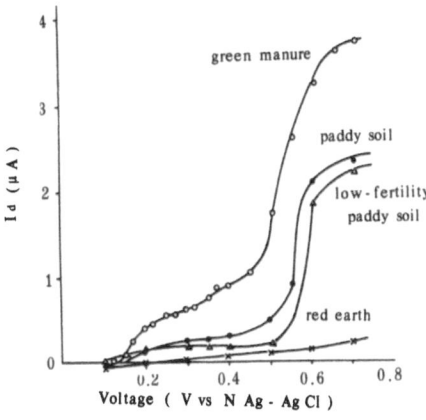

Fig. 2 I-V curves of Fe^{2+} and Mn^{2+} at different pH

TIME OF READING

It is seen from Fig. 3 that the change in reading after 3 minutes is not great. Therefore, a 3-minute reading is taken.

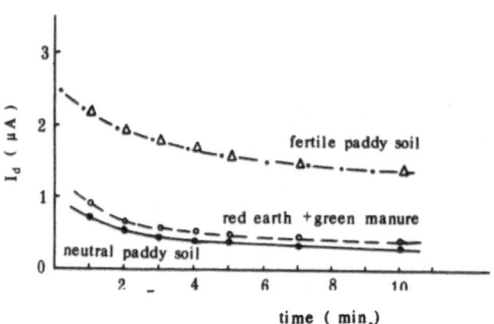

Fig. 3   Current readings at different time

PRECISION

Data shown in Table 1 indicate that the current reading is proportional to the concentration of solution. The relative errors among different determinations are within the range of 6%.

Table 1   Relationship between Id and concentration*

Solution	Concentration	Id	
		$(\mu A/cm.2)$	Ratio per unit conc.
MnSO$_4$	1×10^{-4}M	14.5	0.97
	2×10^{-4}M	28.7	0.97
	3×10^{-4}M	44.8	1.00
Decomposition products of green manure	1:4 dilution	1.96	0.98
	1:2 dilution	3.92	0.98
	Original solution	8.01	1.00
Soil + green manure	1:4 dilution	1.87	1.06
	1:2 dilution	3.56	1.01
	Original solution	7.03	1.00

*Supporting electrolyte: 0.5 M NH$_4$-acetate

SOME EXAMPLES OF PRACTICAL APPLICATIONS

In Table 2 are shown some of the results of field determinations. It is seen that the stronger the reduction in soil (the lower the redox potential), the larger is the amount of reducing substances as expressed by current density.

Table 2 Amount of reducing substances in some soils

Soil type	Redox status	Eh (mV)	I_d (μA/cm^2)	
			0.35(V)	0.7 (V)
Young reddish	Oxid.	450	0.08	0.77
paddy soil	Mod.red	110	1.83	12.5
	Str.red.	20	7.03	20.2
Reddish	Weakly red.	400	0.39	3.11
paddy soil	Mod.red.	150	1.41	5.27
	Str.red.	−10	11.7	16.3
Alluvial	Oxid.	560	0.06	0.21
paddy soil	Weakly red.	370	0.37	4.38
	Mod.red.	150	4.47	9.34
Mud	Ordinary	−	1.23	4.73
	With rusty water	−	8.99	15.2
Dry-farming soil		450	−	1.16
Forest soil		440	−	1.53

CONCLUSION

The features of the proposed method are: (1) The measured value represents the quantity of the reducing substances rather than their relative intensities; (2) It can be used for measurements in situ; (3) It is simple and rapid in operation. We are therefore of the opinion that it is a useful method for characterizing the oxidation-reduction status of soils.

REFERENCES

(1) Yu Tian-ren (ed.), Electrochemical Properties of Soils and Their Research Methods, Chapter 11. Science Press, Beijing (1976). (in Chinese)
(2) Yu Tian-ren, Liu Wan-lan, Studies on oxidation-reduction processes in paddy soils. III. Effect of redox conditions on the growth of rice plants. Acta Pedologica Sinica, 5, 292-304 (1957). (in Chinese with English summary)

(3) Flaig, W., Schrrer, K., Judel, G. K., Determination of redox potential of soil profiles. Z. Pflanzenernahr. Düng. Bodenkd. 68, 203–218 (1955). (in German)

(4) Beilby, A. L., Brooks, W. Jr., Lawrence, G. L., Comparison of purolytic carbon film electrode with wax-impregnated graphite electrodes. Anal. Chem., 36, 22–26 (1964).

(5) Elving, P. J., Smith, D. L., The graphite electrode, an improved technique for voltammetry and chromopotentiometry, Anal. Chem., 32, 1849–1854 (1960).

CHEMICAL EQUILIBRIA OF SULFIDES AS STUDIED
WITH A HYDROGEN SULFIDE SENSOR

Pan Shu-zhen
(Institute of Soil Science, Academia Sinica, Nanjing)

Owing to its theoretical and practical significance, the subject of sulfides in paddy soils has attracted much attention to many soil scientists[1,2] For the determination of sulfides in soils, chemical methods are generally used. But with these methods it is unavoidable to disturb the chemical equilibria originally existing in the soil during extraction and the subsequented termination, and it is also not possible to distinguish the molecular hydrogen sulfide from the ionic HS^- and S^{2-}. In recent years there have appeared two papers pertaining to the study of sulfide in submerged soils with a sulfide-selective electrode[3,4]. But it is necessary to make some assumptions in converting the measured sulfide ion into hydrogen sulfide, and thus the obtained results are still indirect.

In 1973 Ross[5] mentioned the principle of the hydrogen sulfide sensor. But until now no such practical sensor is available, nor is research work on natural systems with such sensor.

In the present work, we tried to make a study on the chemical equilibria of sulfides in paddy soils, with a self-made hydrogen sulfide sensor, in conjunction with the use of a sulfide-selective electrode.

EFFECT OF pH

In the equilibria controlling the solubility of ferrous sulfide and the dissociation of hydrogen sulfide, the pH of the solution should play a dominant role. In Fig. 1 are shown the results with a neutral paddy soil regulated to different pH. It is to be noted that:
(A) the calculated pH_2S from pS^{2-} value agrees fairly well with the actually measured pH_2S, with a difference of less than 0.5 unit. Inasmuch as the two sets of measurements are based on different principles and techniques, the results should be considered reliable. (B) The dependence of pS^{2-} on pH follows a nearly straight line, with a change of 1.2 unit of pS^{2-} for one unit of pH change. (C) In the pH range of 4-8 the pH_2S also changes nearly linearly with the change in pH, with a slope of about 1. (D) At a pH of 5 the pH_2S may attain a value of 5.7, if organic matter is added to the soil and submerged.

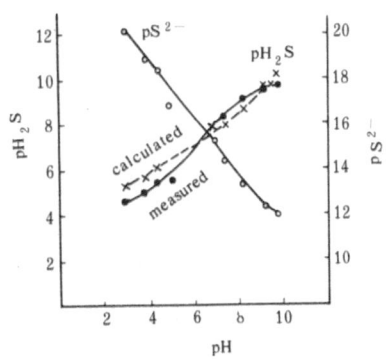

Fig. 1 pH_2S and pS^{2-} as a function of pH for a neutral alluvial paddy soil plus Astragalus

EFFECT OF FERROUS IRON

In order to verify the effect of ferrous iron on the chemical equilibria of sulfides, a sandy paddy soil was incubated after the addition of organic matter and potassium sulfate and then different amounts of ferrous iron were added. From the results shown in Fig. 2 it is apparent that the addition of ferrous iron results in the decrease in the amount of soluble hydrogen sulfide.

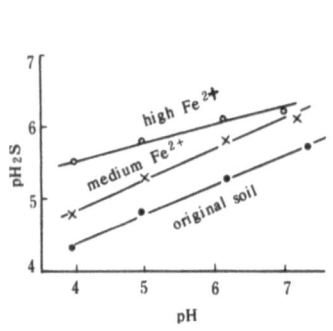

Fig. 2 Effect of Fe^{2+} on pH_2S for a sandy paddy soil (plus K_2SO_4 and O.M.) medium Fe^{2+}: 4 m.e. $Fe^{2+}/100g$ high Fe^{2+}: 8 m.e.

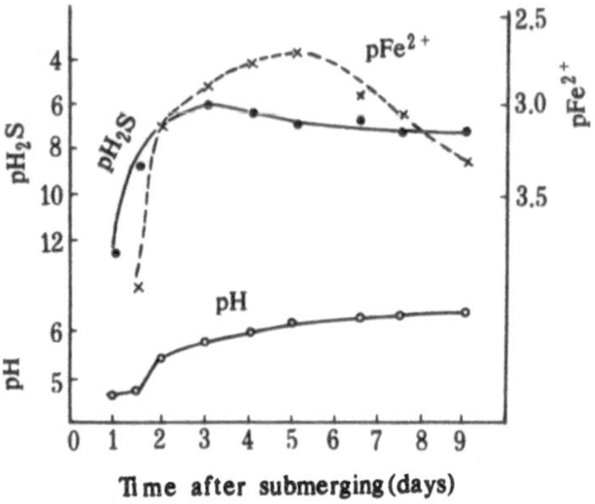

Fig. 3 Dynamics of H_2S and Fe^{2+} in an acid sulfate soil ($35^{\circ}C$)

DYNAMICS

Fig. 3 shows the dynamics of H_2S formation in an acid sulfate soil. It is seen that there is a peak in H_2S concentration after 3-4 days of submergence where the concentration of Fe^{2+} and pH are still not high. It is also worthy of note that the H_2S concentration may maintain a rather high level even in the presence of large amount of water-soluble ferrous iron. The fall in hydrogen sulfide concentration at the latter period is probably due to the precipitation as ferrous sulfied, coupled by the increase in pH.

QUANTITATIVE RELATIONSHIP BETWEEN S^{2-} AND Fe^{2+}

If the concentration of hydrogen sulfide in paddy soils is governed by the association of sulfide ion with hydrogen ion, which is in turn governed by the solubility of ferrous sulfide, the product $(Fe^{2+})(S^{2-})$ will maintain a constant value corresponding to the solubility product, pK=18.4. Some calculations for three paddy soils are shown in Table 1. In the Table neglect of corrections for activity caused by ionic strength and complex formation should introduce an error not exceeding one order of magnitude at most. Thus, the dependence of $pFe^{2+} + pS^{2-}$ on pH, amounting to about 4.5 unit in the pH range of 4-7, is real. Apparently other chemical equilibria, in addition to $FeS \rightleftharpoons Fe^{2+} + S^{2-}$, are involved in governing the soluble ferrous iron and sulfides in paddy soils.

Table 1 Relationship between pH_2S and pFe^{2+}
at different pH (medium
amount of Fe^{2+}added)

pH	Soil no.	pFe^{2+}	pH_2S	pS^{2-}	$pFe^{2+} + pS^{2-}$	
						Mean
	1	0.82	4.8	17.8	18.6⎫	
3.9-4.1	2	1.04	5.3	17.9	18.9⎬	18.8
	3	1.07	5.1	17.9	19.0⎭	
	1	0.86	5.3	16.2	17.1⎫	
4.9-5.2	2	1.11	5.5	16.1	17.2⎬	17.2
	3	1.07	5.3	16.2	17.3⎭	
	1	0.92	5.8	14.5	15.4⎫	
6.1-6.4	2	1.85	5.7	13.9	15.7⎬	15.5
	3	1.30	5.5	14.1	15.4⎭	
7.2	1	1.45	6.1	12.7	14.2	14.2

* Calculated from pH_2S and pH

CONCLUSION

The hydrogen sulfide sensor affords a new tool for the investigation
of the chemical equilibria of sulfides in paddy soils. This tool enables
us to make further studies in the field pertaining to the dynamics of
hydrogen sulfide, which is important to practical agronomists.

REFERENCES

(1) Yu Tian-ren, Liu Zhi-guang, Oxidation-reduction processes in paddy
    soils and their relationship with the growth of rice plant. Acta
    Pedologica Sinica, 12,380-389 (1964). (in Chinese with English summary)
(2) Mitsui, S., Inoganic Nutrition, Fertilization and Soil Amelioration for
    Lowland Rice. Yokendo, Tokyo (1954).
(3) Allam, A. I., Sulfide determination in submerged soils with an ion-
    selective electrode. Soil Sci., 114, 456-467 (1972).
(4) Ayotade, K. A., Kinetics and reactions of hydrogen sulfide in solution
    of flooded rice soil. Plant & Soil, 46, 381-389 (1977).
(5) Ross, J. W., Raseman, J. H., Krueger, J. A., Potentiometric gas sensing
    electrodes. Pure Appl. Chem., 36, 473-489 (1973).

DETERMINATION OF REDOX POTENTIAL (Eh)
BY DEPOLARIZING CURVE METHOD

Liu Zhi-guang
(Institute of Soil Science, Academia Sinica, Nanjing)

Redox potential as an index of the intensity factor of the oxidation-reduction status of a soil has long been widely used in soil research. But in recent years it has been known that the true equilibrium potential is very difficult to be determined rapidly(1). Although attempts such as selection of electrode materials and treatment of electrode surface have been made, the problem has not been solved as yet. Therefore, it is nacessary to search for other measuring ways, if an accuracy of several milivolts is required.

Yu Tian-ren and Li Sung-hua(2) in 1957 applied the depolarizing curve method used by plant physiologists(3) to characterize the redox status of paddy soils. In the present work, based on the same principle, the author made an attempt to determine the Eh in paddy soils rapidly and accurately.

EQUIPMENT

The improved measuring equipment is shown in Fig. 1. For convenience the polarizing voltage was obtained directly from an adjustable constant voltage source. A large-area silver-silver chloride electrode was used as the auxiliary electrode during polarization. A saturated calomel electrode was used as the reference electrode for Eh measurements. The working electrode was a platinum electrode.

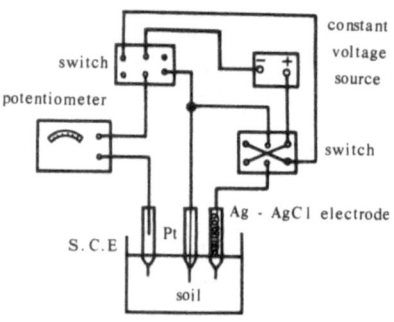

Fig. 1 Equipment for measuring
the depolarization curve

SELECTION OF POLARIZING CONDITIONS

In order to ascertain the measuring conditions, some experiments on polarizing voltage and time, and number of polarizing cycles were made. It was found that the higher the polarizing voltage, the longer the polarizing time, and the more the number of polarizing cycle, the longer the depolarizing time is required. For the sake of unity in operation, and also considering the Eh range commonly encountered in paddy soils, we usually polarized the platinum electrode for 1 minute under a polarizing voltage of $\pm$ 600 mV. In cases where the Eh of the soil is very high or very low, a polarizing voltage of $\pm$ 800 mV should be used.

PROCEDURE

Adjust the polarizing voltage to 600 mV. Connect the positive terminal to the platinum electrode in the soil and polarize for 1 minute. Switch off the polarizing circuit, connect the platinum electrode and the saturated calomel electrode to a potentiometer, and measure the change of the potential during depolarization for 10 minutes. Measure the depolarization curve after cathodic polarization in the same way.

DEPOLARIZING CURVES

In summarizing the results for more than 100 samples, the depolarizing curves may be divided into three classes:

Class 1. The potential 10 minutes after anodic polarization is nearer to the equilibrium potential than that after cathodic polarization. The equilibrium potentials of this class are mostly within the range of 350–650 mV. It may be considered as the oxidation state of soils.

Class 2. The potentials after anodic or cathodic polarization are both near to the equilibrium potential. The equilibrium potentials of this class are in the range of 0–200 mV, reflecting the ordinary reduction state of soils.

Class 3. The difference between the potential after anodic polarization and the equilibrium potential is rather large, but the potential after cathodic polarization shows a comparatively small difference from the equilibrium potential. This class is generally encountered when the Eh of the soil is negative in sign, reflecting the strongly reducing state of soils.

EQUILIBRIUM Eh VALUE FOUND BY EXTRAPOLATION

Depolarization can not be proceeded to the end within a time interval of 10 minutes. Reasoning theoretically, if the electrode is completely depolarized, the two depolarizing curves should intersect each other and have an intersection point just conforming to the equilibrium Eh value. It was observed that within a certain time range there was a semi-logarithmic relationship between the measured Eh value and the depolarizing time. Therefore, the Eh value in mV was plotted against the logarithm of time in second, and the point of intersection of the two straight lines was taken as the equilibrium Eh value. Three examples are shown in Fig. 2. A comparison between the extrapolated Eh values and the actually measured equilibrium Eh values is given in Table 1.

259

Table 1 Comparison of equilibrium Eh obtained by extrapolation
with those actually measured

Redox status	Sample No.	Eh (mV)		
		Measured* (A)	By extrapolation (B)	(A) – (B)
Strongly reducing	8	–123	–112	–11
	32	– 85	– 79	– 6
	4	– 42	– 35	– 7
	18	– 20	– 11	– 9
			mean	8
Reducing	11	5	0	5
	10	32	31	1
	7	35	36	–1
	21	46	46	0
	20	66	69	3
	23	78	78	0
	33	90	92	–2
	3	133	130	3
	19	133	131	2
	26	136	129	7
	24	172	177	–5
			mean	3
Oxidizing	12	380	372	8
	31	448	437	11
	27	554	550	4
	30	615	624	–9
	29	640	632	8
			mean	8

*Steady reading after 48 hours

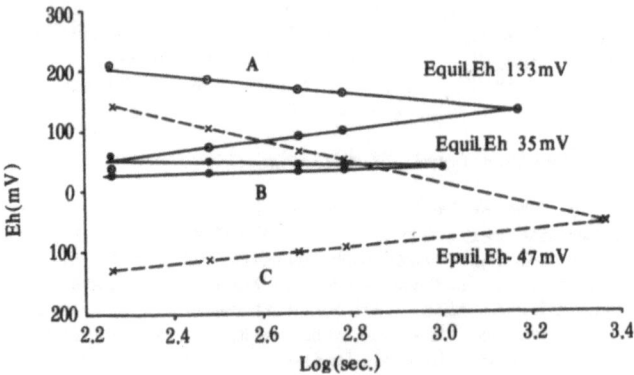

Fig. 2 Eh determined by extrapolation

It is seen from the table that with this method it is possible to get the sought Eh value of the soil with a difference of generally not more than 10 mV from the equilibrium potential, and the needed measuring time is considerably shortened.

CONCLUSION

In conclusion, it is considered that although the mechanism of polarization and depolarization at the platinum electrode is not clear yet, from the practical point of view the method of depolarization may be used advantageously for the rapid measurement of oxidation-reduction potential of soils, if an accuracy of 10 mV is required.

REFERENCES

(1) Yu Tian-ren, Zhang Xiao-nian, Electrochemical Methods and Their Applications in Soil Research, Chapter 11. Science Prese, Beijing (1980). (in Chinese)
(2) Yu Tian-ren, Li Sung-hua, Studies on oxidation-reduction processes in paddy soils. I. Conditions affecting redox potential. Acta Pedologica Sinica, 5,97-110 (1957). (in Chinese with Russian summary)
(3) Gerrel, I. A., Effect of Azotobacter on oxidation-reduction potential of plant tissue. Reports of Academy of Sciences of USSR, 78, 1041-1043 (1951). (in Russian)

A PRELIMINARY STUDY ON PROPERTIES OF ORGANO-MINERAL COMPLEX OF PADDY SOILS IN TAIHU LAKE REGION

Fu Ji-ping, Zhang Jing-shen
(Institute of Soil Science, Academia Sinica, Nanjing)

INTRODUCTION

Soil Structure is an important factor in determining soil fertility, and is closely related to the quantity and properties of the organo-mineral complex of the soil.

The separation of soil particles without modifying the organic and inorganic components is a major difficulty in the study of natural soil organo-mineral complexes. Ultrasonic vibration has been used for dispersion and fractionation of complexes by many workers[1-6].

Because the complexed and non-complexed organic materials usually exist in soil together, the non-complexed or free organic material should be removed by means of densimetric methods[7-10]. The material removed in the light fraction is supposed to be undecomposed or partly humified residues of plants and other soil organisms with the remaining organic material being found chiefly in clay-organic complexes[11-13]. Such a method is used to fractionate organo-mineral complex in order to make sure of the characteristics of paddy soil in different fertility. The purpose of the present work is to study the properties of organo-mineral complex of some representative paddy soils in the Taihu Lake region in relation to soil structure.

Three representative types of paddy soil in the Taihu Lake region were used for experiment: the permeable paddy soil, bleached paddy soil and gleyed paddy soil. For the purpose of comparison the cultivated horizon of a high-fertility soil and a low-fertility soil with about the same clay content for each type was sampled. The clay minerals are predominantly illite and montmorillonite, with a small amount of vermiculite and transitional minerals.

METHODS

1. Destruction Percentage of Structure and Degree of Dispersion

Soil clods about 0.5 cm in size were soaked for 24 hours, and then wet-sieved. The destruction percentage of structure was calculated as weight of <0.25 mm aggregates/weight of sample x 100. The degree of dispersion was obtained from the percentage of micro-aggregates with diameter less than 2 micron divided by the percentage of clay.

2. Separation and Fractionation of Organo-Mineral Complex

The soil was dispersed with ultrasonic vibrator without dispersing agent for 20 minutes, and then fractionated for complexes of different size.

3. Determination of Organo-Mineral Complexation

The heavy fraction of the complex was separated with heavy liquid (mixed solutions of HgI_2 and KI with specific gravity of 2.0). We called the percentage of organic carbon of this fraction in the total carbon of the whole soil as the degree of organo-mineral complexation (DC). The degree of additional complexation (DAC) after application of organic manure was obtained from the equation:

$$DC = \frac{HW \times HC}{SW \times SC} \times 100$$

$$DAC = \frac{(MHW \times MHC) - (HW \times HC)}{SW \times (MSC - SC)} \times 100$$

Where:

H: Heavy fraction

S: Soil (check or original)

M: Manured soil

C: Organic carbon

W: Weight of sample

4. Characterization of Combining Form of Humus

The combining forms of humus were characterized according to their ease of extraction. The humus was extracted with (A) 0.1 N NaOH solution, (B) 0.1 N NaOH + 0.1 M $Na_4P_2O_7$ solution and (C), same as (B) with ultrasonic vibration. The unextracted residual portion (D) was regarded as most tightly combined with the mineral part.

RESULTS AND DISCUSSION

1. Quantity and Quality of Micro-Aggregates

Table 1 shows that paddy soils with high fertility are characterised by comparatively high content of water-stable micro-aggregates in the range of 1–0.5 and 0.5–0.25 mm in size, while most low-fertility soils have a high content of micro-aggregates less than 0.25 mm in size. This seems to imply that the water-stability of aggregates in low-fertility soils is relatively poor, as is evidenced by the high destruction percentage and degree of dispersion shown in Table 2. It is seen from the Table that except for the permeable type of paddy soil, the destruction percentage of the low-fertility soil is higher by 53–95% than the corresponding high-fertility soil. The degree of dispersion after soaking for 24 hours and shaking for 5 minutes is also higher for the low-fertility soil.

2. Organic Matter in Aggregates

It was found that about 80–90% of the soil organic matter were combined with the clay to form organo-mineral complex, with a correlation coefficient of 0.969 between the quantity of complexed organic matter and the organic matter of the whole soil. The organic matter is chiefly concentrated in the aggregates. Table 3 shows that the organic matter content of the micro-aggregates is significantly correlated with that of the whole soil. The

Table 1 Content of water-stable aggregates in paddy soil

Soil type	Fertility	Water-stable aggregate (%)			
		1-0.5mm	0.5-0.25mm	1-0.25mm	< 0.25mm
Bleached	High	14.5	9.5	24.0	46.4
	Low	5.7	5.2	10.9	78.3
Permeable	High	15.4	8.7	24.1	33.6
	Low	11.2	8.0	19.2	33.1
Gleyed	High	13.3	9.6	22.9	42.2
	Low	11.4	9.5	20.9	64.1
Low-yield Gleyed	High	14.6	9.1	23.7	40.9
	Low	11.3	8.7	20.0	64.6

Table 2 Destruction percentage and degree of dispersion of the structure of paddy soil

Soil type	Fertility	Destruction percentage (%)			Degree of dispersion (%)		
		0	1 min.	5 min.	0	1 min.	5 min.
Bleached	High	33.1	45.6	63.9	3.5	6.1	10.9
	Low	64.7	80.2	87.4	7.3	10.5	14.1
Permeable	High	21.5	35.6	51.9	1.7	5.1	7.7
	Low	18.2	35.7	56.7	2.0	5.2	9.1
Gleyed	High	29.2	49.2	61.9	3.8	6.3	11.7
	Low	45.0	59.1	75.7	5.0	8.1	12.0
Low-yield Gleyed	High	33.2	48.1	66.0	0.3	1.4	4.5
	Low	50.8	63.0	71.1	4.6	5.6	8.9

correlation coefficients for the three fractions of the micro-aggregates were statistically significant at 1% level.

The organic carbon content decreases with the decrease in size of the water-stable micro-aggregate, and is higher for high-fertility soils than for low-fertility soils.

Table 4 shows that the degree of organo-mineral complexation differs from soil to soil, and is affected by agricultural practice such as cultivation and manuring. The degree of additional complexation of organic matter increases with the fertility of the soil.

Table 3   Carbon content of water-stable micro-aggregates

Soil type	Fertility	Organic carbon (%)			
		Original soil	1-0.5 (mm)	0.5-0.25 (mm)	0.25-0.06 (mm)
Bleached	High	1.57	1.94	1.76	1.43
	Low	1.12	1.84	1.52	0.95
Permeable	High	1.82	1.97	1.77	1.77
	Low	1.60	1.76	1.61	1.61
Gleyed	High	2.01	2.60	2.29	2.00
	Low	1.61	2.27	1.93	1.56
Low-yield gleyed	High	1.85	2.48	2.25	1.96
	Low	2.09	3.03	2.76	2.19

Table 4   Degree of organo-mineral complexation of the
soil and micro-aggregates

Soil type	Fertility	Degree of complexation (%)				Degree of additional complexation (%)
		Original soil	1-0.5 (mm)	0.5-0.25 (mm)	0.25-0.06 (mm)	
Bleached	High	84.1	83.5	84.7	86.7	84.4
	Low	83.9	79.9	84.9	89.5	0
Permeable	High	83.5	83.8	87.6	84.2	20.9
	Low	91.9	90.9	90.7	89.4	0
Gleyed	High	84.6	83.5	84.3	83.0	92.5
	Low	82.6	77.5	81.4	85.9	0

The destruction percentage of soil structure is negatively related with the complexed organic matter of the whole soil and also of the micro-aggregates 0.06-0.25 mm in size, with correlation coefficients of -0.869 and -0.834 (n = 6) respectively, but is not correlated with organic matter content of the original soil. This seems to imply that it is chiefly the organic matter complexed with the mineral part that is important to the increase of the water-stability of micro-aggregates.

3. Combining Form of Humus in Heavy Fraction of Complex

Table 5 shows the combining form of humus in the heavy fraction of the organo-mineral complex. It is seen that more than 80% of the soil organic matter are in the heavy fraction. The 0.1 N NaOH- extractable fraction

Table 5 Combining form of humus in heavy fraction
of complex

Soil Type	Ferti- lity	Organic matter (%)			Combined humus				A:D
		Whole soil (W)	Heavy frac- tion (H)	H:W	NaOH extra. (A)	$Na_4P_2O_7$ extra. (B)	Ultra- sonic extra. (C)	Resi- dual (D)	
Blea- ched	High	2.79	2.31	82.8	49.3	4.3	4.8	41.6	1.19
	Low	1.85	1.63	88.1	46.0	3.7	5.5	44.8	1.03
Permea- ble	High	3.13	2.55	81.5	47.0	5.1	2.0	45.9	1.02
	Low	2.75	2.48	90.2	42.3	5.2	4.5	48.0	0.88

Table 6 Apparent cation-exchange capacity of humus of organo-
mineral complex less than 2 microns in size

Soil type	Fertility	Original complex		After H_2O_2- treatment		Apparent C E C of humus (m.e./100g)
		Humus (%)	C E C (m.e./100g)	Humus (%)	C E C (m. e./100g)	
Bleached	High	3.93	53.9	0.65	47.7	189
	Low	3.53	57.5	0.64	49.3	284
Permeable	High	3.68	41.5	0.81	39.3	77
	Low	3.36	45.6	0.73	41.6	152

accounts for 42-49% of the combined humus, and the percentage is slightly
higher for the high-fertility soil than for the low-fertility soil. The
ratio NaOH- extractable /residual is also relatively higher for the high-
fertility soil. Since the NaOH- extractable fraction of humus is the most
active part among the four fractions, it plays an important role in the
formation of the organo-mineral complex of soils.

4. Apparent Cation-exchange Capacity of Humus and
Viscosity of Organo-Mineral Complex

As calculated from the decrease in C E C and in humus content after
treating with H_2O_2, the apparent cation exchange capacity of humus
of the organo-mineral complex less than 2 micron in size, is comparatively
higher for the low-fertility soil (Table 6), and the consumed exchange sites
during the formation of organo-mineral complex are more for the high-fertility soil.

Table 7 shows that the viscosity of the complex is comparatively higher
for high-fertility soils, and decreases after treatment with H_2O_2.

Table 7 Viscosity of the organo-mineral complex
less than 2 microns in size

| Soil type | Fertility | Viscosity (centi-poise at 30°C) | |
		Original	After H_2O_2 -treatment
Bleached	High	0.935	0.852
	Low	0.906	0.834
Permeable	High	0.951	0.870
	Low	0.874	0.838

CONCLUSIONS

(1) Soil organo-mineral complex is the foundation of soil quality and closely influences the soil fertility.

(2) About 80-90% of organic matter in paddy soils are associated with clay fraction to form clay-organo complexes.

(3) The degree of additional complexation of organic matter increases with the fertility of the soil.

(4) The higher relative viscosity of the colloidal complex shows a better aggregation occurring in the fertile soils.

REFERENCES

(1) Emerson, W.W., Determination of the contents of clay-sized particles in soil. J. Soil Sci., 22, 50-59 (1971).
(2) Genrich, D.A., Bremner, J.M., A reevaluation of the ultrasonic-vibration method of dispersing soils. Soil Sci. Soc. Amer. Proc., 36 (6), 944-947 (1972).
(3) North, P.F., Towards an absolute measurement of soil structure stability using ultrasound. J. Soil Sci., 27, 451-459 (1976).
(4) Turchenek, L.W., Oades, J.M., Fractionation of organo-mineral complexes by sedimentation and density techniques. Geoderma,21,311-343 (1979)
(5) Parasher, C.D., Lowe, L.E., Isolation of clay size organo-mineral complexes from soil of the lower fraser valley. Canad. J. Soil Sci., 50 (3), 403-407 (1970).
(6) Watson, J.R., Parson, J. W., Studies of soil organo-mineral fractions. 1. Isolation by ultrasonic dispersion. J. Soil Sci., 25 (1), 1-8 (1974).
(7) Ford, G.W., Greenland, D.J., Oades, J.M., Separation of the light fraction from soils by ultrasonic dispersion in halogenated hydrocarbons containing a surfactant. J. Soil Sci., 20 (2), 291-296 (1969).
(8) Fu Ji-ping, Zhang Shao-de, Chu Jin-hai, The use of heavy liquid to determine the degree of organo-mineral complexation in soil. Turang Feiliao, 4, 40-42 (1978). (in Chinese)

(9) Martin Richter, Ichiro Mizuno, Santiago Aranguez, Susana Uriarte, Densimetric fractionation of soil organo-mineral complexes. J. Soil Sci., 26 (2), 112-123 (1975).

(10) Turchenek, L.M., Oades, J.M., Size and density fractionation of naturally occurring organo-mineral complexes. Trans. 10th Int, Congr, Soil Sci., V. 2, 65-72 (1974)

(11) Fu Ji-ping, Zhang Jing-sheng, Effect of green manure on the properties of colloidal complexes of clayey warp soil. Acta Pedologica Sinica, 15 (1), 83-92 (1978). (in Chinese with English summary)

(12) Greenland, D.J., Interactions between clays and organic compounds in soils. Part I. Mechanisms of interaction between clays and defined organic compounds. Soils Fert., 28 (5), 415-425 (1965).

(13) Schnitzer, M., Khan, S.U., Humic Substances in the Environment. Marcel Dekker Inc., New York (1972).

DETERMINATION OF LIME POTENTIAL OF PADDY
SOILS WITH TWO ION-SELECTIVE ELECTRODES

Wang   Jing-hua
(Institute of Soil Science, Academia Sinica, Nanjing)

Schofield[1] in 1955 introduced the concept of "lime potential", i.e., pH-0.5pCa. Since then, some papers have been published on the factors affecting its determination. For the calculation of pCa the generally used method is to analyse chemically the Ca concentration in the solution and then compute in terms of the activity coefficient. By this method it is inevitable to subject to some errors due to the uncertainty in the calculation of single ion activity, to say nothing of the tediousness in analytical operations. Krupsky[2] determined the pH and pCa with a glass pH electrode and a calcium-selective electrode respectively. But the use of a reference electrode may still cause some errors due to the troublesome liquid-junction potential when the salt bridge is in contact with the charged soil particles.

In the present work, lime potential is determined by insertion of a pH glass electrode and a calcium-selective electrode directly into the soil system.

PRINCIPLE OF THE METHOD

Due to the variability of the ratio between pH and pCa in soil systems, it is not possible to know the pH-0.5pCa of the soil by means of a calibration curve with solutions of known pH and pCa. Other means must be sought for.

According to the Nernst equation:

$$E_{Ca} = E_{Ca}^{o'} + 0.5 \ S \ \log a_{Ca} + E_j \tag{1}$$

$$E_H = E_H^{o'} + S \ \log a_H + E_j \tag{2}$$

$$E_{Ca} - E_H = (E_{Ca}^{o'} - E_H^{o'}) + 0.5 \ S \ \log a_{Ca} - S \ \log a_H$$

$$= (E_{Ca}^{o'} - E_H^{o'}) + S \ (pH - 0.5pCa) \tag{3}$$

$$pH - 0.5pCa = \frac{(E_{Ca} - E_H) - (E_{Ca}^{o'} - E_H^{o'})}{S} \tag{4}$$

Where $E_{Ca} - E_H$ is the potential difference between the calcium electrode and the pH electrode, which can be measured with a potentiometer, S the Nernst slope.

The $E_{Ca}^{o'}$ and $E_H^{o'}$ are the "standard potential" of the calcium electrode and the pH electrode respectively. We found them by determining the potentials in solutions of known pH or pCa, and then extrapolating to pH=0 or pCa=0.

In practical work, pH and pCa of the soil were also determined in the usual manner.

COMPARISON BETWEEN PADDY SOIL AND ITS PARENT SOIL

Here are given three cases:

(A) For a weakly acid loessial soil with pH-0.5pCa of about 4, the lime potential rises by about 1.5 unit when developed into paddy soil. The corresponding pH are 5.7 and 6.9 respectively (Fig. 1)

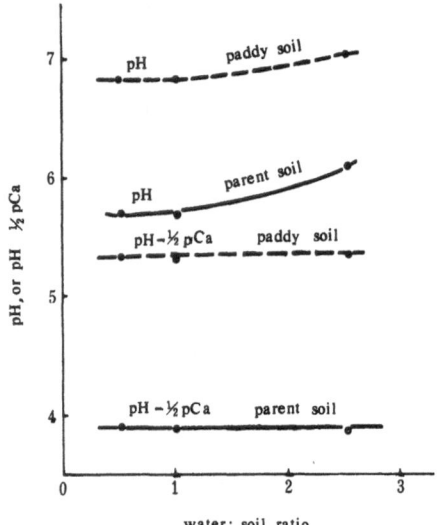

Fig. 1 Comparisons of pH and pH-$\frac{1}{2}$pCa between a loessial paddy soil and its parent soil

(B) For a strongly acid red earth with pH-0.5pCa of about 3, the lime potential rises by about 2 unit after long-term cultivation for rice. The corresponding pH are about 5.2 and 6.7 respectively (Fig. 2).

(C) In Fig. 3 it is shown that for a lateritic soil with pH-0.5pCa of about 2.3, the rise of lime potential is about 1 unit when paddy soil is developed. The difference in pH is small.

Obviously, the change in lime potential after cultivation for rice depends on the lime potential of the original parent soils, the age of paddy soils, and especially such agricultural practice as liming.

IN RELATION TO ADSORPTION AND DESORPTION OF CALCIUM

When a paddy soil is immersed in a $CaCl_2$ solution, adsorption or desorption occurs depending on the concentration of $CaCl_2$ solution and the calcium

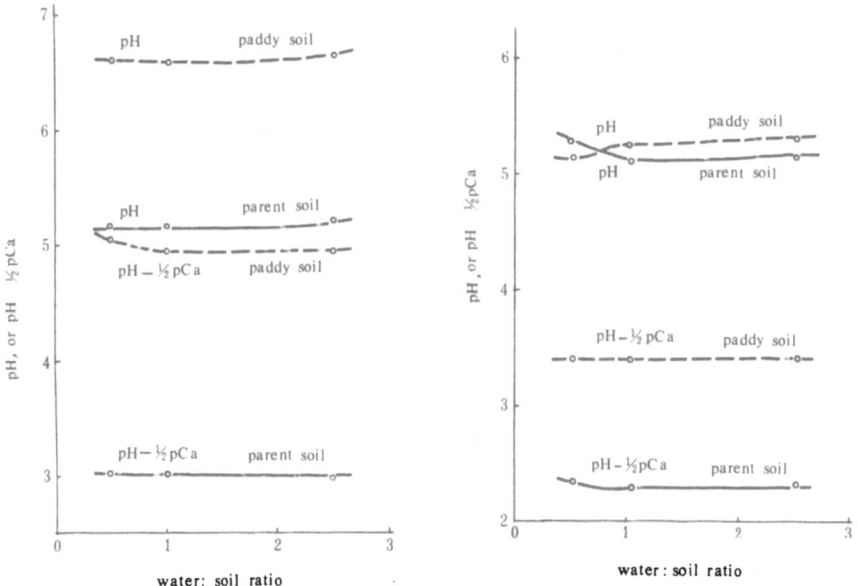

Fig. 3 Comparisons of pH and pH- $\frac{1}{2}$pCa

Fig. 2 Comparisons of pH and pH- $\frac{1}{2}$pCa between a lateritic paddy soil and its
between a reddish paddy soil and its   parent soil derived from granite
parent red earth

status of the soil. At the concentration where there is neither adsorption
nor desorption (Δ pCa=0) the $CaCl_2$ concentration should match the equilibrium solution of the soil in water. Fig. 4 shows that for a loessial
paddy soil with a lime potential of 5.34 and its parent soil with a lime
potential of 3.92, the corresponding zero ΔpCa point are 2.93 and 3.56
respectively.

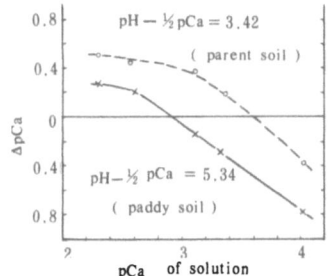

Fig. 4 Adsorption and desorption of calcium by loessial paddy
soil in $CaCl_2$ solution (water:
soil=0.5:1)

271

LIME POTENTIAL AS AN INDEX OF SOIL ACIDITY

From Table 1 it is seen that the $\triangle$(pH–0.5pCa) values between the paddy soil and its parent soil are larger than the $\triangle$pH values. The reason for this is that it reflects the difference in pH plus pCa, instead of pH alone. Thus, we consider that in addition to the theoretical soundness and the technical convenience, the lime potential has the further advantage that it can be used as a more distinct index of the acidity of soils as compared with the conventional pH.

Table 1   Comparison between $\triangle$pH and $\triangle$(pH–0.5pCa)
for paddy soil and its parent soil*

Soil type	pH			pH–0.5pCa		
	Parent soil	Paddy soil	$\triangle$ pH	Parent soil	Paddy soil	$\triangle$(pH–0.5pCa)
Lateritic	5.12	5.23	0.11	2.29	3.40	1.11
Reddish	5.15	6.56	1.41	3.02	4.93	1.91
Loessial	5.71	6.83	1.12	3.91	5.32	1.41

* Water: soil = 1 : 1

REFERENCES

(1) Schofield, R.K., Taylor, A.W., Measurements of the activities of bases in soils. J. Soil Sci., 6,137–146 (1955).
(2) Krupsky, N.K., Alexandrova, A.M., Gubareva, D.N., Electrometric methods for determination of ions activity in soils. X. Electrometric method to determine lime potential in soils. Agrochemistry, 3, 133–138 (1975). (in Russian)

THE AVAILABILITY OF MICRONUTRIENTS IN PADDY SOILS AND ITS
ASSESSMENT BY SOIL ANALYSIS INCLUDING RADIOISOTOPIC TECHNIQUES

K.G. Tiller

(Division of Soils, CSIRO, Glen Osmond, South Australia)

INTRODUCTION

The micronutrients are those elements which are present in small amounts in soils yet essential for the growth of plants. The alternative expression "trace elements" arose from earlier difficulties in analytical procedures which often resulted in analyses being recorded as "trace". "Trace elements" is still a useful description of those elements occurring in small amounts without connotation of their physiological requirement. The micronutrients of importance to plants are currently thought to be Cu, Zn, Fe, Mn, Co, B, Mo, Cl and possible V.[21]

As reviewed by Rahaja et al.[26] field responses of paddy rice to zinc as spray treatments, and copper as soil treatments, were first reported in the early 1950's in India. Since then, zinc deficiency in lowland rice has been widely recognized[37] in many countries and has stimulated much research in all parts of the world but especially in Asia and particularly in the Philippines at the International Rice Research Institute whose annual reports are a valuable source of information on rice nutrition. Zinc deficiency is now a major problem in millions of hectares of wet-land rice especially in Philippines, India, Pakistan, Indonesia and Sri Lanka.[13,14] The soils at risk are commonly neutral to alkaline, unless permanently flooded when pH is less critical, and have low available Zn and high organic matter content.[49] In recent years, Mo deficiency of paddy rice has also been recognized[25] in addition to the established deficiencies of Cu, Zn, Mn and Fe.

In this paper I will stress the micronutrient metals and especially Zn since it is now considered the main limitation to rice production after N and P.

ASSESSMENT OF AVAILABILITY AND GENERAL CONSIDERATIONS

The problem of discussing availability of soil nutrients to plants and its assessment arise from two sources, firstly, the complex chemistry of flooded soils and, secondly, from the limitations of the different concepts of availability.

Chemical reactions

The reactions of micronutrients in soils may be simply represented as shown in Figure 1, with examples of the kind of complexes which may form in solution and their relative stability, given in Table 1.

In this model, the equilibrium between different ionic species in solution and thus between solid and solution phases is determined by the complexing ligands present and their concentration. These data illustrate the important role of pH through the possible formation of hydroxy species (e.g. $ZnOH^+$) and through the effectiveness of organic ligands as determined by their respective acid dissociation constants. The role of organic ligands in controlling metal complexing in soil solution is now well established and partially explains the demonstrated importance of organic matter in soils.

The relevance of metal compounds or high affinity adsorption in controlling micronutrient solubility in soils is a continuing source of debate in soil chemistry. To a degree, the problem is more one of philosophy or of definition, than one of chemistry. One should, perhaps, admit the possibility of micronutrients being associated with the solid phase both in strongly adsorbed forms and as compounds. Probably the most thorough investigation of the possibilities of Zn compounds in flooded soils was carried out by the International Rice Research Institute[11] who concluded that zinc silicates and sulphides may occur in many soils under the appropriate conditions. Other compounds such as Zn phosphates were thought to occur transitorily. On the other hand the solubility of Zn in flooded soils following the peak of biological activity, may be controlled by a pH dominated, high affinity sorption reaction as proposed by Tiller et al.[38] for non-flooded soils. This view was adopted because the low concentrations of Zn in the soil solution of some Australian flooded rice soils were very similar to those found in Zn-poor upland soils[2] in which Zn sulphides are unlikely and zinc silicate is too soluble to account for Zn solubility.[22] In practice it will remain difficult to resolve between those two mechanisms because of assumptions which have to be made about the degree of complexing in soil solution at very low Zn concentrations or alternatively because of the difficulty of determining ionic species at low concentrations. Total concentrations of Cu and Zn in some Australian flooded soils stabilized at values of 5µg/L or less (Table 2) following the early post-submergence peaks of concentration.[2] However the resolution of the relative importance of different mechanisms controlling solubility in flooded soils may be only of academic interest because of the influence of many factors which determine the uptake of zinc by rice.

These factors which determine micronutrient uptake by rice have been considered in many detailed studies and reviews and are related to the following,[9,10,11,12,13,14,20,23,24].

(i) redox potential, which may determine concentrations of Fe and Mn in solution and thus may influence uptake of Zn by nutrient antagonism, (ii) pH changes, (iii) variation in organic matter status, which during anaerobic decomposition can generate organic acids and bicarbonate ions, both of which interfere with the absorption of Zn by rice, (iv) phosphate levels, which can increase the severity of Zn deficiency and (v) management decisions concerning crop rotations and fertilizer practice which control organic matter, zinc inputs, nutrient interactions, cultivar chosen and water policy especially with respect to continuity of submergence. Detailed discussions of these factors are well treated in the reviews quoted and are not discussed here.

Concepts of availability

Although the term availability is widely used to describe the ability of soils to supply the nutrient requirements of plants, it lacks an accepted, exact definition and is sometimes used in different ways by different research groups. Earlier, availability or available nutrient was often considered in terms of a "solubility" which determined plant uptake and which hopefully could be rated by different chemical extractants. A subsequent development was the concept of discrete but related chemical components of

Table 1 Formation constants of some micronutrient complexes

Ligand (L_i^{n-})	Approx. log K_1, $\mu=0$		
	Cu	Zn	Mn
Cl-[1]	0.5	0.5	0
OH-[1]	6.5	5.0	3.5
$SO_4=$[1]	2.5	2.5	2.5
acetate[1]	2.0	1.5	1.0
citrate[1]	7.5	3.0	2.0
valate[1]	8.5	5.0	3.0
fulvate[2]	8.5	2.5	4.0
non-dialysable[3] organic fraction	5.5	4.5	no data

[1] Data from Sillen and Martell[32,33]
[2] Data from Schnitzer and Skinner[29,30]
[3] Separated from soil solution of a soil A horizon with chemical treatment[7]
μ = ionic strength

Table 2 Range in metal concentrations in soil solutions of six flooded Australian rice soils ($\mu g/ml$)

Metal	Maximum values found during 24 week's flooding	Values after flooding for 14 weeks
Zn	0.05 - 3.0	0.001 - 0.015
Mn	5 - 25	2 - 8
Fe	0.2 - 9.0	<0.05 - 1
Cu	not determined	<0.001 - 0.003
Cd	"	<.00001 - 0.0002
Pb	"	<0.002

Data from solutions sampled from potted soils in a glasshouse experiment (Beckwith, Tiller and Suwadji, unpublished data)

275

the soil, called chemical pools,[46] thought to be in quasi-equilibrium with each other. The various pools, such as water-soluble, weakly adsorbed, strongly adsorbed etc provided a better basis of interpretation of different chemical procedures in terms of our knowledge of soil chemistry and also provided an improved framework of interpretation in relation to short- and longer term needs of crops. Problems still remained with respect to the quantification and prediction of nutrient uptake by the plant, and especially rates of supply by the soil.

The complementary approach which stresses the overall chemical inter-actions between solid and solution phases of soils, now has fairly wide acceptance, especially since it provides an input into equations which pre-dict diffusion rates in soils. This approach, often referred to as the Q-I (quantity-intensity) approach, is especially appropriate to many micronu-trients because of their low concentrations and often high affinity for soil surfaces. Some means of quantifying this affinity (Q/I) through factors related to the reaction equilibrium constant such as distribution coeffic-ients or adsorption isotherm slope are important for such nutrients.

For these reasons the prediction of micronutrient availability based on the total amount of available nutrient (Q) in the soil will have severe limitations and could be completely without success unless the tests are restricted to soils in which the chemical affinity factor is effectively constant. The constancy of chemical affinity for the nutrient in question is sometimes achieved by restricting the tests and the derived critical values to defined soil types or landscape units. Good correlations may also be achieved on varied soil types by manipulation of the key soil factors which control the chemical equilibrium, e.g. organic matter status, redox poten-tial, liming to a fixed pH, and which may be achieved regionally by common management decisions. The "success" of different procedures to predict micronutrient uptake by plants or yield response to fertilizers, will be often related to the samples chosen in relation to the breadth of soil and management factors, rather than the superiority or otherwise of the chemical procedure chosen. The likelihood that success will be low when using soil test procedures which remove most of the available nutrient, or lower still if total nutrient content is recovered (e.g. by HF), is exemplified by some Australian experience. Some soil groups which are most closely identified with a particular micronutrient deficiency, e.g. Co deficiency in animals on palehumults*, Zn deficiency in cereals on pellusterts*, and Mo deficiency in legumes on some palexeralfs*, may have relatively high levels of the micro-nutrients concerned.[44] Thus a negative correlation between "available" micronutrient and uptake by plants may be found. Good correlation between soil tests which remove much of the total labile nutrient, and plant uptake or fertilizer response, will be hardest to achieve in soils with greatest affinity for the nutrient.

Measurement of the equilibrium concentration in solution (I) presents several problems. In non-flooded soils the separation of the soil solution is sufficiently difficult on a routine basis that compromise procedures based on neutral salt solutions are often substituted. Sometimes the analysis of the low concentrations of micronutrients is also difficult. It is indeed fortuitous that the agricultural importance of zinc is matched by its partic-ularly high sensitivity by atomic absorption spectrophotometry! Since the amount of micronutrient in soil solution may well only represent the short-term requirement of growing plants, the rate of solution or desorption be-comes a critical factor, yet is not a component of normal test procedures.

---

*Soil descriptions after "Soil Taxonomy" (Soil Survey Staff 1975).

Although many investigations of the prediction of availability have the equilibrium model as their guiding philosophy, there are experimental limitations to its use on a routine basis. In practice, many investigators have compromised by choosing extractants which evaluate the total amount of labile nutrient (or a value correlated closely with it) and have attempted to simulate values equivalent to an intensity variable e.g. soil solution concentration, by the mathematical use of modifying parameters such as pH, organic matter etc which may relate to the affinity of the nutrient for soil surfaces. Alternatively, the tests have been restricted to soils with similar affinity, as discussed earlier. Some of the most successful examples of soil tests fall into the latter category. In the case of rice soils in particular, an additional restriction may also needed to be added, viz the large increase in interfering ions which appear in solution on flooding e.g. Fe^{II}, Mn^{II} and HCO_3^-, which tend to obscure the predictability of soil tests for Zn and Cu.[27]

SOIL TESTS FOR PADDY SOILS:EXPERIENCES AND PROBLEMS

Several considerations could influence one's attitude towards the development of soil tests for the assessment of micronutrient availability in paddy soils. These are the overall effect of submergence on micronutrient uptake by rice plants, the changes in the total amount (solid and solution phase) of labile micronutrient resulting from flooding, and the variation in concentration of micronutrients in soil solution with time and soil properties. Ultimately, one's confidence in soil tests as a factor in farm management decisions must as always depend on field experiments which determine yield variation in relation to added micronutrient fertilizers.

Effect of flooding on micronutrient uptake by rice plants

When a soil is flooded, various soil and management factors (summarized earlier) interact to influence the availability of micronutrients to rice. It is important to note the tendency of alkaline soils to adjust to lower pH values and for acidic soils to become more alkaline on flooding. These effects may have a major influence on metal solubility but interactions arising from enhanced concentrations of organic acids and bicarbonate, manganous and ferrous ions could possibly have overriding effects. Commonly, flooding leads to reduced Zn availability[11] but increased uptake of Zn on both acidic and alkaline soils[5,8] has been reported: Beckwith et al.[2] found either increased, decreased or no change in Zn uptake depending on the soil. However for all acidic soils, the increase in uptake of Zn and Cu was usually relatively less than the change in dry matter production (Table 3), that is, concentration in the rice plant decreased. In two of the most acidic soils, Cu uptake was appreciably increased by flooding during the first 6 weeks but not thereafter. The greatest increases in uptake on flooding, relative to dry matter production, occurred for Mn especially, and Zn and Cd in those soils which were initially alkaline.

Changes in availability on flooding are related to the pH of the soil before flooding (that is, the dry soil) but in a poorly defined way with the possibility of positive or negative effects near neutrality. Although such effects would seem to militate against the use of air-dry soil samples in tests for predicting micronutrient availability in rice soils, successful applications of soil testing (reviewed later) encourage this approach.

Variation of soil solution concentrations with duration of submergence

The application of soil testing for the prediction of zinc deficiency, in particular, has to contend with the observation[11] that the symptoms of Zn deficiency often have their maximum expression when Zn concentrations in solution are highest and that the disappearance of symptoms can coincide with decreasing Zn concentrations in solution. The reported rapid decrease in Zn concentration from a maximum during the first week or so after flooding has been established by IRRI[9,11] but is illustrated here (Figure 2) using data from our laboratory for some representative Australian rice soils (Beckwith, Tiller Suwadji, unpublished data). The peaks for Mn and Fe concentrations in soil solution develope later than zinc, the rates of solubility being dependent on temperature, organic matter and other factors[24]. The relative kinetics of the appearance and disappearance of water-soluble Zn, Mn and Fe stress the need for caution in attempting to simulate Zn deficiency conditions by soil tests based on water-soluble metals.

Variation of trends in soil solution concentration as a result of changes in soil characteristics

Studies at IRRI have established the influence of soil properties such as pH and the levels of organic matter on the function of Zn solubility in rice soils. The studies of Beckwith et al.[2] confirmed the importance of organic matter status of flooded rice soil but proposed the need to consider the kind of organic matter involved. Their investigations were based on paired soils which differed mainly with respect to position in the rotation sequences sometimes used in rice culture in the rice-growing areas of N.S.W. (Australia). Soils sampled immediately following 2 or 3 years of cereal crops including rice, were compared with nearby soils (within 50 metres) which had several years of legume pasture following the cereal-growing phase. The 'pasture' soils always had a higher content of organic matter. Appreciable differences in the trends of soluble metals were found in soils of different management history . Data of the Leeton soil is given as an example in (Figure 3). Added straw generally increased Fe and Mn concentrations and decreased Zn concentrations as predicted from published reports. Concentrations of Fe and Zn in solution were appreciably increased in the 'pasture' soil of higher organic matter content, compared with the 'cereal' soil; the reverse was true for Mn.

Both the added straw, and the increased organic matter following the pasture cycle, were found to affect significantly the metal uptake on some soils when flooded. As expected, this interaction of flooding and increased organic matter sometimes resulted in decreased uptakes of Cu and Zn in the early stages of growth (up to 6 weeks) for both sources of organic matter. The uptake of Cu and Zn was not decreased in the later (major) stages of growth on the more organic-rich soils. In fact for two soil types, Zn and Cu uptakes (flooded) were appreciably higher on the more organic-rich 'pasture' soils during this later growth stage than the paired 'cereal' soils, yet addition of rice straw had no effect on either soil. The mechanism is not clear but enhanced solubility through additional complex formation seems probable. Hence the evaluation of organic matter as a modifier of soil test values used to predict micronutrient availability may, in some farming systems, deserve consideration of its quality as well as quantity.

Table 3   Effect of flooding on uptake of
metals as function of pH[1]

| SOIL NO. | pH Shift on Flooding | | D.M. Yields | [2]Ratios of Values $\frac{Flooded}{Upland}$ at Final Harvest | | | | |
	orig. Soil[3]	after 10 wks submergence		Cu uptake	Zn	Mn	Fe	Cd
445	5.3	6.8	1.07	1.02	0.72	1.17	2.69	0.11
451	5.4	6.8	1.11	0.43	0.61	0.34	2.16	0.05
44	5.4	6.6	0.77	0.74	0.43	0.60	0.91	0.10
450	5.7	7.0	1.00	0.52	0.55	0.65	1.97	0.09
444	5.8	7.0	0.90	0.42	0.35	0.73	1.72	0.02
449	6.3	7.5	1.02	0.27	0.45	1.79	1.71	0.37
448	6.6	7.0	0.75	0.26	0.36	0.68	0.91	0.13
46	7.1	7.0	0.77	0.52	0.63	5.3	1.88	0.68
446	7.3	7.2	1.63	0.43	1.94	17.2	1.39	1.2
447	7.5	7.4	0.90	0.37	0.76	10.2	1.14	1.05
42	8.4	7.4	0.88	0.52	0.89	9.6	1.51	0.61

[1] Beckwith, Tiller & Suwadji (unpublished data)
[2] Values of underlined uptake ratios are 80% of ratios of DM yields or greater
[3] Original soil pH measurements in (1:5) soil:water suspension

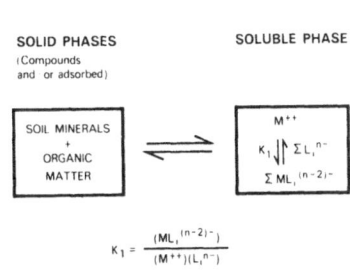

SOLID PHASES (Compounds and or adsorbed)   SOLUBLE PHASE

SOIL MINERALS + ORGANIC MATTER ⇌ M^{++} $K_1 \downarrow\uparrow \Sigma L_1^{n-}$ $\Sigma ML_1^{(n-2)-}$

$$K_1 = \frac{(ML_1^{(n-2)-})}{(M^{++})(L_1^{n-})}$$

Fig. 1   A Simple model of Micronutrient

Reactions in soil

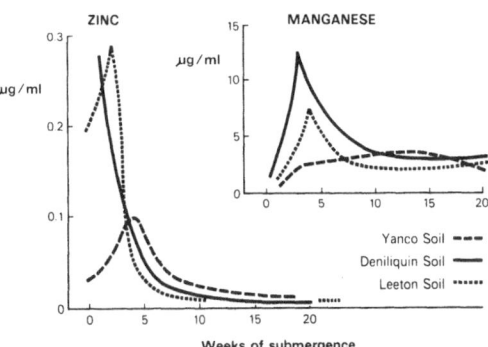

ZINC

μg/ml

MANGANESE

μg/ml

Yanco Soil ---
Deniliquin Soil —
Leeton Soil ······

Weeks of submergence

Fig. 2   Changes in Zn and Mn in soil solution following
submergence of 3 Australian rice soils

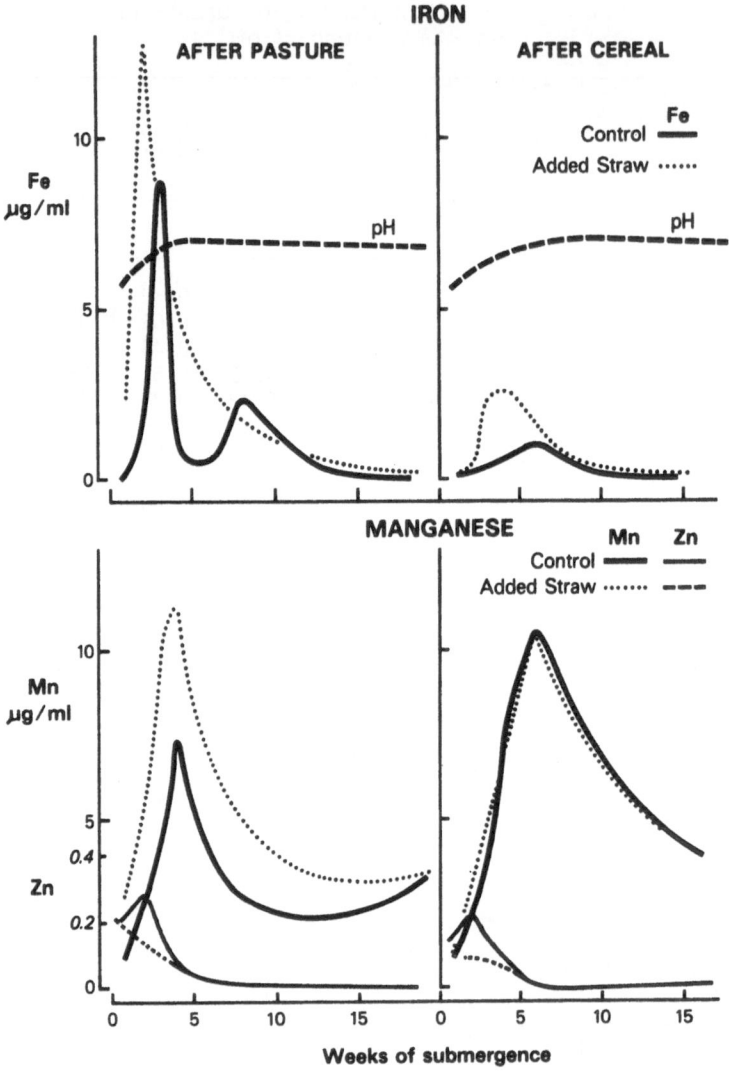

Fig. 3   Water - soluble metals in flooded soils
as affected by crop rotation

The total amount of available micronutrient in soils with
special reference to radioisotopic procedures

The total available content of micronutrient can be viewed as the sum of
the chemical components that are responsive to soil chemical equilibria and
consist collectively of amounts in solution, adsorbed, in compounds (e.g.
ZnS, $ZnCO_3$), and isomorphously substituted in those host minerals such as Mn
and Fe oxides which are themselves sensitive to conditions in paddy soils.
Estimates of total available content provide important reference values but
may have little predictive value in their own right.

The radioactive labelling of the various forms of available micronutrient provides a better means of estimating the sum of those available or
labile forms through isotopic dilution procedures (Figure 5). The technique
has limitations due to irreversible reactions that may take place when the
radioisotope comes in contact with soil surfaces. In our experience, these
reactions are aggravated when the nutrient has a high affinity for the surface and when non-radioactive carrier is used in conjunction with the radioactive label. For these soils of high affinity, radioisotopic procedures
based on soils suspended in neutral salts, as proposed for phosphorus by the
E-value procedure[28] are restricted in their application. If conditions are
not chosen, or manipulated, so as to ensure reversibility of the equilibria
concerned, incorrect and sometimes ridiculous, values will result. Sometimes
suitable conditions are achieved by lowering pH or by introducing complexing
agents into the soil suspension. In fact the chemical environment of the
plant root encourages the desorption of strongly bound forms of nutrients by
such mechanisms. For these reasons the L-value procedure based on the growing plant, which was originally proposed by Larsen[17] for studies of phosphorus, provides a more reliable reference procedure for the estimation of
total amount of available nutrient. The plant acts effectively as a
'pipette' in the radioactively labelled soil. Plants should be the final
"arbiter" of how much available nutrient is in a soil[19]. We have found
that this procedure using carrier-free radioisotopes can provide reliable
values for Zn, Cd, Ni, and Mn at least, and on a large range of soil
types.[38,39,40,41,42].

We have used these radioisotopic procedures to check the suitability of
different chemical extractions in relation to the uptake by plants. Using
radioactively labelled soils under non-flooded conditions, specific activities of zinc in extractants and in growing plants were compared, and showed
that complexing agents and Mg and Ca salts extracted Zn from the same chemical components as did the plants but dilute hydrochloric acid did not. The
latter extractions were done on moist soil taken from the pots in which the
plants were growing. Hence in later experiments, Tiller et al.[42] found
that E-values based on EDTA extraction of labelled soil agreed well with
L-values based on isotopic dilution in growing rice plants under both flooded
and upland conditions (see Table 4). The corresponding r^2 were 0.84^{xxx} and
0.85^{xxx}, respectively.

The L-values provide a reference against which different procedures can
be compared as well as different conditions for one procedure. For example,
a single extraction by 0.1M EDTA[3] (pH6, 7 days) of soils which had been
flooded several months, removed 78% of the corresponding L-values for flooded
soils ($r^2=0.93^{xxx}$) using rice as test plant. The reagents using 0.01M EDTA
and shorter equilibration times, removed about half as much as the 0.1M EDTA
reagent. For upland soils, less of the corresponding L-value was removed by
0.1M EDTA than was the case with flooded soils (Table 4). The Zn_L values
were also used as a reference to assess the variable proportions of total
available zinc that is removed by some well established soil test procedures.

The results are given in Table 4. As might be expected for the range of reagents, and their concentrations used, a variable proportion of the total content of available zinc was removed and the correlation as shown by r^2 generally increased as the proportion removed increased. Accordingly, some of the reagents shown should not be considered as estimates of total available zinc even though they may well be satisfactory indicators of zinc stress under some circumstances. In fact, the predictability of a particular chemical procedure may well be inversely related to these correlation coefficients.

Radioisotopic procedures also provide a means of measuring the potential increase in the total available pool that may result from flooding because of mineralization of organic matter and dissolution of minerals such as Mn and Fe oxides which contain Zn. The total pool may increase even though the solubility decreases. Our investigations showed that Zn_L values for flooded soils only somewhat exceeded Zn_L values of the corresponding non-flooded soils. This trend was paralleled by extraction with 0.1M EDTA, both in its own right and as a basis for the E-value procedure. The relation between values for flooded and non-flooded soils are given in Table 5.

The marked increase in soluble Mn resulting from oxide dissolution under reduced condition is a well established feature of paddy soils. We have also used radioisotopic procedures using ^{54}Mn, to validate some chemical procedures used to assess different chemical components of Mn in soils (Table 6). The mild hydroquinone plus $Ca(NO_3)_2$ procedure of Jones and Leeper[15] corresponded closely to the Mn_L values for non-flooded soils whereas the procedures based on dithionite or hydroquinone plus EDTA[1] agreed well with the Mn_L value of flooded soils. A commonly used procedure to evaluate available Mn_L levels, viz. hydroquinone plus ammonium acetate, gave intermediate values between L-values for flooded and non-flooded conditions and could thus either over - or underestimate Mn status, depending on the soil conditions being studied. Isotopic techniques thus can provide a means of checking independently the specificity of chemical techniques used in availability studies.

However, it needs to be re-stressed, that all these estimates of total labile or available pool of nutrients, such as the L-value, even though reproducible and probably closely identified with the sum of those chemical components which are accessible to plant roots, are generally of little intrinsic value when used alone to predict Zn deficiency in rice. The studies of Stewart et al.[36] on Zn deficiency of rice in Philippines and the experiments of Tiller and Wassermann[39], Tiller et al.[38] and Beckwith et al.[2] support this view. Account should be also made of variable soil affinity for the nutrient or the L-value tests restricted to soils of similar affinity.

THE STATE OF THE ART

Following more widespread recognition of zinc deficiency in the field during the 1960's, it was soon found that zinc uptake by rice plants was unrelated to the total zinc content of the soil[47]. Other investigations around the same period[9] reported that neither total Zn nor that extracted by 0.1N HCl was related to the incidence of Zn deficiency. Sometimes an inverse relation existed because of the strong influence of pH and organic matter content of the soil.

Subsequently the method of Trierweiler and Lindsay[45] based on EDTA and $(NH_4)_2 CO_3$ and air-dried soils, was found to be appreciably better in relating soil properties to the incidence of Zn deficiency of paddy soils[10,49].

Table 4 Zn_L as a reference for other methods of assessing availability[1]

$$Zn_X = b_1 \, Zn_L + b_0 \quad [2]$$

Soil Condition for 'L' Value Measurement[3]	Method (X)	b_1	r^2	no. of soils
Flooded (Rice)	Zn_E using EDTA	1.03	0.84[xxx]	11
Upland (Rice)	Zn_E using EDTA	1.14	0.85[xxx]	11
Flooded (Rice)	0.1M EDTA	0.78	0.93[xxx]	11
Upland (Rice)	0.1M EDTA	0.61	0.59[xx]	11
Upland	0.1M HCl	0.63	0.65[xxx]	20
Upland	0.05M HCl	n.s.	n.s.	20
Uplant (Rice)	0.05M HCl	0.26	0.39[x]	11
Flooded (Rice)	0.05M HCl	n.s.	n.s.	11
Upland	0.01M EDTA-NH_4OAc, pH7	0.35	0.53[xxx]	20
Upland	0.01M EDTA-$(NH_4)_2CO_3$, pH8.6	0.41	0.57[xxx]	20
Upland (Rice)	0.005M DTPA-TEA	n.s.	n.s.	11
Flooded (Rice)	0.005M DTPA-TEA	0.90	0.41[x]	11
Upland	0.005M DTPA-TEA	0.26	0.25[x]	20
Upland	Dithizone-NH_4OAc	0.22	0.43[xx]	20
Upland	1.0M $MgCl_2$	0.11	0.34[xx]	20

[1] After Beckwith et al(2), Tiller et al(38) and Tiller (unpublished data).
[2] Significance of regressions; [xxx]$P<0.001$; [xx]$P<0.01$, n.s. = not significant.
[3] Subterranean clover was test plant for 'L' value where rice is not indicated.

Table 5   Estimates of available zinc following upland (U) and flooded (F) conditions

$$Zn_U = b_1 \, Zn_F + b_0$$

Method	b_1	r^2
Zn_L	0.83	0.95[xxx]
Zn_E (EDTA)	0.86	0.85[xxx]
0.1M EDTA	0.74	0.77[xxx]

Data after Beckwith et al(2)

Generally, Trierweiler and Lindsay's recommended critical value of 1.4µg/g, as developed for corn, proved also to be a reasonable approximation for rice soils unless pH or organic matter content was high.

In another investigation on Philippines soils, Stewart et al.[36] found that several extractants, viz DTPA, $MgCl_2$, and NH_4 acetate were all well correlated with each other and with the Zn contents of young rice plants. This proposal that several extractants may equally well suffice for soil testing for Zn deficiency in rice soils, was supported by further reports[12] that the EDTA-$(NH_4)_2CO_3$ procedure correlated well with a 0.05N HCl procedure. The HCl procedure was preferred because of greater simplicity and economy in routine operations. A working limit for the definition of Zn risk in paddy soils was set at 1.0µg/g although, consistent with previous experience with EDTA-$(NH_4)_2CO_3$, deficient soils of high pH and organic matter status could exhibit higher values.

Meanwhile Giordano and Mortvedt[8] could not recommend DTPA as an extractant of Zn in rice soils because of the lack of agreement between Zn uptake and amount of Zn extracted by DTPA. On the other hand, Gangevar and Chandra[6] investigated critical limits of several extractants with respect to the onset of Zn deficiency of flooded rice and found that NH_4 acetate (pH4.6), dithizone, $MgCl_2$, and DTPA all correlated significantly with Zn in rice plants but a procedure using a dilute acid reagent (0.05N HCl +0.025N H_2SO_4) did not. The critical levels ranged from about 0.5µg/g for the NH_4 acetate to about 2µg/g for DTPA. Sedbury et al.[31] carried out similar comparisons of several common extraction procedures used to predict micronutrient availability and found that none correlated with Zn concentration in and uptake by rice on 25 flooded rice soils. The NH_4 acetate (pH4.6) reagent gave the best correlation for upland rice. Soepardi[34] and colleagues conducted a survey of Zn deficiency in rice soils of Java and found that both 0.05N HCl and DTPA usefully separated soils that responded to Zn application from those that did not in glasshouse experiments.

Such varied experiences would not, at face value, indicate an encouraging future for soil testing in relation to Zn deficiency in paddy rice. Apparent conflicts may well reflect the kinds of soils chosen, especially with respect to the range of properties represented which critically effect Zn deficiency. Much also depends on how the soil tests are assessed, whether by Zn concentrations in the rice plants alone, whether in the glasshouse alone or supported by field evidence and whether leaf symptoms and yield responses to added Zn fertilizer were considered.

It is clear from IRRI experience that soil tests, of whatever kind used, will generally be more useful when considered in relation to other critical factors relating to the soil and the plant. Of the soil factors pH and organic matter are important, but so also are management factors such as water control and fertilizer use. For the plant, micro-nutrient composition in relation to rice variety requires consideration because of variable tolerance to zinc deficiency.

Consideration of suitable soil factors as discussed above may substitute partially for the changes in chemical environment they cause and thus affect changes in Zn solubility or affinity for soil surfaces. Another approach to soil testing which is quite suited to routine operations, is to simulate the flooded conditions of soils during the test procedure. This method was used by Tiller et al.[43] by incubating in test tubes, small samples of soil covered by water containing either glucose, sulphate ions or neither. After 3 weeks, the suspensions, together with controls which had not been incubated under water, were extracted with DTPA and EDTA reagents. Whereas the enhanced levels of Fe and Mn extracted showed that reduction had occurred during incubation, the amounts of Cu and Zn extracted by DTPA were appreciably reduced by the simulated flooding procedure (Table 7). Zinc measured in this way was correlated ($r^2 = 0.79^{xx}$) with the zinc concentration of young rice plants grown on flooded soils in pots in the glasshouse. Further investigations are needed to clarify whether the initial promise of this approach is justified.

Table 6   Confirmation of some chemical extractions
of soil Mn, using radioisotopic procedures

Soil No.	Total Mn	L-value Flooded	L-value Non-flooded	Hydroquinone-EDTA	Hydroquinone-NH$_4$OAc	Hydroquinone +Ca(NO$_3$)$_2$
1	700	370	130	430	310	120
2	410	180	76	190	140	88
3	150	27	13	22	12	17
4	960	870	460	840	690	530

All mean values in µg/g soil

Table 7   DTPA extraction of metals following simulated
flooding of soils in tubes. (examples from
Tiller et al([43])

Soil No	Treatment	pH	Metals extracted by DTPA (µg/g) Cu	Zn	Mn	Fe
42	"Flooded"	7.55	0.29	0.15	50	20
	Control	8.00	0.56	0.25	5	1
46	"Flooded"	7.00	0.04	0.49	6.5	115
	Control	7.05	0.29	0.79	2	3
448	"Flooded"	7.45	0.08	0.42	140	120
	Control	6.35	0.60	1.65	24	13
451	"Flooded"	7.00	≦0.02	0.15	80	245
	Control	5.50	1.2	1.2	54	25

Procedure: 5g soil in test tube; added 10ml solution to "flooded" treatment; store without disturbance 3 wks; add 10ml H$_2$0 to control and then 10ml double strength DTPA-TEA to all tubes, shake 2 hours; centrifuge, filter and analyse

FINAL REMARKS

Generally the methods of assessment of micronutrients for production of paddy rice by soil testing has relied on the procedures developed for crops on dryland agriculture where they have been found very useful[4]. Despite the limitations of soil tests, they can be a valuable aid to crop management. Unfortunately the limitations may be greater in the case of rice soils because of the sensitivity of their chemical conditions to management decisions. The observation has been sometimes made that Zn-deficient soils, including rice soils, usually contain adequate amounts of available Zn. The real problem relates more appropriately to the adequacy of the tests to predict crop response to added Zn fertilizers. General experience suggests that several chemical reagents commonly in use may be equally suitable for the prediction of availability and that the success in practical application varies appreciably from region to region in relation to local variation in soil types. Clearly the incorporation of other soil or management factors will often improve the precision of the tests. Local conditions may militate against the use of some kinds of procedures e.g. areas dominated by alkaline and especially calcareous soils may be unsuited to the use of dilute acid and areas of acidic soils high in Fe and Mn may cause unsatisfactory results with dilute complexing agents because of reagent saturation.

Soil testing has tended towards the use of those reagents and extraction conditions that remove amounts of Zn etc. which are intermediate between that in soil solution and the total available pool. This compromise recognizes the inadequacy of each component alone from the point of view of supply of nutrient to a crop and the restraints of the equilibrium between the two, but also recognizes the practical problem of analysis in routine laboratory operations.

Progress in the development of better soil tests will proceed together with our understanding of the soil chemistry involved and the mechanism of nutrient uptake by plant roots and with the realization that finally all tests gain acceptance though successful calibration against the field performance of crops.

REFERENCES

[1] Beckwith, R.S. (1955). Metal Complexes in Soils. Aust. J. Agric. Res. 6, 685-698.
[2] Beckwith, R.S., Tiller, K.G. and Suwadji, E. (1975). The effects of flooding on the availability of trace metals to rice in soils of different organic matter status. In "Trace Elements in Soil - Plant - Animal Systems" pp. 135-149. (Academic Press Inc : New York).
[3] Clayton, P.M. and Tiller, K.G. (1979). Chemical methods for the determination of the heavy metal content of soils in environmental studies. CSIRO Aust. Div. Soils Tech. Pap. No. 41, 1-17.
[4] Cox, F.R., and Kamprath, E.J. (1972). Micronutrient soil tests. In "Micronutrients in Agriculture". (Soil Sci. Soc. Am.:Madison).
[5] d'Souza, T.J. and Mistry, K.B. (1978). The uptake, distribution and metabolic fate of ^{59}Fe, ^{58}Co, ^{54}Mn and ^{65}Zn in plants and their mobility and availability to crops in typical black and laterite soils. In "International Symposium on the Use of Isotopes and Radiation in Research on Soil-Plant Relationships". (Proc. Symp. Colombo 1978) IAEA., Vienna.

(^{6}) Gangevar, M.S. and Chandra, S.K. (1976). Estimation of critical limit for zinc in rice soils. Commun. Soil Sci. Pl. Analysis 7, 295-310.

(^{7}) Geering, H.R. and Hodgson, J.F. (1969). Micronutrient complexes in soil solution:III. Characterization of soil solution ligands and their complexes with Zn^{2+} and Cu^{2+}. Proc. Soil Sci. Soc. Am. 33, 54-59.

(^{8}) Giordano, P.M. and Mortvedt, J.J. (1972). Rice response to zinc in flooded and non-flooded soil. Agron. J. 64, 521-524.

(^{9}) I.R.R.I. (1968). Annual Report Int. Rice Res. Inst., Los Banos, Philippines.

(^{10}) I.R.R.I. (1969). Annual Report Int. Rice Res. Inst., Los Banos, Philippines.

(^{11}) I.R.R.I. (1970). Annual Report Int. Rice Res. Inst., Los Banos, Philippines.

(^{12}) I.R.R.I. (1972). Annual Report Int. Rice Res. Inst., Los Banos, Philippines.

(^{13}) I.R.R.I. (1975). Annual Report Int. Rice Res. Inst., Los Banos, Philippines.

(^{14}) I.R.R.I. (1976). Annual Report Int. Rice Res. Inst., Los Banos, Philippines.

(^{15}) Jones, L.H.P. and Leeper, G.W. (1951). Available manganese oxides in neutral and alkaline soils. Plant and Soil 3, 154-159.

(^{16}) Katyal, J.C. and Ponnamperuma, F.N. (1974). Zinc deficiency: A widespread nutritional disorder of rice in Agusan del Norte. Philippine Agriculturist 58, 79-89.

(^{17}) Larsen, S. (1952). The use of P^{32} in studies on the uptake of phosphorus by plants. Plant and Soil 4, 1-10.

(^{18}) Lindsay, W.L. and Norvell, W.A. (1978). Development of a DTPA soil test for zinc, iron, manganese and copper. Soil Sci. Soc. Am. J. 42, 421-428.

(^{19}) Loneragan, J.F. (1975). The availability and absorption of trace elements in soil-plant-animal systems and their relation to movement and concentrations of trace elements in plants. In "Trace Elements in Soil-Plant-Animal Systems" pp109-134. (Academic Press Inc.: New York).

(^{20}) Mikkelsen, D.S., and Kuo S. (1977). Zinc fertilization and behaviour in flooded soils. C.A.B. Special Publ. No. 5.

(^{21}) Nicholas, D.J.D. (1975). The functions of trace elements in plants. In "Trace Elements in Soil-Plant-Animal Systems" pp181-198 (Academic Press Inc:New York).

(^{22}) Norvell, W.A., and Lindsay, W.L. (1970). Lack of evidence of $ZnSiO_3$ in soils. Proc. Soil Sci. Soc. Am. 34, 360-361.

(^{23}) Patrick, W.H. and Mikkelsen, D.S. (1971). Plant nutrient behaviour in flooded soils. In "Fertilizer Technology and Use." 2nd. Edit. (Soil Sci. Soc. Am.: Madison).

(^{24}) Ponnamperuma, F.N. (1972). Chemistry of submerged soils. Adv. Agron. 24, 29-88.

(^{25}) Quidez, B.G. (1978). Effects of nitrogen, phosphorus, potassium, copper, molybdenum and zinc on the growth of rice inorganic soils. Phillipp J. Crop Sci. 3, 203-206.

(^{26}) Rahaja, P.C., Yawalkar, K.S., and Singh, R. (1959). Crop response to micronutrients under Indian conditions. Indian J. Agron. 3, 254-263.

(^{27}) Rahmatullah, F.M., Chaudry, F.M., and Rashid, A. (1976). Micronutrient availability to cereals from calcareous soils. II. Effect of flooding on electrochemical properties of soils. Plant and Soil 45, 411-420.

(28) Russell, S.R., Rickson, J.B. and Adams, S.N. (1954). Isotopic equilibria between phosphates in soil and their significance in the assessment of fertility by tracer methods. J. Soil Sci. 5., 86-105.

(29) Schnitzer, M. and Skinner, S.I.M. (1966). Organo-metallic interactions in soils : 5. Stability constants of Cu^{++}, Fe^{++}, and Zn^{++} fulvic acid complexes. Soil Sci. 102, 361-365.

(30) Schnitzer, M. and Skinner, S.I.M. (1967). Organo-metallic interactions in soils. 7. Stability constants of Pb^{++}, Ni^{++}, Mn^{++}, Co^{++}, Ca^{++}, and Mg^{++} fulvic acid complexes. Soil Sci. 103, 247-252.

(31) Sedbury, J.E., Miller, B.J., and Said, M.B. An evaluation of chemical methods for extracting zinc from soils. Commun. Soil Sci. Pl. Analyses. 10, 689-701.

(32) Sillèn, L.G. and Martell, A.E. (1964). "Stability Constants of Metal-Ion Complexes." Special Publ. No. 17. (The Chemical Society, London).

(33) Sillèn L.G. and Martell, A.E. (1971). "Stability Constants of Metal-Ion Complexes." Supplement No. 1 Special Publ. No. 25. (The Chemical Society, London).

(34) Soepardi, G. (1979). Zinc deficiency in paddy soils in Java, Indonesia. Isotope-Aided Studies in Micronutrients in Rice Production with Special Reference to Zinc Deficiency. (Proc. Res. Coord. Meeting, Vienna 1979). IAEA, Vienna.

(35) Soil Survey Staff (1975). "Soil Taxonomy." Agric. Handbook No. 436. U.S.D.A., Washington.

(36) Stewart, J.W.B., Friaz, Betty V., and Lapid, F.M. (1972). Studies on the micronutrient cation status of some Philippine paddy soils with special emphasis on zinc. In "Isotopes and Radiation in Soil-Plant Relationships including Forestry" pp.539-548. (Proc. Symp. Vienna) I.A.E.A., Vienna.

(37) Tanaka, A., and Yoshida, S. (1970). Nutritional disorders of the rice plant in Asia. Int. Rice Res. Inst. Tech. Bull. 10.

(38) Tiller, K.G., Honeysett, J.F., and de Vries, M.P.C. (1972). Soil zinc and its uptake by plants. II. Soil chemistry in relation to prediction of availability. Aust. J. Soil Sci. 10, 151-164.

(39) Tiller, K.G. and Wassermann, P. (1972). Radioisotopic techniques and zinc availability in soil. In "Isotopes and Radiation in Soil-Plant Relationships Including Forestry" (Proc. Symp. Vienna) I.A.E.A., Vienna.

(40) Tiller, K.G. and Wassermann, R. (1973). The effect of flooding on the availability of Zn and Mn to rice. Z. für Pfl. u. Bkde. 136, 57-67.

(41) Tiller, K.G. (1975). Isotope derived criteria for the measurement of soil and fertilizer micronutrient availability. In "Isotope-aided micronutrient studies in rice production with special reference to Zn deficiency." (Proc. Res. Coord. Meeting, Vienna, 1974).

(42) Tiller, K.G. (1979). Applications of isotopes to micronutrient studies. In "International Symposium on the Use of Isotopes and Radiation in Research on Soil-Plant Relationships (Proc. Symp. Colombo 1978). I.A.E.A. Vienna pp. 359-372.

(43) Tiller, K.G., Suwadji, E. and Beckwith, R.S. (1979). An approach to soil testing with special reference to the zinc requirement of rice paddy soils. Commun. Soil Sci. and Pl. Analysis 10, 703-715.

(44) Tiller, K.G. (1981). Micronutrients. In "Soils : An Australian Viewpoint" (CSIRO,: Melbourne) In Press.

(45) Trierweiler, J.F. and Lindsay, W.L. (1969). EDTA-ammonium carbonate test for zinc. Proc. Soil Sci. Soc. Am. 33, 49-54.

(46) Viets, F.G. (1962). Chemistry and availability of micronutrients in soils. Agric. Food Chem. 10, 174-178.

(^{47}) Yoshida, S., and Tanaka, A. (1969). Zinc deficiency of the rice plant in calcareous soils. <u>Soil Sci. Plant Nutr</u>. <u>15</u>, 75-80.

(^{48}) Yoshida, S., Forno, D.A. and Bhadrachalam, A. (1971). Zinc deficiency of the rice plant on calcareous and neutral soils in the Philippines. <u>Soil Sci. Plant Nutr</u>. <u>17</u>, 83-87.

(^{49}) Yoshida, S., Ahn, J.S., and Forno, D.A. (1973). Occurrence, diagnosis, and correction of zinc deficiency of lowland rice. <u>Soil Sci. Plant Nutr</u>. <u>19</u>, 83-93.

A PRELIMINARY STUDY ON SPECIFIC ADSORPTION OF Cu ION
BY PADDY SOILS OF SOUTHERN JIANGSU PROVINCE

Wu Mei-ling

(Institute of Soil Science, Academia Sinica, Nanjing)

Since the sixties research work on specific adsorption of polyvalent ions especially heavy metals by soils has been an active field in soil chemistry. The principal carriers of specific adsorption in soils are oxides and hydrous oxides of iron, manganese, aluminum and silicon of the clay and organic matter (1-3). The enrichment of heavy metals caused by such specific adsorption plays an important role in controlling the concentration of these elements in soil solution and sea-water (4,5). The mobility and availability of some minor elements to plants are also affected by this specific adsorption (6).

The present work is a preliminary study on the specific adsorption of Cu^{2+} ions by some paddy soils.

SOILS AND METHODS

Soil samples were collected from paddy soils of southern Jiangsu Province derived from loess-like material or lake sediments. Some properties of the soils are shown as follows: pH, 6.2-6.8; Organic matter, 1.9-3.8%; Free iron oxides (Fe_d), 1.5-3.0% (Fe_2O_3); Amorphous iron oxides (Fe_o), 0.67-1.71% (Fe_2O_3); Clay content, 22-30% (occasionlly $< 20\%$ or $> 40\%$).

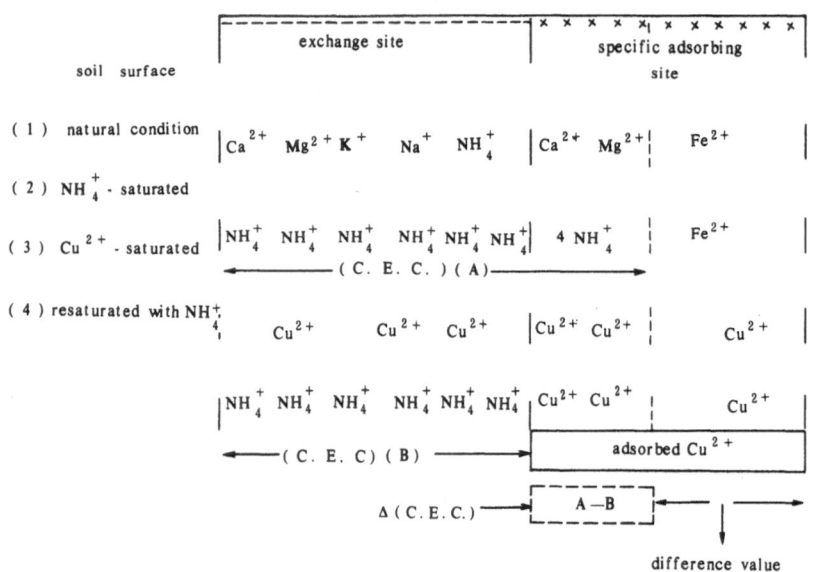

Fig. 1 Chemical behavior of adsorbility of cations by the soil

Methods: since the specifically adsorbed Cu is generally unexchangeable, the "exchange site" and the "specific adsorbing site" were used as index of exchange capacity and specific adsorption capacity of the soil respectively. Their chemical behavior is illustrated in Fig. 1.

(1) Under natural conditions most of the exchange sites are occupied by Ca^{2+}, Mg^{2+}, K^+, Na^+, NH_4^+; the specific adsorbing sites are occupied partly be Ca^{2+} and Mg^{2+}, partly by Fe^{2+} (Soil samples submerged in water previously).

(2) If leached with NH_4Cl solution to determine the CEC, not only the exchange sites, but also the Ca^{2+} and Mg^{2+} on the specific adsorbing sites are occupied by NH_4^+, while Fe^{2+} on the specific adsorbing sites remains unexchanged.

(3) When samples are treated with 1 N $CuCl_2$ solution, both sorts of site are occupied by Cu^{2+}.

(4) If the above sample is treated with NH_4Cl solution again, the NH_4^+ can replace the Cu^{2+} on the exchange sites, while Cu^{2+} on the specific adsorbing sites remains unexchanged.

The method is as follows: Four portions were taken for each sample. Two portions were saturated with NH_4Cl (pH 6.6), the free NH_4Cl were removed by ethanol and then replaced with 0.1 N HCl to determine the CEC(A). Other two portions were treated with 1 N $CuCl_2$ solution (pH 5.5) and then treated with NH_4Cl solution. In the 0.1 N HCl leachate, the amounts of NH_4^+ (CEC,B) and Cu^{2+} were determined separately. The difference between (A) and (B) was taken as ΔCEC. The "difference value" is the amount of adsorbed Cu subtracted by ΔCEC. Some of the results are shown in Table 1.

Table 1 Δ(C E C) and adsorbed Cu for some paddy soils

Sample No.	(C E C) (m.e./100g)	Adsorbed Cu (m.e. /100g)	Sample No.	Δ(C E C) (m.e./100g)	Adsorbed Cu (m.e. /100g)
11	2.22	7.84	118	3.42	8.16
12	1.78	8.42	119	2.88	8.58
13	2.58	8.40	23	5.53	8.00
14	2.04	4.54	27	5.43	13.0
16	2.77	7.80	31	2.68	4.24
17	2.12	8.70	36	0.81	6.96
18	1.66	8.64	74004	2.78	7.65
19	1.99	6.14	74005	2.52	7.54
112	0.94	3.94	74006	3.21	11.2
113	1.18	4.60	74007	3.36	11.6
117	3.10	9.44	74008	2.18	6.68

RESULTS AND DISCUSSION

The adsorbed Cu not exchanged by NH_4^+ but dissolved in 0.1 N HCl may be regarded as the specifically adsorbed Cu. It was found that this portion did not correlate with the clay content, pH or organic matter content of the original soil. There was some correlation between this portion and the free iron oxides content of the soil ($p < 0.05$), but $r^2 = 0.25$. In the following, we shall distinguish this protion into two parts, namely the ΔCEC and the "difference value."

The ΔCEC represents a part of the specific adsorbing sites of the soil. Under natural conditions, it acts as cation-exchange site. However, these sites retain the Cu^{2+} ions so tightly that their ability of acting as exchange site may be lost after treating with Cu^{2+}, thus reflecting as the decrease in CEC of the soil. In Fig. 2 it is shown that this ΔCEC is closely correlated with the content of free iron oxides or amorphous iron oxides. Table 2 shows that this ΔCEC and the amount of adsorbed Cu decrease remarkably (with one exception) after removing the free iron oxides with sodium dithionite, and the decrease in ΔCEC caused by the removal of iron oxides shows significant correlation with the amount of removed iron oxides (Fig. 3). This seems to imply that free iron oxides play an important role in acting as specific adsorbing site as expressed by ΔCEC. On the other hand, it was found that there is no relationship between the change in ΔCEC caused by the removal of organic matter and the removed amount. The amount of adsorbed Cu is decreased to an approximation to ΔCEC after the removal of organic matter (Table 3). All these seem to imply that the specific adsorption of Cu as expressed by ΔCEC is chiefly caused by free iron oxides.

Table 2 Changes in amount of adsorbed Cu and Δ(C E C) after the removal of free iron oxides

Sample No.	Adsorbed Cu (m.e./100g)		Δ(C E C) (m.e./100g)	
	Before removal	After removal	Before removal	After removal
12	8.42	2.79	1.78	0.26
13	8.40	2.22	2.58	0.41
17	8.70	2.23	2.12	0.08
19	6.14	2.19	1.99	1.00
112	3.94	2.96	0.94	1.27
23	8.00	4.22	5.53	1.24
74005	7.54	2.40	2.52	0.77
74007	11.6	3.39	3.36	0.60

Table 3 Changes in amount of adsorbed Cu, Δ(C E C) and difference value after the removal of organic matter

Sample No.	Adsorbed Cu (m.e./100g)		Δ(C E C) (m.e./100g)		Difference value (m.e./100g)	
	Before removal	After removal	Before removal	After removal	Before removal	After removal
12	8.42	2.02	1.78	2.03	6.64	−0.01
13	8.40	1.26	2.58	1.44	5.82	−0.18
17	8.70	1.52	2.12	1.34	6.58	0.18
19	6.14	1.47	1.99	1.31	4.15	0.16
112	3.94	1.36	0.94	1.29	3.00	0.07
23	8.00	1.18	5.53	1.29	2.47	−0.11
74005	7.54	1.15	2.52	1.44	5.02	−0.29
74007	11.6	1.87	3.36	1.26	8.24	0.61

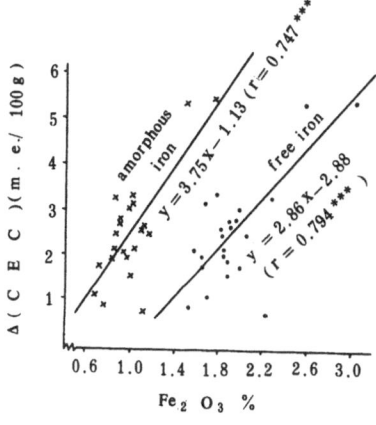

amorphous (or free) iron

Fig. 2 Relationship between (CEC) and amorphous or free iron of the soil

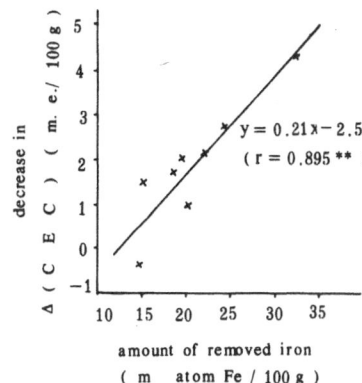

amount of removed iron (m atom Fe / 100 g)

Fig. 3 Relationship between decrease in (CEC) after removal of free iron and the amount of removed iron

The free iron oxides (Fe_d) of the soil are actually composed of a variety of ferric oxides with different crystallinity and of amorphous iron oxides. It is generally considered that amorphous iron oxides (Fe_o) is the least crystallized and the most active part of the free iron oxides. Therefore, the difference between the amount of free iron oxides and that of amorphous iron oxides $(Fe_d - Fe_o)$ may be used as an index of crystalline free iron oxides. It is probable that the fraction Fe_o is more important in inducing the specific adsorption of Cu than the fraction Fe_d-Fe_o, as can be judged by the statistical data shown in Table 4.

Table 4 Correlations between Δ(C E C) and different forms of iron oxides

No. of samples	y	x	r	a	b	p
22	(C E C)	Free iron oxides (Fe_d)	0.794	−2.88	2.86	<0.001
22	(C E C)	Amorphous iron (Fe_o)	0.747	−1.13	3.75	<0.001
22	(C E C)	$Fe_d - Fe_o$	0.466.	−0.20	3.03	<0.05

In addition, the "difference value" is another expression of specific adsorbing sites in the soil. The results in Table 3 show that the "difference value" practically disappeared after the removal of organic matter, namely the amount of adsorbed Cu decreased to about the same as ΔCEC. Besides, there is some correlation between the decreased amount of adsorbd Cu caused by the removal of organic matter and the amount of removed organic matter (r=0.738, p<0.05, n=8). For samples where the free iron oxides has been removed the relationship between the "difference value" and the amount of residual organic matter is very significant (Fig. 4). It seems that the organic matter may play some part in affecting the "difference value". However, there is no significant correlation between the "difference value" and the organic matter content for various soils. It appears that this problem is rather complicated and more detailed work remains to be done.

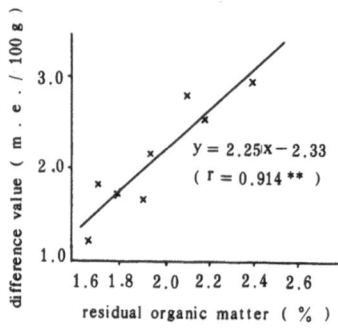

Fig. 4 Relationship between
difference value and
residual organic matter
(samples after removal
of free iron)

294

REFERENCES

(1) Forbes, E.A., Posner, A.M., Quirk, J.P., The specific adsorption of divalent Cd, Co, Cu, Pb and Zn on goethite. J. Soil Sci., 27, 154-166 (1976).

(2) McBride, M. B., Retention of Cu^{2+}, Ca^{2+}, Mg^{2+} and Mn^{2+} by amorphous alumina. Soil Sci. Soc. Amer. J., 42, 27-31 (1978).

(3) Schnitzer, M., Khan, S.U., Reaction of humic substances with metal ions and hydrous oxides. In "Humic Substances in the Environment", Marcel. Dekker Inc., New York, 203-252 (1972).

(4) Jenne, E.A., Controls on Mn, Fe, Co, Ni, Cu and Zn concentration in Solids and water. In "The significant role of hydrous Mn and Fe oxides". Adv. Chem., Ser. 73, 337-387 (1968).

(5) Krauskopf, K. B., Factors controlling the concentration 13 rare metal in sea-water. Geochim. Cosmochim. Acta, 9, 1-32 (1956).

(6) Hodgson, J. F., Chemistry of the micronutrient elements in soil. Adv. Agron., 15, 119-159 (1963).

STUDIES ON ACCUMULATION OF STRONTIUM-90 IN PADDY SOILS

Kinjiro Kawase
(Prof. Emeritus, Dr. Agriculture, Tokyo, Japan)
Eizo Yokoyama
(Niigata University, Niigata City, Japan)

INTRODUCTION

About 80% of Strontium-90, a fallout of nuclear tests, which has settled on the earth is concentrated in the region bounded by north latitudes 30 and 60° of the North Hemisphere.[1] Japan is located in this region. According to the meteorological results of analysis, ^{90}Sr concentration in the precipitation in 1970 was about 73 mCi/Km² in Tokyo on the coast of the Pacific Ocean, and almost twice the amount, 143 mCi/Km², in Akita on the coast of the Japan Sea. Basically, ^{90}Sr is the most important component of the radioactive fallout with respect to the soil-plant ecosystem.

As a result of the studies of the conditions of ^{90}Sr accumulation in each of the horizons of the various types of paddy soils along the Japan Sea coasts of Niigata Prefecture, it has been found that in 1967-1971, ^{90}Sr, concentration was definitely higher than that in the precipitation, and its concentration changed intricately by such factors as terrain, parent materials, soil classes, Eh, (Redox condition), influent and effluent irrigation water. This paper will report the changes stemming from the factors.

METHOD

The paddy soil samples were taken from 11 different locations in Niigata Prefecture. The soil types were 5 strong gley soils, one gray soil, 3 gray brown soils, one ando soil, and one gravel soil. At each location, forms and conditions of 3 - 5 layers of pedogenic horizons were noted. The thickness of each horizon was 10-20 cm. For each of the samples, the following were examined -- as physical composition -- mechanical texture, field volume weight, three phase distribution, and -- as chemical composition —— pH (H₂0), pH(KCl), exchange acidity , base exchange capacity, exchangeable base, percentage of base saturation, total carbon, total nitrogen, hot HCl soluble Fe_2O_3, absorptive coefficient of P_2O_5, drying effects, and available SiO_2. 90Sr was extracted by the NaOH-HCl method, and its concentration determined by the Science and Technology Agency radioactive strontium analysis method. Analytical procedure is shown in Fig. 1.

RESULTS

The results obtained are shown in Table 1 in which soil layer number, year of sampling, location, and depth of sampling, concentration of ^{90}Sr

Soil Sample (50-200 g)

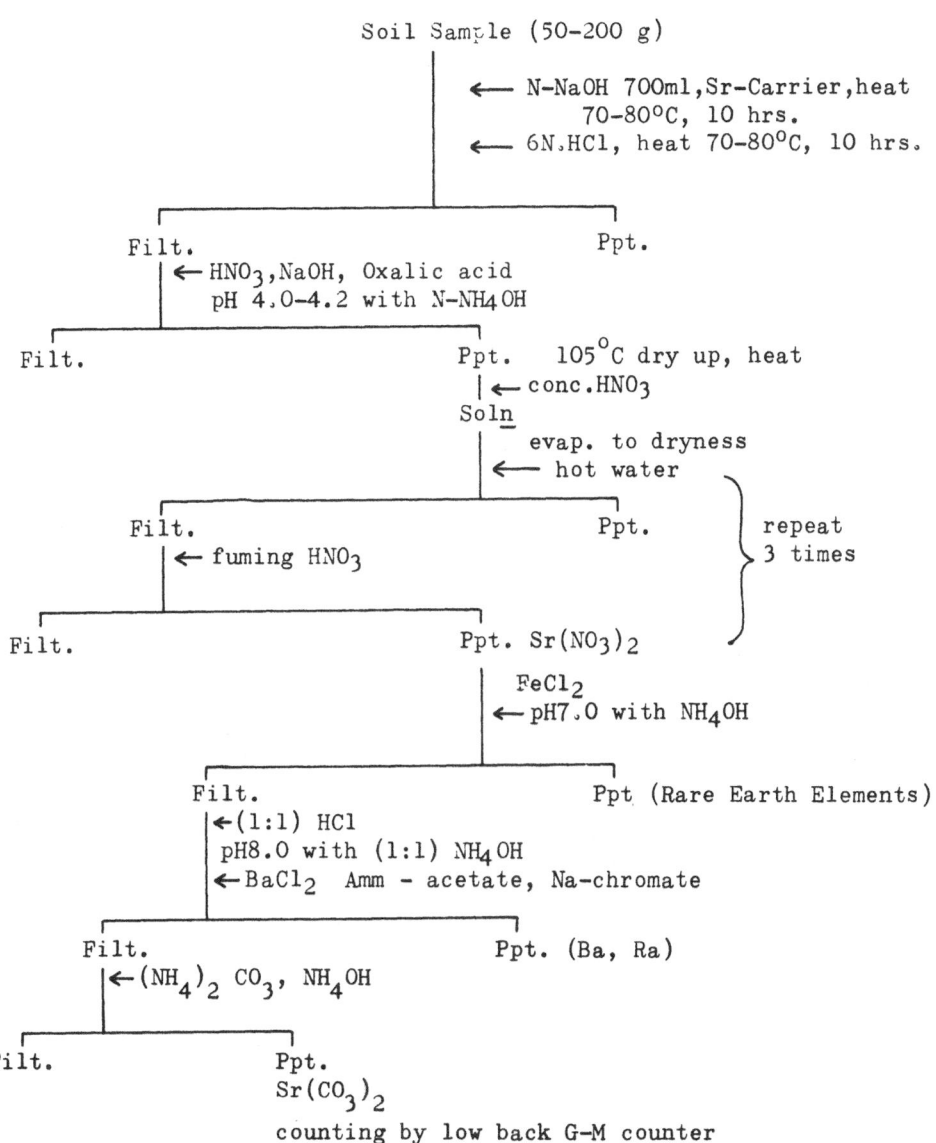

Fig. 1 Analyt. Procedure

at each horizon, in addition to the total concentration at each location
are given. In Table 2, index numbers of the horizons of ^{90}Sr concentration VS soil layers have been shown. In Table 3-13 are shown the physical
and chemical composition of each paddy soil samples. These tables were
distributed at SPS (1980.10.22). Therefore, they were omitted here.

DISCUSSION

Referring to Table 1, No.1 and No.2 were obtained at Gosen-Shi in
April, 1967, No.1 is gray brown soil with a clay content of 25.6 - 28.0%
and its soil class is SiC; it is an oxydative paddy soil whose whole section is mottled with rusty and Mn mottles, ^{90}Sr concentration is 162 m Ci*
(at depth 50 cm); it is 60% in No.I layer, 22% in No.II, and 82% of the
total is found in the range 0 - 25cm; and the extent of transition to or
penetration into the lower layers is not appreciable.

No.2 is gray brown soil. Its soil class is CL and its clay content
is 22.6-26.5%; it is an oxydative paddy soil in which rusty and Mn mottles
are found scattered throughout. ^{90}Sr concentration is 201mCi (depth 50cm);
it is 46% in No.I layer, 37% in No.II layer, and 83% is found in 0-20cm.

At this layer, well-water irrigation is rendered; nevertheless, the
accumulation of ^{90}Sr is relatively great.

No.3 was obtained at Muramatsu-Cho in April, 1967. The paddy soil
type is ando soil. Its soil class is SiC, containing 31.5 - 34.3% clay
and only a small amount of rusty mottles. No.I layer contains 9.61% humus
and its absorptive coefficient of P_2O_5 is 1.551 - 1.756. It is an oxydative
paddy soil whose parent material is volcanic ash. ^{90}Sr concentration is
316 mCi (depth 50 cm), which is the highest concentration among the samples
taken from the 11 locations. It is about 2.2 times that of ^{90}Sr (143 mCi)
found in the precipitation at Akita (at the end of 1970). It is 43% in
No.I layer, 36% in No.II layer, 11% in No.III layer, and 10% in No.IV
layer. Even in 38 - 50 cm of No.IV layer, 32 mCi still remains, and the
value of No.IV layer is also substantial. This probably results, because
the parent material is volcanic ash and its retentive power of ^{90}Sr is
weak. A plausible explanation for the maximum value shown is that the
annual rainfall, 2,500mm, is especially heavy even in Niigata Prefecture,
and the paddy soil is further contaminated by the influent irrigation
water.

No.4, 5 and 6 were obtained at Hatano-Cho in April, 1968. This town
is located on the island of Sado in the Japan Sea, 43 Km from Niigata City.
No.4 is gley soil and its soil class is HC; its clay content is 47.0 - 55.1
with increasing amounts in the lower layers. It is a reductive paddy soil in
which rusty mottles are found scatter up to No.III layer. ^{90}Sr concentration is 115 mCi (depth 70cm). It is 33% in No.I layer, 60% in No.II layer,
and in 0 - 25cm, it is 93%.

---

* ^{90}Sr concentration should be expressed as mCi/Km^2, but in this
  paper Km^2 will be omitted for simplicity.

Table 1 ^{90}Sr concentration of paddy soils (Unit mCi/Km^2)

Soil layer No.	Sampling Year	Location	Layer No. Depth (cm)	I	II	III	IV	V	Total
1	1967	Gosen-Shi	50	97	35	14	16		162
2	1967	Gosen-Shi	50	93	73	21	4	10	201
3	1967	Muramatsu-cho	50	137	114	33	32		316
4	1968	Hatano-Cho	70	38	69	6	2		115
5	1968	Hatano-Cho	55	55	28	39	32		154
6	1968	Hatano-Cho	50	48	29	4			81
7	1969	Kariwa-Mura	82	19	98	9	2	1	129
8	1969	Kariwa-Mura	51	64	12	5	1		82
9	1971	Itakura-Cho	45	52	45	6			103
10	1971	Itakura-Cho	49	67	49	9			125
11	1971	Itakura-Cho	65	47	32	3	1		83

Table 2
Index numbers of horizons of paddy soil ^{90}Sr concentration

Horizon No. / Soil layer No.	I	II	III	IV	V
1	60	22	8	10	
2	46	37	10	2	5
3	43	36	11	10	
4	33	60	5	2	
5	36	18	25	21	
6	59	36	5		
7	15	76	7	2	1
8	78	15	6	1	
9	51	44	6		
10	54	40	7		
11	56	39	4	1	

There is but slight transition to No.III layer, because its soil class is HC. No.5 is gravel soil and of soil class LC, containing 25.1 - 31.4% clay; the content increases with lowering of the layers. Rusty mottles are found scattered in all layers. Below 55cm, it is an oxydative paddy soil of great, medium, small gravelly layers containing Mn mottles. ^{90}Sr concentration is 154 mCi (depth 55 cm), 36% in No.I layer, 18% in No.II, 25% in No.III layer, 21% in No.IV layer, and in 28 - 50cm of No.IV layer 32 mCi is indicated. Below 55cm there are gravel layers, and the upper layers are of fine sandy soil, thus the water dissolves it and penetrates into the lower layers. No.6 is strong gley soil and soil class LC; its clay content is 39.6 - 42.1% which becomes higher as layers get lower. No.6 is a reductive paddy soil containing scattered rusty mottles. ^{90}Sr concentration is 81 mCi (depth 50 cm) and it is the lowest among the samples taken at the 11 locations. It is 59% in No.I layer, 36% in No.II layer, and 95% is found in 0 - 29cm, and it decreases as the layers get lower. There is a difference of almost 90% in the concentration from samples No.4, 5, 6 which are taken from the same location, Hatano-Cho. The difference probably arises from the complicated combined effects of irrigation water and underground water unique to the paddy soil.

No.7 and No.8 were obtained at Kariwa-Mura in November, 1969. No.7 is strong gley soil and its soil class is SCL with a clay content of from 17.7 to 19.9 for No.I and No.II layers, and it is the lowest among the samples taken from the 11 locations. It is a reductive paddy soil containing only a slight amount of rusty mottles. ^{90}Sr concentration is 129 mCi (depth 82 cm); 15% in No.I layer, 76% in No.II layer, and 91% is found in 0 - 23cm, and sectionally, No.II layer shows a high concentration amounting to 97.5 mCi. No.8 is strong gley soil and its soil class is HC containing from 36.3 to 52.4% of clay, higher in the upper and lower in the lower layers. It is a reductive paddy soil containing practically no rusty mottles. ^{90}Sr concentration is 82 mCi (depth 51cm). 78% is in No.I layer, 15% in No.II layer, and 93% is found in 0 - 31cm. No.I layer is of HC soil class and contains 52.4% clay; and the highest percentage is found in this layer, since its water permeability is poor. However, ^{90}Sr is washed away along the layer and its concentration is found to be low.

No.9, 10, and 11 samples were taken from Itakura-Cho in April, 1971. No.9 is gray brown soil with soil class LC, containing 32.0 - 39.9% clay, and it is found more abundantly in No.II and No.III layers. Moreover, the soil is an oxydative paddy soil rich in rusty mottles. ^{90}Sr concentration is 103 mCi (depth 45cm), 51% in No.I layer, 44% in No.II layer, and 95% is in 0 - 20cm, and there is less tendency to penetrate into the lower layers. No.10 is strong gley soil and its soil class is as follows: No.I layer is LC, No.II layer and the others following are HC. The clay contents are 26.2% and from 48.7 to 53.2% for No.I layer and for No.II to No.IV layers, respectively, and it is higher at the lower layers.

This is a reductive paddy soil containing only a small amount of rusty mottles at No.II layer. ^{90}Sr concentration is 125 mCi (depth 49cm), 54% in No.I layer, 40% in No.II layer, and 94% is found in 0 - 23cm. Soil class is HC, and there is less tendency to penetrate into the lower layers, since it is of reductive nature. No.11 is gray soil with soil

300

class of LC and contains 27.3 - 34.5% clay, it exists abundantly in No.II layer. Rusty mottles are found scattered in all layers. It is an oxydative paddy soil with a clay film formation at No.III layer and the lower. ^{90}Sr concentration is 83 mCi (depth 65cm) which is low. Since it is microtopographically located at an elevated position, there may occur a considerable loss in concentration by washing. 56% in No.I layer, 39% in No.II layer, and 95% in 0 - 19cm; and there is only a slight tendency or diffusion into the lower layers. Despite its being oxydative, it shows but slight tendency for diffusion into the lower layers. This may be associated with the clay minerals.

REFERENCE

(1) M.W. Meyer, J.S.Allen and L.T. Alexander,
    Strontium-90 on the earth's surface, IV. Health and Safety
    T.I.D.24341 (1968), Nature, 219,584 (1968).

RESIDUAL BHC IN PADDY SOILS AND ITS CONTAMINATION TO RICE

Zhang Shui-ming, Ma Xing-fa, Li Xun-guang*
(Institute of Soil Science, Academia Sinica, Nanjing)

At present, hexachloro-cyclohexane (BHC) is still one of the dominant pesticides commonly used in China. Whether a long term application of BHC would lead to the contamination of soil and crop is still a problem in dispute. The present article deals mainly with BHC remaining in paddy soils and its contamination to rice.

MATERIALS AND METHODS

(1) The plowed layer samples of paddy soils, 0—20cm in depth, were collected from Jiangsu, Sichuan, Guangdong, Hunan, Jilin and Henan provinces where BHC had been applied by the years of sampling.

(2) Pot experiment on the uptake of BHC residue by rice was carried out with paddy soils collected from Jiangsu Province.

(3) All samples were analysed by means of gas chromatography(1).

RESULTS AND DISCUSSION

1. Residual BHC in Paddy Soils

Analyses of 140 samples showed that BHC remaining in the plowed layer of paddy soils ranged from 0.0214 to 1.47ppm, with a mean of 0.293 ppm. This means that in spite of its continuous usage BHC remaining in paddy soils is not very high. In most of the samples, residual BHC is less than 0.5 ppm (Fig. 1).

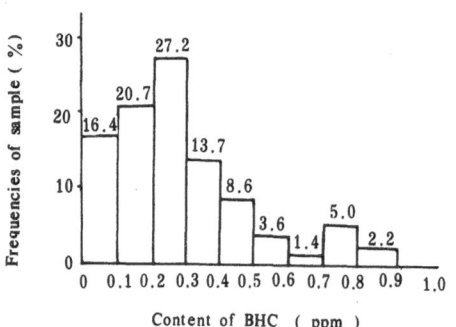

Fig. 1 Amount of BHC remained in soil

---

*Co-workers in this research also include: Cai Dao-ji and Li Xiao-ping.

The low content of BHC residue after its long term application in paddy soils is probably caused by its rapid degradation(2-5). A supplementary experiment showed that all the four BHC isomers were degraded rapidly in submerged soils. The rates of degradation of the isomers were in the order: $\gamma -> \alpha -> \beta$-and δ-BHC. 85.5% of γ-BHC were degraded in 7 days, and 92.1% in 14 days. For α-BHC there was a 3-day lag phase, and for β-and δ-BHC a 7-day lag phase respectively in the degradation. After the lag phase, they were all degraded rapidly. More than 90% of these three isomers were degraded in 28 days in a water-logged soil, and nearly all of the four isomers disappeared in 2 months under the same condition (Fig. 2).

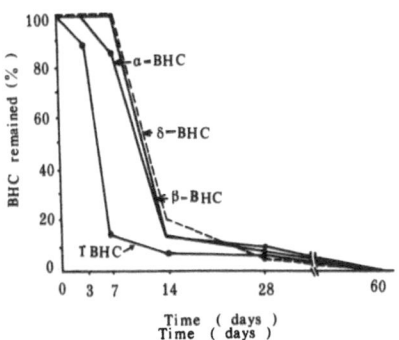

Fig. 2 Rate of degradation of
BHC in water logged soil

It was also observed that the rate of degradation in the treatments applied with 1% of milk vetch was faster than that in the control.

If we take Q_n as the amount of BHC residues accumulated, Q the yearly dosage of BHC, and K the rate of the residue remaining in soil per year, then,

$$Q_n = \sum_{n=0}^{n} OK^n = Q\ (1+K+K^2+...+K^n) = \frac{1-K^n}{1-K}\ Q \qquad (1)$$

The value of K depends on the rate of degradation and the migration of pesticides in soil. Since more than 95% of the BHC were degraded in 2 months on account of the rapid degradation in water-logged soils, the K value should be less than 0.05. Then,

$$Q_n < 1.05Q \qquad (2)$$

This means that the increase of BHC in soil should be less than 5% after many years if equal quantity of pesticide is applied every year. Generally, it does not lead to the severe accumulation of BHC. The amount of BHC residue in a soil is mainly governed by the dosage applied in that year.

2. The Relative Stability of the Four BHC Isomers

BHC consists of α, β, γ and δ isomers. Their stabilities in soil are

Table 1 shows that the content of β-BHC in soils is highest, amounting to 33.4-48.6% (mean = 44.4%) of the total, while those of others are α-BHC 25.8-49.8% (mean = 35.0%), δ-BHC 7.3-14.8% (mean = 13.0%), and γ-BHC only 4.4-19.2% (mean = 7.6%). However, it is difficult to judge the relative rate of dissipation from the amount of each isomer remaining in the soil, for the contents of these isomers in the original BHC powder were unequal (α about 67%, β-8%, γ 15%, δ 7.5%). Instead of using the ratio of α/γ, β/γ and δ/γ to indicate the relative persistence of each isomer in soil by some authors, we think it more appropriate to use the following equation:

$$S = \frac{q_i}{Q_t P}$$

Where S is the index of relative persistence, P the % of each isomer in BHC powder, q_i the amount of each isomer remaining in soil, and Q_t the total residual BHC in soil. The data for 140 samples of paddy soils are shown in Table 2. It is seen that the S for the four isomers is in the order: $\beta_s <$ $\delta_s \gg \gamma_s$ or α_s implying that β-and δ-BHC are more stable than α-and γ-BHC in soils.

Table 1   Content of the four BHC isomers in soil

Range (ppm)	No. of samples	Mean value of BHC content (ppm)	% of each isomer			
			α	β	γ	δ
0.0-0.1	23	0.0426	38.6	34.8	19.2	7.3
0.1-0.2	29	0.144	25.8	48.6	13.7	11.8
0.2-0.3	38	0.244	32.8	45.6	7.7	14.0
0.3-0.4	19	0.347	34.6	46.0	6.7	12.9
0.4-0.5	12	0.452	33.1	45.5	7.4	13.8
0.5-0.6	5	0.520	41.1	39.4	6.5	13.1
0.6-0.7	2	0.646	39.5	42.4	8.0	10.3
0.7-0.8	7	0.756	37.1	45.3	4.4	13.3
0.8-0.9	3	0.840	49.8	33.4	6.9	9.7
> 1.0	2	1.346	34.1	46.2	4.9	14.8
0.0214-1.47	140	0.293	35.0	44.4	7.6	13.0

Table 2   Index of relative persistence of BHC in paddy soil

Range (ppm)	No. of samples	Index of relative persistence (S)			
		α	β	γ	δ
0.0-0.1	23	0.575	4.35	1.28	0.99
0.1-0.2	29	0.384	6.07	0.91	1.57
0.2-0.3	38	0.490	5.70	0.513	1.87
0.3-0.4	19	0.516	5.74	0.445	1.72
0.4-0.5	12	0.495	5.73	0.494	1.84
0.5-0.6	5	0.613	4.92	0.433	1.75
0.6-0.7	2	0.590	5.30	0.534	1.38
0.7-0.8	7	0.553	5.66	0.294	1.77
0.8-0.9	3	0.742	4.17	0.46	1.30
> 1.0	2	0.510	5.77	0.326	1.97
0.0214-1.47	140	0.522	5.54	0.507	1.73

3. Uptake of Residual BHC by Rice

Pot experiment shows that residual BHC in paddy soils collected from 6 counties in Jiangsu Province is 0.134 to 0.506 ppm (Table 3). The content of BHC in brown rice lies between 46-84 ppb. Its percentage of uptake is 11.4-35.6%, with a mean value of 18.5%. The percentages of uptake for each isomer are different. In brown rice, the uptake of α and γ-BHC are rather high, amounting to 22.8 and 26.6% respectively, while that of β-and δ-BHC are 12.8 and 13.8% respectively. The correlation between the BHC content in brown rice and that remaining in soil is not significant (correlation coefficient: r = 0.647), but the uptake of γ-BHC in brown rice increases with the increase of residual γ-BHC in soil, with a correlation coefficient of 0.925.

Table 3   Uptake of residual BHC in brown rice

Locality	BHC isomer in soil (ppb)					BHC isomer in brown rice (ppb)					Rate of uptake (%)				
	α	β	γ	δ	Total	α	β	γ	δ	Total	α	β	γ	δ	Total
Wujiang	46	50	20	16	134	23	15	7	3	48	50	30		35	18.8 35.6
Xiangshui	19	98	16	17	150	23	15	7	4	49	121	15.3	43.7	23.6	32.6
Wuxi	143	67	35	60	305	24	13	8	5	50	16.8	19.4	22.8	8.3	16.4
Danyang	63	190	23	35	311	23	12	7	4	46	36.6	6.3	30.5	11.4	14.8
Jiangning	170	184	45	58	457	34	24	11	15	84	20.0	13.0	24.5	25.8	18.4
Jurong	235	132	49	90	506	27	14	10	7	58	11.7	10.6	20.4	7.8	11.4
Mean value	112.7	120.7	31.3	46	310.5	25.7	15.5	8.3	6.3	55.3	22.8	12.8	26.6	13.8	18.5

REFERENCES

(1) Editorial Group of the Analysis of Environmental Contamination, Aanlysis of Environmental Contamination. Science Press, Beijing, 313-327 (1980). (in Chinese)
(2) Kohnen, R., Haider, K., Jagnow, G., Investigations on the microbial degradation of lidane in submerged and aerated moist soil. Environmental Quality and Safety Supplement, V. III, Pesticides, 222-225 (1975)
(3) Sethunathan, N., Microbial degradation of insecticides in flooded soil and in anaerobic cultures. Residue Rev., 47, 134-165 (1973).
(4) Suzuk, M., Yamato, Y., Watanabe, T., Persistence of BHC (1, 2, 3, 4, 5, 6-Hexachlorocyclohexane) and dieldrin residues in field soils. Bull. Environ. Contam. Toxicol., 14 (5), 520-529 (1975).
(5) Castro, T.F., Yoshida, T., Effect of organic matter on the biodegradation of some organochlorine insecticides in submerged soils. Soil Sci. Plant Nutr., 20 (4), 363-370 (1974).

DECOMPOSITION OF PLANT MATERIALS IN RELATION TO THEIR
CHEMICAL OCMPOSITION IN PADDY SOIL

Shi Shu-lian, Lin Xin-xiong, Wen Qi-xiao
(Institute of Soil Science, Academia Sinica, Nanjing)

It has been well known that soil organic matter is one of the most important natural factors determining the productivity of soil. As soil organic matter is derived from plant materials, information on the quantity and quality of soil organic matter formed from plant materials under given conditions,is the most importance for establishing a rational cropping system. This paper presents some preliminary experimental results of the decomposition and humification of plant materials in paddy fields. Results of pot experiments on the effect of chemical composition of some green manures on their nitrogen availability are also presented.

AMOUNT OF HUMUS FORMED IN RELATION TO THE CHEMICAL
COMPOSITION OF PLANT MATERIALS

13 plant residues including straws of cereal and leguminous crops and various kinds of green manure crops were examined. They were dried at $< 80^{\circ}C$ and passed through a 0.42 mm seive. Each kind of plant material was added to soil with very low organic matter content at a rate of 2% (W:W), mixed and transfered into a carborundum tube. A series of carborundum tubes were buried in the plough layer of paddy field. At intervals samples were taken for analyzing carbon content[1].

Fig. 1 shows the percentage of plant carbon remained at different time after decomposition commenced. It indicates that under favourable conditions, decomposition of plant residues proceeded most rapidly in the 1st month, and went down in the 4th month. It suggests that nearly all the easily decomposable chemical constituents were decomposed within the initial 3 months. Because of the differences in the content of decomposable chemical constituents of plant residues, the largest difference in decomposition rate among various plant residues was at the initial period of decomposition. Although the difference in % of carbon remained in soil per unit wieght of added plant carbon among various plant residues decreased as the decomposition went on, it still could be noticed even in the third year (Table 1). Accordingly, it may be concluded that the chemical composition of plant residues not only affects on their initial decomposition rate, but also on the amount of humus they formed.

Since plant residues enter into soil yearly, it is preferable to use the humification coffeicient as an index denoting the contribution of plant residues to soil organic matter content. Results of decomposition experiments conducted under field condition show that humification coefficient varied greatly according to the different kinds of plant matreials (Table 2). It is correlated significantly with the lignin content, and (% of lignin). [% (water soluble fraction + organic solvents soluble fraction + carbohydrate)]$^{-1}$;as well; but it is not correlated with the water souble fraction content and the C:N ratio (Table 3).

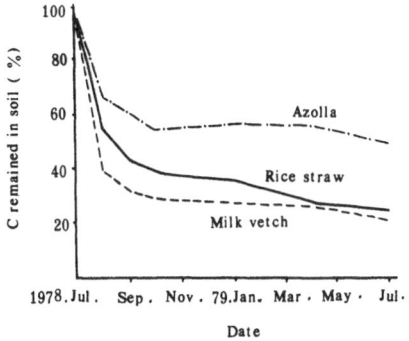

Fig. 1 D composition of various plant materials in paddy field

Table 1 Percentage of plant carbon remained in soil at different decomposition period

Material	Year		
	1st	2nd	3rd
Milk vetch	18	16	13
Common water hyacinth	24	20	17
Rice straw	23	22	18
Azolla	43	40	33

Table 2 Chemical composition and humification coefficient of plant materials

Material	C:N	Lignin (%)	Humification coefficient
Rice roots	39.3	17.4	0.50
Azolla	11.2	20.2	0.43
Aesbania	24.5	11.8	0.37
Sum crotalaria	28.5	15.3	0.36
Wheat roots	49.3	20.7	0.32
Wheat straw	104	19.9	0.31
Vetch	11.8	13.5	0.30
Common vetch	14.7	11.5	0.27
Common water hyacinth	16.3	10.2	0.24
Rice straw	61.8	12.2	0.23
Broad bean stalk	12.6	8.7	0.21
Corn-stalk	51.0	13.2	0.19
Milk vetch	14.8	8.6	0.18

Table 3 Correlation coefficients between humification coefficient and chemical constituents of plant materials (n = 13)

Chemical constituent	Correlation coefficient (r)
C:N	0.005
% of water soluble fraction	−0.369
% of lignin	0.660*
% of lignin). [% (water sol. frac.+ org. solv. sol. + Carbohydrate)] -1	0.750**

* $P < 0.05$
** $P < 0.01$

AVAILABILITY OF NITROGEN OF PLANT MATERIALS WITH
DIFFERENT CHEMICAL COMPOSITION

Pot experiments were carried out to investigate the availability of nitrogen of some green manures with different chemical composition 15N-labelled milk vetch, azolla and common water hyacinth were used as testing materials. Rice, buckwheat and barley were used as testing crops successively(2).

Table 4 shows that green manures with different chemical composition differ from each other not only in their nitrogen availability during the first growing season after they have been incorporated into soil, but also in their residual nitrogen availability during succeeding growing seasons. Among plant materials tested, nitrogen originated from milk vetch is most available both during the first and succeeding growing seasons, and the availability of residual nitrogen originated from azolla seasons is the smallest.

Table 4 Availability of nitrogen of some green manures

Material	Recovery by 1st crop (%)	Nitrogen availability ratio**		
		2nd crop	3rd crop	Closed* incubation
Milk vetch	37.6	2.62	3.00	1.61
Azolla	25.4	1.39	1.04	0.81
Common water hyacinth	20.2	1.71	1.63	1.11

*Soil samples taken after 3rd crop.
**Nitrogen availability ratio

$$= \frac{\% \text{ of residual } ^{15}N \text{ originated from manure assimilation by plant}}{\% \text{ of native soil organic nitrogen assimilated by palnt}}$$

Table 5 shows the distribution of residual ^{15}N in soil fraction. As can be seen from table 5, while only about 2-4% of residual ^{15}N derived from milk vetch or common water hyacinth is found in the light fraction before the 2nd cropping, nearly 50% of residual ^{15}N originated from azolla remains to be decomposed and at the end of the 3rd growing season ,^{15}N originated from azolla in light fraction amounts to 41% of the total residual ^{15}N. These results demonstrate clearly that the high lignin content of azolla is responsible for the low availability of its residual nitrogen.

Table 6 shows the relative distribution of soil nitrogen and residual labelled nitrogen in heavy fraction, significantly more ^{15}N is found in the amino-acid and unidentified fraction of hydrolyzates as compard with native soil nitrogen. As it is generally acknowledged that nitrogen in the amino-acid is easy to be absorbed by crops. Nevertheless, no difference in the relative distribution of ^{15}N in heavy fraction among green manures tested

Table 5 Percentage distribution of applied ^{15}N in soil fractions

Treatment	Distribution (%)			
	After 1st crop		After 3rd crop	
	LF*	HF*	LF	HF
Milk vetch	2	98	4	96
Azolla	48	52	41	59
Common water hyacinth	4	96	3	97

  * LF-Light fraction (SP. gr,< 1.80).
 ** HF-Heavy fraction.

can be observed although the residual nitrogen availability of them differs considerably from each other.

Table 6 Distribution of soil N and residual labelled N in soil(% of the total ^{14}N and ^{15}N in HF respectively)

Treatment	Nitrogen distribution (%)					
	NH$_4$ -N		Amino-acid		Unidentified N	
	^{14}N	^{15}N	^{14}N	^{15}N	^{14}N	^{15}N
	After 1st crop					
Azolla	31.2	19.4	28.6	36.5	17.7	28.5
Milk vetch	30.9	20.2	35.1	41.8	8.5	17.8
Common water hyacinth	32.2	23.2	33.5	39.3	12.3	22.3
	After 3rd crop					
Azolla	30.6	16.8	16.6	23.1	24.5	39.6
Milk vetch	28.3	17.6	26.6	32.2	13.2	24.4
Common water hyacinth	31.7	23.8	31.0	38.9	17.2	20.7

There is significant difference among manure-derived nitrogen in heavy fraction as to their extractability, 31-35% of the nitrogen derived from milk vetch can be soluble in 0.1N NaOH and 0.1M Na$_4$P$_2$O$_7$ solution, while only 19-20% of the nitrogen derived from azolla is extrated (Table 7).

It seems that different newly formed humus will be produced from plant materials with different chemical composition.

309

Table 7   Distribution of ^{15}N in humus fractions
(% of total ^{15}N in the HF)

Treatment	Distribution (%)					
	HA		FA		Humin	
	1*	2*	1	2	1	2
Milk vetch	14.7	12.8	20.4	18.7	65.0	68.5
Azolla	7.8	6.4	11.9	12.4	80.3	81.2
Common water hyacinth	10.3	8.6	14.2	13.5	75.5	78.0

* 1— soil samples taken after 1st crop; 2—soil samples taken after 3rd crop

REFERENCES

(1) Lin Xin-xiong, Cheng Li-li, Shi Shu-lian, Wen Qi-xiao, Decomposition of plant residues as effeted by their chemical composition and rotation system. Acta Pedologica Sinica, 17(4), 319-327(1980). (in Chinese with English summary)
(2) Shi Shu-lian, Wen Qi-xiao, Liao Hai-qiu, The availability of nitrogen of green manures in relation to their chemical composition. Acta Pedologica Sinica, 17(3), 240-246(1980). (in Chinese with English summary)

SOME PHYSICAL PROPERTIES OF PADDY PROFILES IN TAIHU LAKE REGION

Yu De-fen, Yao Xian-liang

(Institute of Soil Science, Academia Sinica, Nanjing)

Farmers living in the Taihu Lake region used to consider that good structure is one of the most important factors for the high productivity of paddy soils. High-yield paddy soils are readily permeable and easily managed, but low-yield soils are not. In order to study the nature of paddy soil profiles relating to their permeability in that region, we chose more than 10 representative profiles in fields grown milk vetch in spring and measured some of their physical properties both in situ and in laboratory.

SOILS AND METHODS

Most of the paddy soil profiles under study are derived from loessial lacustrine deposits. The texture of various profiles is rather uniform with a clay content (<0.001 mm) of about 25% in cultivated horizon and slightly higher in 50-100 cm horizon.

Total porosity, aeration porosity and water-holding porosity of soil were measured on undisturbed soil samples collected with a volume-weight ring of 100 cm^3 in capacity. Total porosity was obtained from specific gravity and volume weight. Aeration porosity was calculated as the difference between total porosity and water-holding porosity.

The infiltration rate was measured with a hard plastic tube of 4.4 cm in diameter and 13 cm in height. Ten tubes were imbedded into each horizon for determination, and then the steady infiltration rate after flooding the soil was recorded. The following formula was used for correcting temperature variations:

$$K_{10} = \frac{K_t}{0.7 + 0.03 \ t^\circ C}$$

where

K_t —— actually measured infiltration rate (mm/min);

K_{10} —— infiltration rate at $10^\circ C$;

t —— water temperature at the time of measurement;

0.7 and 0.03 —— empirical constants.

Hardness of soil was measured with a penetrometer, and mean value was taken from 20 measurements for each horizon.

RESULTS AND DISCUSSION

1. Distribution of Porosity in Profile

It is seen from Fig. 1 that the porosity in cultivated horizon is relatively high, but it decreases markedly from plowpan downwards. The aeration porosity in the cultivated horizon of high-yield paddy soil with good structure is relatively high, generally 7-10% or more, but about 2-4% in other horizons below. In most cases the aeration porosity in cultivated horizon of low-yield soils with poor structure is less than 5%, and even less than 1% in plowpan and percogenic horizons. It seems that the distribution of aeration porosity in profile is closely correlated with the good permeability of high-yield paddy soils and the liability of water-logging for low-yield soils.

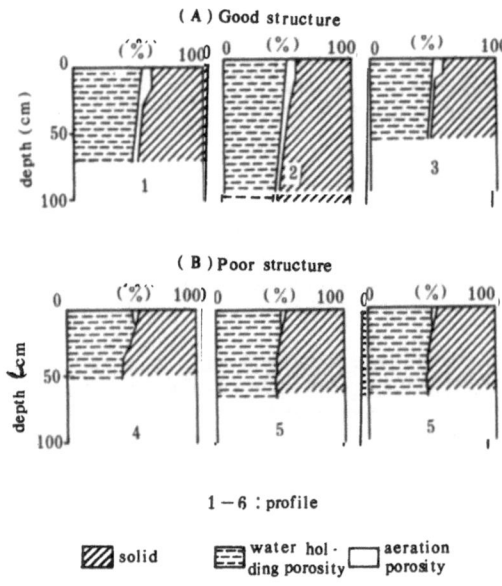

Fig. 1 Distribution of different
porosities in paddy soil
profiles

2. Infiltration Rate of Various Horizons in Profile

The infiltration rate of soil is affected by aeration porosity.
Fig. 2 shows that the infiltration rate in spring is highest in cultivated horizon and lowest in plowpan. The rate is about 1-12 mm/min in cultivated horizon, 0-0.5 in plowpan and 0.5-2.5 in percogenic horizon.
The rates in the three horizons of the paddy soils with good structure are

generally >3, >0.1, and >0.5 mm/min respectively. But, the rate in
plowpan of low-yield paddy soil is nearly zero. Autumn plowing is an
important factor influenced infiltration rate. The rate is rather low
in plowpan when the field was planted with milk vetch without plowing.
But the rate becomes relatively high when the soil was planted with wheat
or rape owing to autumn plowing and the formation of some crackings in the
plowpan caused by dehydration. Consequently, the soil as a whole is still
permeable to certain extent, although the structure in the plowpan is very
dense. It is beneficial to the improvement of permeability of the plowpan
if the field is plowed in autumn and frozen in winter.

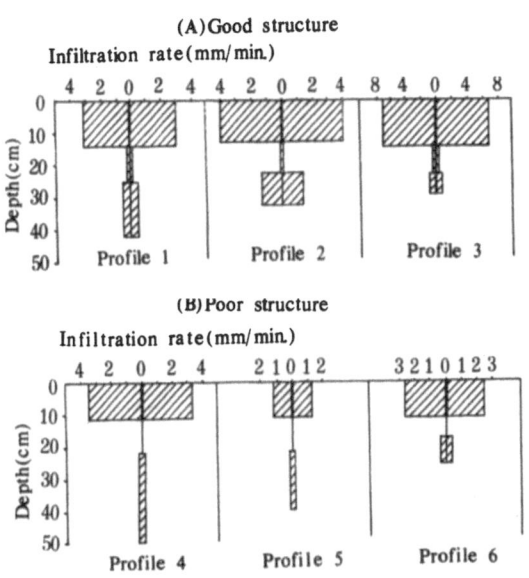

Fig. 2 Infiltration rate of paddy
        soil profiles

3. Hardness in Various Horizons

    Under the influence of intensive cultivating operations, the structural
units in cultivated horizon are mostly clods formed by artificial crushing
rather than aggregates formed by natural weathering. Therefore, the brittle-
ness for the air-dried clods is an important index of soil structure[*]. In
Fig. 3 it is shown the relationship between hardness of soil and its moisture
content under natural conditions. It is seen that the hardness is lowest
(2-4 kg/cm^2) in cultivated horizon, highest (4-10 kg/cm^2) in percogenic

---

[*]Yao Yian-liang, Yu De-fen, Preliminary investigation of structural characteristics of fertile paddy soils. Acta Pedologica Sinica, 15 (1),
1-12 (1978). (in Chinese with English summary)

horizon, with substratum (3-7 kg/cm^2) intermediate. However, it may be
on account of the dehydration that the hardness is slightly higher in
cultivated horizon of well-structural soils. The hardness of percogneic
horizon or substratum depends mainly on its moisture or clay contents.
Of course, the hardness in cultivated horizon is very changeable, depending chiefly on the moisture content of soil.

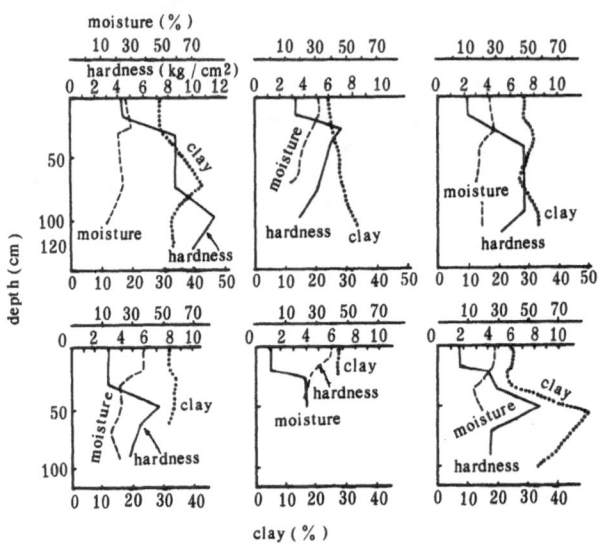

Fig. 3 Hardness, clay content and
moisture content of paddy
profiles

Factors affecting the permeability and hardness include size and
quantity of pore space, clay content and moisture content, etc. Statistical data show their correlations in the Table 1.

From the above it is apparent that (A) it is necessary to increase
the aeration porosity in order to increase the infiltration rate or
decrease the hardness of soil; (B) a adequate aeration porosity is an
important criterion of fertile clayey paddy soils.

CONCLUSIONS

Based on the present work, we consider that during the growing
period of dry-farming crops, the high-yield paddy soil should have the
following aeration porosities:
    cultivated horizon: 7 or 10%,
    plowpan: 2-4%,
    percogenic horizon: 3-5%,
and the following infiltration rates:

cultivated horizon: 3 mm/min,
plowpan: 0.1 mm/min,
percogenic horizon: slightly higher than plowpan.

Table 1 Correlations of some physical factors
         in paddy soils

Y	X	r	p	R	N
I	Pa	0.652	<0.01	Y=0.832x-1.326	30
H	Pa	-0.455	<0.01	Y=7.804-0.642x	42
H	M	-0.595	<0.01	Y=14.582-25.706x	42
H	Dv	0.688	<0.01	Y=14.202x-13.455	42
H	C	0.403	<0.01	Y=0.342+0.175x	56

I—Infiltration rate (mm/min); H—Hardness
(kg/cm^2); Pa—Aeration porosity (V%);
M—Moisture content (W%); Dv—Volume weight
(g/cm^3); C—Clay content (%)

HYDROLOGIC STATUS OF PADDY FIELDS
DURING RICE-GROWING PERIOD

Zhang Bing-gang
(Guangdong Institute of Soil Science, Guangzhou)
Chen Zhi-xiong
(Institute of Soil Science, Academia Sinica, Nanjing)

The hydrologic status of a paddy field reflects the liquid pressure of the moving soil water. Experiences in China show that the drainage condition of paddy field is one of the important factors affecting the growth and yield of rice. In order to drain the paddy field rationally so as to control the water permeability to a proper range, a perfect knowledge of the hydrologic status of the whole field is required. For, the movement of soil water is determined by Hydraulic gradient and hydraulic conductivity of the soil while the former factor is controlled by the water pressure within the whole soil.

It is well known that the flow rate is controlled by layer of poorest permeability alone as water flows through a stratified soil. In vertically downward flow, if the soil layer with poor permeability lies by the upside, then a soil layer with negative pressure may appear within the profile[*]. For most of the paddy fields there is generally a compact plowpan with low water permeability, thus the presence of a soil layer with negative pressure in a submerged field is quite possible.

In the following, based on the results of tensiometer measurements, we shall discuss the position of the layer with negative pressure, the magnitude of the negative pressure and factors affecting it.

HYDROLOGIC STATUS OF PADDY FIELDS
AT THE PERIOD OF SUBMERGING

Measurements conducted on mid-gleyed clayey paddy soils in Zhujiang Delta showed that the hydraulic status was different in various parts of a field. It is seen from Fig. 1 that near the drainage ditch the layer with negative pressure is very distinct and the negative pressure is of greater magnitude (Fig. 1 a). With increasing distance from the ditch, the magnitude of negative pressure becomes smaller (Fig. 1 b). At a certain distance beyond the ditch, water pressure remains positive throughout the whole profile (Fig. 1 c).

---

[*] Shunsuke Takagi, Analysis of the vertical downward flow of water through a two-layered soil. Soil Sci., 98-103 (1960).

The above difference is apparently related to different flux in various parts of the field. It is conceiveable that the flux near the drainage ditch is relatively high. And the layer with negative pressure may appear wherever the flux is high enough.

Another necessary condition for the formation of a layer with negative pressure is the presence of a poor permeahility layer lying above it.

Table 1 Some physical properties of a paddy field (Guangdong)

Horizon	Depth (cm)	<0.01mm particle (%)	Volume weight (g/cm³)	Hydraulic Conductivity (cm/hr.)
Cultivated	2-7	62.7	$1.26^{\pm}0.05$	$0.747^{\pm}0.541$
Plowpan-1	15-20	64.8	$1.33^{\pm}0.02$	————
Plowpan-2	25-30	75.3	$1.39^{\pm}0.02$	$0.016^{\pm}0.022$
Illuvial	40-45	69.5	$1.31^{\pm}0.03$	$0.095^{\pm}0.071$

It is seen from Table 1 that the hydraulic conductivity of the plowpan is only about one-sixth of the lower-lying layer. Should water flow through such a stratified soil profile, the water pressure in various layers would undergo some changes. It is only when the water pressure gradient in the plowpan is increased and that in the lower-lying layers decreased, that water will flow with a steady speed across the various horizons of whole profile. If the water pressure at the lower border of plowpan becomes less than zero pressure, a layer with negative pressure results, as is shown in Fig. 1 a and 1 b.

In Fig.1a, it is also worthy to be noted that the magnitude of layer with negative pressure is larger for the field with a lower water table (Fig. 1a-2, 1a-3) than for those with a higher water table (Fig. 1a-1). This may be related to the horizontal flow of water.

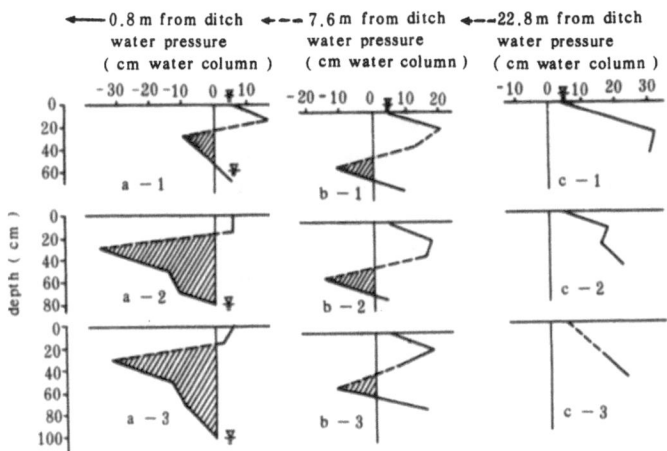

Fig. 1 Water Pressure under different drainage conditions for a paddy field

317

What is the practical significance of the negative pressure layer?

It is a well-known fact that the prerequisites for a good drainage system are a low water table in the drainage ditch and a proper distance between ditches. To improve the drainage of a paddy field, it seems reasonable to dig a drainage ditch at the places where there is no negative pressure layer.

Inasmuch as the negative pressure layer is a reflection of low water permeability of the plowpan under certain flux, a plausible reasoning seems to be that it is possible to use the magnitude of the negative pressure layer at a definite position as a qualitative index of permeability of a plowpan.

It is not clear yet as to whether the negative pressure layer has some bearings on rice growth. However, it is generally observed that rice grown near a drainage ditch often ripens in advance and has better grains. Whether this is related to fast percolation or to the change in oxidation-reduction conditions of the soil deserves to be further studied.

HYDROLOGIC STATUS OF PADDY FIELDS AT THE PERIOD OF SUMMER-DRYING

In China, summer-drying is a commonly used agricultural practice for paddy field. Farmers usually dry the soil slightly till small cracks appear in it, or moderately or heavily till large cracks appear, depending on soil conditions and the growth condition of rice plant. Some measurement on a clayey alluval paddy soil with a tensiometer showed that the suctions at a depth of 15cm for the slightly, moderately and heavily dried fields were about 270, 460 and 650 cm water column respectively as is shown in Fig. 2. It is suggested that these figures can be used as a semi-quantitative index of the degree of summer-drying for the same type of soil mentioned above.

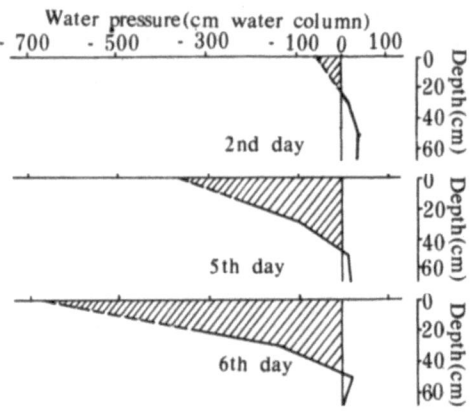

Fig. 2 Change in water pressure
during summer drying of
a paddy field

CONCLUSION

The hydrologic status of paddy soils differs both from that of
unsaturated arable soils and of completely saturated gleysol. The
frequent presence of layer with negative pressure makes it impossible
to study the hydrologic status of a paddy profile by means of a piezometer,
which can only register positive pressure. A tensiometer provides a
useful tool for such purpose.

RELEASE OF NITROGEN FROM DECOMPOSING LEGUME ROOTS AND NODULES

John S. Waid, Azizah Chulan*
(Department of Microbiology, La Trobe University,
Bundoora, 3083, Australia)

At present there is a considerable world-wide research effort under way to boost biological nitrogen fixation. The areas of research thrust[1] include (a) the improvement of legume inoculation (rhizobial technology), (b) the modification of farming systems and the creation of new legume cultivars so that legume green manures, forage legumes and legumes intercropped with non-legumes are used in agriculture more extensively and more efficiently, (c) the use of the techniques of genetic engineering to fashion and improve hosts as well as symbiotic N_2-fixing micro-organisms, (d) efforts to harness to the needs of agriculture symbiotic combinations, such as the <u>Azolla</u>-<u>Anabaena</u> or actinomycete (<u>Frankia</u>)-nodulated angiosperm associations, which hitherto have not been exploited to the full nor have the conditions under which such associations fix N_2 optimally been determined and, (e) investigations to understand and to use biological N_2-fixing associations with grasses, grain cereals and other non-legumes.

Research on biological N_2-fixation has become important because the cost of industrially-fixed N fertilizer is high, not only in terms of cash expenditure by the farmer, but also in terms of fossil fuel consumption and the consequent depletion of non-renewable resources. For example, Phillips <u>et</u> <u>al</u>.[2] have shown that about 40% of the energy consumed in the production of the corn crop in the USA is used to manufacture the N fertilizer applied to this crop.

Biologically-fixed N is undoubtably an attractive alternative to industrially-fixed N because it is cheaper and uses solar energy instead of fossil fuel. Consequently the pace of research has quickened and there is a considerable and ever-increasing annual output of publications on every conceivable aspect of biological N_2-fixation. And an ample number of field-scale experiments have demonstrated the benefits of legumes to companion or suceeding non-legume crops in terms of their enriched N nutrition. Yet there is surprisingly little information on the mechanisms whereby fixed N is transferred from living or decomposing legumes to the soil or to companion crops. There are, for example, very few publications on the microbiology or chemistry of the decomposition of legume roots and root nodules[3].

It has been claimed by Wittwer[1] that "Biologically fixed nitrogen is slow-release nitrogen. Losses from nitrification and denitrification, which now approximate 50 to 65% of the fertilizer nitrogen, could be reduced by 50%." This claim requires careful examination before it can be substantiated because the organic N in decomposing legume residues with low C-to-N ratios could be prone to rapid mineralisation to NH_4^+ and NO_3^- with consequent losses of N from the rooting zone in soil either as NH_3 through volatilisation or as N_2 or N_2O through denitrification of NO_3^-, or through leaching of NO_3^-. Indeed the biologically-fixed N might well be as vulnerable as industrially-fixed N to various mechanisms of loss from the soil system. Therefore it is

---

*Present address: Faculty of Science and Environmental Studies, Universiti Pertanian Malaysia, Serdang, Malaysia.

important that we discover how efficiently legume-fixed N can be conserved in plant residues or in soil organic matter before its eventual transfer to non-legumes. One way to conserve legume-fixed N might be to breed legumes which release N slowly perhaps by selection for legumes with polyphenol or tannin-rich tissues resistant to microbial decay. Indeed many of the problems which may have to be faced could well prove to be analagous to those investigated during the last three decades when slow-release N fertilisers were formulated and tested.

Legumes are not necessarily cheap crops to grow because to fix N they may require provision of fertilizer P and treatment with pesticides and so forth, thus making demands on non-renewable or difficult-to-renew mineral, energy and soil resources. The farmer makes an investment when he sows and cultivates a legume crop & if he expects a return in the form of biologically-fixed N_2 it is up to research scientists to ensure he is not disappointed.

EXPERIMENTAL

We have investigated one minor aspect of this problem namely the release of N during the decomposition of the roots and nodules of the tropical leguminous cover-crop <u>Centrosema</u> <u>pubescans</u> Benth. This legume is used as a ground-cover in plantations where it has proved to be very effective in enriching or maintaining the amount of N in the ecosystem.

In a laboratory study, described in full detail elsewhere[4] we measured the rates of decomposition of radicles, laterals and nodules of roots of <u>C. pubescens</u> and the accumulation of total N and mineral N in the soil.

To do this samples (20g) of an air-dry, sieved (<2mm) silty loam (pH 4.9) with or without (controls) the incorporation of 200 mg root segments of <u>C. pubescens</u>, were moistened to 50% moisture-holding capacity at 0.33 bar (33 kPa). The root portions were healthy, freshly-harvested, clean samples of nodules, laterals or radicles. Soil samples were incubated aerobically at 30°C for 16 weeks. The radicles, laterals and nodules respectively contained 1.65, 3.46 and 6.91 mg N per sample and their C to N ratios were 16.5, 7.0 and 4.2.

The root fragments disappeared rapidly and were difficult to recover by hand-picking. Less than 40% and 10% of the initial dry weight of roots remained as recoverable fragments after 1 and 16 weeks respectively; nodules decomposed more rapidly than laterals and laterals more quickly than radicles. At 16 weeks the total N contents of the controls had not changed but soils treated with radicles, laterals and nodules respectively showed gains over the controls from week 1 onwards and by week 16, these amounted to 8.2, 17.1 and 31.8% N. Concentrations of mineral N increased in all soils from week 4 onwards and by week 16, the amounts of mineral N that had accumulated in nodule, lateral and radicle-treated soils were 1.62, 1.16 and 1.14 times greater that in the controls (201.1 µg min $N.g^{-1}$ o.d.). The mineral-N ($NH_4^+ + NO_3^-$-N) at 16 weeks represented 37, 19 and 32% and NO_3^--N 2, 5 and 7% of the total accumulated N in soils treated with nodules, laterals and radicles respectively.

Under the conditions of this experiment, visible fragments of legume roots and nodules disappeared rapidly and all of the legume root N and all but an unaccounted for 5.6% (which may represent experimental or analytical errors) of the nodule N was transferred to the soil without any apparent net loss or gain of volatile N.

Our results confirm that legume root and nodule residues incorporated in soil disappear rapidly and that their disappearance is associated with an increase in the concentration of mineral N in soil. But we were not tracing the origin of the mineral N by use of labelled N so we are unable to say whether the mineral N was derived from the decomposing residues or more labile components of the organic matter of the soil used or both.

There were no apparent gains or significant losses of N from the soils but we do not know to what extent the artificial system used minimics the field situation nor were we measuring losses or gains using labelled N. Our experiments suggest that more detailed investigations of the fate of biologically-fixed N might be used to advantage to ensure the N gained is transferred to non-leguminous crops.

ACKNOWLEDGEMENTS

John Waid thanks the Nanjing Institute of Soil Science, Academia Sinica for the opportunity to participate in this symposium, and the Australian-China Council and the Australian Research Grants Committee for financial support. Azizah Chulan thanks the Australian Development and Assistance Bureau and the Universiti Pertanian Malaysia for their support of her research studies in Australia and Malaysia respectively.

REFERENCES

(1) Wittwer, S.H., Nitrogen fixation and agricultural producitivity. BioScience, 28, 555 (1980).
(2) Phillips, R.E., Belvins, R.L., Thomas, G.W., Frye, W.W., Phillips, S.H., No tillage agriculture. Science, 208, 1108-1113 (1980).
(3) Waid, J.S., Decomposition of roots. In Biology of Plant Litter Decomposition (Dickinson, C.H., Pugh, G.J.F., eds), Volume 1, Academic Press, London, 175-211 (1974).
(4) Chulan, A., Waid, J.S., Loss of nitrogen from decomposing nodules and roots of the tropical legume Centrosema pubescens to soil. In Nitrogen Cycling in South East Asian Wet Monsoonal Ecosystems. (Wetselaar, R., Simpson, J., Rosswall, T. eds) Australian Academy of Sciences, Canberra (in press).

INVESTIGATION ON ECOLOGICAL DISTRIBUTION OF FUNGI IN PADDY SOILS

Hao Wen-ying, Yao Hui-qin, Xu Yue-rong
(Institute of Soil Science, Adademia Sinica, Nanjing)

INTRODUCTION

Paddy soils are derived from various kinds of parent materials under wet cultivation over different parts of China. They developed through repeated and alternate processes of oxidation and reduction. They are usually badly aerated and possess a low oxidation-reduction potential during the growing period of rice[1]. As a rule, soil microflora varies with the changes of soil properties[2,3]. Fungi are generally considered as obligate aerobes. Nevertheless, fungi survive even in permanently water-logged and poorly aerated soil, such as permanently submerged paddy soils[4], muds and marshes[5,6]. This article aims to present the results of investigation on the ecological distribution of fungi in different types of paddy soils in China.

MATERIALS AND METHODS

Soil samples were collected from the plowed layer of paddy soils derived from the following parent materials in different localities:

1) Loess deposit near Nanjing
2) Lacustrine deposit in the Taihu Lake district   } East China
3) Lacustrine deposit in northern Jiangsu Province
4) Quaternary red earth and alluvial deposit
   along the Ganjiang River (a tributary of the   } Central China
   - Changjiang River)
5) Lateritic red earth and alluvial deposit   } South China
   along the Zhujiang River
6) Alluvial deposit along the tributaries of   } West China
   the Changjiang River in Sichuan Province
7) Alluvial deposit along the Liaohe River and   } Northeast China
   the Mudan River

According to soil-water regime, the paddy soils are grouped under two categories. The first is the temporarily sumberged paddy soil with better aeration (abbreviated henceforth as T.P.S.). It is usually under rice-wheat rotation. The second is the permanently submerged paddy soil which is permanently water-logged or wet caused by high underground water level (abbreviated henceforth as P.P.S.). Both types of soils and the corresponding upland fields and virgin soils are taken for comparison. The soil fungi were isolated by means of dilution plate method on Martin's medium. After identification, the frequency of occurrence and the relative number of each genus and/or species in different kinds of soils were estimated.

RESULTS AND DISCUSSION

1. Number of Fungi in Paddy Soils

Generally, 10^3–10^5 fungal propagules were found in one gram of paddy soil (Fig. 1).

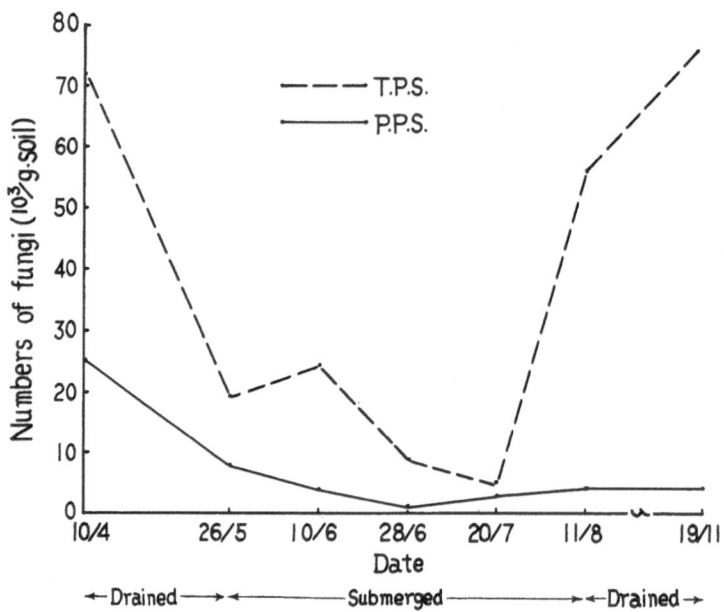

Fig. 1 Numbers of fungi in different paddy soils

Fig. 1 shows that the amount of fungi in paddy soils fluctuates with the changes in soil-water regime. In T.P.S., fungal counts are always larger than those in P.P.S.. Drainage in T.P.S. improves soil aeration, and this in turn accelerates the growth of fungi, whereas in P.P.S., fluctuation of fungal number is insignificant.

2. General Characteristics of Mycoflora in Paddy Soil as
Compared with Upland Field and Virgin Soil

Comparing the paddy soils in such regions as yellow brown soil, red soil and lateritic soil with corresponding upland fields and virgin soils, we found that soil mycoflora varied with the differences in soil properties as a result of different mode of utilization (Table 1).

Table 1 shows that Coniothyrium, Emericellopsis and Westerdykella** are found in paddy soils only, the frequencies of their occurrences in these soil regions being 50-95. %, 25-40 % and 30-100 % respectively. But in upland field or virgin soil, these genera rarely appear. It may be presumed that the mycoflora of paddy soil is characterized by those fungi.

Table 1 Frequency of the common fungi isolated in different types of soils

Soil region	Mode of utilization	1*	2*	3**	4*	5*	6*
Yellow-brown earth	Virgin soil	0	0	0	91	100	91
	Upland field	0	0	0	42	83	33
	Paddy soil	95	40	30	55	100	45
Red earth	Virgin soil	0	0	0	+	+	+
	Upland field	0	0	0	100	50	0
	Paddy soil	74	25	33	84	100	67
Lateritic red earth	Virgin soil	0	0	0	71	100	57
	Upland field	0	0	0	100	100	67
	Paddy soil	50	+	100	75	100	100
Paddy soils de-veloped on alluvial deposits	(Guangdong)	0	25	75	0	100	75
	(Jiangxi)	50	25	50	100	75	0
	(Hubei)	94	0	0	63	94	63
	(Jiangsu)	42	67	33	75	100	84
	(Liaoning)	100	25	100	0	100	50

1*	Coniothyrium	4*	Aspergillus
2*	Emericellopsis	5*	Penicillum
3**	Westerdykella	6*	Trichoderma

** Cleistothecia globose, smooth, 160-370 μ in diameter, superficial or embedded, black, wall membranaceous carbonaceous. Asci numerous, subglobose to elliptical, 10.6 x 12.3 , many-spored (c. 32 spores). Ascospores globose to short cylindrical, one-celled, dark gray, 2.4-3.0 x 2.4-4.7μ, not ornamented by spiral bands. It is closest to Pseudeurotium multisporum (Saito & Minowa) Stolk.

3. Mycofloral Characteristics of Different Types of Paddy Soils

Table 2 demonstrates the differences in mycoflora between two types of paddy soils. The Coniothyrium occurs more frequently in T.P.S. than in P.P. S., whereas the occurrences of Emericellopsis and Westerdykella are on the contrary. They appear more frequently in P.P.S. than in T.P.S.. This is especially true for Westerdykella whose frequencies of occurrence in P.P.S. derived from lacustrine deposit in northern Jiangsu Province, the Taihu Lake district in southern Jiangsu Province and Quaternary red earth are 82 %, 44 % and 67 % respectively, being significantly higher than that in the corresponding T.P.S.,

Fig. 2 shows the relative numbers of these genera. The relative number of Emericellopsis and Westerdykella are also higher in P.P.S. than in T.P.S.. The relative number of Emericellopsis in P.P.S. in different regions is about 3-10 times greater than that in T.P.S.; while Westerdykella is about 2-8 times greater in P.P.S. than in T.P.S.. As for the T.P.S. in Guangdong Province, Westerdykella is even much greater than that in P.P.S. in other regions. This seems to be due to the long term flooding of the field as a result of 2 or 3 rice crops every year. On the other hand, the relative number of Coniothyrium is usually greater in T.P.S. than in P.P.S..

Table 2 Frequency of occurence of some fungi isolated
in different paddy soils

Soil		Coniothy- rium	Emericel- lopsis	Westerdy- kella
			%	
Derived from loess	T.P.S.	100	29	0
deposit near Nanjing	P.P.S.	75	50	0
Derived from lacustrine deposit in Taihu Lake	T.P.S.	100	20	10
district	P.P.S.	44	70	44
Derived from lacustrine deposit in northern	T.P.S	64	55	27
Jiangsu Province	P.P.S.	39	79	82
Derived from quaternary red earth and alluvial	T.P.S.	83	18	41
deposit in Jiangxi Prov.	P.P.S.	50	0	67
Derived from lateritic red earth and alluvial	T.P.S.	0	10	73
deposit in Guangdon Prov.	P.P.S.	--	--	--
Derived from alluvial deposit in	T.P.S.	92	23	39
Northeast China	P.P.S.	--	--	--

4. Ecological Distribution of Penicillium spp. in Different Paddy Soils

Among soil fungi, Penicillium is the most widely distributed genus.
Its frequency of occurrence and relative number are much larger than other
genera in almost all soils. It is also found in about 90 % of paddy soils.
However, the species of this genus differ greatly among different types of
paddy soils (Table 3).

The data in Table 3 indicate that Penicillium stipitatum and P. spi-
culisporum occur only in P.P.S.. Althouth P. vermiculatum is rather frequently found in both paddy soils, both the frequency of occurrence and relative number of this species in T.P.S. are much laess than that in P.P.S..

The large amount of fungi in T.P.S., especially under drained condition, was probably due to the improvement of soil aeration. On the contrary, Emericellopsis was on the decrease as a result of soil reclamation
by changing from P.P.S. to T.P.S.(4).

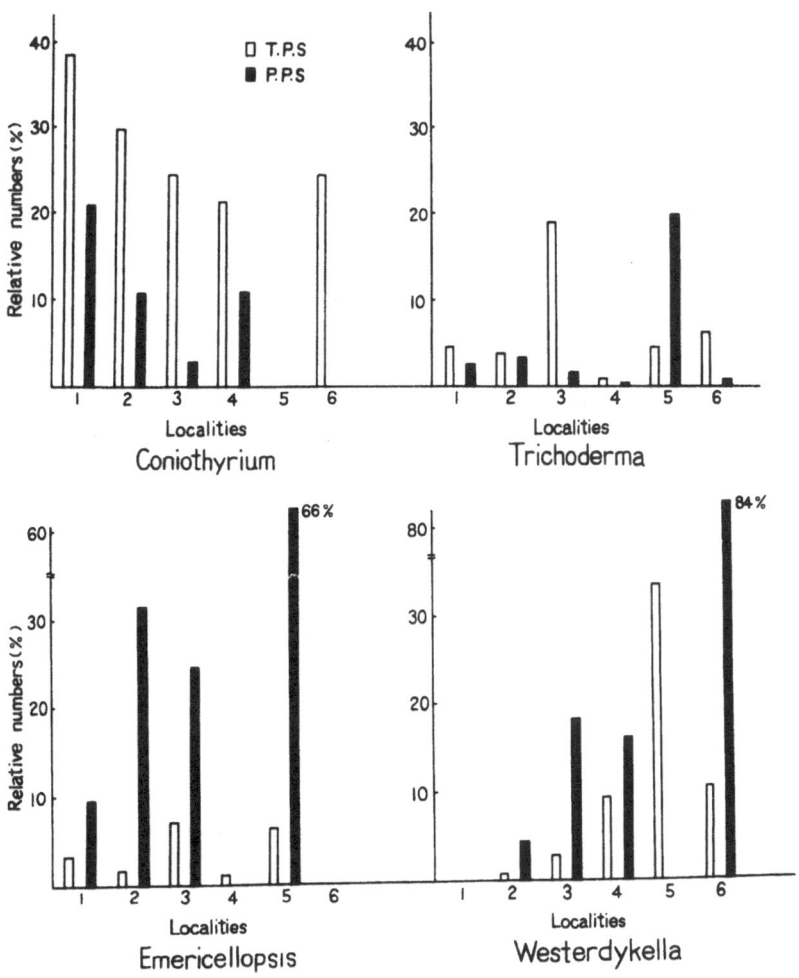

Fig. 2 Relative numbers of fungi in paddy soils
in different localities

Previous pure culture studies in the laboratory revealed that most of the isolates tested could grow even in atmosphere containing 0.5 % oxygen(3). Therefore, we would infer that soil aeration might play an important role in fungous growth, but it could not be considered as a limiting factor for the growth of all fungi. Other edaphic factors induced by water-logging might also influence the survival of soil fungi.

From the data mentioned above, the characteristics of mycoflora in paddy soils derived from various kinds of parent materials in different localities are similar in occurrence. In addition to paddy soil, the above genera and Penicillium spp. existing typically in P.P.S. also occur in other gleyed soil or soil with high moisture content and rich in organic matter. For

instance, P. stipitatum and P. vermiculatum were also found under mangrove in South China, in humic yellow soil in the Xizang (Tibet) Autonomous
Region, and in deep layer of chernozem in Northeast China. Westerdykella
was found in mangrove swamp as well. Emericellopsis occurs in humic ye-
loow soil apart from paddy soils, and it even constitutes 81 % of the total
fungous flora in phospho-calcic soil in the Xisha Islands(7). In other
countries, Orpurt, Backus(8). Davidson(9) et al. also observed that Emericellopsis and Westerdykella were found only in wet soils. The data mentioned
above also revealed that soil fungi are ubiquitous. They exist wherever soil
condition is suitable for their survival. As mentioned above, the fungi existing typically in both types of paddy soils, could characterize the soil
properties as a whole. Survival of fungi in relation to microenvironment and
the role of fungi in transformation of organic matter in paddy soil remains
a matter to be investigated.

Table 3 Distribution of Penicillium spp. in
different types of paddy soils

Species	P.P.S.		T.P.S.	
	Frequency of Occurence	Relative Number %	Frequency of Occurence	Relative Number
P. stipitatum	52.7	15.9	0	0
P. spiculisporum	61.2	10.4	0	0
P. vermiculatum	95.0	21.0	45.5	8.3
P. funiculosum	61.2	27.2	72.5	66.5
P. oxalicum	61.2	6.0	27.4	2.5

REFERENCES

(1) Institute of Soil Science, Academia Sinica, Nanjing (ed.), Soils of
    China. Science Press, Beijing, 351-359(1978). (in Chinese)
(2) Cao Zheng-bang, Hao Wen-ying, You Chang-fen, Gu Xi-xian, The microbiological
    characteristics of rice soils. 1. A study of the number and activity of
    microorganisum in the fundamental types of rice soils at varying levels
    of fertility in eastern and central China. Acta Pedologica Sinica, 7,
    218-226(1959). (in Chinese with Russian summary)
(3) Hao Wen-ying, Cao Zheng-bang, You Chang-fen, The microbiological proper-
    ties of paddy soils. 2. The relationship between microflora and soil
    fertility. Acta pedologica Sinica, 9, 1-8(1961). (in Chinese with English
    summary)
(4) Hao Wen-ying, Yao Hui-qin, Xu Yue-rong, Ecological distribution of
    fungi in submerged paddy soil. Acta Pedologica Sinica, 17, 346-354
    (1980). (in Chinese with English summary)
(5) Apinis, A. E., Chester, C.G.C., Ascomycetes of some salt marshes and
    sanddunes. Trans. Brit. Mycol. Soc. 47(3), 419-435(1964).
(6) Puhg, G. J., The fungal flora of tidal mud- flats. In "The Ecology
    of Soil Fungi, an International Symposium" (ed. by Parkinson, D., Waid,
    J. S.), Liverpool University Press, 202-208(1960).
(7) Soil Expedition of the Institute of Soil Science, Acadamia Sinica,
    Soils and Guanos of Xisha Islands. Science Press, Beijing, 35(1977).
    (in Chinese with English summary)

(8) Bachus, M.P., Orpurt, P.A., A new <u>Emericellopsis</u> from Wisconsin, with notes on other species. Mycologia, 53(1), 64-83(1961).

(9) Davidson, D. E., Christensen, M., <u>Emericellopsis</u> <u>stolkia</u> sp. nov, from saline soils in Wyoming. Trans. Brit. Mycol. Soc. 57(3), 385-391(1971).

MECHANISMS OF BACTERIAL IRON-REDUCTION IN FLOODED SOILS

J. C. G. Ottow*
(Institut für Bodenkunde and Standortslehre, Universitat Hohenheim
27 Emil Wolff Strasse, D-7 Stuttgart-Hohenheim-70, Germany)

One of the most characteristic features of flooded soil is the rapid
and intensive reduction of the Fe(III)-oxides. Within a few weeks, dissolved Fe(II) may reach values up to a few thousand ppm. Generally, the
amount and rate of Fe(II)-formation depends on the content and type of organic matter, the initial pH of the soil, and the composition of the operating redox systems (O_2, NO_3^-, Mn-IV-oxide as well as ratio of amorphous to
crystalline iron-III-oxides) at the time of submergence(1-4).

The higher the content of easily decomposable organic matter, and the
lower the amount of nitrate, as well as Mn(IV)-compounds, the more intensive
is the accumulation of Fe(II). Iron reduction is beneficial to rice growth
because it increases the solubility of phosphorus, potassium, silicon, and
molybdenum in neutral to alkaline soils, but it may be toxic to rice ("bronzing"), particularly in acid soils or in Oxi -and Ultisols that are low in
nutrients(5).

Usually, Fe(II) in solution does not accumulate until the trapped oxygen
has respired, the nitrates are reduced, and most of the Mn(IV)-compounds are
transformed into their soluble Mn(II)counterparts(Table 1). These clear successive processes have led to the assumption that soil reduction proceeds in
the sequence predicted by the thermodynamics of the involved redox systems.
Such an interpretation holds that the reductive transformations are caused by
a decrease in the redox potential (Eh) and the accumulation of reducing substances as the result from microbial activity. This mechanism of iron reduction would be basically chemical in nature.

In a direct biochemical mechanism, Fe(III)-compounds are thought to act
as specific electron acceptors during metabolic energy conservation (ATP
synthesis)(2). This latter mechanism would be entirely enzymatic and linked
to specific reductases of certain bacteria.

Most soils contain a spectrum of different iron(III)-oxides ranging from
completely amorphous to highly crystalline (goethite, hematite). Crystalline
Fe(III)-oxides are more stable and exothermous than their noncrystalline
counterparts(7). These thermodynamic considerations have resulted in the
assumption that the lower the degree of crystallinity, the higher the extent
to which the pedogenic iron oxides may be reduced(1). If Fe(III) is indeed
acting as an electron acceptor, then the amorphous Fe(III)-compounds, rather
than the crystalline Fe-oxides, should be dissolved. These hypotheses are

---

* Present address: International Rice Research Institute, P. O. Box
933, Manila, Philippines

Table 1 Sequence of redox processes and their thermodynamics (at standard conditions) in flooded soils containing sufficient amounts of easily decomposable organic matter

Succession of processes	Eh(mV) at beginning of process [1]	Redox systems involved [2]	Eo (mV) [3]
Respiration: $O_2 \rightarrow H_2O$	> 400	$O_2 + 4H^+ + 4e \rightleftharpoons 2H_2O$	+814
NO_3^- reduction	+600 to +300	$2NO_3^- + 12H^+ + 10e \rightleftharpoons N_2 + 6H_2O$	+741
Formation of $Mn(II)$ [4]	+500 to +220	$MnO_2 + 4H^+ + 2e \rightleftharpoons Mn(II) + 2H_2O$	+410
Production of $Fe(II)$ [4]	+400 to +180	$Fe(OH)_3 + 3H^+ + e \rightleftharpoons Fe(II) + 3H_2O$	-185
S^{2-}-accumulation	+100 to -200	$SO_4^{2-} + 10H^+ + 8e \rightleftharpoons H_2S + 4H_2O$	-214
CH_4 formation	-150 to -280	$CO_2 + 8H^+ + 8e \rightleftharpoons CH_4 + 2H_2O$	-244

[1] Data from Aomine (1962), Ponnamperuma (1972), and Yoshida (1978)

[2] Modified after Ponnamperuma (1972). Most of these redox systems should be regarded as electromotively sluggish (nitrate, MnO_2) or even inactive $\underline{/Fe(OH)_3, SO_4^{2-}\_/}$

[3] Eo (Em) = standard electrod potential at 50% reduction and a pH of 7.0

[4] H_2 is evolved during fermentation, usually before and during active iron-reduction. Together with CO_2, formate, and acetate, it acts as a substrate for methane production

the object of this presentation.

MATERIALS AND METHODS

To test the postulates outlined above, we used model experiments with
Fe_2O_3-powder (hematite) or selected soils (with high Fe_o/Fe_d-ratio) under a
given set of standard conditions of pH, particle size (sieved between 63-
125μ m), and available energy (2% glucose). For iron-reducing bacteria,
common nitrogen-fixing bacteria (Bacillus polymyxa and Clostridium butyricum)
were chosen because they develop well under anaerobic conditions, lowering the
Eh(8)

1. Model Experiments

Half-gram samples of Fe_2O_3-powder (Merch) or soil were weighed into
150-x-25 mm boiling tubes, saturated with 25 ml of a 2%-glucose solution, and
mixed and adjusted to pH 7-7.2 (1 N NaOH) before sterilization (1 min at
121°C). In parallel tubes, 0.25 mM KNO_3 and/or 0.1 g of MnO_2 powder (sieved,
Merck) were added. Each sample was inoculated either with 0.7 ml of a spore
suspension obtained from Bacillus polymyxa (nitrate reductase positive =
nit^+) or with a strain of Clostridium butyricum (nit^-)(8,9). The tubes were
homogenized and incubated anaerobically (N_2/CO_2 = 9/1 30°C). At regular
intervals three   tubes were taken and analyzed for pH, Eh, Fe(II),
Mn(II), and glucose(9). The rH (reduction intensity) was calculated according to the formula:

$$rH = \frac{Eh(mV)}{29} + 2 \text{ pH } (30°C).$$

At rH = 0, the environment is completely reduced. At rH 42.6, entirely
oxidizing conditions exist(10).

2. Differentiation of Fe-Compounds

Because acid ammonium oxalate-soluble Fe (Fe_o) only approximates the
amount of noncrystalline Fe oxides in soil, the use of Fe_d-Fe_o gives only
a rough estimate of the content of crystalline Fe(11-13). Oxalate extractable
Fe (Fe_o) was determined, according to Schwertmann(14), and the total amount
of active iron (Fe_d) by the method of Mehra and Jackson(15). The extracted
Fe was analyzed quantitatively by ASS (Philips Unicam S90A).

3. Model Experiments With ^{59}Fe-Labeled Iron Oxides

Both crystalline hematite (Fe_2O_3 - powder, Merck) and goethite (Ward's
Co., Rochester, U.S.A.) were sieved and labeled by radiation in a nuclear
reactor(16). In these experiments, 470 mg of a highly amorphous gley soil
(Fe_o/Fe_d = 0.78) was mixed with 30 mg of either ^{59}Fe-labeled hematite or
goethite and inoculated as described. All samples were examined for Fe_o,
Fe_d, and [59]Fe(16).

RESULTS

1. Sequence of Reduction by B. Polymyxa (nit+)

In Fig. 1, the development of iron reduction, rH, and glucose consumption as the result of B. polymyxa S55 (nit+), is shown. In this, and the following graphs, the uninoculated control values are omitted because pH (7.2-6.9) rH (19-21), Fe(II)-production (1-2 ppm), and manganese reduction(Mn(II)= 5-15 ppm) changed only to a minor extent. From Fig. 1 it is apparent that entirely reducing conditions (rH = 0) were established by B. polymyxa within 2 to 3 days, while glucose consumption and ironreduction still proceeded. Obviously, completely reduced conditions are reached several days before maximal Fe(II) production. If nitrate is added (Fig. 2), both the drop in rH and the production of Fe(II) are retarded. However, as soon as nitrate has been utilized, iron reduction starts at completely reduced conditions(rH = 0). If MnO_2 is added (Fig. 3), iron reduction becomes suppressed throughout the incubation period. Again, the rH dropped within the completely reduced level (rH = 2-3) and remained at this level while Mn(II)-production continued. In the presence of both nitrate and MnO_2 (Fig. 4), intensive manganese reduction begins only after nitrate has been largely dissimilated. Under these conditions iron reduction remains negligible.

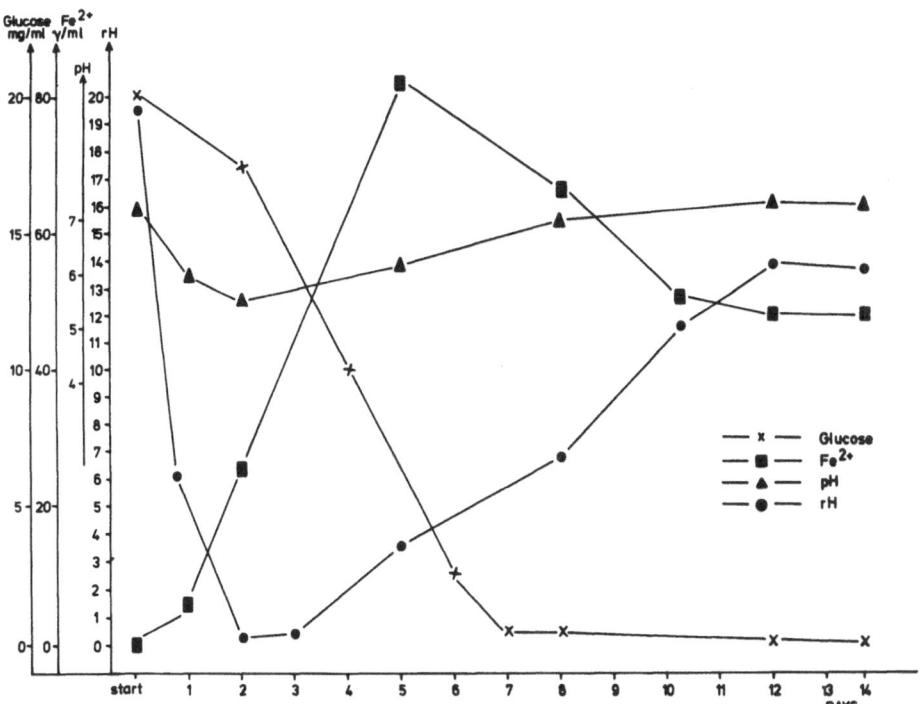

Fig. 1 Fe(II)-formation with **Bacillus polymyxa** S55 (nit+) in relation to glucose utilization and development of pH and rH under anaerobic (N_2/CO_2=9/1; 30°C) standard conditions

333

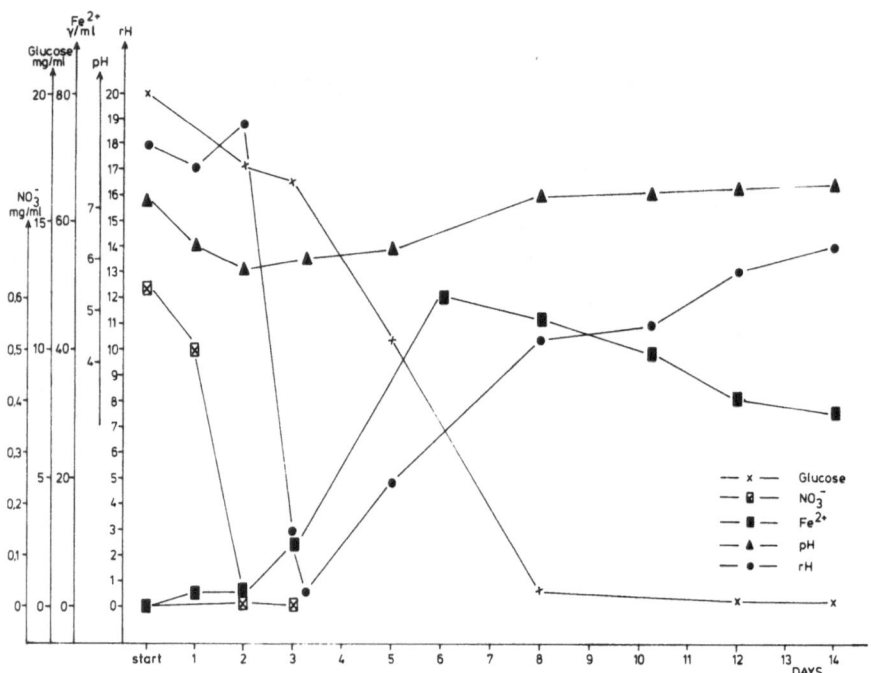

Fig. 2   Influence of nitrate (KNO₃) on iron reduction, glucose
          utilization, pH, and rH with B. polymyxa S55 (nit⁺)

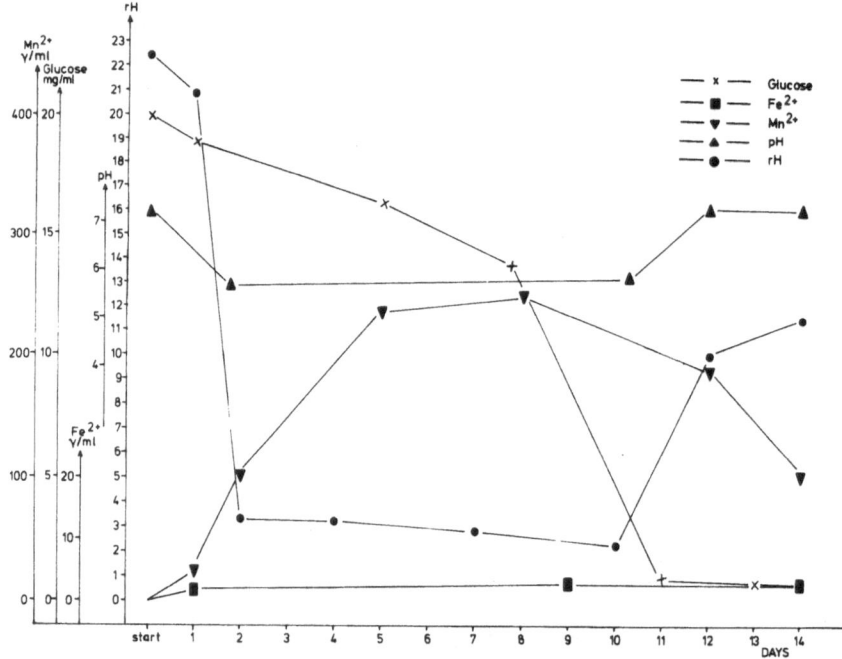

Fig. 3   Influence of manganese dioxide (MnO₂) on iron reduction,
          glucose utilization, pH, and rH with B. polymyxa S55 (nit⁺)

334

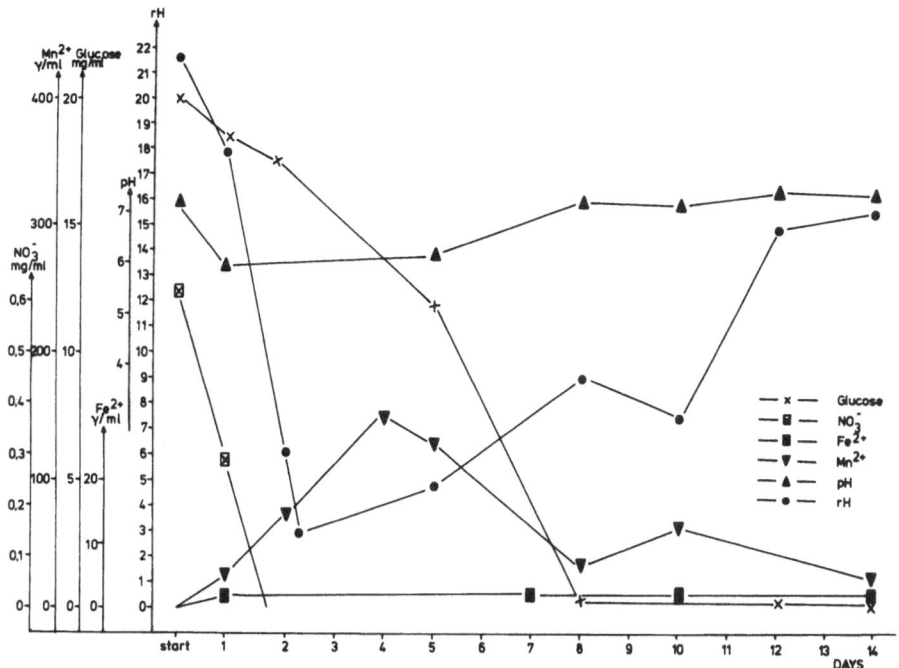

Fig. 4   Influence of nitrate and MnO_2 on iron reduction, glucose
utilization, pH, and rH with B. polymyxa S55 (nit$^+$)

Apparently, the sequence of reductive transformations as affected by
B. polymyxa nit$^+$ proceeds at highly reduced conditions (rH = 0-3) as predicted by the thermodynamics of the redox systems involved (see Table 1).

2. Sequence of Reductive Processes With C. Butyricum (nit$^-$)

C. butyricum is physiologically related to B. polymyxa, but differs from
the latter in its inability to reduce nitrate (lacking nitrate reductase).
If Fig. 5 (iron reduction at standard conditions) is compared to Fig. 6 (with
nitrate), the influence of nitrate on Fe(II)-formation as observed with B.
polymyxa S55 could not be confirmed. Nitrate is not attacked at all, although
completely reduced conditions (rh = 0) are established within 2 or 3 days.
The addition of MnO_2 (Fig. 7), however, suppresses iron reduction significantly
in a comparable environmental situation. If nitrate and MnO_2 are added simultaneously (Fig. 8), only MnO_2, but not nitrate, affects iron reduction considerably. Again, nitrate remains unchanged throughout the experiment at
reduced conditions (rH = 0). Neither Mn(II)-production nor Fe(II)-accumulation
seems to affect nitrate in the medium.

335

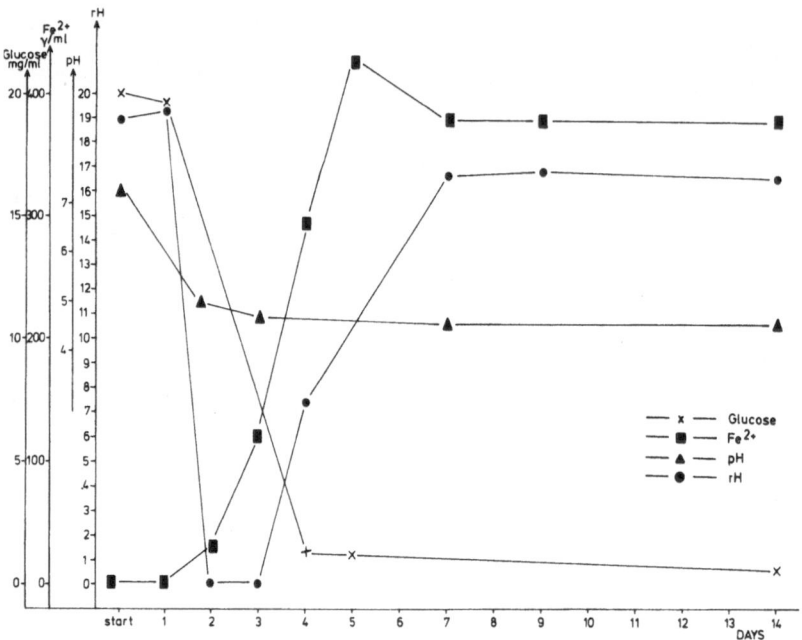

Fig. 5 Iron-reduction with <u>Clostridium butyricum</u> S22a (nit⁻) in
relation to glucose consumption, pH, and rH under anaerobic
standard conditions ($N_2/CO_2=9/1$; $30^{\circ}C$)

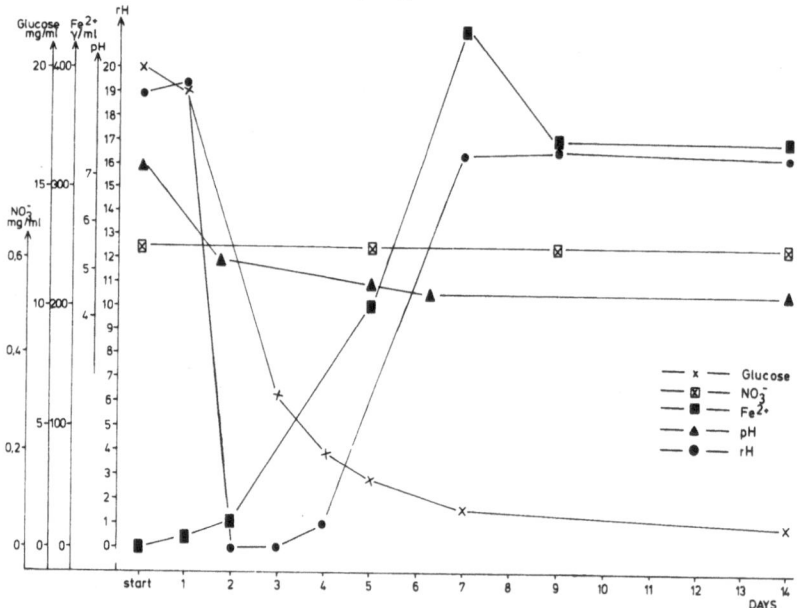

Fig. 6 Influence of nitrate (KNO_3) on iron reduction, glucose
utilization, pH, and rH with <u>C. butyricum</u> S22a (nit⁻)

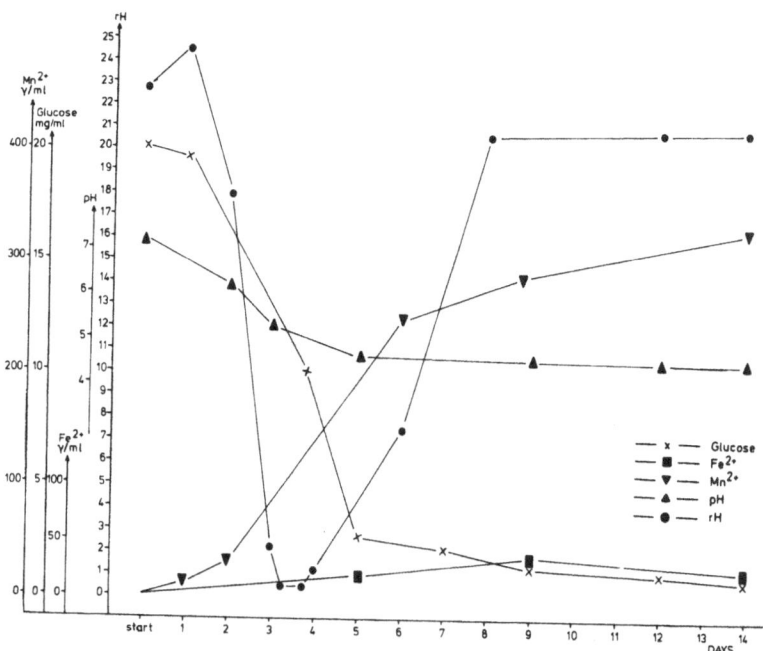

Fig. 7   Influence of MnO₂ on iron reduction, glucose utilization, pH and rH with C. butyricum S22a (nit⁻)

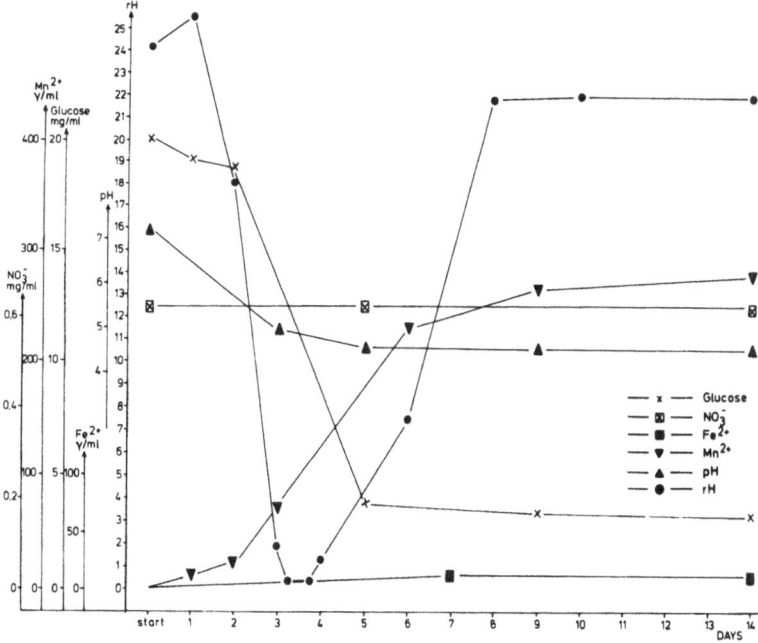

Fig. 8   Influence of nitrate on the formation of Mn(II), Fe(II), glucose utilization, pH, and rH with C. butyricum, S22a

In summary, it is clear that the sequence of reduction processes with a nit⁻ bacterium like C. butyricum is different from that observed with B. polymyxa S55 (nit⁺). Undoubtedly, nitrate is not reduced in preference to Mn (IV) or Fe(III), although completely reduced conditions are established. Additionally, NO_3^- is not transformed by the accumulation of Mn(II) or by Fe(II)-formation.

3. Preferential Reduction of Amorphous Iron(III)-Oxides

The question whether amorphous Fe(III)-oxides are reduced in preference to the more stable and exothermous crystalline counterparts is answered in Tables 2 and 3. In Tables 2 and 3, the formation of Fe(II) form amorphous gley material (highly oxalate soluble: $Fe_o/Fe_d = 0.78$) mixed with either ^{59}Fe-labeled hematite or goethite is demonstrated.

From both tables, the following general conclusions can be drawn. First, the total amount of free iron oxides (Fe_d), particularly the amount of non-crystalline $Fe(Fe_o)$, decreases with increasing iron reduction. Compared to the Fe_o fraction, the amount of crystalline oxides (Fe_d-Fe_o) decreases only slightly. Consequently, the Fe_o/Fe_d of both mixtures diminishes with time. Second, the percentage of ^{59}Fe in the total amount of active iron oxides (Fe_d) increases with time in both soils. This percentage should remain constant, however, if the mixed amorphous and crystalline Fe oxides were reduced randomly. Third, the percentage of [59]Fe(II) dissolved from crystalline hematite (Table 2) or goethite (Table 3) is considerably lower than the amount reduced from noncrystalline gley soil material. These percentages should be in the same range, if the Fe oxides are attacked at random.

DISCUSSION

The results of the present experiments have wide-ranging effects on the interpretation of reductive transformation in flooded soils. First, these data make clear that the apparent sequence of redox processes (Table 1) is not governed by the thermodynamics of the various redox systems involved, but exclusively by the presence of specific microbial enzymes that catalyze these energy-requiring, reductive transformations. If a rapid and intensive decrease in the redox level (pH – Eh relationship) is responsible for the sequential reduction of nitrate, Mn(IV)- and Fe(III)-oxides in flooded soils, this sequence should be established whenever the Eh passes from a positive (about 400 to 700 mv to a low negative (about -300 to -350 mv) redox potential. This generally accepted view is rejected by the present studies.

At the microsite, a reducing organism will prefer that inactive external oxidized compound as an electron acceptor, which requires less activation energy, provided the organism is equipped with the corresponding enzymes. This economic principle rather than the pH-Eh relationship of the redox couple involved ensures the highest gain in potential energy which may be used for energy-conserving reactions (ATP-synthesis). With a facultative anaerobic organisms, like B. polymyxa (nit⁺), energy conservation with external electron acceptors will undoubtedly proceed in the sequence $O_2/H_2O \longrightarrow NO_3^-/NO_2^- \longrightarrow$ Mn(IV)/Mn(II)$\longrightarrow$ Fe(III)-amorph./Fe(II) $\longrightarrow$ Fe(III)-cryst./Fe(II), because this organism is equipped with all enzymes required to catalyzed the redox transformation essential to lower the activation energy and to drive the electromotively sluggish reactions. However, with an obligate anaerobic

Table 2 Fe(II)-formation from noncrystalline gley soil material mixed with ^{59}Fe-labeled hematite in model experiments

Days	Fe_d	Fe_o	Fe_{d-o}[1]	Fe_o/Fe_d	$^{59}Fe_d/Fe_d$ - total %	Non-labeled Fe(II) in % from gley soil (470 mg soil = 4.1 mg Fe)	^{59}Fe(II) in % from ^{59}Fe-hematite (30 mg Fe-hematite = 20.9 mg Fe)
	mg Fe/tube						
Start	25.1	4.10	21.0	0.16	83.5	0.5	0
4	23.1	2.13	20.97	0.10	90.0	42.9	0.11
8	21.8	1.92	19.88	0.08	92.2	70.1	0.45
13	21.2	1.60	19.60	0.08	92.5	70.8[2]	0.62

[1] Fe_{d-o} (= crystalline part) corresponds practically to the total amount of Fe added as hematite powder

[2] Highest Fe(II) in solution was measured after 13 days of incubation (anaerobic at 30°C)

Table 3 Fe(II)-formation from noncrystalline gley soil material mixed with ^{59}Fe-labeled goethite in model experiments

	Fe_d	Fe_o	Fe_{d-o} 1)	Fe_o/Fe_d	$^{59}Fe_d/Fe_d$-total %	Nonlabeled Fe(II) in % from gley soil (470 mg gley soil = 4.1 mg Fe)	^{59}Fe in % from ^{59}Fe-goethite (30 mg goethite = 0.96 mg Fe)
		mg Fe/tube					
Start	14.10	3.20	10.90	0.23	70.7	0.3	0
4	12.25	2.60	9.75	0.21	79.5	29.9	2.1
8	11.80	2.05	9.75	0.17	81.9	61.4	2.3
13	11.05	1.76	9.26	0.16	84.2	70.1	3.22

1) Fe_{d-o} = amount of crystalline Fe

bacterium, such as C. butyricum (lacking cytochromes, as well as nitrate reductase), only Mn(IV)-and Fe(III)-compounds will act as suitable electron acceptors for some specific reductase(s) ("mangani" - and/or "ferri-reductases")(9,16-18) and the sequence of transformations will be restricted to Mn(IV) ——— Fe(III)-amorph. ——— Fe(III)-cryst. The latter reductases are probably linked to $NAD/NADH_2$ or $FAD/FADH_2$ type of enzymes and thus, must have relatively low operational redox potentials (around E_o' of -220 to 320 mV). These reductases will not function until the Eh of the microenvironment has been decreased to the appropriate level by metabolic activtiy. This prerequisite in redox level may in fact explain why the environment of a growing and actively metabolizing bacterial population becomes entirely reduced (rH = 0)before managenese or iron are attacked significantly (Fig. 1 and 5). The observed establishment or traverse of certain "critical" redox levels in flooded soils (Table 1) is responsible for the misleading view that a definite electron activity (pE or Eh) rather than the induction of specific electron-transferring enzymes is thought to be responsible for the overall sequence of reductive transformations.

Second, these results impair with the role and significance of the soil Eh. Although there is no doubt on the pH-Eh level of a soil as a measure of the proton and electron "pressure" of that particular biosystem, its role as a causative agent has been largely overestimated. This may be illustrated by the relationship

$$pFe^{(TT)} = pE + 3pH - 17.9$$

which has been established in several freshly flooded soils containing sufficient amounts of easily decomposable organic matter(1,19). From this relationship, it may be concluded that Fe(II)- formation is caused by an increase in electron activity (pE - negative logarithm of electron activity and thus another expression of Eh) and high pH of the submerged soil. Such an interpretation implies a chemical reduction mechanism rather than a direct biochemical one. In fact, a low Eh and a low pH may only affect the solubility (and thus mobility) of the physiologically linked Fe(II)-formation. The drop in Eh is just an expression of an actively metabolizing microbial population. As soon as the aveilable amount of energy has been exhausted, Eh will increase again(1), although the Fe(II) concentration in solution remains at a relatively high level for a long time. This behavior may explain the numerous papers that failed to find any relationship between the redox level of the soil and the amount of Fe(II) produced(20,21).

The Mechanism of Electron Transfer at the Microsite

In any soil (either flooded or with a locally reduced partial O_2-pressure), drop in Eh and the production of Fe(II) are the result of the metabolic activity of numerous facultative and obligate anaerobic bacteria(17,22,23) that use amorphous Fe(III)-compounds as an electron acceptor in preference to the crystalline Fe-forms during biological oxidation and energy conservation:

$$\text{Organic matter} \xrightarrow[\text{ADP} + P_i]{\text{dehydrogenases}} H^+ + e + ATP + \text{products}$$

Iron(III) oxides as H-acceptor (hydration):

$$\text{>Fe} \text{——} OH + H^+ + e \xrightarrow{\text{reductase(s)}} Fe(II) + H_2O$$

At a suitable anaerobic microsite, it is hard to visualize how iron-reducing bacteria actually manage to use amorphous rather than crystalline Fe(II)-oxides as a final electron acceptor, especially since an intimate contact between cell surface and hydrated Fe-compounds must be regarded as an essential prerequisite for the successful transfer of electrons[16,18]. Even if a selective transfer of electrons to amorphous Fe(III)-compounds is mediated by certain unknown "carriers", it remains inconceivable how this specific transfer continue while iron dissolution proceeds. An active metabolizing population using noncrystalline Fe(III)-oxides in their direct surrounding should switch to crystalline forms in the close vicinity as soon as the amorphous compounds become exhausted. Such a view, however, is not consistent with the data presented in Tables 2 and 3. These tables show that very little ^{59}Fe(II) is dissolved from crystalline material. Apparently, the electron-transferring reductases are specific enough to select between the amorphous and crystalline Fe(II)- compounds, probably by the differences in the standard free energy of formation (ΔGf^{o}). This is quite remarkable, because the difference in the ΔGf^{o} for hematite (-177.7 kcal$\cdot$mol^{-1} and amorphous oxides (-166.5 kcal$\cdot$mol^{-1} is relatively small [24] Such a preferential reduction of amorphous Fe(III)-compounds among a natural consortium of various iron oxides has indeed been demonstrated. Thus, in flooded soils a gradual decrease in the ΔGf^{o} from values around -162 kcal$\cdot$ kcal$\cdot$mol^{-1} "Fe(OH)$_3$" at the start of submergence to values close to -177 kcal$\cdot$mol^{-1} "goethite" after 10 weeks of incubation has been measured[20].

REFERENCES

(1) Ponnamperuma, F.N. The chemistry of submerged soils. Adv. Agron. 24, 29-96 (1972).
(2) Ottow, J.C.G., and H. Glathe, Pedochemie und Pedomikrobiologie hydromorpher Boden; Voraussetzung und Ursachen der Eisenreduktion. Chem. Erde 32, 1-44 (1973).
(3) Yoshida, T., Microbial metabolism in flooded soils. In: Soil Biochemistry, Vol. 3 (E.A. Paul and A.D. McLaren, ed.), Marcel Dekker, Inc. New York-London, p. 83-133 (1975).
(4) Yamane, L., Electrochemical changes in rice soils. In : Soils and Rice. Int. Rice Res. Inst., Los Banos, Philippines, p. 381-399 (1978).
(5) Tanaka, A. and S. Yoshida, Nutritional disorders of the rice plant in Asia. Int. Rice Res. Inst. Tech. Bull. No. 10, 51 O. (1970).
(6) Aomine, S. A review of research on redox potentials of paddy soils in Japan. Soil Sci. 94, 6-13 (1962).
(7) Scheffer, F., E. Welte, and F. Ludwing, Zur Frage der Eisenoxydhydraten im Boden. Chem. Erde 19,51-64 (1957).
(8) Hammann, R., and J.C.G. Ottow, Reductive dissolution of Fe$_2$O$_3$ by sac-charolytic clostridia and Bacillus polymyxa under anaerobic conditions. Z. Pflanzenernaehr. Bodenkd. 137, 108-115,(1974).
(9) Munch, J.C., and J.C.G. Ottow, Modelluntersuchungen zun Mechanismus der bakteriellen Eisenreduktionen in hydromorphen Boden. Z. Pflanzenernaehr. Bodenked. 140, 549-562,(1977).
(10) Jacob, H. E., Redox potentials. In: Methods in microbiology, Vol. 2 (J.R. Norris & D.W. Ribbons, ed.) Acad. Press, New York-London, p. 91-123, (1970).
(11) McKeague, J.A., J.H. Day, Dithionite-and oxalate-extractable Fe and Al as aids in differentiating various classes of soils. Can. J. Soil Sci. 46, 13-22, (1966).
(12) Borggaard, O.K. Selective extraction of amorphous iron oxides by EDTA from a mixture of amorphous oxides, goethite and hematite. J. Soil Sci. 27, 478-487, (1976).

(13) Kodama, H., J.A. Mc Keague, R.J. Tremblay, J.R., Gosselin, and M.G. Townsend, Characterization of iron oxide compounds in soils by Mossbauer and other methods. Can J. Earth Sic. 14, 1-15, (1977).

(14) Schwertmann, U., Differenzierung der Eisenoxide des Bodens durch Extraktion mit saurer Ammoniumoxalatlosung. Z. Pflanzenernaehr. Bodenkd· 105, 194-202, (1964).

(15) Mehra, O.P., and M.L. Jackson, Iron oxide removal from soils and clays by a dithionite-citrate system buffered with sodium bicarbonate. clays Clay Miner. 7, 317-327, (1960).

(16) Munch, J. C. and J.C.G. Ottow, Preferential reduction of amorphous to crystalline iron oxides by bacterial activity. Soil Sci. 129-15-21, (1980).

(17) Ottow, J.C.G., Selection, characterization and iron-reducing capacity of nitrate reductase less (nit⁻)mutants from iron-reducing bacteria. Z. Allg. Mikrobiol. 10, 55-62, (1970).

(18) Munch, J.C., Modelluntersuchungen zu den Mechanismen der bakteriellen Reduktion Von Eisen(III)-Oxiden in hydromorphen Boden. Diss. University of Hohenheim, Stuttgart-Hohenheim, Germany, In: Hochschulsamml. Natuwiss. Biol. (Freiburg) 6,1-132, (1980).

(19) Ponnamperuma, F.N., Physicochemical properties of submerged soils in relation to fertility. Int. Rice Res. Inst. Res. Paper Ser. 5, pp. 1-32, (1977).

(20) Breemen, N. Van, The effect of ill-defined ferric oxides on the redox characteristics of flooded soils. Neth. J. Agr. Sci. 17, 256-260, (1969).

(21) Gotoh, S. and W.H. Patrick, Transformation of iron in a waterlogged soil as influenced by redox potential and pH. Soil Sci. Soc. Amer. Proc. 38, 66-71, (1974).

(22) Ottow, J.C.G., The distribution and differentiation of iron-reducing bacteria in gley soils. Zbl. Bakt. Abt. II 123, 600-615, (1969).

(23) Ottow, J.C.G. and H. Glathe, Isolation and identification of iron-reducing bacteria from gley soil. Soil Biol. Biochem. 3, 43-55, (1971).

(24) Schuylenborg, H. Van, Sesquioxide formation and transformation. In: Pseudogley and Gley (E. Schlichting and U. Schwertmann, ed.), Verlag Chemie, Weinheim, Germany, p. 93-102, (1972).

INVESTIGATION ON NITROGEN-FIXING
BACTERIA IN RICE RHIZOSPHERE

Qiu Yuan-sheng, Zhou Shu-ping, Mo Xiao-zhen
(Guangdong Microbiology Research Institute, Guangzhou)
You Chong-biao
(Institute of Application of Atomic Energy, Chinese Academy of
Agricultural Sciences, Beijing)
Wang Da-si
(Institute of Microbiology, Academia Sinica, Beijing)

It has been confirmed in many laboratory and field experiments that the nitrogen-fixing activity in rice rhizosphere is rather high. There has been a number of different nitrogen-fixing bacteria isolated from rice rhizosphere, and more than 10 genera were reported [1-3]. Their N_2-fixing activities were also extensively investigated. By means of ^{15}N tracer technique, it was demonstrated that the nitrogen fixed in rice rhizosphere might be absorbed rapidly by rice plant. Therefore, the nitrogen-fixing bacteria associated closely with the rice root may be different from other free-living nitrogen-fixing bacteria in paddy soil.

In 1977, two strains of N_2-fixing bacteria (A-15 and E-26) with high capability of nitrogenase activity were isolated from rice root in Guangdong Province. They were identified tentatively as Alcaligenes faecalis and Enterobacter cloacae respectively[4].

Strain A-15 is an aerobic Gram-negative organism. It is rod in young culture and short rod in older culture with peritrichous flagella, containing no lipid bodies in cell. Oxidase positive and litmus milk alkaline. It can use a variety of organic acids such as malic, lactic, succinic, benzoic etc, but not sugars as its carbon source; and will dix nitrogen under microaerophilic conditions.

In nitrogen-free liquid culture, it could grow and fix N_2 by lowering the partial pressure of oxygen, and no pellicle was formed. It grew well on nitrogen-free malate semi-solid medium containing 0.2% agar on which a layer of white pellicle was formed. Its acetylene reduction activity varied with the incubating time. The peak of nitrogenase activity occurred after 16-20 hours of incubation at $33^{\circ}C$, but declined rapidly thereafter (Fig. 1). The maximum acetylene reduction activity ranged from 500-700 nmol C_2H_4/ml culture/hr, and its specific activity was 70-90 nmol C_2H_4/mg cell protein/min. In pure culture, the nitrogenase activity of A-15 was rather high, but the duration of N_2-fixation was shorter. After 22 hours of incubation, in each 100 ml culture medium about 0.25 g of organic acid was consumed while total nitrogen was increased by about 5 mg.

The efficiency of N_2-fixation was approximate to 20 mg of nitrogen fixed per gram of malic acid consumed. The capability of N_2-fixation of A-15 was also demonstrated by using ^{15}N tracer technique, and the amount of nitrogen fixed was 30 μg of N/ml culture after adding $^{15}N_2$ for 37 hours of incubation, its specific activity was 200-250 nmol NH_3/mg cell protein/hr.

344

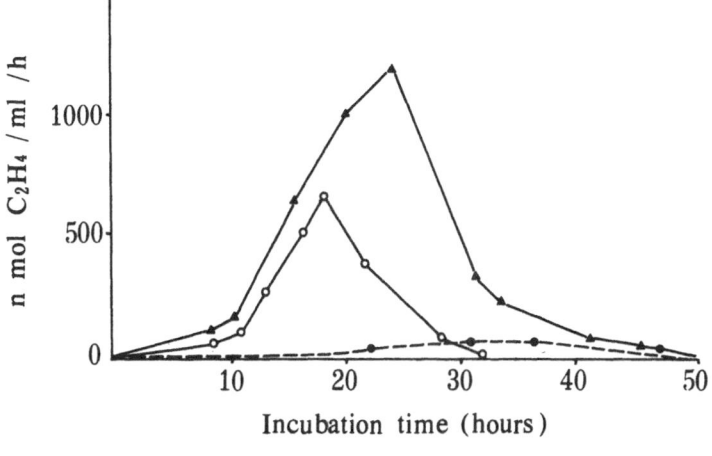

Fig. 1 Nitrogenase activity of E-26 and A-15
in pure and mixed culture

Strain E-26 is a facultative anaerobic, Gram-negative rod with peritrichous flagella. Gas and acid were produced from fermentation of glucose and sucroce. Nitrogen fixation carried out significantly only under anaerobic conditions (Fig. 2). Its nitrogenase activity was up to 200-300 nmol C_2H_4/ml culture/hr, and the specific activity was 30-40 nmol C_2H_4/mg cell protein /min. The acetylene reduction rate of E-26 was lower than that of A-15, but its duration of N_2-fixation lasted longer than that of the latter. After 72 hours of incubation, 4 mg of N was fixed in each 100 ml culture. Its efficiency of N_2-fixation was 7.5 mg nitrogen fixed per gram of sucrose consumed. It is indicated in Fig. 1 that although the nitrogen fixation activity of E-26 was not very high in separate culture, it did raise significantly in culture mixed with A-15, and the acetylene reduction activity of the mixed culture was up to 1300-1500 nmol C_2H_4/ml culture/hr. The total nitrogen increased by about 13 mg per 100 ml culture, but the efficiency of N_2-fixation was not increased obviously because of the increase of carbon source consumption. These two strains spread widely and simultanously in the rice rhizosphere, and their concordance in N_2-fixation is remarkable.

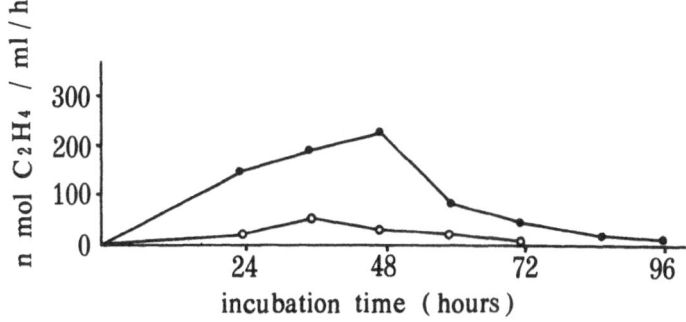

Fig. 2 Nitrogenase activity of E-26 under aerobic
and anaerobic conditions

By using the ^{15}N tracer technique it was found that both strains inoculated to rice seedlings could grow well and fix N_2 associated with rice without adding extra carbon source, and the N_2-fixation in mixed culture was higher than that in separate culture. It was also found that the amount of nitrogen fixed by the bacteria associated with normal young seedlings was more than that of the seedlings without seed residues. The fixed nitrogen of the former was 7-11 μg/ plant, and the latter was only 2-3 μg/ plant during 48 hours of incubation (Table 1). It was shown in another experiment that the fixed nitrogen increased significantly with the addition of extra carbon source, and it could be assimilated rapidly by rice. After 69 hours of inoculation, the enrichment of ^{15}N in rice leaves was significantly higher than the background, and about 1/3 of the nitrogen fixed by the bacteria was translocated into the roots and the leaves (Table 2). The results mentioned above revealed that the association of A-15 and E-26 with rice can well set up. The rice could provide carbon source for the growth and N_2-fixation of the bacteria, while a portion of nitrogen fixed by the bacteria was absorbed rapidly by rice. The amount of nitrogen fixed by the association between bacteria and rice depends greatly on the carbon source supplied. Therefore, as a potential nitrogen source for rice cultivation, this associative N_2-fixation is worthy of consideration.

Table 1 Amount of nitrogen fixed by bacteria associated with rice (without extra carbon source)

	Inoculant	Sample	Atom % ^{15}N excess*	Total N (mg)	N_2 fixed (μg)	Mean N fixed (μg/plant)
Normal	A-15	Medium	2.205	1.26	310	
		Root	0.083	3.65	33	
seedling		Leaf	0.015	15.05	25	
		Total			358	7.5
	E-26	Medium	2.377	1.42	377	
		Root	0.030	3.15	10	
		Leaf	0.013	16.50	23	
		Total			410	8.5
	A-15 & E-26	Medium	2.907	1.44	467	
		Root	0.092	4.65	47	
		Leaf	0.010	21.40	23	
		Total			537	11.2
Seedling	A-15	Medium	0.907	0.50	50	
		Root	0.037	2.70	11	
without		Leaf	0.017	11.15	17	
		Total			78	2.2
seed	E-26	Medium	0.970	0.70	75	
		Root	0.048	2.80	15	
residue		leaf	---	11.50	--	
		Total			90	2.5
	A-15 & E-26	Medium	1.120	0.74	92	
		Root	0.114	2.85	36	
		Leaf	---	9.80	--	3.6
		Total			128	

*Abundance of ^{15}N = 9.31 atom % in the gas

Table 2 Distribution of nitrogen fixed by bacteria
in rice plant and medium(with
extra carbon source)

Carbon source	Inoculant	Part analyzed	Atom % ^{15}N excess*	Total N (mg)	N$_2$ fixed (μg)	Distribution (%)
0.5% malate	A-15	Medium	5.529	0.50	300	68.7
		Root	1.580	0.71	125	28.6
		Leaf	0.056	1.93	12	2.7
1 % sucrose	E-26	Medium	1.591	1.05	187	79.6
		Root	0.506	0.58	33	14.0
		Leaf	0.073	1.83	15	6.4
0.25% malate	A-15 & E-26	Medium	5.035	0.53	290	54.4
and		Root	2.701	0.77	232	43.5
0.5% sucrose		Leaf	0.057	1.87	11	2.1

*Abundance of ^{15}N = 9.31 atom % in the gas

RETERENCES

(1) Soil Microbiology. In "The International Rice Research Institute
Annual Report for 1970", 47-59(1971).
(2) Balandreau, J.P., Rinaudo, G., Oumarov, M.M., Dommergues, Y.R.,
Asymbiotic N$_2$ fixation in paddy soil. In "Proceedings of the
1st International Symposium on Nitrogen Fixation"(ed. by Newton,
W.E., Hyman, C.J.), Washington State University Press, 611-628
(1975).
(3) Osamu, D. C., Watanabe, I., Fixation of dinitrogen-15 associated with
rice plants. Appl. Environ. Microbiol. 39, 554-558(1980).
(4) Buchanan, R.E., Gibbons, N.E.(ed.), Bergey's Manual of Determinative
Bacteriology. Williams & Wilkins Company, 275,325(1974).

BIOLOGICAL NITROGEN FIXATION ASSOCIATED WITH WETLAND RICE

I. Watanabe

(The International Rice Research Institute, Los Baños, Laguna, Philippines)

Wetland rice can be grown continuously with reasonable yield levels without N fertilizers. Long-term fertility trials in temperate and tropical regions have shown that about 50 kg N/ha is absorbed by every crop of rice grown without addition of N fertilizers (Koyama and App 1979). Quantitative assessment of this N gain is not yet satisfactory, but biological N_2 fixation undoubtedly contributes to the N enrichment in paddy fields.

Sen (11) indicated the presence of heterotrophic N_2 fixing bacteria in the rice root, but the significance of his suggestion was neglected after the research of De (4), who showed the importance of blue-green algae as N_2 fixing agents in the floodwater of wetland rice soils. Döbereiner and Campelo (5) paid attention to the presence of N_2 fixing bacteria in or on the root of tropical graminous plants. The development of a sensitive acetylene reduction technique opened the way for study of N_2 fixation associated with a number of non-nodulating plants. Rinaudo and Dommergues (10) and Yoshida and Ancajas (17) found that some N_2 fixation (acetylene reduction) was associated with the wetland rice root. In the Soil Microbiology Laboratory of the International Rice Research Institute, a series of experiments have elucidated further N_2 fixation associated with wetland rice. This paper reports the outline of the IRRI research achievement.

NITROGEN BALANCE STUDIES

Measurement of N balance by Kjeldahl analyses is the simplest and most natural way of quantifying biological N_2 fixation in soil-plant systems. Rice plants were grown in pots in continuously flooded soil low in total N (0.08%) for several croppings. To eliminate phototrophic N_2 fixation by blue green algae, the surface of some pots was covered with black cloth (Table 1). These results suggest that the presence of rice is necessary to detect significant N gain and that significant N gain is obtained by heterotrophic N_2 fixers. The roles of rice to stimulate the N gain would be: 1) the supply of organic matter from rice, and 2) removal of inorganic N from the soil. The removal of N would reduce the chance of soil N losses and provide a low N condition, which would be favorable for N_2-fixation around the rice root.

ACETYLENE REDUCTION ASSAYS

Great caution is required to evaluate the results of acetylene reduction activity (ARA) assays, because the assay gives only indirect and probably a semiquantitative evaluation on N_2 fixation.

In a flooded soil, the water restricts the transfer of acetylene to the soil and return of the formed ethylene to the atmosphere. Measures were taken to overcome these obstacles during field assay of wetland rice using plastic bags (9). Through the gas transport system existing in wetland rice, acetylene is quickly transported to the root and the evolved ethylene is effectively, although not completely, recovered.

To avoid difficulties associated with slow gas transfer in flooded soil, rice plants grown in the field were transferred to water culture and acetylene reduction assays were conducted and ethylene content in water and gas phases was analyzed (13) and factors affecting N_2-fixation were examined. Steps were taken to avoid the contamination of blue green algae, which were often very active in N_2 fixation (6, 16). The results of field (in situ) acetylene reduction assays in the Philippines (14, 15, 16) and Thailand (3) are summarized as follows: 1) acetylene reduction assay value associated with normally growing wetland rice fell in the range of 0.2 - 1.0 mmol C_2H_4/m^2/day, 2) the highest activity per plant appeared near heading stage, and differences among rice varieties were noticed; the differences varied with the growth stage of rice.

$^{15}N_2$ GAS INCORPORATION

To directly prove the N_2 fixation associated with rice, plants were enclosed in a chamber and $^{15}N_2$ gas was provided (7). To avoid the dilution of $^{15}N_2$ by unlabelled dinitrogen in flooded soil, rice plants grown in the field were transferred to water culture and enclosed in the chamber. The results of two trials for 7 days are shown in Table 2. The results are summarized as follows: 1) a considerable ^{15}N was found in the root as well as in the outer leaf sheaths and basal nodes, 2) the transfer of fixed ^{15}N to growing tissue seems to be low during 7 days exposure to $^{15}N_2$, and 3) the value of $^{15}N_2$ incorporation was almost the same as that estimated by ARA.

Recently, Yoshida and Yoneyama (19) also confirmed $^{15}N_2$ incorporation in rice grown in soil. In rice-soil system, fixed ^{15}N were found more in soil than in rice. In contrast to the Ito et al (7) data, ^{15}N content in the panicle comprised a large fraction (about 30% of ^{15}N in plant).

Eskew and App (unpublished) developed more sophisticated techniques, which enclosed the rice plant and soil, and permitted negligible leak of N_2. Most of the fixed ^{15}N was found in soil and the transfer to the growing parts was small during the exposure. When the plant was grown until harvest after removal from the enclosure at the heading stage, a considerable amount of ^{15}N was found in the grain.

CONTRIBUTION OF ROOT AND BASAL PORTION OF SHOOTS

It has been generally considered that heterotrophic N_2 fixation is associated with the rice root or the rhizosphere. ARA assays of the plant from which roots were removed and ^{15}N feeding experiments revealed that the basal portion of shoots, including outer leaf sheaths and stems, is also important site of N_2 fixation. ARA associated with rootless plants, depending on the varieties and growth stage, accounted for 10-100% of ARA of intact plants. The population of N_2 fixing bacteria and in vitro ARA were concentrated to the basal portions of shoots that were in the soil and floodwater (Table 3). It is not recommended, therefore, to express ARA of the whole intact plant on the basis of root weight of plant.

Table 1   Nitrogen balance studies at IRRI (Source: (1))

	Pot surface	Crop N removal	Soil N change	Misc. input	Positive balance
			mg N/pot		
Experiment 1 (6 crops)					
Planted	light	1175	− 604	27	544
Planted	dark	1148	− 961	27	160
Fallow	light	0	193	24	169[ns]
Experiment 2 (4 crops)					
Planted	dark	997	− 795	21	181
Fallow	dark	0	− 243	0	− 243

ns = not significant

Table 2 $^{15}N_2$ incorporation into rice plants in growth chamber for 7 days (Source: (7))

Samples	^{15}N content (atom % excess)	Apparent N_2 fixing rate (μmol N_2/day)	Percent to total
IR26			
Root	0.71	5.4	47
Outer leaf sheath	1.57	5.0	43
Inner leaf sheath	0.052	1.0	8
Leaf blade	0.003	0.07	< 1
Young panicle	0.005	0.015	< 1
Latisail			
Root	0.50	1.8	28
Outer leaf sheath	0.50	3.1	46
Basal node	0.25	0.41	6
Inner leaf sheath	0.028	0.55	9
Leaf blade	0.031	0.49	8
Young panicle	0.017	0.14	3

POPULATION OF N$_2$ FIXERS

Azospirillum, _Beijerinckia_, and _Enterobacter_ have been reported as N$_2$ fixing inhabitants of rice root. The most-probable-number techniques were used to count N$_2$ fixing bacteria on or in the rice root, by using semisolid agar medium either with malate or glucose (14). The presence of N$_2$ fixing bacteria was detected by acetylene reduction. The number of aerobic N$_2$ fixing bacteria on glucose medium was always 10 - 100 times higher than the number on malate medium and was nearly the same as the number of aerobic heterotrophic bacteria grown on tryptic soy agar.

The colonies on tryptic soy agar were tested on semisolid glucose medium supplemented with yeast extract or casamino acid (12). Those bacteria required a small amount of yeast extract or casamino acids for nitrogenase activity. As Table 4 shows, large portions of the heterotrophs were active in N$_2$ fixation (2). The majority of these bacteria were close to _Pseudomonas_, and had been previously designated as _Achromobacter_-like (12), but their exact taxonomic position is not yet identified.

Other groups of N$_2$ fixing bacteria found from rice roots, stems and outer leaf sheath were acid-gas forming Enterobacteriaceae (close to _Klebsiella_) and _Azospirillum_ (mostly close to _A_. _brasiliensis_). These three major groups of N$_2$ fixing bacteria isolated from rice root and stems were inoculated on germ-free culture of rice. The rice plants were grown for 8 weeks in flood sand-vermiculite culture with 18 mg N/pot and then subjected to ARA assays (Table 5). _Azospirillum_, or the mixture of _Azospirillum_, _Pseudomonas_-like, and Enterobacteriaceae, gave the highest activity and the MPN count of _Azospirillum_ on malate medium was the lowest. But the activity was much lower than that associated with field grown rice plants. The interaction between N$_2$-fixing bacteria and non-fixing microflora is suggested.

DRYLAND RICE VERSUS WETLAND RICE

Yoshida and Ancajas (18) showed that the ARA of dryland rice roots was less than the wetland rice roots. The field and water culture ARA assays conducted by Barraquio and Watanabe (unpublished) also showed that ARA activity associated with dryland rice was almost negligible as compared with that by wetland rice. The population of N$_2$ fixing bacteria was measured for the rice grown in both conditions (Fig. 1). The number of N$_2$ fixing bacteria was much higher in or on wetland rice roots and basal shoots than that associated with dryland rice. Particularly, the number of N$_2$ fixing bacteria associated with the basal portion of shoots was much higher in wetland rice. The basal portion of shoots beneath the floodwater are in a suitable environment for the development of N$_2$ fixing bacteria.

The favorable condition of submergence to the development of aerobic N$_2$ fixing bacteria was also confirmed by comparing the incidence of N$_2$ fixing bacteria in the heterotrophs from roots of wetland plants (rice, wild rice, _Monochoria_) and dryland plants (_Cassia_, _Panicum_, _Zea_ and others). The fraction of N$_2$-fixing bacteria to total aerobic heterotrophic bacteria was several percentage or less on the roots of dryland plants (Table 4).

Table 3   In vitro ARA and N_2 fixing population of
stems and outer leaf sheath (Source: unpublished)

Samples	5 hr ARA C_2H_4 nmole.g^{-1} fw	Heterotroph 10^{-6} g^{-1} fw	N_2 fixing bacteria Glucose	Malate
Stems				
0 - 1 cm*	84	–	1	1
1 - 2 cm	84	–	0.2	0.6
2 - 3 cm	30	–	0.2	0.07
3 - 4 cm	5	–	0.002	0.005
Outer leaf sheath				
0 - 1 cm*	27	970	1	1
1 - 2 cm	32	150	2	2
2 - 3 cm	15	150	2	2
3 - 4 cm	15	130	1	0.8
4 - 5 cm	6	40	0.1	0.2

*From the lower position

Table 4   Occurrence of combined –N requiring N_2 fixing
bacteria in cultivated and wild rice grown in
flooded soil (Source: (2))

	Samples	Total heterotrophic bacteria g^{-1} fw	Percentage of N_2 fixing bacteria
Oryza sativa			
Latisail	Histosphere	9.2×10^7	84
	Stems	3.0×10^7	49
IR26	Histosphere	4.9×10^7	92
	Outer leaf sheath	9.7×10^7	15
Khao Lo	Histosphere	2.7×10^8	75
	Stems	5.5×10^7	59
Oryza australiensis			
	Histosphere	7.3×10^6	75
	Stems	2.8×10^7	9
	Rhizomes	2.4×10^7	49
O. punctata			
	Histosphere	8.7×10^7	90

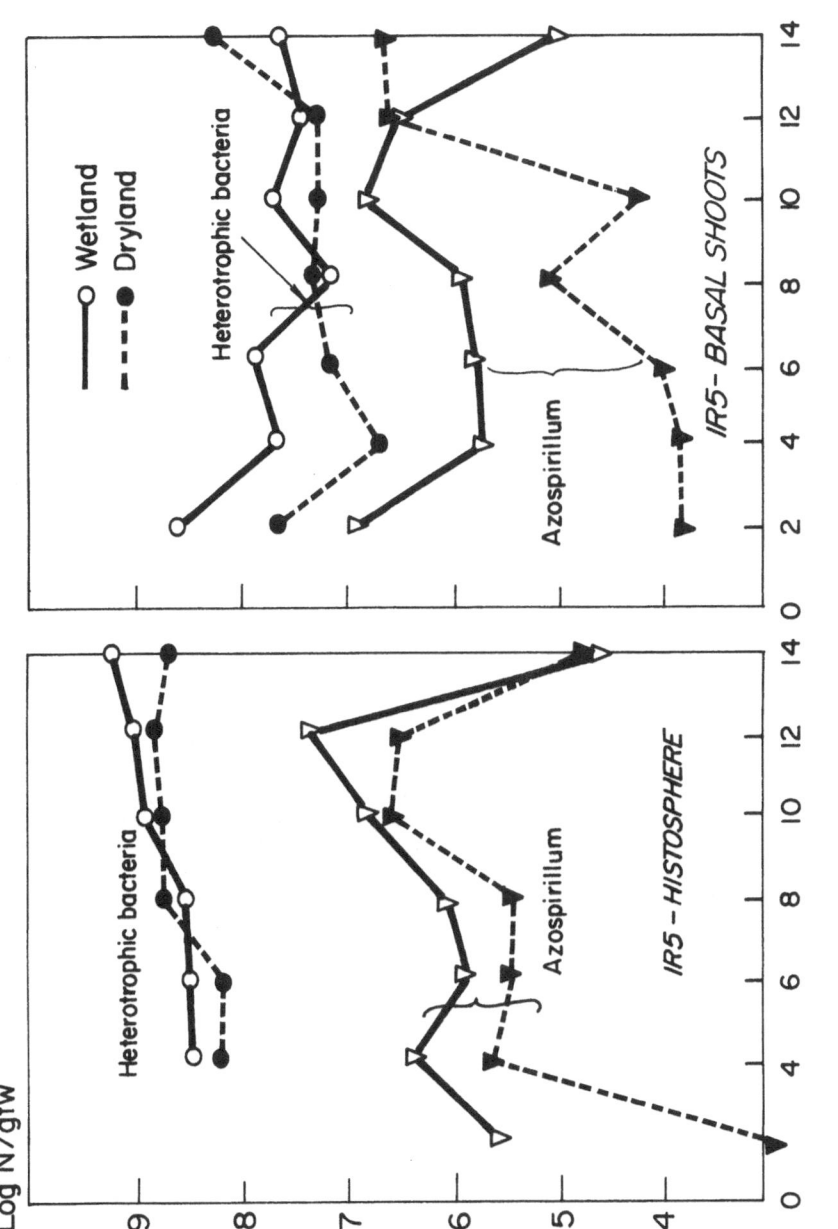

Log N/gtw

Wetland

Dryland

Heterotrophic bacteria

Azospirillum

IR5 – HISTOSPHERE

Heterotrophic bacteria

Azospirillum

IR5- BASAL SHOOTS

Week

Fig. 1   Comparison of microbial population between wetland
and dryland rices (1980)

353

Table 5   The population of N_2-fixing bacteria and their activity
inoculated to rice plant (Source: unpublished)

Treatment	MPN g^{-1} fresh root	ARA (20 h) nmole C_2H_4 per plant
Noninoculated	0	200 ± 40**
a. Pseudomonas-like	1×10^7	250 ± 70
b. Enterobacteriaceae	1.9×10^6	560 ± 250
c. Azospirillum*	4×10^5	1500 ± 470
Mixture of a,b,c	9×10^6	1900 ± 370

*On malate medium. **Average and standard error of 3 replicates

SIGNIFICANCE OF ASSOCIATIVE N_2 FIXATION

Although quantitative assessment of associative N_2 fixation with wetland rice is not satisfactory, the contribution of this process to N_2 fixation in the paddy soil and rice system is not so high as suspected previously (18), but it significantly adds to fixation by N_2 fixing agents (blue green algae, heterotrophic N_2 fixers on crop residue and soil organic matter). Little is known about organic matter supply from the rice plant to these bacteria, the energy efficiency of this association, and the mechanism of nitrogen transfer to host plants. Results so far are controversial regarding the rapidity of the transfer of the fixed N_2 to rice plants. A better combination of rice cultivars and bacteria strains would be one way to increase associative N_2 fixation. At this moment, data are still fragmental. Even with such combinations, active N_2 fixation would result in less production due to the competition of photosynthate between energy consuming N_2-fixation process and the growing overground parts.

ACKNOWLEDGMENT

This research was supported by the United Nations Development Programme and conducted in cooperation with A. App, M. Alexander, O. Ito, W. Barraquio, M. de Guzman and D. Cabrera.

REFERENCES

(1) App, A., Watanabe, I., Alexander, M., Ventura, W., Daez, C., Santiago, T., De Datta, S. K., Nonsymbiotic nitrogen fixation associated with the rice plant in flooded soil. Soil Sci. (in press). (1980).

(2) Barraquio, W. L., Watanabe, I., Occurrence of aerobic nitrogen fixing bacteria in wetland and dryland plants. Soil Sci. Plant Nutr. (in press). (1981).

(3) Cholitkul, W., Tangcham, B., Sangtong, P., Watanabe, I., Effect of phosphorus on N_2-fixation as measured by the field acetylene reduction technique in Thailand long term fertility plots. Soil Sci. Plant Nutr. 26,291-299 (1980).

(4) De, P. K., The problems of nitrogen supply of rice. I. Fixation of nitrogen in the rice soil· under waterlogged condition. Indian J. Agric. Sci. 6,1237-1245 (1936).

(5) Döbereiner, J., Campelo, A. B., Non-symbiotic nitrogen fixing bacteria in tropical soils. Plant Soil special volume, 457-470 (1971).

(6) Habte, M., Alexander, M., Nitrogen fixation by photosynthetic bacteria in lowland rice culture. Appl. Environ. Microbiol. 39,342-347 (1980).

(7) Ito, O., Cabrera, D. A., Watanabe, I., Fixation of dinitrogen-15 associated with rice plants. Appl. Environ Microbiol. 39,554-558 (1980).

(8) Koyama, T., App, A., Nitrogen balance in flooded rice soils. Nitrogen and Rice, IRRI, pp. 95-103 (1979).

(9) Lee, K. K., Watanabe, I., Problems of acetylene reduction technique applied to water-saturated paddy soils. Appl. Environ. Microbiol. 34, 654-660 (1977).

(10) Rinaudo, G., Dommergues, Y., Validity of estimating biological nitrogen fixation in the rhizosphere by acetylene reduction method. Ann. Inst. Pasteur (Paris) 121,93-99 (1971).

(11) Sen, M. A., Is bacterial association a factor in nitrogen assimilation by rice plant? Agric. J. India 24,229-231 (1929).

(12) Watanabe, I., Barraquio, W. L., Low levels of fixed nitrogen required for isolation of free living N_2-fixing organisms from rice roots. Nature 277,565-566 (1979).

(13) Watanabe, I., Cabrera, D. A., Nitrogen fixation associated with the rice plant grown in water culture. Appl. Environ. Microbiol. 37,373-378 (1979).

(14) Watanabe, I., Barraquio, W. L., De Guzman, M., Cabrera, D. A., Nitrogen fixing (acetylene reduction) activity and population of aerobic heterotrophic nitrogen fixing bacteria associated with wetland rice. Appl. Environ. Microbiol. 37,813-819 (1979).

(15) Watanabe, I., Lee, K. K., Alimagno, B. V., Seasonal change of N_2-fixing rate in rice field assayed by in situ acetylene reduction technique. I. Soil Sci. Plant Nutr. 24,1-13 (1978a).

(16) Watanabe, I., Lee, K. K., De Guzman, M., Seasonal change of N_2-fixing rate in rice field assayed by in situ acetylene reduction technique. II. Soil Sci. Plant Nutr. 24,465-471 (1978b).

(17) Yoshida, T., Ancajas, R. R., Nitrogen fixation by bacteria in the root zone of rice. Soil Sci. Soc. Amer. Proc. 35,156-157 (1971).

(18) Yoshida, T., Ancajas, R. R., Nitrogen fixing activity in upland and flooded rice fields. Soil Sci. Soc. Am. Proc. 37,42-46 (1973).

(19) Yoshida, T. and Yoneyama, T., Atmospheric dinitrogen fixation in flooded rice rhizosphere determined by the N-15 isotope technique. Non-symbiotic Nitrogen Fixation Newsletter 8(1),4-17 (1980).

STUDIES ON THE PREPARATION OF DRIED INOCULUM OF BLUE-GREEN ALGAE AND ITS APPLICATION IN PADDY FIELD

Huang You-xing, Fang Guang-ru, Yan Yu-zhou,
Wang Ting, Su Guo-feng
(Soil and Fertilizer Research Institute, Jiangsu Academy of
Agriculture Sciences, Nanjing)

The nitrogen-fixing blue-green algae (<u>Anabane</u> <u>azotic</u>) possesses the ability of fixing nitrogen and carbon. Certain amount of amino acids, polypetides and carbohydrates are excreted from metholic process during its growth. Soil nitrogen and organic matter are increased through the mineralization of the dead algal mass so as to enhance the fertility. Result obtained from many investigators(1-3) indicated that the nitrogen-fixing blue-green algae might be reared in paddy field for the purpose of improving nitrogen status. Recently, progress on this aspect has been made in our country, and the inoculation of algal mass to paddy field has been practised in large-scale in Hubei Province(4,5).

Fresh algae is generally used for multiplication and inoculation nowadays. A huge amount of inoculum would be required for extension application. In Jiangsu Province, the period suitable for the multiplication of the blue-green algae ranges from the middle of May to the later part of June, and the time for inoculation algal mass to the paddy field of the second crop is one month later sometime around the end of July to the first part of August while high temperature prevails. The periods of multiplication and inoculation are not synchronous. However, the blue-green algae can maintain its vitality after desiccation(6). Taking advantage of such unique character, the use of dried algal mass as the inoculum might be a practicable approach to meet the requirement of large-scale application.

This paper describes the experiments on large scale preparation of dried inoculum of blue-green algae and the exploration of effective measures in improving its survival nitrogen-fixing activity and floating character in the paddy fields are presented.

MATERIALS AND METHOD

The strain Shui-sheng 1042(7) of <u>Anabane</u> <u>azotic</u> obtained from the Institute of Hydrobiology, Academia Sinica, was cultivated in Shui-sheng III medium at 30°C under artificial illumination of 5000 Lux of 14-hour day length. Acetylene reduction method was employed in the determination of nitrogen-fixing activity. Results were calculated by peak height ratio method(8). Soil organic matter and total nitrogen were determined by Tiurin method and Kjeldahl method(9) respectively. Plant samples were digested with H_2O_2(9) and their nitrogen and phosphorus contents were determined by routine methods.

1. Experiments on the Proper Time for Collecting Dry Algae Culture

Each 50 ml of culture solution sterilized in a 200 ml flask was inoculated with 1 ml of fresh algae and incubated at room temperature. Three bottles of fresh algae were collected every other day. After water being leached, they were air-dried at room temperature and weighed. A sample of 20 mg of algae powder from each batch was inoculated into 30 ml of culture medium in a 150 ml flask and cultured under illumination in an incubating room. And then the nitrogen-fixing activity was determined 4 days after.

2. Experiments on the Selection of Absorptive Materials for Preparing Dry Algae Inoculum

For each 1000g of fresh algae (sun-dried wt, 30g) was thoroughly mixed with the following absorptive materials: 250g rice hull, 250g wheat hull, 50g foamed plastic grain, and 400g vermiculite respectively, and pure algae powder was used as a check. The mixture was sun-dried and broadcast evenly over the water surface just as the transplanted seedlings had resumed, the application rate of algae powder was 37.5 kg/ha (pure algae powder).

3. Experiment on the Effect of Nitrogen-fixing Algae on Paddy Yield

Experiment was carried out on the paddy field with double cropping rice. Both small plots design and field comparison were employed simultaneously. All of the plots and fields for comparison were fertilized as practised by the local farmers, but in addition, a sufficient amount of superphosphate was added, 60 kg/ha of fresh algae were applied to each plot (45 kg/ha for large field for comparison) respectively. Samples of rice plants and soil samples were collected and analysed after harvest. When covering all over the water surface, the algae mass was hoed into the soil. Yield and agronomic characters of rice were determined after harvest.

RESULTS AND DISCUSSION

1. The Proper Time of Collecting Fresh Algae for the Preparation of Dry Algae Inoculum

The growth curve of blue-green algae might be divided into three stages: lag phase, log phase, and stationary phase(10). Bubbles began to appear in 15 minutes after the fresh algae were introduced into culture medium. Nitrogen-fixing activity could be detected obviously within the first hour (Fig. 1). The acetylene reduction ability might reach 6.41 nM C_2H_4 per gram of dried algae inoculum per hour. The nitrogen-fixing activity within 1 to 10 hours was increased linearly with time. Regression was very significant ($p < 0.01$). In the mean while, the dry weight of algae mass remained nearly unchanged. It was illustrated from the fact that during the "lag phase" only the induced enzyme was produced, and no significant increment in the number of cell was observed. In 24 hours after inoculation, the inoculum proceeded to the stage of logarithmic growth, and its physiological activity was greatly enhanced. For instance, the dry weight of strain "Shui-sheng 1042" was increased from 4.62 mg to 26.2 mg per bottle within three days at this stage. The rate of cell division (K) reached 0.58 (K) per day. The dry weight increased up to 43.0 mg on the fifth day, but the rate of cell division lowered down to 0.25 in the meanwhile. During the 7th to 11th day the rate of increase in dry weight dropped rapidly and the rate of cell division merely maintained around 0.1 (Fig. 2, Table 1). After the inoculum proceeded to the stationary phase, there was a little increase in dry weight.

The dried inoculum prepared from different stages of growth possessed considerably different recovering ability in the paddy field. The nitrogen-fixing activity of the dried inoculum prepared in the log phase was 217.7 and

Table 1 Rate of multiplication of blue-green algae and their
variations in nitrogen-fixing activity after the
recovery of dried inoculum prepared
at different dates

Time (days)	Dry algae yield (A) (mg/flask)	Rate of cell division (K)	Nitrogen-fixing activity after recovery (B) (nMC$_2$H$_4$/mg/hr.)	(A x B) (nMC$_2$H$_4$. flask/hr)
0	4.6			
3	26.2	0.58	218.0	5700
5	43.0	0.25	195.0	8380
7	52.6	0.10	233.0	12300
9	63.4	0.09	69.5	4410
11	74.5	0.06	54.0	4020
15	80.9	0.04	37.3	3020

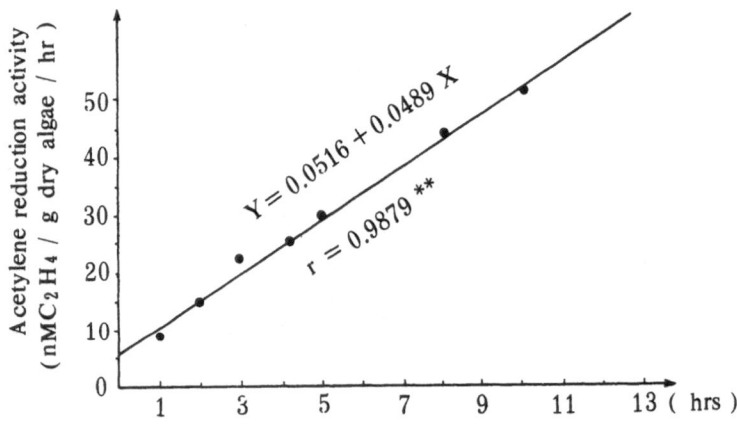

Fig. 1 Acetylene reduction activity of fresh
algae in culture solution

214.1nM C$_2$H$_4$ per mg dried inoculum per hour when incubated in the recovery
culture for 4 and 5-7 days respectively. The recovering ability of dried
inoculum prepared from algae of more than 7 days old reduced very markedly
(Table 1, Fig. 2). Thus, both the yield of algae mass (A) and the ability
of recovery (B) should be considered as important factors in determining the
optimum time for the preparation of dry inoculum, that is to say, a maximum
value of A X B would be most promising. Results of this experiment suggested
that the suitable time for collecting algae should be in the linearly growing
phase.

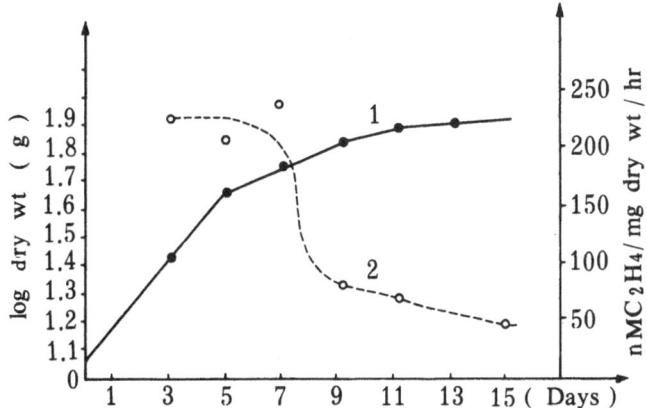

Fig. 2 Recovering ability of dried inoculum of
nitrogen-fixing blue-green algae at different stages
of growth 1 log dry wt; 2. nitrogen-fixing activity

2. Effect of Temperature on the Activity of Nitrogen-fixing
   Blue-green Algae During the Preparation of Dried Inoculum

The nitrogen-fixing activity was 25.01 and 21.01 nM C_2H_4 per mg dry
inoculum per hour for inoculums ventilated dried at temperatures of 40°C
and 50°C respectively. The nitrogen-fixing activity dropped down to 17.6
and 13.2 nM C_2H_4 per mg dry inoculum per hour for those dried at 60°C and
70°C respectively (Table 2). It might be concluded that the optimum temperature for the preparation of dried inoculum would be close to 40°C.

Table 2 Effect of drying temperature on the recovering
ability of algae

Temperature 0°C	40	50	60	70
Nitrogen-fixing activity nMC_2H_4/mg/DW/hr	25.01	21.01	17.16	13.32

3. Selection of Absorptive Materials

The effect of various absorptive materials on the recovering ability
and the multiplication rate of the dried inoculum, when used in paddy field,
were quite different (Figs. 3,4). The rice hull was most promising, it
induced the highest rate of multiplication. The algae mass of dried ino-

culum absorbed with rice hull covered 66 and 90% of the water surface after 2 or 3 days respectively and the entire water surface 6 days later. In the meantime, the fresh weight of algae mass reached to 12.8 t/ha. The second promising one was wheat hull, in this case, fresh weight of algae mass re ached to 10.13 t/ha after 6 days, as compared with 6.75 t/ha of the pure inoculum treatment.

The recovering ability of dried inoculum was very low, when foamed plastic grain and vermiculate were used as absorptive materials. The fresh weight of algae mass was 9.0 t/ha, and 6.75 t/ha, respectively on the 10th day after it was introduced on to the paddy field.

In this experiment, we mixed 50 kg of fresh algae mass with 12.5 kg rice hull or wheat hull, air-dried the mixture and applied it onto the paddy field, a covering of 13 and 29% of the water surface was resulted. In comination with the canopy of rice plants the total shading area might be extended to 40%. This was favorable to the multiplication of the algae mass for the growth of weeds would be inhibited. The hulls of rice and wheat can be obtained everywhere, and are very cheap. Hence, they are considered to be the most suitable absorptive materials.

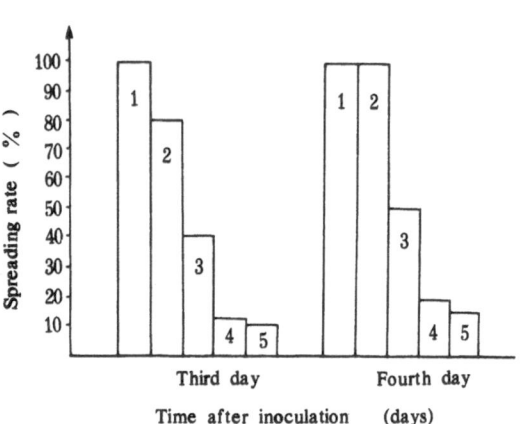

Fig. 3 Spreading rate of dried inoculum prepared with different adsorptive materials. 1. rice hull; 2. wheat hull; 3. pure algae powder (control); 4. vermiculite; 5. foamed plastic grain

4. Effect of Bule-green Algae on the Rice Yield

The merit blue-green algae would be attributed to the increase of nitrogen supply. In field experiment carried out during 1977-1978, the nitrogen and phosphorous contents of the rice plants grown on the treated plots were significantly higher than those of the controls (Table 3).

The yields of treated plots were 9.3% and 9.1% higher than those of the check plots respectively (Table 4).

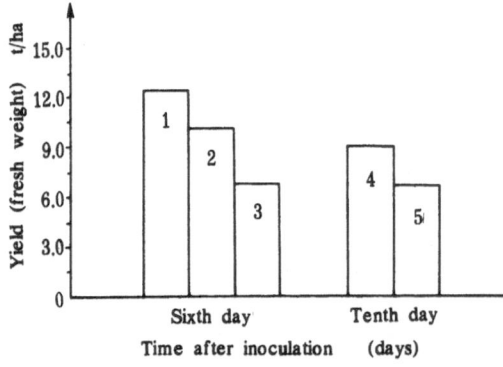

Fig. 4 Yield of fresh algae cultured from dried inoculum prepared with different adsorptive materials. 1. rice hull, 2. wheat hull; 3. pure algae powder; 4. vermiculite; 5. foamed plastic grain

Table 3 Effect of inoculation of nitrogen-fixing blue-green
algae on nutrient contents of rice plant

Year	Treatment	N %	P_2O_5 %
1978	Control	0.84	0.13
	Inoculated	1.01	0.15
	Increase	0.17	0.02
1977	Control	0.94	0.31
	Inoculated	1.02	0.34
	Increase	0.08	0.03

Table 4 Effect of inoculation of nitrogen-fixing blue-green
algae on rice yield

Treatment	Yield (kg/ha)	Field increases (%)	Plot yield (kg/ha)	increases (%)
Control	5870		4780	
Inoculated with algae	6420	9.3	5210	9.1

REFERENCES

(1) Agarwal, A., Blue-green algae to fertilize Indian rice paddies. Nature (London), 279, 181(1979).
(2) El-Nawawy, A. S., Hamdi,B.Y.A., Research on blue-green algae in Egypt. In "Nitrogen Fixation by Free-living Micro-organisms" (ed. by Stewart, W.D.P.), Cambridge University Press, London, 219-228(1975).
(3) Section of Experimental Algal Ecology, Institute of Hydrobiology, Academia Sinica, Studies on the large-scale algalization of the late rice field by the inoculation of nitrogen-fixing blue-green algae. Acta Hydrobiologica Sinica 6, 299-310 (1978). (in Chinese with English summary)
(4) Li Shang-hao, Studies on the nitrogen fixing blue-green algae using as a fertilizer on late rice cultivation. Lanzao Gongzuo Tongxun, 7, 1-13 (1980). (in Chinese)
(5) Institute of Soil Science, Academia Sinica, Nanjing (ed.), Soils of China. Science Press, Beijing, 434-435(1978). (in Chinese)
(6) Zeng Ji-mian, Ye Qing-quan, Li Shang-hao, Liu Fu-rui, Wang Li-mei, A method for conservation of the nitrogen fixing blue-green algae for inoculation. Acta Hydrobiologica Sinica, 4, 452-455(1959). (in Chinese with English summary)
(7) Institute of Hydrobiology, Academia Sinica, Incubation and application of nitrogen-fixing blue-green algae. Lanzao Gongzuo Tongxun, 1, 13-15 (1975). (in Chinese)
(8) Laboratory of Nitrogen Fixation of Shanghai Institute of Plant Physiology, A simplified procedure for the determination of acetylene reduction by the nitrogen-fixing system. Acta Botanica Sinica, 16, 382-384 (1974).

(in Chinese)
(9) Jiangsu Branch, Chinese Academy of Agricultural Sciences, Method
     of Soil and Fertilizer Analysis. Shanghai Science & Technology
     Press, Shanghai, 2-3, 164-165 (1960). (in Chinese)
(10) Tamiya, H., Watanabe, A., Experiments in Phycology. Akihiko-Hattori,
     Kyoto, 195-197(1965). (in Japanese)

PRELIMINARY STUDIES ON PROCESS OF NITROGEN EXCRETION BY AZOLLA

Liu Zhong-zhu, Chen Bing-huan
(Soil and Fertilizer Research Institute, Fujian Academy of
Agricultural Sciences, Fuzhou)

Song Wei
(Institute of Application of Atomic Energy in Agriculture,
Chinese Academy of Agricultural Sciences, Beijing)

INTRODUCTION

In China the cultivation of azolla has a long-standing history, and
the research of symbiotic nitrogen fixation by azolla and anabaena has
been carried out for nearly a century. But it is just in the recent
years that scientists have turned their attention to the exploration on
the excretion of nitrogen by azolla.

In the course of studies on the nitrogen fixation by azolla, nitrogen
was found in nitrogen-free culture solution, judging from which one could
imagine the existence of nitrogen excretion by azolla(1-3). In recent
years, Peter(4) has put forward the problem of nitrogen excretion by azolla
as a theoretical subject. He discovered that anabaena isolated from Azolla
caroliniana could excrete 50% of N fixed in the form of NH_4^+, but the normal
Azolla caroliniana could not excrete any nitrogen. Brill(4) once mentioned
that wild strain of Azolla mexicana could excrete 20% of N fixed in the
form of NH_4^+, but up to now we have not seen any detailed reports on the
experiments yet. The problem of nitrogen excretion by azolla bears significance not only in theoretical aspect but also in practical application.
It is of great significance to a further study on whether azolla can excrete
nitrogen or not, and on the possibility of its utilization by rice.

Using ^{15}N tracer technique, we conducted experiments to investigate
the process of nitrogen excretion by two species of azolla and its utilization by rice plants.

MATERIALS AND METHODS

1. Materials for the Experiments

For azolla, Azolla imbricata (Roxb) Nakai strain "Shipping Green" and
Azolla filiculoides Lam were used. For rice, Agr. No. 1 (Oryza sativa
Subsp. Keng) were used.

2. Conditions for the Experiments

The experiments were conducted under artificial light with intensity
of 20,000 lux, 12 hrs. per day and at the temperature of 25°C.

3. Assay of ^{15}N

$^{15}N_2$ was prepared from $(^{15}NH_4)_2SO_4$ and $(^{15}NH_2)_2CO$ according to the method described previously(5) and purified by liquid nitrogen trap. The ratio of $^{15}N_2:O_2$ in the mixed gas was 8:2. The mixed gas was injected into the experimental chamber by the method of draining water. Azolla was cultivated under $^{15}N_2$ for a certain time according to the experiment project; and then the amount of nitrogen fixed was measured by Kjaldahl method, and the abundance of ^{15}N, with mass spectrometer. Two purified mixed gases, with ^{15}N abundance of 92.44% and 10.1% respectively were used in our experiments.

4. Culture Solution and Methods

1) The following two culture solutions were employed:

A. Yoshida's nitrogen-free culture solution, with its composition as follows (mg/l): $NaH_2PO_4.2H_2O$ 40, K_2SO_4 40, $CaCl_2$ 40, $MgSO_4.7H_2O$ 40, $MnCl_2.4H_2O$ 0.5, $Na_2MoO_4.2H_2O$ 0.1, H_3BO_3 0.2, $ZnSO_4$ 0.01, $CuSO_4.5H_2O$ 0.01, Fe-citrate.

B. Zhejiang Agr. nitrogen-free culture medium No.6302 (mg/l): $CaSO_4.2H_2O$ 171.7, KH_2PO_4 41.6, $MgSO_4.7H_2O$ 41.6, KCl 22.8, EDTA-Na_2 3.7, $FeSO_4.7H_2O$ 2.8, Na_2MoO_4 0.3, H_3BO_3 o.s.

2) Water sealing equipment: A white porcelain dish (22 cm in diameter, 9 cm in height) was covered with a glass dish (19 cm in diameter). Inject the mixed gas of $^{15}N_2$ into it by draining water to form a water sealing chamber for growing azolla. Put azolla underwater into the chamber. The culture solution is 1100 ml and $^{15}N_2$ abundance is 10.1%.

Two methods for cultivating rice seedling and azolla were employed: 1) Rice seedling and azolla were growing in different air conditions but the same culture medium, namely that azolla was growing inside the chamber with an atmosphere tagged with ^{15}N while rice seedling was under the ordinary atmosphere. However, the rice roots can uptake the excrements from the culture solution growing azolla simultaneously. 2) Both rice seedling and azolla were growing in the same ^{15}N tagged atmosphere.

RESULTS AND DISCUSSION

1. Nitrogens Excretion by Azolla

Azolla was cultivated in the $^{15}N_2$ environment for 1,2 or 3 days respectively, then removed to the ordinary atmosphere for continual cultivation of 3 more days. The culture solution was changed once a day and ^{15}N-content of the solution and azolla were analyzed separately. The results are shown in Fig. 1.

It may be seen from Fig. 1 that azolla can excrete a part of fixed N_2 into the culture solution. Each gram of azolla can excrete 0.85, 0.93 and 4.33 μg NH_3 on the first, second and third day of cultivation respectively. After azolla was removed from $^{15}N_2$ environment into the ordinary atmosphere, it continued to excrete the fixed nitrogen. This process, according to our observation, did not come to an end after a couple of days.

2. Evidence for Nitrogen Excretion of Azolla

Some scientists have considered that the nitrogen in the culture solution is derived from the renewal of organs, decomposition of senescent abscised roots as well as the results of activities of microorganisms, but not from the physiological nitrogen excretion. In order to clarify the above-mentioned question, we have conducted the following experiments:

1) Comparison of N-content between culture solution and senescent abscised roots. In the above experiment, while changing the culture solution every day, we collected the abscised roots and determined the total nitrogen content of the culture solution and abscised roots.

The result reveals that the total nitrogen content of the culture solution growing azolla was generally higher than that of the abscised roots; the biomass and nitrogen content of abscised roots decreased as the number of times in renewal of the culture solution increased (Fig. 2). The nitrogen content of culture solution has no clear correlation with it. Such facts may indicate that the nitrogen present in the culture solution does not come from the falling roots to the full extent.

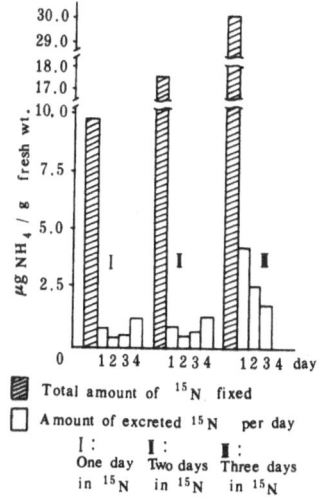

Fig. 1 Dynamics of $^{15}N_2$ fixation and ^{15}N excretion by Azolla imbricata

We took two kinds of ^{15}N-labelled azolla roots (the cutting roots of A. imbricata strain "Shipping Green" 0.6 g and the abscised roots of A. filiculoides Lam. 0.9 g) and dipped them respectively in Yoshida's culture solution for 24 hours. Then we detected the nitrogen content as well as the abundance of ^{15}N of the solution and the roots. According to these data, the extract ability of N were calculated (Table 1).

The results show that at $25^{\circ}C$, the extract ability of N of those roots are 22-31%, therefore, the nitrogen excreted by the abscised roots of azolla in the culture solution is only small in quantity.

Judging from the experiments, it may be concluded that the nitrogen in the culture solution is derived mostly from the excretion by azolla in the course of its growth. The nitrogen come from the decomposition of abscised roots is small in quantity, and nitrogen released by the abscised roots is even smaller.

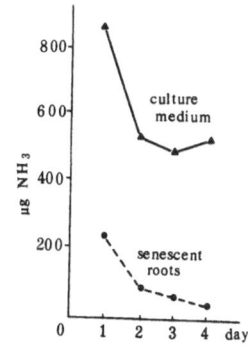

Fig. 2 Nitrogen contents in culture medium excreted by Azolla imbricata and in senescent roots of Azolla imbricata

365

Table 1 Content of ^{15}N in root and its extract

| | Root | | Root extract | | |
	Content of N (μg NH$_3$)	^{15}N abundance(%)	Content of N (μg NH$_3$)	^{15}N abundance(%)	Rate of extract(%)
Senescent root of A. fil-iculoides	1070	0.463	347	0.458	30.8
Cut root of A. im-bricata	480	0.772	384	0.483	22.2

2) Evidence for nitrogen excretion by sterilized azolla. After steril-ization(5), azolla was cultivated under sterilized condition for one week. The culture solution being examined by a microscope, no microorganisms were discovered. Then the azolla was removed into $^{15}N_2$ chamber to be cultivated for 2 days and the content of nitrogen excreted was analyzed (Table 2).

Table 2 ^{15}N fixation and excretion by sterilized azolla

| Azolla* | | Culture medium | | Total ^{15}N fixed |
^{15}N (μg NH$_3$)	abundance (%)	^{15}N (μg NH$_3$)	abundance (%)	(μg NH$_3$)
98.2	0.430	25.9	0.601	124.1

* Fresh wt. 5.75g

The results show that in the axenic condition azolla still has the capacity of nitrogen excretion. Furthermore, the amount of ^{15}N excreted into the solution by sterilized azolla are higher than those by the non-sterilized one. Why does sterilized azolla have a greater ability to excrete nitrogen is a matter still to be investigated. From our experi-ments, one can see that the vegetative mass of sterilized azolla is thinner than that of non-sterilized one, and the cover area of the former is 1/3 larger than that of the latter with the same weight. Whether the greater capability of N excretion of the latter is due to the increased area of azolla with thinner vegetative mass in contact with water remains to be fur-ther studied. But the above-mentioned facts shwo that the nitrogen excreted by azolla is not a result of the activity of microorganisms and algae in the circumstances.

3. Utilization of Nitrogen Excreted from Azolla by Rice

The experiment for studying the utilization of nitrogen excreted from azolla by rice consisted of three treatments. 1) The sterilized rice seed-lings and azolla were cultivated in the same atmosphere containing $^{15}N_2$ for 4 days. 2) Azolla was cultivated under $^{15}N_2$ but the top of sterilized rice seedling was exposed to the ordinary atmosphere while their roots were grown in the same solution together. 3) As a check, the rice seedlings were culti-vated in the same condition as treatment 2, but azolla was absent. After 4 days of cultivation, the total content of nitrogen and abundance of $^{15}N_2$ of azolla, rice and culture solution were analyzed (Table 3).

Table 3   Utilization of excreted N from
azolla by rice-seedlings*

Treatment No.	Wt. of azolla (g)	Total ^{15}N fixed by azolla (μg NH_3)	^{15}N in rice seedlings (μg NH_3)	^{15}N in culture medium (μg NH_3)
I	4.4	160	8.54	11.1
II	4.4	652	8.38	17.6
CK	0	0	0.94	0

*wt. of rice seedling 1.1-1.15 g

It is clear from Table 3 that the nitrogen excreted into the culture solution can be promptly utilized by rice. From treatment 2 and the check, one can see that ^{15}N in the rice seedlings is absorbed mainly from the excretion of azolla. Under the conditions of sterilized rice seedlings, the nitrogen fixing microorganisms remained in/on the rice roots only play a minor role in nitrogen supply to rice seedlings.

CONCLUSION

The increase of nitrogen in the culture solution growing azolla is mostly due to the excretion of azolla vegetative mass by its physiological process. The decomposition of abscised roots only gives a small quantity of nitrogen, and the direct releasing of nitrogen from abscised rotts is even smaller. Azolla in the sterilized solution can still excrete nitrogen. A portion of nitrogen is excreted on the same day as it is fixed and the rest excretes on the days followed. Rice seedlings can absorb and utilize the nitrogen excreted by azolla in the ambient solution.

REFERENCES

(1) Chen Ju-ming, Chen Qiang, The mineral nutrient and fertilizing technique for azolla(1964, ms.). (in Chinese)
(2) Zeng Din, Preliminary exploration on the growth, multiplication and nitrogen fixation of azolla. Journal of Amoy University, 3, 114-138 (1954). (in Chinese)
(3) Wenzhou Institue of Agricultural Sciences, Azolla(1966, ms.). (in Chinese)

(4) Peters, G.A., The Azolla <u>Anabaena</u> <u>azollae</u> symbiosis. In "Genetic Engineering for Nitrogen Fixation" (ed. by Hollaender, A.), Plenum Press, New York, 231-258(1977).
(5) You Chung-biao, Hu Tie-min, The tracing technique and dissociation of nitrogen compound in the biological samples. Yuanzineng, 11, 995 1000(1965). (in Chinese)

ON ECOLOGICAL DISTRIBUTION OF BACTERIA IN PADDY SOILS

Yin Rui-ling, Hao Wen-ying
(Institute of Soil Science, Academia Sinica, Nanjing)

It is the low O_2 content of soil air, low oxidation-reduction potential and the accumulation of reducing substances in paddy soil[1] that govern the growth of soil microbes as well as alter soil bacteria both in number and in composition. This article deals with the ecological distribution of bacteria in the plowed layers of paddy soils in different localities in China, with the corresponding upland fields taken for comparison. The influence of water-logging on bacterial flora in different paddy soils of China has also been investigated.

NUMBER AND COMPOSITION OF BACTERIA IN PADDY SOILS

Most of the soil bacteria are aerobic or facultatively anaerobic, both capable of existing under submerged soil condition. The results obtained from dilution plate method on nutrient agar show that amount of bacteria in plowed layer of paddy soils is about 10^7 per gram soil on average, and is by no means less than that in the corresponding upland fields. Since the paddy soils are usually richer in organic matter and plant nutrients (2), they contain a larger number of bacteria than that in the soils of adjacent upland fields (Table 1).

Table 1 Numbers of bacteria in paddy
soils and in upland fields

Soil zone	Mode of soil utilization	Bacterial number ($\times 10^4$/g dry soil)
Lateritic red earth (Guangdong)	Paddy soil	3300
	Upland field	1050
Red earth (Jiangxi)	Paddy soil	2420
	Upland field	2480
Yellow brown earth (Jiangsu)	Paddy soil	1280
	Upland field	1030

The species of bacteria in both paddy soils and upland soils collected from Guangdong, Jiangxi and Jiangsu provinces are very similar to each other. The genera commonly present are: Bacillus, Arthrobacter, Pseudomonas, Alcaligenes, Flavobacterium, Serratia, Chromobacterium and Micrococcus. Among them, Bacillus, the predominant genus, contains such species as B. firmus, B. megaterium, B. pumilus, B. circulans, B. subtilis, B, cereus, B. mycoides and B. sphaericus. However, the component of the bacterial species in paddy soil and the corresponding upland soil differs greatly. According to the data obtained from nutrient agar, with equal part of malt agar, it is shown that B. megaterium is the predominant spore-forming bacteria in paddy soils (Table 2).

Table 2 The species of Bacillus in paddy soils and in upland fields in different localities (% of isolates)

Soil zone	Mode of soil utilization	B. megaterium	B. subtilis	B. cereus	B. sphaericus	B. pumilus
Lateritic red earth (Guangdong)	Paddy soil	70.3	6.4	11.5	4.3	4.2
	Upland field	50.3	19.3	19.0	4.4	6.2
Red earth (Jiangxi)	Paddy soil	61.6	18.6	8.2	2.9	8.5
	Upland field	31.9	47.0	12.9	4.7	3.1
Yellow brown earth (Jiangsu)	Paddy soil	49.2	30.9	3.3	4.4	12.3
	Upland field	31.4	65.5	0.7	0.2	2.3

It constitutes 50% of the total Bacillus in paddy soils in different localities, and is usually more than that in the corresponding upland fields while B. subtilis and B. cereus occur principally in the upland fields. The similarities on the occurrence of those Bacillus spp. in the paddy soils from different localities also reveals that there is no evidence of zonality in the distribution of soil bacteria.

NUMBER OF BACTERIA IN DIFFERENT TYPES OF PADDY SOILS

According to the regime of water-logging, the paddy soils may be divided into two categories: the temporarily submerged paddy soil (T.P.S.) and the permanently submerged paddy soil (P.P.S.). The number of bacteria differs greatly in these two categories of paddy soils in East China, Southwest China and Northeast China. It is often the case that there is less number of bacteria in P.P.S. than in T.P.S. (Fig. 1), ranging from 4 to 10 millions per gram of dry soil in P.P.S. and 10 to 17 millions per gram of dry soil in T.P.S.. Nevertheless, the number of bacteria changes markedly in T.P.S. as a result of the variation of soil water regime. It increases significantly after drainage under dryland crops and decreases under submerged condition. Although seasonal changes in bacterial number also occur in permanently flooded paddy soil, its fluctuation is to a less extent than that under alternatively wet and dry conditions.

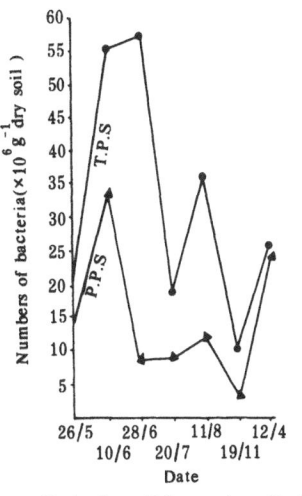

Fig. 1 Changes of bacterial
numbers in P.P.S. and
in T.P.S.

BACTERIA FLORA IN DIFFERENT TYPES OF PADDY SOILS

The percentage of the isolates of various bacteria also differs in
T.P.S. and P.P.S.. The results we have obtained in the same fields in
three successive years, during waterlogging period show that Bacillus constitutes approximately 80% of the total isolates. Under dry farming the
percentage of Gram negative rods raised up to more than 90% while Bacillus decreased suddenly (Figs. 2,3). However, the spore-forming bacteria were always predominant in P.P.S.. Apart from B. megaterium, B.
firmus was the most predominant species on nutrient agar plate, constituting more than 60% of the total colonies of P.P.S., whereas the
amount of Pseudomonas spp. was found much higher in T.P.S. (Table 3).
Table 3 also shows that both kinds of bacteria vary with different soil
water regime. The relative number of B. firmus was greater in P.P.S.,
while the range of fluctuation was narrow (Fig. 4). But in T.P.S. it
fluctuated greatly according to the water regime of soil. On the contrary,
Pseudomonas spp. was very rare in P.P.S.. In T.P.S. it dropped in July
and August as the soil was flooded, but raised up in the drained season
(Fig. 4,5).

Laboratory experiments indicated that B. firmus grew vigorously both
under anaerobic and aerobic conditions (Figs. 6,7). The growth rates under
both conditions were nearly the same. In Pseudomonas spp. the growth
rate under aerobic condition was up to 5-10 times larger than that under
anaerobic condition. It is therefore believed that the fluctuation of
both species in soil might be related to their adaptation to low O_2 content
in atmosphere. But apart from this, the various reducing substances and
intermediate products[3] formed during the anaerobic decomposition in P.P.
S. might also have influence to some extent on the normal growth of soil
microbes.

371

Table 3 The common bacteria in different types of
paddy soils (% of isolates)

Locality	Pseudomonas spp*	B. firmus	B. megaterium	B. cereus
South China				
P. P. S.	11.7	67.5	11.3	0.0
T. P. S.	28.9	0.0	2.2	60.0
Southwest China				
P. P. S.	8.3	50.2	16.6	8.3
T. P. S.	38.9	8.3	27.7	0.0
East China				
P. P. S.	4.9	41.0	14.8	4.9
T. P. S.	53.1	19.8	8.6	0.0
Northeast China				
P. P. S.	0.0	55.3	12.1	5.1
T. P. S.	86.6	0.0	0.0	13.4

* Including Alcaligenes

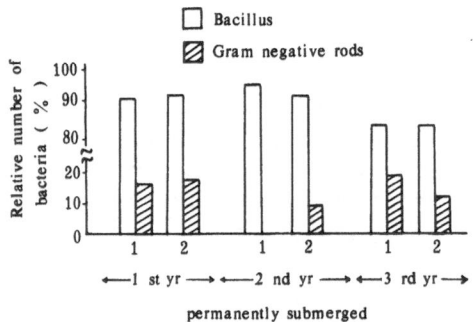

Fig. 2 Changes of the relative
number of bacillus and gram negative
rods in P.P.S.

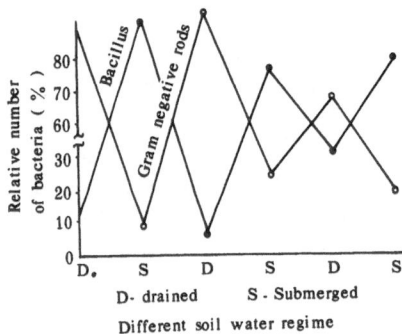

Fig. 3 Changes of the relative
number of bacillus and gram negative rods in T.P.S.

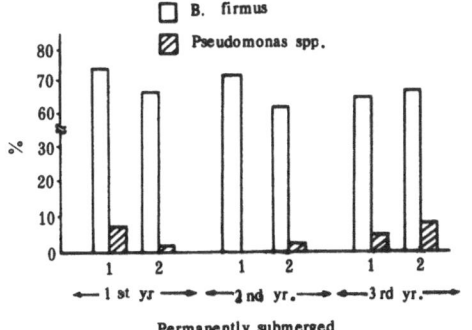

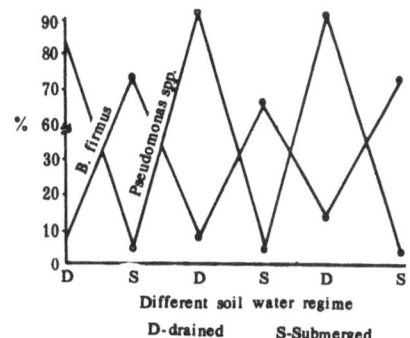

Fig. 4 The percentage of B. firmus
and pseudomonas spp. in P.P.S.

Fig. 5 The percentage of
firmus and pseudomonas spp. in
T.P.S.

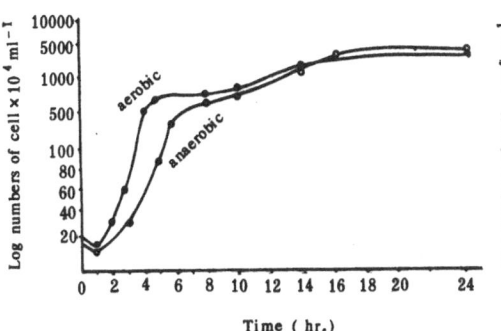

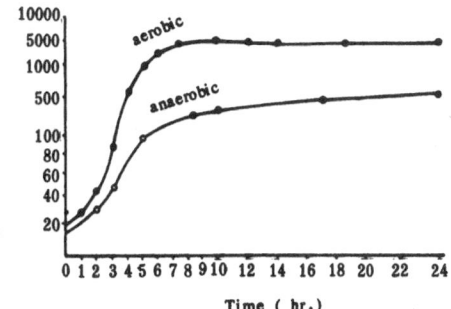

Fig. 6 Growth curve of B. firmus
under aerobic and anaerobic conditions

Fig. 7 Growth curve of
pseudomonas spp. under aerobic
and anaerobic conditions

Influence of continuous rice-cropping on the changes of bacterial composition

In recent decades, the acreage of double cropping (rice-wheat, or
rice-milk vetch) decreases and that of triple cropping (rice-rice-upland
crop) prevails in the Taihu Lake area, Jiangsu Province. As a result,
the secondary gleying process develops and the release of available
nutrient retards under a rather persistently waterlogging condition[4].
It has been observed that the bacterial flora in paddy soils also varied
with such a change of farming system. The relative number of B. firmus
increases but that of Pseudomonas spp. decreases (Figs.8,9). And the
whole composition of bacteria in soil under triple cropping system
becomes similar to that in P.P.S.. Such change in bacterial composition
might be taken as a signal for the change of soil properties.

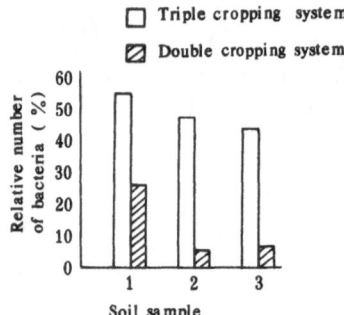

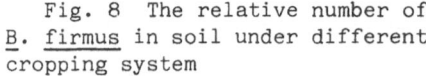

Fig. 8 The relative number of
B. firmus in soil under different
cropping system

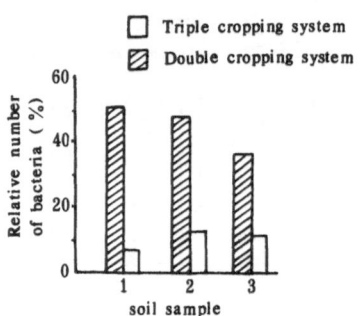

Fig. 9 The relative number of
pseudomonas spp. in soil under
different cropping system

REFERENCES

(1) Yoshida, T., Microbial metabolism of flooded soils. In "Soil Bioche-
    mistry,V.3" (ed. by Paul, E.A., Mclaren, A.D.) Marcel Dekker, Inc., New
    York, 83-122(1975).
(2) Institute of Soil Science, Academia Sinica, Nanjing (ed.), Soils of
    China. Science Press, Beijing, 361-363 (1978).
(3) Takijma, Y., Studies on organic acid in paddy field soil with re-
    ference to their inhibitory effects on the growth of rice plants.
    Parts 1 and 2. Soil Sci. Plant Nutr., 10(5), 204-224 (1964).
(4) Liu Yuan-chang, On the change of soils in properties under triple-crop
    system with double cropping of rice and the maintenance of their
    fertility (1978, ms.). (in Chinese)

IRON TOXICITY IN RICE PLANTS AND NITROGENASE ACTIVITY
IN THE RHIZOSPHERE AS RELATED TO POTASSIUM APPLICATION

G. TROLLDENIER
(Büntehof Agricultural Research Station, Hannover,
Federal Republic of Germany)

Introduction

Microbial transformations in flooded soils differ significantly from those in aerated upland soils. In a fertile upland soil about 50 per cent of the total volume consists of
pores. More than one third of the pore volume is filled with
air, the other part contains water. These soil conditions are
most suitable for root development of upland crops. The presence of oxygen and the predominance of oxidation processes are
characteristic for microbial transformation in upland soils.
When the air is expelled by flooding, the environment of soil
organisms changes drastically. As the diffusion of oxygen in
water is very low, aerobic microorganisms rapidly exhaust oxygen
from the soil solution. After oxygen has been consumed, facultative anaerobes may thrive subsequently which are later followed by strictly anaerobic organisms.

Oxygen status of the rhizosphere and iron toxicity

Rice differs from many other plants in so far as its roots
are largely independent of environmental oxygen. Rice roots obtain oxygen by intercellular air channels from shoots and surface roots. Oxygen may even be excreted into the rhizosphere, as
indicated by the brownish colour of the root surface (Fig. 1).
The excreted oxygen is utilized at least in part by rhizosphere
organisms. In a densely populated rhizosphere oxygen demand
exceeds the supply, thus causing anaerobiosis even on the root
surface. As a consequence excessive uptake of manganese and iron
are not prevented by oxydation and precipitation at the root
surface. In soils with high concentrations of ferrous iron the
plants are forced to take up high amounts of iron which injure
the plant and cause a widespread physiological disorder known
as iron toxicity (Tanaka and Yoshida, 1970).

Oxygen status of the rhizosphere and nitrogen fixation

Maintenance of the oxidizing power of rice roots or low
oxygen consumption by rhizosphere organisms, respectively, are
important not only for preventing iron toxicity but also for
active nitrogen fixation in the rhizosphere. Enumerations of
aerobic and anaerobic nitrogen fixing bacteria revealed a predominance of aerobic organisms in the rhizosphere (Balandreau
and Hamad-Fares, 1975). They constitute a large portion of the
total bacterial population (Trolldenier, 1977 b). For example,
at one date 8 million aerobic and 1 million anaerobic nitrogen
fixers per gram fresh root were counted on the roots of the
variety IR 20 in addition to 18 million total saprophytic
aerobic bacteria. At another sampling date, 255 million total

bacteria, 59 million aerobic nitrogen fixers, but only 23.000 anaerobic nitrogen fixing bacteria were observed on the roots of the variety IR 8.

As nitrogenase is inactivated by oxygen, it is understandable that aerobic nitrogen fixers fix nitrogen more efficiently at subatmospheric oxygen tension. This can be visualized when soil crumbs from the root zone are placed on a nitrogen-free agar medium and incubated at different oxygen concentration. Colonies of nitrogen fixers on plates incubated at 5 and 10 per cent oxygen reach a greater diameter than colonies incubated at atmospheric oxygen tension. This illustrates the importance of an oxidized rhizosphere in a submerged soil for efficient nitrogen fixation. On the other hand, a habitat well aerated as the rhizosphere of rice grown under upland conditions is less suitable for nitrogen fixation as shown by Yoshida and Ancajas (1973).

The inhibitory effect of atmospheric oxygen concentration on nitrogen fixation is revealed when roots are incubated at different O_2 pressures. Table 1 shows that intermediate oxidative conditions in the rhizosphere are more favourable for nitrogen fixation than excessive aeration or absence of oxygen.

Crucial for the ability of aerobic nitrogen fixers to compete successfully with other organisms seems to be the excretion of oxygen (and nitrogen) by rice roots resulting in a distinct oxygen tension in the rhizosphere of healthy rice roots. In a submerged soil a gradient in redox potential exists from the oxygenated root surface to the reduced soil distant from the roots (Trolldenier, 1977 a). Along this gradient optimum conditions are provided for aerobic as well as for anaerobic nitrogen fixers.

Effect of potassium on general microbial activity
in the rhizosphere

A series of studies revealed that potassium deficiency in contrast to nitrogen deficiency enhances microbial numbers and activity especially at low oxygen levels in the rhizosphere (Trolldenier, 1979 a). This effect is caused by more root excretions due to an accumulation of low-molecular compounds in potassium deficient plants. At potassium deficiency oxygen consumption in the rhizosphere is faster than at a balanced nutrition. The higher oxygen consumption results in higher denitrification.

Effect of potassium on iron toxicity

The microorganisms in the rhizosphere of lowland rice counteract the oxidizing power of rice roots. Oxidizing power is highest with complete nutrition and lowest - due to the high number of microorganisms - when potassium is deficient (Trolldenier, 1977 a). The decrease in oxidizing power of rice roots is reflected by their black discolouration caused by precipitation of ferrous sulphide (Fig. 2). The black roots did not have any oxidizing power. Accordingly, plants low in potassium

Table 1　Effect of oxygen concentration on acetylene reduction
of roots from plants grown in solution culture
(TROLLDENIER, 1977 b)

| | Ethylene formed (nmol/g fresh root) at | | |
	21 % O_2	3 % O_2	0 % O_2
After　3 hours	1.4	100.3	61.8
After 14 hours	11.9	654.8	549.0

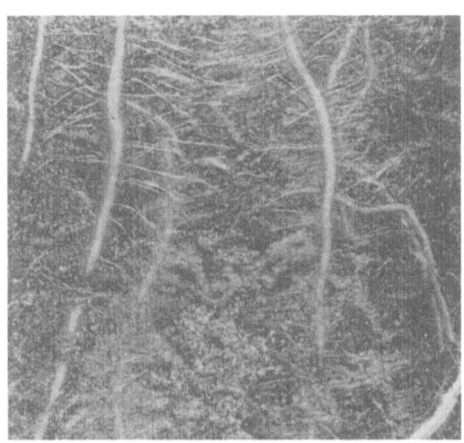

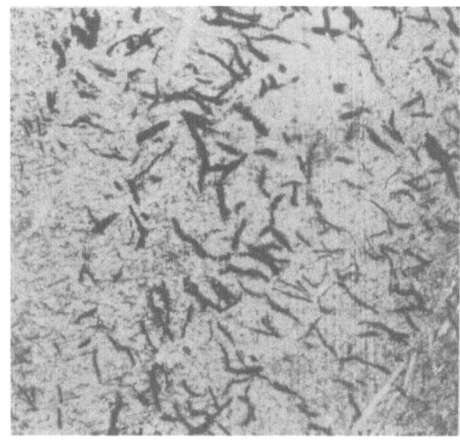

Fig. 1

Healthy rice roots with
sufficient oxidizing power
indicated by brownish colour

Fig. 2

Black discolouration of lateral
roots indicating loss of oxidizing power due to K deficiency

take up more iron than plants better supplied with K; this is often resulting in iron toxicity (Table 2).

Effect of potassium on nitrogenase activity in the rhizosphere

The loss in oxidizing power of rice roots by potassium deficiency implies a ceasing in the proliferation of aerobic micro-organisms, which abundantly colonize healthy brownish rice roots. It seems probable that also the nitrogenase activity of the predominating microaerophilic nitrogen fixing bacteria is affected. First evidence was obtained in a long-term fertilizer trial. Nitrogenase activity was lowest in the treatment with no fertilizer. The application of N, P or K significantly stimulated nitrogenase activity (Trolldenier, 1975). Although combined nitrogen at a concentration of 10 ppm and more depresses nitrogenase activity, this critical threshold is exceeded in the rhizosphere probably only in the first stages of growth and immediately after top dressing. The overall beneficial effect of fertilizer nitrogen on nitrogen fixation under field conditions may be due to the fact that nitrogen dressings increase the capacity for nitrogen fixation by promoting plant growth.

Whether phosphorus has a specific or a more general effect similar to that of mineral nitrogen has still to be elucidated.

The beneficial effect of adequate potassium application on nitrogenase activity has been confirmed in a pot experiment with two different levels of nitrogen and potassium and two different water regimes (Trolldenier, 1979 b). In one half of the pots the soil remained flooded also during interim fallow. In the other pots the soil was exposed to dry fallow and was flooded again before transplanting. After flowering, nitrogenase activity was measured in situ by acetylene reduction assay. The plants of each pot were covered with a plastic bag which reached into the water thus providing a gas-tight volume. The air in the bags was replaced by a mixture of acetylene and air at a ratio of 1 : 4. For the evaluation of the volume of the bag, propane was added, as well. Gas samples were removed in the beginning and after 24 hours for analysis by gas chromatography.

Table 3 shows the dry matter weight of the shoots at harvest and the amount of ethylene formed per pot. Yield was influenced by the water regime as well as by the fertilizer supply. The yield mean values of all four treatments after wet fallow were above those of the dry fallow treatments, but a significant difference could only be observed at N_2K_2. The low difference of N fertilizer levels had no effect on yield, whereas a considerable effect on yield was obtained by higher K application.

The nitrogenase activity was characterized by a considerable deviation of the single values as can be seen in the very high coefficients of variation. Accordingly, the nitrogenase

Table 2 Potassium and iron content in the dry matter of shoots
57 days after transplanting (TROLLDENIER, 1977 a)

Exchangeable K (me/100 g soil)	Shoot Dry weight (g)	K (% in dry matter)	Fe (ppm in dry matter)
0.08	16.9	0.45	520
0.15	20.7	0.93	400
0.25	25.3	1.41	380
0.40	27.7	2.00	330

Table 3 Influence of water regime and fertilizer application
on shoot dry matter and nitrogenase activity
(ethylene formation) (TROLLDENIER, 1979 b)

Previous fallow	Fertilizer level	Shoot DM g/pot	CV[1]	Ethylene formation μmol/24 h/ pot	CV[1]
Dry	$N_1 K_1$	151.1 a[2]	12.7	10.5 a	24.8
	$N_2 K_1$	164.0 a	18.5	6.8 a	70.1
	$N_1 K_2$	208.9 bc	7.2	27.6 a	26.9
	$N_2 K_2$	210.9 bc	4.3	24.8 a	63.3
Wet	$N_1 K_1$	161.0 a	20.1	12.6 a	75.5
	$N_2 K_1$	174.1 ab	25.1	14.9 a	36.6
	$N_1 K_2$	244.9 cd	11.5	115.3	47.8
	$N_2 K_2$	253.3 d	16.9	43.3 a	43.8

[1] CV = Coefficient of variation

[2] In a column, means not followed by a common letter are significantly different at 5 % level

activity is only significantly higher in the N_1K_2 treatment
with wet fallow than in all the other treatments, although
the K_2 treatments also testify a higher nitrogenase activity,
in particular after wet fallow. It can, moreover, be observed
that the mean values of the ethylene formation are higher
after wet fallow than after dry fallow. A significant difference
is, however, not obtained, for the deviation is too high.

A complete analysis of variance reveals, however, in the
F tests significant influences of the factors water regime, N
fertilizer levels and K fertilizer levels (Table 4). Nitrogenase activity is with a 1 per cent level of error higher
after wet fallow. The higher N fertilizer level has a slightly
negative influence on nitrogenase activity.

Table 4   F tests of the analysis variance on the influence of
water regime and fertilizer application on nitrogenase
activity (ethylene formation) (TROLLDENIER, 1979 b)

Source of variation	F values
Wet/dry (water regime)	12.8 **
N fertilization	5.3 *
K fertilization	24.2 ***
Wet/dry x N fertilization	3.6
Wet/dry x K fertilization	8.8 **
N fertilization x K fertilization	4.9 *

Significantly different at 5 %*, 1 %** and 0.1 %*** level,
respectively

The F test also reveals a strong positive influence of K
fertilizer supply on nitrogenase activity. This influence can
even be observed when taking the different dry matter production
into account.

The higher nitrogenase activity after wet fallow with adequate potassium supply might be due to lower redox potential in
the soil providing optimal oxygen tension on the root surface.
However, further studies are necessary to elucidate the effect
of soil conditions on nitrogen fixation.

REFERENCES

(1)   Balandreau, J., Hamad-Fares,I.,Importance de la fixation
d'azote dans la rhizosphère du riz. Bull. Soc. Bot. de
France 122, 109-119 (1975)

(2)   Tanaka, A., Yoshida, S., Nutritional disorders of the rice
plant in Asia. Int. Rice Res. Inst., Manila, Philippines
(1970)

(3)   Trolldenier, G., Influence of fertilization on atmospheric
nitrogen fixation in rice fields. Proc. 11th Colloq. Int.
Potash Inst. Bornholm/Denmark, pp. 287-292 (1975)

(4)   Trolldenier, G., Mineral nutrition and reduction processes
in the rhizosphere of rice. Plant and Soil 47, 193-2o2
(1977 a)

(5)   Trolldenier, G.,Influence of some environmental factors on
nitrogen fixation in the rhizosphere of rice. Plant and Soil,
47, 203-217 (1977 b)

(6)   Trolldenier, G., Effects of mineral nutrition of plants and
soil oxygen on rhizosphere organisms. In: Soil-borne plant
pathogens. Academic Press London, New York, San Fransisco,
Schippers, B., Gams, W., pp. 235-240 (1979 a)

(7)   Trolldenier, G., Nitrogenaseaktivität in der Rhizosphäre
von Sumpfreis in Abhängigkeit von der Mineralstoffernährung.
Mitteilg. Dtsch. Bodenkundl. Gesellsch., 29, 339-344 (1979 b)

(8)   Yoshida, T., Anjacas, Rosabel, R., Nitrogen-fixing activity
in upland and flooded rice fields. Soil Sci. Soc. Amer.
Proc. 37, 42-46 (1973)

PRINCIPLES OF CLASSIFICATION AS APPLIED TO PADDY SOILS

Klaus W. Flach and Oliver W. Rice, Jr.
(Soil Conservation Service, U.S. Department of Agriculture,
Washington, D.C., and Broomall, Pennsylvania)

Soils used for paddy have been difficult to classify in older systems of soil classification, not only because of the dominantly genetic orientation of those systems but also because of the strong effects of management on soil properties, the complexity of soil patterns due to land shaping, and the strong influence of water management and water quality on crop response.

Many different soils have been cultivated to paddy, probably as many as those cultivated to any other crop. Moormann (3) states that rice is grown on soils in all 10 orders recognized in Soil Taxonomy (5). Soils in 8 suborders are of major importance in rice growing, soils in 6 suborders are of local importance, and those in 14 suborders are of minor importance. He discussed soils in 43 suborders that are used for rice growing.

Many but not all "non-paddy" soils are dramatically changed by wet cultivation of rice that includes artificial flooding and shaping of the soil surface.

The use of the term "paddy soil" may cause some to conclude that all "paddy soils" have the same characteristics and properties and, hence, similar response to management. This obviously, is far from the truth; hence the term "soils for paddy" is used in this paper to emphasize that such soils have a great diversity of properties and management requirements.

Many soils used for rice are naturally wet. Such soils do not change much when rice culture is started. If well-drained soils are prepared for paddy, the amount of water that is lost through the soil must be reduced. This is accomplished by puddling, destroying the structure of the surface layer, and, in some soils, creating a dense layer, sometimes called a plow pan or traffic pan, immediately below the surface layer.

Flooding results in a moisture regime that changes the redox potential, nutrient cycles, and soil flora and fauna. The most obvious change in the soil within a short time results from the change in redox potential. Mottling appears in the surface soil shortly after saturation. Other visible changes that develop more slowly can be redistribution of mineral and organic colloidal material within the profile.

The use of irrigation water that carries minerals and salts can result in accumulations of organic materials, salts, and mineral solids. In some farming systems the organic-matter level in the surface layer is reduced to very low level, and in some areas the irrigation water causes significant changes in soil temperature.

Clearly, if we are to be successful in transferring experience, classificatic must reflect the properties of a soil after it has been modified by rice culture. To predict how a soil that has not been used for paddy in the past will respond to rice culture, classification must also relate to the properties of the soil before it has been modified.

We believe that Soil Taxonomy (5) can be amended to develop taxonomic classes that will be useful in classifying soils used for paddy and that modern soil survey methodology based on this taxonomy can be fully responsive to the specific needs of rice culture. This requires, however, that detailed attention be paid to soil phases that adapt definitions of taxa to local needs and to the map units that translate abstract taxonomy to the complex three-dimensional real world.

THE PLACE OF SOILS USED FOR PADDY IN TAXONOMIC SYSTEMS

A system for classifying soils used for rice is an integral part of any general system of soil classification. It is tempting to create a separate system for classifying "paddy soil" that would consider only soil properties important to growing paddy rice. Such a narrow "technical system" would make it impossible to relate soils used for paddy rice to other soils and to determine, for example, what soils in the world that are not now used for rice can be advantageously used for growing rice. More important, a system that relies too heavily on current technology and on soil properties now recognized as being important to the cultivation of one crop, rice, may not be flexible enough to meet new requirements and may be rapidly outdated.

A strong argument can be made for a taxonomic system with a strong genetic bias. Without such a bias, relationships between soils cannot be readily understood; the properties and the classification of a given soil cannot be predicted from environmental factors.

Base saturation, for example, is recognized as an important soil property that is used as a diagnostic criterion in many soil classification systems. We can afford to determine it only on relatively few soil samples. Yet, base saturation can be predicted from our knowledge of soil genesis. Hence, in a practical soil survey the soil scientist who uses a genetic approach has to make only relatively few measurements to verify his prediction rather than making thousands of determinations on randomly selected individual soil samples.

Finally, a system for classifying soils used for rice must fit into a hierarchial system of soil classification. There are thousands of different kinds of soils, each with hundreds of properties. We cannot remember them all, but we can remember and relate the taxa in an orderly hierarchial system. Neither can we convey information about broad principles or broad geographic areas if we cannot generalize with relatively few taxa at an appropriate level.

THE CLASSIFICATION OF STRONGLY MODIFIED SOILS IN SOIL TAXONOMY

From the beginning, the creators of Soil Taxonomy (5) have emphasized that soils must be classified as they are, not as they may have been at one time or what they may be at some time in the future (4). Hence, they developed a system that relies on observable and measurable soil properties. To avoid frequent changes in the classification of a soil, diagnostic criteria were selected that are not easily changed by soil use and certainly not as a function of one cycle or a few cycles of cultivation.

The classification of some soils may change, however, either as the result of more or less drastic intentional changes by man, such as deep plowing or terracing, or of unintentional changes, such as erosion or flooding. Alfisols whose argillic horizon has been incorporated in the plow layer may become Entisols, and Mollisols that have lost their epipedon through erosion may become Alfisols or Inceptisols. In some classes in Soil Taxonomy, it has been found advantageous to adopt definitions that preserve the classification of a preexisting condition. Examples are Aquic suborders that have a high water table "unless artificially drained" or Aridisols that are "dry unless irrigated." Some taxa and diagnostic criteria have been established specifically to reflect man-induced soil properties. Good examples are the anthropic epipedon, the agric horizon, and the sulfuric horizon.

Overriding considerations on which induced changes should be recognized in classification are the probable permanence of the change and the residual effect of preexisting soil properties. Obviously, we do not want to change the classification of a soil just because a farmer decides not to irrigate in a given year. The decision to retain a classification that reflects wetness although the soil has been drained is more difficult to defend since most drainage systems, once established, have some degree of permanence. Yet, it would seem unrealistic to ignore the prominent morphological features induced by poor soil drainage.

In applying the logic of Soil Taxonomy to soils used for paddy, it is probably desirable to use the more permanent induced changes that are reflected in observable and measurable soil properties at the appropriate level and to recognize the more transient and superficial features at the phase level.

ADAPTING TAXONOMIC SYSTEMS TO LOCAL NEEDS

It has been pointed out that systems of soil classification must be able to address broad and basic relationships between soils. It is equally important that these systems be able to address factors that are important to soil use and management at the local level.

In Soil Taxonomy (5) this flexibility is provided at the soil series level and, although not formally a part of the classification system, at the soil phase level.

At the series level, pragmatic criteria adapted to local conditions are introduced in the classification system. Consequently, the series has the necessary homogeneity to allow detailed and specific transfer of technology from one site to another. This is not to say that higher categories of Soil Taxonomy cannot be used for technology transfer. At the family level in particular, specific statements about soils can be made but these statements must be broader and less specific than those that apply at the series level. In general, if technology is to be transferred at the family level, the judgment of a soil scientist is needed to verify if, and with what reservations, a given soil management practice can be applied to all soils of a family. Soil series, on the other hand, are defined narrowly enough that the land manager can use soil information without the assistance of a soil scientist. The development of a well-documented framework of soil series, however, requires a well-structured and experienced soil survey organization, which is not yet available in many parts of the world.

We do not know to what extent current series criteria are adequate for classifying soils used for paddy. Soil series criteria are being used successfully in transferring specific management recommendations for rice within the United States, but rice culture in the United States is relatively inextensive and largely restricted to naturally poorly drained soils that require relatively simple management (1).

Phases are functional groupings created to serve specific purposes within individual soil survey areas. They are most commonly used to subdivide series, but they can also be used to subdivide taxa at any level of the classification system. Any property or combination of properties can be used to differentiate phases. The choice of properties and limits is determined by the importance of the property to the purposes of the survey and by how consistently the phase criteria can be applied. Texture of the surface soil, slope, stoniness, differences in materials below the actual soil, and similar criterial have been used as phase criteria for dryland crops. In some places, phases have been used to recognize differences in soil temperature and in the soil moisture regime that may be caused by differences in aspect. Other phase criteria may be desirable for soil surveys in rice-growing areas. Phases, for example, can be used to separate soils that respond differently to irrigation water of differing qualities or that have a different susceptibility to puddling.

Ultimately, the decision whether to recognize an attribute of land as a phase criterion depends on pragmatic considerations of convenience and usefulness. If, for example, otherwise similar soils respond differently to irrigation water of a given quality and that irrigation water supply is likely to be stable for a long time, a subdivision of the soil property responsible for the different response can be used as a phase criterion. If the quality of the irrigation water changes frequently, the effect of irrigation water on soil performance in better discussed in the map unit description.

Relating Soil Taxa to Units of Land

Soil taxa are conceptual units that have no exact counterparts in the real world. The areas that a soil scientist delineates on a map consist inevitably of parts of several taxa. Some map units consist of similar soils. If so, the map unit is named after the most representative kind of soil, and management recommendations for the representative soil usually apply to the whole map unit.

Other map units consist of contrasting soils. They are called soil complexes. Rice paddies that have been constructed by means of cut and fill usually are complexes. Management recommendations for complexes must be developed from a knowledge of the individual soils and how these soils in combination affect management of the land.

In dealing with soil complexes we must distinguish sharply between the transfer of management technology, which must be based on narrowly defined pure taxonomic units, and the application of this knowledge to pieces of land, which must consider how the components of the complex affect the management of whole fields. No useful knowledge about a soil can be developed if a field experiment is conducted on a soil complex, and no knowledge of soil use can be applied if the effect of all the soils in a field on the management of this field is not fully considered.

Although soil patterns may be complex and intricate, highly detailed mapping is not justified if the different soils can be identified by someone without technical training in soil science or if a cultivator cannot separately manage the smallest of the delineations. The size of the smallest practical management unit is an important consideration in choosing the level of detail of a soil map.

Excessively detailed soil maps may be too expensive to make and impractical to use. With such maps, recommendations on the most effective management of contrasting soils in a field must be made jointly by a soil scientist and a soil management expert. They must not be left to the farmer. Our goal is to design map units and interpretations of the greatest usefulness to the land manager.

SOIL SURVEYS AND LAND CLASSIFICATION

Modern systems for interpreting soil surveys and other natural resource surveys for land evaluation have not been developed to the extent of soil classification system nor have they been used as widely. Yet, the ultimate usefulness of soil surveys depends on such systems. The absence of a more or lees universal approach has hampered the transfer of information from area to area within a country and from country to country. The reason for the slow development of such systems has to do with the diversity of objectives of land classification systems and, in many places, the absence of up-to-date quantitative information about soils that can be related to defined soil taxa. In other words, not enough data about soil performance are being collected and many of the data available are not related to defined kinds of soil. Two recent developments in this area are worthy of attention.

As one aspect of the application of soil survey information, in 1978 the National Cooperative Soil Survey of the United States started a program of assisting users of soil surveys to rate the soils they use in terms of soil potential.

Soil potential gives the rating or rank of soils in such a way that users of soil surveys can readily determine the relative quality of a soil for a particular use compared with other soils in a given area. Yield or performance level, the relative cost of applying available technology to minimize the effects of any soil limitation, and the adverse effects of any continuing limitations on social, economic, or environmental values are considered.

A systematic procedure is used to identify for each soil mapping unit the soil limitations to be overcome, the locally feasible measures for overcoming these soil limitations, the performance level of the soil, and the limitations continuing after corrective measures have been applied. The procedures are equally applicable to agricultural and nonagricultural uses. Soil potential ratings developed for a number of uses provide an excellent analysis that can be used in allocating resources.

A second modern approach is a comprehensive product of the Food and Agricultural Organization of the United Nations. It is called A Framework for Land Evaluation (2). The word "Framework" used in its name indicates that it is not an evaluation system but a set of principles and concepts that can be used to construct an individual land development project or a regional or national evaluation system. It was prepared with the recognized need for all systems, whether local or national, to have comparable and compatible components. It stresses that land must be evaluated in terms relavant to the physical, economic, and social context of the area, including, for example the location of markets, acceptable systems of land tenure, and availability of resources, but not strictly economic and social characteristics. It provides for three levels of intensity that correspond to the objectives. It uses as its starting point a soil survey or similar survey but requires that the inclusion of other aspects of the environment. It encompasses the Land-Capability Classification system and the Soil Potential Rating system used by the USDA Soil Conservation Service.

REFERENCES

(1) Flach, K. W., and D. F. Slusher. Soils used for rice culture in the United States. Pages 199-214 in Soils and Rice, International Rice Research Institute, Los Banos, Philippines (1978).

(2) Food and Agricultural Organization of the United Nations. Soils Bull. 32. A framework for land evaluation. Rome. 72 p. (1976).

(3) Moormann, F. R. Morphology and classification of soils on which rice is grown. Pages 255-272 in Soils and Rice. International Rice Research Institute, Los Banos, Philippines (1978).

(4) Smith, G. D. Lectures on soil classification. In Pedologie, Special No. 4. Belgian Soil Sci. Soc., Ghent (1965).

(5) U.S. Department of Agriculture, Soil Conservation Service, Soil Survey Staff. Soil taxonomy: A basic system of soil classification for making and interpreting soil surveys, Agric. Handb. No. 436, U.S. Government Printing Office, Washington, D.C. (1975).

CLASSIFYING RICE SOILS ON THE BASIS OF THEIR PHYSIOLOGICAL CHARACTERISTICS

Hou Guang-jiong

(Southwestern Agricultural College, Chongqing)

SALIENT ASPECTS OF THE GENESIS OF RICE SOILS

1. Composite Forms of Structure Aggregates

With only a few exceptions, the rice soils usually manifest themselves by relatively big structural aggregates, which can only be partly pulverized by plowing during the growing season of upland crops usually in rotation with rice. The current literature consequently always deals with the structural forms of rice soil as being one of strictly hydrogenic in nature. Practically little has been said about the physicochemical reactions that are intervening in structure-forming processes. Questions thus arise as to what role the soil parent material plays in the genesis of rice soils, or why similar types of structural aggregates may differ widely in their rate of slaking in water. But till now there still remains much to be debated.

Recently, a simple test carried out by the writer and his colleagues, Qing Chang-le, Zhou Yuan-fang and Long Zai-zhong gave us the hint that structural aggregates are formed in paddy soils throught two different agencies: the one is in close connection with the energy levels at which the thinly water-coated colloidal particles attract each other electrically to form primary aggregates, and the other is in connection with the hydro-pneumatic reactions that take place in the whole solum in concordance with irrigation and drainage practices. In Table 1. tests were carried out on four soil samples, two of which were siallites, and the other two, allites. The discrepancy in "liquid limit" figures was found in each sample through a comparison between Hou's "Plasticity Range Method"[1-4] and Atterberg's classical "Plasticity Limit Method". The two samples of siallitic soil, being characterized by a greater divergency in liquid limit figures, indicate that the potential water-holding capacity (this can be estimated by the "Plasticity Range Method") is in reality a measure of the capillary potential of their primary aggregates. The higher zeta-potential of amphoteric colloids tends to swell up energeticly on wetting throughout the cross section of their huge, hydromorphic aggregates. The above discussion may well explain why siallitic soil aggregates usually slake faster than those of allitic soil. The thermodynamic changes in the water-holding capacity of primary aggregates should also be emphasized as a factor affecting swelling, thereby, structural changes. We may even say that clusters or granular-like forms of soil structure are actually present even in soil suspensions (water suspension, of course), otherwise there would be no differences in plastic figures as shown in Table 1. Since the structure features of taxonomic importance are related to a greater extent with types of clay minerals, the necessity of taking primary aggregates into consideration and thereby the colloidal behaviour characteristic of parent materials should be emphassized in classifying rice soils.

Table 1 Effects of quality of soil colloids on soil plasticity
(Discrepancy in "lower plastic point" figures obtained
by the use of "Atterberg's method" versus "Hou's method.)

Soil family name	"Lower plastic point" value		Numerical difference between two readings
	by" Atterberg's method"	by" Hou's method"	
Feixianguan	32.6	28.8	3.8
Shaximiao	33.4	25.0	8.4
Jiaguan	30.8	28.9	1.9
Laoweng	25.8	25.6	0.2

2. Soil Heat-water Regime Under Normal Drainage Condition

Tests has been made on the diurnal fluctuations in the temperature
of surface soils at the depths of 5 cm in order to see how thermal changes
in the water-holding capacity in different soil types affect soil heat-
water regime and whether or not the overlying irrigation water layer
interferes with the heat transfer of solar radiation. Data gathered so
far lead us to the belief that rice soils do gove very quick response to
periodical changes in air temperature, provided the thickness of the
irrigation water layer at the testpits are kept unchanged. In addition
to this, a new fertility index of rice soil, designed to characterize the
relative power of temperature stabilization, has been suggested. It
seems very likely that an increase in sticky range of siallitic soil,
accompanied by a simultaneous increase in air temperature, naturally
results in an enlargement of heat capacity of those primary aggregates
as a whole, and thereby, a stronger buffer power to changes in air temperature. On the contrary, an allitic soil shows only slight increase in
sticky range, hence a lower buffer power toward atmospheric temperature
changes. The fertility index finally adopted represents a diurnal
summation of hourly deviation from the air temperature. The practical
significance of this index as shown in Table 2 is found to be not only in
concordance with the grades of soil productivity, but also with their
behaviour under cultivation practices. Conditions seem to be more complicated when the soil heat-water regime is related to cultivation practices.
The need of a longer duration of uniform temperature may be explained as
a means of excluding untimely thermal transformations in the form of
primary aggregates. Thus in the case of allitic paddy soils, a delay
in late autumn plowing, especially when the rice field is newly drained,
will cause the formation of water-tight clods very resistant to plowing
implements. A delay in digging drainage channels in fields of these
soils is also proved to be a vain attempt. The reason may be that at a
lowering of soil temperature below the activation range of its own, the
water-adsorbing power of primary aggregates is lessened, and simultaneously, the primary aggregates disintegrate and disperse. The deter-
iorated internal drainage of this soil does not show the least sign in
its outlook, and may last for years long with low rice yields. Besides,
hard clod formation may also be encountered in siallitic soils due to
unproper cultivation, but then the clods are destroyable throughtilling
practices. The above discussions may be sufficient to show the applica-
bility of soil temperature determination tests in soil classification
studies.

Table 2 Soil physiological index as a guide to rating potential
fertility of soil in single field plots

Soil family	Fiedld no.	Soil condition	depth (cm)	Relative temp.					Soil physiol. index**	Soil grade***
				10am	12am	14pm	16pm	18pm		
Shaximiao		well-	5	105	104	104	104	106		
	2	drained	15	105	98	94	100	104	2	1st
(siallitc)			25	119	98	90	94	98		class
	3	Poorly	5	107	104	110	106	109		
			15	108	97	98	96	105	6	2nd
		drained	25	112	103	97	94	102		class
	1	Not well-	5	104	107	105	109	112		
			15	112	112	98	102	110	8	2nd
		drained	25	117	111	99	99	105		class
Suining	2	Highly	5	111	111	101	109	113		
(siallitic			15	116	111	99	105	112	12	3rd
allite)		dispersed	25	122	114	99	104	109		class
	4	Timely	5	105	107	104	107	109		
		water-	15	108	105	98	96	105		1st
		logged	25	110	107	96	98	100	5	class

* Taking the temperature of air one meter above field surface as 100,
  calculate the soil temperature records into %
** This index is calculated in order to show the relative power of
  stabilizing soil temperature
*** 1st class-index value 1-5, 2nd class-index value 6-10, 3rd class-
  index value 11-15

EVIDENCES OF SOIL PHYSIOLOGICAL BEHAVIOUR

The term "Soil Physiological Behaviour" is suggested by the writer to
denote those inherent properties of soils which are changing from time to
time under the joined action of their environmental factors, such as daily
or seasonal weather changes, the physiological processes taking place in
plants, and human activities. The colloidal part of soil, like the protein-
eous materials in plant, is the sent of its physiological activities. To
say more precisely, the soil is designated by its metabolic vigor and self-
regulating ability toward heat-water regime.

We can now say with certainty that a fertile soil with higher metabolic
vigor usually gives relatively high yields with lower expense of fertilizing
materials. In addition to this, we are also convinced of the presence of
stabilizing ability toward its shifting heat-water regime, as is stated in
the foregoing paragraphs. What is in urgent need now is some indices which
would be responsible for reflecting the multiple interactions between the
soil and its environs as a whole.

Preliminary tests were devised by the writer in 1963, and were reported by Yu Yuan-chu, Luo You-fang[5] and Pang Bang-yu. In these tests, soil
types are differentiated on the basis of the time and number of tillering of
crops growing on them. This is recommended as a routine procedure for regis-
tering the physiological characteristics of a definite type of rice soil.
For the sake of clarity, only two textural subgroups of the siallitic
purple soil are compared to their "tillering behaviour". Conclusions may
finally be drawn as follows. The normal hydrogenetic or submorphic soil
(tendency to form prismatic structure, but not glei horizon) of loamy textured behaves much better in keeping physiological harmony with crop growth
than the slightly glyied soil of sandy loam texture does Experimental data
reveal that the loamy textured soil, due to its higher content of siallitic
colloids, has a stronger stabilizing power toward soil temperature. The
latter mentioned condition is requisite for maintaining vigorous microbiological activities, hence a normal crop growth.

SUGGESTIONS TO A SYSTEM OF RICE SOILS WITH PARTICULAR
REFERENCE TO THEIR PHYSIOLOGICAL BEHAVIOUR

Soils are quite different from nutrient solutions in Their use as a
medium for plant growth. The former in their natural state can last for
very long time if they are skillfully managed. The latter, however, has to
be renewed at very short time intervals if high crop yields are to be obtained. The fundamental purpose of soil classification, therefore, is to
ascertain the mechanism of physiological processes that are incessantly
taking place in the soil. Soils, especially the rice soils, differ widely
in their colloidal composition, and thus give rise to the importance of
differentiating parent materials on the basis of their colloidal behaviour.
Rice soils also offer specially diversified dispersed phase. They respond
quickly to changes in weather conditions, run-off from surrounding sloping
lands, irrigation water of different source, and the manifold types of
groundwater flow, even when they are overlaid by irrigation water layer.
It is this heat-water regime that makes crop yields to be questioned to a
greater extent by seasonal climate than by the fertility potential of soil
types. The classification system of rice soils as suggested below may
fill in the gap still laid between the macroscopic and microscopic factors
of soil formation.

The tentative system outlined for classifying rice soils consists of
six categories.

1st category: SOIL GENUS

Here disturbance coming from regional drainage conditions or regional
climate is taken into consideration. A prevailing coldness in the soil is
harmful to the normal processes of both soil physiology and plant growth.
A good illustration of the effect of regional drainage on soil physiological behaviour was reported by Zhang Xian-wan[6].

2nd. category: SOIL ORDER

Under this category, the rice soils are differentiated into 9 subgroups: oily-looked, calmy, tough, soft, stiff, brittle, idle, chilly, and
poisonous. Each of them denotes one basic phase of soil physiological
behaviour, particularly those acknowledged in farm practices. Assessments

of composite soil structure with reference to their textural influences are to be made. The details have been given in one of my earlier papers[7].

3rd category: SOIL FAMILY

Results of both field and laboratory studies on soil parent materials are to be systematized here for an elucidation of the nature of soil colloids.

4th category: SOIL CLASS

Textural and structural profiles are to be studied here on the basis of collective determination of their heat-water regime. The subgroups are titled with the name of hydrogenetic structure types which are described elsewhere[8].

5th category: SOIL SPECIES

This category is devoted to meet the demands of rice growers, and of course also agronomists. Items denoting the proper choice of fertilizer species, crop varieties, rotation systems, etc, must be given.

6th category: SOIL VARIETY

As we are generally aware, rice soils are submitted to all sorts of land leveling or land reform for better harvests. "Ill soils", especially those induced by harmful surface deposits, local brine water, and uplifted groundwater table are to be differentiated.

We have made arrangements to bestow soil surveyors with field equipments in order to facilitate the adoption of the above-stated classification system. Field experiments, however, are still to be promoted for a comprehensive study of representative rice soils.

REFERENCES

(1) Hou K.C., Chang S.W., Studies on soil plasticity. Pochvovedenie, 5, 17-24 (1956). (in Russian)
(2) Hou K.C., Chang S.W., Nature of exchangeable bases adsorbed on clay colloids in relation to soil plasticity. Trans. 4th Int. Congr. Soil Sci., 136-137 (1949).
(3) Hou K.C., Huang J.H., Method of determining soil plasticity curves. Trans. 4th Int. Congr. Soil Sci. 44-46 (1949).
(4) Hou K.C., Tao C.T., Differentiation of soil types by the use of plasticity method. Trans. 4th Int. Congr. Soil Sci., 339-342 (1949).
(5) Yu Yuan-chu, Luo You-fang, Studies on the "tillering behaviour" of various types of purple soils (1965, ms.). (in Chinese)
(6) Zhang Xian-wan, Paddy soils in Chengdu plain and their classification. In this Proceedings.
(7) Chu L.T., Ma Y.T., Sung T.C., Hou K.C., The nomenclature of the various horizons of paddy soils. Special Soils Publication, China, Series B, 4, 85-91 (1938). (in Chinese with English summary)
(8) Hou Guang-jiong(ed.), Essentials of the Characteristics of Agricultural Soils in China. Chapt. 3 (1980, in press). (in Chinese)

THE RECENT TREND OF PADDY SOIL CLASSIFICATION IN JAPAN

TAKESHI MATSUI

(Regional Planning Consultation, Co. Ltd., Japan)

INTRODUCTION

In Japan, several schemes for paddy soil classification have been proposed, but the general agreement has not yet been reached.
Kamoshita(1940)[1] was the pioneer of this field. He originally introduced
the modern concept in pedology into the paddy soil classification, and
established the following soil types : Bog soils, Half-bog soils, Meadow
soils, Gray lowland soils and Brown lowland soils according to the catenary
sequence based on the differences in groundwater table. This scheme for
paddy soil classification and its variations have been applied to various
scientific and administrative soil survey projects in Japan, for example, to
the Land Classification Survey sponsored by the National Land Agency, and to
the Fertilizer Reclaiming Program by the Ministry of Agriculture and
Forestry.

However, subsequent progress in studies of the genesis and classification of paddy soils has raised some questions against Kamoshita's system.
The most important one was that effects of the irrigation water on the paddy
soil morphology were not taken into account in his system. Then, Uchiyama
(1949)[2] proposed a new classification scheme by introducing these effects.
Kanno(1956)[3] developed this trend, and proposed an unique classification·
scheme based on the composite effects of irrigation- and ground-water to the
paddy soil morphology. The basic concept of the hydrosequence consist of the
Ground-water, Intermediate and Surface-water types of inorganic paddy soils
was thus established. Concurrently, Yamazaki(1960)[4] devised a similar
scheme on the basis of the hydrosequence in paddy soils independently. As a
result of the determinations of redox potentials of numerous paddy soil horizons with reference to their morphological features, he distinguished several
diagnostic paddy soil horizons, and set up ten soil families according to the
horizon sequences and grouped them into two great soil groups, such as, Ground-
water type and irrigation-water type.

Following aforesaid trend, the Group of Japanese Pedologists (GJP, 1969)[5]
devised a preliminary classification system of Japanese paddy soils according
to the author's proposal (Matsui, 1966)[6].
This system was based upon the multicategorical morpho-genetic principle.
That is to say, Category 1 (genetic soil type) consists of the Surface-water
and Stagnant-water gley-like lowland paddy soils, similar upland ones and
hydromorphic paddy soils (Bog soils and Gley soils). Category 2 (soil subtype)
is composed of transitional soil into the other soil types. Category 3 is
identified by the preceding natural soil types before reclamation for paddy
fields, suggested by the proposal of paddy soil classification by the Soil
Survey Group, Soil Inst. Acad. Sci., China (1959)[7].
Category 4 is divided by the differences in permeability of solum which may
be estimated by the gross texture and structure of the solum. Category 5 is
classified by the intensities of paddy soil formation, such as, mottle-
forming process, gleyzation, bleaching, etc. The lower categories are divided
by the lithology of parent materials and the texture and content of organic
matter of top soil, as well as the depth of solum.

Kyuma and Kawaguchi (1966)[8] proposed a new great soil group,
"Aquorizem" for only Surface-water gley-like paddy soils which was character-
ized by mottling formation,and regarded the other paddy soils as "antraquic"
(Dudal, 1965)[9] subgroups of the equivalent original soil group, such as,
Bog soils and gley soils. Mitsuchi(1969[10], 1974[11]) proposed a new process
"grayzation in addition to the aforesaid paddy soil formations. Upon driving

after drainage, the gleyed horizons which were formed during water-logged stage, turn gray with some rusty mottles. He set up several genetic horizons, and proposed a two-dimentional classification scheme for paddy soils. Along the hydrosequence axis, Surface-water paddy soils, Fluctuated ground-water ones and Stable ground-water ones are arranged. While along the permeability-sequence axis, Brown lowland paddy soils, Gray lowland ones and Hanging-water gley lowland ones form in line. Recent study on the genesis and classification of Chinese paddy soils in Tai-lake basin by Xu, Q. et al (1980)[12] seems to be influenced by Mitsuchi's idea.

Along aforesaid trend, the author tried to revise his multicategorical classification system for paddy soil classification (Matsui, 1966) as follows:

THE PLACE OF PADDY SOILS IN A COMPREHENSIVE CLASSIFICATION SYSTEM FOR JAPANESE SOILS

Recently the author proposed a new comprehensive classification system for Japanese soils independent on any administrative purposes (Matsui, 1978)[13]. This system is affected by the Russian and European trends of soil classification. Thereafter he modified this system in some parts including the classification of paddy soils (Matsui and Nagatsuka, 1980)[14].

This new soil classification system consists of the following multi-categorical units.

Phylum : Classified on the basis of differences in the megascopic hydromorphic conditions (terrestrial, semi-terrestrial and aquatic). Anthropogenic soils including paddy soils is added at this level.

Class : Classified on the basis of differences in the degree of soil zonality. Paddy soil is placed at this level because of its peculiar intrazonal character.

Order : Classified on the basis of differences in main soil formers (phytogenic, lithogenic, pydrogenic, etc.). In the case of paddy soils, the preceding soil phylum before reclamation for paddy fields (terrestrial, semi-terrestrial and aquatic) is taken into account. But the aquatic paddy soil order is regarded as a synonym of natural aquatic soils, because the morphology of these paddy soils is not substantially different from that of natural aquatic soils (Peat soils, Muck soils, Meadow soils, Gleys and Immature polder soils).

Genetic soil type : Classified on the basis of differences in the basic horizon sequence forresponding to the specific landscape. In the case of paddy soils, differences in hydromorphism (Surface-water type and Stagnant-water one) are taken into account.

Surface-water type is characterized by the presence of Bg horizon, in which iron and manganese mottlings are illuviated as the effect of leaching by irrigation water. While stagnant-water type is characterized by the presence of surface gley horizon underlain by Bg horizon. This surface gley horizon has been formed by the effect of stagnant-water in the upper part of solum on account of its very slow permeability.

Soil subtype : Classified on the basis of the valiability towards the other genetic soil type. In the case of Surface-water upland paddy soils, typic and immature subtypes are classified. The former is characterized by the presence of illuvial mottled Bg horizon, while the latter by no evidence of mottling-formation, which is regarded as the most diagnostic feature of Surface-water paddy soils.

In the case of Surface-water lowland paddy soils, typic, bleached, Brown warp soil-like and fluctuated ground-water subtypes are identified. Typic subtype is characterized by the presence of illuvial Bg horizon. Bleached one defined by the contrast of bleached A2(e)g horizon and illuvial B2irg horizon, which is caused by the strongly acting reduction-leaching and subsequent oxidation-illuviation processes. Brown warp soil-like subtype is not effected by the mottling formation, instead it is characterized by the oxidation through the profile. It can be correlated with Kamoshita's Brown low-

land soil and Uchiyama's Brown oxidized type. Fluctuated ground-water subtype is the synonym of Kanno's Intermediate type. It is characterized by the
presence of the BgG horizon which includes gleyed mottles and specific haloed
tubular mottles (Yamazaki, 1960).

As mentioned before, soils derived from aquatic soils cannot be differentiated from the equivalent natural aquatic soils, so these paddy soils are
named by the way that the name of the equivalent natural aquatic soils is succeeded by - (hyphen) paddy soils, for instance, Typic gleys-paddy soils,
Lowmoor peat soils-paddy soils, etc.

LOWER CATEGORIES OF PADDY SOIL CLASSIFICATION THAN SOIL SUBTYPE

Categories lower than soil subtype are classified into soil species, subspecies and variety in descending order. Each categorical unit is identified
by the following criteria.

Soil species : Classified on the basis of differences in the intensity of the
main pedogeneses. As mentioned before, main pedogeneses in paddy soils are
mottling formation, bleaching, gleyzation and hanging-gleyzation. Then this
unit should be set up as follows;

Strongly mottled	Strongly bleached
Moderately mottled	Moderately bleached
Weakly mottled	Weakly bleached
Strongly gleyed	Strongly hanging-gleyed
Moderately gleyed	Moderately hanging-gleyed
Weakly gleyed	Weakly hanging-gleyed

Soil subspecies : Classified on the basis of differences in the lithology,
nature of organic matter and gross texture in paddy soils derived from semi-
terrestrial (lowland) soils. But in the case of paddy soils derived from
terrestrial (upland) soils, differences in preceding genetic soil types
should be taken as diagnostic criteria, because in these paddy soils, the
inherent morphological, physical and chemical characteristics of preceding
soil body may be considered to surpass lithology, organic matter and gross
texture to make up the morphological features of paddy soil profiles (Soil
survey Group, Soil Inst. Acad. Sci. China, 1959).

1) Criteria for paddy soils derived from semi-terrestrial soils

Lithology	Nature of organic matter
Quartzose	Peaty
Arkose	Mucky
Micaceous	Marshy
Mafic-mineralized	Humic

	Gross texture
Expanding lattice mineralized	Coarse grained
Non-expanding lattice mineralized	Medium grained
Allophanic	Fine grained
Calcareous	Gravelly
Salty	

2) Criteria for paddy soils derived from terrestrial soils

Preceding genetic soil type
Brown forest soils
Yellow-brown forest soils
Red and yellow soils
Volcanogeneous Kuroboku (Ando) soils
Non-volcanogeneous Kuroboku soils
Pseudogleys
Stagnogleys
Lithsols
etc.

394

Soil variety :
1) Criteria for paddy soils derived from semi-terrestrial soils.

The criteria of soil variety are modes of superposition of parent materials combined with the depth of solum. Modes of superposition are represented as follows;
Parent material in upper part of solum / that in lower part e.g.
Medium grained humic allophanic / Fine grained expanding lattice mineralized
Depth of solum is classified as follows;

Very shallow	:	0 - 15cm
Shallow	:	15 - 30cm
Medium	:	30 - 50cm
Thick	:	50 - 100cm
Very thick	:	100cm

2) Criteria for paddy soils derived from terrestrial soils

The criteria are lithology of preceding genetic soil type combined with the depth of solum.

CONCLUSION

The author reviewed the recent trend of paddy soil classfication in Japan, placed paddy soils in his comprehensive classification system for Japanese soils, and proposed a revised multi-categorical classification system for Japanese paddy soils.

Paddy soil class, which is a member of the Anthropogenic soil phylum, is divided into soil orders on the basis of differences in the preceding soil phylum before reclamation for paddy soils (terrestrial, semi-terrestrial and aquatic soil orders). But the aquatic paddy soil order is regarded as a synonym of natural aquatic soils, because the morphology of these paddy soils is not substantially different from that of equivalent aquatic soils (Peat soils, Muck soils, Meadow soils, Gley and Immature polder soils).

Genetic soil type is classified on the basis of differences in hydromorphism (Surface-water type and Stagnant-water one).
Soil subtype is classified by the valiability towards the other genetic soil type, i.e. typic, immature, bleached, Brown warp soil-like and fluctuated ground-water subtypes.

Soil species is classified by the intensity of main pedogeneses in paddy soils, i.e. mottling formation, bleaching, gleyzation and hanging gleyzation.

Soil subspecies is classified by lithology, nature of organic matter and gross texture of solum in paddy soils derived from semiterrestrial soils. But in the case of upland paddy soils, differences in preceding genetic soil type should be taken as criteria instead of lithology, organic matter and gross texture.

Soil variety is classified on the basis of differences in the local variation of soil materials, e.g. modes of superposition, and the depth of solum.

REFERENCES

(1) Kamoshita, Y., The soil types of the Tsugaru Plain, Aomori Prefecture, Nippon. Joru. agric. Expt. Station, Tokyo, Vol. 3, 401-420, (Japanese with English summary) (1940).

(2) Uchiyama, N., Morphology of paddy soils. 185pp, (japanese), Chikyushuppan, (1949).

(3) Kanno, I., A scheme for soil classification of paddy fields in Japan with special reference to mineral paddy soils. Bull. Kyushu Agric. Expt. Station, 4, 261-273, (1956).

(4) Yamazaki, K., Studies on the pedogenetic classification of paddy soils in Japan. Bull. Toyama Pred. Agric. Expt. Station, Special Publication, 1, 1-105, (Japanese), (1960).

(5) Group of Japanese Pedologists, On the 6th Approximation of Paddy Soil

Classification. Pedologist, 13(2), 105-111, (Japanese), (1969).

(6) Matsui, T. A proposal on a new classification system of paddy soils in Japan. Pedologist, 10(2), 68-87, (1966).

(7) Soil Survey Group, Soil Inst., Acad. Sci. China Genesis and classification of rice soils in South China. Acta Pedologica Sinica, 7(1), 28-41 (Chinese), (1959).

(8) Kyuma, K., Kawaguchi, K. Major soils of Southeast Asia and the classification of soils under rice cultivation (Paddy soils). The Southeast Asia Studies (Kyoto University), 4, 100-122, (1966).

(9) Dudal, R., Problem of genesis and classification of paddy soils. Geography and classification of soils of Asia, 189-192, (Russian)(1965), Moscow: Nauka. Cited from Kyuma and Kawaguchi, (1966).

(10)Mitsuchi, M. On the two kinds of soil sequence, i.e. hydrocatena and permeability of the solum, in paddy soils of lowland area - in relation to the genesis and classification of lowland paddy soils - . Pedologist, Vol. 13(1), 2 - 13, (Japanese with English summary), (1969).

(11)Mitsuchi, M. Pedogenic characteristics of paddy soils and their significance in soil classification. Bull. National Inst. Agric. Sciences (Japan), Ser. B, 25, 29 - 115, (Japanese with English summary), (1974).

(12)Xu, Q. et al The genesis and classification of the paddy soils, Tai-lake Basin, Jiangsu Province, China. Acta Pedologica Sinica, 17(2), 131-142, (Chinese with English summary), (1980).

(13)Matsui, T. A tentative scientific classification and systematics of Japanese soils. Pedologist, 22(1), 56-70, (Japanese), (1977).

(14)Matsui, T., Nagatsuka, S. A comprehensive classification system for Japanese soils. 24th International Geographical congress (1980, Tokyo), Main Session, Abstracts, 1, 292-293, (1980).

PADDY SOILS IN CHENGDU PLAIN AND THEIR CLASSIFICATION

Zhang Xian-wan
(Department of Soil Science, Chengdu Branch of Academia Sinica)

Chengdu Plain is situated at the upper reaches of the Changjiang River in the subtropical zone, with a height of 450-750 m above the sea level and a gradient of 4‰. It covers an area of about 7500 square kilometres, with an average annual rainfall of about 1000 mm. It is generally used for a rotation of rice and winter dryland crops. As devoted mainly to rice plantation this plain is characterized by its high elevation and vast expanse of rice fields.

GENETIC CHARACTERISTICS OF PADDY SOILS IN CHENGDU PLAIN

Chengdu Plain is a complex alluvial fan (Fig. 1) with Holocenic sediments lying on the eroded yellow earth of Quaternary period. Both the sections of these sediments and the groundwater level very with the distances from the river source. Here the Qingbai River runs through from the west to the east, dividing the plain into two parts: the Minjiang River deposit system and the Tuojiang River deposit system. The former is of grey colored loamy deposit, while the latter slightly brownish grey colored clay loamy deposit. Sample analysis shows that both consisted of swelling clay minerals with good arability and favorable water and nutrient supplies.

In the middle part of the plain, the parent material of the eroded Quaternary yellow earth usually appears as a belt of secondary terrace along the river banks, or as substratum overlain by modern alluvial deposits. In either case, the presence of the yellow earth layer in soil profiles and its wide variation in depth exert profound effects on the soil water regime and soil fertility as well.

On the edge of the plain, although the date and sources of the Pleistocene formation remain under debate(1), yet there now can be differentiated three types of yellow earth: 1) The acidic yellow earth is a clay in which we can see gravels in different sizes distribute throughout the whole horizon, indicating that it is of glacial origin, and acid in reaction, belonging to the sediment of Ya'an age in the Middle Pleistocene epoch. 2) The alluvial yellow earth has an significant well-sorted gravel layer, neutral in reaction, showing that it was once transported and redeposited thereafter and belongs to the upper Pleistocene series. 3) The carbonate concretion-bearing yellow earth, with slight alkaline reaction is a yellow earth macerated with calcic water, its age being still in argument, Records on pH, total exchangeable bases and available phosphorus content of these soils (Table 1) show also an increase in turn from the first one to the third. As a consequence, the fertility as well as its productivity of these types of yellow earth varies just in the same way as in their pH value.

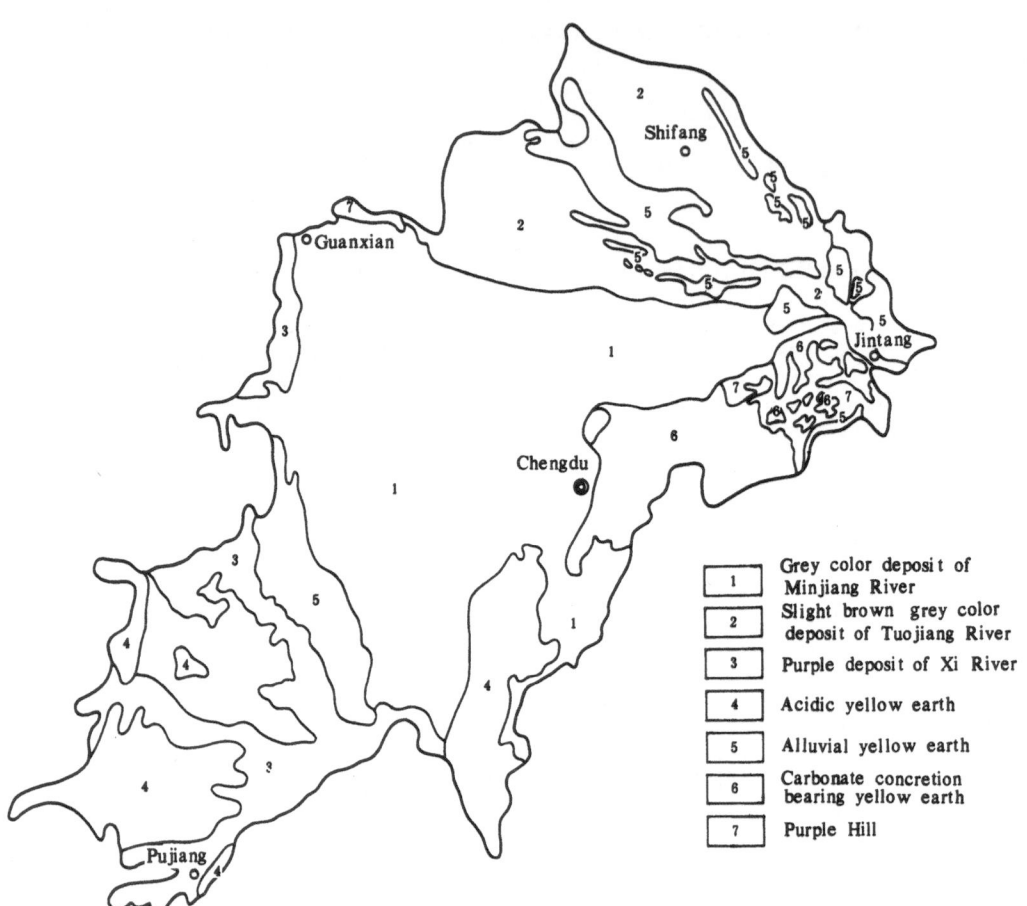

Fig. 1 Map of geographical distribution of paddy soils on Chengdu
      Plain

The fertility of soils,when regional characteristics are taken into
consideration, must be looked upon as a reaction product of regional hydrological factors. Owing to the abundance of water, the greater gradient
and the lighter texture, the water percolation in soil profile of paddy
soils in Chengdu Plain is uniformly, and a homogenous original colour is
kept in the profile with less differentiation. Except in the lower waterlogged region, gleization is not significant and accumulation of salts is
little(2). On the other hand, when comparing the Minjiang River basin with
the Tuojiang River basin, hydrological differences are, however, to be noticed. The upper streams of the Minjiang River get their water supply mainly
from melted snow of the mountainous source area, which covers far much greater area that the Tuojiang River and thus can maintain relatively rich water
supply, forming a steady flow of infiltration water and a higher ground water
level of about 1-3 m underground. The coldness of snow-water, the swiftness
of current and the side-flowing of ground water toward the lower part of the
alluvial fan bring about troubles, such as high ground water table, cold
subsoils, depressed growth vigour of rice plants to the farmers of the Minjiang

398

Table 1 Properties of three types of yellow earths

Parent material	PH	Available phosphorous (P_2O_5mg/100g)	Total exchange-able bases (m.e./100g)	Clay (%)	Silt (%)
Acidic yellow earth	6.0-6.5	1.3	8.97	20-40	30-50
Alluvial yellow earth	7.0-7.8	2.75	25.15	20-40	40-50
Carbonate con-cretion bearing yellow earth	7.8-7.9	3.23	52.41	60-74	14-21

River basin, especially in the top of the alluvial fan. On the contrary, the Tuojiang River watershed shows more abrupt changes and much greater water level fluctuations. As a result, the ground water table of soil in the Tuojiang River basin is usually low, about 3-5 m underground and the internal drainage of soil is generally fair. In addition, the good cultivation tradition of economical use of water in irrigation contributes to higher records on crop yields in this drainage area. The only trouble is that floods are not infrequently encountered in most parts of this area. It can be seen from the statements above that hydrological factors are of greater importance to the determination of the fertility level of rice soils in Chengdu Plain.

CLASSIFICATION OF THE PADDY SOILS IN CHENGDU PLAIN

Turning now to the problem of classification of rice soils, it is true that without a preliminary study of geography, a reasonable grouping of rice soils can hardly be fulfilled. This is because most of the megascopic genetical factors are in fact the main key to a deep look into soil genesis and classification from the standpoint of soil fertility implications.

A system for classifying paddy soils is thus suggested after having had a thorough understanding of the geographical features of Chengdu Plain. The skeleton of the proposed system is now outlined below.

1. The variation of water regimes in paddy soils is the leading indicator of the development of paddy soil.

Since paddy soils are the special types of hydromorphic soils formed under artificial submersion with periodical wetting and drying cultivation, their genetic courses, features and characteristics are mainly affected by the variations of soil water regimes. Soil scientists in our country have by degrees enriched their knowledge of the variations of soil water regimes. On the basis of what they have classified for the paddy soils(3-5), the present author is of the same opinion that the soil water regimes may still be classified as submorphic, percolatic, primary hydromorphic, hydromorphic and glei types, but they have been given some new conception. By submorphic type we mean that the water moves slowly and homogenously in the soil profile, thereby it keeps a original colour and a good loose structure bearing only rust spot and streak in the profile. The percolatic type means that water percolates quickly and homogenously, nutrients are eluviated; but the primary hydromorphic type indicates that the water percolates slowly, alternating the wet markedly with the dry. The colour of the profile also keeps in original form and a significant prismatic structure can be formed. When discussing the hydromorphic type, we refer neither to ground water podzol nor to illuviation, but a process through which both leaching and illuviating are developed in the subsoil owing to the wet alternating with the dry. As to gleitype, it indicates that as soil is in a waterlogged condition, the free iron is changed into reducing form.

2. The hydrogenetic profile characteristics are no doubt a major factor to different types of paddy soils. Since they are the reaction product of this periodically immersed soil and its environment. There is, however, something to be clarified as to the intensity and scope of the environmental factors which are active in soil-forming processes. They can also be taken as criterion for a rational system of soil classification. For example, the paddy soil in the Minjiang River basin predominated with a lower temperature must be differentiatedl from that in the Tuojiang River basin, which has a favourable heat-water regime.

The thermo-water dynamics present in different texture profiles are also worthy to be noticed. Fig. 2 shows the different water status in a submorphic grey loamy soil and a hydromorphic grey clay loamy soil. We can see that the moisture content of loamy soil is usually higher than that of the other, especially in the rainless winter season. The loamy soil can regulate its moisture from gaseous and capillary water and simultaneously give a high yield to the wheat crop. Fig. 3 shows the temperature fluctuations of the same two soils, hydromorphic clay loamy soil has a higher temperature either in winter or summer season due to the less water content in the profiles.

3. Studies on the relative fertility of different parent materials as well as hydromorphological features of soil profiles show that the quality of soil mineral colloids affected the productivity of paddy soil. Thus in the case of Holocenic sediment, its high productivity over Pleistocene yellow earth can only be due to its better colloidal behaviour. Therefore, we should pay more attention to the differentiation of parent materials.

4. In the course of our field studies our interest was also centered on the ecological index of soil-rice plant inter-relationships, quite a good coordination was found between the rice plant and soil type characteristics(6).

Table 2 gives an observation on cross section of the wheat stem growing on the same soils mentioned above, showing that the wheat stem on hydromorphic grey clay loamy soil has a more thickness of cortex and a larger number of vascular strand than that on submorphic grey loamy soil. The quality of rice plant, ramie and tobacco was found similar in these two soils. Therefore, the adaptability of crops and their physiological and biochemical characteristics may be regarded as the comprehensive indicators of the morphology and fertility of paddy soils.

For the present, an attempt is made to formulate the systems of paddy soils in Chengdu Plain as seen in Table 3.

Soil subtype: Water regimes may be regarded as factors to subdivide soil into subtypes.

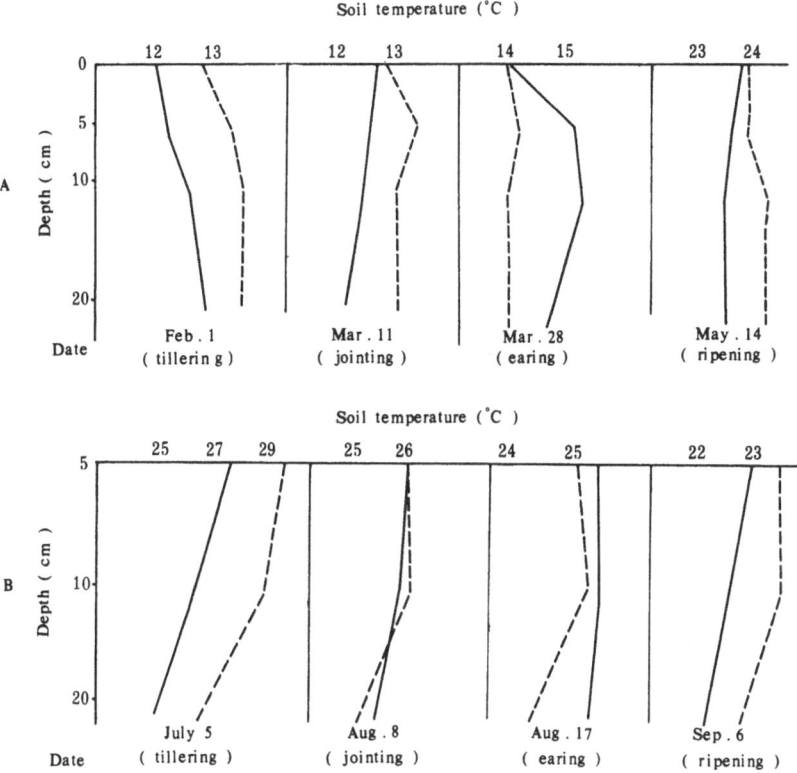

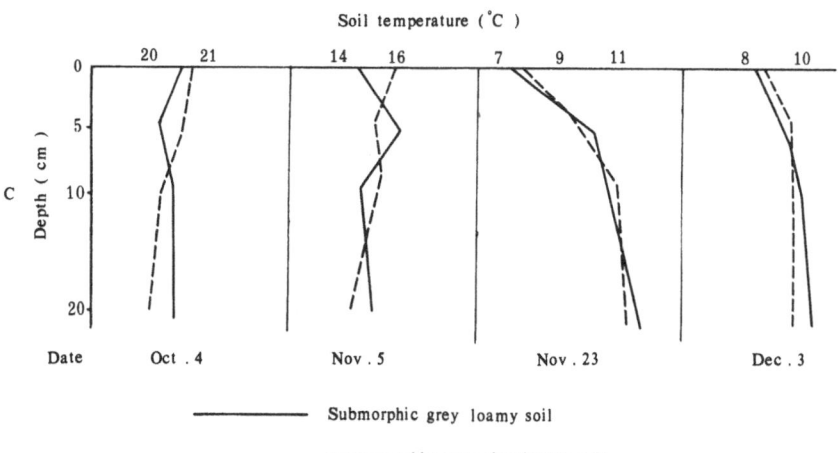

———————— Submorphic grey loamy soil

– – – – – – Hydromorphic grey clay loamy soil

Fig. 2 Soil temperature change on two paddy soils in four season (in 1961-1962)
A. Each growing period of wheat (winter & spring);
B. Each growing period of rice (summer);
C. Bush vetch field (autumn)

1. In late autumn

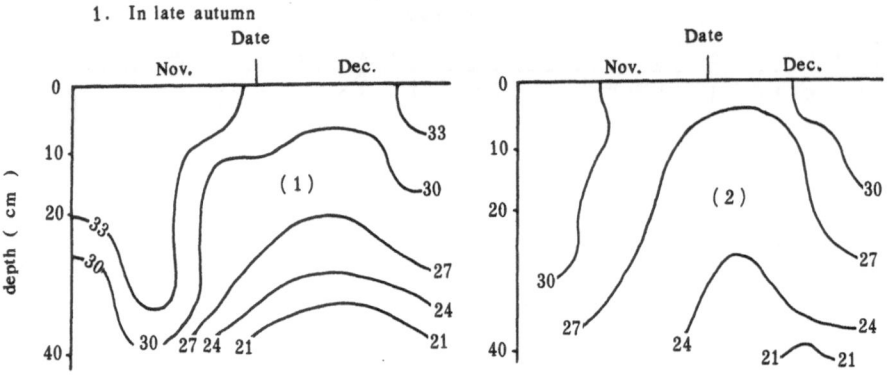

2. In growing period of wheat (Nov.- Apr.)

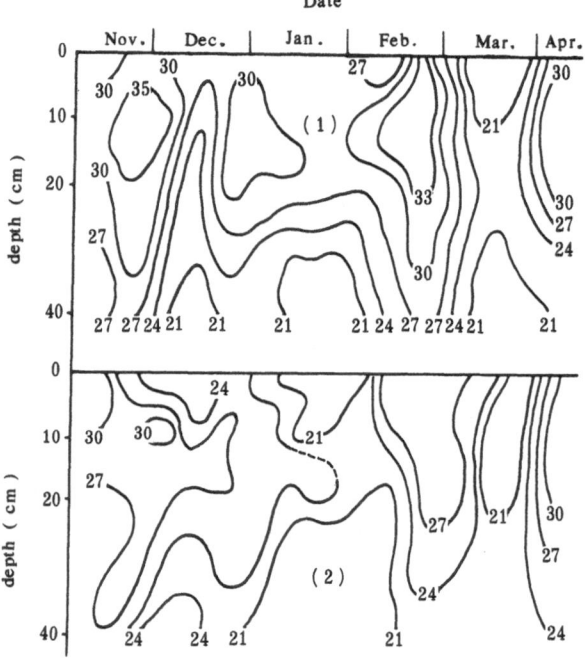

Fig. 3 Moisture isogram of two paddy soils: 1. Submorphic grey loamy soils; 2. Hydromorphic grey clay loamy soil (in 1961-1962)

Table 2 Anatomic characters of stem of wheat on two
paddy soils in Chengdu Plain

Knob locality	Paddy soil	Knob Dia. lengh (cm)	(cm)	Number of large vesculas strand	Number of small vesculas strand	Thickness of epidermis (μ)	Thickness of lower epidermis (μ)	Thickness of stem wall (μ)
1	Submorphic grey loamy soil	6.2	3.74	32.8	14.2	1.67	6.01	70.5
	Hydromorphic grey clay loamy soil	4.8	4.04	35.4	14.0	1.85	8.08	75.3
2	Submorphic grey loamy soil	11.6	4.22	30.6	15.1	1.15	5.68	50.4
	Hydromorphic grey clay loamy soil	9.7	4.38	35.2	14.6	1.28	6.25	63.6

Soil family: Soil belongs to this class of same parent materials.

Soil species: Soils are grouped according to texture, texture profiles, etc. which affected the thermo-water status directly.

Soil variety: Based on the ecological and plant physiological evidence of fertility of various soil types, soil species are subdivided into varieties.

REFERENCES

(1) Faculty of Hydrogeology, Chengdu Institute of Geology, Engineering-geological properties of Chengdu clay. Bulletin of Chengdu Institute of Geology, 1(1960). (in Chinese)
(2) Zhang Xian-wan, Li Dao-chun, Studies on genetic types of waterlogged rice soil in Chengdu plain, China. Turang Tongxun, 11(1979). (in Chinese)
(3) Chu L.T., Ma Y.T., Sung T.C., Hou K.C., The nomenclature of the various horizons of paddy soils. Special Soils Publication, China, Series B, 4, 85-91(1938). (in Chinese with English summary)
(4) Cao Sheng-geng, Yao Yu-cheng, Delimitation, nomenclature of genetic horizons of paddy soils and their characteristics. Turang Zhuanbao, 36, 179-205(1964). (in Chinese)
(5) Xu Qi, Lu Yan-chun, Zhu Hong-guan, The genesis and classification of the paddy soils, Tai-lake basin, Jiangsu Province, China. Acta Pedologica Sinica, 17, 120-132(1980). (in Chinese with English summary)
(6) Zhang Xian-wan, Li Dao-chun, The relationship between cropping system formation and soil type characteristics in grey sediments in Chengdu Plain(1963, ms.).

Table 3 Classification system of paddy soils in Chengdu Plain

Subtype	Soil family	Soil species	Soil variety
Submorphic	Grey color Alluvial deposit	You Sha Tu(loam)	Da Tu You Sha(slight clay loam)
		Sha Tu(loamy sand)	Bai Yan Sha(unfertile)
		Da Tu(clay loam)	You Sha Da Tu(fine sand clay loam)
		Bai Shan Ni(clay loam/loamy clay)	Huo Bai Shan(fertile)
		Cao Tian(water-logged)	Er Cao Tian(less water-logged)
Percolatic	Slight brown Grey color deposit	You Sha Tu(loam)	Ban Sha Ni(fine sand loam)
		Sha Tu(loam sand)	Shi Tou Tian(graveling)
		Ni Tian(clay loam)	Huang Ni Di Ni Tian (yellow earth underlying)
		Cao Tian(water-logged)	Fei Cao Tian(fertile)
Primary hydromorphic	Acidic yellow earth	Bai Shan Ni(Fe-eluviate)	Huo Bai Shan(fertile)
		Huang Ni Da Tu (structural)	Shan Xue Da Tu(fertile)
		Tie Gan Zi Ni(Fe-coating)	
		Si Huang Ni(Fe-illuviate	
		Lan Ni Tian(water-logged)	
Hydromorphic	Alluvial yellow earth	Huang Ni Tian (structural)	Shan Xue Huang Ni(fertile)
		Bai Shan Ni(Fe-eluviate)	Huo Bai Shan(fertile)
		Si Huang Ni(Fe-illuviate)	
		Tie Gan Zi Ni(Fe-coating)	
		Cao Tian(water-logged)	
Gley	Carbonate Concretion bearing Yellow earth	Pao Huang Ni (structural)	
		Huang Ni Tian (slight Fe-illuviate)	
		Lan Ni Tian(water-logged)	

404

THE APPLICATION OF CLUSTER ANALYSIS TO THE
MATERIAL CLASSIFICATION OF PADDY SOILS IN TAIHU LAKE AREA

Liu Duo-sen   Lu Yan-chun
(Institute of Soil Science, Academia Sinica, Nanjing)

Numerical taxonomy of soils is a form of expression for their material classification. The present study is an attempt to carry on material classification for paddy soils in the Taihu Lake area using cluster analysis as the basic method in numerical taxonomy.

Thirty samples were collected from plowed horizons of paddy soils in the area of the Taihu Lake. To identify the soils, seven items of indices were taken from each sample, i.e. contents of organic matter, total nitrogen, total phosphorus, total potassium, coarse silt, clay and cation exchange capacity.

The 210 original data mentioned above were standardized by the method of standard deviation. Each of the 30 paddy soils may be indicated by a 7 dimension vector based on the standardized values of above mentioned indices.

With a comparison between the statistical values of affinity obtained from the "character difference" and Euclidean distance of the 7 dimension vector of the 30 samples, 435 pairs of corresponding values of distance of the pedons were calculated. Further analysis has confirmed that so far as the classification of paddy soil in the Taihu Lake area is concerned, the result obtained from the "character difference" is some what better than that from Euclidean distance. The "character difference" between pedons i and l is

$$d_{il} = \sum_{j=1}^{m} \left| x_{ij} - x_{lj} \right|$$

$$(i, l = 1,2,\ldots,n)$$

In the equation, i and l are the coded numbers of samples. j is the coded number of indices, n is the total number of samples, m is the total number of indices, while x_{ij} and x_{lj} are the standardized values of index j of samples i and l respectively.

The cluster analysis of the samples was carried on respectively using minimum, maximum and mean square distances as the distances between groups of soils. Comparison of the results obtained from the three methods has showed that the minimum distance with higher concentration of space and lower sensitivity is not a suitable method for soil classification of these samples due to minimal variations between pedons in a smaller distribution area; while the methods of mean square distance and maximum distance are better than minimum distance method, and the mean square distance is the best one. The mean square distance D_{rk}^2 between groups G_r and G_k is

$$D_{rk}^2 = \frac{1}{n_r \, n_k} \sum_{i\epsilon G_r, l\epsilon G_k} d_{il}^2$$

In the equation, n_r and n_k are the numbers of samples included in groups G_r and G_k respectively, d_{il} is the distance between samples i and l.

Using the "character difference" distances of absolute values as the distance between samples and mean square distance as the distance between groups, a dendrogram of the relationships among 30 paddy soils (Fig. 1) is obtained. In accordance with Fig. 1,5 material patterns are divided as follows:

1) Pattern of low fertility;
2) Pattern of medium fertility with coarse silty texture;
3) Pattern of medium fertility with fine silty texture;
4) Pattern of high fertility;
5) Pattern of high potential fertility.

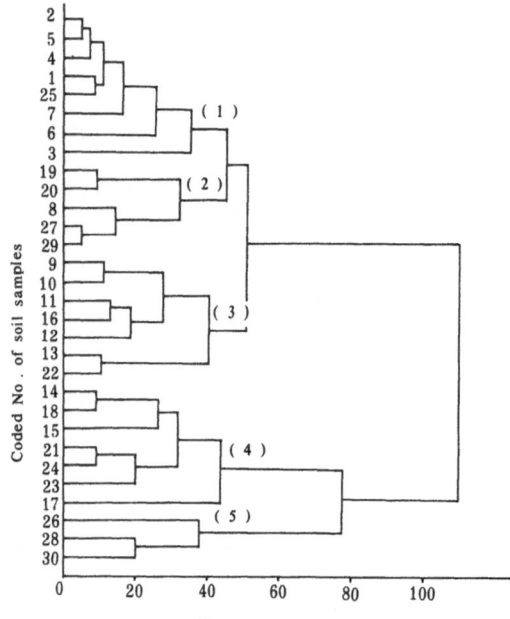

Fig. 1 Dendrogram of the relationships among 30 paddy soils in Taihu lake area
Patterns:
(1) Low fertility;
(2) Medium fertility with coarse silt;
(3) Medium fertility with fine silt;
(4) High fertility;
(5) High potential fertility

The topographical conditions and the means and standard deviations of the observal values of material classification indices are arranged in Table 1. The paddy soils in one material pattern have roughly the similar characteristics of fertility.

A linkage graph of the material patterns (Fig. 2) is obtained on the basis of the 5 material patterns treated with the method of graph theory. The relationships among the material patterns may be directly expressed in Fig. 2. In Fig. 2, the real line denotes the minimum spanning tree of the linkage graph, it indicates the minimum figure of general affinity among the 5 material patterns. The figure implies that the patterns of lower fertility

Table 1 Means ± standard deviations of indices values of different material patterns of paddy soils in Taihu Lake area

Material pattern	Topography	Nos of soil samples	O.M. (%)	Total N (%)	Total P_2O_5 (%)	Total K_2O (%)	Coarse silt (%)	Clay (%)	CEC (m.e./100g)
1. Low fertility	Hill	8	1.48±0.302	0.086±0.0441	0.049±0.0193	1.39±0.221	52.4±4.29	14.1±3.41	9.48±1.819
2. Medium fertility with coarse silt	Plain*	5	2.10±0.438	0.120±0.0208	0.133±0.0456	1.87±0.240	56.3±3.61	10.9±3.42	12.56±1.671
3. Medium fertility with fine silt	Plain*	7	2.16±0.694	0.116±0.0230	0.122±0.0539	1.50±0.147	45.3±5.93	19.7±3.79	14.77±2.400
4. High fertility	Plain*	7	2.68±0.406	0.156±0.0185	0.169±0.0264	1.91±0.498	40.5±5.22	23.6±4.52	19.23±1.847
5. High potential fertility	Depression	3	4.13±0.194	0.230±0.0089	0.189±0.0794	2.04±0.235	47.8±5.70	12.6±5.00	18.61±1.777

*Parent materials of soils in plain include loessal and alluvial deposits

may be cultivated and promoted directively to the most fertile pattern of paddy soils under certain conditions.

The properties of plowed horizon are influenced, to a great extent, by cultivation, fertilization and the application of river mud or soil transported from other place, though their properties are generally related to their genetic characteristics. Therefore, the patterns of material classification are not the same with the types of genetic classification, but they may also partially reflect its position in genetic classification. For the side bleached paddy soils in hilly area and the waterlogged paddy soils in the depression by the lake, their positions in the material classification are the same with those in genetic classification, they belong to patterns of low fertility and high potential fertility respectively; similarly, for the stagnating paddy soils in higher plain and permeable paddy soils in lower plain, they have roughly the same patterns both in material and genetic classifications, and belong to patterns of medium fertility with fine silty texture and of high fertility; while for the percolating paddy soils developed on alluvial deposit, it varies irregularly between medium and high fertility patterns, and no correlation with the genetic classification has been found.

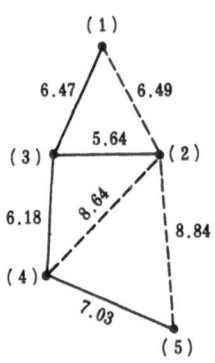

Fig. 2 Linkage graph of
5 material patterns
of paddy soils in
Taihu lake area
Patterns:
(1)Low fertility;
(2)Medium fertility
    with coarse silt;
(3)Medium fertility
    with fine silt;
(4)High fertility;
(5)High potential
    fertility

Different genetic types of paddy soils may get some similarities in their properties, through anthropic activity of cultivation, and many even become of the same pattern of material classification, which indicates the significant effect of human activity on the directive cultivation and formation of soils.

EVOLUTION AND DEVELOPMENT OF ALLUVIAL PADDY SOILS IN ZHUJIANG DELTA

Lu Fa-xi, Zhu Shi-qing, Yuan Cai-ting,
Shen Dao-ying, Yang Yuan-ying, Luo Lian-xiang
(Guangzhou Institute of Soil Science)

Zhujiang Delta covers a total area of about 10,000 square kilometres. As shown on the attached map, the margins of the Delta are surrounded by rolling hills. Along the river courses and coastal areas of the Delta occurs alluvial paddy soil at different stages of development with a total area of 400,000 hectares, known as an important agricultural region in South China (Fig. 1)(1,2).

The present article deals with the evolution and development of alluvial paddy soils in the Delta region. The formation of paddy field in the polders is rather recent in history. It comes into being just soon after the construction of the dam(3,4). Table 1 shows that the elevation of paddy field, which formed in the 19th century, still ranges 0.2-0.7 meter below the sea level. It is the dam that prevents the paddy soil from being flooded by the sea tide

Table 1 Formation factors of different paddy soils

Soil type	Elevation above sea level (m)	History (years)	Ground water table (cm)
Strongly gleyed	−0.7-0.2	150	0-30
Weakly gleyed	−0.2-0.4	150-200	40-60
Hydromorphic	0.4-1.0	200-400	70-100
Hydromorphic	0.2-5.0	400-1000	80-100

It is the usual farm practice in this area that large quantities of silty and clayey deposits of the river courses as high as 225-300 tons per hectare are dressed as mud manure on paddy field for every 2-3 years. Each dressing has added a layer of mud about 1.5-2.0 cm in thickness to the land surface. Table 1 indicates the relations between the duration of culture, the rising of land elevation and the lowering of underground water in soil profile. It also indicates that the formation of a well-developed hydromorphic paddy soil requires more than 2 centuries under such a system of soil management(2,5).

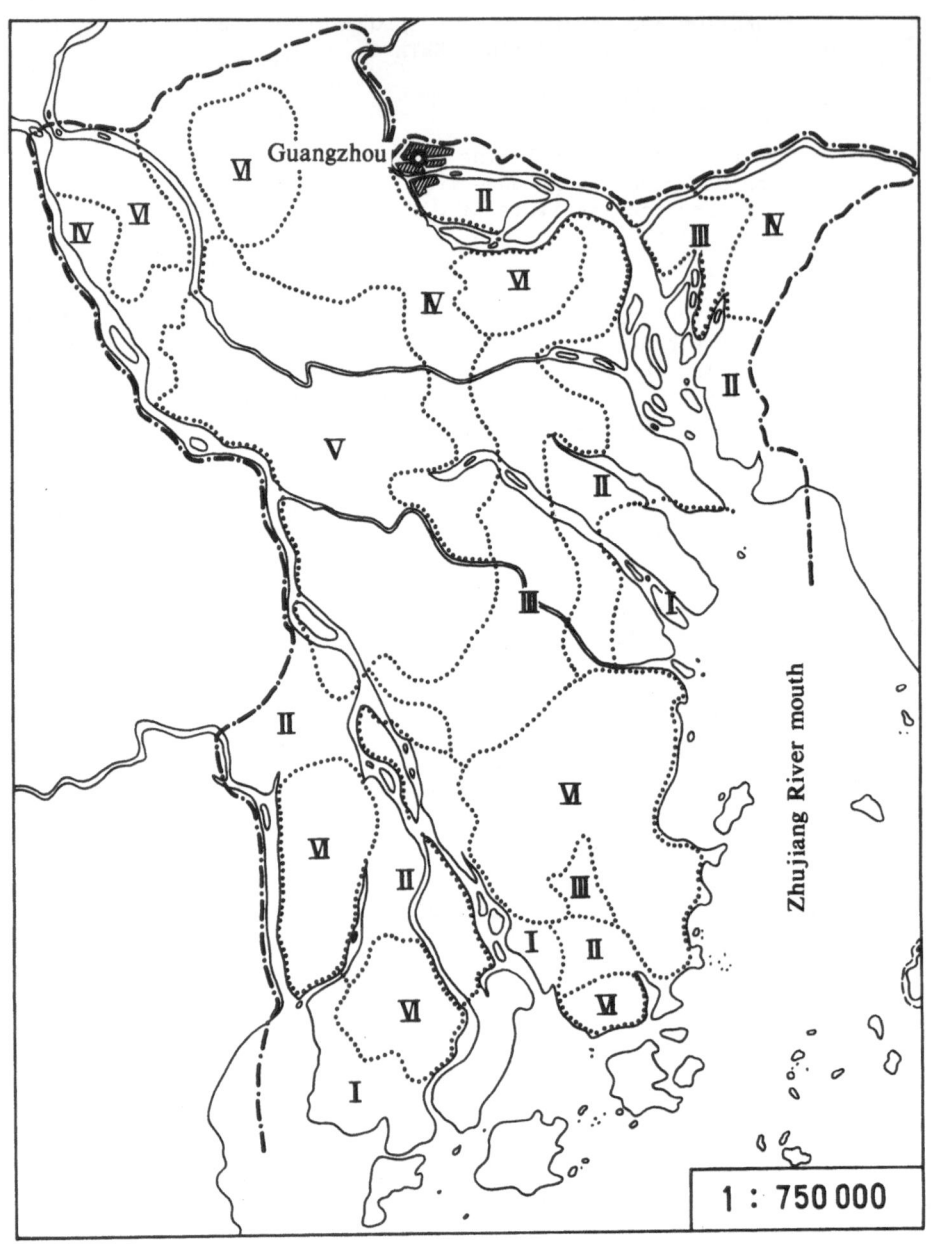

Fig. 1 Schematic map of Zhujiang Delta
    I. Low level polder;
   II. Intermediate level polder;
 III. High level polder;
  IV. Old polder;
   V. Mulberry orchard combined with fishery;
  VI. Rolling hill

Fig. 2 shows the process of desalinization of alluvial paddy soils
from three different origins. The great quantity of rainfall in this
region facilitates the desalinization process of the salty alluvial materials along the development of strongly gleyed paddy soils to a hydromorphic
paddy soil. The effect of gleization in the surface soil layer usually
disappeared in about 2 decades cultivation (from AG to A), A plowpan layer
(P) is formed in 30-40 years, and subsequently the reducing illuvial horizon
(W), redoxing illuvial horizon (WB) and oxidizing illuvial horizon (B) appear
in the soil profile. The development of a typical hydromorphic paddy soil
(Fig. 3 last profile, upper part) usually requires about more than 200 years
under the above-described measures of soil management. Under good fertilization with proper irrigation and drainage, both weakly gleyed paddy soil and
hydromorphic paddy soil can give high annual yield of about 10.5-12.00 tons
of grain per hectare for double cropping of rice(1,2).

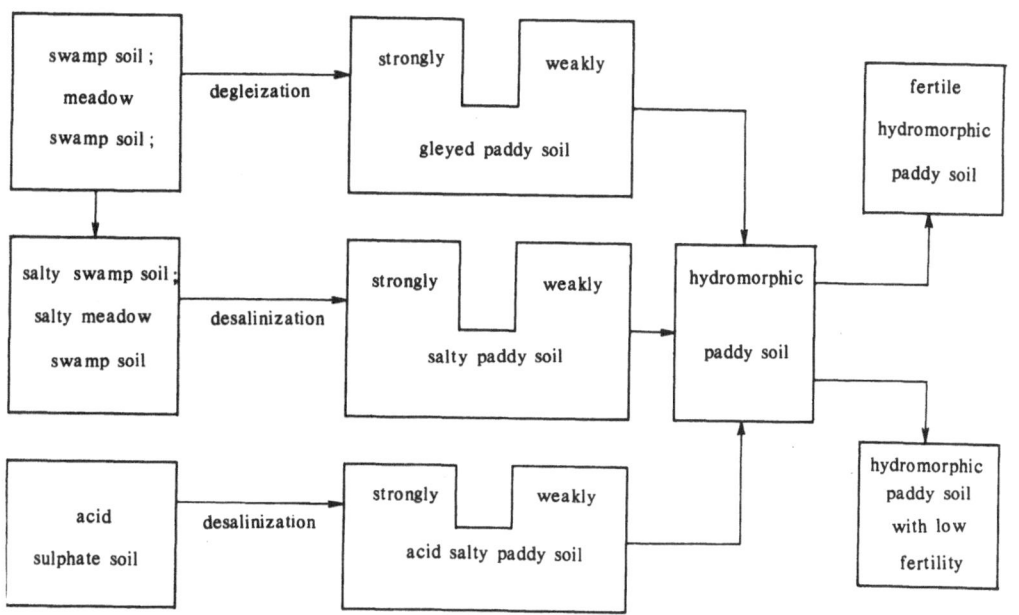

Fig. 2 Evolution and development of alluvial
paddy soils

In the newly emerged coastal area from mangrove swamp, salty paddy
soils are developed from acid sulphate soil. However, the area of acid
salty paddy soil is very small.

Table 2 gives some chemical composition of the alluvial paddy soils
in Zhujiang Delta. With the exception of salty types, the content of
soluble salts in the soils is harmless to any crops. Nitrogen and
phosphorus have to be applied for an adequate yield of rice. At present,
K-nutrition of the soil seems sufficient. The higher content of the organic

411

matter in soils ranks the paddy soil of the Delta area as one of the fertile soil types in China(6).

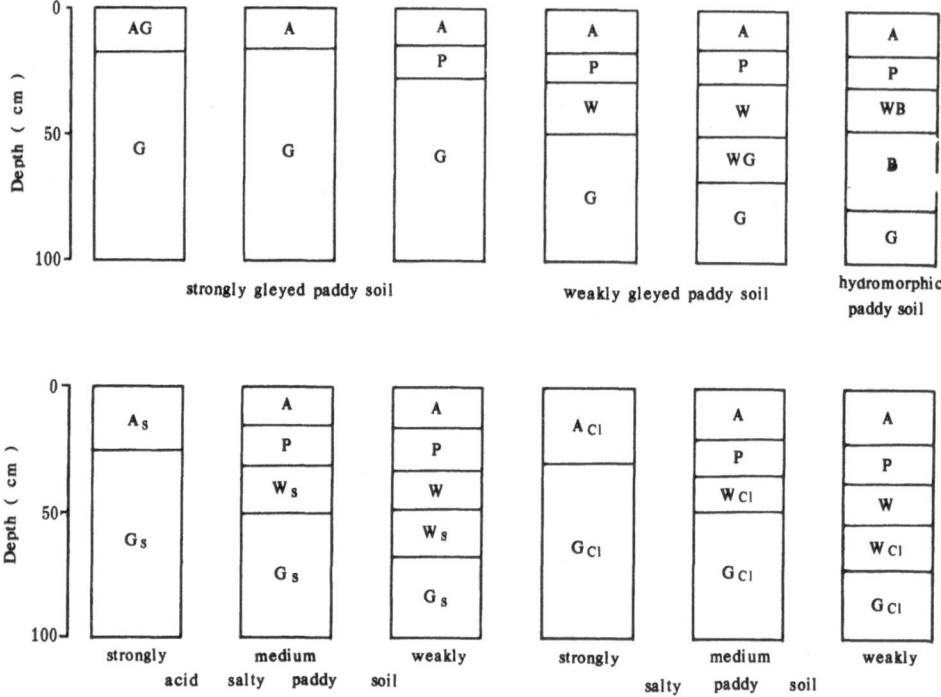

Fig. 3 Morphological characteristics of different paddy soils

Table 2 Chemical composition of different paddy soils

Soil type	Organic matter (%)	N (%)	P (%)	K (%)	Available K (ppm)	Total salt (%)
Fertile hydromorphic	3.67	0.223	0.058	1.76	80	—
Low fertile Hydromorphic	2.15	0.126	0.039	1.72	71	0.059
Strongly gleyed	2.69	0.161	0.048	1.91	97	0.086
Weakly gleyed	2.48	0.177	0.059	1.93	160	0.014
Salty	2.46	0.160	0.045	1.83	191	0.014–0.775
Acid salty	3.69	0.186	0.042	1.94	198	0.238–0.930
Salty swamp soil	2.43	0.132	0.071	1.86	250	0.789

REFERENCES

(1) Tang Tai-yee, Report on the soil survey of Panyu County, Zhongshan County, Nanhai County, Sanshui County and Dongguan County, Guangdong, China(1930, ms.). (in Chinese)

(2) The Soil Common Survey and Land Utilizing-planning Committee of Guangdong Province, The agricultural soil archives in Zhujiang Delta (1962, ms.). (in Chinese)

(3) Lu Fa-xi, et al. A discussion on the agricultural division of soil in Zhujiang Delta(1964, ms.). (in Chinese)

(4) The Editorial Department of Agricultural Archives on Zhujiang Delta, The agricultural archives of Zhujiang Delta (1967, ms.). (in Chinese)

(5) Gong Zi-tong, Chen Zhi-cheng, The soils of Zhujiang Delta. Turang Zhuanbao, 36, 69-129(1964). (in Chinese)

(6) Zhongnan Research Department for Soil Science, on the basic characteristics and the practices for fertility improvement of some main paddy soils in Zhujiang Delta(1964, ms.).(in Chinese)

ON THE CLASSIFICATION OF BASIC CATEGORIES
OF PADDY SOILS IN TAIHU LAKE REGION

Liu Yuan-chang
(Institute of Soil Science, Academia Sinica, Nanjing)

Though great progress has been made on genetic classification of paddy soils in recent years, there is divergence of understanding on the conception of paddy soil. The difference of viewpoints is not only reflected on the classification of the soil, but also on the implication of various categories in classification. In our opinion, paddy soil is a kind of anthropo-aquatic soil. Its formation and development are not only affected by the natural conditions, but also by human activities of cultivation. The soils used for rice cultivation under submergence and possessing a plowed horizon, plowpan and percosubmergic horizon are regarded as paddy soils.

Formerly , in the study of the classification of paddy soils, more attention was paid to the classification in high level categories than to that in the low level categories. Two basic categories are popularly used in soil classification system, i.e. soil species and soil series. The basic categories are the foundation of soil classification. The features and properties used as the criteria for the classification in basic category level should be relatively stable, while the more variable factors should be considered as the criteria in the even more lower category level. At the same time, the criteria for the classification in these categories should be as concrete as possible. Only in this way is it possible to avoid the confusion of the same soil with different names or different soils with the same name, and to be conducive to classification and mapping of soils. It is evident that the properties and features of a profile in the depth of 0.5 m can directly affect the growth of crops, determine the variation of the features of certain diagnostic horizons, and simultaneously influence the means for the promoting agricultural productivity. Therefore, some characteristics of the profile in the depth of 0.5 m are used as criteria for the classification of basic categories of paddy soils in the Taihu Lake region. The basic categories used in the classification in the Taihu Lake region differ from both the soil species and soil series. The term "soil species" is used in the classification of paddy soil in this region and "soil variety" as its supplementary category. According to the characteristics of paddy soils in Taihu Lake region, in case the textural profile is homogeneous, the difference of texture of the soils is used as the main criterion for the division of the taxa in soil species level, while in case the textured profile is heterogeneous, the location and thickness of specific horizons in profile are adopted.

In the system of classification of paddy soils in the Taihu Lake region, five categories are adopted, they are great soil group, sub-group, family, species and variety. According to the relationship among the patterns of soil solum structure, water regime, eluviation and illuviation, the paddy soils in this region are divided into five great soil groups, i.e. permeable, side-bleached, stagnating, waterlogged and percolating paddy soils. The representative soil species of the five great soil groups are described as follows.

Permeable paddy soil: There is no impeding layer in the profile in the depth of 0.5 m or 1.0 m with homogeneous texture. The soil is generally permeable, but its permeability depends on athe texture of the upper soil layer in depth of 0.5 m, and is better in loam than in clay. Based on the difference in texture, the soils are subdivided into three soil species (Table 1).

Table 1 Division of soil species of permeable paddy soil with homogeneous profile

Soil species	Coarse silt (0.05-0.01 mm) %	Physical clay (<0.01 mm) %	Clay (<0.001 mm) %
Light clay	30-40	60-70	> 30
Heavy loam	30-50	45-60	25-30
Medium loam	40-60	30-45	20±

Side-bleached paddy soil: This soil widely spreads on the terraced fields in hilly regions. The development of the soil is affected by both the topography and the soil layer covered on the original soil by anthropic transportation in the process of building the terraced fields, which induce strong eluviation of the soil. Therefore, according to the thickness of the whitish bleached horizon, the soils are subdivided into three soil species (Table 2).

Table 2 Division of soil species of side-bleached paddy soil

Soil species	Thickness of whitish bleached layer (cm)
Thin layer	<30
Medium layer	3 30-50
Thick layer	> 50

Stagnating paddy soil: This soil is distributed mostly in plains and fuvial terraces. Due to highly weathered parent material and apparent eluviation and illuviation in profile, a bleached horizon (whitish bleached horizon) and illuviation horizon of clay, iron and manganese (stagnating horizon) are formed in this soil. For subdividing the soils in soil species level, two factors should be taken into consideration, i.e. the texture of the profile in the depth of 0.5 m and the location of the whitish bleached horizon in profile. The thickness of the whitish bleached horizon is often related to the texture; the more clayey the texture is, the thinner the bleached horizon is. At the same time, the location of the bleached horizon in profile is affected by human activities. Based on the location of the

415

bleached horizon in the profile, three soil species are subdivided (Table 3).

Table 3 Division of soil species of stagnating paddy soil

Soil species	Location of bleached horizon in profile (cm)
Upper level	<30
Middle level	30-50
Lower level	>50

Waterlogged paddy soil: This soil mainly spreads in polder area. The degree of degleization of the soil depends on the microrelief which is related to the deposition of the parent material and the application of river mud as manure after the reclamation of the soil, Firstly, with the microrelief changing from high to low, the loamy or clayey layer gets thicker progressively and the surface gleization of the soil becomes apparent. Secondly, with soil surface lifting from the centre to the side area of the polder field due to human activities such as the application of river mud, the thickness of soil layer resulted from anthropic activity is increasing from the centre to the side area of the polder field, and the soil texture becomes lighter. Based on the texture and the thickness of loamy or clayey layer in profile, the soils are subdivided into three soil species (Table 4,5).

Table 4 Division of soil species of water-logged paddy soil with homogeneous textured profile

Soil species	Coarse silt (0.05-0.01 mm) %	Physical clay (<0.001 mm) %	Clay (<0.001 mm) %
Medium loam	45-60	30-45	<20
Heavy loam	35-45	45-60	20-25
Light clay	<35	>60	>30

Table 5 Division of soil species of water-logged paddy soil with heterogeneous textured profile

Soil species	Thickness of heavy loam or clay layer (cm)
Thin-layer	<30
Medium-layer	30-50
Thick-layer	>50

416

Percolating paddy soil: This soil widely spreads in the alluvial plains of the Changjiang and other rivers. Based on the soil texture and the location of interlayer of sand and clay in profile, the soil species are divided as shown in Tables 6 and 7.

Table 6 Division of soil species of percolating paddy soil with homogeneous profile

Soil species	Coarse silt (0.05-0.01 mm) %	Physical clay (<0.01 mm) %	Clay (<0.001 mm) %
Medium loam	45-60	30-45	< 20
Heavy loam	35-45	45-60	20-25
Light clay	> 35	> 60	> 30

Table 7 Division of soil species of percolating paddy soil with heterogeneous profile

Soil species	Location of interlayer of sand and clay in profile (cm)
High position	< 30
Middle position	30-50
Low position	> 50

Soil tilth is a comprehensive manifestation of a series of tilling characteristics of the soil including soil structure. It is related not only to the texture, but also to the amount and properties of organic matter in soil. Meanwhile, tilth is frequently affected by water regime. The variation of water regime is generally the important conditions for the formation of soil structure. At the same time, tilth is a variable factor which can be changed apparently in several years under the influence of agricultural practice. For example, the settling and compacting characters of whitish bleached paddy soil can be improved by the application of large amount of organic manure in three or four years. Another example is that the stiffness of the soil due to the long term submergence in the Taihu Lake region can be improved by drainage or cultivation of upland crops. Therefore, the soil tilth together with the amount of Fe-organic coatings, the presence of the gley layer and the degree of stiffness (resistance to compression of dry soil clod) are regarded as the creteria for the division of soil varieties.

Take heavy loamy permeable paddy soil as an example, three soil varieties are subdivided on the basis of the criteria mentioned above (Table 8).

According to analytical data of the Wuxian Institute of Agricultural Sciences, Jiangsu, the nitrogen supplying capacities of these three varieties of soil are quite different (Table 9). Therefore, different amounts of nitrogen fertilizer should be applied and different management

417

should be adopted on the three soil varieties respectively.

Table 8 Division of soil varieties of heavy
loam permeable paddy soil

Soil varieties	Fe-organic coatings	Surface gley layer	Modulus of rupture (kg/cm^2)	Fertility
1	+++	−	<26	High
2	++	− +	26–30	Moderate
3	−	++	> 30	Low

Table 9 Supplying capacity of nitrogen of different
heavy loamy permeable paddy soil

Soil varieties	O. M. (%)	Total N (90)	NH_4-N in tillering stage of rice (ppm)	Mineraliza- tion rate
1	3.14	0.163	39.7	2.44
2	3.13	0.158	33.5	2.12
3	2.21	0.115	20.5	1.78

CHARACTERISTIC FEATURES OF PADDY SOILS OF JAPAN

Masanori Mitsuchi
(National Institute of Agricultural Sciences, Japan)

A long continuation of wetland rice cultivation often causes characteristic changes in soil morphology and properties, including surface gray
coloring and the development of iron and manganese accumulation horizons.
It should be noted that it is only in well-drained soils that the characteristic changes take place under wet cultivation of rice. The characteristic
features of well-drained paddy soils are closely related to their peculiar
type of water regime. It is almost unthinkable without perpetual irrigation by men that well-drained soils are kept submerged for about ninety
consecutive days, with resultant charcteristic water regime, i.e. simultaneous submergence and percolation followed by oxidation after drainage.

In this paper, considerations are restricted to those paddy soils that
have acquired such characteristic features. These artificially altered
soils correspond roughly, though not exactly, to Aquorizem of Kyuma and
Kawaguchi (1), and anthraguic and hydroferric units of Moormann and Van
Breemen (2).

It has been widely recognized that artificial irrigation intensifies
the aquic moisture regime by raising groundwater level and often brings
about an important re-formation of soils. However this kind of changes
will not be considered here because it is not characteristic and can be
caused by natural rise of groundwater table as well. On the other hand,
extremely ill-drained (peraquic) soils do not undergo any significant
changes through cultivation of rice.

LOWLAND PADDY SOILS OF GOOD EXTERNAL DRAINAGE

Changes under cultivation of rice are most pronounced in well-drained
lowland soils, or what we call Brown Lowland Soils, which correspond to
Fluvents of US soil taxonomy (3), Brauner Auenboden of West Germany (4),
and Brown Alluvial Soils of England (5). But the changes vary in kind as
well as in degree with the permeability of soils (internal drainage)(6).

1. Highly Permeable Lowland Paddy Soils

In highly permeable paddy soils derived from lighter-textured and/or
loosely packed Brown Lowland Soils, a distinctly developed plowpan plays
a rate-determining role in percolation, and enhances the separation between water saturated upper profile and unsaturated (and therefore aerobic)
lower profile.

One of the conspicuous features is the gray coloring accompanied by
the segregation of iron into rusty mottles in the upper part of the profiles
due to seasonal surface gleying, while lower profiles retain their original
brown color. Fig. 1 shows that gray coloring has spread to 34 centimeters deep in Ogawa and 40 centimeters deep in Sekijo paddy soils. Adjacent unirrigated soils are dull-brown colored (10YR 4/2-3) up to the sur-

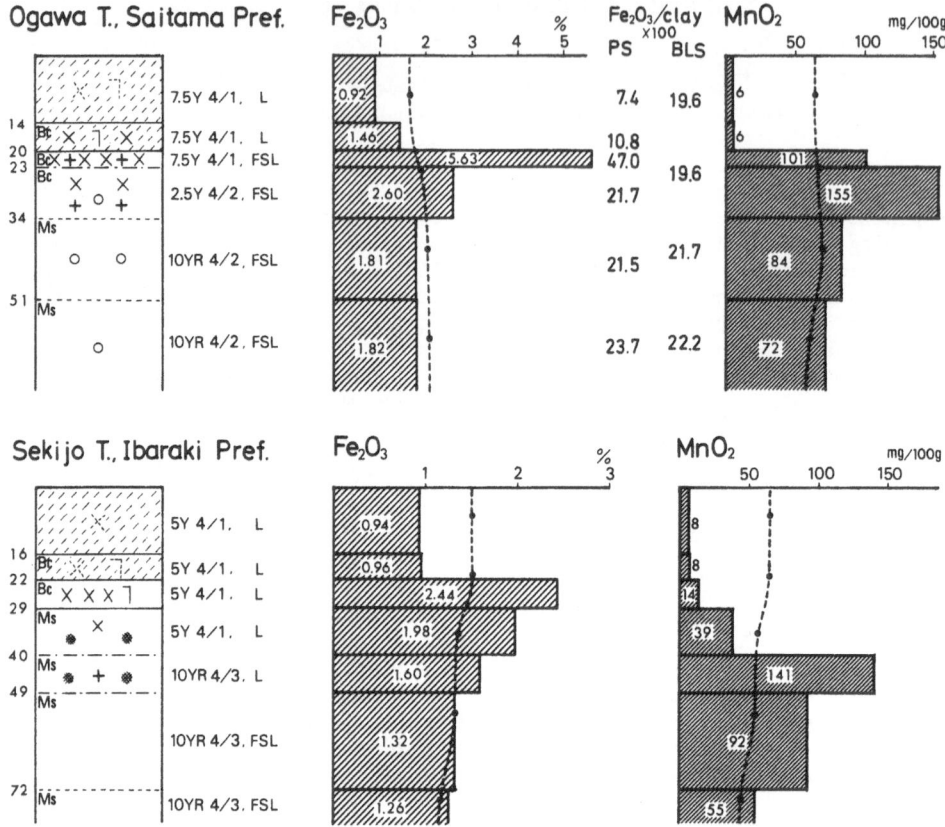

Fig. 1 Highly permeable lowland paddy soils derived
from Brown Lowland Soils (Fluvents)
(dotted line; adjacent unirrigated soils)

Legend for Figs. 1 to 4

Rusty mottles

    shape   X   thread-like ferruginous mottles along root channels (ferran)

                     ∫   tube-like ferruginous mottles along root channels (neoferran)

                     ⌐   film-like ferruginous mottles on ped-faces and cleavage planes

                     ✳   cloud-like ferruginous mottles in the matrix

                     ●   spot-like manganiferous mottles in the matrix

                     +   thread-like manganiferous mottles along root channels

    abundance   X few, XX common, XXX many, XXXX abundant

    distinctness   X⌐∫ prominent-distinct, ⋉⌐∫ clear-faint

Structure   Bc; blocky, Pr; prismatic, Ms; massive

Organic matter   ▨ >10%, ▧ 10-5%, ▦ 5-2%, ☐ < 2%

face with no rusty mottles. Blocky structures in the gray colored upper
horizons indicate the alternation of wetting and drying.

Another characteristic feature is the horizon differentiation between
the eluviated Ag and the illuvial Bg horizons, which is indicated by vertical distribution patterns of rusty mottles and iron and manganess contents.
The Bg horizons show a further separation between the iron-and manga-
nese-illuvial subhorizons. This could be further confirmed by comparing
the iron and manganese oxide contents with those of adjacent unirrigated
soils (dotted lines). Divalent iron and manganese leached from the upper
profile seem to precipitate completely upon a direct oxidation in aerobic
subsurface soils which contain enough oxygen in pore spaces, giving rise
to pronounced (often indurated) accumulation horizons.

2. Slowly Permeable Lowland Paddy Soils

In slowly permeable paddy soils derived from finer-textured and/or
closely packed Brown Lowland Soils, a saturated percolation occurs throughout profiles. Under such condition, gray coloring extends down to more
than 50 centimeters or even to nearly one meter. Two examples, Tadotsu
and Higashichichibu paddy soils, are given in Fig. 2. Their neighbouring
unirrigated soils are dull-brown colored (10YR 3-4/2.5-3) throughout.

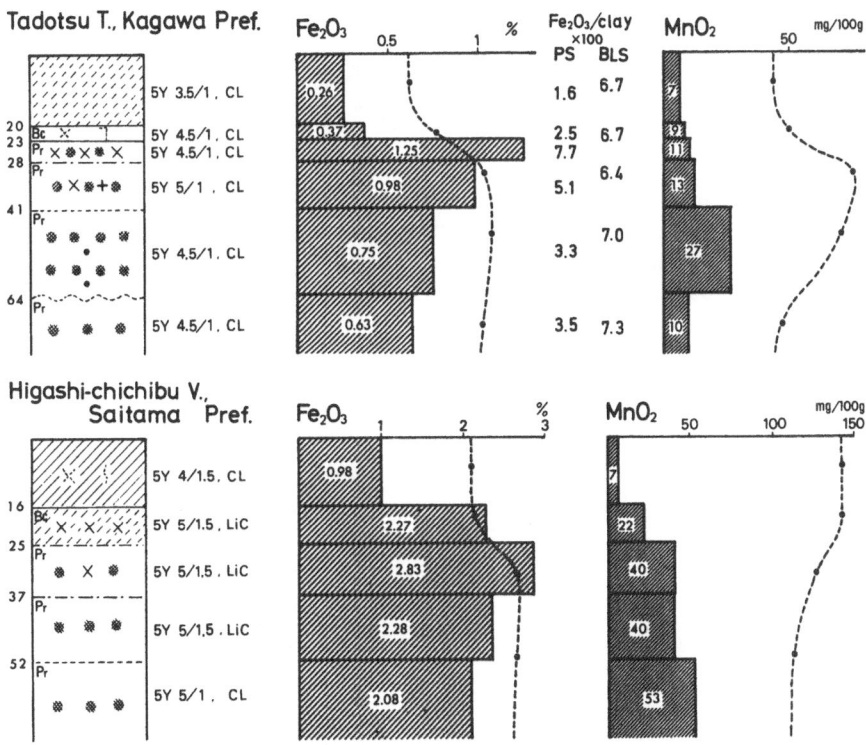

Fig. 2   Slowly permeable lowland paddy soils derived
from Brown Lowland Soils (Fluvents)
(dotted line; adjacent unirrigated soils)

421

In the gray colored surface horizons, rusty mottles are dominated by thread-like type along root channels and film-like type on cleavage planes indicating a seasonal surface gleying, while gray colored lower horizons resemble in morphology the marmorierter Horizont of Pseudogley (7), with low-chroma mottles spreading along the voids (ped-faces and channels) and cloud-like rusty mottles in the matrix. Observation in thin sections shows that there are virtually no difference in size distribution of soil particles between gray colored ped-faces and matrix (Photo. 1). This suggests that many of the gray colored ped-faces are neither clay skins nor flood coatings (gleyans)(8). Weak birefringence that ped-faces occasionally exhibit (e.g. Photo. 1-3) seems to indicate the rearrangement of soil particles by pressure. It seems that the gray colored lower profiles have been formed through the loss of iron along the walls of voids by 'reductive-eluviation', and in-situ segregation of iron in the matrix. Here the writer would like to refer to the gray colored lower horizons tentatively as 'pseudogleyed' horizons, which appear to be equivalent to what is called 'percosubmergic' horizons by Chinese soil scientists (9). Reducing substances in percolating water are considered to play an important role in 'pseudogleying' of lower horizons (10).

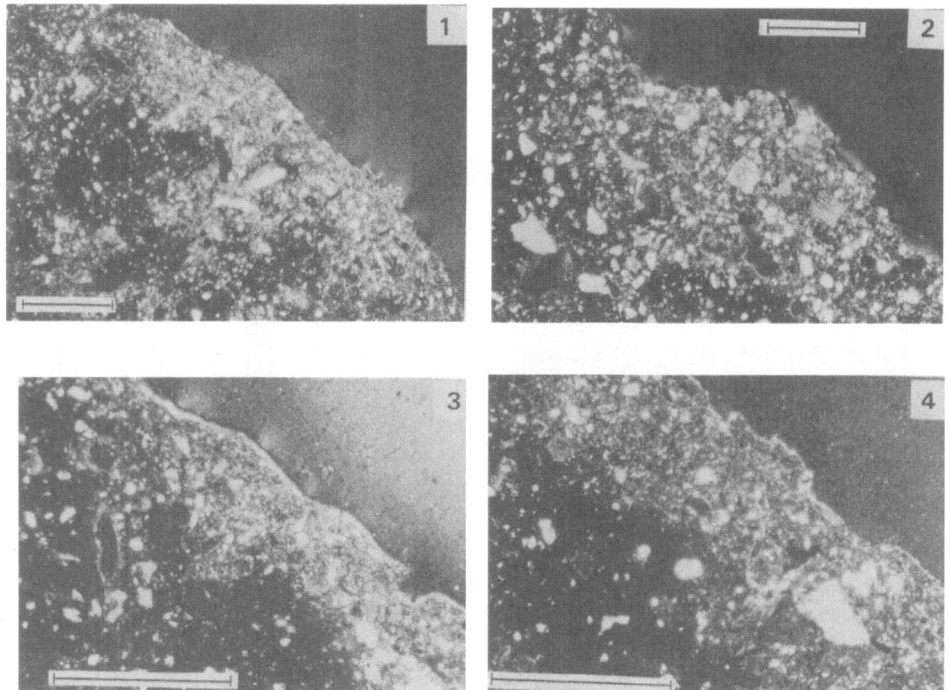

Photo. 1 Photomicrographs around gray colored surfaces of blocky and prismatic structures of the 'pseudogleyed' horizons of lowland paddy soils (scale; 1 mm)
1. Higashichichibu V., Saitama prefecture (37-52 cm)
2. Tadotsu T., Kagawa prefecture (41-64 cm)

3. and 4. Nagano C., Nagano prefecture (30-68 cm)

Eluviation-illuviation process of iron is also evident from vertical distribution patterns of rusty mottles as well as iron oxide contents (Fig. 2). In these soils iron accumulation commonly takes place in the transitional zone between the upper gleyed and the lower pseudogleyed horizons. According to Wada and Matsumoto (11), ferrous ions eluviated from the Ag are tentatively adsorbed in subsurface soil by exchange reaction instead of being oxidized directly, and wait off-season when they leave exchange sites and precipitate. However, a comparison of iron oxide contents with those of adjacent unirrigated soils (dotted lines) indicates that the development of iron-illuvial horizons is less marked than in highly permeable paddy soils. This is probably because the effect of accumulation is offset by the loss of iron occurring in subsurface soils. The comparison also reveals that iron and manganese have been leached from the lower 'pseudogleyed' horizons.

UPLAND PADDY SOILS

Cultivation of wetland rice is also practised on Pleistocene terraces and piedmonts as long as irrigation water can be made avairable. Brown Forest Soils, Red-Yellow Soils and Andosols are the major soils occurring there.

As will be mentioned later, many of the upland paddy soils of Japan are considerably drained because of their favorable physical properties. Therefore the same kind of pedogenic process as in well-drained lowland paddy soils is taking place in many of these soils. Two examples, the one derived from Red-Yellow Soil, and the other from Andosol, are presented in Figs. 3 and 4. Pedogenic characters are essentially the same as those of lowland paddy soil. By the way it is interesting to note that cobalt behaves in the manner similar to manganese (Fig. 3).

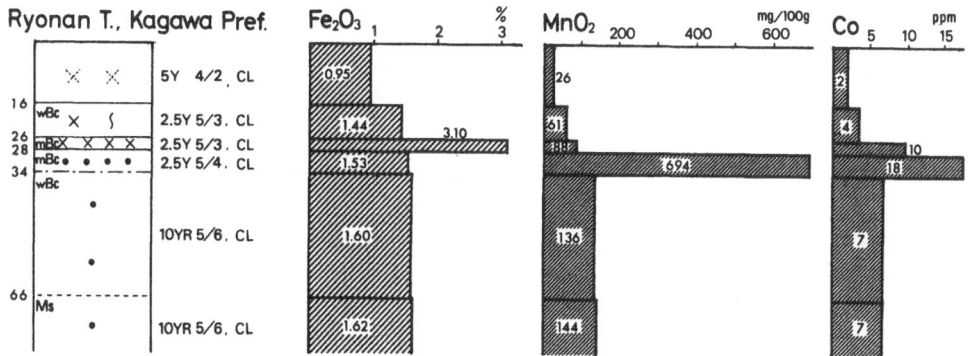

Fig. 3 Upland paddy soil derived from Yellow Soil

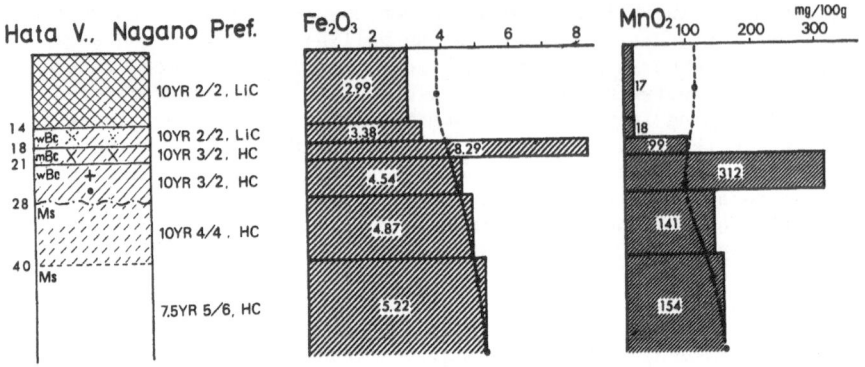

Fig. 4 Upland paddy soil derived from Andosol
(dotted line ; adjacent unirrigated soil)

Generally speaking, however, the changes under rice cultivation are
rather weak in upland paddy soils and the characters of original soils
tend to persist for many years. Probably this is due both to a low organic matter status and to an advanced stage of 'aging' or crystallization
of iron oxides, both being unfavorable for microbial reduction and mobilization of iron oxides.

FACTORS CONTRIBUTING TO CHARACTERISTIC FEATURES
OF JAPANESE PADDY SOILS

As the survey of paddy soils of Tropical Asia proceeds, it has been
gradually realized that aquorizemic or anthraquic soils are seldom encountered in Tropical paddy soils, particularly those of vast deltaic regions,
and rice cultivation can be regarded in most cases as a type of land use.
However 'paddy soil forming process', which leads to the formation of aquorizemic soils, had been considered, errouneously but naturally, as a quite
common process in paddy soils by many Japanese soil scientists. Why is
that? This is the question to be discussed here.

Japan is a mountainous country with alluvial lowlands, the major rice
growing area, accounting for about 15 percent of the land (Fig. 5). Rivers
originating from mountainous regions quickly run into the ocean after dis-
charging coarser materials on small-scaled alluvial lowlands. Frequent
downpours in catchment areas may add to a large transportation capacity.
Fig. 6 shows a geomorphological outline of the alluvial lowland along the
R. Kurobe and its vicinity, Toyama prefecture, in which the lowland consists entirely of alluvial fan. The distance from the apex to the river
mouth (13 kilometers) is too short for the river to form a flood plain or
delta downstream. This geographical situation is by no means exceptional,
but rather common in alluvial lowlands of Japan. According to the result
of the National Land Survey sponsored by Economic Planning Agency, alluvial fans plus natural levee, the well-drained parts, amount to as large
as two third of the total alluvial lowlands with deltas and polders, the
poorly drained land, comprising the remaining one third (Fig. 5). This
situation is also reflected in the particle size distribution of lowland

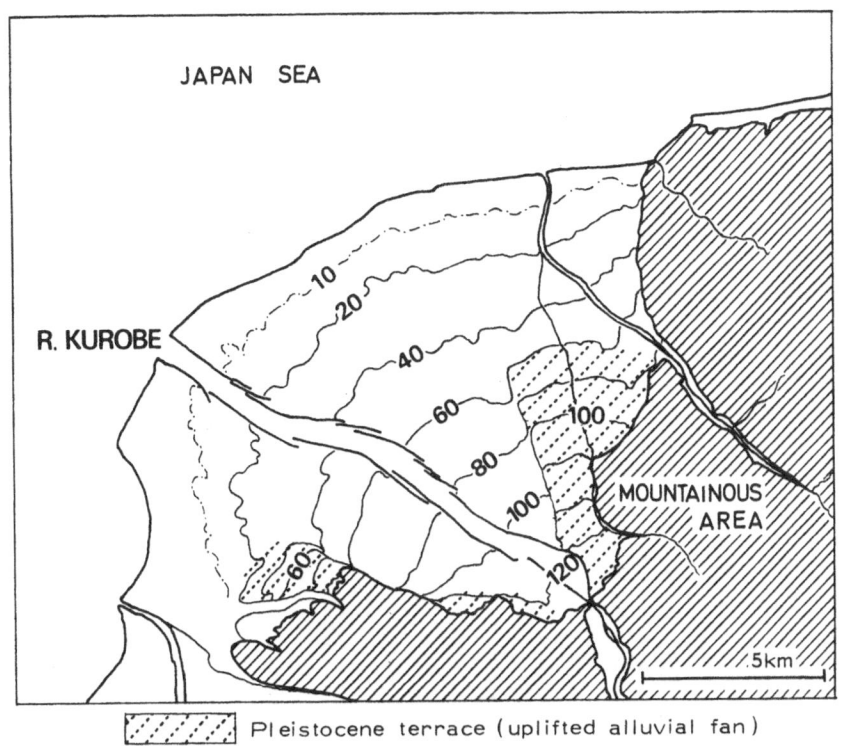

Fig. 5 Alluvial lowland along the R. Kurobe, consisting entirely of alluvial fan (Kaizuka, 1977)

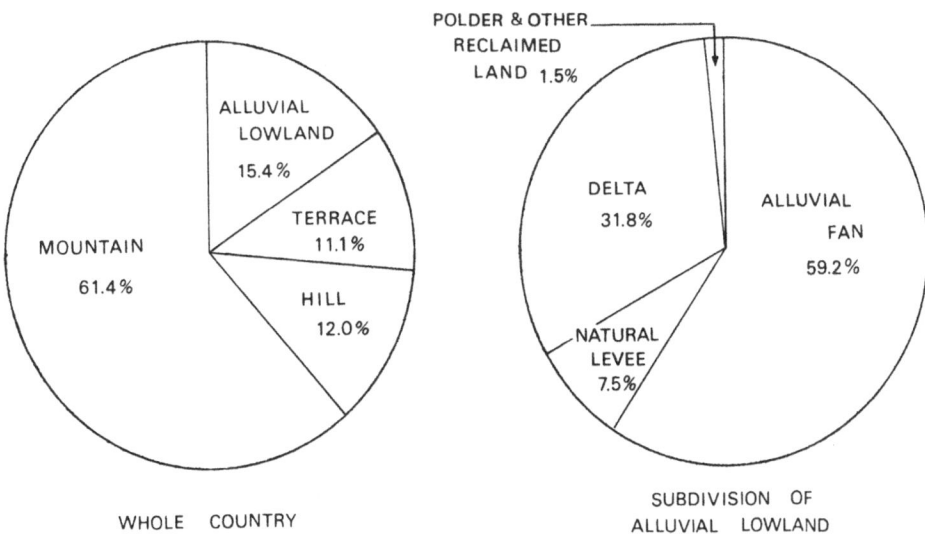

Fig. 6 Geomorphological make-up of Japan archipelago

soils given in Fig. 7, which shows that 81 out of 125 samples plotted, about two third of the total, have medium to coarse textures. Therefore soils well-drained, both externally and internally, prevail in alluvial plains of Japan.

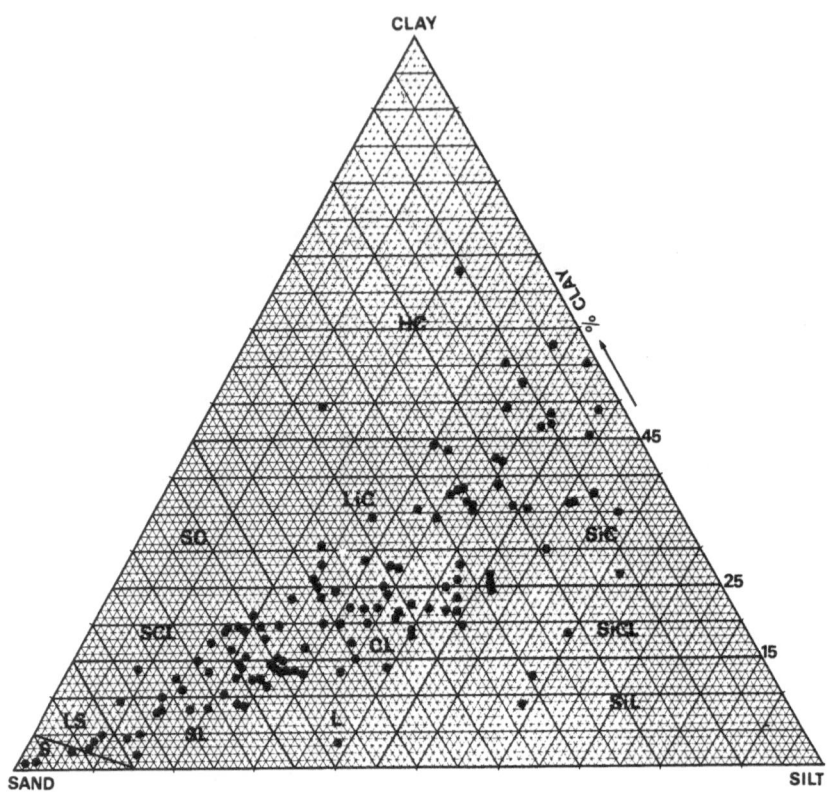

Fig. 7 Particle-size distribution of lowland
paddy soils of Japan

Turning back to Fig. 5 again, it is noteworthy that the Pleistocene terraces on both sides of the plain are also former alluvial fans. Again this is not uncommon in Japan. Therefore upland soils of Japan are in average lighter textured relative to those of other part of the world. In addition Andosols widely occurring there are by nature highly permeable to water. Consequently many of the upland soils are drained to a considerable degree.

On the other hand, the geomorphological character of alluvial plains is inevitably accompanied by flash floods and shifting of river course. In order to safeguard the rice cultivation against flood damages and secure a stable supply of water, there had been no other choise but to control rivers and introduce artificial irrigation systems. It was in 16th century that the techniques for river control and irrigation scheme were basically established. This situation enabled farmers to increase soil productivity through artificial means. Since early Yedo era (17C), being encouraged by the then feudal government, the heavy application of organic manures has taken root in rice cultivation of Japan (12). Traditional heavy application of fertilizers has provided a favourable condition for a strong gleying by stimulating microbial activity.

Geomorphological character of alluvial lowlands and Pleistocene terraces, together with the intensive managment of soil and water, may have contributed largely to a widespread occurrence in Japan of the soils that have acquired features specific to paddy soils.

It seems quite natural that Japanese soil scientist (14) was the first to find the 'degradation' phenomenon triggered by severe leaching of iron from the plough layer and the measures for its amendment. Troublesome practice of fertilizer top-dressing at various growth stages of rice plant is related at least partly with the effort to minimize the loss of nutrient elements by leaching.

REFERENCES

(1) Kyuma, K., Kawaguchi, K., Major soils of southeast Asia and the classfication of soils under rice cultivation, The Southeast Asian Studies, 4; 290-312 (1966).
(2) Moormann, F.R., van Breemen, N., Rice: Soil, Water, Land , 185p., Int. Rice Res. Inst., Los Banos, Philippines (1978).
(3) Soil Survey Staff, SCS, USDA, Soil Taxonomy, a basic system of soil classification for making and interpreting soil surveys, Agriculture Handbook, No. 436, 754p. (1975).
(4) Mückenhausen, E., Entstehung, Eigenschaften und Systematik der Böden der Bundesrepublik Deutschland, 148p., DLG-Verlag-GmbH (1962).
(5) Avery, B.W., Soil classification in the soil survey of England and Wales, J. Soil Sci., 24; 324-338 (1973).
(6) Mitsuchi, M., Pedogenetic characteristics of paddy soils and their significance in soil classification, Bull. Nat. Inst. Agr. Sci., B25; 29-115 (in Japanese with English summary) (1974).
(7) Blume, H.P., Zum Mechanismus der Marmorierung und Konkretionsbildung in Stauwasserboden, Z.f. Pflanzenernähr. Bodenkunde, 119; 124-134 (1968).
(8) Brammer, H., Coatings in seasonally flooded soils, Geoderma, 6; 5-16, (1971).
(9) Xu Qi et al., The paddy soil of Tai-hu region in China, Nanking Inst. Soil Sci., Academia Sinica (in Chinese with English summary) (1980).
(10) Okazaki, M., Wada, H., Some aspects of pedogenic processes in paddy soils, Pedologist, 20; 139-150 (1976).
(11) Wada, H., Matsumoto, S., Pedogenic process in paddy soils, Pedologist, 17; 2-15 (1973).
(12) Kaizuka, S., Geomorphology of Japan, 234p., Iwanami-shinsho, Iwanami Pub. Co. (in Japanese) (1977).
(13) Yoshida, T., Japan's agricultural technology——Its characteristic and outlook, Farming Japan, 12; 10-20 (1978).
(14) Shioiri, M., Studies on the degradation of paddy soils and measures for its amendment (in Japanese), Nogyo oyobi Engei(Agriculture and Horticulture), 20; 39-40 (1945).

ON THE CLASSIFICATION AND USE OF SOME DARK CLAYEY PADDY SOILS IN CHINA

Huang Rui-cai, Wu Shan-mei
(Nanjing Agricultural College)

On the 1:5,000,000 scale soil map of the world, sheet VIII-3 (FAO/ UNESCO, 1977)a large extent of land surrounding the big lakes and along their tributaries and some other scattered areas south of the Changjiang River valley in the middle part of China are delineated as vertisols denoted by Vp, 66 3a on the map (1). These dark clayey soils have been used to grow rice for centuries. As early as 1950, H. Oakes and J. Thorp also suggested that the Sajong black soils (i.e. black soils with lime concretions (2), some of which have been used as rice fields at present) of the Huaihe River Plain in Central and East China have properties similar to the Grumusols in Texas of the United States (3). Now Grumusols in Texas have been considered as typical Vertisols (4). Thus, it seems that Vertisols are quite extensive in China. However, we are not familiar with this interpretation and correlation and with some reservations (5). There is urgent need to gather further scientific information and promote international exchange of views about this problem.

This paper deals with nine profiles of the dark clayey paddy soils from three areas. Profiles No. 1-5 inclusive of the north Huaihe River Plain are located in one area (Area A) of East China representing the monsoon-temperature region (6,7). Profiles No. 6 and 7 of southern Jiangsu are located in another area (Area B) of East China representing the monsoon-subtropical region (8-10). Profiles No. 8 and 9 are located in one area (Area C) of Hainan Island representing the tropical region with distinct wet and dry seasons (11).

Field observations and laboratory studies were made in order to afford data for classifying the nine profiles according to the criteria of a typical Vertisol and elucidate their soil fertility status in relation to utilization in agriculture.

The profile characteristics of the nine dark clayey soils in the three different areas are given below.

Profiles No. 1 and 2 are very dark gray (5Y 3/1, moist) to olive gray (5Y 4/2, moist) in the surface and subsurface layers. Profile No. 3 is dark grayish brown (2.5Y 4/2, moist); profiles No. 4 and 5 range from brownish gray (2.5Y 5/2, moist) to dark grayish brown (2.5Y 4/2, moist) in the upper layers. Profiles No. 6 and 7 are very dark grayish brown (2.5Y 2/2, moist) to black (2.5YR 1.7/1, moist) in the upper layers, but the substrata are light brownish gray (10YR 6.5/2, moist) to dark grayish brown (2.5Y 4/2, moist) in colour. Of these two profiles, water table varied between 50 and 65 cm below ground surface last year. Bog iron ore occurs at a depth of about 80-100 cm below the surface. Profiles No. 8 and 9 are dark gray (N 4/ , moist) to black (N 2.5/ , moist) in the surface and subsurface layers. Basaltic bed rock occurs within 1 meter from the surface of the soil.

Except the substrata of profiles No. 6 and 7 of southern Jiangsu (12) all the soil layers of the nine profiles are quite heavy to very heavy in texture.

The pH values of the five profiles of the north Huaihe River Plain are above 7, some reaching 8.2 in the substratum; those of the two profiles of southern Jiangsu vary from 6.5 to 7.6, most of them being above 7; those of the two profiles of Hainan Island are between 5 and 6.7.

The structure of the surface soils of all the nine profiles are granular to blocky. The subsoils of five profiles of the north Huaihe River Plain are blocky to angular blocky, while those of the two profiles of southern Jiangsu are prismatic. The subsoils of the two profiles of Hainan Island are structureless in autumn just after the harvest of rice.

When dry, all the upper layers of the nine profiles crack to a great extent. The dark clayey paddy soils of Hainan Island might crack to as much as 8-10 cm wide and more than 35 cm deep during the dry season.

Particle size distribution, expansion and shrinkage, settling volume and time of settling, organic matter content, total N and C:N, total and available P and K, cation exchange capacity, X-ray diffraction analysis and electron microscopy, and micromorphology of eight profiles (excluding profile No. 9) were investigated in the laboratory. The results are given in Tables 1-4, Fig 1-3 and Pl. 1.

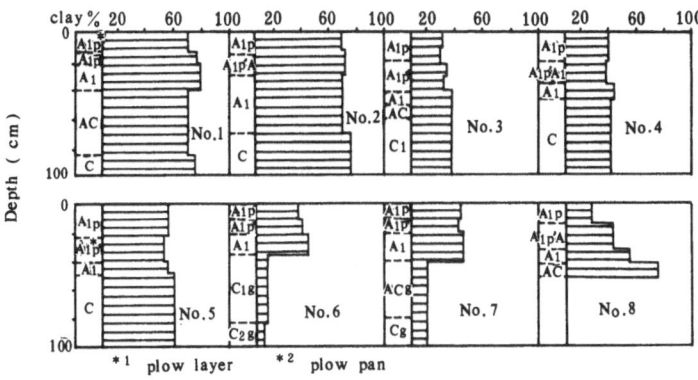

Fig. 1 Distribution with depth of clay

The clay contents ($<$0.002mm) are more than 35 to over 70% (Table 1, Fig. 1) (6). The values of expansion and shrinkage are approximately proportional to the clay contents of the different soil profiles (Table 2) (13,6). The differences in the nature of soil colloids of different profiles are revealed in the study of settling volume and time of settling (Table 3) (14-16). Profiles No. 1 and 2 of the north Huaihe River Plain showed abnormal results. In the soil-water suspension of the AC or C layers of these two profiles, after the coarser soil particles and aggregates were settled, thick layers of floccules appeared overlying the sediments. For instance, with the 70-100 cm layer of profile No. 2 the

Table 1 Particle size distribution and soil texture*

Profile No.	Depth (cm)	Soil separates (%)			Soil texture
		2-0.05mm	0.05-0.002mm	<0.002mm	
Area A — Paddy soils of North Huaihe River Plain					
1	0-110	1.02	27.23	71.75	Heavy clay
2	0-100	1.09	28.60	70.31	Heavy clay
3	0-107	8.32	56.61	35.07	Silty clay loam
4	0-95	2.51	55.87	41.62	Silty clay
5	0-98	2.76	55.14	42.10	Silty clay
Area B — Paddy soils of Southern Jiangsu					
6	0-39	1.56	58.06	40.38	Silty clay
7	0-47	1.30	56.64	42.06	Silty clay
Area C — Paddy soil of Hainan Island					
8	0-50	4.04	50.49	45.47	Silty clay

Ma, T.S; Wu, S.M. By pitette method
Fong. M; Wu. S.M. Zhuang. G.L. (fractionation of clay)
* Average value of each profile (or section of Prof. no. 6 & 7)

Table 2 The coefficients of expansion and shrinkage and ESI*

Profile No.	Depth(cm)	Coef. of expansion (%)	Coef. of shrinkage (%)	ESI (%)**
Area A — Paddy soils of North Huaihe River Plain				
1	0–110	35	40	75
2	0–100	27	36	63
3	0–107	22	28	50
4	0–95	24	37	61
5	0–98	25	36	61
Area B — Paddy soils of Southern Jiangsu				
6	0–39	18	31	49
7	0–47	15	25	40
Area C — Paddy soil of Hainan Island				
8	0–50	23	35	58

Xu, S.L. By Keen-Raczkowski's method

* Average value of each profile (or section of prof. no. 6 & 7)
** The expansion and shrinkage indices

431

floccules amounted to 14 ml as against a volume of 14 ml of sediments underneath in the 100 ml-graduate. Moreover, it took 62 days for the supernatant liquid to become clear. The general tendency was that the settling volumes (cc/g) increased with the clay contents in the soil profile; but the time of settling of the suspension to reach a state of clear supernatant liquid was less regular. This test may give some information on the suitability of the flooded soil to the transplanting of rice seedlings in respect to compactness and fluidity of the soil. However, it is doubtful that any semi-quantitative results may be obtained with this test for settling volume whenever there are too many floccules to show the real volumes of the soil particles and aggregates.

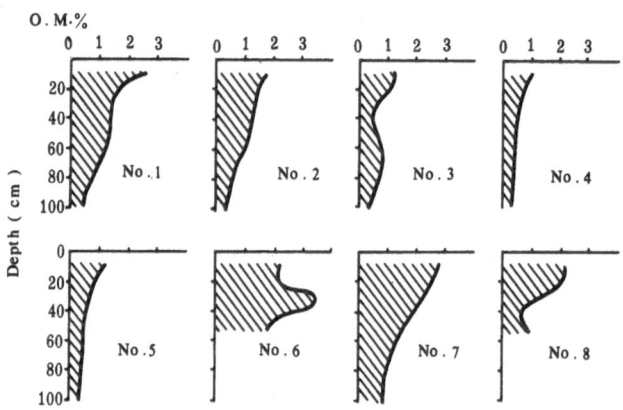

Fig. 2 Distribution with depth of organic matter

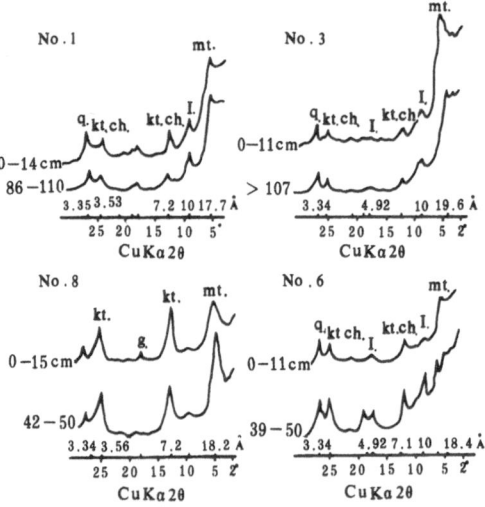

Fig. 3 X-ray diffraction patterns of
clay particles

432

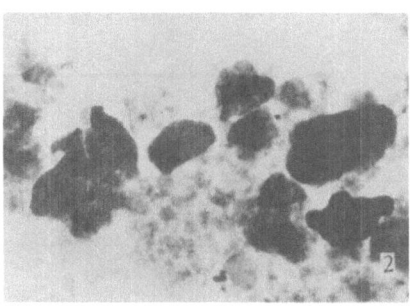

1) Montmorillonite (X15,000).
No. 5 41-98cm

2) Hydrous mica, Montmorillonite.
Kaolinite. No. 3 0-11 cm

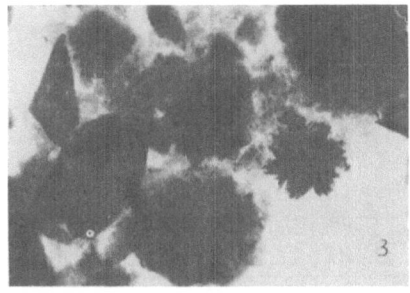

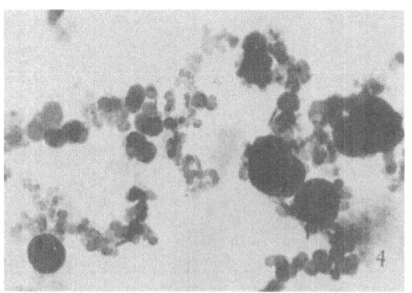

3) Goethite (X15,000). No. 6
39-83 cm

4) Hematite (X20,000). No. 8
42-50 cm

Plate 1 The kinds of clay minerals of different soil profiles

All the eight dark clayey paddy soils analyzed are originated from or closely related to the hydromorphic soils of either the past or recent times. However, except profile No. 6 of southern Jiangsu the O.M. contents of the different layers of the profiles are below 3%, mostly below 1.5% in the plowpan of the soil (Fig. 2). The N contents are roughly proportional to the contents of O.M. in soil, the C:N varying from 6.3 to 18. There is a tendency that C:N ratios increase from the humid temperate to the humid subtropical and tropical regions. It is worth mentioning that in profile No. 8 of Hainan Island the O.M. could not be removed with H_2O_2, probably due to its chemical combination with the mineral grains of the soil.

The total and available phosphorus of all the eight profiles are low to extremely low (Table 4). Phosphatic fertilizers must be applied if high yields of rice and other crops are expected.

Available potassium contents are very variable (Table 4). Profiles No. 1 and 2 of the north Huaihe River Plain have plenty of available K, being above 200 ppm in the whole depth of profile, while profile No. 8 of Hainan Island has low contents of around 20 ppm throughout the profile.

Cation exchange capacity of all the eight profiles are not high according to the diagnostic character of a Vertisol (Table 4). They vary from less than 10 m.e. 100 g soil to 30 m.e. 100 g soil. They are roughly proportional to the contents of clay and O.M. in the soil.

433

Table 3 Settling volume and time of settling

Profile No.	Depth (cm)	Settling volume (cc/gm)			Time of settling
		Total	Sediments	Floccules	
Area A - Paddy soils of North Huaihe River Plain					
1	0-14	1.90	1.75	0.15	13 days
	14-21.5	2.00	1.80	0.20	23 days
	86-110	3.40	1.60	1.80	52 days
2	0-15	1.58	1.40	0.18	103 days
	15-30	1.70	1.70	--	6 days
	30-70	2.20	1.40	0.80	47 days
	70-100	2.80	1.40	1.40	62 days
3	0-11	1.30	1.30	--	24 hrs.
	11-22	1.45	1.35	0.10	20 days
	42-107	1.40	1.30	0.10	11 days
4	0-20	1.55	1.50	0.05	20 days
	20-33	1.80	1.60	0.20	13 days
	43-95	1.90	1.40	0.50	31 days
5	0-15	1.55	1.30	0.15	43 days
	15-26	1.50	1.25	0.25	43 days
	41-98	1.70	1.40	0.30	31 days
Area B — Paddy soils of Southern Jiangsu					
6	0-11	1.80	1.70	0.10	20 days
	11-24	1.70	1.60	0.10	20 days
	39-50-83	1.25	1.20	0.05	19 days
7	0-9	1.70	1.65	0.05	20 days
	9-22	1.60	1.55	0.05	20 days
	>47	1.50	1.45	0.05	19 days
Area C — Paddy soils of Hainan Island					
8	0-15	1.30	1.25	0.05	13 days
	15-36	1.55	1.40	0.15	103 days
	42-50	1.50	1.45	0.05	47 days

Huang, R.C. and Ma, T.S. By Middleton's method, modified

X-ray diffraction examinations (Fig. 3) show that profiles No. 1,3,4
and 5, especially No. 3 of the north Huaihe River Plain have more conspicuous
amounts of montmorillonite and hydrous mica with traces of chlorite, kaolinite and quartz in the surface soils. Profile No. 6 of southern Jiangsu
has greater amount of hydrous mica, less amount of montmorillonite and
traces of chlorite kaolinite and quartz. Profile No. 8 of Hainan Island
has dominant amounts of both kaolinite and montmorillonite with only traces
of gibbsite and quartz (17).

Electron microscopic photographs are taken to show the kinds of clay
minerals in the representative soil layers of the different profiles (Pl. 1).
The shape of the montmorillonite is thin and fluffy; the size of the

kaolinite is very small. There are also some iron oxides such as goethite and hematite in profile No. 8.

All the eight profiles of dark clayey paddy soils have a thick col-loidal-like soil matrix in which are embedded minute skeletal grains. These soils have a porphyropeptic fabric. Parallel alignments of the cells of cellulose were seen in the partly decomposed plant tissues under the microscope with low magnification. There are different kinds of mineral grains including quartz. In profile No. 8 of Hainan Island quartz is the only dominant mineral. The humus looks like minute grains scattering in the web-like constitution of the soil matrix. Hydrous iron oxides are mixed with the humus in a colloidal state. Some of them are condensed toward the wall of the pores and root channels, and others tend in various degrees to form stains and minute concretions in almost all of the eight profiles (18). With crossed Nicols there was seen flecked extinction in the thin sections of the soil of most profiles on rotation of the stage of polarized microscope.

Table 4 Chemical properties and nutrient contents

Prof. No.	pH[*]	Total N[**] (%)	Total P[***] (ppm)	Avilable P[***] (ppm)	Avilable K[***] (ppm)	CEC[****]
Area A — Paddy soils of North Huaihe River Plain						
1	7.4-8.2-7.7	0.12	190.3	1.0	227.3	27.3-30.6
2	7.1-8.2	0.082	239.8	1.5	219.9	—
3	7.0-8.1-7.7	0.061	123.8	0.6	107.0	16.5-22.4
4	7.3-7.9	0.048	176.3	0.7	143.7	—
5	8.0-7.7	0.062	182.0	0.9	131.0	19.3-21.4
Area B — Paddy soils of Southern Jiangsu						
6	7.0-7.9	0.16	356.0	2.9	122.9	20.5-6.9
7	6.5-7.6	0.12	174.8	1.3	70.2	17.7-12.1
Area C — Paddy soil of Hainan Island						
8	5.0-6.7-6.0	0.072	164.5	1.2	20.9	8.9-15.4

Li, J.J., Sun, W.L., Sun, F.L.

   * From A to C horizens; by pH meter 1:1 soil & water
  ** Plow layer, by the Semi-micro Kjeldahl method
 *** Average value of each profile, by Olsen's method
 *** Average value of each prafile, by flame photometer after NH_4OAc extraction
**** A and C respectively, by NH_4OAc method

Primary aggregation is faint or invisible in the thin sections of the soil. Except in the plow layers roots and root channels or pores are very few.

Clay skins (cutans) are seen on the surface of structural units of several profiles especially profiles No. 6 and 7 of southern Jiangsu. Micromorphologically, the extent of reorientation of clay caused by the differential movements of the soil materials as a result of expansion and contraction in contrast to the orientation of clay due to illuviation calls for further study (19).

Conclusively speaking, five points must be considered in preference to all other details in the classification of the dark clayey paddy soils in respect of Vertisols (20, 21). They are: 1) a climate of alternative wet and dry seasons, 2) heavy texture with great expansion and shrinkage, 3) haploidization or simple in appearance of the soil profile, i.e. an A_1 (gray to black in colour) C profile, 4) no eluviation and illuviation except for lime, and 5) water regime of the soil profile especially influenced by irrigation for growing rice.

Considering the above-stated five points and the results of study of the nine profiles of dark clayey paddy soils of the three different areas(2), we come to know that all the nine profiles deviate much from Houston black clay—the type Vertisol in all specifications (22, 23). Nevertheless, each of the profiles has some of the important characteristics of a Vertisol. Irrespective of the other schemes of soil taxonomy especially that of the classification of paddy soils as an independent soil order or great soil group, the nine soil profiles in question are tentatively classified(according to the U.S. Soil Taxonomy, 1975 (20) into the following soil families:

Profiles No. 1 and 2.... Aquic Pelludert, very fine, mixed, mesic. They are wet gray to black expanding clay soils of the monsoon-temperate climate.

Profile No. 3 .... Entic Chromudert, clayey, mixed, mesic. It is a bright brownish expanding clay soil of the monsoon-temperate climate.

Profiles No. 4 and 5 .... Aquic Chromudert, clayey, mixed, mesic. They are wet brownish expanding clay soil of the monsoon-temperate climate.

Profiles No. 6 and 7 .... Vertic Fluvaquent, clayey, mixed, thermic. They are wet dark clayey soils of stratified flood and lacustrine deposits with very high water table in the monsoon-subtropical climate. At some season in most years they become dry enough to show deep and wide cracks as in Vertisols.

Profiles No. 8 and 9 ... Aquic Pelludert, clayey, kaolinitic and montmorillonitic, hyperthermic. They are wet dark clayey expanding soils from residual and slope deposits of basaltic origin in the tropical climate.

Thus, from the present study, we can see that some of the dark clayey paddy soils are Vertisols while others are hydromorphic soils intergrading to the Vertisols at subgroup level. We should work out more precise standards to define Vertisols.

The coverage of dark clayey paddy soils in the north Huaihe River Plain is not large. The Sajong black soils of that region have been used to grow upland crops for centuries (2). Near the villages a rotation of rice and wheat has been adopted since 1949. For continuous rice and green manure the annual yield might reach 6,000 to 8, 250 kg/ha of rice on more fertile soils (profile No. 1), but the annual yield of continuous rice without green manure was only 2,250 to 3,000 kg/ha on less fertile soils (profile No. 4).

436

In southern Jiangsu the dark clayey paddy soils located in the diked lowlands could grow only one season of rice each year, with the fields remaining flooded all the year round and the annual yield of rice being only 750-1,500 kg/ha. After improving the drainage systems, single cropping of rice has been changed to the rice-wheat or rice-rice-barley rotation, and the total annual yield has been increased to 5,250 kg/ha of rice and 3,000 kg/ha of wheat per year, or 3,750 kg/ha of 1st rice crop, 3,000 kg/ha of 2nd rice crop and 1,500 kg/ha of barley (profile No. 6).

The dark clayey paddy soils of Hainan Island grew only one crop of rice before the 1960s. More fertile soil produced about 3,750-4,500 kg/ha of rice each year. Now some of the rice fields have been changed to grow double crops of rice each year through better water conservancy facilities. With enough fertilizers the annual yield of rice may be further increased by two to three times.

For the dark clayey paddy soils of the three areas under study, there exist some soil problems and means of improvement in common, such as 1) heavy texture and unfavorable soil structure (24), 2) low nutrient contents of P and K and minor elements such as Mn and Zn (5), 3) excess of soil water after rice harvesting and during the rainy spring. Adequate drying, deep plowing, suitable cultivation, application of balanced fertilizers and good drainage, such are the effective means of improvement to obtain sustained high yields of rice and other crops.

REFERENCES

(1) FAO/UNESCO, Soil Map of the World. Sheet VIII-3 (1977).
(2) Thorp, J., Geography of the Soils of China. Published by the National Geological Survey of China (1936).
(3) Oakes, H., Thorp, J., Dark-clay soils of warm regions variously called Rendzina, Black Cotton Soils, Regue and Tirs. Soil Sci. Soc. Amer. Proc., 15, 347-354 (1951).
(4) Soil Survey Staff, U.S.D.A., Soil Taxonomy. 375-382 (1975).
(5) Institute of Soil Science, Academia Sinica, Nanjing (ed.), Soils of China. Science Press, Beijing, 549-552 (1978). (in Chinese)
(6) Zhang Jun-min, Genesis and classification of the Shajiang Black Soil (1979 , ms.). (in Chinese)
(7) Kovda, V.A., The Natrual Environment and Soils of China. Science Press, Beijing, 132 (1960). (in Chinese translation)
(8) Xu Qi, Geographical distribution and genesis of the "White Soil" of the middle and lower Yangtze River. Acta Pedologica Sinica, 10, 44-54 (1962). (in Chinese with Russian summary)
(9) Soil Survey Committee, Liyang County, Jiangsu, Soil survey report of Liyang County (1959, ms.). (in Chinese)
(10) Soil Survey Committee, Yixing County, Jiangsu, Soil survey report of Yixing County (1959, ms.). (in Chinese)
(11) Han Yuan-beng, Wu Yu-wen, Huang Yuan-lue, Zheng Xiang-sheng, The Hainan Island. Guangdong People's Publishing House, Guangzhou, 15-23 (1976). (in Chinese)
(12) Xu Qi, Lu Yan-chun, Zhu Hong-guan, Genesis and classification of paddy soils in Taihu Lake region (1979, ms.).
(13) Huang Rui-cai, Laboratory Direction of Soil Science. Published by Nanjing University, 14-15, 28-29 (1950). (in Chinese)

(14) EPPS Committee, Examination of the Physical Properties of Soils. Chongqing Branch, Publishing House of Scientific & Technical Information, Chongqing, 546-556 (1979). (in Chinese translation)

(15) Middleton, H.E., Byers, H.G., The settling volume of soils. Soil Sci., 37, 15-27 (1934).

(16) Olmstead, L. B., The sedimentation volumes of soil from certain major soil groups. Soil Sci. Soc. Amer. Proc., 4, 89-93 (1939).

(17) Thorez, J., Practical Identification of Clay Minerals (A handbook for teachers and students in clay mineralogy) Dison, Belgique, 2-3 (1976).

(18) Richardson, J. L., Hole, F.D., Mottling and iron distribution in a Glassoboralf-Haplaquoll hydrosequence on a glacial moraine in northwestern Wisconsin. Soil Sci. Soc. Amer. J., 43, 552-558 (1979).

(19) Cao Sheng-geng, Jin Guang, Micromorphological diagnosis of the characteristics of the soil fertility of rice paddy soils (1979, ,s.). (in Chinese)

(20) Buol, S.W., Hole, F.D., McCracken, R.J., Soil Genesis and Classification. Iowa State University Press, Iowa, 218-225 (1973).

(21) Dudal, R., Dark clay soils of tropical and subtropical regions. FAO Agric. Dev. Pap., 83, 71-98 (1965).

(22) Kunze, G. W., Templin, E. H., Houston black clay, the type grumusol. II. Mineralogical and chemical characterization. Soil Sci. Soc. Amer. Proc., 20, 91-96 (1956).

(23) Templin, E. H., Mowery, I. C., Kunze, G. W., Houston black clay, the type grumusol. I. Field morphology and geography. Soil Sci. Soc. Amer. Proc., 20, 88-90 (1956).

(24) Deng Shi-qin, Xu Meng-xiong, Texture of paddy soils and its regulation in Taihu Lake region. Turang, 5, 175-177 (1979). (in Chinese)

ON THE GENESIS, CHARACTERISTICS AND UTILIZATION OF THE ACID SULPHATE PADDY SOIL IN CHINA

Wei   Qi-fan
(Institute of Soil Science, Academia Sinica, Nanjing)

Acid sulphate soil occurs in large areas of the tropic and subtropic regions of the world.   In China, it scatters over the estuaries and bays at the river mouths south of 25° N-latitude (Fig. 1), but is more concentrated at the coastal areas of western Guangdong Province.   Acid sulphate soil has a total acreage of about 67 thousand hectares in China. Most of them have been utilized for rice and are denominated as acid sulphate paddy soil or acid salty paddy soil in soil classification(1,2).

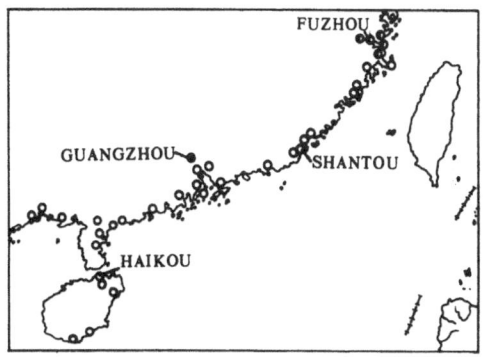

o Acid sulphate paddy soil
● Acid sulphate paddy soil formed by
   manuring with mangrove residue

Fig. 1 Distribution of acid sulphate paddy soil in China

GENESIS AND CHARACTERISTICS

Acid sulphate soil is developed on the sea beach where previously covered by mangrove forest.   The mangrove was later overlain by the continuous deposit of alluvial material from the rivers.   Profile of acid sulphate soil is thus formed with a surface layer of alluvial material and a buried soil layer rich in organic matter.   As the coastal land emerges from sea water, the acid sulphate soil is usually used for rice.   Land levelling, ᵢ cultivation and fertilization gradually improve the soil property. The acid sulphate paddy soil is classified into three local soil types, i.e. severe acid sulphate paddy soil, medium acid sulphate paddy soil and moderate acid sulphate paddy soil, according to the degree of mellowing (Table 1).

439

Acid sulphate paddy soil becomes more mellowed as cultivation goes on. The severe type is usually cultivated for about 10–20 years only, with a thin surface layer of less than 25 cm thick. Thickening of surface soil, lowering of acidity and decreasing of sulfide in buried organic layer characterize the degree of mellowing. However, even on the acid sulphate paddy soil of moderate type, conditions are still inadequate for healthy growth of rice (Table 1).

Table 1   Classification of acid sulphate paddy soil

Local soil type	Depth of buried organic layer(cm)	Total S in surface layer (%)	pH surface layer	Water soluble Al^{3+} (m.e./100g)	Toxicity to rice
Severe type	25	0.3	2.0–2.5	1	Severe
Medium type	50	0.2	3.0	0.3–0.6	Medium
Moderate type	50	0.15	3.5	0.1	Moderate

Acid sulphate paddy soils have a layer of dark colored subsoil about 20–40 cm thick, containing organic matter 3.2–6.6 %, with H_2S odor. Although this layer has been highly humified. the decayed organs of mangrove are still perceivable (1). The contents of total sulphur and water soluble sulphur for two typical profiles of acid sulphate paddy soil are shown in Fig. 2. Notice that the content of total 5 in the thick buried organic layer of a severe acid sulphate paddy soil can be as high as 2.5 % and water soluble S 0.6 %.

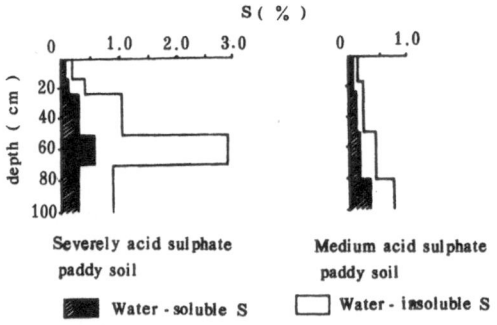

Fig. 2 Content of insoluble S and water soluble S in acid sulphate paddy soils

Oxidation of sulphide proceeds rapidly after the field has been drained from a submerged condition. The final oxidation product of sulphide is sulphate, and mainly may be in the form of aluminium sulphate. However, when soil clods are exposed in air yellow speckles which perhaps are jarosite will be observed on its surface.

Table 2 pH of acid sulphate paddy soil

Depth (cm)	Submerged in summer	Drained in winter	Air-dried
0–14	5.5	3.1	3.1
14–26	5.5	2.9	2.8
26–50	5.5	2.6	2.3
50–70	5.5	3.0	2.2
70–100	7.0	5.1	3.0

Table 2 shows that under submerged condition the soil acidity is not very strong, when field has been drained pH value is quickly lowered. The strongly acid reaction of acid sulphate paddy soil is mainly induced by high concentration of active Al^{3+} up to 0.37 m.e./100g soil in surface horizon and 23 m.e./100g in buried organic horizon in a severe type (Table 3). Owing to the low redox potential of acid sulphate paddy soil, especially in the buried organic horizon, exceedingly high figures of water soluble Fe^{2+} and Mn^{2+} have also been found(1–4,6).

Table 3 Composition of water- soluble cations in acid sulphate paddy soil

Soil type	Depth (cm)	Horizon	pH	Water-soluble cation (m.e./100g)			
				H^+	Al^{3+}	Fe^{2+}	Mn^{2+}
Severe type	0–14	Surface	3.1	0.07	0.37	0.61	0.02
	50–70	Buried organic	2.2	1.48	23.17	2.44	0.53
Medium type	0–16	Surface	3.5	0.02	0.05	0.04	0.02
	80–95	Buried organic	2.5	0.50	13.76	3.58	0.43

RECLAMATION OF ACID SULPHATE PADDY SOIL

As shown in Table 1, toxicity of acid sulphate soil hinders plant growth in all three types. Rice plant has been stunted in growth even in the moderate type. Cultivation usually fails without appropriate measures of reclamation.

Reclamation of acid sulphate soil through rice plantation begins with field levelling, followed by the construction of proper irrigation and drainage ditches. Subsequently the soluble toxic components are washed and drained out. Amelioration of acid sulphate paddy soil is well accomplished by the application of mud, manures together with chemical fertilizers. Liming has been proved very effective to check the acid toxicity(7,8).

Serious deficiency of available phosphorus has been observed in some acid sulphate paddy soils. For example, those acid sulphate soils that occur in the coastal area of western Guangdong contain only traces of available phosphorus as determined by current extractants. Rice crop has practically failed without the application of phosphatic fertilizer(5).

Table 4   Effect of P-fertilizer on yield of rice on an acid
sulphate paddy soil

Treatment	Yield (t/ha)	
	1st year	2nd year
CK	0.15	---
Superphosphate, 0.6 t/ha	2.8	2.3
Morocco apatite*, 1.5 t/ha	3.2	3.0
Guizhou apatite**, 1.5 t/ha	2.5	2.5

Data from Wang Zhen-rong
* $CaCO_3$ content 11%
** $CaCO_3$ content 3%

Table 4 gives the results from field experiment laid in years 1964-65 on an acid sulphate paddy soil at the coastal land, western Guangdong. All plots received adequate amount of $(NH_4)_2SO_4$ as nitrogen fertilizer. Results show the necessity of phosphatic fertilizer. Further more, due to the strong soil acidity, rice plant has good response in yield to powdered rock phosphate. As shown in Table 4, loosely crystallized (Morocco) has a better effect than micro-crystallized apatite (Guizhou). The high content of $CaCO_3$, 11 %, in Morocco rock surely plays an important role in the reclamation of the acid sulphate paddy soil. Table 4 also shows that the powdered rock phosphate has a better after effect for succeeding crop than that of superphosphate. Utilization of powdered rock phosphate has been extended in this area.

REFERENCES

(1) Gong Zi-tong, Zhou Rui-rong, Genesis of strongly acid saline paddy soil of Southern Guangdong. Acta Pedologica Sinica, 12, 183-191(1964). (in Chinese with English summary)
(2) Huang Ji-mao, A study on the chemical charateristics of strongly acid saline paddy soils in the coastal areas of Guangdong Province. Acta Pedologica Sinica, 6, 114-122(1958).(in Chinese with Russian summary)
(3) Zhang Jun-min, Shi Hua, Gong Zi-tong, Wei Qi-fan, A Preliminary study on the coastal acid sulphate soils in western Guangdong Province. Turang Tongbao, 1, 19-22(1958). (in Chinese)

(4) Moormann, F. R., Acid sulfate soils (cat-clays) of the tropics. Soil Sci., 95 (4), 271-275 (1963).

(5) Hesse, P.R., Phosphorus relationships in a mangrove-swamp mud with particular reference to aluminium toxicity. Plant Soil, 19 205-218 (1963).

(6) Sahrawat, K.L., Iron toxicity to rice in an acid sulfate soil as influenced by water regimes. Plant Soil, 51(1), 143-144 (1979).

(7) Tanaka, A., Navanero, S. A., Growth of the rice plant on acid sulfate soils. Soil Sci. Plant Nutr., 12 (3), 23-30(1966).

(8) Jordan, H. D., The utilization of saline mangrove soils for rice growing. Proc. Third Interafr. Soils Conf., 327-331(1959).

SOME CHEMICAL PROPERTIES OF SOILS FROM THE TAOYUAN PREFECTURE, CHINA, IN PARTICULAR THEIR FERTILITY

Hideo Okajima, Hiroki Imai
(Faculty of Agriculture, Hokkaido University, 060 Japan)

Taoyuan prefecture lies between North latitudes 28°24'26" and 29°24'13" and East longitudes 110°50'36" and 111°36'32". It has an area of 4,441 sq. km. The annual rainfall is 1,460 ml and the average temperature is 16.5°C. The maximum temperature is 40.6°C and the minimum is −15.8°C.

The proportion of mountains to rivers and crops is 7, 1, and 2 respectively. Arable land for crops is 90,000 ha, including 70,000 ha of lowland paddy fields. The main soil groups in the arable area are the purple soils, the red soils and the alluvial soils. Top soils from several locations were taken in October, 1979, to investigate their fertility. Sampling sites are plotted in Figure 1.

Sample No. 1 is of red soil for upland crops. Sample No. 2 −paddy soil− was taken from near the site of sample No. 1. Sample No. 3 is one of the typical alluvial paddy soils. Paddy yields in this location are the medium there. Sample No. 5 is a purple soil for upland crop; it covers extensive areas. Sample No. 6 −paddy soil− was taken from near the site of sample No. 5.

Sample No. 4 is alluvial paddy soil which was taken from the trial plot for maximizing paddy yield conducted by Mr. Li and experimental farm researchers. A heavy application practice of plant nutrients has been properly conducted there, resulting in high yields.

CHEMICAL PROPERTIES OF SOILS

The general and chemical properties of the soils are given in the Tables. Sample No. 1, red soil, is acidic with low nitrogen, low carbon, low CEC and low base saturation. Available phosphorus is extremely low.

Table 1 General properties of soils

Sample No.	pH H$_2$O	pH KCl	Soil texture	Clay minerals*	Remarks
1	5.2	4.2	HC	Ka>I>Q	upland (red soil)
2	6.3	5.7	HC	I>Ka>Q	paddy
3	6.3	5.2	HC	Q>I>Ka>14Å	paddy
4	7.4	6.9	HC	Q>I>Ka>14Å	paddy
5	7.8	6.9	LiC	I>Q>Ka>14Å	upland (purple soil)
6	6.5	5.7	LiC	Q>I>Ka>14Å	upland

* X-ray analysis
  (ka: kaolinite, I: illite, Q: Quartz, Å: 14 Å minerals)

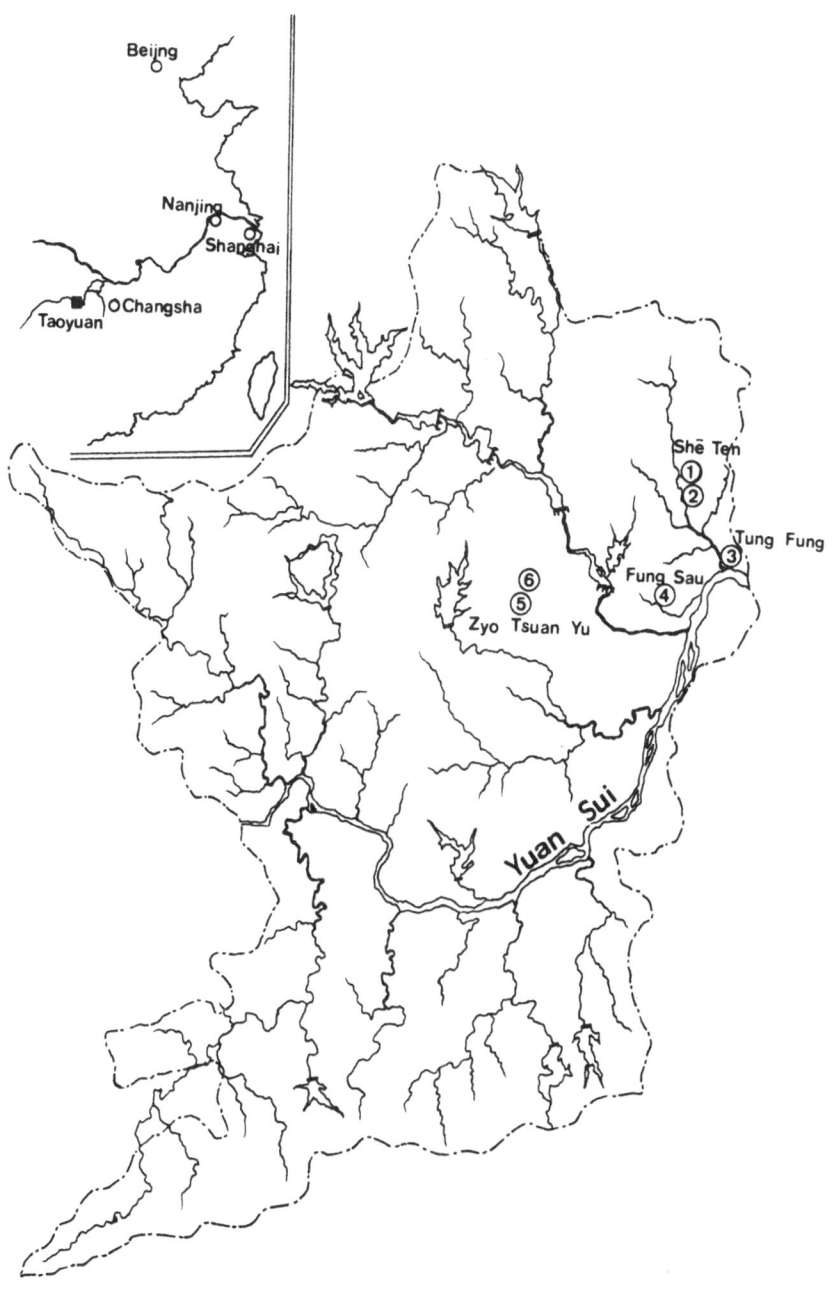

Fig. 1 Sampling Sites in Taoyuan

445

Table 2 Cation exchange capacities, exchangeable
bases and C, N contents

Sample No.	CEC (me/100g)	Exch. bases (me/100g)				Base saturation (%)	C (%)	N (%)
		Ca	Mg	K	Na			
1	8.3	0.76	0.42	0.13	0.08	17	0.59	0.12
2	10.9	8.54	0.82	0.20	0.17	89	2.15	0.25
3	11.6	7.98	1.46	0.23	0.14	85	1.71	0.21
4	13.8	18.40	2.27	0.47	0.15	154	2.41	0.27
5	17.2	31.30	0.52	0.26	0.09	187	0.83	0.14
6	9.2	7.77	0.50	0.20	0.17	95	1.63	0.20

Table 3 Available phosphorus and phosphorus
absorption coefficient

Sample No.	Available P(P_2O_5 mg/100g)			P Absorp. coef. (P_2O_5 mg/100g)
	Bray 1	Bray II	Truog	
1	0.34	0.22	0.44	641
2	0.90	2.58	1.94	747
3	1.44	7.81	2.14	739
4	5.86	31.0	10.9	868
5	0.90	15.7	9.19	1058
6	1.93	12.3	2.78	641

0.1 N HCl soluble elements are also low. Especially, potassium and
zinc are at nearly critical values for plant growth (Table 4).   X-ray dif-
fractions determined that kaolinite is predominate in clay minerals; this
corresponds to their low CEC. No. 1 soil may be really low in fertility.

Table 4 0.1N – HCl soluble elements

Sample No.	Ca	Mg	K	Na	Mn	Zn	Cu	Al
	←———me/100g ———→				← ——— mg/100g ———→			
1	0.32	0.49	0.10	0.03	1.3	0.12	0.37	87.3
2	7.66	1.06	0.19	0.10	9.9	0.53	1.05	98.7
3	7.83	2.07	0.21	0.11	13.5	0.67	1.00	94.5
4	28.2	7.06	0.60	0.17	35.6	1.18	0.64	47.1
5	43.4	1.20	0.16	0.07	9.7	0.15	0.11	19.8
6	7.31	0.22	0.22	0.16	16.8	0.51	0.69	61.8

Sample No. 2 –paddy soil– shows different characteristics, although
it was taken from near the site of sample No. 1.   There is an increase in
pH and exchangeable bases such as calcium, with a consequent increasing

of base saturation. There is also an increase in carbon and nitrogen. There is potassium and phosphorus increase to great extent. It may be said that soil fertility increased with rice cultivation. Kaolinite gives up its dominant place to illite.

Sample No. 3 -alluvial paddy soil-shows nearly the same character in chemical as No. 2; a quartz occupies the top position in clay minerals while 14 A mineral appears.

Sample No. 5, purple soil, derived from purple shales, shows a typical calcareous character with high pH and with free calcium carbonate. It is low in organic matter and nitrogen but high in available phosphorus. Truog-P is 9 mg per 100 g soil. The CEC is relatively high, reflecting a dominance of illite content and the existence of 14 A minerals. Exchangeable magnesium and potassium are relatively low, while zinc is at an especially critical content for plant growth (1 - 2) as shown in acid soluble matters (Table 4). Sample No. 6 shows nearly the same properties as No. 3, though it was taken from a site very near of No. 5.

All the elements measured in sample No. 4 are sufficient for plant growth. This may be a genuine example of a man-made fertile alluvial paddy soil. It contains free calcium carbonate.

Available irons for plants, measured by Morgan's method, are low in upland soils No. 1 and No. 5, while high in all the other paddy soils (Table 5). Free oxidized irons, however, are abundant in the upland soils. It appears that iron compounds are likely to be changed to less available forms in the upland locations due to their oxidative nature. Free oxidized iron contents of the paddy soils in Taoyuan are not particularly low, but they may, more or less, be degraded. It has been recognized, generally, that degraded soil is characterized by the eluviation of iron, manganese, phosphorus, potassium, magnesium, etc. from its surface to a lower horizon.

Table 5 Iron status and calcium carbonate contents

Sample No.	Available Fe, Morgan[*] (mg/100g)	Free oxidized Fe, Asami & Kumada(mg/100g)	Free $CaCO_3$ (%)
1	0.59	2100	–
2	23.7	1080	–
3	74.4	829	–
4	39.5	862	2.18
5	0.44	907	3.40
6	59.3	684	–

* pH4.8

Available phosphorus in all the paddy soils except No. 4 is low, raging from 1.94 to 2.78 mg per 100 g soil for Truog-P and 2.58 to 12.3 mg for Bray 11-P. These figures, however, do not always mean that the phosphorus response is high there. For instance, Miyake recently indicated that in high temperature conditions such as a tropical region, the rice plant is able to grow in a low available phosphorus (Bray II-P) when the capacity factor of the phosphorus supplying power is large enough (3).

447

CONCLUSION

In general, the parent materials of the paddy soils are derived from the surrounding elevated areas. Sample No. 2 was taken from the foot of an elevated upland field, red soils, while sample No. 6 was taken from the foot of elevated upland purple soils. It is hard, however, to find out their characteristic properties as reflecting their sampling sites or their parent materials.

Far from it: both the soils are very much like the typical alluvial paddy soil, No. 3. It can be said that at these sites a paddy soil has a tendency to converge to an uniform character. One of the reasons may be lengthy human activity in rice cultivation. Sample No. 4 may offer evidence of how large are the effects of rice cultivation on the chemical properties or the fertility of soil.

Soils investigated in Taoyuan may be amerable to improvement for crop production. The main limiting factors for plant growth may be insufficiency of some plant nutrients and probably a good response to fertilization may be expected in a proper management. There is a possibility, however, that some other factors such as related troubles with low CEC or the minor elements will affect the way in which fertilizer dosages are increased.

ACKNOWLEDGEMENT

The authors wish to express their gratitude to Dr. Li Ching-kwei, deputy director of the Institute of Soil Science, Academia Sinica, for his valuable suggestions on the selection of the soils investigated, to Mr. Shi San, deputy secretary-general, Academia Sinica for his efforts in arrangements for visiting Taoyuan prefecture,to Mr. Liu Shin-tei, chairman, Revolution Committee, Taoyuan prefecture, for his kind offices in the survey, and also to Dr. S. Tamura, professor emeritus, Tokyo University, Japan, team leader, visiting party to China. The field trip was guided by Mr. Kui Kuo-chao, chief, Agricultural Bureau, Taoyuan prefecture, whose co-operation is gratefully acknowledged.

REFERENCES

(1) Tanaka, A., Shimono, K., Ishizuka, Y., Zinc deficiency as the cause of the "Akagare", J. Sci. Soil and Manure, Japan, 40, 415-419 (1969)
(2) Yokoi, Y., Kikuchi, S., Studies on zinc deficiency in the corn plants in Tokachi district, Hoku-Nō, 44, 11-27 (1977)
(3) Miyake, M., On the phosphorus response in the paddy fields, Indonesia, Dr. Thesis, Hokkaido University, Japan, 1-169 (1980)

448

ON THE CHARACTERISTICS OF LEACHING AND ACCUMULATION OF CHEMICAL ELEMENTS IN TWO EUTRIC PADDY SOILS IN TAOYUAN COUNTY, NORTHERN HUNAN PROVINCE

Chen Zhi-cheng, Jiao Jian-ying, Zhao Wen-jun
(Institute of Soil Science, Academia Sinica, Nanjing)

The leaching and accumulation of chemical elements in the process of soil formation play an important part in revealing soil development and soil fertility. Based on the data obtained from the chemical determination of major and minor elements in two eutric paddy soil (No. s-48, s-76), and in comparison with the adjacent leached limestone soil (No. s-49) and leached purplish soil (No. s-47), a discussion is here made on the role of leaching and accumulation of the chemical elements in the formation of eutric paddy soils.

MATERIALS AND METHODS

Two eutric paddy soils were collected from Tao yuan County of Hunan Province and had been used for double cropping rice. Profile No. s-48 was derived from the marlite in the lower valley of the hills. This kind of soil was poorly drained belonging to eutric paddy soil of reducing type. The contrasting profile No. s-49, leached limestone soil, was derived from the same parent rock on the hillside. Soil No. s-76 was derived from calcareous purplish sandy shale in the upper valley of the hills under better drainage conditions belonging to redox eutric paddy soil. The contrasting profile No. s-47, leached purplish soil, was derived from the same parent rock. The clay content ($<1\mu$) and pH value of these soils are given in Table 1[1].

Table 1 pH value and clay content of soils

Group	Depth (cm)	Horizon	pH	Clay($<1\mu$) (%)
Paddy soil (No. s-48)	0-13	A	6.9	25.8
(from marlite)	13-22	Pg	7.2	23.2
	22-31	Bg	7.3	23.2
	31-45	G	7.4	25.0
Leached limestone	0-15	A	7.1	33.6
soil (No. s-49)	15-30	AB	6.9	42.6
	30-60	B	6.6	41.6
Paddy soil (No. s-76)	0-10	A	7.7	15 .4
(from purplish	10-20	P	8. 0	15.5
sandy shale)	20-35	Bg1	8.5	13.4
	35-80	Bg2	8.5	11.6
Leached purplish	0-10	A	6.0	19.6
soil (No. s-47)	10-25	AC	6.6	12.0

Micro elements, except Cr, were extracted by aqua regia-HClO$_4$ digestion. Samples for Cr analysis was digested by HNO$_3$-HF. Except Mo which was determined colorimetrically by sodium sulfocynate, atomic absorption spectrophotometry was used for all other microelements[2].

The rate of leaching loss during soil formation was calculated on the basis of parent rock, taking the content of TiO$_2$ as an index[3].

RESULTS AND DISCUSSION

The results obtained from a paddy soil (No. s-48) derived from the weathering product of marlite show that, besides Ca, Na, Mg, and P, which were severely leached, Mn, K, Si, Fe, and Al were leached in different degrees (Fig. 1). As compared with the leached limestone soil (No. s-49), the leaching of Ca and P was slightly reduced, but the leaching of Mn, Si, K, and Na was increased to some extent. The Fe and Al, which showed an apparent accumulation in the leached limestone soils, also showed some leaching loss from the paddy soil. Among the trace elements, Mo was leached heavily, Ni, Zn and Co were also leached to a certain degree, but, on the contrary, Cu and Cr were accumulated (Fig. 2). As compared with the profile No. s-49, the leaching loss of Mo, Ni, Zn and Co was increased, while the accumulation of Cu and Cr decreased considerably.

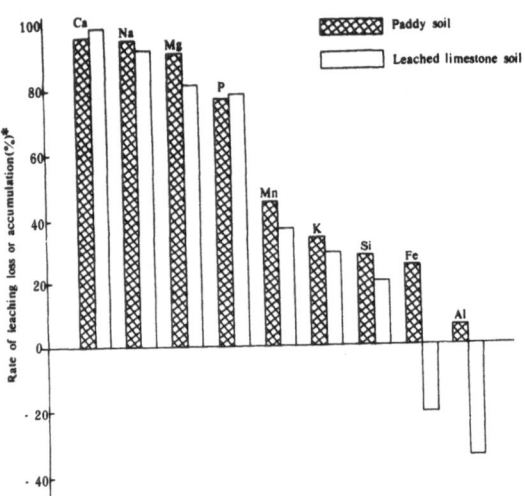

*Weighted average for the whole profile
Fig. 1 Comparison of leaching and accumulation of
major elements between paddy soil and
leached limestone soil

In the paddy soil (No. s-76) originated from weathering product of calcareous purplish sandy shale, besides the heavy leaching loss of Na, P and Mn, the leaching loss of Si, Al, K and Fe also occurred to some extent, but as for Ca and Mg, considerable amount was accumulated (Fig. 3). As compared with the leached purplish soil (No. s-47), except a considerable

accumulation of Ca and Mg, the leaching loss of P in the cultivated horizon and plowpan was decreased (Table 2).   But there was a significant

Table 2 Comparison of phosphorous content and its
leaching loss among different horizons

Profile No.	Depth (cm)	Horizon	P_2O_5 (%)	Leaching loss (%)
s-76	0-10	A	0.14	+37
	10-20	P	0.14	+39
	20-35	Bg1	0.10	+46
	35-80	Bg2	0.09	+61
s-47	0-10	A	0.08	+63
	10-25	AC	0.12	+21

increase in the leaching rate of Fe, K, Na, and also Al, with Si and Mn
to a less extent.   Among the trace elements, the leaching loss of Cu was
obvious, and that of Zn, Ni, and Cr was only to a certain extent.   As
for Co, a little amount was accumulated (Fig. 4).   As compared with
profile No. s-47, the leaching loss of Zn, Ni and Cr was increased, and
the accumulation of Co decreased to some extent, but an obvious increase
in leaching loss of Cu was observed.

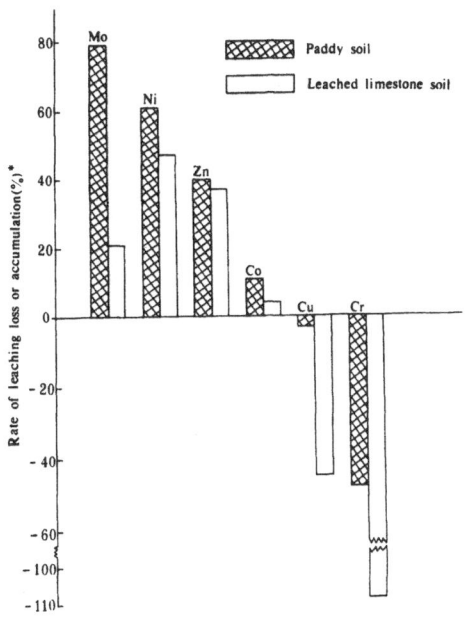

Fig. 2 Comparison of leaching
and accumulation of
microelements between
paddy soil and leached
limestone soil

*Weighted average for the whole profile

451

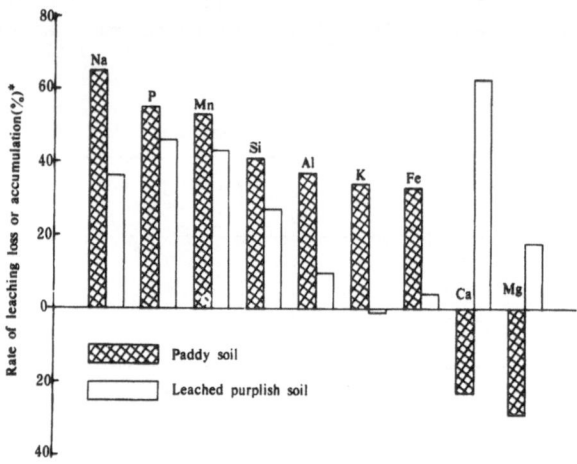

*Weighted average for the whole profile

Fig. 3 Comparison of leaching and accumulation of major elements between
paddy soil and leached purplish
soil

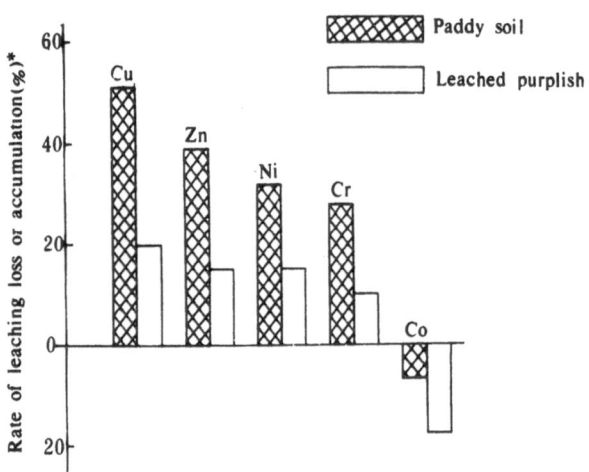

*Weighted average for the whole profile

Fig. 4 Comparison of leaching and accumulation of
micro elements between paddy soil and
leached purplish soil

According to the results mentioned above, when compared with upland
soils derived from the same parent rock, the characteristics of leaching
and accumulation of chemical elements in two eutric paddy soils are as
follows:

1) Some mobile or weakly mobile elements were decreased in leaching

as Ca and P in profile No. s-48, or accumulated to some extent, as Ca and Mg in profile No. s-76. This may be due to the application of fertilizers and irrigation or being brought in from seepage water[4].

2) Most of other elements showed an increase of leaching in different degrees. The increase in leaching of Na, Mg, K, Si, Fe, Mn and Mo may be caused by waterlogging-reduction. In consideration of pH of soils, the increase in leaching loss of Al, Cr, Ni, Co, Cu and Zn is unlikely to be waterlogging-reduction, but perhaps caused by the loss of clay through the differentiation of particles in the process of transportatior(5,6). Of course the loss of clay particles may in turn bring about the leaching loss of Na, Mg, K, Si, Fe, Mn and Mo.

To sum up, it seems that in eutric paddy soils the leaching loss of chemical elements through the solum is not great. Although the leaching loss of some elements are aggravated due to the loss of clay particles in the process of transportation and water logging-reduction, the supply of nutrient elements through fertilization and irrigation makes the leaching loss of some elements slackened.

REFERENCES

(1) Comprehensive Survey Group for Agricultural Modernization of Taoyuan County of Hunan Province, Academia Sinica, Soils and Their Appropriate Utilization, Selected Reports of Comprehensive Survey in Taoyuan County, Hunan Science & Technology Press, Changsha, 74-112 (1980). (in Chinese)
(2) Research Group of Microelements, Institute of Soil Science, Academia Sinica, Nanjing, Methods for Analysis of Trace Elements in Soils and Plants. Science Press, Beijing, 375-412 (1979). (in Chinese)
(3) Perelman A. I., Geochemistry of Epigenetic Process. Science Press, Beijing, 224-234 (1975). (in Chinese translation)
(4) Institute of Soil Science, Academia Sinica, Nanjing (ed.), Soils of China. Science Press, Beijing, 465-469 (1978). (in Chinese)
(5) Li Qing-kui, Cui Zheng (ed.), Proceedings of Symposium on the Research of Trace Elements. Science Press, Beijing, 86-156 (1964). (in Chinese)
(6) Brooks, R. R., The Distribution and Mobilization of Minor Elements in Soils, Geobotany and Biogeochemistry in Mineral Exploration. Harper & Row Publishers, 80-90 (1972).

EFFECTS OF NEOTECTONIC MOVEMENT ON DEVELOPMENT
OF PADDY SOILS IN CHANGJIANG DELTA

Lu Jing-gang
(Zhejiang Agricultural University, Hangzhou)

GENERAL DESCRIPTION

By the neotectonic movement we mean in the main geological tectonic movement during the Quaternary period. We have found that it affected the soil forming factors and soil developing processes quite markedly. We have already published some papers in this respect [1,2]. Based on that, this report further discusses the effects of neotectonic movement on the development of paddy soils.

According to characteristics of the neotectonic movement, Changjiang Dalta may be divided into eight regions as shown in Fig. 1[3,4]. Due to the differences of land forming age and the effects of neotectonic movement, the topography and parent material of soils are different, resulting in the distribution of different natural soils in this region. At present, however, most of the soils situated on the plains and hills in this area are in the paddy soil order as they have been under wet rice cultivation for many years. But the development and characteristics of these paddy soils were also affected by the neotectonic movement.

Moreover as this region is an old agricultural area, human activity—soil improvement, land leveling, cultivation and fertilizing often emphasize or conceal the effects of neotectonic movement. But all of these are superposed on top of the neotectonic movement which is very important as it markedly influenced the soil characteristics.

EFFECTS OF NEOTECTONIC MOVEMENT ON SOIL FORMING FACTORS

The land surface forming age of this region was closely associated with the neotectonic movement (Fig. 2). To the west and southwest of the Taihu Lake there is a strongly uplifted area (mountainous area), where three denudational surfaces of different height were formed in the Tertiary period. Owing to strong erosion only young soils could develop on these surfaces. They are usually kept as forest land. In this region most paddy soils belong to the surface water type (not influenced by ground water) and contain a relatively large amount of gravels.

The lower hill area in the periphery of the mountains was weakly uplifted or relative stable during the neotectonic movement. Here, a wide spread of paddy soils is distributed. In the southern part of this region paddy soils developed from red soils whose parent material was Quaternary red earth (Q_2). While in the northern part paddy soils developed from yellow-brown earth whose parent material was Quaternary Xiashu loess (Q_3)

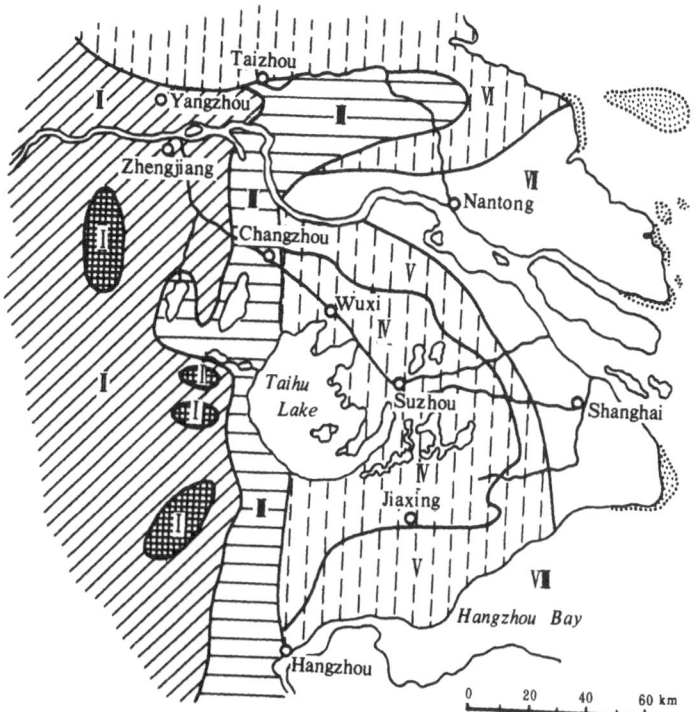

Fig. 1 Relations between the characteristics of neotectonic
movement and geomorphology and soil distribution in
the Changjiang Delta
I. Strongly uplifted mountain area;
II. Weakly uplifted hilly area;
III. Relative stable lower hill area;
IV. Subsided Taihu Lake depression;
V. Subsided external edge of Taihu Lake depression;
VI. Subsided northern coastal sandy spit area;
VII. Strongly subsided debouchment;
VIII. Strongly subsided Hangzhou Bay

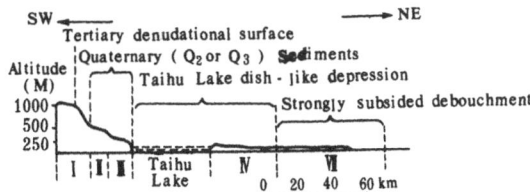

Fig. 2 Geomorphologic section in the Taihu
Lake area showing the relation between
the characteristics of neotectonic
movement and soil distribution
I.II.III.IV.VII: Neotectonic divisions
whose discription is similar to Fig. 1

455

Most of these paddy soils in the hilly area also belong to the surface water type. Only at the edge of the hills or the bottom of shallow valleys can bleached paddy soils be found.

The eastern part of Changjiang Delta is a vast plain of neotectonic subsidence where paddy soil type also changes with the soil developing age (Table 1). Immature and salty paddy soils spread in the coastal region of the plain. Farther inside from the coast is a plain netted by many rivers and canals. The paddy soils on this water abundant plain belong to periodical submergic, gley or bleached type. And the paddy soils of bog type lie in the center of dish-like depression.

The neotectonic movement also affected soil development by changing the vegetation or other factors. For instance, some acid sulfate soils can be found in the southeastern region of the Taihu Lake area, some of them have been improved for rice cropping. But these paddy soils are still strongly acidic and retain some original properties of the soil from which it developed.

In additon, there is no peat or muck distributed east of the line from Taicang to Caojing. This indicates that moor vegetation had not grown there. But peat or muck layer is often found in the soil profile west of that line. All of these are controlled by the extent of neotectonic movement and land forming age.

SOIL PROFILE OVERLAPPING AS A RESULT OF NEOTECTONIC MOVEMENT

Besides, owing to the effects of neotectonic movement, the overlapping of soil profile can usually be found. This is especially common in the vast eastern plain.

Soil profile overlapping can be divided into two types: the first may be called "parent materials overlapping" which can be traced to the overlapping of geological sediments prior to the development of the present soil; the other results from a new soil developing process with some diagnostic characteristics being superposed on an older soil profile. The latter may be called 'developmental overlapping' (Fig. 3). Both types of overlapping mentioned above could be caused by the neotectonic upward or downward movement and often appear in the same soil profile.

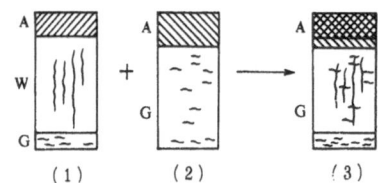

(1)   (2)   (3)

Fig. 3 A hypothetical sketch of
       soil profile development
       under overlapping
       (1)Developed under well-
          drained process;
       (2)Developed under poorly
          drained process with high
          water table;
       (3)Overlapped profile

Table 1 Developing conditions of paddy soils with different forming age from the sea coast to inland (from east to west)*

No.	Site	Soil forming age		Soil local name	Soil type	Soil profile	Calcic reaction	Brief account of soil development
		Approximate years	sequence					
No.1 Shanghai	Nicheng Commune, Nanhui County about 1 km from coast	< 800	1	loamy sand	Immature paddy soil	A(P)C	Strong in whole profile	With the initial differentiation of P horizon
No.2 Shanghai	Rice Seedproduction Farm of Huanglu Commune, Nanhui County	< 800	2	sandy loam	Immature paddy soil	APC	strong below the depth of 25 cm	With a distinct P horizon and some colloidal coating on the surface of soil structure in lower part of profile
No.3 Shanghai	Huinan Brigade, Nanhui Commune, Nanhui County	800-820	3	yellow clayey soil	Periodical submergic paddy soil	APW (B)	Strong below the depth of 60 cm	B horizon is slightly differentiated, weak columnar structure and some Fe-Mn concretions containing calcic precipitations in the subsoil
No.4 Shanghai	By the Office of Zhuangiuo Commune, Shanghai County	130-1600	4	ancient coastal upland soil	Periodical submergic paddy soil	APW (B)	No reaction above the depth of 100 cm	The differentiation of B horizon is still not obvious. But colloidal coating on the surface of soil structure may be found in all lower layers below P

457

No.	Location	Detail	Years	No.	Soil type	Paddy soil	Profile	Reaction	Description
No.5	Shanghai	Xujing Brigade, Xujing Commune, Qingpu County	5000–6000	5	yellow mottled soil	Periodical submergic paddy soil	APWBBG	React above the depth of 40cm which may have been from the recent sediments of debouchment	There is a distinct B horizon and columnar structure in the lower part of the profile. A large amount of rust streaks. rust spots are in both W and B horizon. Many bluish grey root pore space through out the whole profile even to the muck layer below the depth of 100cm
No.7	Shanghai	About 500 m south of Qingpu Agricultural Research Institute	> 6000	6	bluish grey clay	Gley paddy soil	AP(W)G	No reaction above the depth of 150cm	Differentiation of soil horizons is very distinct. Below the depth of 40cm is the grey layer. In the upper part of the gley layer there are some red rust spots, but the lower part of it belongs to the typical G Horizon
No.1	Suzhou	Wuxian Institute of Agricultural Sciences	> 6000	7	red rust spot clayey soil	Periodical submergic paddy soil	APWB	Same as above	Every horizon and the structures in it are all differentiated distinctly. Appreciable amount of Fe_2O_3 has mobed downward and precipitated in the subsoil

*(1) Signs used in soil profile: A–plough layer, P–plowpan, W–periodical submergic horizon.
    B–illuviated horizon, G–gley horizon

 (2) Paranthesis indicates that this particular horizon can not be seen distinctly

Comparatively speaking, the overlapping due to changing soil forming process is more difficult to identify than the first type in the field, yet it affects the characteristics and properties of the soils greatly. As both types of the soil profile overlapping may occur more than once, it can therefore lead to a very complicated profile.

In association with the overlapping of paddy soil profiles in this area, four basic kinds of morphologically distinct soil layers should be mentioned:

1) Yellowish soil layer containing many Fe-Mn concretions, rust spots and rust streaks which are produced in either neotectonic uplift relatively stable environment with good drainage.

2) Black soil layer containing muck or peat produced in the environment of moorland which was developed in formal dish-like depression.

3) Bleached soil layer, white in color with most of the Fe, Mn etc removed; product of a water bleaching process.

4) Greyish blue soil layer, a product of gleization due to the high watertable.

Each of the four soil layers mentioned above is relatively stable in color and characteristics. They indicate a specific environmental influence which the respective soils have undergone. But under the influence of neotectonic movement, the different soil layers often overlap in one soil profile. In this region paddy soil profile can be found with these four soil layers overstratumed one another without any order.

Sometimes certain anomalous features may overlap in one layer, forming some "irregular phenomena". For example, the black muck soil layer formed in the environment with poor drainage. But in this very layer columnar structure, a product of alternate wetting and drying process, may also be found. In the white soil layer developed under the bleaching process, there is often a large amount of Fe-Mn precipitation which is usually not associated with the bleaching process. Moreover, the yellowish mottling layer which was formed in an oxidizing environment, can be found under the ground water level where the soil environment is reductive.

All of these "irregular phenomena" result from soil profile overlapping. And the environment in which these soils developed has changed because of the tectonic upward or downward movement.

CONCLUSION

In short, neotectonic movement can affect soil formation, determine the type of soil developing process (salinization, meadow formation, moor formation, bleaching or zonal soil developing process) and the degree of soil development. Although those soils have become paddy soils after a long time of rice culture, but according to the morphological characteristics and properties, it can still be identified that they had been affected by the neotectonic movement. Therefore, the role of neotectonic movement should not be overlooked in the research of paddy soils of Changjiang Delta.

REFERENCES

(1) Lu Jing-gang, Relations between the neotectonic movement and the formation and distribution of red soils of hilly area in Zhejiang Province. Acta Pedologica Sinica, 13, 161-169 (1965). (in Chinese with Russian summary)
(2) Lu Jing-gang, The meanings of neotectonic movement on the formation and the classification of red soils of hilly area. Colloquium on Soil Geography and Soil Classification. Zhejiang Science & Technology Press, Hangzhou, 102-108 (1979). (in Chinese)
(3) Yu Zhe-ying, The neotectonic movement of Changjiang Delta (1959, ms.). (in Chinese)
(4) Chen Ji-yu, Yu Zhi-ying, Yun Cai-xing, Topographical development of Changjiang Delta. Acta Geographica Sinica, 25, 201-220 (1959). (in Chinese with Russian summary)

CHARACTERISTICS OF GEOGRAPHICAL DISTRIBUTION OF FERTILE PADDY SOIL IN FUJIAN PROVINCE

Zhu He-jian, Guo Cheng-da, Lin Zhen-sheng, Chen Zhen-gao
Tan Bing-hua, Li Quan-bao
(Fujian Teachers' University, Fuzhou)

Being located along the southeast coast of China, Fujian Province belongs to south and middle subtropical zone, mild in climate and abundant in rainfall. About 80% of the total area of the province are covered with mountains and hilly land. Its topographical features are complicated and varied, lower in the southeast and higher towards the northwest. It can be divided into three categories, namely, the northwest mountain area, the south hilly land and the coastal plain. The fertile paddy soils are distributed mainly over the coastal plain and the valley basins in the hilly area. The annual yield per ha of rice in the area under study was more than 12 tons for the past three years. According to the different level in agricultural management, an order can be listed as follows: coastal plain > south hilly land > northwest mountain area. Differences of hydrothermal conditions in these regions are shown in Table 1. The temperature of mountain area is lower than that of the coastal plain and hilly land, while the rainfall, the humidity and the temperature range are higher in the mountain area. However, both the temperature and the humidity of coastal plain and hilly land are similar, only with lesser rainfall in the coastal plain.

Table 1 Hydrothormal condition in different regions

Types	Represent-ed county	Anual average tempera-ture (^{o}C)	Annual tempera-ture range (^{o}C)	Annual precipita-tion (mm)	Annual relative humidity (%)
Northwest mountain area	Jianyang	18.2	21.4	1780.3	83
Coastal plain	Putian	20.2	17.3	1264.4	78
South hilly land	Xianyou	20.2	17.7	1422.1	78

Note: The items listed above are mean values of 10 years

The regional differences mentioned above can affect the properties and development of fertile paddy soils. Thus various geographical characteristics will be necessarily shown in different indices. A total of 45 soil samples collected from 109 high yield sites in tnese regions were analyzed. The results of our investigation are as follows.

REGIONAL DIFFERENCES IN ORGANIC MATTER AND NITROGEN

The regional differences of fertility of fertile paddy soils can be shown in its physical and chemical properties, especially in the indices of organic matter and nitrogen. The order of these indices is as follows: northwest mountain area > coastal plain > south hilly land (Table 2).

Table 2 Variation of organic matter and nitrogen in the fertile paddy soils of different regions

Types	Number of soil samples	Organic matter (%)	Total N (%)	Hydrolyzable N (ppm)
Northwest moun-tain area	13	3.2	0.20	120
Coastal plain	22	2.5	0.15	80
South hilly land	10	2.0	0.13	70

The organic matter and the nitrogen differ in various regions by reason of different hydrothermal conditions. In the mountain area, because of lower temperature and more rainfall, the absorptive power of rice root, the mineralization of organic matter, and the circulation of nitrogen are all lower than those in the coastal plain and the hilly land. Besides, the field management level is lower in this region. In this way, the contradiction is further sharpened between the nutrient requirement by rice and their supply by soil. Higher level of organic matter and nitrogen are required to meet the needs of higher rice yield thereby. In the coastal plain, the hydrothermal condition is, however, much better for rice growth, the absorptive power of rice root is relatively stronger and the mineralization of organic matter is quicker. Consequently, the indices of organic matter and nitrogen in such region may be lower than those in the mountain area. Moreover, geographical position is another factor influencing the fertility of fertile paddy soil. The fertile paddy soil of the hilly land and the coastal plain is situated in the upper-middle and down stream regions respectively. The nutrients lost from the former position are mostly deposited in the latter plain. Therefore, the content of organic matter and nitrogen in the coastal plain is higher than those in the hilly land.

REGIONAL DIFFERENCES OF HUMUS COMPOSITION

Humus composition may vary with different locations. In fertile paddy soils, the content of humic acid and the ratio of humic acid to fulvic acid are higher in the hill-plain than those in the mountain area (Table 3).

As shown in Fig. 1a, within the hill-plain region, numerical values of the humus components show an increasing tendency from the south to the north. The peak appears at Putian site where the maximum yield was obtained in the province. Again in this Fig. 1b, a comparison was made among three sites located at the same latitude of about 27°N, but in different distances from the sea. It also shows that the content of humic acid and the ratio of humic acid to fulvic acid tend upwards from the

462

east to the west. Apparently, humus composition of the soils varies greatly with the distance from the sea.

Table 3 Humus composition of fertile
paddy soils in different regions

| Types | % of total carbon | | Humic acid: Fulvic acid |
	Fulvic acid	Humic acid	
Hill-plain	28.9	8.4	0.30
Mountain area	28.7	12.5	0.43

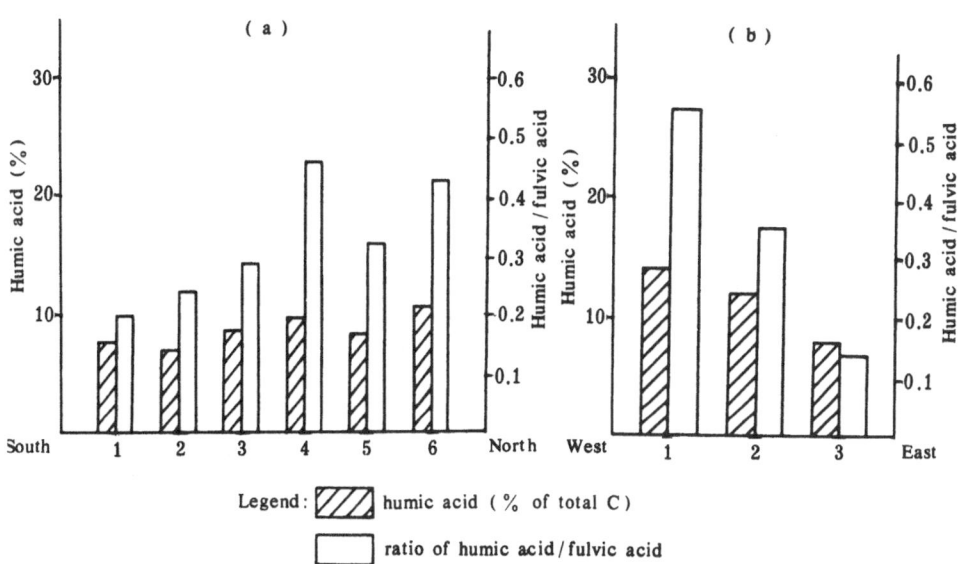

Legend: ⬚⬚⬚ humic acid (% of total C)

☐ ratio of humic acid/fulvic acid

Fig. 1 Variation of humic acid and humic acid/fulvic acid of the fertile paddy soils in different regions

Location (a)	Location (b)
1. Longhai (24°27'N)	The distance from the sea
2. Xiamen (24°27'N)	1. Jianning 275 km
3. Nanan (24°57'N)	2. Jianou 127 km
4. Putian (25°26'N)	3. Fuan 32 km
5. Changle (25°58'N)	
6. Fuzhou (26°04'N)	

REGIONAL DIFFERENCES OF SOIL PHYSICAL PROPERTIES

The fertile paddy soil of coastal plain derived from alluvial and marine deposit has more clay content. The content of microaggregates and the rate of structural maintenance are also high, though that of organic matter is usually of medium level only. In the mountain area, the fertile paddy soil contains more organic matter but less clay, so both the content of microaggregates and the rate of structural maintenance are less than those in coastal plain. In the hilly land, the clay as well as the organic matter is less in content. Therefore, the soil in this region are characterized by their lowest content of microaggregates (Table 4).

Table 4 Some physical properties of fertile paddy
soils in different regions

Types	Microaggregates (1-0.005mm) (%)	Rate of structural maintenance* (%)	Clay ($<$0.001mm) (%)
Northwest mountain area	26.2	37.4	17.5
Coastal plain	27.1	39.3	21.1
South hilly land	16.8	22.2	13.6

*Rate of structural maintenance

$$= \frac{(1\text{-}0.01\text{mm microaggregates}) - (1\text{-}0.01\text{mm mechanical particles})}{1\text{-}0.01\text{mm microaggregates}}$$

Since there exist obviously differences in fertility of fertile paddy soils, it is necessary to define different standards for various regions in accordance with local conditions. A standard for evaluating soil fertility in one area can not be used in the other. Besides, the measures for maintaining high soil fertility should be various according to different regions. In order to improve soil fertility and increase soil nutrient content, fertilization should be emphasized in hilly and plain fields, whereas in mountainous fields, the availability of nutrients from the soil potential fertility should be of the first consideration and water control is an important approach to it.

On the other hand, despite of the r gional difference in soil fertility, fertile paddy soils in all regionsunder investigation still share some common characteristics, namely, soil conditions favorable for the growth of rice including the balance of water, nutrient, air and heat elements in soil fertility. They are interpreted as follows.

1) Deep plowing, deep furrowing, reasonable water control and rotation between dry farming and waterlogging planting are effective practices in creating certain characteristics of soil profile and forming a soil solum favorable for rice growing and high grain yield, i.e., plowlayer thicker than 15 cm, most of old sole-layers being destroyed with new thin one formed, mottling layer (30-50 cm) well developed with clear mottle and moderate ground-water table, as shown in Table 5. A plot may get low

yield only because of high ground-water table with high ferrous content
even though its physical and chemical properties are as good as those
of high yield field. Thus, the gley horizon of high yield fields should
be below 60 cm.

Table 5 The characteristics of paddy field
of different yields

| Types | Organic matter (%) | Total N (%) | Total P₂O₅ (%) | Total K₂O (%) | Ferrous (ppm) | Mechanical composition (%) | | | Depth of ground-water table (cm) | Yield of rice (ton/ha) |
						> 0.02 (mm)	0.02-0.002 (mm)	< 0.002 (mm)		
Fertile paddy field	3.9	0.22	0.14	2.1	150	43.3	17.3	39.4	below 80	12
Low-yield paddy field	3.0	0.21	0.10	2.4	500	46.3	17.1	36.6	40-60	6

2) The appropriate physical properties of fertile paddy soil is
typified by water conserving capacity and good permeability in addition
to loose, soft plow layer suitable for cultivation. When irrigated to
form a water layer of 3 cm in depth, the soil will keep the water through
3-5 days, making the soil conditions always refreshed. In general, such
soil conditions deal with several factors especially the texture and the
structure of soil. Among the soil samples from sandy loam to clay under
analysis, 80% are from sandy clay loam to loamy clay. With regard to
microaggregates, the separate of 0.05-0.25 mm is in the great proportion
(Table 6) which can be regarded as a reference indication for the assessment.

Table 6 Microaggregates of cultivated
horizon of fertile paddy soils
(45 Samples)

| Aggregate size | Composition(%) | |
(mm)	Range	Mean
1 -0.25	10-39	22.6
0.25-0.05	17-45	31.5
0.05-0.01	15-37	24.8
0.01-0.005	5-15	8.3
0.005-0.001	3-16	9.9
< 0.001	1-6	2.9

3) Under certain level of cultivation and management, a particular
requirement of plant nutrients is in line with a high yield, therefore,
a nutrient index should be recommended for fertile paddy soil. Although
it may vary with different cultivation and management, there still exists
a certain criterion for high yield soil. Of course, a high yield may be
attained in soil of low fertility through better management and higher
fertilization, though unsteadily. On the other hand, soil of high fertility may bring about a low yield through poor management and unreasonable
fertilization.

MICROMORPHOLOGICAL FEATURES AS THE INDICATION OF THE FERTILITY OF PADDY SOILS

Cao Sheng-geng, Jin Guang
(Institute of Soil Science, Academia Sinica, Nanjing)

There are numerous types of paddy soils which occur extensively in large areas in China. Each of them has its specific characteristics of soil fertility. The Chinese peasants through long practice of production have well recognized the fertility of paddy soils in the field, and many Chinese soil scientists have done a series of researches for characterizing the various types of the paddy soils[1,2]. Present paper reports that micromorphological features can be used as additional criteria for the judgement of soil fertility. About 180 thin sections of paddy soils of different fertility were examined using a polarizing microscope, and the relationship between the micromorphological features and the characteristics of soil fertility was studied.

FERRIC CONCENTRATIONS

The diffusion organo-ferrans, the diffusion ferrans and the ferric rings around plant roots are three main factors which show specific significant relationship to the fertility of paddy soils.

The diffusion organo-ferrans are only found on the walls of vughs and in the adjacent soil matrix in the cultivated horizon of highly fertile paddy soils[3,4](Fig. 1). They are composed of the organo-ferrans and the diffusion spots of organo-ferric substances. The organo-ferrans are 10-50μ in thickness, and very dark reddish brown (2.5YR 2/2, 2/3, 2/4) to brownish black (5YR 2/2) in color, while the diffusion spots are reddish brown (2.5YR 4/8) to bright reddish brown (2.5YR 5/8) in color. Sometimes, crystallized fibrous aggregates appear on the surface of organo-ferrans. Judging from the optical characteristics, they may be a kind of goethites.

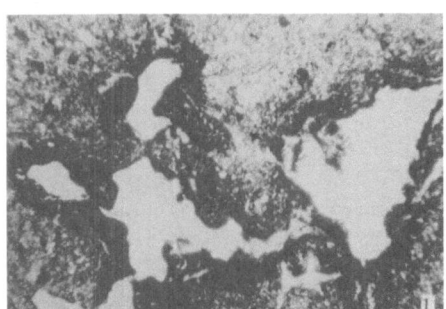

Fig. 1 Diffusion organo-ferrans.
The white areas are the
vughs. Plain light, x 25

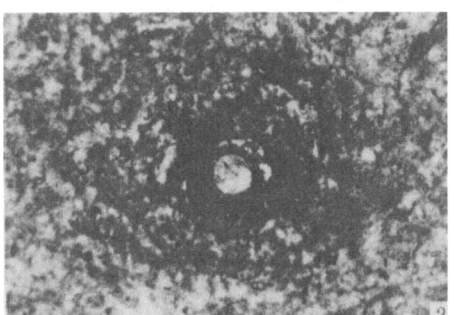

Fig. 2 A ferric ring around plant
root with thicker halo.
Plain light, x 70

The diffusion ferrans, which are corresponding to rust spots in macromorphological feature, are reddish brown to brown in color with a hue ranging from 5YR to 10YR. The reddish brown diffusion ferrans with a hue 5YR indicate a more fertile paddy soil,while paddy soils bearing brown diffusion ferrans with hue 7.5YR-10YR are usually less fertile. When the thin sections are observed under microscope the reddish brown diffusion ferrans are similar to the diffusion spots of organo-ferric substances of the diffusion organo-ferrans, but on closer examination it will be seen that the color of the diffusion spots tends to be more reddish and the hue usually falls on 2.5YR. In addition, the amount and concentration of diffusion ferrans also bear a positive relationship with soil fertility.

Besides the coloration, the thickness of diffusion halos on the ferric rings may also serve as an indication for soil fertility. Fertile soils appear thicker diffusion halos (Fig. 2).

VOIDS (PORES)

Voids in different types of paddy soil differ in shapes, sizes and amounts. Few root pores (less than 1%) and fine fissures are found in the cultivated horizon of low fertile paddy soils. About 2% of root pores in the area of thin section of the cultivated horizon exist in medium fertile paddy soils. In highly fertile paddy soils, not only more root pores (up to 4-5%), but also some larger root vughs and chambers can be found.

The porosity of cultivated horizons is increased with increasing soil fertility. For example, the porosity of a less fertile paddy soil derived from red earth is 7.5%, the medium fertile paddy soil is 10%, and the highly fertile one is 19.9%.

PARTICLE SIZE

Paddy soils, either excess of or lack of skeleton grains, are usually low in fertility and poor in tillage. For excess of sandy particles and lack of soil matrix, and a close arrangement of skeleton grains, the soils are easy to settle and compact, which are usually characteristic of paddy soils derived from granitic materials. The soils with large amount of silty skeleton grains are also apt to settle and compact (Fig. 3). On the contrary, a heavy clayey paddy soil is lack of skeleton grains and contains soil particles less than 0.01mm in diameter, which is also difficult to till.

FABRIC OF SOIL MATRIX

In the submerged paddy field, the cultivated horizon of the soil is usually structureless. However, there exists a coagulating matrix in fertile paddy soils (Fig. 4). A distinctive appearance of such matrix may be regarded as an indication of fertile paddy soils.

OPTICALLY ORIENTED CLAYS

Paddy soils are usually cultivated and levelled under flooded condition before transplantation of rice seedlings. However, in some paddy soils, such as the purplish paddy soils of heavy clayey type, poorly drained paddy soils in marsh areas, the fine particles remain suspended as paste over a rather long time after tillage. Consequently, the rice seedlings cannot stand well in such a pasty surface. Under crossed nicols, large amount of striated oreintation of clays can be found (Fig. 5)

CARBONATE CONCENTRATIONS

The presence of carbonate nodules and particularly large amount of cryptocrystalline carbonate interflorescences in the matrix of paddy soils derived from non-calcic materials makes the micromorphological features in low productive anthropogenic calcareous paddy soil (Fig. 6). They are induced by long-term of over liming[5].

If a thin layer of cryptocrystalline carbonate concentrations beneath the cultivated horizon is present, and the calcitans on the walls of pores and fissures are formed, the paddy soil should be a strong carbonatized one.

SULFATE CONCENTRATIONS (JAROSITES)

They are found in the low productive acid sulfated paddy soils on the coastland of south China. The soil fertility is closely related to the depth of buried mangrove layer in soil profile[6]. As a result, the various forms and amounts of jarosites are formed according to the soil development. Their color and form are similar to the ferric concentrations as observed under plain light (Fig. 7). But jarosites can be recognized by tneir lower value of coloration ($<$4) under plain light and by their light gray color (5Y 8/2) or pale yellow color (5Y 8/3) under reflected light. In addition, they are yellow (5Y8/6, 5Y8/7) in color as observed under visual condition.

PEDORELICTS (REMAINS OF PARENT MATERIAL)

In the soil matrix of cultivated horizon of incipiently developed infertile paddy soils derived from the red earths, the "raw soil" blocks originated from red earths can be still observed (Fig. 8). Usually they are 0.05-0.3mm in diameter with subangular shape and distinctive boundary. As further development of the paddy soil goes on, "raw soil" blocks become rounded and diffused, and finally disappeared.

The micromorphological features of paddy soils as indication of soil fertility varies with soil types. For example, the high productive paddy soils derived from the lacustrine deposit of Taihu Lake area are characterized by the diffusion organo-ferrans, the reddish brown diffusion ferrans, the reddish brown ferric rings. However, these formations can not be always observed in highly anthropogenic mellowed paddy soils derived from the gray alluvial material of Zhujiang Delta, where vughs and coagulating matrix characterize the micromorphological features of fertile soils. Present paper illustrates that micromorphological study of soils may serve as one of the criteria for the judgement of the fertility of paddy soils.

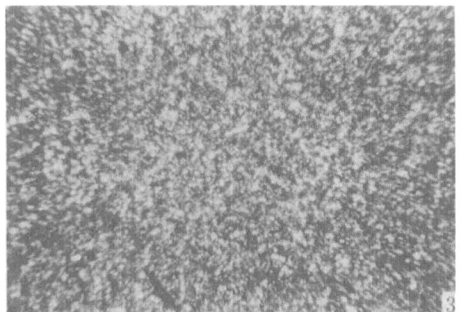

Fig. 3 Excess of silty skeleton
grains. Plain light,
x 25

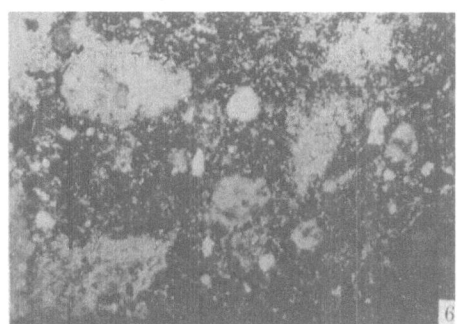

Fig. 4 Coagulating soil matrix.
The white areas are
quartzs and fine
fissures. Plain light,
x 80

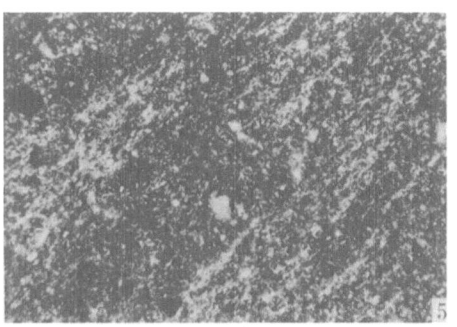

Fig. 5 Striated orientation
of clays in a heavy
clayey purplish
paddy soil.
Crossed nicols, x 80

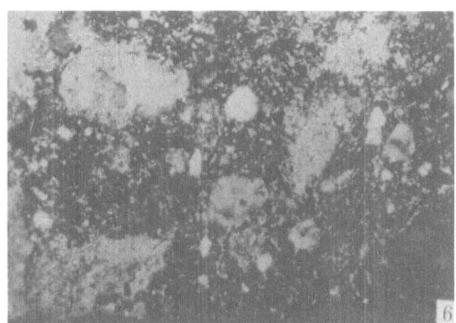

Fig. 6 Carbonate nodules and
interflorescences in
an anthropogenic
calcareous paddy soil.
Crossed nicols, x 80

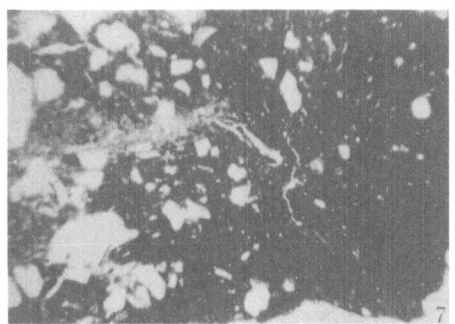

Fig. 7 Jarosite spots and
nodule (lower
right). Flain
light, X 25.

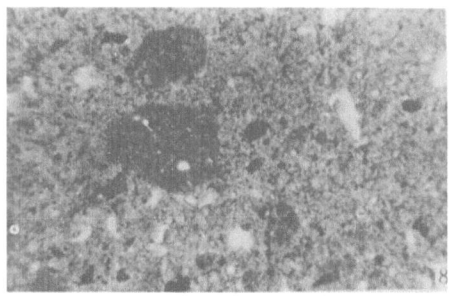

Fig. 8 "Raw soil" blocks-
pedorelicts. Plain
light, X 25

REFERENCES

(1) Cao Sheng-geng, Yao Yu-cheng, Nomenclature of genetic horizons of paddy soils and their characteristics. Turang Zhuanbao, 36, 179-205 (1964). (in Chimese)

(2) Institute of Soil Science, Academia Sinica, The Environmental Conditions of Soils for High Yield of Rice. Science Press, Beijing (1961). (in Chinese)

(3) Cao Sheng-geng, The peculiarity of soil formation of the paddy soils derived from red earth in Jiangxi, China. Acta Pedologica Sinica, 12(2), 155-163(1964). (in Chinese with English summary)

(4) Gu Xin-yun, Li Shu-qiu, On the characteristics of Fe-organic coating in paddy soils. In this Proceedings.

(5) Shi Hua, Hou Chuang-qing, Formatiom and amelioration of paddy soils with calic horizon in Guangxi Province. Turang Tongbao, 3, 5-10 (1961). (in Chinese)

(6) Gong Zi-tong, Chen Zhi-cheng, The soils of the Zhujiang Delta. Turang Zhuanbao, 36, 69-129(1964). (in Chinese)

DIFFERENTIATION OF IRON OXIDE IN PADDY SOILS IN TAIHU LAKE REGION

Xu Zu-yi, Chen Jia-fang
(Institute of Soil Science, Academia Sinica, Nanjing)

In the genesis of paddy soils the differentiation of iron oxid is one of the main causes for forming specific morphological profiles. In this paper, based on the contents of total iron, free iron oxide and amorphous iron, we shall discuss the differentiation, degree of freeness and activity of iron in some types of paddy soils in the Taihu Lake region.

Seven profiles with 36 samples were used for the study. The topography where the soil distributed, the parent material and the horizon sequence of the profiles are given in Table 1. Where

A   plowed layer
P   plow pan
L   bleached horizon
W   percogenic and submergenic horizon
B_g B horizon influenced by gleyization
C   parent material

Table 1 Some information of the soil used for study

Profile No.	Location	Topography	Parent material	Horizon sequence					
1	Jiang ning	Rolling land	Xia shu Loess	A	P	W	B_g	C	
2	Li yang	Rolling land	Xia shu Loess	A	P	L	B_g	C	
3	Jin shan	Plain	Loess like	A	P	L	W	C_1	C_2
4	Wu xi	Plain	Loess like	A	P	L	W	C	
5	Kun shan	Plain	Lacustrine	A	P	W	B_g	C	
6	Ma qiao	Plain	Alluvium form Changjiang River	A	P	W	B_g	C	
7	Jia xing	Plain	Alluvium from Qian tang River	A	P	W	B_g	C	

Total iron of the soil was digested with mixed acids of $HClO_4$ and HF. Free iron oxides and amorphous iron were extracted with dithionitecitrate-bicarbonate solution(1) and Tamm's solution respectively. Then the iron in solution was determined by colorimetric method. The percentage of free iron oxide (Fe_d) with respect to total iron (Fe_t) is termed degree of freeness of iron oxide, and is used as an index of weathering strength. The term of activity of iron oxide is designated as the percentage of amorphous iron with respect to free iron oxide (Fe_o/Fe_d)(2).

The w horizon, as a consequence of the differentiation of iron and

manganese compounds of the soil, is an important genetic horizon of typical
paddy profile. Now, based on the variation of iron content of different
horizons in the profile, we try to discuss the relationship between the
differentiation of iron and the formation of this horizon. For this, we
group together the genetic horizons with similar content of total iron or
free iron oxide in the same profile, and calculate the mean values and
standard deviation. In Tables 2 and 3 shown that eluviation and illuviation
of iron are obvious in profiles 4 and 2. Profile 3 shows the same trend.
the differentiation of iron among different horizons in profiles 1,5 and 6
is not obvious. For these three profiles the coefficients of variation

Table 2 Differentiation of total iron in the paddy profile

Profile No.	Horizon	M ± S	Coefficient of Variation
1	A P W Bg C	5.08±0.08 5.46	0.02
2	P L A Bg C	3.62±0.26 4.48±0.42 5.80	0.07 0.09
3	A P L W C_1 C_2	4.81±0.18 6.48±0.54	0.04 0.08
4	L A P C W	3.70 5.15±0.54 7.23	0.10
5	A P W Bg C	4.58±0.07	0.02
6	A P W Bg C	5.15±0.22 6.21	0.04
7	A P W Bg C P	4.72±0.11 5.21	0.02

of total iron content among various horizons excluding the C horizon are
all smaller than 0.04. However, their W horizons are distinct morphologically.
It is supposed that the formation of W horizons in these profiles is due to
local differentiation, in which the iron diffuses or transports from the
interior of structural units to their exterior or the surface of pore space,
and then forms rusty spots or streaks there under intermittent oxidation-
reduction conditions, thus showing the peculiar morphological features of W
horizon. This mode of W-horizon formation is different from that of eluviation-illuviation process, as is the case in profiles 4 and 2.

The degree of freeness of iron oxide does not show significant variation
among different horizons in same profile for profiles 1,2,3, with a coefficient of variation less than 0.06, while for other profiles there are some
variations (Table 4). The degree of freeness in highest in C horizon for
profiles 1,3,6, in Bg horizon for profiles 2,5, and in W horizon for profiles
4,7, while the value is lowest in A horizon for profiles 1,2, in P horizon for
profiles 3,5,7,in L horizon for profile 4, and in Bg horizon for profile 6.

Therefore, it seems that the degree of freeness of iron does not reflect the characteristics of genetic horizons in a given region, although it may be affected by the parent material and topography to a certain extent, as can be seen from Table 4.

Table 3 Differentiation of free iron oxide in the paddy profile

Profile no.	Horizon	$N \pm S$	Coefficient of variation
1	A P W Bg C	2.25 ± 0.10 2.60	0.04
2	P L A Bg C	2.10 ± 0.12 2.69 ± 0.16 3.12	0.06 0.06
3	A P L W C_1 C_2	2.29 ± 0.22 3.10 ± 0.08	0.10 0.02
4	L A P C W	1.12 2.11 ± 0.12 3.56	0.06
5	A P W C Bg	1.12 ± 0.13 1.47	0.12
6	A P W Bg C	1.30 ± 0.06 1.61	0.05
7	A P Bg C W	1.32 ± 0.06 1.61	0.04

Table 4 Degree of freeness of iron oxide ($Fe_d/Fe_t \cdot 100$)

Profile no.	Range	$M \pm S$	Coefficient of variation
1	42.5–47.6	44.9 ± 2.3	0.05
2	53.5–58.6	56.3 ± 2.4	0.04
3	44.4–52.4	47.9 ± 3.0	0.06
4	30.3–49.2	40.6 ± 7.1	0.17
5	22.1–31.5	26.4 ± 4.0	0.15
6	22.7–41.9	28.5 ± 7.6	0.27
7	24.6–34.4	28.6 ± 3.9	0.14

The activity of iron oxide varies markedly among different horizons of the same profile, with coefficient of variation of 0.46-1.08. The value for profile 1, although lower than for other profiles is still as high as 0.24. It has been observed from Fig. 1 that the activity of iron oxide is correlated significantly with the organic matter content of the soil, with a correlation coefficient of 0.822 (n=29, $p < 0.001$), except for the plowed layer in which the activity of iron oxide may perhaps be influenced by frequent alternation of activation and ageing. This seems to imply that organic matter plays an important role in the activation of iron oxide in paddy soils under condition of alternating of irrigation and drainage. This mechanism is quite different from the process of podzolization in which the activity of iron oxide is highest in B horizon(2).

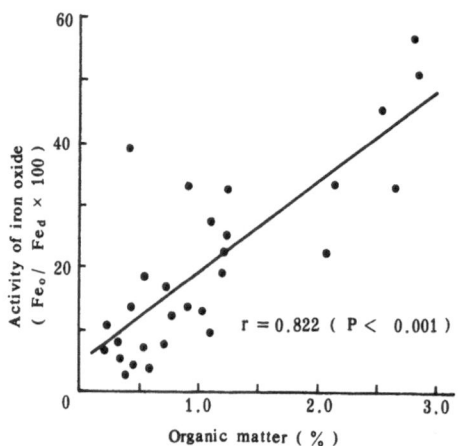

Fig. 1 Effect of the organic matter content of the soil on activity of iron oxide

REFERENCES

(1) Mehra, O.P., Jackson, M.L., Iron oxide removed from soils and clays by dithionite citrate system buffered with sodium bicarbonate. Clays & Clay Mineral, 7, 319-329(1960).
(2) Schwertmann, U., The differentiation of iron oxide in soil by extraction with ammonium oxalate solution. Z. Pflanzenernahr. Dung. Bodenkd., 105, 194-202(1964). (in German)

CHANGES IN CLAY MINERALS IN THE GENESIS OF PADDY SOILS

Zhang Xiao-nian
(Institute of Soil Science, Academia Sinica, Nanjing)

Paddy soil is derived from its parent soil after a long period of submergence and cultivation. What is the relationship between the caly minerals of paddy soil and that of its parent soil? Are there any changes in the clay minerals of the gley horizon due to intensive reduction and leaching? In this paper, based on results of identification of clay minerals of clay fraction ($<2\,\mu$) of some soil samples, we shall try to have a preliminary discussion no these questions.

CHANGES IN CLAY MINERALS IN THE GENESIS OF PADDY SOILS DERIVED FROM DIFFERENT PARENT SOILS

The clay minerals of a paddy soil come primarily from its parent soil, but it is also possible that some changes may take place during the genesis of the paddy soil. We have distinguished three cases in this respect.

1) For parent soils abundant in K-containing minerals, the changes in clay minerals during the genesis of paddy soils are comparatively remarkable due to the process of depotassication. Taking a paddy soil derived from purplish soil as an example, the content of illite is lower, and that of vermiculite and kaolinite higher than its parent soil, although illite is still the dominant clay mineral as in the case of the purplish soil. Chemical analysis shows that the K_2O content of the clay fraction of the paddy soil decreases by 1.5%, and the CEC increases by about 5 m.e./100g as compared with that of the preceding purplish soil (Table 1).

Table 1 Some properties of the clay fraction of a paddy soil and its preceding purplish soil (Jiangxi)

Sample	CEC (m.e./100g)	K_2O (%)	Combined water (%)	Clay minerals
Surface horizon of paddy soil	18.8	3.74	9.9	ill. decrease, kl. & verm. increase
Substratum of paddy soil	18.3	3.64	9.2	
Preceding purplish soil	13.2	5.25	7.3	ill. dominant, with some kl. & verm.

2) For parent soils with medium amount of K-containing minerals, there is a slight change in clay minerals during the genesis of paddy soil. Table 2 shows that for a paddy soil derived from loess the content of K_2O is 0.5%

lower and the CEC 2-3 m.e. higher in the clay fraction than that of the
parent soil. The dominant clay minerals are illite and montmorillonite
with a certain amount of vermiculite and kaolinite in both cases, although
the content of illite is slightly lower in the paddy soil.

Table 2 Some properties of the clay fraction of paddy
soil and its preceding loessial soil (Nanjing)

Sample	CEC (m.e./100g)	K_2O (%)	Combined water (%)	Clay minerals
Surface horizon of paddy soil	42.7	2.67	9.2	ill. slightly lower
Substratum of paddy soil	41.0	2.54	10.2	
Preceding loessial soil	39.3	3.06	10.6	ill. & mt. dominant, with some kl. & verm.

3) For highly weathered laterite with very low content of K-containing
minerals, there is little change in clay minerals during the genesis of the
paddy soil, except the strong leaching of free iron oxides. The dominant
clay minerals are kaolinite in both cases, with some gibbsite and iron
oxides. The CEC is slightly higher for the paddy soil, whereas the SiO_2/Al_2O_3 ratio remains practically unchanged(Table 3).

Table 3 Chemical properties of the clay fraction of a
paddy soil and its preceding laterite(Guangdong)

Sample	CEC (m.e./100g)	SiO_2 (%)	Al_2O_3 (%)	Fe_2O_3 (%)	K_2O (%)	$\dfrac{SiO_2}{Al_2O_3}$	$\dfrac{SiO_2}{Fe_2O_3}$
Paddy soil (subsurface)	9.6	32.94	35.60	11.83	0.29	1.57	7.43
Parent soil	5.2	30.30	33.29	18.14	0.10	1.55	4.46

EFFECT OF GLEYIZATION ON CLAY MINERALS

Two pairs of the cultivated horizon and its corresponding gley horizon
of paddy soils derived from Quaternary red clay were identified for their
clayminerals. No remarkable difference could be found from their differential
thermal curves and X-ray data. All are predominated by ill-crystallized
kaolinite and illite. However, chemical analyses of the clay fraction show
a slightly stronger depotassication and slightly higher CEC for the cultivated horizon (Table 4). Another paddy soil derived from laterite shows a
higher SiO_2/Fe_2O_3 ratio in the clay fraction of the gley horizon due to strong
decrease in Fe_2O_3 content as compared with the cultivated and substratum

476

horizon. The CEC is also lower for the gley horizon. However, K_2O content and SiO_2/Al_2O_3 ratio are about the same (Table 5). X-ray diffraction pattern shows a not very distinct peak of montmorillonite for the clay fraction of the cultivated horizon, while the gley horizon does not. It is not clear whether the difference is due to the destruction of montmorillonite originally existing in the gley horizon or due to the admixture in the cultivated horizon from other sources.

Table 4   Chemical properties of the clay fraction of cultivated and gley horizons of two paddy soils derived from Quaternary red clay

Locality	Horizon	CEC (m.e./100g)	K_2O (%)	Combined water (%)
Zhejiang	Cultivated	25.6	1.95	9.7
	Gley	21.6	2.51	8.9
Hunan	Cultivated	20.9	2.57	10.7
	Gley	18.5	2.74	10.1

Table 5   Chemical properties of the clay fraction of a paddy soil derived from laterite

Horizon	CEC (m.e./100g)	SiO_2 (%)	Al_2O_3 (%)	Fe_2O_3 (%)	K_2O (%)	$\dfrac{SiO_2}{Al_2O_3}$	$\dfrac{SiO_2}{Fe_2O_3}$
Cultivated	24.6	35.74	27.99	16.14	0.17	2.17	5.9
Substratum	20.4	36.28	28.20	18.77	0.15	2.19	5.2
Gley	17.9	40.96	32.99	9.26	0.17	2.11	11.8

In the lower part of some red soils there is frequently a plinthitic horizon intermingled with red and white zones. It is generally considered that the formation of this horizon is also caused by gleyization. We have collected three samples of this horizon from South China, and identified for the mineral composition of the clay fraction of the red and white zones separately. X-ray diffraction pattern (Fig. 1) and differential thermal curves (Fig. 2) did not show obvious difference between the two fractions in the same horizon. The dominant clay minerals are kaolinite in the clay fraction of plinthitic horizon derived from marine deposit (Guangdong). Illite and kaolinite are the dominant clay minerals of plinthitic horizon derived from metamorphic rock (Guangdong). Clay mineral composition of plinthitic horizon derived from Quaternary red clay is predominated by ill-crystallized kaolinite and illite (Jiangxi). Chemical analyses (Table 6) show that the Fe_2O_3 content of the white zone is less than half of the red zone, resulting in a higher SiO_2/Fe_2O_3 ratio. There is no obvious difference in SiO_2/Al_2O_3 ratio. The CEC of the clay

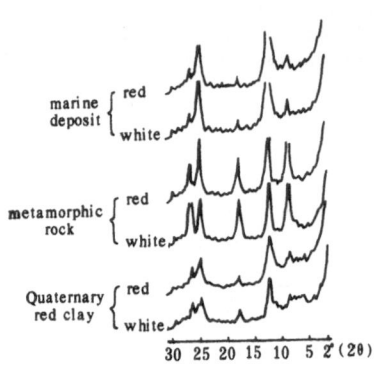

Fig. 1 X-ray diffraction
pattern of clay fraction of
red and white fraction of
three plinthites derived from
different parent materials

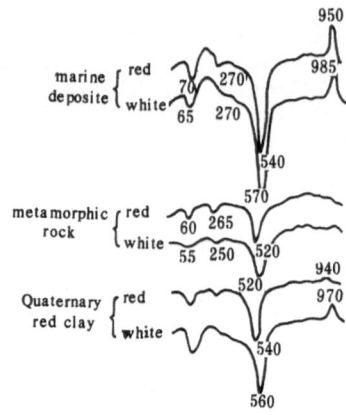

Fig. 2 Differential thermal
curves of clay fraction of red
and while white frations of three
plinthites derived from different
parent materials

Table 6 Chemical properties of the clay fraction of red and
wite zones of plinthites

Locality	Parent material	Zone	CEC (m.e./100g)	SiO_2 (%)	Al_2O_3 (%)	Fe_2O_3 %)	Combined (%)	$\dfrac{SiO_2}{Al_2O_3}$	$\dfrac{SiO_2}{Fe_2O_3}$
Guang-dong	Marine deposit	Red	15.8	45.24	32.20	9.17	12.2	2.38	13.2
		White	17.6	46.95	35.04	3.92	12.1	2.27	31.9
Guang-dong	Metamorphic rock	Red	9.8	40.56	29.38	14.42	8.2	2.34	7.5
		White	13.0	46.77	32.25	4.77	7.8	2.46	25.9
Jiangxi	Quaternary red clay	Red	18.1	39.69	28.17	19.77	–	2.39	5.3
		White	26.7	47.86	33.41	4.06	–	2.42	31.8

fraction is higher in the white, zone, especially for those derived from
Quaternary red caly. The difference in CEC for plinthites derived from marine
deposit and metamorphic rock is relatively small.

From the above it follows that:

(A) In the process of gleyization a large part of free iron oxides in
the clay fraction is reduced and leached, with the resulting formation of a
grayish-blue or gray gley horizon. In some lower horizons of red soils the
free iron oxides may also be segregated into a plinthitic horizon intermingled
with red and white zones by gleyization.

(B) The layer silicate minerals of the clay fraction may also undergo some alterations during intensive gleyization and leaching, but the extent is much smaller than the change in iron oxides.

(C) Morphologically, the change in iron oxides manifests itself very markedly while the alterations of crystalline silicate minerals can only be distinguished by chemical means. It is frequently not possible to identify the changes in mineral composition caused by gleyization by the conventional physical methods, if the resolution power of the instrument is not very high.

CLAY MINERALS OF PADDY SOILS IN TAIHU LAKE REGION

Xu Ji-quan, Yang De-yong, Jiang Mei-yin
(Institute of Soil Science, Academia Sinica, Nanjing)

The Taihu Lake is situated in the northern part of subtropics, with a mild and humid climate. According to the results of primary mineral analysis(1), the minerals in the silt fractions are less weathered within the soil profiles of this region. Moreover, due to the long period of rice culture, these soils suffered deep artificial influence. In the present study the relationship between the mineral composition of clay fraction in paddysoils and the type of geomorphology and parent material was examined, and some aspects about the nature of the bleached horizon were also discussed.

RELATIONSHIP BETWEEN THE CLAY MINERAL COMPOSITION AND THE TYPE OF GEOMORPHOLOGY AND PARENT MATERIAL

Paddy soils of different geomorphological unit in this region are developed on different types of parent materials. This can be distinguished by the composition and some properties of minerals in the clay fraction.

1. Western Hilly Part

The parent materials encountered in this part are mostly Xiashu loess, but weathering products from other rocks such as tuff and limestone can also be seen. The clay content of Xiashu loess always exceeds 30%. Uuder the action of weathering and leaching after sedimentation, Xiashu loess varies in their properties from north to south. Nearby Nanjing it has higher degree of base saturation, with a SiO_2/Al_2O_3 ratio of 2.67-3.10 in the clay fraction ($<1\mu$). The K_2O content and CEC are about 3% and 30 m.e./100g respectively. Judging from the X-ray diffraction spectrum, the clay minerals are poorly crystallized, consisting mainly of hydrous micas, together with some chlorite. Vermiculite occurred in surface and some lower horizons. In the clay fraction of Xiashu loess collected from Liyang County, the SiO_2/Al_2O_3 ratio is 2.67-3.17, CEC, 32-41 m.e./100g, and K_2O 2.2-2.8% (highest in the surface). In addition to abundant hydrous micas, small amount of kaolinite and smectite are present, chlorite occurs in less amount than in Nanjing, but vermiculite almost absent within 1 meter. The mineral composition varies in different layer of the Xiashu loess, hence, the paddy soils developed on its denudation planes may also be dissimillar in mineral composition. Taking vermiculite as an example, it is almost absent in the surface layer of a silty paddy soil collected from Dongshan, Jiangning County, but is present in the plowpan of the same profile; while it is present in the whole profile of a silty paddy soil collected from Qingxiu of the same county; it is found only in the Bg horizon of a surface bleached paddy soil in Shangxing, Liyang County (Fig. la-b).

In a stiff bleached paddy soil influenced by the weathering product of tuff, the upper horizons contain large amount of smectite and kaolinite, thus sharply distinguished from underlying Xiashu loess below 45 cm (Fig. 1c).

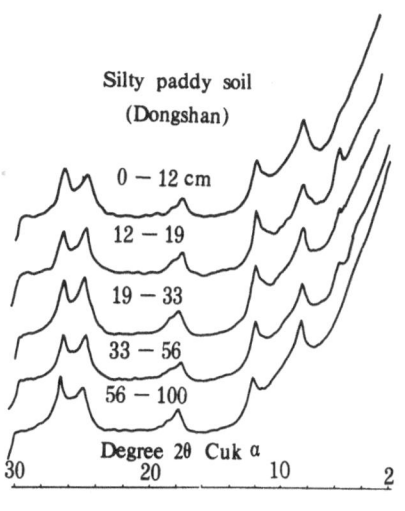

a. Silty paddy soil
   (Dongshan)

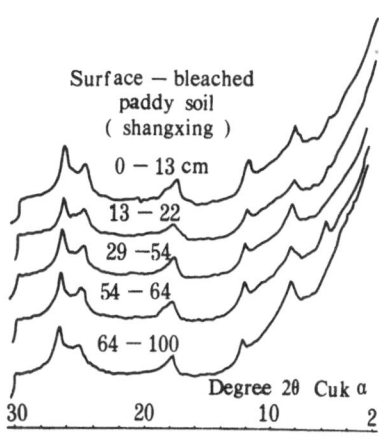

b. Surface-bleached paddy soil
   (Shangxing)

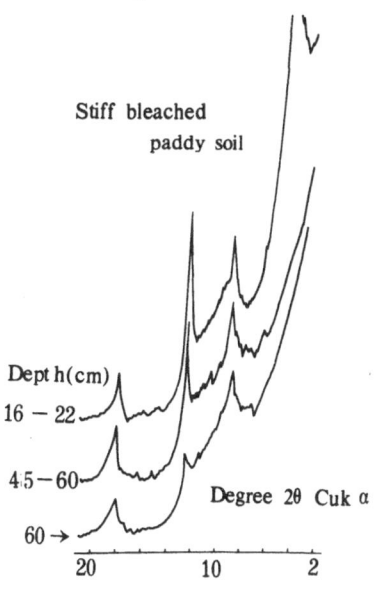

c. Stiff bleached paddy soil
   (Xinchang)

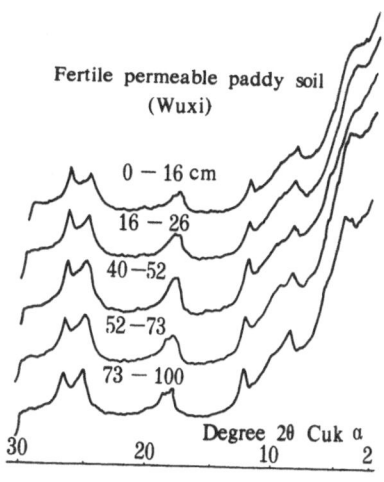

d. Fertile permeable paddy
   soil (Wuxi)

Fig. 1 X-ray diffraction pattern (I)

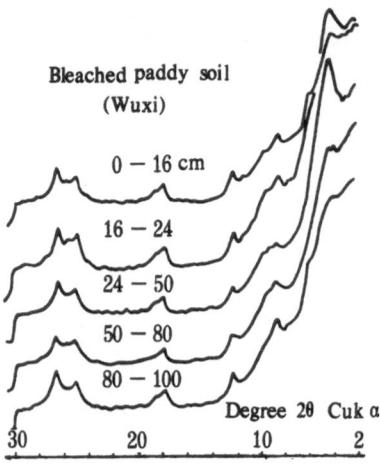

Bleached paddy soil
(Wuxi)

0 — 16 cm

16 — 24

24 — 50

50 — 80

80 — 100

Degree 2θ Cuk α

30   20   10   2

e. Bleached paddy soil
(Wuxi)

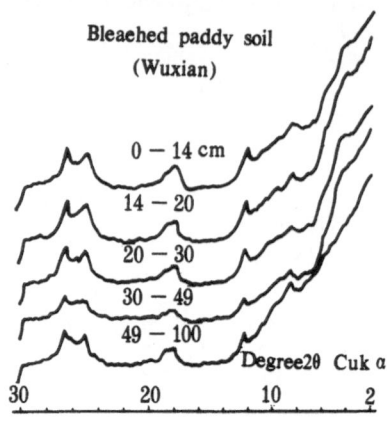

Bleached paddy soil
(Wuxian)

0 — 14 cm

14 — 20

20 — 30

30 — 49

49 — 100

Degree2θ Cuk α

30   20   10   2

f. Bleached paddy soil
(Wuxian)

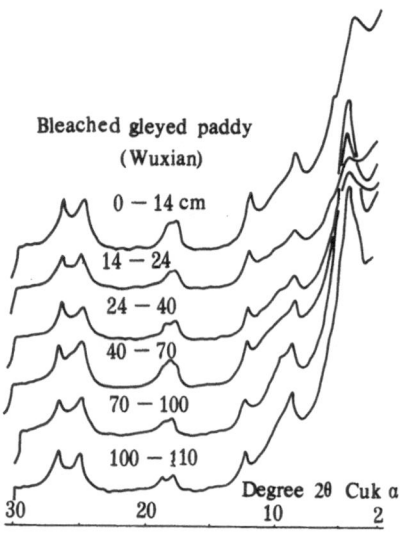

Bleached gleyed paddy
(Wuxian)

0 — 14 cm

14 — 24

24 — 40

40 — 70

70 — 100

100 — 110

Degree 2θ Cuk α

30   20   10   2

g. Bleached gleyed paddy soil
(Wuxing)

Fig. 1   X-ray diffraction pattern (II)

2. Low Plain Along Taihu Lake

In Wuxi and Suzhou regions, in the lower part of paddy soils there is generally a brown substratum, which is heavier in texture than Xiashu loess.

Its clay fraction contains significant amount of smectite. From the clay mineral composition it can be supposed that this brown substratum together with the permeable soils laying above it were originated from Xiashu loess, though the permeable soil is light in texture and contains some smectite.

The content of smectite in lower horizons of the fertile permeable paddy soil in Wuxi County seems to be higher than that of the upper part (Fig. 1d). The decrease of CEC can be explained in terms of increasing amount of free iron oxide. The smectite peaks of a bleached paddy soil from Wuxi County are more intensive than others, especially the bleached horizon (Fig. 1e). However, the clay mineral composition of the bleached paddy soil from Wuxian County is rather uniform throughout the whole profile, including the bleached horizon (Fig. 1f).

It seems that parent material of a gleyed paddy soil with sandy substratum from Wujiang County and a bleached gleyed paddy soil from Wuxing County belong to other types They are alos rich in hydrous micas and smectite, accompanied with some chlorite, but the smectite peak is relatively strong, especially the lower horizons of the latter (Fig. 1g). The decrease in CEC of the clay fraction may also be due to the influence of free iron oxide.

3. Alluvial Plain of Changjiang River

Although all the paddy soils of this geomorphological unit are derived from recent alluvial deposits of the Changjiang River, the clay mineral composition varies slightly due to sorting effect. These soils are dominantly hydrous micas and smectite, associated with chlorite. Smectite tends to increase in the clayey layer, while chlorite tends to increase in sandy soils.

4. Northern Plain of Hangzhou Bay

Paddy soils in this part have some similarities in clay mineralogy with that in alluvial plain of the Changjiang River. In some local districts more micas are found from coarse textured parent materials. As compared with that in Wujiang and Wuxing Counties, the paddy soils in Jiaxing County (i.e. weakly-gleyed soil and permeable paddy soil) contain more hydrous micas. The K_2O content may be as high as 4%. In the lower horizons of a permeable paddy soil the intensity of 10 Å and 17.7 Å reflections varies markedly, suggesting the possibility of variable sedimentary conditions. In addition, the existence of significant amount of chlorite suggests that these soils sedimented more recently, and pedological weathering was relatively weak. The clay mineralogy of a sedimentary paddy soil from Shanghai is similar to the weakly-gleyed paddy soil from Jiaxing County, and is more uniform throughout the whole profile.

MINERALOGICAL CHARACTERISTICS OF THE CLAY FRACTION OF THE BLEACHED HORIZON

The questions about leaching and/or translocation of materials and destruction of clay mineral in the bleached horizon have attracted much attention. Unfortunatly, due to the influence of crystallinity and other factors, only the relative abundance can be estimated by comparision of kinds of clay mineral species from X-ray diffraction patterns. It is now hardly possible to determine the true quantities and to detect lattice

destruction of clay minerals. However the evidence of material translocation and lattice destruction ,ay be estimated roughly from variations in
clay content, mineral type and free iron oxide (Fig. 2). It is valuable in
recognizing the bleached horizon.

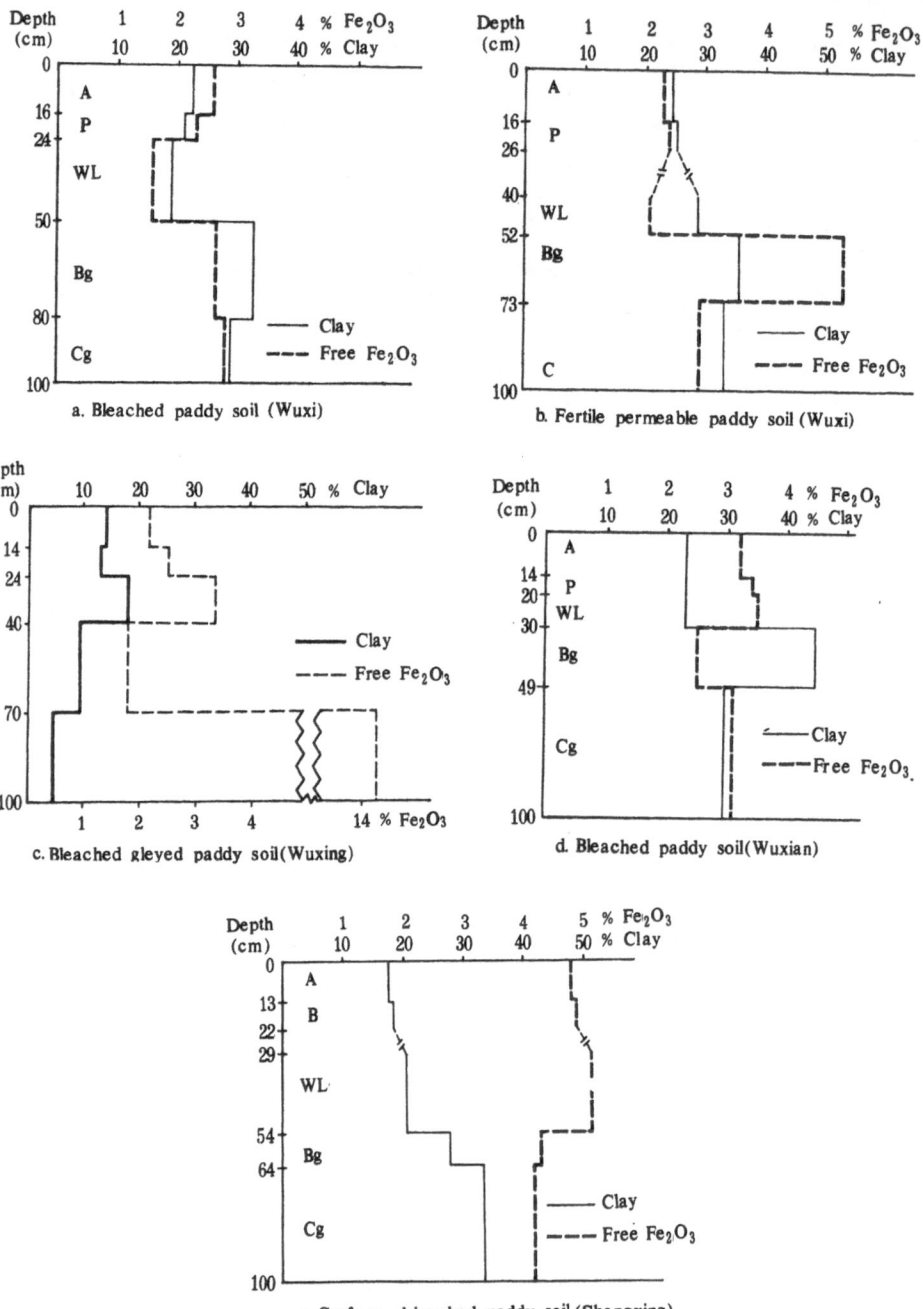

Fig. 2   Distribution of clay and free iron oxid in
         soil profiles

Fig. 2a shows that the contents of free iron oxide and clay are lowest in the bleached horizon of a bleached paddy soil from Wuxi County. This means that leaching is probable, or alternatively, the original sediment might be low in clay content. The highest 17.7 Å peak in this horizon indicates that it is rich in smectite or it is well crystallized. It seems that lattice disruption is improbable, or if any, would be restricted to the disordered portion. Iron leaching in the bleached horizon of a fertile permeable paddy soil (Fig. 2b) and a bleached gleyed paddy soil (Fig. 2c) is more probable, but the clay content in these horizons does not show sharp change as compared with horizons just above or beneath it. The smectite peaks increase steadily with depth, hence, no evidence for clay translocation or destruction can be drawn, As for a bleached paddy soil from Wuxian County (Fig. 2d) and a surface-bleached paddy soil from Liyang County (Fig. 2e), neither the mineralogical composition nor the clay content gives a positive indication of material translocation or lattice disruption.

Several authors have noticed the possibility of dual nature of parent material in the formation of bleached paddy soil(2-3). The texture and clay mineral composition in upper horizons of these five bleached paddy soils differ markedly from those of the underlying horizons. It is very probable that they were deposited in different periods. As distinguished from other bleached soils, the upper horizons of a bleached gleyed paddy soil are heavier in texture than the lower ones. Possibly, it is a type of ancient gleysol(3). The reason for the low CEC in the bleached horzion of some bleached paddy soil from Wuxi County is not known exactly yet. The question as to whether this is related to ferrolysis(4) deserves further study.

REFERENCES

(1) Luo Jia-xian, Preliminary studies on the primary mineral of some paddy soils in Tahu Lake region. In this Proceedings.
(2) Zhou Chuan-huai, The so-called" Bai-ty" paddy soil in Jiangsu Province. Acta Pedologica Sinica, 6,217-227(1958). (in Chinese)
(3) Yu Tian-ren, Xie Jian-chang, Yang Guo-zhi, Gao Zi-qin Chen Jia-fan, Shen Ren-shui, Din Chang-pu, Zhou Qi-kun, Studies on the infertile "white soil" in Tai Lake region. Acta Pedologica Sinica, 7, 42-58(1959). (in Chinese with English summary)
(4) Brinkman, R., Ferrolysis, a hydromorphic soil forming process. Geoderma, 3,199-206(1970).

ON THE CHARACTERISTICS OF Fe-ORGANIC COATING IN PADDY SOILS

Gu Xin-yun, Li Shu-qiu
(Institute of Soil Science, Academia Sinica, Nanjing)

Vast areas of fertile paddy soils occur on Taihu Lake region. At the lower portion of the cultivated horizon of these fertile paddy soils emerges the reddish Fe-organic coating which can be considered as an important morphological criterion for soil fertility. The condition for the formation of Fe-organic coating and its significance on soil fertility have been studied by some scientific workers(1,2). The present article deals with the characteristics of Fe-organic coating of paddy soil by micro-morphological method.

MATERIALS AND METHODS

Samples of permeable paddy soil, bleached paddy soil and immature bleached paddy soil derived from loessial deposity were collected from Taihu Lake region. Soil clods were separated from these samples and the Fe-organic coatings were cut down from the clod surface for investigation. At the same time, a thin layer of the adjacent grey soil was scraped down from the same clod and used as the sample of the soil mass. The separated Fe-organic coating would be inevitable mixed with some soil mass. Furthermore, the reddish brown iron mottlings scraped down from the sample of B horizon, the synthetic humus iron complex and the anthropogenic rusty speckles were used as reference samples for comparison. The morphological characters of the soil samples are shown in Table 1.

Table 1 The morphological feature of samples

Sample	Soil horizon	Morphological features
Permeable paddy soil (Fe-organic coating)	Cultivated horizon	Reddish coating in speckle shape, persist in coloration
Permeable paddy soil (soil mass)	Cultivated horizon	Greyish brown with rusty speckles and streaks
Bleached permeable soil (Fe-organic coating)	Cultivated horizon	Reddish brown coating in speckle shape, turned into rusty color in a few days
Immature bleached paddy soil (soil mass)	Cultivated horizon	Greyish brown and brown soil clods with a few rusty speckles and streaks
Iron rusty speckle	Illuvial horizon	Bluish (dark) grey soil-clods attached with reddish brown speckles and streaks

Methods: The Na-dithionate reduction method was used for the determination of free ferric oxide; iron complex was extracted by Na-pyrophosphate; clay minerals were determined by X-ray diffraction Co $K\alpha$ constituents of organic matter were determined by infrared spectrography; the variaties of iron oxide were determined by Mossboure spectroscopic analysis; electron microscope and eletron microprobe were used to study the coatings and soil micro-morphology.

RESULTS

The X-ray diffraction analysis showed that the composition of clay minerals in permeable paddy soil, bleached paddy soil and immature bleached paddy soil was nearly the same, consisting mainly of hydrated mica accompanied with chlorite, kaolinite, quartz and some iron oxide. Mössbouer spectroscopic analysis showed that in the coating and soil mass, the types of iron oxide in these three soils were virtually identical, also containing hematite, goethite, lepidocrocite, maghemite and some divalent ferrous compounds, but the content of hydrated iron oxide in the reddish coating was higher than that of the soil mass.

As it could be seen from Table 2, the content of SiO_2, Al_2O_3 and TiO_2 in the coating was lower than that in the soil mass, but the total amount of iron and manganese was about double while the difference on the amount of free iron oxide was still greater (Table 3). The organic matter in the Fe-organic coating was 20-30% higher than that in the soil mass, but there was not much difference in organic matter between the Fe-organic coating and the soil mass of the immature bleached paddy soil.

Table 2 Chemical composition of the reddish coating
and the corresponding soil mass(%)

Sample	SiO_2	Al_2O_3	Fe_2O_s	MnO_2	TiO_2
Permeable paddy soil (coating)	61.08	13.53	9.81	0.10	0.72
Permeable paddy soil (soil mass)	66.31	15.35	5.01	0.05	0.83
Bleached paddy soil (coating)	67.82	11.88	8.17	0.13	0.55
Bleached paddy soil (soil mass)	73.91	12.64	4.11	0.05	0.78
Immature bleached paddy soil (coating)	67.31	11.46	9.64	0.08	0.70
Immature bleached paddy soil (soil mass)	73.38	11.68	3.87	0.04	0.80

Table 3 The content of iron and organic matter
in the coating and soil mass (%)

Sample	Soil pH	Organic matter	Free ferric oxide	Fe-organic complex
Permeable paddy soil (coating)		3.66	7.15	0.49
	5.10			
Permeable paddy soil (soil mass)		3.14	1.57	0.13
Bleached paddy soil (coating)		3.21	5.36	0.29
	5.72			
Bleached paddy soil (soil mass)		2.35	1.58	0.15
Immature bleached paddy soil (coating)		2.25	5.05	0.14
	6.00			
Immature bleached paddy soil (soil mass)		2.14	1.34	0.07

The two absorption in 2926 cm^{-1} and 2850 cm^{-1} of the infra-red spectra showed the existence of aliphatic compounds, which appeared mroe clear in the coating material than in the soil mass. This indicated that there was more aliphatic material in the coating. The microphotograph made in situ with a scanning electron microscope showed that the cross-section of the soil mass covered with a coating had the appearance of porous and fibrous micros-tructure and the soil mass had a micro aggregate structure of different sizes (Fig. 1). Analysis of the micro-region with an electron microprobe showed that the content of iron in the coating was higher than that in the soil mass (Fig. 2).

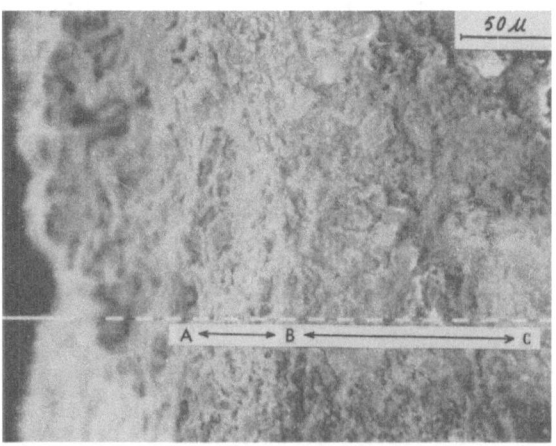

Fig. 1 Scanning electron micrograph cross-section of soil mass with coating

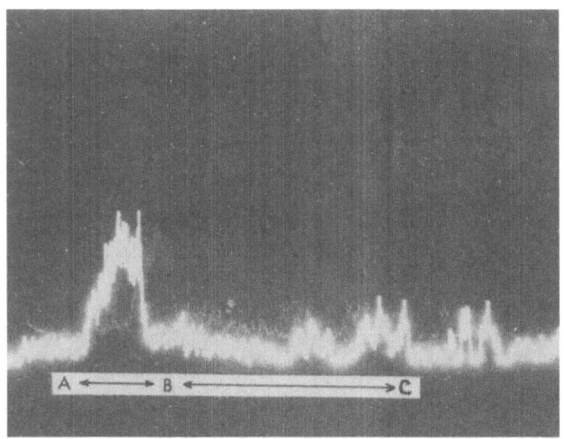

Fig 2 Electron microprobe showing higher
content of iron in coating

The observations made under electron microscope showed that in the
samples of Fe-organic coating, besides the large amount of clay mineral,
there were also some materials: of net-work structure (Fig. 3) and they
were identified with electron diffraction as non-crystaline (amorphous).
These materials did not appear in the soil mass. The net-work structures
had different shapes and on most of them were attached various forms of
iron oxide whose sizes were about 2-7 micron. They either existed separately
or, in most cases, conjugated with clay mineral (Fig. 4). The net-work struc-
true would be destroyed by repeated treatment with hydrogen peroxide (Fig. 5).

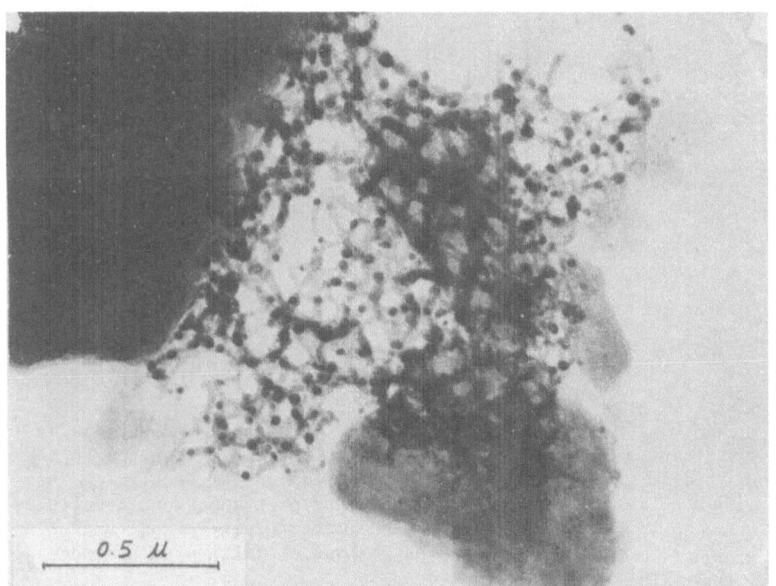

0.5 μ

Fig. 3 T.E.M. photo: Net-work structure of coating
in bleached paddy soil

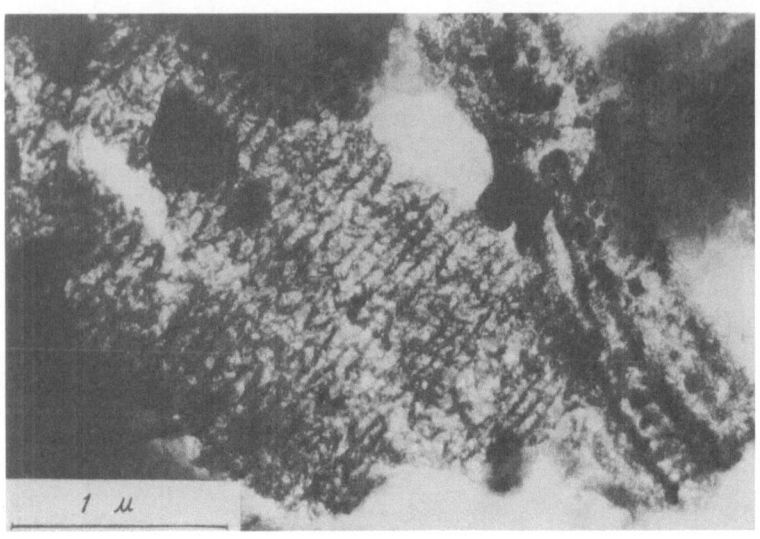

Fig. 4 T.E.M. phote: Net-work structure of
coating in permeable paddy soil

In the coating from the immature bleached paddy soil, very few net-work
structure materials were found. Its main constituents were various forms of
iron oxide. In the iron mottling of B horizon and in the anthropogenic rusty
material, only goethite and hematite existed and no net-work materila was
found. The synthetic humus iron complex was a kind of polymer formed of
numerous globular colloidal materials which were also different from the
reddish coating in soils in the net-work structure (Fig. 6).

DISCUSSION

1) The reddish coating in the cultivated horizon of fertile paddy soil
was an organic-mineral complex consisting of organic matter and various forms
of iron oxide, part of which had an specific net-work structure, whereas the
majority were conjugated with clay mineral. The micromorphological feature of
the reddish coating was different from that of the rusty speckle in B horizon
and the anthropogenic rusty material. The artificially synthesized humus-
iron complex was a porous polymer composed of numerous globular colloidal
materials.

2) The forming process of Fe-organic coating was assumed to be as follows;
during waterlogging period, when the soil was in a permeable condition, a
considerable amount of organic acid was formed in the course of decomposition
of organic matter part of which formed complex with ferrous ions and moved
down with the water. Then it adsorbed on the surface of clevages in subsoil.
When the field was draind, the reducing status of the cultivated horizon was
changed into oxidizing status, part of the Fe-organic comple adsorbed on the
soil clods was oxidized and decomposed, the released ferrous iron which was
then partly converted into ferric oxide. The other part underwent complicated
physico-chemical and biochemical processes and formed a net-work structure
material. The scarlet coating in the lower portion of cultivated horizon was
the outgrowth of these repeated processes.

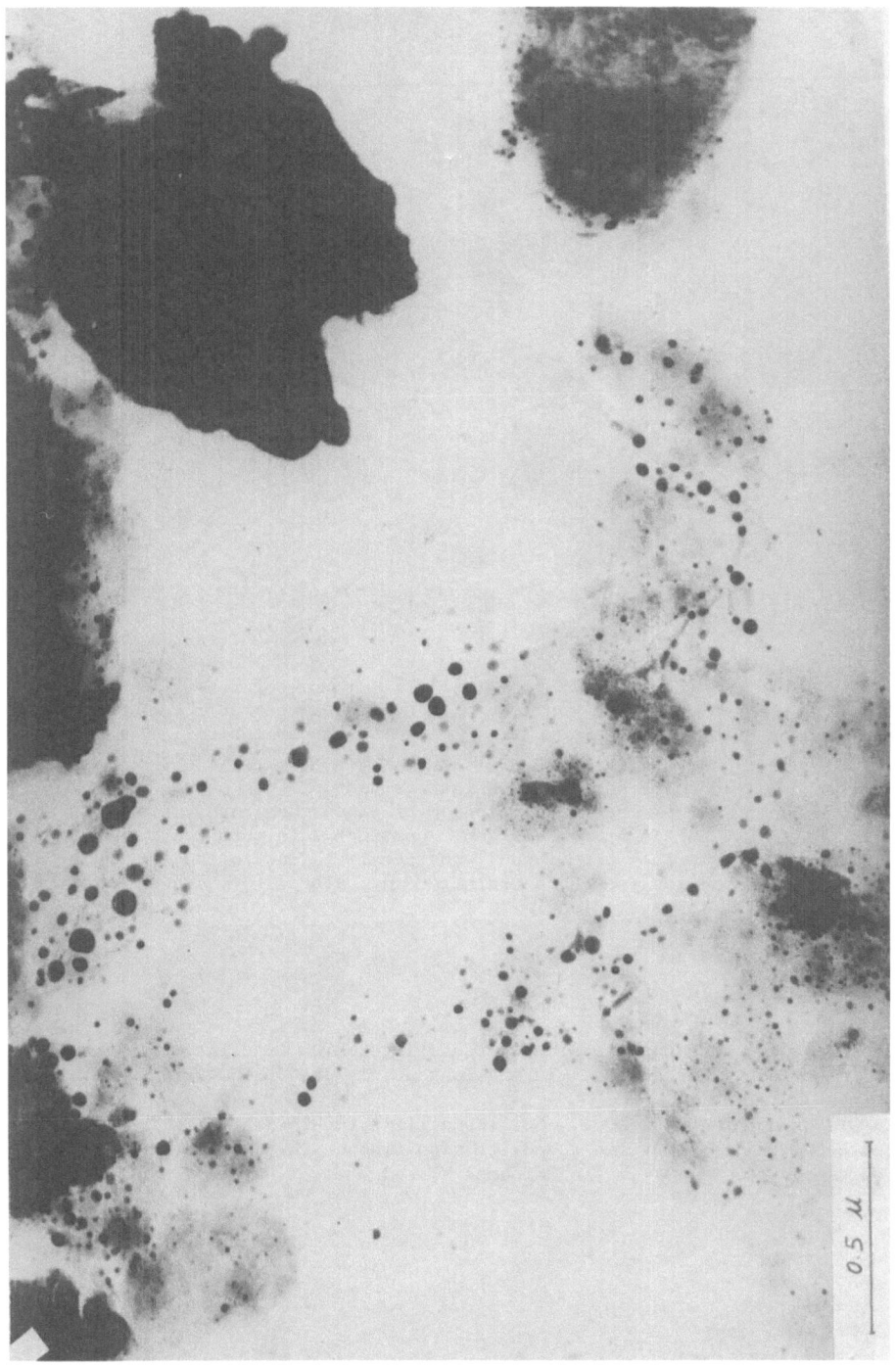

0.5 μ

Fig. 5 T.E.M. photo: Net-work structure of coating destroyed by H_2O_2 in bleached paddy soil

491

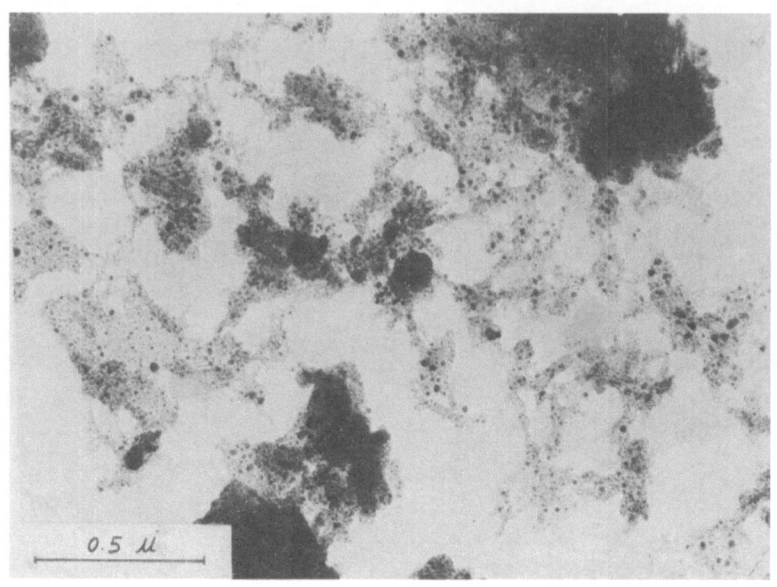

Fig. 6　T.E.M photo: The porous polymer of synthetic
humus-iron complex

3) The Fe-organic coating was formed in non-calcareous soil where drainage
was good and organic matter was ample.　Sometimes, in permeable soil containing
less organic matter, a kind of reddish brown coating was formed in the cultivated
horizon; but, the red color was not as bright as that of the Fe-organic
coating, and when the soil became dry the color would turn into rusty one,
thus, it was termed as "Immature Fe-organic coating".　In this coating,
very few net-work materials were found.　Whether it is due to the lack of
organic matter to provide enough co-ordinated radicals to form a complex
with Fe-ions rmains to be studied.

REFERENCES

(1) Gong Zi-tong, Farmen's experience of distinguishing fertility of paddy
　　soils in Taihu Lake region.　Turang Tongbao, 1, 45-48(1960). (in
　　Chinese)
(2) Xu Qi, Chen Zhi-cheng, 1961.　Soil conditions for formation of iron-
　　humic coating in paddy field and its significance in fertility.　Turang
　　Tongbao, 4, 43-49(1961). (in Chinese)

THE MAGNETIC SUSCEPTIBILITY DISTINCTION OF
PADDY SOILS IN TAIHU LAKE REGION

Yu Jin-yan, Zhao Wei-sheng, Zhan Shuo-ren
(Zhejiang Agricultural University, Hangzhou)

Soil magnetism is a new branch of pedology which makes use of magnetic techniques in the study of soils[1]. The present paper gives some results of an attempt to classify paddy soils of the Taihu Lake region on the principle of soil magnetism.

MATERIALS AND METHODS

The magnetic susceptibility profile of defferent paddy soil types and of their corresponding upland soils were studied by genetic horizons with air-dried soil samples or their monolithes (326 samples from 75 profiles). The volume magnetic susceptibility ($\varkappa$) of soil was examined and measured with a magnetic susceptimeter WCL-3*. The value of $\varkappa$ was converted into value of mass magnetic susceptibility x , in terms of 10^{-6} emu/g by the formula $x = \varkappa/D$, where D is the apparent density or volume weight of the soil.

RESULTS AND DISCUSSION

The experimental results obtained are described as follows.

1. Character of Magnetic Susceptibility
Profile Common to All Paddy Soils

It has been shown that the $\varkappa$ value of hydromorphic soils is lower than that of automorphic soils[2,3]. The $\varkappa$ value of the paddy soils is even much lower than that of the upland soils, probably due to artificial flooding and application of organic fertilizers during a long period of time. It is probable that these practices mentioned above would have caused remarkable reduction, solution, hydration, and eluviation of magnetic minerals in the paddy soils.

A comparison of the magnetic susceptibility profile (m.s.p.) of three groups, including paddy soils and their related upland soils, derived from the same parent material, is given in Fig. 1.

In the first group, there are three yellow mottled soils, all silty loam in texture, developed on alluvial deposit (Fig.1 A), one of which is

---

* Product of Beijing manufactory of geological instruments.

an old paddy soil that has low x' value throughout the profile. The x value of its ploughed horizon is 5 units, and of the underlying horizons as low as 3-2 units. The x value of the ploughed horizon of a young calcareous paddy soil varies between 12-9 units which is much higher than that of the old paddy soil, but still much lower than that of the related upland soil. For the upland soil, the x value varies from 18 to 16 units in the upper part of the profile, and from 15 to 13 units in its underlying layers.

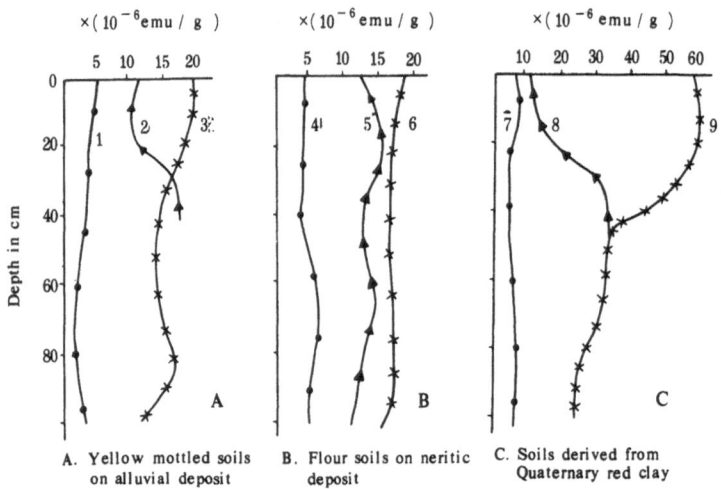

A. Yellow mottled soils on alluvial deposit B. Flour soils on neritic deposit C. Soils derived from Quaternary red clay

Fig. 1   The magnetic susceptibility profile of paddy soils and their corresponding upland soils
1,4,7. Old paddy soil; 2,5,8. Young paddy soil;
3. Upland soil; 6. Upland soil of mulberry grove;
9. Krasnozem of tea plantation

In the second group, three floury soils including an old paddy soil, a young paddy soil and an upland soil of mulberry grove, with texture of coarse silty light loam, derived from nertic deposit have been examined (Fig. 1 B). The x values in the upper part of these profiles are $\sim$4 units, 11-14 units, and 17-15 units respectively.

In the third group, a krasnozem of tea plantation and two paddy soils are examined (Fig. 1 C). All of those soils are derived from the Quaternary red clay on the hill. The x value of the old paddy soil is as low as 6-4 units throughout the profile, and the x value in the upper part of the young paddy soil profile on the red clay is 20-12 units. But the krasnozem of tea plantation has much higher x value, with $\sim$58 units in the ploughed horizon and 30-25 units in the B and C horizons.

It is apparent that 1) the magnetic susceptibility x of paddy soils is by a long way lower than that of their relevent upland soils, 2) the x value of the old paddy soils is lower to a certain amount than that of the young paddy soils of same genesis, and 3) the stronger are the magnetic properties of the related upland soils, the more remarkable is the decrease in the $\varkappa$ value when the soils are put under paddy cultivation (Table 1).

Table 1   Mass magnetic susceptibility of A-horizon
in paddy soils, as compared with upland
soils, developed from the same parent
material

($\varkappa$, 10^{-6} emu/g)

Groups of soils	Old paddy soils	Young paddy soils	Upland soils
Yellow mottled soils	∿ 5	12-9	18-16
Floury soils	∿ 4	14-11	17-15
Soils derived from Quaternary red clay	6-5	17-12	∿ 58

2.   Differentiation of Magnetic Susceptibility
Profile of Various Paddy Soil Types

A clear differentiation of the magnetic susceptibility profile among various paddy soil types will appear, because the soil moisture regime and pattern of soil construction are different, which affect the processes in the soil of oxidation and reduction, crystallization and decrystallization, leaching and accumulation of ferromanganese minerals, and affect also the movement of clay particles which are principal carriers of secondary fer-romagnetic minerals.

Three types of paddy soils, including the permeable, the lessivating, and the waterlogged in the Taihu Lake region(4) have been researched (Fig. 2).

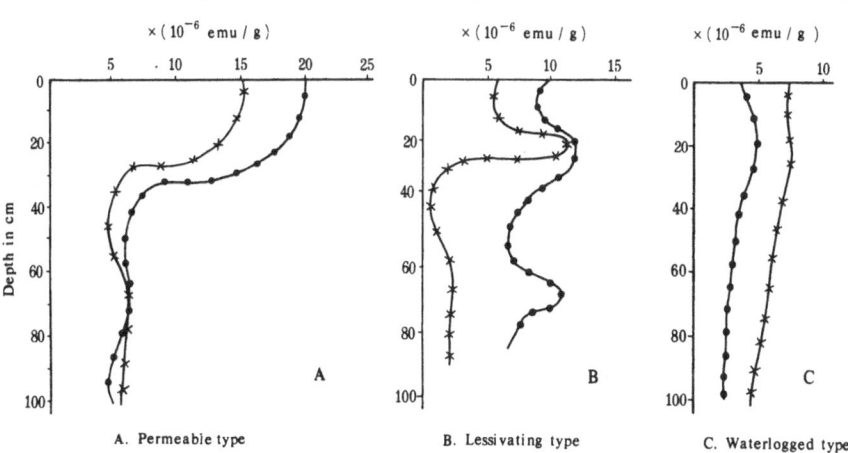

A. Permeable type   B. Lessivating type   C. Waterlogged type

Fig. 2   The magnetic susceptibility profile of
different paddy soil types

495

1) The permeable paddy soils: This type of paddy soils is developed under well-drainage conditions. There is a suitable alternation of dry and wet periods. It is favorable to the replenishment of bases and organic matters, and thus, as denoted in the researches on chernozems and other soils, such conditions would be also favorable to the formation and accumulation of secondary ferromagnetic minerals in the surface layer of soil(5-7). Certainly, it the ploughed horizon of permeable paddy soils, the χ value shows marked increases, being 20-15 units as compared with only about 5 units in horizons beneath it (Fig. 2 A).

2) The lessivating (bleached) paddy soils: This type of paddy soils, as typified by the presence of one or more bleached or albic horizons (L-horizons) in the subsoil, is formed under the influence of the stagnating water in reducing conditions. The ferromanganese materials and clay particles in the L-horizons were leached(8,9). The χ value of the L-horizons is lower than that of the horizons either above or underneath it (Fig. 2 B).

There may be found occasionally, however, in the L-horizons a few ferromanganese concretions with extremely high χ value of 100-300 units; these concretions may bring the magnetic susceptibility of soil mass in these horizons to a higher level than the other horizons.

3) The waterlogged paddy soils: The dominant genesis of this type of paddy soils is gleization, in which free iron and manganese oxides are reduced intensively and the χ value rapidly lowers throughout the profile (Fig. 2 C). But these soils are often characterized by the abundance of organic matter and heavy in texture, and they may be converted into permeable paddy soils after being reclaimed by drainage, then the χ value goes up to a higher level of 16-23 units.

3. The Magnetic Indices of Paddy Soils and Their Diagnostic Significance

Two magnetic indices are suggested by us for paddy soils; The magnetic alteration coefficient of paddy soil (m.a.c.) K_p and the magnetic alteration ratio of paddy soil to upland soil (m.a.r.) K_R. They are expressed in the following formulas:

$$K_p = \chi_{PA}/\chi_{oc}$$

$$K_R = \chi_{PA}/\chi_{uA}$$

Where, χ_{PA} denotes the χ of A-horizon of the paddy soil; χ_{oc}, the χ of C-horizon of the original soil before paddy cultivations; and χ_{uA}, the χ of A-horizon of upland soil on the same parent material.

The magnetic indices of three types of paddy soils in the Taihu Lake region are shown in Table 2.

CONCLUSIONS

The essential conclusions that might be drawn from our preliminary examinations are as follows:

1) Magnetic susceptibility as expressed by χ is much lower in paddy

Table 2　The magnetic indices of the three types of paddy
soils in the Taihu Lake region

Soil types	K_p	K_R
Permeable	0.5-1.1	0.4-0.7
Lessivating	0.4-0.7	0.2-0.3
Waterlogged	0.2-0.4	∿0.2

soils than in the upland soils developed on the same parent materials.

2) Each type of paddy soils in the Taihu Lake region has its own characteristics in "morphological figure" of magnetic susceptibility profile (m.s.p.).

3) Two magnetic indices suggested, may perhaps be used together with the
"morphological figure" of m.s.p., as diagnostic criteria for genetic indenti
fication and classification of paddy soils.

REFERENCES

(1) Yu Jin-yan, Soil magnetism __ a new field of soil science researches
    Turangxue Jinzhan, 4, 1-12(1979). (in Chinese)
(2) Vadyunina, A.F., Babanin, V.F., Magnetic susceptibility of soils in the
    European part of USSR. Trans. 10th Int. Congr. Soil Sci., Moscow, V. 1,
    357-362(1974). (in Russian with English, German and France Summary)
(3) Yu Jin-yan, Zhan Shuo-ren, The Magnetic susceptibility profile of the
    main soil types in China (1980, ms.).
(4) Xu Qi, Lu Yan-chun, Zhu Hong-quan, The genesis and classification of the
    paddy soils in Tai-lake basin, Jiangsu Province, China. Acta Pedologica
    Sinica, 17, 120-132(1980). (in Chinese with English summary)
(5) Mullins, C.E., The magnetic susceptibility of soils and it's significance
    in soil science __ a review. J. Soil Sci., 28, 223-246 (1977).
(6) Henin, S., Le Borgne, E., On magnetic properties of soil and their
    interpretations by soil scientists. Trans. 5th Int. Congr. Soil Sci.,
    Leopoldville, V. 2, 13-17(1954). (in France, with English summary)
(7) Schwertmann , V., Fischer, W. R., Taylor, R. M., New aspects of iron oxide
    formation in soils. Trans. 10th Int. Congr. Soil Sci., Moscow, V. 6(1),
    237-247(1974).
(8) Xu Qi, The geographic regularity of belozems in basin of middle and lower
    reaches of the Yangtse River and their formative processes. Acta Pedologica
    Sinica, 10, 44-54(1962). (in Chinese with Russian summary
(9) Yu Tian-ren, Xie Jian-chang, Yang Guo-zhi, Gao Zi-qin, Chen Jia-fang,
    Shen Ren-shui, Ding Chang-pu, Zhou Qi-kun, Studies on the infertile
    "white soil" in Tai-lake region. Acta Pedologica,Sinica, 7, 42-58(1959).
    (in Chinese with English summary)

EFFECT OF LONG-TERM RICE CULTIVATION ON SOME PROPERTIES OF SODA SOLONCHAKS AT QIAN GORLOS IRRIGATION REGION OF JILIN PROVINCE IN CHINA[*]

Wang Chun-yu, Wang Ru-yong, Zhang Su-jun,
Zhang Xiu-lan, Tian Lin-jie
(Institute of Forestry and Pedology, Academia Sinica, Shenyang)

The Qian Gorlos irrigation region is located at long. $124^{\circ}10'$-$125^{\circ}2'$E and lat. $45^{\circ}0'$-$45^{\circ}8'$N. The growth of rice on saline-alkali soils in this region has a very long history in China. This region is spread widely with soda solonchaks which have been used for rice cultivation and on these soils a grain yield up to 6142.5 kg/ha was obtained. The present paper deals with the effects of long-term rice cultivation on some properties of the soils. The results obtained in 1955-1978 are summarized as follows.

EFFECT OF LONG-TERM RICE CULTIVATION ON THE CONTENT OF WATER-SOLUBLE SALTS IN THE SOILS

Under conditions of long-term rice cultivation, a slower leaching of water-soluble salts occurred in the soda solonchaks. This process was much quicker in the initial stage of rice cultivation, and later, the desalting cultivated horizon with larger depth was gradually formed in soil profile, desalinization reached to 73.7%. Under the condition of rice cultivation, the mineralization rate of the ground water was decreased, or at least unchanged.

Table 1 Effect of long-term rice cultivation on pH,
water-soluble salts and desalinization
rate of soda solonchak

Depth (cm)	Year	pH	Soluble salts (%)	Desalinization (%)
0-5	1955	11.5	1.34	92.5
	1978	8.3	0.10	
0-30	1955	10.5	0.70	73.7
	1978	8.9	0.18	
0-110	1955		0.42	47.4
	1978		0.22	

----

[*] This work has been instructed by Prof. Chen En-feng.

EFFECT OF LONG-TERM RICE CULTIVATION ON
THE EXCHANGEABLE BASES OF SOILS

The content of exchangeable Na of the soda solonchaks was very high, but that of exchangeable Ca and Mg was low. Under long-term rice cultivation and application of large amount of organic manure, the alkalinity was greatly decreased, and conversely, the exchangeable Ca and Mg were greatly increased, up to more than 10 times.

Table 2 Effect of long-term rice cultivation
on the composition of water-soluble
anions of soda solonchak

Depth (cm)	Year	Anions m.e./100g			
		CO_3^{2-}	HCO_3^-	Cl^-	SO_4^{2-}
0-5	1955	8.84	8.97	1.24	0.75
	1978	0	1.10	0.12	0.13
0-30	1955	2.02	6.74	0.39	0.28
	1978	0.48	1.68	0.28	0.22
0-110	1955	0.99	4.21	0.18	0.16
	1978	0.77	2.03	0.15	0.22

Table 3 Effect of long-term rice cultivation
on the composition of water-soluble
cations of soda solonchak

Depth (cm)	Year	Cations m.e./100g		
		Ca^{2+}	Mg^{2+}	Na^+
0-5	1955	0.13	0.11	19.55
	1978	0.24	0.25	0.86
0-30	1955	0.22	0.07	9.15
	1978	0.13	0.22	2.18
0-110	1955	0.11	0.10	5.34
	1978	0.08	0.10	2.80

Table 4 Effect of long-term rice cultivation on the
exchangeable bases of soda solonchak

Depth (cm)	Field	Cations(m.e./100g)		% in exch. cations	
		$Ca^{2+} + Mg^{2+}$	Na^+	$Ca^{2+} + Mg^{2+}$	Na^+ (Alkalinity)
0-5	Virgin soil	1.21	20.9	5.4	92.2
	Paddy field	16.9	1.73	85.4	8.8
0-30	Virgin soil	1.58	21.9	6.7	91.7
	Paddy field	14.8	5.61	70.0	25.3

FFECT OF LONG-TERM RICE CULTIVATION ON THE
CONTENT AND COMPOSITION OF SOIL HUMIC SUBSTANCES

With the alternation of draining and submerging of soils by long-term rice cultivation and application of organic manure, the decomposition and regeneration of organic matters were facilitated and the organic matter was accumulated even in the plowed layer, and its humus content was gradually increased, and the ratio of humic acid to fulvic acid was decreased, but still more than 1.

Table 5 Effect of long-term rice cultivation on the content and composition of soil humic substances

Depth (cm)	Field	Organic C (%)	Humic acid (% of org. C)	Fulvic acid (% of org. C)	Humic ——— Fulvic
0-5	Virgin soil	0.82	14.54	6.08	2.39
	Paddy field	1.43	7.64	6.95	1.09
0-30	Virgin soil	0.71	8.91	3.04	2.93
	Paddy field	1.30	6.62	6.50	1.01

EFFECT OF LONG-TERM RICE CULTIVATION ON
THE PHYSICAL PROPERTIES OF SOIL

The physical properties of the soda solonchaks were very poor. Under wetting, it was dispersed, and when drying, very hard and compact. there-fore, it was an unstructural soil. Under long-term rice cultivation, the soil structure was improved, the volume weight of the soil being decreased, and the porosity of the soil being increased.

Table 6 Effect of long-term rice cultivation on soil physical properties

Field	Depth (cm)	Volume weight (g/cm^3)	Structural coefficient (%)
Virgin soil	0-25	1.50	2.41
Paddy field	0-25	1.35	76.5

EFFECT OF LONG-TERM RICE CULTIVATION ON
THE MICROBIOLOGICAL PROPERTIES

Under long-term rice cultivation on soda solonchaks, the favorable soil conditions were created for the development of microorganisms. The quantity of soil microflora and the biochemical activity of soil were increased greatly.

Table 7 Effect of long-term rice cultivation on the biochemical activity of the soda solonchak

Depth (cm)	Field	Respiration intensity (mg CO_2/100 g/day)	Urease (mg NH_3 -N/g)	Invertase (m.e.H_2SO_4 /g)	Catalase (m.e. $KMnO_4$g)
0-15	Virgin soil	3.0	44	0.06	0.74
	Paddy field	14.0	47	0.54	0.89
15-30	Virgin soil		39	0.14	0.41
	Paddy field		32	0.24	0.83

CONCLUSION

Under long-term rice cultivation on the soda solonchaks, its effects on the soil physical, chemical and microbiological properties were pronounced. Of course, these favorable effects could only be found under proper conditions of irrigation, fertilization, and rice cultivation. Through the decrease of the salinity, the change of the composition of exchangeable bases, the accumulation of organic matters, and the improvement of structure and biochemical activities in the soda solonchaks, the properties of soil profile, especially those of plowed layer were greatly improved; soil layers became looser and more permeable. As a result of long-term rice cultivation, soda solonchaks tended to forming of paddy soils through of the continuous process of desalinization.

REFERENCES

(1) Chen En-feng, Wang Ru-yong, Cheng Tung-liang, Hu Shin-ming, Shin Pin-ruh, The meadow solonetzic solonchak at Kuo-chien-chi irrigated region of Chilin Province. Acta Pedologica Sinica, 5(1), 61-77(1957). (in Chinese with English summary)
(2) Chen En-feng, Wang Ru-yong, Hu Shin-ming, Wang Shun-yu, Cui Lian-wu, The melioration of soda-solonchak at Kuo-chien-chi irrigated region of Chilin Province. Acta Pedologica Sinica, 10(2), 201-215 (1962). (in Chinese with Russian summary)

ON THE TENDENCY OF ORGANIC MATTER ACCUMULATION IN PADDY SOIL UNDER TRIPLE CROPPING SYSTEM IN SUBURBS OF SHANGHAI

Xi Zhen-bang
(Soil and Fertilizer Institute,
Shanghai Academy of Agricultural Sciences)

Under certain ecological environment and agricultural system, the content of soil organic matter generally sustained in a dynamic-balance status. Once the agricultural system changed, this balance broke up instantly, and before the new balance was established, the content of O.M. in soil would change either higher or lower.

Since the development of double rice of triple cropping system in 1964, the multiple crop index of field rose from 160% to about 240%, and the plowing frequency as well as the crop yield per unit area also increased, so much attention has been paid to the changing tendency of soil organic matter(1). This is a complicated problem and demands further research. According to our surveys and experiments made recently, it was found that the organic matter and the total nitrogen in the cultivated horizon of some main paddy soils in the suburbs of Shanghai showed a tendency of accumulation. Take for example the survey of five experimental plots in our Academy. After the development of triple cropping system in 1973, the content of organic matter in the cultivated horizon had increased by $0.56\pm0.14\%$ and the total nitrogen by $0.0096\pm0.0011\%$ ($\overline{X}\pm SD$), as compared with that in 1964 when double cropping system was being practised (Table 1).

Table 1 The change of content of
O.M. and total N in the
cultivated horizon

Years	O.M.(%)	Total N(%)
1964	1.58 ± 0.14	0.096 ± 0.0047
1975	2.14 ± 0.19	0.105 ± 0.0065
Increment	0.56 ± 0.14	0.0096 ± 0.0011

Note: The soil was greyish-brown upland paddy soil.
Samples were taken from five sites in the
same plot in the cultivated horizon in different
years, the depth being 0-18 cm in 1964 and 0-15
cm in 1975. The data ($X\pm SD$) were the average of
5 sites of and the standard difference

According to the statistical data from four main paddy soils in the suburbs of Shanghai, the content of O.M. and N in the cultivated horizon in 1973-1977 had increased by 0.5-0.6% and 0.02-0.03% respectively, as compared with that in 1962-1964 (Table 2).

Why was this accumulative tendency formed? At present there is still no identical viewpoint. Results from preliminary investigation and study showed that the fundamental reason might be: with the development of triple cropping system and the increase of multiple cropping index, the application of fertilizers increased tremendously, which laid a material foundation for the accumulation. According to the statistical data in 1973-1976, the application of chemical N and organic manure were 360 N kg/ha and 263 N kg/ha respectively, an increase of about 4-folds and 1-fold respectively as compared with that in 1963. The balance sheet of nitrogen in the field during 1966-1975 showed that an average of 53 kg of N had been gained by every ha of farmland each year (Table. 3).

Under certain ecological conditions, the C/N of the same type of soil is generally stable (2,3). Thus, the increase of N income implies that carbon and organic matter may also increases. Supposing that the C/N of the soil in Shanghai is 10, then there would be an increase of 914 kg (53 x 10 x 1.724 = 914) of organic matter per ha every year. If this went on for 10 years, it would come to 9140 kg which could make the soil increase its organic matter by 0.4%.

It should be pointed out that the chemical nitrogenous fertilizer accounted for 62% of N income in the N balance sheet(1). This made the triple cropping system feasible, and through increasing the reservation of solar energy and the returning of organic material to the field, the consumed organic matter in the soil could be compensated(4). Results from calculations of storing solar energy in agricultural products and the amount of chemical nitrogenous fertilizers before and after the development of triple cropping system showed that with the increase of applied N fertilizer and grain yield per unit area, in spite of the decrease of energy storing efficiency per unit chemical N (from 1:6 to 1:3), the total amount of solar energy stored in the agricultural products had apparently increased from 66 million K-cal/ha to about 100 million K-cal/ha. Thus, under the present status— about 20% of agricultural products being returned to the fields in the form of organic manure from the increase of the application of chemical N, every ha of farmland would have an increase of 5.3-9.2 million K-cal/ha of biological energy, equivalent to an increase of 1200-2200 kg of organic material per hectare each year (Table 4).

It could be seen that the application of chemical fertilizer produced through the consumption of mineral energy, capable of making the crops store more solar energy and increasing the organic material return to the fields, had an active effect on the main solar energy storing form-organic matter in the soil(4).

On the other hand, under the condition of triple cropping system, due to the shortening of growth duration of each rice crop and also due to the prolonging of waterlogging period by about 40 days, the mineralization condition of the soil organic matter was getting poorer than ever, and was beneficial to the accumulation of O.M.(1,5). In accordance with these, the percentage of recovery of chemical fertilizer and organic manure in the very season was apparently decrease(5,6). The determinations made in the field in recent years showed that the recovery rate of N from ammonium bicarbonate and from organic manure by double cropping rice was $25.6 \pm 8.2\%$ (n = 22, by difference), and $8.1 \pm 6.6\%$ (n = 18, by difference) respectively, both lower than the average level of other districts in China. At the same time, the yield increasing efficiency per unit weight of fertilizer N was also decreased accordingly, say, more than 1/3 (Table 5).

Table 2 The accumulative tendency of O.M. and N in the cultivated horizon of paddy soils in the suburbs of Shanghai

Type of soils	Years	Number of samples	O.M.(%)	Total N (%)	Increment(%) O.M.	Increment(%) Total N
Greyish-blue lowland paddy soil	1973–1977	12	5.10$^+_-$0.61	0.27$^+_-$0.032	0.56*	0.023**
	1962–1964	4	4.45$^+_-$0.20	0.248$^+_-$0.026		
Greyish-brown upland paddy soil or brownish-yellow upland paddy soil	1973–1977	11	2.85$^+_-$0.49	0.158$^+_-$0.019	0.62**	0.020*
	1962–1964	9	2.23$^+_-$0.38	0.138$^+_-$0.013		
Sandy upland paddy soil	1973–1977	10	2.35$^+_-$0.27	0.141$^+_-$0.018	0.63**	0.029**
	1962–1964	5	1.72$^+_-$0.31	0.112$^+_-$0.014		

Note: 1. * and ** indicate a 5% and 1% of significant level respectively by T test

2. Most of the samples in 1973–1977 were obtained from 0–15 cm and 0–18 cm in 1962–1964

Table 3 The balance sheet of nitrogen in the agriculture of Shanghai suburbs (the average in 1966–1975)

N input		N output	
Object	kg/ha/year	Object	kg/ha/year
Chemical fer.	238	Harvest	194
Hog manure	69	Loss	
		(denitrification+NH_3^+volatilization)	
Green manure			
Winter green munure	11.3	Chemical fertilizer	107
Summer green manure	3.8	Organic manure	26.3
Night soil, garbage	4.5	Leaching, runoff	1.5
Rice straw	5.3		
Rapeseed and cotton aeed cake	9.8		
Seeds	7.5		
Irrigation water	9.8		
Precipitation	23.3		
Total	382		329

Note: The above table is calculated by the method described in the reference
(1), loss through leaching and runoff adopted from the data of Suzhou Prefecture and precipitation from the data suggested by Lu Ru-kun

505

Table 4 Promotion of chemical N on solar energy storing in agricultural products

Year	Yield (t/ha)		storing of energy in agricultural product (million K-cal)		The relation of chemical N with energy storing		
	grain	straw	Each year, ha.	20% return to field	Amount of N applied (kg/ha)	Amount of energy stored(million) K-cal/kg N)	Energy consumed by chemical N: energy storing by crop
1964	7.2	8.6	66.4	13.3	115	0.15	1:6.0
1975	10.1	12.1	93.2	18.6	287	0.08	1:3.2
1979	12.2	14.6	113	22.6	411	0.08	1:3.2

Note: The selection of parameter: every hectare of farmland received 11.4 billion K-cal solar energy (Shanghai), every gram of agricultural product stored 4.2 K-cal of energy. The production of every gram of chemical N consumed 25 K-cal energy, The contribution of chemical fertilizer was counted as 50% to the total in increment yield

Table 5 The efficiency of fertilizer N (before and after the development of triple cropping system)

Years	Increase in yield (kg) of rough rice per kg N applied		
	Ammonium sulphate	Ammonium bicarbonate	Organic manure
Around 1964	19.5+2.75 (n=4)	12.9 (n=1)	7.5*
1977-1979	8.95 (n=2)	9.41+4.5 (n=22)	4.45+2.93 (n=18)

Note: * indicates investigated and statistical data

Table 6 Yield of double cropping rice under different fertilizer treatment (1979)

Double cropping rice	Full fertilizer yield (t/ha)	Relative yield(%)				Relative contribution (%) fertilizers to the yield	
		Full fer.	Chemical fer.	Org. manure	Non-fer.	Chemical fertilizer	Organic manure
Early rice	6.5	100	100.1	81.4	77.0	23.1	4.4
Late rice	5.7	100	100.8	82.4	75.2	25.6	7.2

Note: 1. The data given in the table are the averages of 8 sites
2. In the chemical fertilizer plot, 893 kg/ha of ammonium bicarbonate was used; in the organic manure plot hog manure was used; in the full fertilizer plot, ammonium bicarbonate used was similar to chemical fertilizer plot, but supplemented with hog manure similar to the organic manure plot; the non-fertilizer plot was used as a control

The yield increase of organic manure in the very season was low, especially the first season, only an increase of 4-7% over the nonfertilizer plot being observed. So, in order to obtain an increase in yield of the very season, one couldn't help relying on chemical fertilizers (Table 6) (1).

The above-mentioned accumulative tendency of organic matter in soil might be beneficial to the successive increase in crop yield on the soils of low fertility where the content of organic matter was lower than 1.5%, and exert detrimental influence on the waterlogged paddy soils where the content of organic matter had already reached a level of 3-6%. Owing to the increased relying on chemical fertilizer N, the production cost increased significantly. These facts have brought about a situation in which the current agricultural system seems to be in low economic efficiency and have given rise in recent years to such problem as often to be complained by the broad masses.

Practical work has proved that to maintain a high crop yield per unit area, to set up an agricultural system with high effciency and to maintain an optimum content of soil organic matter, the following points must be taken into consideration:

1) To carry out the rotation of rice and upland crops and improve the physical property of soil and the decomposition condition of organic matter.

2) To change the method of returning organic manure, in which the organic manure must be rotted and then applied mainly to upland crops (wheat, barley, naked barley, rapeseed and cotton).

3) To practise according to soil conditions and rationally decrease the multiple crop index (1).

4) To increase the recovery rate of chemical fertilizer in the very season.

REFERENCES

(1) Hseung Yi, Xu Qi, Yao Xian-liang, Zhu Zhao-liang, Effect of cropping system on the fertility of paddy soil. Acta Pedologica Sinica, 17 (2), 101-119(1980). (in Chinese with English summary)
(2) Russell, E.W. Soil Conditions and Plant Growth. 10th edition, Science Press, Beijing, 252-271(1979). (in Chinese translation)
(3) Schmitt, L., Ertel, H.(ed.), A Hundred Years of Successful Fertilizer Practice. Communications of the IIIrd World Fertilizer Congress with All Contribution to the Discussions, 43-64(1959).
(4) Burrows, W.C., Sticker, F.C., Nelson, L.F., Nitrogen fertilization helps plants capture solar energy. Fert. Solutions, 22(2), 82-83(1978).
(5) Chu Chao-liang, Liao Hsian-ling, Tsai Kuei-shihn, Yu Chin-chou, Soil nutrition status under "rice-rice-wheat" rotation and the response of rice to fertilizers in Suzhou district. Acta Pedologica Sinica, 15(2), 126-137(1978). (in Chinese with English summary)
(6) Hsi Chen-pang, Pien Yi-chieh, Kuang An-chi, Liu Te-pen, Liu Ming-ying, Studies of peak nutrient uptake of double cropping rice and the method of basal dressing with volatile nitrogen fertilizer to the whole plough layer. Acta Pedologica Sinica, 15(2), 113-125(1978). (in Chinese with English summary)

STAGNANCY OF WATER IN PADDY SOILS UNDER THE TRIPLE
CROPPING SYSTEM AND ITS IMPROVEMENT*

Li Shi-jun, Li Xue-yuan
(Huazhong College of Agriculture, Wuhan)

Shiyue Production Brigade is located on the north bank of the Changjiang
River at the foot of the Dabie Mountain. The soils developed on granitic
gneiss in situ or washed down from upper stream are of the Yellow-brown earth
type and has been under cultivation for hundreds of years. Paddy soils occupy
about 90% of the total cultivated area in the Brigade, and most of paddy soils
(approximately 70%) are distributed in the bottom of concave topography and
slightly rolling flat land. Before 1955, the soils were given to double cropping system (rice-winter seed crop or green manure) a year. From 1956-1965,
the cropping system was changed to double rice-green manure a year (rice-rice-
milk vetch). From 1966-1972, part of milk vetch was replaced by barley, wheat
or rapeseed. Owing to the improvements in production conditions and farming
technique and the changes in the cropping system, the annual output of crops
has remarkably increased. Since 1972, the area of rice-rice-winter seed cropping has been extended to 72% of the total paddy fields and the average yields
per hectare of double cropping rice have gone up to 10500-12000 kg and that of
wheat up to 1500-2250 kg (or rape-seed 750-1500 kg). However, as a result of
the prolonging waterlogging each year, the properties of soil showed a tendency towards degradation and annual output stayed put or even decreased. In
recent years, the Brigade practised a new system of rice-cotton rotation.
One or two crops of cotton were grown between years of rice culture. The
winter crops following cotton grew much better than those under the continuous
triple cropping system. The rice crop grown the next year after cotton was
also markedly improved.

Since 1977, the soil conditions under the triple cropping system have
been investigated, as well as the effects of cotton crop on soil properties
and succeeding crops.

SOIL FACTORS LIMITING HIGH-YIELDING OF CROPS

Most of the Brigade's soils have a plow horizon of 15-17 cm in thickness and are rich in plant nutrients (Table 1). Since 1972, on a number of
paddy soils located at the bottom of concave topography and slightly rolling
flat land, the winter crops showed pronounced symptoms of soil water stagnancy
and anaerobiosis; the tillering of early rice had to be postponed; and the
demand for N fertilizer much increased without significant benefit on yield
(Fig. 1). Soil factors limiting high-yielding of crops on these soils are
mainly in two aspects.

---

*The project was carried out under direction of Prof. Chen Hua-kui.

509

Table 1 The contents of nutrient in plow horizon (Oct.,1977)

Plot No.	Crop system	pH	Total N (%)	Available P (ppm)	K(ppm)
1	Rice-rice-winter seed crop	6.0	0.215	23	106
2	Rice-rice-milk vetch or winter seed crop	6.2	0.138	39	121
3	Rice-rice-winter seed crop	6.2	0.218	30	149
5	Rice-cotton rotation	6.2	0.153	28	168
6	Rice-rice-milk vetch or winter seed crop	6.2	0.139	37	86
11	Rice-rice-winter seed crop	6.0	0.167	43	110

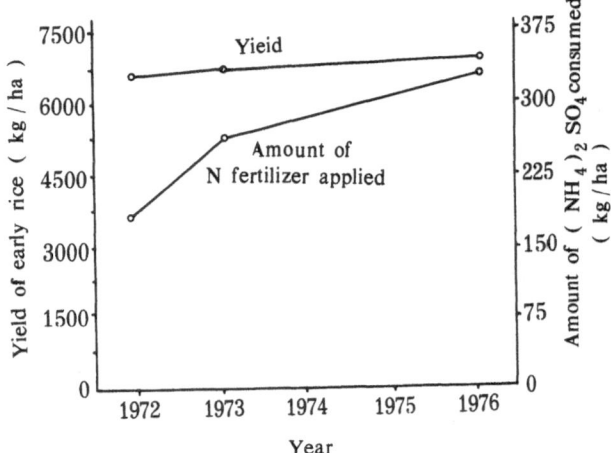

Fig. 1 Relation between the yield of early rice and amount of N fertilizer consumed in the twelfth team

1. Decrement of Soil Non-capillary Porosity and Aeration

Table 2 shows that the non-capillary porosity of soils at a depth of 6-11 cm is apparently related to the cropping system. Soil samples were taken at the time of harvesting late rice before the next plowing. The average value of non-capillary porosity of rice-cotton rotation amounted to 7.0-8.6%, that of continuous rice-rice-winter seed crop amounted to 4.7-5.9%, that of continuous rice-rice-winter seed crop or milk vetch amounted to less than 4.7%, that of continuous rice-rice-milk vetch amounted to 1.8%. The non-capillary porosities at a depth of 22-27 cm varied from 0.0-3.5% which were affected similarly by different cropping systems. Apparently, the low non-capillary porosity of continuous triple cropping system is mainly the result of longer period of waterlogging[1,2] as well as the puddling effect of mechanized plowing and harrowing[3], which are destructive to soil structure. The condition of soil aeration is also affected by differences in topography and soil texture.

Table 2   Effect of cropping system on the aeration
of plow horizon (Oct., 1977)

Cropping system	Non-capillary porosity (%)	Soil aeration
Rice-cotton rotation	7.0-8.6	Satisfactory
Rice-rice-winter seed crop	4.7-5.9	Rather poor
Rice-rice-winter seed crop or milk vetch	<4.7	Poor
Rice-rice-milk vetch	1.8	Very poor

2. Development of Gleying

As shown in Fig. 2, in heavy textured plots located at the bottom of concave topography, a grayish-blue gleying layer often develops below the plow horizon or the plowpan, if the soils are continuously given to triple cropping system. Below the gleying layer, there appears a yellowish brown illuvial horizon. Gleying is most pronounced in rice-rice-milk vetch plots.

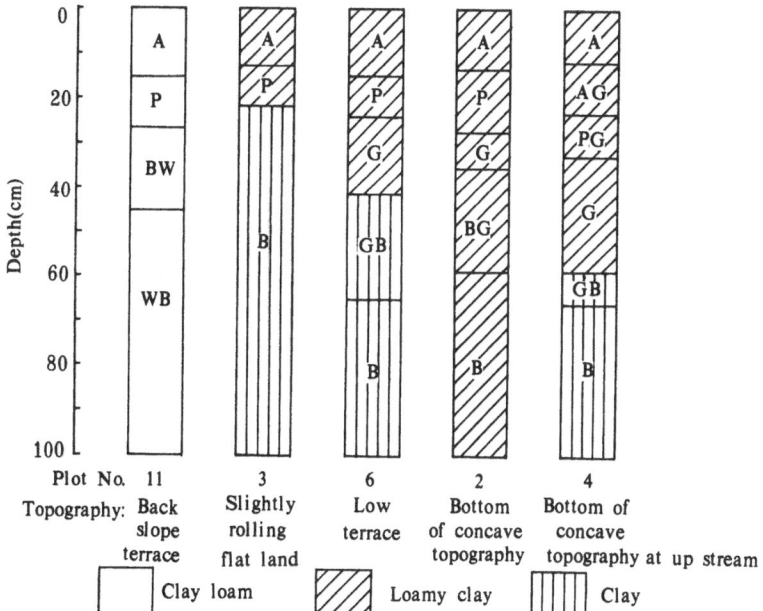

Fig. 2   Effect of topography and texture on gleying
A. plowed layer; P. plowpan; B. illuvial horizon;
BW. B-percogenic horizon; WB. percogenic-B horizon;
AG. glei subplowed layer; PG. glei plowpan; GB. glei
B horizon; BG. B-glei horizon and G gley horizon

As gleying develops, soil aeration deteriorates further. The gleying horizon impedes the downward movement of surface water. Fig. 3 shows that the more developed the gleying, the more lowered the Eh value in the entire soil profile in spring. With the development of gleying by degrees, the curves of Eh value at different depths change from mild variation to sharp ups and downs. Data presented in Fig. 4 show that with the advancement of gleying, there are corresponding lowering of Eh value and increasing of active reducing matter during the early stage of early rice. Gleying also retards the development of root system of early rice, lowers its capability of nutrient absorption, postpones the development of early rice and results in decreased yields under the same cultivation conditions (Fig. 5). Probably, the diminution of absorbing root system is one of the reasons of the increased demand for chemical fertilizers.

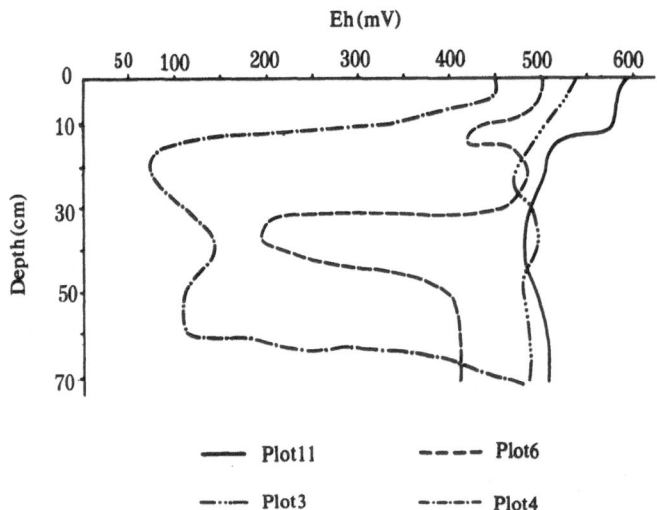

Fig. 3 Eh values in soils of different degree of gleying (Mar. 1979)

Gleying is detrimental to the production of winter crops. For instance, in 1977-1978, under the same cultivation conditions, the yields of wheat were 3045 kg/ha on a gleyed loamy clay plot (plot 2), 3405 kg/ha on a nongleyed but compacted loamy clay plot (plot 3), and 3735 kg/ha on a nongleyed and aeration satisfied clay loam plot (plot 11) located on terrace.

RICE-COTTON ROTATION AS MEANS OF IMPROVING WATER-AIR RELATION AND CROP YIELD

1. Cropping Systems on Comparable Plots

Plot 3 and plot 5 are comparable plots: before 1977, both were planted to continuous fice-rice-wheat (or milk vetch), both were compacted and were similar in other aspects.

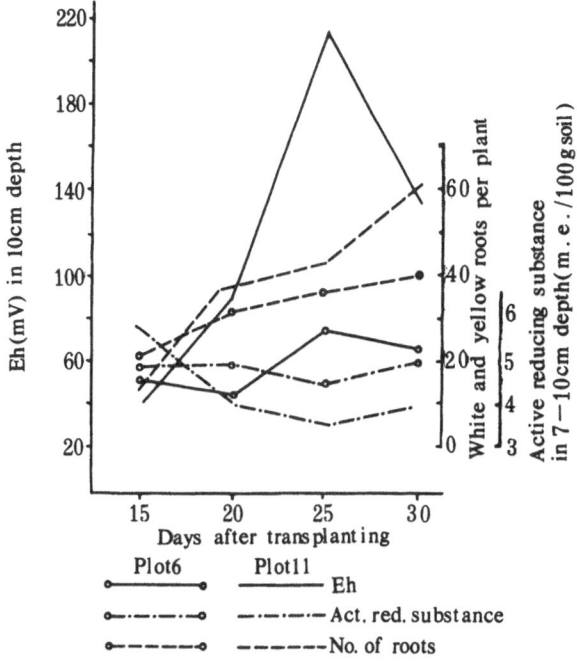

Fig. 4 Effect of gleying on soil properties
and rice roots (1979)

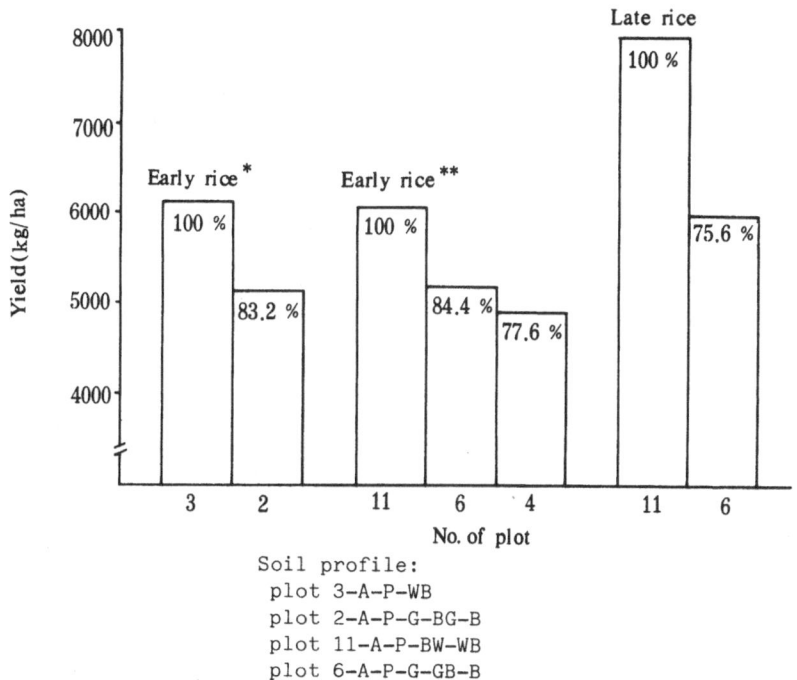

Soil profile:
  plot 3-A-P-WB
  plot 2-A-P-G-BG-B
  plot 11-A-P-BW-WB
  plot 6-A-P-G-GB-B
  plot 4-A-AG-PG-G-GB-B
*The last crop is rapeseed **The last crop is wheat
Fig. 5 Effect of gleying on rice yield (1979)

From the winter sowing of 1976 up to the antumn harvest of 1979, Plot 3 continued the triple cropping system, while plot 5 was changed to two years of cotton and similar winter crop as the plot 3; after this, plot 5 was replanted to rice-rice in 1979. Another plot (plot 1) was selected, in which one crop of cotton was grown after continuous rice-rice-winter crop and after cotton, the plot was replanted to rice-rice again.

2. Improving of Soil Porosity after One or Two Years of Cotton

As shown in Table 3, after one or two years of cotton the non-capillary porosity improved markedly. Soil samples were taken during the harvesting time of late rice, in 1977 and 1978, non-capillary porosities increased to 7.0-8.6%, which were satisfactory for plant growth.

Table 3 Effect of rice-cotton rotation on soil

aeration (soil profile: A-P-WB)

Plot No.	Texture	Cropping system	Year	Depth (cm)	Non-capillary porosity(%)
3	Loamy clay	Rice-rice-winter seed crop	1977	6-11	4.7
				22-27	0.0
5	Loamy clay	Intercepted by 1 year cotton	1977	6-11	8.0
				22-27	3.5
		Intercepted by 2 year cotton	1978	6-11	8.6
				22-27	3.2
1	Clay loam	Rice-rice-winter seed crop	1977	6-11	5.9
				22-27	0.2
		Intercepted by 1 year cotton	1978	6-11	7.0
				22-27	3.3

3. Effect of Two Years of Cotton on Succeeding Crops

1) Effects on soil Eh value, active reducing matter and number of Healthy rootlets (white and yellow). As shown in Fig. 6, Eh values were higher, contents of active reducing matter were lower and the number of healthy rootlets was higher in plot 5 than those in plot 3, as sampled in 15-30 days after transplanting in 1979.

2) Effect on crops yields. As shown in Fig. 7, the yields of early rice, late rice and wheat in plot 5 were 8.6, 19.3 and 22.8 % higher respectively than those in plot 3.

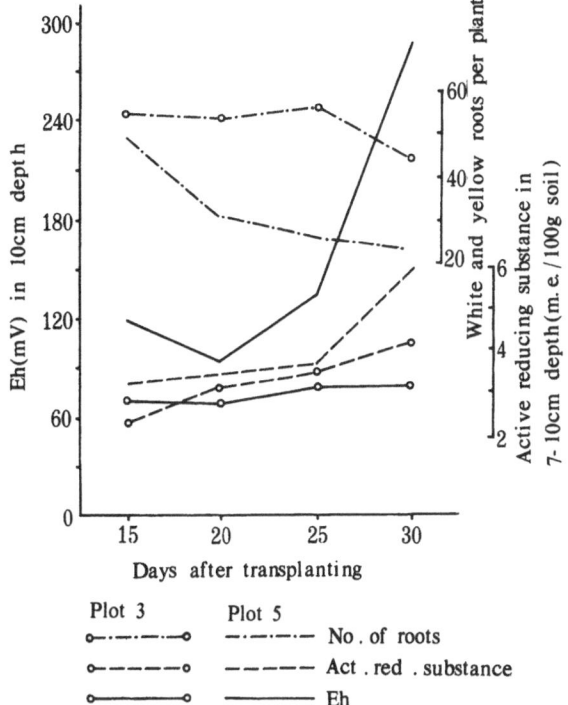

Fig. 6   Effect of cotton on next years soil
properties grown to rice (1979)

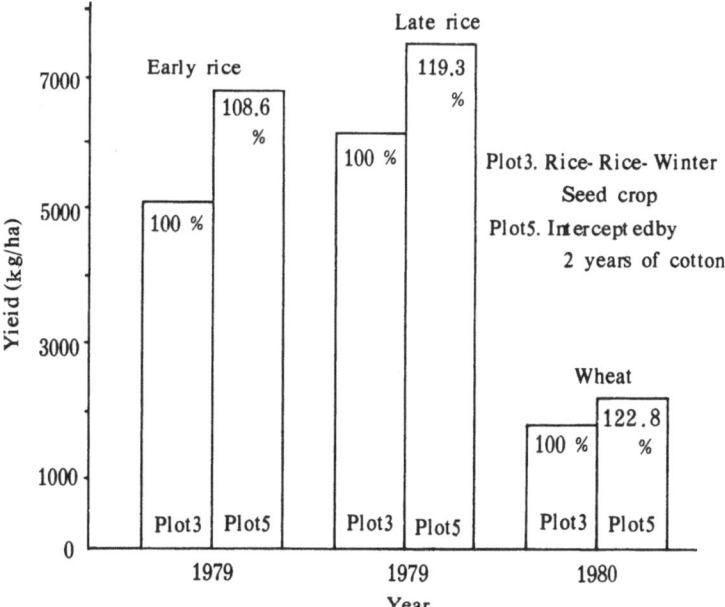

Fig. 7   Effect of rice-cotton rotation on crop yield

REFERENCES

(1) Chen Jia-fang, Wu Mei-ling, He Qun, Liu Bin, Studies on the compacting
    of paddy soils in Snzhou, Jiangsu. Turang, 6,286-290(1975). (in Chinese)
(2) Chen Jia-fang, Zhao Cheng-zhai, Zhou Zheng-du, Characteristics of pore
    space in arable horizon of compact paddy field in Suzhou Prefecture
    Jiangsu. Turang, 3, 81-84(1978). (in Chinese)
(3) Institute of Soils, Fertilizers and Plant Protection, Shanghai Academy
    of Agricultural Sciences, Effect of plowing and harrowing on soil
    structure under various soil moisture conditions. Turang, 2, 52(1978).
    (in Chinese)

DRAINAGE OF PADDY SOILS IN TAIHU LAKE REGION

Cheng Yun-sheng

(Institute of Soil Science, Academia Sinica, Nanjing)

The fertility of paddy soils in the Taihu Lake region is relatively high(1-3). For example, the annual grain output in Wuxi County, Suzhou Prefecture of this region has topped more than 12 tons per hectare. These soils are heavy in texture and rich in nutrients. In the last decade these soils have been cultivated under a system of three crops a year with two crops of rice. Since then the duration of waterlogging has been prolonged, the water table rose and soil aeration worsened. It is now more important to apply field drainage installations to improve the physical conditions of these soils.

The soils in this region contain 20 to 30 % of clay, being less in the upper layer and higher in the lower(2,3). The volume weight of the soil increases gradually from 1.20 g/cm^3 in the cultivated horizon to 1.35 g/cm^3 in the plowpan and to 1.56 g/cm^3 in substratum. With the exception of plow layer where the total pore space is greater than 50 %, the total pore space in lower layers decreases gradually being only 40 % in substratum (Table 1).

Table 1 Volume weight and porosity of different
horizons (mean of 15 profiles)

Horizon	Volume weight (g/cm^3)	Total porosity (%)
Cultivated	1.20±0.07	53.45±2.59
Plowpan	1.35±0.09	50 02±3.37
Subsoil	1.50±0.08	44.14±2.77
Substratum	1.56±0.06	41.78±2.04

Among the different kinds of pores, the distribution of various equivalent pores is given in Table 2. Pores with diameter greater than 0.2 mm in the plow layer occupy 22 % of the total proe space, while those in the plowpan occupy only 11 %; pores with diameter in the ranges of 0.2 - 0.01, 0.1 - 0.05, 0.05 - 0.01 and 0.01 - 0.005 mm in plow layer and plowpan occupy 2 -5 % and 1 - 3 % of the total respectively; pores with diameters smaller than 0.005 mm in the two layers occups 66 % and 81 % of the total respectively. It is evident that the unfavorable distribution of various equivalent proes in the plowpan may lead to poor permeability of air and water. The saturated hydraulic conductivity K is given in Table 3. these values are quite different in various horizons due to different distribution of the various equivalent pores. With the exception of cultivated horizon, where the K value is very high, those in lower horizons are very low. Therefore, if there is no field drainage installation in these soils, the air-water relationship in the soil can not be regulated only by natural soil drainage.

Table 2 Distribution of soil equivalent pores

Diameter of pores (mm)	Cultivated horizon (29)*		Plowpan (28)*	
	%	% of total	%	% of total
0.2	11.48±3.11	22	5.49±2.30	11
0.2-0.1	0.92±0.52	2	0.41±0.11	1
0.1-0.05	1.08±0.34	2	0.54±0.14	1
0.05-0.01	2.64±0.86	5	1.69±0.70	3
0.01-0.005	1.55±0.45	3	1.26±0.53	3
<0.005	34.95±3.08	66	40.92±3.15	81

* Number of samples

Table 3 Saturated hydraulic conductivity (K) of different horizons of a paddy profile

Horizon	K (cm/day), 20°C
Cultivated	1040
Plowpan	1.66
Subsoil	3.10
Substratum	1.00

The results of investigations in the last ten years showed that the tile drainage installed in paddy soils has made effective control of the excess water in the soil(3-5). For example, in Fig. 1 the change in ground water table during dry-farming period indicated that if the precipitation was greater than 30 mm the water table rose rapidly. The depth of the water table after the rain was lower in the plot installed with 110 cm tile than in the plot installed with 80 cm tile and was higher in check plot than in two plots installed with tiles. However, the rise of water table after the rain lowered as the depth of the tile installed increased. When the daily precipitation is over 50 mm the water table will rise up near surface layer(3). After raining the lowering speed of the water table hastened as the depth of the tiles insta led increased. The superiority of the tile drainage is more evident during rainy years(3). Fig 2 shows that after raining the soil water suction during the same period was greater in the plot installed with the tiles than in check plot, and was also greater in the upper layer than in the lower ones. However, the magnitude of the variation in soil water suction with its influence on the depth of soil horizon was in accordance with the variation in depth of the water table. That is to say the lower the tile installed the better the draining of the soil water.

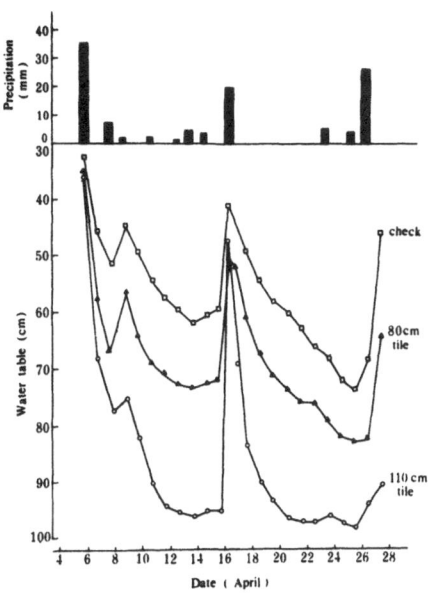

Fig. 1 Variations in depth of the water
table with tile drainage

The results from the experiment using ^{32}P in Table 4 indicated that the root system of barley went down to 40 cm deep in the plot installed with tiles, while that in check plot reached to a depth of only 15-20 cm. This is the reason why crops have well-developed root system and high yield in a field installed with tile drainage. Generally, 10 to 20 % of increase in wheat or barley yield may be obtained(3,4).

The application of drainage in paddy soils is also beneficila to rice crop(6). Although large quantity of irrigation water is applied to rice field, the soil proes have not been entirely occupied by water. Experimental results indicated that the volume weight of paddy soil under submerged conditions was about 1.00 g/cm^3 due to the swelling of the plow layer as the soil body absorbed water. After draining the water content of the soil gradually decreased to a point near field capacity and the volume weight of the soil gradually increased to about 1.20 g/cm^3 . Thus, the three phases of the soil, i.e. solid, liquid and gas, varied greatly(6,7). Under submerged conditions the gaseous phase occupied 1 to 5 %. Fig. 3 shows that at the 8th day after draining the proportion of the gaseous phase increased by over 10 %. The oxidation-reduction potential of the soil increased from negative values before draining to over 500 mV after draining(6). The oxidation-reduction potential of paddy soils is mainly controlled by oxygen status of the soil. The amount of dissolved O_2 in soil solution is directly proportional to the partial pressure of O_2 in the soil air.

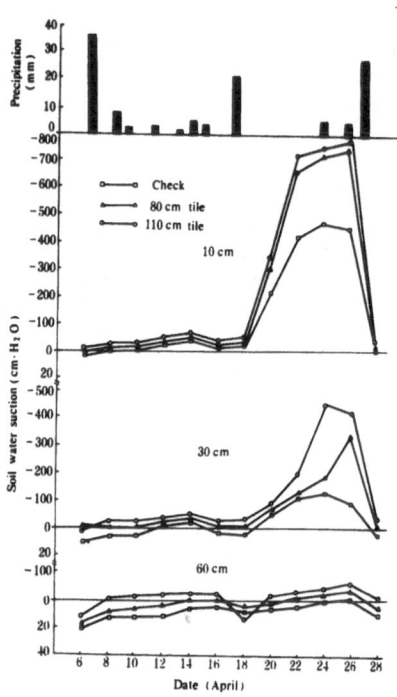

Fig. 2 Variations in soil water
suction with tile drainage

Table 4 Effect of soil drainage on root
activities of barley ($32p$)

Depth (cm)	Tile plot		Check plot	
	Counts/min	%	Counts/min	%
3	1996	41.9	3550	79.5
15	1125	24.2	623	19.4
30	1168	24.5	29	0.9
45	473	9.4	4	0.2

The results from our investigation in Fig. 4 indicated that the amount of O_2 in soil air increased from 5 % before draining to 18 % after draining immediately and then biological activities reduced it to 8 %, while that remained about 5 % in flooding plot. The amount of CO_2 increased markedly at the early period of draining and then dropped to a lower figure——increased from about 8 % to 12 % and then decreased to below 6 %. This amount of CO_2 was lower than in flooding plot. If CO_2 accumulates in excessive amount in the soil it will be unfavorable to plant roots and microorganisms. Thus, the drainage of paddy soil for renewing soil air and preventing the accumulation of CO_2 is a matter of great significance to ensuring a favorable environment for the growth of plants(6).

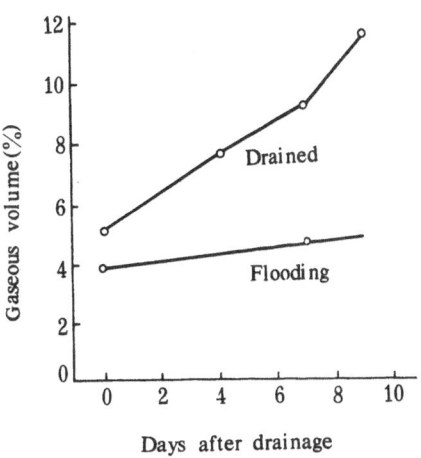

Fig. 3 Effect of drainage on gaseous volume of a paddy soil

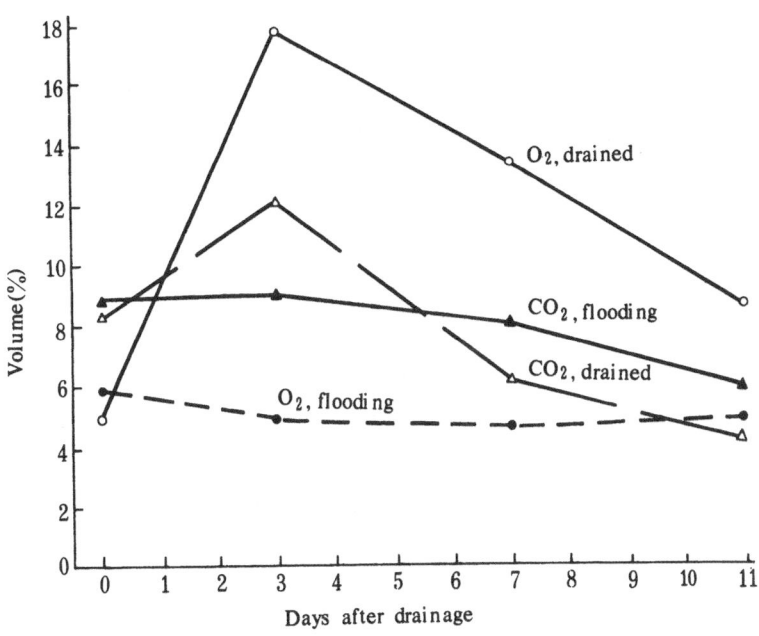

Fig. 4 Effect of draingage on composition of soil air

The drainage of paddy soils may also lead to the increase in dissolved O_2 in soil solution. The results of investigation from a field study are given in Fig. 5. It was evident that the amount of the dissolved O_2 of the percolation water in a plot with drainage increased from 0.65 mg/1 before draining to 1.69 mg/1 after draining, while in check plot the amount of dissolved O_2 remained at 0.6-0.8 mg/1. All of these findings fully indicated that the drainage of paddy soils greatly improved the aeration condition of soils and is favorable for the growth of rice. Generally, 10 % in yield increase of rice can be obtained[6].

The results from the recent investigation by Kunshan Farm Land Drainage-irrigation Institute[5] showed that those fields installed with tile could drain away most of the water within 3 to 4 days after a heavy rainfall. The speed of the drainage was 2 times faster than that of natural soil drainage. Moreover, it is possible to keep the irrigation water in the paddy field until 2-3 days before the harvest. This is favorable not only for rice, but also for the land preparation of the succeeding crop. The net result is a greater potential for crop production all the year round.

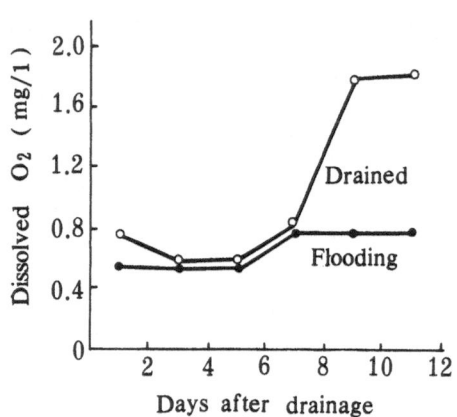

Fig. 5 Effect of drainage on dissolved O_2 content of the percolation water

The depth and spacing of tiles vary with the soil. As shown in Table 5, the depth of tiles for clayey soil, loamy soil and sandy loam should be about 1.1-1.2, 1.0-1.2 and 1.0 m respectively, and the spacing between tiles about 6-10, 15 and 20 m respectively. Therefore, the field tile drainage is an indispensable practice to increase the fertility of paddy soils.

Table 5 Depth and spacing of tiles (m)

Soil	Depth	Spacing
Clayey	1.1 - 1.2	6 - 10
Loamy	1.0 - 1.2	15
Sandy loam	1.0	20

REFERENCES

(1) Institute of Soil Science, Academia Sinica, Nanjing(ed.), Soils of China, Science Press, Beijing, 23-35 (1978). (in Chinese)
(2) Hseung Yi, Xu Qi, Yao Xian-liang, Zhu Zhao-liang, Effect of cropping system on the fertility of paddy soils. Acta Pedologica Sinica, 17(2), 101-119 (1980). (in Chinese with English summary)

(3) Zhao Cheng-zhai, Cheng Yun-sheng, The effect of the tile drainage on rice soils in Suzhou district, Jiangsu. Acta Pedologica Sinica, 15(2), 187-193(1978). (in Chinese with English summary)

(4) Yang Jin-lou, Zhu Ji-cheng, Jiang Su-zhen, The drainage problem of the soils in the suburbs of Shanghai, Acta Pedologica Sinica, 15(1), 95-100(1978). (in Chinese with English summary)

(5) Kunshan Farm Land Hydraulic Station, Jiangsu Province, Effect of the tile drainage on the melioration of soil waterlogging. Turang, 2,45-46(1978). (in Chinese)

(6) Cheng Yun-sheng, Zhao Guo-hua, A preliminary study on the effect of periodical drainage of paddy soils. Acta Pedologica Sinica, 11(3), 275-285(1963). (in Chinese with Russian summary)

(7) Satoru Motomura, Ashara Seirayosakol, Wisit Cholitkul, The changes in some physical and chemical properties of paddy soils under water management. Symposium on water management in rice field, Proceedings of a symposium on tropical agriculture researches, August, 1975. Tropical agriculture research series 9, 101-115 (1976).

IMPROVING RICE SEEDLING EMERGENCE IN FLOODED SOIL
BY USE OF CALCIUM PEROXIDE

S. Yoshida, F.T. Parao
(The International Rice Research Institute)
P.O. Box 933, Manila, Philippines

INTRODUCTION

Direct seeding normally practiced by farmers in the tropics is to broadcast pregerminated seeds on drained surface of puddled soil. Drained condition provides rice seeds with adequate oxygen for normal growth, but it also
favors early and vigorous growth of weeds. Furthermore, seedling emergence
becomes poor and uneven when it rains right after seeding and when the seeds
are covered by water. Pregerminated or presoaked seeds can also be seeded
into water of puddled soil. This, however, often causes floating and overturn
of seedlings, particularly when temperature is high, resulting in low percentages of seedling emergence. It is also a common experience that a direct-
seeded rice crop is more susceptible to lodging than a transplanted rice crop,
presumably because the direct-seeded crop has less degree of anchorage than
the transplanted crop. These problems can be overcome if seeds are placed
below soil surface of flooded puddled field and oxygen is supplied to the
seeds. A new technology - coating rice seed with calcium peroxide - has been
developed in Japan to serve such purpose (Yamada, 1951; Ota and Nakamura,
1970; Mitsuishi, 1975; Nakamura, 1978). We feel that the calcium peroxide
coating technique will have a great potential for rice cultivation in tropical
monsoon climates. The present paper describes the calcium peroxide coating
technique as it is adapted to tropical rice cultivation.

CALCIUM PEROXIDE AS OXYGEN SUPPLIER

Calcium peroxide when in contact with water releases oxygen:
$$2\ CaO_2 + 2\ H_2O \longrightarrow 2\ Ca(OH)_2 + O_2$$
Coated to rice seeds it generates and supplies oxygen to emerging coleoptile,
true leaves, and roots. Rate of oxygen release is dependent on pH of the soil
surrounding the coated seed. When the medium is acid, the rate of oxygen
release is fast, and when alkaline, the rate is slow. Test by Kurosawa (1976)
indicated that pH of soil in the close vicinity of coated seed is increased by
calcium peroxide up to 10. Thus, the rate of oxygen release proceeds slowly.
Beyond 5 mm distance of coated seed, however, calcium peroxide has no effect
on soil pH.

A commercial product of calcium peroxide known as Calper G contains
gypsum as an adhesive. Increased adhesiveness of Calper G has made it easy
to handle the coated seeds and also has made it possible to use a wet paddy
seeder capable of placing seeds below soil surface in flooded fields
(Nakamura, 1978).

COATING RICE SEEDS WITH CALPER G

Dry or pre-soaked seeds are coated with Calper G using a coating machine. Seeds are first immersed in water and floating seeds are discarded. Selected seeds are air-dried or soaked in water for 24 hours, and placed in the coating machine. The seeds are mixed with Calper G at 1:1 ratio in weight. Then, water, equivalent to about 30-35% of the weight of dry seed, is carefully sprayed to moisten the seeds. The coated seeds can be stored for about 10 weeks without any reduction in percentage of seedling emergence. That length of storage period is perhaps sufficient to make full use of one coating machine in preparation of the coated seeds for large hectarages of paddy fields.

LABORATORY EXPERIMENTS

Seedling Emergence of Different Rice Varieties from Flooded Puddled Soil

Varieties differ in their ability to emerge from flooded puddled soil (Fig. 1). Percentages of seedling emergence ranged from 0% for LMN 111 to about 60% for IR8. The emergence capacity of IR8, an indica variety, appeared as good as Reimei and Koshi-hikari, both japonicas. When seeds were coated with calcium peroxide, the rate of seedling emergence was sped up, and percentages of seedling emergence were increased to 70-95%. Improvement of seedling emergence seemed higher in varieties LMN 111 and Peta which appeared more sensitive to submerged condition than other rice varieties. Because the rate of seedling emergence was faster when coated with calcium peroxide, shoot and root length of seedlings were also better (Table 1). Thus, coating rice seeds with Calper G accelerates seedling emergence of rice seeds in flooded puddled soil, and hence subsequently shoot and root growth.

Effect of Temperature on Rate and Percentage of Seedling Emergence

Seedling emergence was slower at lower temperatures (Fig. 2). But coating seeds with calcium peroxide sped up seedling emergence in all temperatures. At the end of the experiment, percentages of seedling emergence from uncoated seeds ranged from 10% for 26/18°C to 40% for 35/27°C. When seeds were coated with calcium peroxide, they were increased to 60% for 26/18°C to 80-90% for higher temperatures. Thus, coating seeds with calcium peroxide insures high percentages of seedling emergence from flooded puddled soil at high temperatures.

Effect of Seeding Depth on Percentage of Seedling Emergence at Different Temperatures

Increased seeding depth adversely affects seedling emergence. To test effect of seeding depth on seedling emergence, the coated and uncoated seeds were placed 1, 2, 3, and 5 cm below the surface of puddled soil.

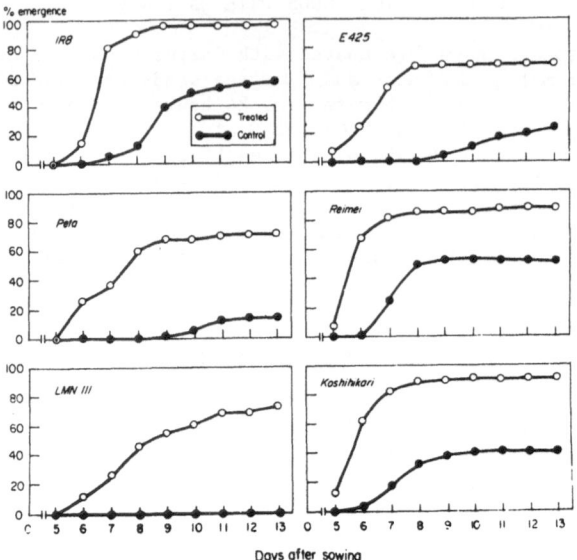

Fig. 1   Effect of calcium peroxide seed coating on seedling emergence of 6 varieties
from flooded soils at 29/21°C. (Seeding depth: 1 cm; water depth: 5 cm)

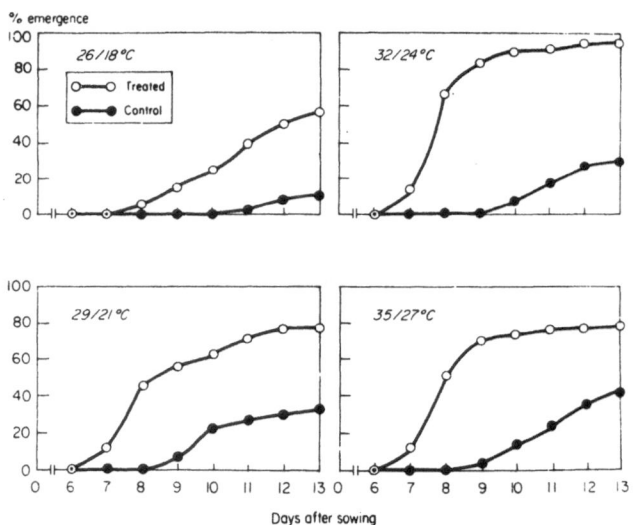

Fig. 2   Effect of calcium peroxide seed coating on seedling emergence of IR36
from flooded soils at different temperatures. (Seeding depth: 1 cm;
water depth: 5 cm)

Table 1   Effect of calcium peroxide seed coating on
seedling growth of 6 varieties at 29/21°C*

Variety	Group	Shoot length (cm)		Root length (cm)	
		Control	Treated	Control	Treated
IR8	Indica	20.2	29.5	8.8	13.0
Peta	Indica	12.6	24.3	4.3	7.7
LMN 111	Indica	0	23.8	0	7.5
E425	Indica	15.5	23.6	5.0	9.6
Reimei	Japonica	25.2	25.2	7.0	7.2
Koshihikari	Japonica	23.5	27.4	8.8	8.9

*2 weeks after sowing

Table 2   Percentage of seedling emergence of CaO_2-coated
seeds under different temperatures

Condition of seed	Temper- ature (°C)	Treated				Control			
		Soil depth (cm)							
		1	2	3	5	1	2	3	5
Dry	26/18	82	65	3	0	0	0	0	0
	29/21	93	53	32	0	3	0	0	0
	35/27	77	32	12	0	3	0	0	0
Presoaked	26/18	82	60	12	0	7	0	0	0
(48 hrs)	29/21	80	60	3	0	5	0	0	0
	35/27	70	30	0	0	15	0	0	0

Counted 22 days after seeding; water depth was 5 cm. Dry seeds were selected
seeds after immersing in water and removed of floated ones then air-dried
before coated. Presoaked seeds were dry seeds soaked for 48 hours, then
coated

Coating rice seeds, presoaked or dry, with Calper G improved seedling emergence up to a soil depth of 3 cm compared with the control (Table 2). The highest percentage of seedling emergence was obtained at 1 cm soil depth. At 2 cm, percentages of emerged seedling ranged from 30-32% for 35°/27°C to 60-65% for 26°/18°C. Apparently, adverse effects of increased seeding depth is more conspicuous at a high temperature, 35°/27°C. Very few seedlings emerged at 3 cm soil depth, and no seedlings came out at 5 cm soil depth.

Those results indicate that seeding depth is very critical. To get maximum benefit from Calper G, seeds must be placed at around 1 cm below soil surface.

FIELD EXPERIMENTS

A Comparison Among Different Methods of Seedling Establishment

Four methods of seedling establishment were compared to study effectiveness of calcium peroxide coating in wet land direct seeding (Table 3).

Fields were prepared and puddled in a usual way. Nitrogen, phosphorus, and potassium were applied at rates of 125 kg N/ha, 50 kg P_2O_5/ha, and 50 kg K_2O/ha, respectively. Machine seeded refers to the treatment where coated or uncoated seeds were sown by a wet paddy seeder. Broadcast flooded implies that coated or uncoated seeds were broadcast into flooded puddled soil before soil particles were fully settled. In the broadcast drained, only uncoated seeds were broadcast on to the soil surface after the field was drained. This is a common practice in the Philippines. The fourth treatment was transplanting of 20-day old seedlings at 20 x 20 cm spacing.

A uniform weed control measure was adopted for direct seeded plots. Twenty-five kg/ha of 2,4-D IPE granules were applied about 7 days after seeding. In addition, 2,4-D amine at a rate of 0.8 kg/ha a.i. was sprayed later when necessary. In the transplanted crop, weeds were removed by hands.

Coating seeds with calcium peroxide insured sufficient numbers of seedlings to emerge from flooded puddled soil. As a consequence, grain yields of wetland direct-seeded crops were comparable to those of the broadcast-drained and the transplanted plots. When uncoated seeds were placed in flooded puddled soil, however, percentages of seedling emergence were extremely low and uneven, resulting in low yields.

Effect of Seeding Density on Grain yield of Wetland Direct-Seeded Crop

The use of calcium peroxide adds extra cost to the farmer. Since the amount of calcium peroxide required is proportional to the amount of rice seeds to be used, it is important to examine whether seeding density can be reduced. Three seeding rates, i.e., 30, 50, and 70 kg/ha were examined in this experiment (Table 4).

Table 3   A comparison of different planting methods, 1980, IRRI*

Methods of sowing	Calcium peroxide coating	Number of emerged seedlings (no./m^2)	Grain yield (t/ha)
Machine seeded**	With	174	5.5
	Without	8	0.6
Broadcast, flooded	With	216	5.4
	Without	30	2.7
Broadcast, drained	Without	259	5.6
Transplanted***		(25)	5.4

*Variety IR36; seeding rate was 50 kg/ha
**Row spacing - 30 cm
***20 x 20 cm spacing

Table 4   Effect of seeding density on grain yield of IR36 in wet land direct seeding, 1980, IRRI

Seeding rate (kg/ha)	Number of emerged seedlings/m^2	Grain yield (t/ha)
30	157	4.20
50	223	3.94
70	273	4.24

Seedling emergence ranged from 157 plants/m^2 at 30 kg seeds/ha to 273 plants/m^2 at 70 kg seeds/ha. Since the emerged seedlings were evenly distributed in the field and IR36 is a high tillering variety, difference in the initial plant density did not affect grain yields among three seeding rates. In traditional direct seeding practice, seeding rates of 100 to 120 kg/ha are commonly used. The present experiment indicates that by coating rice seeds with calcium peroxide, seeding rate can be reduced to one-third or one-fourth of the current practice. Compared with the number of plants per square meter used in transplanted system, it might be even possible to reduce the seeding rate to 10 kg/ha provided that even distribution of seeds in the field is insured. By reducing seeding rate, we can reduce costs of both calcium peroxide and seeds. Thus, one important question in the wetland direct seeding will be how to insure the even distribution of seeds in the field. At present, tillering capacity of a variety is considered not important in the direct seeding rice cultivation. This is because the current seeding rate is so high that tillering does not contribute much to yield. When seeding rate is reduced, however, use of high tillering varieties might become imperative even in the direct seeding rice cultivation as in transplanted rice.

REFERENCES

(1) Kurosawa, T., Effect of seed coating with calcium peroxide on seedling stand in the mechanized direct-sowing rice culture on the paddy field. 2. Oxygen diffusion rate and acidity of surrounding soil of the coating seed. Rep. Tohoku Br. Crop Sci. Soc. Japan 17, 42-43 (1976).
(2) Mitsuishi, S., Study on the direct underground sowing in the submerged field of rice. Special Bulletin Ishikawa Pref. College of Agric. No. 4 (1975).
(3) Nakamura, Y., Studies on the direct underground sowing machines of rice seed in submerged paddy fields. Special Bulletin Ishikawa Pref. College of Agric. No. 7 (1978).
(4) Ota, Y., Nakayama, M., Effect of seed coating with calcium peroxide on germination under submerged condition in rice plant. Proc. Crop Sci. Soc. Japan 39, 535-536 (1970).
(5) Yamada, N., Calcium peroxide as an oxygen supplier for crop plants. Proc. Crop Sci. Soc. Japan 21, 65-66 (1951).

RICE FERTILIZATION AND WASTEWATER NUTRIENTS RECYCLING

Paolo Sequi
(C.N.R. Institute for Soil Chemistry, Pisa, and
Chemical Institute of the University of Udine, Italy)

In the opinion of ecologists, modern agriculture is generally considered as a primary cause of pollution, due to the possible loss from the soil of the added agrochemicals. This commonplace is widespread, though serious proofs of its validity are often lacking.

As a matter of fact, soil represents the hearth of natural defenses in the fight against pollution. Available data for total biomasses, e.g., show that at least two thousand kilograms of living organisms are present in a normal soil, with respect to an average of about one hundred kg of humans and two hundred kg of animals who live above the soil in a densely populated country, say Italy (1). Most chances for equilibrium of environment come from such an astonishing amount of soil biomasses, from their specialization, and the high ratio between soil and human (i.e. polluting) biomasses (2, 3).

Agriculture can be actually defined as the distortion of natural equilibria which is carried out in order to produce food and fiber. A cultivated field can be considered as an <u>artificial</u> ecosystem. Strictly speaking, agriculture is in contrast with the goals of environmental conservation in the same manner as other human activities, such as building of houses or towns. On reconsidering human activities from a more realistic point of view, however, agriculture always is potentially damaged from environmental pollution rather than compromising environmental equilibria. We must realize that soil, in fact, assumes the role of the natural cleaner of environment.

ENVIRONMENT AND AGRICULTURE: THE MODEL OF RICE CULTIVATION

Rice cultivation can excellently exemplify how agriculture is a main depolluting rather than contaminating activity. Since in developed countries water quality is often poor, and rice needs water for growth, analytical determinations carried out on in- and out-flowing water can easily give an idea of the overall action of the paddy soil cultivation on the general pattern of pollutants exchange through ecosystems. A long-term survey program has shown in Italy (4-6) that concentration of undesired chemicals in water is substantially lowered after usage for paddy soil irrigation, in spite of the concentration effects of evapotranspiration.

Some examples from the impressive mass of available findings are reported in Table 1-3. Data concerning the balance of nutrients (Tab.1) refer to the average data from two rice fields, Aramino and Biandrate, and are calculated from one-year average analytical determinations of nutrient contents in in-flowing (input) and out-flowing water (output). Although manure and chemical fertilizers were added to the two rice fields (e.g. 30 metric tons per hectar of farmyard manure, 180 kg/ha calcium cyanamide and 210 kg/ha ammonium sulphate for the Biandrate field), the amount of nutrients released with out-flowing water was always smaller than that supplied by irrigation.

Data for selected heavy metals, analogously calculated, are reported in Tab.2. As expected, soil generally retains substancial amounts of metals; data here not reported show that an appreciable proportion is then absorbed by the crop. Manganese, as known, is poorly retained under reduced conditions and in some instances its content was increased after contact of water with the soil, although comparable figures are shown from the average data. Tin is the only metal whose content is higher in out-flowing water, presumably due to the use of agrochemicals like "brestan 60" (tin triphenyl acetate), a pesticide commonly used in paddy fields for the control of growth of algae. However, one could be more amazed for the high level of tin in in-flowing water than for the increase of tin content caused by a well-defined agronomic practice.

The water content of selected pesticides (Tab.3) is generally decreased after irrigation of the rice fields. An exception concerns propanil, a commonly used herbicide, but it could be perhaps justified to be more worried about parathion level in in- flowing water than about the level of propanil in out-flowing water. Anyhow, content of pesticides was always low, except for Ordram, a herbicide arising out of rice fields upriver from those considered. No doubt, its level requires a constant attention in order to pursue a best environmental quality, although authors suggest that volatilization and degradation processes occurring in paddy fields may operate efficiently enough to be fully reassuring from an ecological point of view. Polychlorinated biphenyls (PCB), finally, are not pesticides, but chemical pollutants arising at present in the Italian environment from electrical industries (7). The PCB level in samples of some Italian surface waters is much higher than those proposed as acceptable (8), and data reported in tab.3 confirm the widespread presence of such pollutants. The soil input of PCB is not attributable only to irrigation water; e.g. 3.76 g/ha/yr of PCB come to Aramino soil from rain water (5). So, the balance of 0.63 g/ha of PCB retained and possibly degraded in soil (Tab.3) may prove to be inadequate and both the reality of the attribution to soil of the specific role of scavenger and the general deterioration of environmental quality in developed countries become more apparent.

Many findings are also available on the balance of pathogens in Italian irrigation water-paddy soil systems. Data reported in Tab.4 are referred to the average data from two years of experimentation in a rice field (Briona) different from those previously cited, irrigated with water where industrial and town sewage were discharged. The findings need no careful comments, as they are self-explanatory. Total bacterial counts

Table 1 Soil retention of nutrients from irrigation water (5, 6)

	Input kg/ha	Output kg/ha	Balance kg/ha
N	149	55	94
P_2O_5	24	13	11
K_2O	222	132	90
CaO	4691	3329	1362
MgO	1341	865	476

Table 2 Soil retention of some metals from irrigation water (5, 6)

	Input kg/ha	Output kg/ha	Balance kg/ha
Chromium	1.44	0.53	0.91
Copper	1.18	0.58	0.60
Lead	0.53	0.23	0.30
Manganese	7.47	7.35	0.12
Nickel	2.37	1.23	1.14
Tin	0.58	1.17	-0.59
Zinc	3.09	1.22	1.87

Table 3 Soil retention of some pesticides from irrigation water (5, 6)

	Input g/ha	Output g/ha	Balance g/ha
DDT (**)	0.35	0.08	0.27
DDE (**)	0.02	n.d.	0.02
Dieldrin	0.35	0.16	0.19
Alpha-HCH	0.73	0.82	-0.09
Beta-HCH	0.16	0.16	–
Gamma-HCH (lindane)	1.98	1.17	0.81
Delta-HCH	0.08	0.05	0.03
Parathion	8.49	0.15	8.34
Methylparathion (**)	0.05	n.d.	0.05
2,4,5-TP	2.74	1.47	1.27
Ordram (molinate)	1054	663	391
Propanil (*)	n.d.	7.97	-7.97
PCB	6.13	5.50	0.63

n.d. = not detected. Some figures refer only to the Aramino (*) or Biandrate (**) rice-field, as not detected in the other field.
Some abbreviations: HCH = hexachlorociclohexane; TP = trichlorophenoxy-
propionic acid; PCB = polychlorinated biphenyls.

show no spectacular differences between in- and out-flowing water, whilst numbers of pathogens (total coliforms, fecal coliforms, fecal streptococci) is dramatically reduced after passage on rice fields.

DISPOSAL OF WASTEWATER BY APPLICATION TO THE SOIL

Industrialized countries at present face the apparently conflicting needs of primary production, environmental quality, and energy requirements. Each problem can be classified with others as matter of finite resources: both the finite nature of food production or depolluting capacity of ecosystems and the limited supply of known energy sources are a reality beyond dispute. Since disposal of organic wastes represents a major environmental problem, recovery of available resources seems to be the more logical fate of such wastes. As is well known, anaerobic digestion of organic wastes yields energy under easily utilizable forms, like the gaseous product methane. Application to the soil of organic wastes, or organic residues after fermentation, is a practice which allows an ultimate recovery of energy.

Many chemical pollutants of wastewater are plant nutrients. The recycling of nutrients from wastewater into soil via land treatment is a promising process, as it involves problems of environmental quality, plant nutrition, and energy conservation. In fact, major pollutants in wastewater normally are nitrogen, phosphorus, oxygen-demanding materials and trace elements. Nearly all of them are simultaneously the major components of fertilizers and soil amendments required for crop production. No doubt, recycling of wastewater is no less complicated than other methods of disposal. Its advantage is that the treatment produces rather than consumes resources.

As known, treatment of wastewater consists of primary, secondary and sometimes tertiary processes, with possible energy recovery from anaerobic digestion. They originate several kinds of semiliquid sludges and a final liquid effluent, all to be suitably discharged. Land is the more obvious placing of sludges and effluents, but application to the soil must be performed both in a controlled quantitative manner, to achieve the removal of pollutants, and at the opportune period of the year, to introduce the supply of organic matter and nutrients in a well-timed occasion during crop cultivation. So, land disposal requires storage of large amounts of products to be applied to the soil at the proper time. However, systems have been developed where storage is unnecessary, i.e. wastewater is applied to the soil in continuous. Such systems are often suitable for wastewater withouth previous treatment. They are the infiltration systems, where wastewater is renovated by the soil inorganic components, microorganisms, and plants, as it moves through the soil profile, and the overland flow system, where most of the water flows over a relatively impermeable soil surface and the renovative action essentially depends on microbial and plant activity (Fig. 1).

Paddy soils are specifically involved in a possible application of

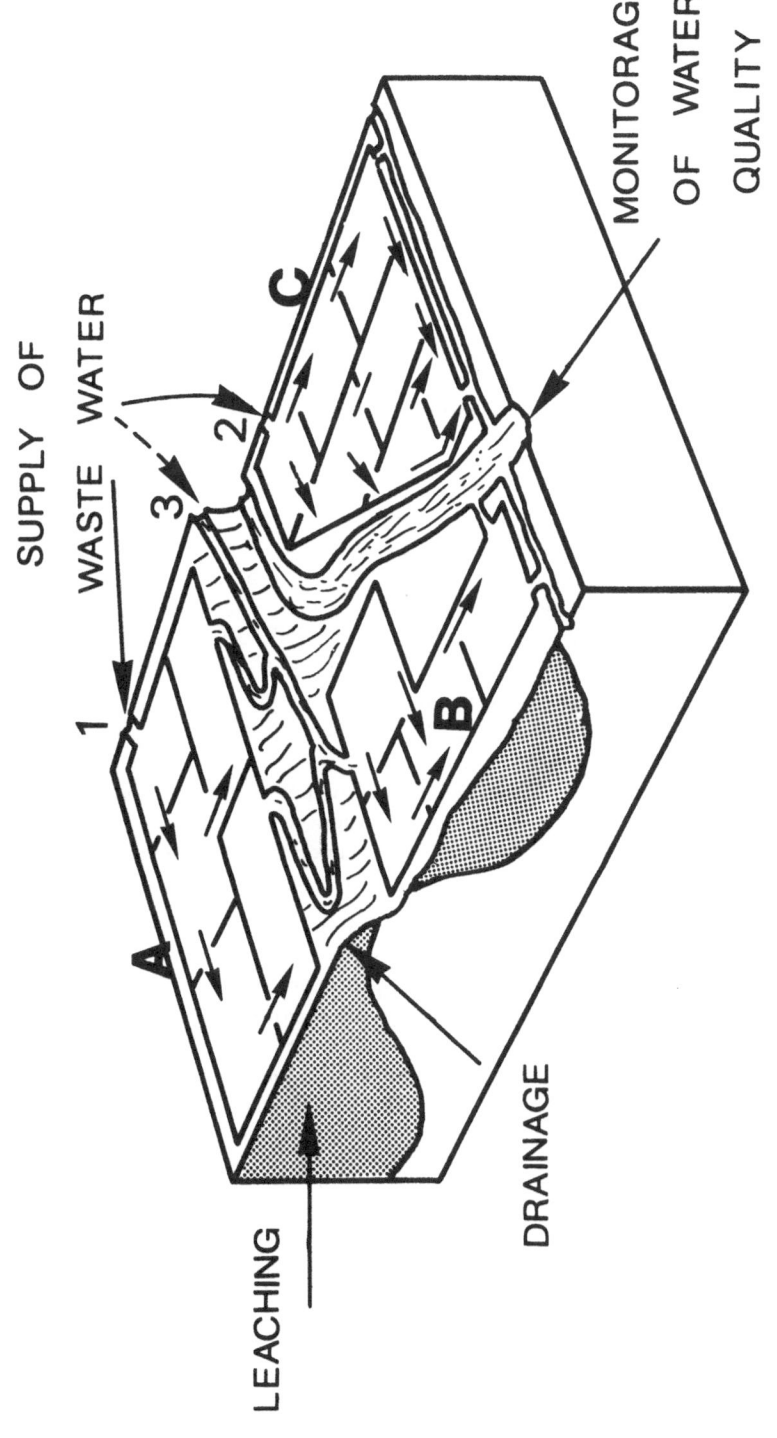

SUPPLY OF WASTE WATER

MONITORAGE OF WATER QUALITY

LEACHING

DRAINAGE

Fig. 1 A possible arrangement of rice-fields in overland flow disposal of wastewater

the overland flow system to wastewater treatment. A slight slope of fields highly facilitates the system operativeness. Flowed water must be collected in channels where its quality can be easily monitored to decide further use, which can vary from groundwater recharge to industrial application.

RICE FERTILIZATION WITH WASTEWATER: SOME BENEFITS AND RISKS

As stated before, many chemical pollutants of wastewater are plant nutrients. The available amounts of total organic waste are impressive: in Italy, for instance, sewage sludge is reported to be yearly 1.8 million tons of dry matter, containing about 5 per cent nitrogen and 1-1.2 per cent phosphorus (9). The amount of animal manure is even more striking: it accounts for 57.2% of total wastes produced in the country. It can be calculated that about 7,000 kg of animal manure per year and per hectar of agricultural utilized surface are available in Italy (9).

Although such figures are probably overestimated, they can give an idea of the order of magnitude of availability and, of course, of problems concerning disposal of animal wastes in the country. However, the geographic distribution of livestock plays an essential role in influencing both availability and problems of disposal. For instance, nutrients in pig, poultry and cattle manure do not include in Italy excessive amounts of nutrients, if referred to the total agricultural surface. However, in a district of the Po valley between Reggio and Mantova (the geographic region including Pegognaga, Suzzara, Cà del Bosco, Guastalla and Novellara, about 20 X 20 km) nutrient amounts rise up to 330 kg nitrogen, 190 kg phosphate, 250 kg potassium, and 900 g copper per year and per hectar of agricultural surface (Tab. 5). It is obvious that such a large availability of nutrients involves favourable perspectives to agriculture, but entails drastic and compulsory solutions to the disposal problems.

Apart from the general effects of the application of wastewater to the soil, which produces a well known improvement of physical and chemical fertility (10), it seems important to recall shortly how applied nutrients can be utilized by the crop. A very high percentage of nitrogen applied with wastewater is removed in overland flow systems. About two thirds of nitrogen are removed by the rice crop; the balance seems to be withdrawn by denitrification. Phosphorus removal is higher than foreseeable, taking into account the reduced amount of trivalent iron compounds in paddy soils. Since ferrous oxides are prevalent under reduced conditions, the high adsorption capacity of reduced soils is ascribed to their large specific surface area; the resulting phosphorus compounds are presumably easily available to plant roots. In any case, aluminum ions are added sometimes to wastewater to decrease phosphorus solubility; aluminum phosphates are then periodically plowed under the soil.

The removal of trace elements is often quantitative and attributed to the surface organic mat, where most of the heavy metals are bound. As known, organic matter exerts a tremendous influence both on physical and chemical properties of soil; in some instances it has been shown that

536

organic matter can a decisive factor in determining properties like specific surface area or cation-exchange capacity, also by masking the presence of inorganic soil constituents (11). The occurrence of a surface organic mat in paddy-soils may lead to the attribution of soil properties mainly, if sometimes not solely, to the organic matter, expecially in the interaction between soil and substances carried out by flowing water. The specific involved interaction include both a marked influence of metal on soil organic matter (12, 13) and the inverse influence of organic matter on behaviour of metals, generally affecting their availability (14, 15).

Heavy metals, depending on their properties, can be essential, useful, indifferent or toxic to plants and/or the following food chain. Significative amounts of toxic metals are often present in wastewater and can limit the volumes applicable. Lead is the more common heavy metal limiting the application (15).

A further problem often arises from variability of concentration of heavy metals (and obviously nutrients) in wastewater. The level of toxic metals, e.g. cadmium, may range from 2 to 1500 ppm, with a variation of the order of one thousand times (Tab. 6). A similar problem occurs with animal slurries, because the percentage of water in stall-feeding systems, where livestock are concentrated in narrow spaces, is very variable due to the addition of difficultly foreseeable amounts of washing waters which increase the volume of wastewater. So, not only heavy metals, but also major constituents may vary considerably; an example of variation of the chemical composition of a pig slurry used for continuous addition to the soil in Italy is given in Tab. 7. High and sudden variations of nutrient and pollutant levels make difficult to plan a controlled addition of wastewater to irrigation water. It is recommended to check water quality very often, for at least one or two indicator parameters, if possible by means of a monitoring station operating in continuous.

Eutrophication phenomena may also occur, particularly causing an anomalous growth of algae; also, the use of pesticides on the rice crop may be heavily affected by the presence of organic matter in water. No doubt, however, the major problem in using the overland flow system, in order to achieve both the improvement of water quality and rice fertilization, is the presence of pathogens. It has been suggested that rice grown in the presence of wastewater should be used for consumption only after some processing step that would definitely eliminate any possibility of pathogen transfer to humans. However, such precaution may prove to be inadequate. Technical measures to prevent any risk during wastewater disposal are the main problems in planning fertilization of rice with wastewater.

The more obvious precaution is the use of waste water only after biological (secondary) treatment and disinfection by chlorine or ozone (16). However, a field disposition can be arranged where wastewater is applied directly to grass fields (water-tolerant forages) and only out-flowing water reach rice fields; in Fig. 1, grass fields should be represented in this case by the area A, and rice fields by the area B. This system could be very suitable in practice, because overland flow on grass

Table 4 Soil removal of pathogens from irrigation water (8)

	Input	Output	% Removal
Total bacterial biomass (*)	19,000	14,000	26
Coliform bacteria (**)	114,000	5,500	95
Fecal coliforms (**)	13,600	60	99,5
Fecal streptococci (**)	7,700	450	94

(*) CFU (Colony Forming Units at 37°C)/ml
(**) MPN (Most Probable Number)/100 ml

Table 5 An estimate of nutrients in wastewater from livestock production

	N kg/ha	P_2O_5 kg/ha	K_2O kg/ha	Cu g/ha
Italy Total	29	14	28	50
District RE-MN	330	190	250	900

Table 6 Variability of heavy metal levels in European sewage sludge (15)

Element	ppm
Cd	2 - 1500
Co	2 - 260
Cr	20 - 8800
Cu	50 - 8000
Hg	0,1 - 55
Mn	75 - 2500
Ni	15 - 5300
Pb	50 - 3000
Zn	700 - 50000

("grass-filtration") is reported to be very effective in the removal of some pollutants like BOD_5 and heavy metals, but leaves most nutrients, expecially phosphorus, in out-flowing water (17-19). The residence time of water should be greater, up to twenty times, in the rice fields, so that suitable flooding times, water velocities, and possibly the opportune ratio between the surfaces cultivated with grass and rice should be consequently provided. A crop rotation can be also planned where waste-water is applied to grass fields enhancing soil fertility and rice is cultivated only after one or more cycles of the first crop.

A fascinating possibility could be represented by a rotation aquaculture-rice. Aquaculture allows growth in succession of phytoplancton-molluscs and brine shrimp-fin fish (trouts, etc.), seaweeds-lobsters, benthic algae-omnivorous fish; small crustaceans easily grow using as food source detritus and feces produced in the system. The area gives many products of commercial value and many be used also as recreational. The reduction of polluting agents is very efficient (Tab. 8) and can permit further cultivation of rice without any sanitary problems.

Table 7   Chemical composition of a pig slurry

	Minimum values (%)	Maximum values (%)
Dry matter	0.85	7.11
Organic matter	0.56	5.81
Soluble salts	0.07	0.36
Total N	0.15	0.46
Organic N	0.05	0.20
Total P_2O_5	0.07	0.41
Soluble P_2O_5	0.006	0.07
Total K_2O	0.14	0.35
Soluble K_2O	0.025	0.31
Ca	0.03	0.20
Mg	0.015	0.095
Na	0.025	0.08
Cl	0.04	0.125
SO_4	0.002	0.045
BOD_5 (mg/kg)	2415	8190
pH	6.6	7.4

Table 8   Reduction of wastewater pollutants by aquaculture (20)

	BOD (%)	COD (%)	Coliforms (%)
Natural wetlands	80.1	43.7	86.2
Artificial wetlands	91 - 95.7	61 - 86.8	90 - 99.7
Macrophytes	35.4 - 97	60.9	0 - 96.1
Invertebrates	69 - 93	55 - 67	72.1 - 99.9
Fish	96.7	-	99.9

REFERENCES

(1) Sequi, P. Sostanza organica del terreno: significato, qualità, fun-
zioni. L'Italia Agricola, 111 (12), 51-68 (1974).

(2) Cervelli, S., Nannipieri, P., Sequi, P., Interactions between agrochemicals and soil enzymes. In: Soil Enzymes (R.G. Burns Ed.) pp.
251-293, Academic Press (1978).

(3) Sequi, P., La lotta contro l'inquinamento del terreno. In: Il
disinquinamento in Italia (in press).

(4) Corbetta, G., Leonzio, M., Indagini sulle condizioni ecologiche
delle risaie italiane. Il risicoltore (1976).

(5) Ente Nazionale Risi, Inquinamento delle acque del terreno e delle
risaie del comprensorio Dora Baltea-Ticino-Po. ENR, Milano (1978).

(6) Tinarelli, A., Leonzio, M., Indagini sulle caratteristiche fisiche e
chimiche delle acque di irrigazione nel comprensorio Dora Baltea-Ticino-Po. Ente Nazionale Risi, Quad. n.12 (1979).

(7) Puccetti, G., Leoni, V., PCB and HCB in the sediments and waters of
the Tiber estuary. Marine Pollution Bull., 11, 22-25 (1980).

(8) Ente Nazionale Risi, Esami microbiologici delle acque di irrigazione
e di scolo delle risaie nel comprensorio Dora Baltea-Ticino-Po. ENR,
Milano (1978).

(9) Fenoglio, G.P., Farneti, A., Bassetti, A., Colanzi, G., Fermentation-
-hydrolysis process for the utilization of organic materials. In:
Raw materials research and development. Studies on secondary raw
materials. Part IV, vol. II. Comm. Europ. Commun., DG XII (1979).

(10) Sequi, P., Guidi, G., Somministrazione di prodotti di rifiuto per il
miglioramento del terreno. L'Italia Agricola, 115 (5), 55-72 (1978).

(11) Sequi, P., Aringhieri, R., Destruction of organic matter by hydrogen
peroxide · in the presence of pyrophosphate and its effect on soil
specific surface area. Soil Sci. Soc. Am. J., 41, 340-342 (1977).

(12) Sequi, P., Guidi, G., Petruzzelli, G., Influence of metals on
solubility of soil organic matter. Geoderma, 13, 153-161 (1975).

(13) Giovannini, G., Sequi, P., Iron and aluminium as cementing substances of soil aggregates, II. J. Soil Sci., 27, 148-153 (1976).

(14) Sequi, P., Cervelli, S., Sostanza organica e inquinamento del terreno. L'Italia Agricola, 112 (10), 57-81 (1975).

(15) Sequi, P., Petruzzelli, G., Il riciclaggio dei rifiuti: pericolo per
l'inquinamento del terreno. L'Italia Agricola, 115 (10), 55-72
(1978).

(16) Hunt, P.G., Lee, C.R., Land treatment of wastewater by overland flow
for improved water quality. In: Biological control of water pollution (J. Tourbier, R.W. Pierson Jr. Eds.) pp. 151-160, University
of Pennsylvania Press (1976).

(17) Scott, T.M., Fulton, P.M., Removal of pollutants in the overland
flow (grass filtration) system. Prog. Wat. Tech., 11, 301-313 (1979).

(18) Jenkins, T.F., Martel, C.J., Pilot scale study of overland flow land
treatment in cold climates. Prog. Wat. Tech., 11, 207-214 (1979).

(19) Lee, C.R., Peters, R.E., Overland flow treatment of a municipal
lagoon effluent for reduction of nitrogen, phosphorus, heavy metals
and coliforms. Prog. Wat. Tech., 11, 175-183 (1979).

(20) Tortell, Ph., The utilisation of waste nutrients for aquaculture.
Prog. Wat. Tech., 11, 483-498 (1979).

THE STUDY OF NITROGEN DISTRIBUTION AROUND RICE RHIZOSPHERE

Liu Zhi-yu,   Qin Sheng-wu
(Institute of Soil Science, Academia Sinica, Nanjing)

It is interesting to note that the soil around plant root forms a special nutrient environment, in which the mobility and distribution of nutrient ions directly reflect the plant-soil relationship. However, in this regard only a few attempts were recorded in the literature, before the more sensitive methods of measuring the amount of microzonal nutrient had been established(1). In recent decade, many investigations reported the change of micro-regional nutrients around the rhizosphere(2,3), but little concerning rice rhizosphere under waterlogged condition(4,5).

Present paper deals with the variation of nitrogen concentration in the microzone of soil phase which is closely contacted with the functional root surface.

MATERIALS AND METHODS

Rice seedlings were grown in a perspex frame to get a root layer in uniform size and shape. Generally, the rice formed a plane root layer in one month, with a length of about 13 cm and width of 4 cm just like the frame (Pl. 1). The seedlings were then transferred into water culture or soil culture according to requirements.

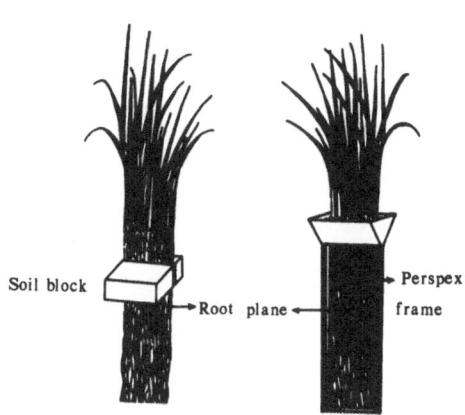

Plate 1 Illustration of root plane
and soil block

A neutral clay loamy paddy soil collected from the Taihu Lake area
was used, it contains organic matter 2.68%, total N 0.165%, available
P 9.1 ppm (by 0.5 M NaHCO₃ extractant) and exchangable K 9.1 mg per 100 g
soil (by N NH₄OAc extractant). The soil was ground and passed through
100 mesh sieve, and then mixed uniformaly with $(^{15}NH_4)_2SO_4$ (containing 13.3
atm.% ^{15}N), thus the rate of fertilizer-N in soil was about 0.01 g N/100g. The
soil was packed into a perspex box ($2 \times 4 \times 4$ cm³) provided both sides of
the box with a movable glass plate. The soil bulk density in the box was
1.1 g/cm³ in air dried weight. During the experiment, the two soil boxes
were moistened and the root plane was fitted between the two opposite sides
of the boxes. In order to prevent the root from penetrating into the soil
block, a piece of permeable nylon membrane was put between each phase of
the root plane and soil blocks while the water and nutrient ions could pass
through the membrane readily.

The intact rice plant with two soil boxes was put into a pot, with
15 cm diameter and 19 cm height. The pot was filled with soil or quartz
sands and cultured under a submerged condition. At the end of each growing
period, e.i. 6, 12, and 18 days the plant was took out from the pot carefully.
Then, the soil boxes were removed from root plane, and immediately immersed
into liquied nitrogen. The frozen soil blocks were sliced parallel to the
root-soil interface, with section thickness of about 1 mm. Successive sections were prepared up to 10 mm or 40 mm from each soil block. Total N and
^{15}N in each section were determined. Similar tests were made for unexploited
soil as blanks, all results were subtracted by the data obtained from the
blanks.

RESULTS AND DISCUSSION

1. Distribution of Total N and Fertilizer-^{15}N

In all cases, after the soil blocks were contacted with root layer
in each growing period of 6,12 and 18 days, a gradient of N concentration
appeared, and showed a negative correlation with the distance from soil-
plant contact phase in both soil blocks. This depletion of nitrogen was
mainly caused by the decrease of fertilizer- ^{15}N. The concentration of
fertilizer N in the soil near the root plane was reduced by 30-70%. The
relationship between the distance from the root plane to the rate of
depletion of fertilizer N in soil is given in Fig. 1. It shows that within
a distance of 10 mm there was a rapid fall in fertilizer-N, and then the rate
of ^{15}N depletion decreased gradually beyond 10 mm distance, indicating that a

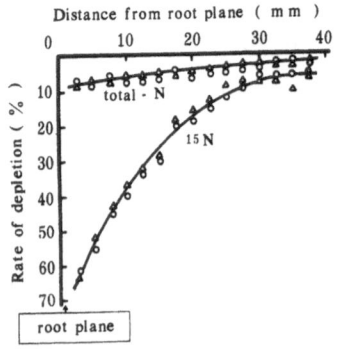

Fig. 1 Distribution profile of
total N and fertilizer-^{15}N
(o and Δ denote measurement on
opposite soil blocks)

542

main uptake zone occured within 10 mm. Results obtained from two opposite soil blocks were very much the same. But these results were different from the results Harmsen and Jager had got(6). They worked with upland crops and found an accumulation of nitrogen adjacent to plant root.

2. Variation and Distribution of N-depletion in the Soil

The amount of depletion of fertilizer ^{15}N in the soil varies with the growth of rice, the uptake of fertilizer- ^{15}N by rice is shown in Table 1. When the soil blocks contacted with root plane from 6 to 12 days, the rice plant uptook form 98 to 404 μg ^{15}N and the depletion of fertilizer- ^{15}N in soil slices within 10 mm varied from 113 to 221 μg. After 12 days, the growth of rice plant appeared stunted. The uptake of ^{15}N by the rice in the period of 12-18 days only increased from 404 t0 606 μg. In the meantime, the depletion rate of fertilizer ^{15}N in soil blocks within a 10 mm distance also declined. Within first 6 days, the amount of ^{15}N uptake by rice plant was almost in consonance with the depletion of fertilizer- ^{15}N in the soil within 10 mm, and yet in the growing period of 12-18 days the uptake of ^{15}N by rice surpassed the amount of depletion in 10 mm soil. Evidently, the movement of available nitrogen in the soil around rhizosphere toward the plant root had compensated for the depletion of ^{15}N in soil areas near by the root.

Table 1 Relationship between uptake of ^{15}N by rice
and the depletion of ^{15}N in soil*

Time of uptake (days)	^{15}N depletion ($\mu g/g$ soil) with distance (mm)			Mean ^{15}N uptake (μg/plant)
	0-2.5	7.5-10	Total amount within 10 mm	
6	43.8	12.8	113	98
12	65.1	45.6	221	404
18	84.4	68.9	304	606

* Mean values for replication

The distribution of depletion zone in soil blocks, as shown in Fig. 2, is conformable with the rate of depletion. Depletion of fertilizer- ^{15}N in first 6 days was confined to 10 mm, beyond this limit the concentration of ^{15}N was same as unexploited soil block, at 12 days the limit extended to about 40 mm and at 18 days over 40 mm. Farr et al. worked on K-depletion in soil rhizosphere, showed a similar result(2). Of course, the distribution of depletion zone as mentioned above, only gives a relative value, because the extension of depletion zone mainly indicates the mobility of available nitrogen in soil

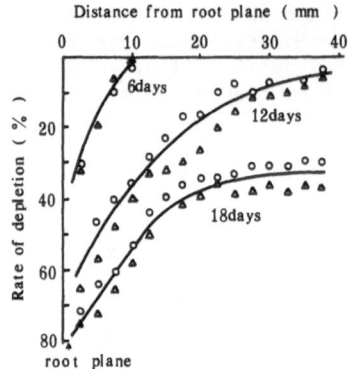

Fig. 2 Variation of depletion
rate of ^{15}N at different time of
rice growth

3. Effect of Root Position on the Variation
of N-content in the Soil

Comparison of the depletion of fertilizer-N in soil blocks, at the root
plane 4-6 cm and 9-10 cm below the root node showed that the concentration of
fertilizer-N varied significantly between the two positions (Fig. 3). The
variation is also affected by the growth of rice or the time of contact.
Within 6 days the depletion rate of upper layer was the same as lower layer.
However, in 12 days the rate of ^{15}N-depletion in the upper layer with a distance of 2.5 mm apart from the contact, was 27 % greater than the lower

layer, and within the distance of 10 mm
was 25 % greater, and in 18 days the differences for 2.5 mm and 10 mm distance
were up to 30 % and 32 % between two
layers respectively. It is believed that
the rice plant at the beginning of the
experiment was just at initial tillering
stage, the upper layer, i.e. 4-6 cm below
root node, had more active roots, and
uptook nitrogen quite vigorously, which
affected the amount of depletion and distribution of fertilizer-N in soil blocks
markedly. Subsequently, it was also
related to the weight of active root, the
upper layer was 13.2 mg dry wt. per cm^2
and lower only was 7.4 mg. These results
indicated that the nitrogen variation in
rhizosphere was directly relevant to both
of activity and quantity of absorbing
root per unit area.

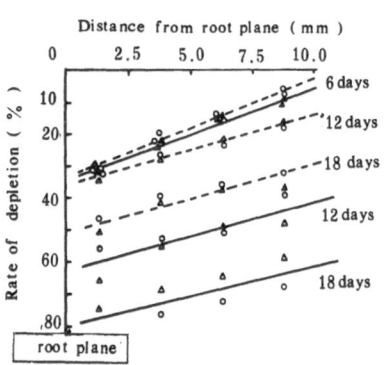

Fig. 3 Influence of root portion
on ^{15}N-depletion
----upper portion (4-6 cm)
——lower portion (8-10 cm)

4. Effect of Temperature on the Depletion of
Fertilizer-N in Soil

The content of rhizosphere nitrogen was markedly affected by the temperature. Under an average of 17°C at the growing period the depletion zone of the soil blocks was limited within 10 mm after 6 days, and the rate of depletion was 34.1 % in 2.5 mm, and 2.7 % in 10 mm. However, when the root plane was contacted with soil blocks at an average of 27°C, the effect of ^{15}N depletion was beyond 10 mm, and the rate of depletion was 48.3 % in 2.5 mm and 28.8 % in 10 mm (Fig.4).

Evidently, a higher temperature may promote the uptake of nutrients and water by rice plant, as well as the mobility of nutritive ions in soil. An increasing rate of ion transfer usually affects the accumulation of nitrogen to root surface. But in this experiment the temperature effect on rice plant to accelerate the uptake of nitrogen more than on the rate of transfer of available nitrogen in the soil. The increased rate or transpiration of plant at a higher temperature induced no accumulation of fertilizer-N on plant root. It gives the evidence that the transfer of nitrogen by mass flow is rather less significant.

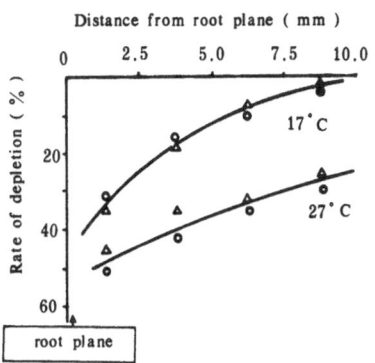

Fig. 4 Influence of temperature on ^{15}N-depletion (6 days)

5. Effect of Soil Volume on the Variation
of N-content in Soil

The relationship between the volume of soil, which contained a same concentration of fertilizer-N, and the amount of N-depletion in soil blocks, was investigated (Table 2). In this experiment the volume of each block was doubled, thus, the experiment was carried out in two treatments, the first one in a soil volume of 128 cm3, and the second 256 cm^3. The rate of depletion in 256 cm^3 soil bulk was about 14 % lower than in 128 cm^3 ones. Similarly, the depletion zone of the former soil blocks was limited to 18 mm and the latter to 25 mm. Obviously, within a certain extent, a greater supplying capacity of available nitrogen may reduce the depletion rate of fertilizer-N in rhizosphere and reduce the area of depletion zone in soil.

545

Table 2 Effect of soil volume on the variation of
available N content in soil*

Distance from root plane (mm)	Soil bulk, 128 cm³		Soil bulk, 256 cm³	
	Content of N (μg/g soil)	Rate of depletion (%)	Content of N (μg/g soil)	Rate of depletion (%)
0–2.5	4.67	62.6	6.43	48.6
2.5–5.0	5.79	53.7	7.58	39.4
5.0–7.5	6.99	44.2	8.72	30.2
7.5–10	7.73	38.1	9.45	24.4
30	11.6	6.7	11.6	6.7
Initial soil	13.5	–	13.5	–

* Mean values for replication

CONCLUSION

Present paper reveals that there appears a depletion zone of nitrogen
in rice rhizosphere. This depletion is mainly induced by the decrease of
fertilizer-N. The extension of depletion zone in soil blocks is in consonance with the rate of depletion. The depletion rate of fertilizer-N
is greater in soil blocks contacted with upper root layer than with the
lower. Moreover, and increased temperature at the growing period promotes
the rate uptake of nitrogen by rice plant and meantime the limit of depletion
zone of fertilizer-N in soil blocks is expanded. However, no accumulation of
N-ion has been found near root surface at a higher temperature, while both the
rate of ion transfer in soil and the transpiration of plant are accelerated.
It gives the evidence that the transfer of nitrogen by mass flow is rather
less. Generally speaking, it is shown that the microregional variation of
nitrogen in rice rhizosphere is both affected by nitrogen uptake of plant
and supply of soil, and the uptake of rice plant plays more important role.

REFERENCES

(1) Barber, S.A., Nutrient flux at soil-root interface. Trans. 11th Int.
    Congr. Soil Sci., 3, 43-48 (1978).
(2) Farr, E., Vaidyanatnan, L.V., Nye, P.H., Measurement of ionic concentration
    gradient in soil near roots. Soil Sci., 107, 385-391(1969).
(3) Bhat, K.K.S., Nye, P.H., Baldwin, J.P., The concentration profile in the
    rhizosphere of roots with root hairs in a low-P soil, Plant Soil, 44,
    63-72(1976).
(4) Chino, M., Application of electronprobe X-ray micro-analysis to the
    localization of chemical elements with and around rice roots grown in
    soils under submerged condition. Japan Agric. Res. Quart., 11(3), 129-
    135 (1977).
(5) Yu Tian-ren, Ho Chun, Chiang Pei-fan, Suan Chia-siang, Sie Chian-chang,
    Electrical conductivity of paddy soil in relation to their fertility.
    Acta Pedologica Sinica, 7(3-4), 145-158(1959).(in Chinese with English
    summary)
(6) Harmsen, G.W., Jager, G., Determination of quantity of C and N in the
    rhizosphere of young plants. Nature London, 195, 119-1120(1962).

FERTILIZER MANAGEMENT OF PADDI SOILS WITH PHYSICAL CONSTRAINTS

H. R. von Uexkull
(East and S.E. Asia Program, The Potash Institutes, Singapore)

INTRODUCTION

Rice is the only cereal crop that finds its optimal environment under flooded soil conditions. For this reason the physical, chemical and biological environment of the rice plant, and especially it's rhizosphere is totally different from that of other food grain crops. It is not only different, it is also far more complex.

Under upland conditions the effects of applied fertilizer nutrients can in most cases be largely interpreted in the light of soil chemistry and plant physiology. Nutrient interactions - where they are observed, are usually easily explained by the principles of plant nutrition.

In the case of the flooded rice, water logging results in profound changes in soil chemistry. The biosphere also undergoes changes, which in turn interact with plant nutrients added as fertilizer. Under certain conditions, for example, the addition of soluble phosphate may greatly stimulate the growth of algae. Vigorous growth of algae results in a depletion of dissolved CO_2 in the water (during day time) which in turn causes a rapid increase in the pH to a level where volatilization of NH_3 may become a problem. Addition of P to a deficient soil can therefore result in decreased efficiency of applied nitrogen, thereby masking a physiological effect of P. Similarly the application of P and K on soils rich in above elements can stimulate biological fixation of Nitrogen, resulting in apparent PK responses that in a physiological sense are primarily responses to (biologically fixed) nitrogen. Until recently the interaction of chemical fertilizer with the biosphere had received very little attention. A better and deeper understanding of the problems involved may contribute toward more efficient use of fertilizer and could possibly be of particular interest for farmers operating under conditions of environmental and economic constraint, where only moderate levels of fertilizer can be used.

PATTERNS OF FERTILIZER RESPONSE

General

Nitrogen is universally the most deficient plant nutrient for rice. A great part of the breeding work has been for "nitrogen responsiveness" and most of the fertilizer research has been done with nitrogen fertilizer. By comparison very little work has been done with the other nutrients. This concentration on nitrogen fertilizer has possibly contributed to the fact that until recently very little attention has been paid to biological nitrogen fixation.

Normally it can be assumed that nitrogen will be the most limiting nutrient and that responses to the other nutrients can only be expected once yields have been increased through the application of nitrogen.

Similarly it could be expected that where each of several nutrients, when applied alone, significantly increased the yield, such nutrients applied in combination would produce positive interaction effects. A review of a great number of fertilizer experiments carried out in several countries of Asia and Africa showed however, that only a minority of them falls into such "normal" pattern.

"Normal" and "Abmormal" Response Patterns

An example for a "normal" response pattern is given in Fig. 1.

In the above case all nutrients performed poorest when applied alone and best when combined. One would expect that most experimental results would fall into such pattern. This is, however, not the case. On the contrary, in the majority of the cases the response pattern is rather different as is shown in Fig. 2 and 3.

An evaluation of 3.146 simple fertilizer trials carried out under the auspices of the UNDP/FAO-Fertilizer Programme in the Philippines from 1977-80 in 14 provinces showed a similar trend. In the overwhelming majority of the fertilizer trials there was a negative interaction between N on one side and P and K on the other. Only on soils where both P and K were very deficient, there was a positive interaction between nitrogen and PK and vice versa as shown in Table 1.

Fertilizer Application and Nitrogen Fixation

The results presented in Fig. 1-3 and Table 1 suggest that the application of phosphorous and potassium alone or in combination can stimulate biological nitrogen fixation, even on soils where the rice plant itself may not require fertilizer P or K (for the yield levels reached in the experiments.) On the basis of the results obtained at IRRI, Chandler (1979) stated that ...."there is evidence that when phosphrous - and, to a lesser degree potassium, is applied and nitrogen is not, the result is an increase in the amount of nitrogen available to the rice plant." Recent N-balance studies by Koyama and App (1979) suggest that about 20-50 kg. N/ha./crop of wetland rice must be supplied from sources other than soil, water and fertilizers, although much higher figures have been recorded.

Although it has been known for a long time that atmospheric nitrogen is fixed and made later available to the rice

YIELD INCREASE
OVER CONTROL, kg./ha.

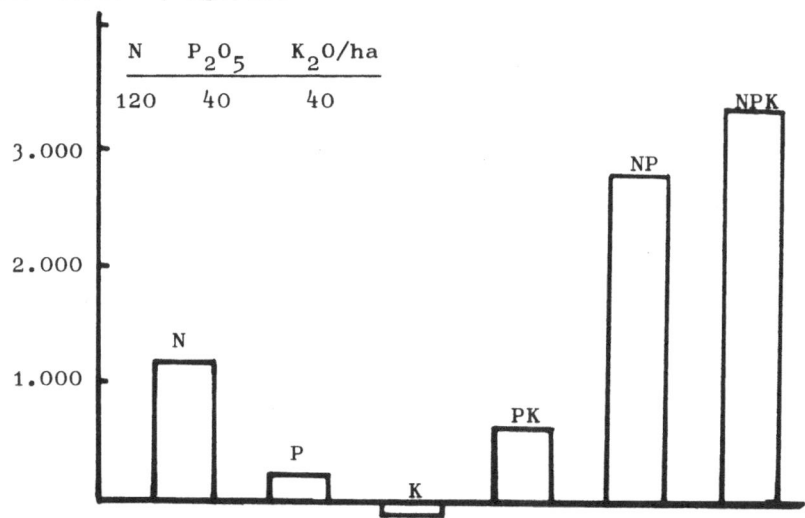

Fig. 1   Example for a "normal" fertilizer response
pattern. Louisiana Clay Loam, Laguna, Philippines
(IRRI, 1980)

Table 1   N-PK - Interaction Effects on Soils with Different
Soil Fertility in the Philippines
Calculated from FAO-Data, 1980

Soil Fertility Status	PK-Effect, kg./ha.		N-Effect, kg./ha.	
	without N	with N	without PK	with PK
1. High P, high K	553	204	982	633
2. Low, P, high K	627	294	1.007	674
3. Med. P, low K	622	513	894	785
4. Very low P,very low K	1.053	1.268	1.061	1.276

A applied at 70 kg./ha., P_2O_5 and K_2O at 30 kg./ha.

1.   Averages from 272 experiments in Albay Province

2.   Averages from 143 experiments in Antique Province

3.   Averages from 291 experiments in Mindoro Oriental Province

4.   Averages from 95 experiments in Querino Province

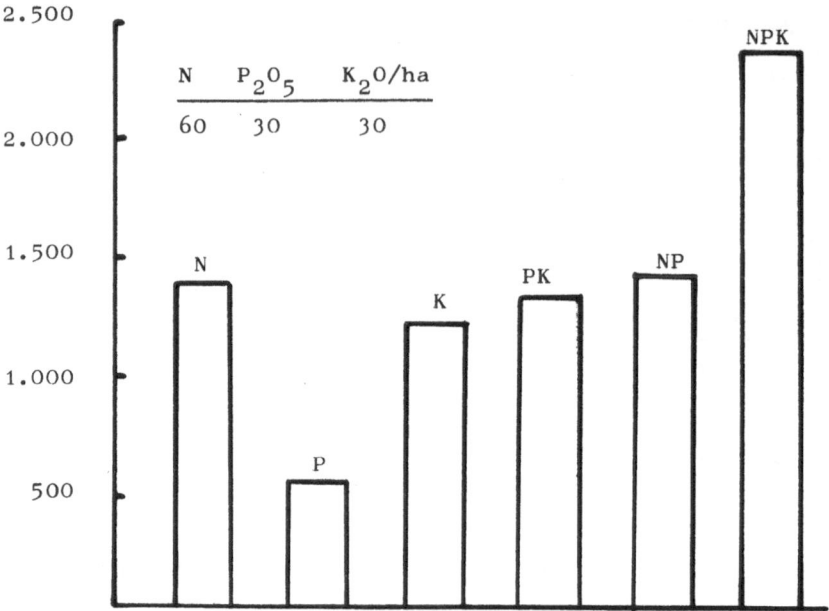

Fig. 2   Example for an "abnormal" fertilizer response
          pattern. Four-year averages from 6 cropping seasons,
          1976-1979. Titabor, N.E. India. (All India Coordi-
          nated Rice Improvement Project, 1979)

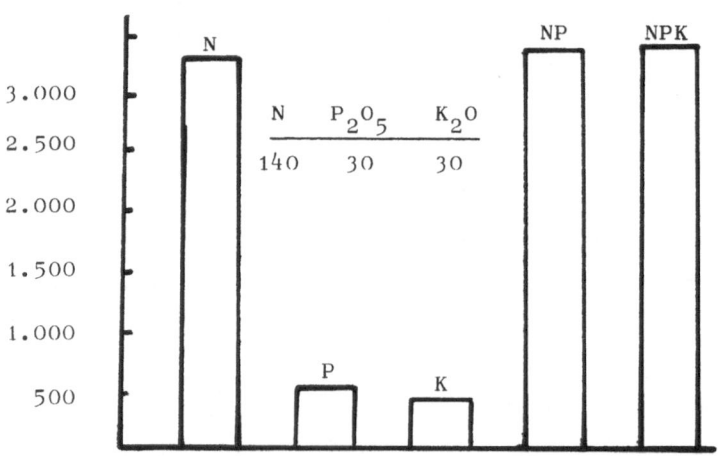

Fig. 3   "Abnormal" fertilizer response pattern.
          1973-1978 dry season results from the IRRI
          long-term fertility experiment

plant both in the soil (by hetotrophic bacteria) and in the irrigation water (mainly by phototropic algae) (De, 1936 De et al 1950; Bunt 1961; Konishi et al 1961; Yoshida et al 1973; Venkataraman 1975; Watanabe 1977), very little attention has been paid as to the interactions between nutrients, the micro-organisms in the soil and the water and the rice plant.

Legume fixation of nitrogen is at a maximum only when the available soil nitrogen is at a minimum and when at the same time phosphorous and potassium supply is adequate, (Tisdale & Nelson, 1975; Griffith, 1980). The stimulating effect of phos-phorous (and potassium) on the growth of algae is well establish-ed (Moore 1966; Davidson 1967; Singh 1977; IRRI 1980.) Srini-vasan (1979) investigated the effect of organic and inorganic fertilizer on the growth of algae.

Phosphate and, to a lesser extent, potash stimulated algae growth whereas nitrogen clearly depressed it. Venkataraman ('1979) assumed that nitrogen application reduced phosphate avail-ability to plants and organisms which according to him could account for the depressing effect of nitrogen on the growth of algae. Watanabe and Lee (1975) analysed the population of var-ious kinds of N_2-fixing micro-organisms in the plots of the long-term fertility experiment of the International Rice Research Institute and found no consistent differences between fertilized and non-fertilized plots, inspite of the fact that nitrogen fixation was considerably higher in the non-fertilized plot. Nitrogenase activity in algae in fertilized plots (40kg. N/ha + 30kg. each of P_2O_5 and K_2O/ha. just before transplanting and 20kg. N/ha. at panicle initiation) was low up to 21 days after transplanting. Most of the early nitrogenase activity found in fertilized plots was therefore ascribed to the activity of he-terotrophic bacteria in the soil or in the root zone. By con-trast, in the non-fertilized plots photosynthetic N_2 fixation predominated.

While phosphorous deficiency and low pH are probably the factors most commonly limiting the growth of blue-green algae (and their nitrogen fixation capability), potassium is more likely to affect root growth and the micro-flora in the rhizo-sphere. K-deficincy results in increased root-secretion of low-molecular weight carbohydrates that attract saprophytic micro-organisms. Presence of larger numbers of such micro-organisms causes a sharp drop in the O_2-pressure that in turn could affect the activity of the N-fixing micro-flora. (Trolldenier, 1977, 1979.)

Research in China (Liu, 1979) has shown that azolla pro-pagates rapidly in soils containing 21-87 ppm of available phosphorous but grows poorly on soils containing 2.1 - 8.7 ppm. It was also shown that P was most effective during the warmer periods of the year (spring, summer and autumn) whereas K was most effective in winter.

In Vietnam 1 kg. of P_2O_5 produces about 2-3 kg. of biolo-gically fixed N and 1 kg. of K_2O about 1 kg. of N. At the same time K is said to improve the efficiency of P, (Tuan and Thuyet, 1979). In India addition of phosphorous is considered essential to azolla's rapid multiplication, while application of potash proved to be beneficial, (Singh, 1979).

While the link between phosphorous and nitrogen-fixing micro-organisms is very clear-cut, the effect of potassium is still obscure. Kawasaki et al (1980) reported that the top : root ratio increased with a decrease in the K-concentration of the nutrient solution and that this trend was more pronounced with Indica type of rice than Japonica rice. The beneficial effect of K could therefore also be an effect on root development.

Apart from phosphorous and potassium, elements like calcium and molybdenum can increase availability of nitrogen to the rice plant. This can be either a direct effect or an indirect effect through the stimulation of better root growth.

Fertilizer Application and Nitrogen Losses

Recovery of nitrogen fertilizer by the rice plant is often rather poor. This is particularly the case where nitrogen is broadcast to the surface of the soil or applied after transplanting into the irrigation water.

Nitrification of ammonia nitrogen in the oxydized soil layer, followed by di-nitrification has been identified as one mechanism leading to nitrogen loss. (Mitsui 1954, Watanabe and Mitsui 1979.) A second mechanism which may explain many of the field observations is dependent upon ammonia volatilization into the atmosphere. (Bouldin and Alimagno 1976.) The potential for NH_3 losses becomes appreciable once the pH of the irrigation water exceeds 7.4 . Algae growth may increase the pH drastically during the day because of utilization of dissolved CO_2 for photosynthesis.

Bouldin and Alimagno showed that losses of NH_3 by volatilization from the water surface following broadcast application could amount to 30-60% of the applied N. In a similar study Mikklesen et al (1977) observed nitrogen losses from top-dressed fertilizer (applied 10 days after transplanting) of about 20%. The bulk of NH_3 was lost within the first 3 days. As could be expected, it was also observed that water pH values and NH_3 losses were higher during the dry than during the rainy season. The study also confirmed the numerous earlier findings that placement or incorporation of nitrogen minimized losses in comparison with top-dressed applications.

While phosphorous and to a lesser extent potassium - stimulate micro biological activity in the irrigation water and in the rhizosphere, above elements at the same time may help to create an environment conducive to increased losses of fertilizer ammonia, especially where, the N-containing fertilizer is applied into the irrigation water which is still a rather common practice among farmers in S. E. Asia.

DISCUSSION

With most soils suitable for paddy cultivation already in use, future increases in rice production will have to come more and more on the existing land through yield-rising practices, such as the intelligent use of fertilizer. The need for the adoption of yield-rising practices comes at a time when fertilizer prices are increasingly affected by high costs of energy. In the past, most research was directed toward maximizing yield with little attention being paid as to the efficiency of the fertilizer being used.

A review of past experimental data and of recent work on nitrogen fixation and nitrogen losses as affected by fertilizer use suggests the possibility to develop more efficient fertilizer usage techniques for paddy rice, especially on soils with no firmly secured source of irrigation water. Existing evidence permits us to assume that:-

a. Biological nitrogen fixation can play an important role as a source of nitrogen for rice.

b. Nitrogen fixation is promoted by applications of phosphorous, potassium and possibly some other elements.

c. High concentration of nitrogen in the soil solution and the irrigation water depress biological nitrogen fixation.

d. A high concentration of phosphorous in the irrigation water often favours growth of algae which cause the pH of the water to rise to levels where volatilization of NH_3 becomes a problem.

e. The superior effect of deep placed nitrogen or of slow release nitrogen is probably not only due to lower denitrification and volatilization losses, but also due to less interferance with biological fixation.

f. The lower the level of applied nitrogen fertilizer, the more important it will be to maximize biological nitrogen fixation.

g. Where phosphorous is deficient for the rice plant, it should be incorporated into the soil before transplanting.

h. Where phosphorous is to be applied in order to stimulate growth of bluegreen algae at least some of it should be applied into the irrigation water.

i. Where potassium is primarily needed for the plant, it should be applied in split dosages before transplanting and at maximum tillering.

j. Where potassium is needed primarily to stimulate N-fixation, it should be incorporated into the soil as a basal dressing before transplanting.

Currently the level of nitrogen used in rice culture depends largely on the degree of water control. Where water control is "good" to "perfect" very high levels of nitrogen (up to 180 kg. N/ha.) are often used with economic advantage. On the other hand in areas with uncertain water supply, hardly any fertilizer is used.

Based on our present knowledge and experience it may be possible to develop fertilizer management techniques suited for a wide range of different conditions. For the sake of simplicity we classify them into 3 groups or categories.

Group I - Optimal conditions (perfect water control)

Yield target: 6 - 10 t. grain/ha.
N-fertilizer input: 90 - 180 kg. N/ha.

Under conditions of perfect water control, maximization of yield is done by making full use of chemical fertilizer with no consideration given to biological N-fixation. Growth of algae in the irrigation water is undersirable as this may lead to nitrogen losses. P and K should be applied to feed the rice plant but not any other organisms. Most of the past fertilizer work has been done under "optimal" conditions.

Group II - Sub-optimal conditions (marginal water control)

Yield target: 4 - 6 t. grain/ha.
N-fertilizer input: 45 - 70 kg. N/ha.

Fertilizers used and techniques of application should be compatible with biological N-fixation. Slow release forms of N, deep placement of N are practiced not only to reduce N losses due to di-nitrification and volatilization, but also to interfere less with the process of biological N-fixation. P and K are not only applied to feed the plant, but also to stimulate biological N-fixation.

Group III - High risk conditions (no water control = rainfed paddies)

Yield target: 3 - 5 t. grain/ha.
Nitrogen input: 30 - 45 kg. N/ha.

Most of the paddi rice acreage in S.E. Asia falls into this category. As water supply depends on natural rainfall, fields frequently dry out. As a basal N-dressing is too risky, most farmers apply it into the standing water or do not apply any fertilizer. For this group of farmers a "low risk" technology has to be developed that would permit them to maximize yields with a minimum of risks involved. For this, fertilizer techniques totally different from those used under Group I or Group II conditions will be needed.

To minimize danger of nitrogen loss and to maximize nitrogen fixation, fertilizer nitrogen is only applied at panicle initiation, once it is clear that there is enough water for a good crop. Nitrogen broadcast by that time will be efficiently utilized as there is an adequate canopy to prevent volatilization and there is also a sufficiently developed surface root system to intercept nitrate nitrogen before it moves down into the reduced soil layer. Phosphate and potash, on the other hand, will not be lost even if the soil dries out. They can therefore safely apply before planting in order to stimulate nitrogen fixation.

Phosphate applications may have some times to be split, with some of the phosphate applied into the water to feed bluegreen algae. All potash would have to be incorporated into the soil for maximum efficiency.

To substantiate some of the assumptions behind the suggested practices, a number of experiments have been initiated. Unfortunately most experiments are not yet harvested resp. evaluated. Preliminary data are available from 2 experiments in the Philippines and the results are shown in Fig. 4 and 5.

In the experiment "A" (Fig. 1) ordinary prilled urea was ompared with sulphur coated urea in the presence and absence of P and K. It was assumed that slow release, sulphur coated urea would interfere less with biological N-fixation and thus there would be more pronounced effect of P and K on yield. This appears to be the case.

The aim of experiment "B" was to test the effect of timing of N-application in the presence and absence of P and K. The basal dressing with P and K was to feed the plants and to stimulate N-fixation. If P and K stimulate N-fixation, a late top-dressing of N should be more effective than a basal dressing. The preliminary results obtained seem to bear out this point. (Fig. 2)

SUGGESTIONS FOR FUTURE RESEARCH

The data presented suggest the possibility to develop agronomic techniques that could be used by farmers operating under conditions of environmental constraint caused by insecure supply of irrigation water.

There appear to be many possibilities to develop fertilizer and other agronomic practices that would be compatible with or that would even stimulate N-fixation by (a) azolla, (b) blue-green algae, (c) in the rhizosphere, (d) by legumes. Depending on the main source of biologically fixed N, different fertilizer techniques may be needed. As most biologically fixed

YIELD INCREASE
OVER CONTROL, kg./ha.

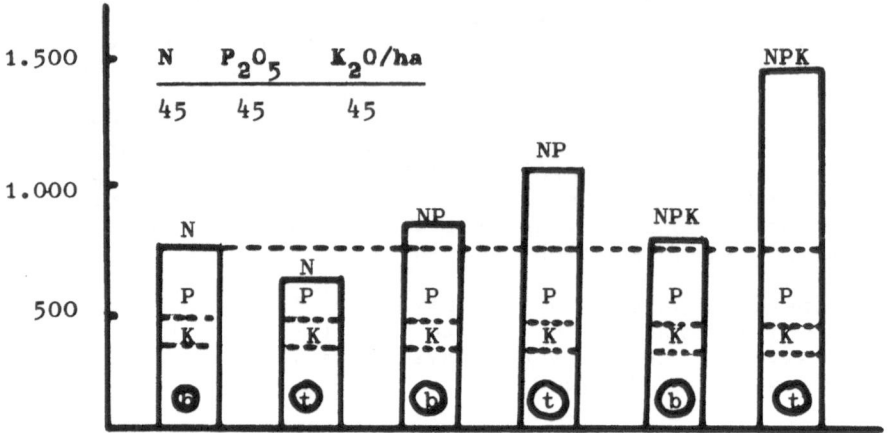

Fig. 4   Effect of nitrogen timing on grain yield in the
          presence and absence of P and K. (v. Uexkull, prel.
          results, unpublished)
          b = all Nitrogen as basal application before transpl
          t = all N as top dressing, 40 days after transpl

YIELD INCREASE
OVER CONTROL, kg./ha.

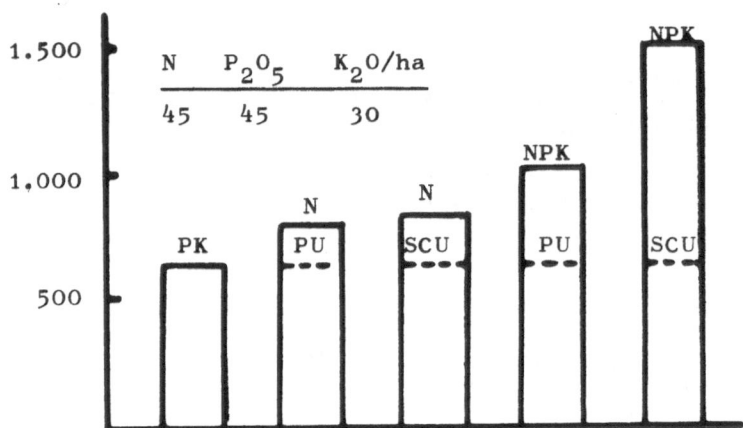

Fig. 5   Effect of prilled urea vs. sulphur coated
          urea in the presence and absence of P and K.
          Var. IR 42.   Preliminary data
          PU = Prilled Urea
          SCU= Sulphur coated urea

N will become available to the rice plant only during the second half of the active growing period, farming systems relying heavily on biologically fixed N will face problems related to low nitrogen supply to the rice plants up to panicle initiation. Tillering in particular may be affected.

In the past one major target of the plant breeder was nitrogen responsiveness. A variety very responsive to fertilizer nitrogen may not be the one to interact most effectively with N-fixing organisms. Different nitrogenase activities have been associated with different varieties suggesting that some varieties might be able to develop more specific and effective links with N-fixing organisms. (Rinaudo et al 1975; Lee and Yoshida, 1977.) A new criteria for breeding could be the ability to tiller at lower nitrogen levels. Interactions of differently spaced varieties with fertilizer rates, time of application and different methods of applications could yield valuable data as to how to maximize yields per unit input. We have to keep in mind that we have to increase food production per unit area at a time when input costs are on the increase. We must therefore maximize yields and at the same time maximize efficiency of inputs.

CONCLUSION

There is a large - so far untapped potential to increase yields on rainfed rice paddies. Fertilizer techniques developed for rice paddies with controlled irrigation are usually too risky for most farmers who operate under conditions of environmental constraint. Evidence presented in this paper suggests the possibility to develop techniques that would minimize the risk and at the same time increase efficiency of the applied fertilizer. To get full benefit from such techniques, further research will be needed.

REFERENCES

(1) All India Coordinated Rice Improvement Project. Progress Report. Section 2, Agronomy. Kharif 1979.
(2) Bouldin, D.R. and B. V. Alimagno. NH_3 volatilization from IRRI paddies following broadcast applications of fertilizer nitrogen. Terminal Rpt. of D. R. Bouldin as visiting scientist at IRRI. 1976.
(3) Bunt, J. S. Nitrogen fixing blue-green algae in Australia rice soils. Nature. 192:479-80. 1961.
(4) Chandler, R. F. Rice in the Tropics. Westview Press, Boulder, Colorado. p. 40. 1979.
(5) De. P. K. The problem of the nitrogen supply of rice. I. Fixation on nitrogen in the rice soils under waterlogged conditions. Ind. J. Agric. Sci. 6:1237-1245. 1936.
(6) De, P. K., M. Sulaiman. Fixation of nitrogen in rice soils by algae as influenced by crop, CO_2 and inorganic substances. Soil Sci. 70:137-151. 1950.

(7)   Dewson, R. S.   Potential for nitrogen fixation by micro-
      organisms in rice paddies.   Int. Rice Comm. Newsletter
      16:1-10.
(8)   Griffith, W. K.   Effects of potassium on nitrogen fixa-
      tion.   In: Potassium for Agriculture.   Potash & Phosphate
      Institute: 132-139.   1980.
(9)   International Rice Research Institute.   Annual reports
      for 1972-1977.   Los Banos, Philippines.   1973-1978.
(10) International Rice Research Institute.   International
      Network on Soil Fertility and Fertilizer Evaluation for
      Rice.   (Insfer.)   Rep. on the International Long-Term
      Fertility Trial in Rice.   1980.
(11) International Rice Research Institute.   Research High-
      lights for 1979.   Los Banos, Philippines p. 81.   1980.
(12) Kawasaki, T.   Response of rice by types to the level of
      nutrients in culture solution.   Paper presented at the
      annual meeting of the Soc. of the Sci. of Soil and
      Manure.   Tottori, Japan.   (In Japanese) 1980.
(13) Konishi, C., Seino, K.   Studies on the maintenance mech-
      anism of paddy soil fertility in nature.   Bull. Hokuriku
      Agric. Expt. Sta. 2:41-136.   (In Japanese) 1961.
(14) Koyama, T., App, A.   Nitrogen balance in flooded rice
      soils.   In: Nitrogen and Rice International Rice Res.
      Inst. Los Banos, Philippines. p. 95-104.   1979.
(15) Liu Chung Chu.   Use of azolla in rice productions in
      China.   In: Nitrogen and Rice Int. Rice Res. Inst.   Los
      Banos, Philippines. 375:394.   1979.
(16) Mikkelsen, D.S., De Datta, S.K., Obcemea, W.N.   Factors
      affecting ammonia volatilization losses from flooded envir
      ment of rice.   Paper presented at the Int. Rice Res. Conf.
      April 18-22, 1977.   IRRI. Los Banos, Philippines. 1972.
(17) Mitsui, S.   Inorganic nutrition, fertilization and soil
      ametioration.   Yokendo Press, Tokyo. pp. 107.   1954.
(18) Moore, A.W.   Non-symbiotic nitrogen fixation in soil and
      soil-plant systems.   Soils and Fertilizers 29:113-128. 1966.
(19) Rinaudo, G., Hamad-Fares, I., Dommergues, Y.R.   Nitrogen
      fixation in the rice rhizosphere: Methods of measurement
      and practices suggested to enhance the process.   In: Bio-
      logical Nitrogen Fixation in Farming Systems of the Tropics.
      Ed. A. Ayanaba and P.J. Dart.   John Wiley & Sons, Chichester,
      N.Y. Brisbane, Toronto. p. 313-322. 1975.
(20) Singh, P.K.   Use of Azolla in India.   In: Nitrogen and Rice.
      Int. Rice Res. Inst. Los Banos, Philippines. 406-417.   1979.
(21) Srinivasan, S.   Algae multiplication and fertilizer prac-
      tices.   Int. Rice Res. Newsletter 4:3.   1979.
(22) Trolldenier, G.   Influence of fertilization on atmospheric
      nitrogen fixation in rice fields.   Proc. 11th Colloq. Int.
      Potash Inst., Bornholm, Denmark. p. 287-292.   1975.
(23) Trolldenier, G.   Nitrogenaseaktivitaet in der Rhizosphere
      von Sumpfreis in Abhaengigkeit von der Mineralstoffernae-
      hrnug.   Mitt. Dt. Bodenk. Ges. 29:339-344.   1979.
(24) Tuan, D.T., Thuyet, T.Q.   Use of Azolla in rice production
      in Vietnam.   In: Nitrogen and Rice.   Int. Rice Res. Inst.
      Los Banos, Philippines. 395-404.   1979.
(25) UNDP/FAO.   Fertilizer demonstration and pilot scheme distri-
      bution 1978/79.   Rept. to the Govt. of the Philippines.
      1980.

(26) Venkataraman, G.S.   The role of blue-green algae in tropi-
     cal rice cultivation.   In: Nitrogen Fixation by Free-
     Living Micro-organisms.   Ed. W.D.P. Stewart. p. 207-218.
     Cambridge Univ. Press.   1975.
(27) Watanabe, I. Biological nitrogen fixation in rice soils.
     Paper presented at the Symposium "Soils and Rice" at the
     Int. Rice Res. Inst. Los Banos, Philippines. Sept. 20-23.
     p. 289-305.   1977.
(28) Watanabe, I, Lee, K.K.   Non-symbiotic nitrogen fixation
     in rice and rice fields.   In: Biological nitrogen fixation
     in farming systems of the tropics.   Ed. A. Ayanaba and
     P. J. Dart. John Wiley & Sons.   Chichester, New York,
     Brisbane, Toronto. p. 292-305.   1975.

RATIONALIZATION OF FERTILIZER APPLICATION TO RICE IN A COOL ENVIRONMENT

Akira Tanaka* and Shinichiro Sekiya**
(Faculty of Agriculture, Hokkaido University* and Hokkaido
National Agricultural Experiment Station**, Sapporo, Japan)

INTRODUCTION

In 1870s people in the Japan mainland started to migrate to and settle in the Hokkaido Island (42-45° N), where the winter is very cold and long. As these people were accustomed to eat rice as the staple food, they brought seeds with them from the mainland, and tried to grow rice plants, but could not produce much rice, because varieties and cultural methods, which they used, were not adapted to the climatic condition in Hokkaido. By improvements of variety and cultural method during the last 100 years, however, rice has now become one of the most important crops in Hokkaido.

Total area planted to rice in Hokkaido was extremely small till 1900, then started to increase rapidly, continued to increase until 1930, decreased considerably from 1940 to 1950 due to shortage of labor during the Pacific War, increased rapidly again between 1955 and 1969 due to improvement of irrigation facilities and also to favorable prices of rice to farmers by a government policy, and then decreased abruptly from 1970 due to a government regulation to reduce rice production because of surplus of rice in Japan (in 1979 only 64% of rice fields was permitted to plant rice in Hokkaido)(Fig. 1).

The average rice yield per unit field area in Hokkaido was kept almost constant until 1930s (Fig. 1). This does not mean that the yield was kept stagnant: The yield in old rice areas increased consistently; but rice cultivation expanded to the areas where environmental conditions were less favorable and the yield in such areas was lower; thus the average yield was kept constant. After 1950 the average rice yield increased continuously, and is now more than 5 t/ha of brown rice.

The rice yield fluctuated remarkably from year to year. During 1900s it was 2.46 t/ha in 1907, and was only 0.21 t/ha in 1902. The fluctuation still remains even at present; during 1970s the yield was 5.36 t/ha in 1978, and was 2.73 t/ha in 1971. The fluctuation was due mainly to the fluctuation of weather conditions from year to year. There is a very close correlation between the temperature during the growing season and the rice yield. For example, the yield decreases abruptly when the mean temperature in July and August is lower than 19° C (7). Thus, how to minimize the damages caused by low temperatures has been one of the most important issues of rice technology in Hokkaido.

There are two types of low temperature damages, i.e. destructive damage and delayed growth damage: The destructive damage is abnormalities of reproductive organs which are caused by low temperatures below about 15° C during the development of sexual organs, and results in an increase of sterility percentage; and the delayed growth damage is an extension of total growth duration caused by low temperatures at various growth stages, especially during vegetative growth, and results in a failure of complete ripening of grains before frosts come in autumn (1)

Results of researches to improve technology to obtain consistently high rice yields by minimizing the effect of low temperature damages, especially through rationalization of fertilizer application, are described in this paper.

VARIETAL IMPROVEMENT

The first variety which was successfully grown fairly extensively in the central part of Hokkaido was Akage, which was selected by a farmer from an old farmer's variety. Pure line selection from local varieties and cross-breeding began in 1900 and 1913, respectively, by agricultural experiment stations, and about 100 varieties (mostly results of cross breeding) have so far been released.

One of the major objectives of rice breeding was early maturity. This characteristic was very important to make the rice plant to complete ripening before frosts come in autumn, especially in years with cool weather under which the delayed growth damage was expected. Generally, the shorter the duration, the lower is the yield. Thus, consistence of yield regardless of the weather conditions prevailing between one year and the next, rather than high yields in good years, was considered to be important in rice breeding in early days. It is necessary to flower before the middle of August to complete ripening before frosts come in autumn. This means that the panicle primordia should be initiated by the middle of July when the daylength is more than 14 hours. Because of this reason, varieties should be photo-non-sensitive. Varieties with such a characteristic are, however, frequently sensitive to temperature. Thus, it was necessary to obtain varieties which had a relatively stable growth duration regardless of temperature conditions.

By introduction of low-temperature protected nursery beds in 1940s as will be described later, however, the season avilable to rice growth was artificially expanded, and it became possible for breeders to seek varieties with potential to produce a consistent high yield by making the growth duration longer.

Although the season available for rice growth become longer, the chances of the destructive damage does not decrease because the damage is caused by a few-day cool spell below a critical temperature in July when sexual organs are differentiating. Thus, breeding to increase resistance to the destructive damage is still necessary.

As nitrogen application is indispensable to obtain high rice yields, varieties with high nitrogen response are necessary. Rice plants become more susceptible to the low temperature damages by nitrogen application because flowering becomes late and uneven, and also the sexual organs become more susceptible to the destructive damage. Thus, varietal difference in nitrogen response is of particular importance.

Fig. 2 (left) illustrates the changes of various traits from old to more recent recommended varieties in Hokkaido (11): Akage or Bozu-6, which were leading varieties many years ago, have tall stature, droopy leaves, and fewer panicles. On the contrary, more recent varieties, such as Mimasari and Yu-kara, have short stature, an erect growth habit, and many panicles. Thus, the modern varieties have an ability to utilize solar energy by photosynthesis at higher efficiencies, especially when the leaf area index is large due to an ample supply of nitrogen. They produce high yields, and respond to nitrogen application at high levels (Fig. 2, right). In the breeding of nitrogen responsive varieties, lodging resistance and also blast resistance were, of course, very important. Leading varieties in Hokkaido at present are Ishikari, Tomoyutaka, and Kitahikari, which are offsprings of Yukara.

561

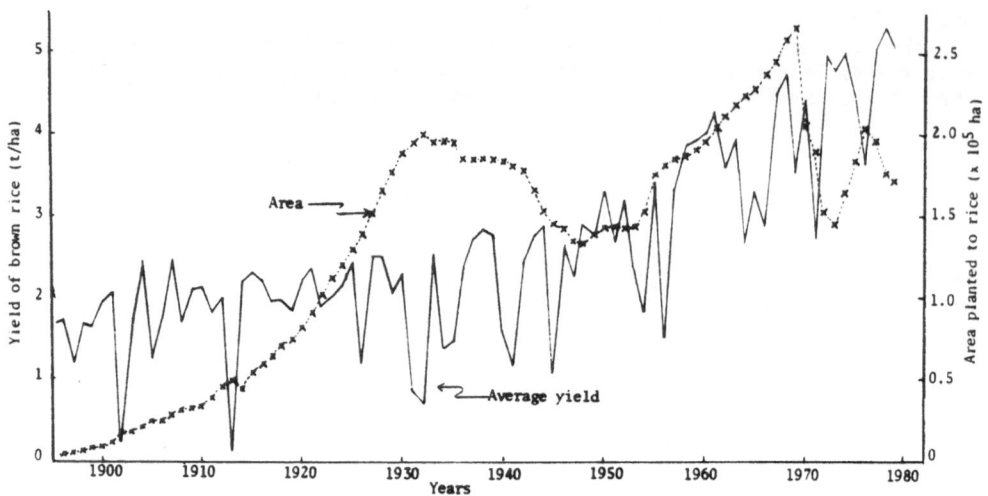

Fig. 1   Changes of area planted to rice and
average yield of rice in Hokkaido

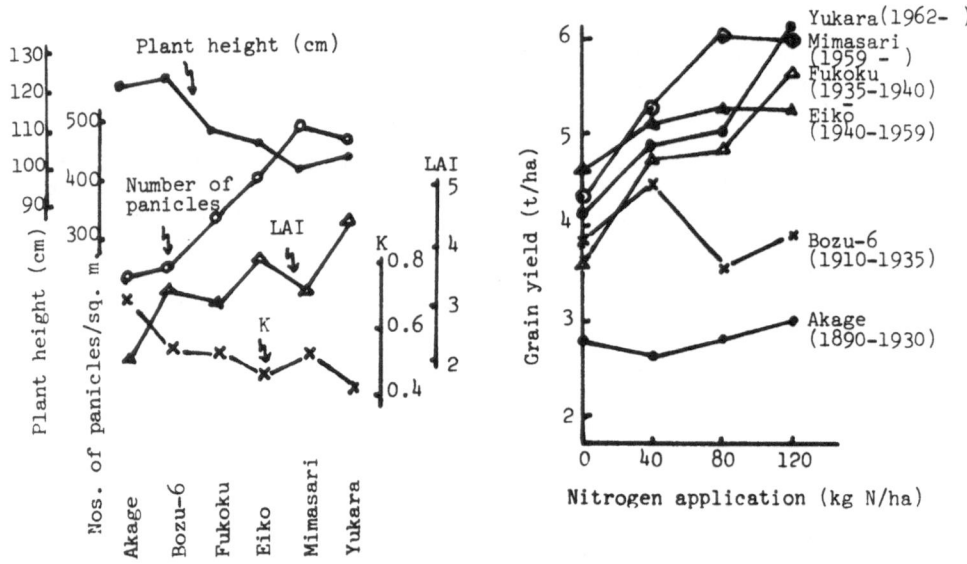

Fig. 2   Comparison of various traits (left) and nitrogen response (right) among old and new varieties in Hokkaido

Growth pattern is another important characteristic. Growth patterns of rice plants in Hokkaido and warmer areas in Japan were compared in early 1950s. It was revealed that in Hokkaido the growth rate at early stages was slower due to lower temperatures, and the panicle primordia were initiated before the maximum tiller number stage, while it occurred after the stage in warmer areas(2). Thus, it was necessary in Hokkaido to increase the growth rate at early stages, and also to increase the tillering rate to move the date of the maximum tiller number stage earlier.

By comparing dry matter production at different growth stages among old and new varieties, it was demonstrated that in very old varieties the dry matter production is small at all growth stages; in early stages of the varietal improvement breeders concentrated to increase the growth rate at the early growth stages, though which the growth pattern was improved, a considerable total dry matter production was secured, and the yield was increased even if the temperature was low during early growth stages; and in the more recent varietal improvement breeders attempted to increase the dry matter production after flowering by improving the plant type as described earlier, though which the yield increased by the increases of total dry matter production as well as of harvest index. The objective of breeders at present is to increase the dry matter production at both early and after flowering, although it is extremely difficult to accomplish.

NURSERY BED TECHNOLOGY

In Hokkaido the season when the temperature is above a critical is very short. The frost free period is only 160 days in the central part of Hokkaido. Since planting is made in spring when the temperature is below 12° C, establishment of seedlings tends to be slow, and this results in delay of whole growth cycle, and frequent failure of complet ripening. Thus, how to make the establishment faster is an important technique.

The method to establish plant stands in the main field changed dramatically in the past (Fig. 3). In early years the only method was to transplant seedlings which were raised in submerged nursery beds in the open air.

In 1900 a direct sowing method was invented by a farmer. By using a simple hand operated equipment seeds were sown on the surface of paddled submerged soil at 16 hills by one operation. This method became popular among farmers because the income of farmers increased considerably since by this method it became possible to cover a larger area with their own labor although the yield per unit field area did not increase. By the introduction of this method the area planted to rice in Hokkaido increased very rapidly, and by 1930 almost all areas with environmental conditions suitable for rice cultivation had been converted to the rice field.

In 1935 a method to raise seedlings in low temperature protected upland nursery bed, was introduced. In early days the bed was surrounded by wooden boads at the sides and covered with oiled papers at the top, and the seeds were sown on a soil which was kept under a ordinal upland moisture condition. Later, vinyl film tunnels or houses were used to cover the bed. By this method it became possible to sow seeds much earlier in spring, the establishment after transplanting became much faster, and the yield increased and stabilized remarkably.

In 1960s, however, labor became more and more scarce and expensive. Thus, about 1970 machine transplanting was introduced. Various machines were developed, various methods to raise seedlings which were suited for each type of machines were developed, and the machine transplanting became very popular among farmers (94% in 1980).

Studies on the process of seedling establishment demonstrated that after

563

transplanting it takes about one week before the seedlings absorb an appreciable amount of nutrients from the soil in the main field; they start to absorb nitrogen and potassium; there is no appreciable absorption of phosphorus until two weeks after transplanting. Thus, it is important to hasten the absorption of nutrients, especially phosphorus, to make a more quick seedling establishment (5).

The seedlings raised by the submerged nursery bed are more slender and lower in carbohydrate content, have weaker roots, and take more time to establish than those raised by the upland nursery bed.

When seeds are sown thin and seedlings are kept in the low temperature protected upland nursery bed for more than 40 days, a few tillers develope on the nodes at lower positions of the main stem. These seedlings are very active to establish, and the tillers exist at transplanting produce a large panicle. Thus, it was recommended to use "mature seedlings" than "ordinal seedlings" (Table 1), although a larger area of nursery beds is required to raise the mature seedlings.

When an adequate amount of fertilizers is applied to the low temperature protected upland nursery bed, the seedlings are high in contents of both nutrient elements and starch, more active in the vegetative vigor, and capable of developing new roots more quick so that they can establish satisfactory even under low temperatures (4). Thus, it was recommended to apply 30 g each of N, P_2O_5 and K_2O per m^2, and to topdress 13 g N per m^2 3 to 5 days before transplanting. It was also advised to maintain the soil pH ranging from 4.5 to 5.0, because if it is higher the roots become weak, and the seedlings become more susceptible to "Murenae", a disease caused by fusarium (6).

In very cool areas it was recommended to top-dress phosphorus fertilizers to the nursery bed just before transplanting, and to transplant the seedlings with their roots covered with a mixture of soil and fertilizers.

While uprooting the seedlings from the upland nursery bed, a portion of roots are cut. The roots, which are present on the rice plant at transplanting, are able to absorb nutrients when the plant is transferred to the main field; these roots promote the initiation and development of new roots and the establishment. The greater the length of these roots, the more active the nutrient uptake and development of new roots; there is, however, no more increase in these activities when the roots are longer than 3 cm (4). Thus, it was recommended to keep the 3 cm roots of seedlings as much as possible when the seedlings are uprooted.

For mechanical transplanting, it became necessary to make the seedlings suitable for machines. Various types of transplanting machines were manufactured, and various methods to raise seedlings suited to each machine were introduced (Table 1). Generally speaking, the seeds are sown on a tray. The tray has about a 60cm x 30cm surface area and a 2-3 cm depth. A mixture of soil and fertilizers is put into the tray, and the seeds are sown on it and are covered with an adequate material. These trays are placed in an incubator, kept at about $30°$ C until germination starts, and then these are transferred to a vinyl house. After seedlings are grown for 20-35 days, the tray on which seedlings are growing are carried to the field, and fed to a transplanting machine, and 4-5 seedlings are picked up in a bunch by a device of the machine at a time from the tray, and the bunch is transplanted to a hill. Four to twelve rows can be planted at a time depending upon the size of machine. One type of tray has many compartments in a tray, and a few seeds are sown to each compartment, and the seedlings in a compartment will be transplanted as a hill. Another type of tray has holes at the bottom, and the tray is placed on a fertilized soil in a nursery house. In this case the roots of seedlings elongate throughout the holes into the soil below the tray and absorb nutrients from the soil. The shorter the duration in the tray, the more convenient because smaller seedlings are easier to handle with a machine and also it requires less number of trays to cover a unit main field area.

However, the longer the duration, the better for more quick establishment of the seedlings. By experience, mature seedlings are becoming more popular among farmers.

FERTILIZER APPLICATION TO THE MAIN FIELD

A long term N, P, K trial (1926-1967) conducted by the Hokkaido National Agricultural Experiment Station demonstrated that grain yield varied considerably from year to year. In years of high grain yields the yield of -P plot was reasonably high, but in years of low yields it was extremely low. Conversely, the yield of -N plot was relatively low in years of good crop yields, but relatively high in years when most crops were poor (12). Thus, it can generally be said that the application of higher rates of phosphorus and lower rates of nitrogen is recommendable in areas where rice plants are liable to the low temperature damages.

Optimum amount of nitrogen application changes with varieties as mentioned earlier. It also changes from a year to the other depending upon the weather condition: It was frequently experienced that with a given variety the optimum amount was 120 kg N/ha in some years, while in other years the highest yields were obtained without nitrogen application. Thus, it was almost impossible to recommend an optimum amount of nitrogen application.
However, due to improvements of varieties and cultural methods the situation has become somewhat better. A following recommendation can be made: 60 kg N/ha at planting, and an additional top-dressing of 20 kg N/ha after panicle initiation if the weather condition of the year is favorable. This recommendation is a conservative one which is not to obtain a very high yield in good years, but to obtain consistent relatively high yields in any year.
By nitrogen application at a shallow layer of the soil, the seedling starts to absorb nitrogen earlier and the growth at early stages is promoted, but the loss of nitrogen by denitrification is more; whereas by nitrogen application at a deep layer, the rice plant absorb more nitrogen at later growth stages and produce more yield, but a delay of ripening may occur (3). Thus, applications of 1/3 of nitrogen at the shallow layer (5 cm) and 2/3 at the whole layer are recommendable.
It was considered that nitrogen top-dressing was not safe in Hokkaido. The reason was as follows: In warmer areas the top-dressing at the panicle initiation stage gives positive effect to the grain yield because the nitrogen absorbed at this growth stage promotes only the development of panicle-primordia since no tiller is growing at this growth stage; whereas in Hokkaido, if nitrogen is top-dressed àt this growth stage, the nitrogen absorbed promotes development of tillers rather than the development of panicle-primordia, the flowering becomes late and uneven, and results in a decrease of grain yield. However, by the improvement of cultural method, the growth pattern has been changed, the panicle primordia are initiated after the maximum tiller number stage even in Hokkaido; and the top-dressing gives generally a positive effect. It was further demonstrated that nitrogen top-dressing at booting has more chances of good effects by making the ripening better.
There is a believe that improvement of soil fertility by application of compost is necessary to obtain a very high yield because nitrogen is supplied to rice plants slowly but steady when soil fertility is high. On a soil whose fertility had been built up, rice plants absorb a large amount of nitrogen slowly for a long period and produce a very high yield; whereas, even on a poor soil it is possible to make the amount of nitrogen absorbed large by a heavy nitrogen application at planting, but the yield is low (9). However, it is almost impossible to use a large amount of compost at present because there is no animal in rice farms and also no labor to produce compost. Thus,

Table 1 Seedling raising methods of various types

Age of seedlings	Type of tray (size in cm)	Amount of seeds (ml per m² or tray)	Amount of fertilizers (g per tray)		Duration in nursery (days)	Weight of seedling (g per 100 plants)	Number of trays required per ha
			Basal (N, P$_2$O$_5$, K$_2$O, each)	Top-dressing (N)			
		Low temperature protected upland nursery beds					
Ordinal	Ordinal bed	300/m²	30/m²	13 (per m²)	30–40	3.0–4.5	100 m²
Mature		220/m²	30/m²	4.5 x 2 + 9 (per m²)	45	6.0	200 m²
		Nursery beds for mechanical transplanting					
Young	Box-mat (58x28x3)	400	1	1	20–25	1.0–1.2	200
Medium	Box-mat (58x28x3)	200	1	1 x 2	30–35	2.0	340
Medium	Frame (59x27x18 74 slits; 0.7x12)	150	0*	1 x 2	35	2.5	340–430
Mature	Pots 62x31x2.5 (448 pots with 1.6 cm diameter)	2-3 seeds per pot (60 per bed)	0.5*		35	3.0–4.5	500–550

* 30 g/m² to the soil below the tray

Table 2   Effect of phosphorus application at graded levels on the yield of rice plants on a volcanic soil

Year	Treatments			
	P-0	P-100	P-250	P-500
	Phosphorus application (kg P$_2$O$_5$/ha)			
1968 - 70	0	100	250	500
1971	0	100	100	100
1972 -	0	0	0	0
	Yield of brown rice (t/ha)			
1968	3.4	4.7	4.8	4.3
1970	4.2	6.4	6.4	5.9
1971	2.4	5.8	5.8	5.3
1973	3.7	6.4	6.4	6.4
1975	4.6	6.1	6.2	6.1
1977	2.6	6.0	6.2	6.3
1979	1.8	5.5	6.3	6.1

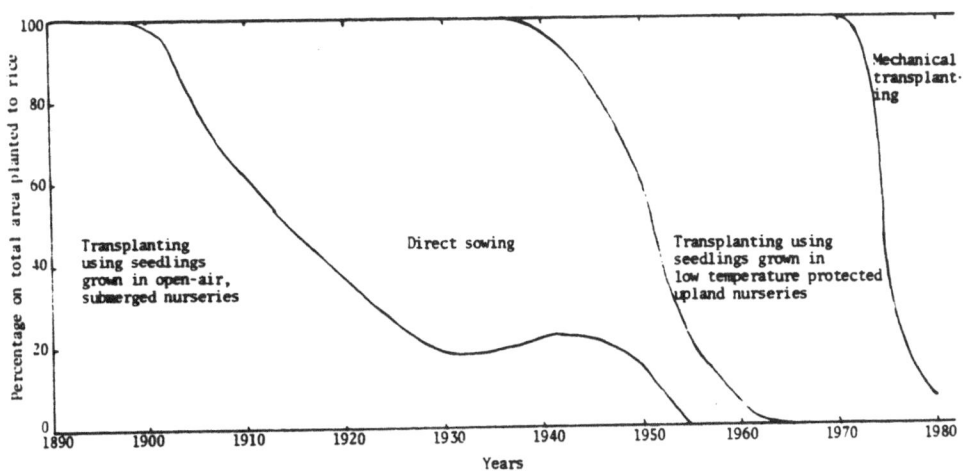

Fig. 3.   Changes of planting methods of rice in Hokkaido

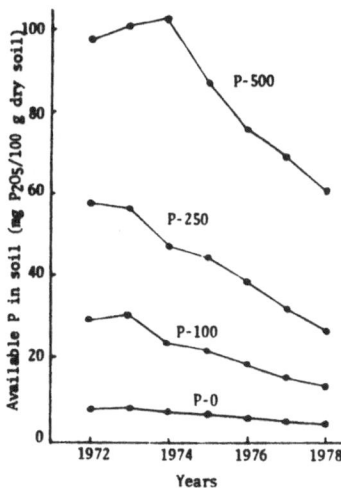

Fig. 4.   Available phosphorus in the soil at successive
years at the plots which received graded levels
of phosphorus before discontinuing phosphorus
application (P-0~ P-500 indicate the treatments
shown in Table 2)

question arises as to whether it is possible to produce a very high yield by using only chemical fertilizers: By top-dressing an adequate amount of nitrogen each at several growth stages it is possible to make the rice plant to absorb a large amount of nitrogen with the process similar to that on a high fertility soil, and to produce a very high yield (10). However, there remains a question how to top-dress nitrogen so many times with a minimum labor.

Phosphorus application promotes the establishment of seedlings, makes the flowering earlier, and the grain yield higher. These effects are more apparent in years with an unfavorable weather.

An experiment was conducted on a volcanic ash soil which was low in phosphorus supplying power. In this experiment graded amounts of phosphorus were applied for three years (1968-70), in the next year (1971) 100 kg P_2O_5/ha was applied uniformly to the plots which received phosphorus in previous years, and then from 1972 no phosphorus was applied to all plots (Table 2). During the early stage of this experiment (1968-70) the yield at the P-0 plot was reasonably high, and a yield increase was observed only upto the P-100 plot. In 1971 the yield at the P-0 plot became apparently lower indicating that the available phosphorus in the soil was already depleted. When no phosphorus was applied after 1972, the plot from which the maximum yield was obtained shifted from the P-100 plot to the P-250 plot after 6 years.

Farmers in Hokkaido have been applying a large amount of phosphorus every year, and soils on which rice has been grown for many years are already high in phosphorus at present. Thus, it is necessary to find out whether phosphorus application is necessary or not to a given soil.

Available phosphorus in the soil measured by the Bray No. 2 method using the soil kept under submerged condition (at the maximum tiller number stage) was higher with larger amount of phosphorus application and decreased with years of successive rice plantings when no additional phosphorus was applied (Fig. 4). By combining the informations from Table 2 and Fig. 4, it can be said that about 30 mg P_2O_5/100 g dry soil is the critical level of available phosphorus in the soil to obtain a maximum yield (8). From the decreasing trend of available phosphorus in the soil without phosphorus application, it can be speculated that for example, if the available phosphorus is 100 mg P_2O_5/100 g dry soil and rice is planted on the soil annually, it takes about 15 years till phosphorus application become necessary.

REFERENCES

(1) Ishizuka, Y., Shimazaki, Y., Tanaka, A., Satake, T., Nakayama, T., Rice growing in a cool environment. Food & Fertilizer Technology Center, Taipei, Taiwan. p. 98 (1973).
(2) Ishizuka, Y., Tanaka, A., Studies on ecological characteristics of rice plant grown in different localities, especially from standpoint of nutrio-physiological character of plant. J. Sci. Soil and Manure, Japan. 27, 1-6, 95-99 (1956).
(3) Ishizuka, Y., Tanaka, A., Studies on placement of fertilizers in rice plant. J. Sci. Soil and Manure, Japan. 33, 88-92 (1962).
(4) Ishizuka, Y., Tanaka, A., Nutrio-physiology of the rice plant. Yokendo Co. Ltd., Tokyo pp. 307 (1963).
(5) Miyake, M., Hoshi, S., Comparison of rice culture systems in Hokkaido. Res. Bull. Hokkaido Nat'l Agr. Expt. Stat. 75, 53-59 (1960).

(6) Nishikata, T., Konno, S., Naganuma, Y., Studies on the outbreak of physiological damping-off of rice plant, so-called "Murenae", in frame nursery. Part 1. Influence of soil reaction on the outbreak of "Murenae". Res. Bull. Hokkaido Nat'l. Agr. Expt. Stat. 66, 17-32 (1954).

(7) Sasaki, T., Nakayama, T., The relationship between the yields of paddy rice and weather conditions in the Abashiri district of Hokkaido. Nogyogijutsu 22, 12 (1967).

(8) Sekiya, S., Kogano, K., Miyazaki, N., Shiga, H., Studies on phosphorus application to the rice plant in cool area. Part 9. Effect of phosphorus accumulated in the soil. Abst. the 1980 Meeting. Soc. Sci. Soil and Manure, Japan. 26, 100 (1980).

(9) Shiga, H., Kakimoto, A., Doi, Y., Awazaki, H., Miyake, M., Sekiya, S., Kataoka, T., Environment and characteristic of rice plants in a high yielding paddy field in Hokkaido. Res. Bull. Hokkaido Nat'l Agr. Expt. Stat. 99, 30-40 (1971).

(10) Shiga, H., Miyazaki, N., Sekiya S., Time of fertilizer application in relation to the nutreint requirement of rice plants at successive growth stages. Proc. Inter. Seminar on Soil Environment and Fertilizer Management in Intensive Agriculture. Soc. Sci. Soil and Manure, Japan. 223-229 (1977).

(11) Tanaka, A., Yamaguchi, J., Shimazaki, Y., Shibata, K., Historical changes in plant types of rice varieties in Hokkaido. J. Sci. Soil and Manure, Japan. 39, 526-534 (1968).

(12) Yamaguchi, N., Shiga, H., Miyake, S., The results of a 41 years N, P, K trial on rice at Kotoni. Hoku-No 38, No. 7, 1-14 (1971).

TRANSFORMATION AND DISTRIBUTION OF ORGANIC AND INORGANIC FERTILIZER NITROGEN IN RICE AND SOIL SYSTEM

Huang Dong-mai, Gao Jia-hua , Zhu Pei-li
(Soil & Fertilizer Research Institute, Jiangsu
Academy of Agricultural Sciences, Nanjing)

Combined application of chemical fertilizer with farm manure is popularly practised in Chinese farming. In order to clarify the characteristics of nitrogen supply in this fertilization system, pot experiments were conducted in the greenhouse at the soil and Fertilizer Research institute , Jiangsu Academy of Agricultural Science.

MATERIALS AND METHODS

Four paddy soils were used in present experiments: (A) A heavy clay loam derived from the lacustrine deposit of Taihu Lake. (B) An acid clay loam with bleached subsoil derived from highly bleached loess, distributed on the hilly land nearby Nanjing. (C) A heavy clay loam derived from moorland soil, northern Jiangsu Province. (D) A saline soil of sandy loam texture, strongly calcareous, derived from the alluvial deposit along the Huanghe River, northern Jiangsu Province.

Some chemical properties of soils used are shown in Table 1.

Table 1 Some chemical properties of the soils
used in the experiment

Soil	pH	O.M.(%)	T.N(%)	C:N
A	6.5	2.43	0.133	10.6
B	6.0	0.86	0.054	9.2
C	7.0	2.91	0.150	11.3
D	8.5	0.73	0.058	7.2

Crotalaria, drilled with $(^{15}NH_4)_2SO_4$ was planted(1). The tops of the plant were harvested at the beginning of flowering and the air-dried plant materials, containing 6.5% of ^{15}N in total nitrogen were pulverized into fine powder, and then used as the source of ^{15}N labelled organic manure. $(NH_4)_2SO_4$ tagged by 10.6% ^{15}N of the total nitrogen was applied as inorganic labelled nitrogen fertilizer.

Greenhouse experiments with pot, in size of 15 x 20 cm, containing 3 kg of air-dried soils, were conducted for rice plant. All treatments received adquate supply of phosphorus and potassium fertilizer. The rate of nitrogen application was 0.24 g N per pot for both organic and inorganic fertilizers and no nitrogen for check pots. The treatments included:

1) ^{15}N labelled crotalaria.
2) 1/2 ^{15}N labelled crotalaria + 1/2 $(NH_4)_2SO_4$[*].
3) 1/2 ^{15}N labelled $(NH_4)_2SO_4$+ 1/2 crotalaria[*].
4) ^{15}N labelled $(NH_4)_2SO_4$.
5) Check (No nitrogen applied).

Rice seedlings, planted on soil No. A, C and D, were harvested at maturity. For the acid clay loam, soil No. B, the nitrogen trans-formation was investigated at different stage of rice growth.

Nitrogen in soil and plant samples was determined by semimicro Kjeldahl method and the ^{15}N abundance by mass spectrography.

RESULTS AND DISCUSSION
1. The Contribution of Different Nitrogen Sources to the Total Nitrogen Uptake by Rice

Results showed that in general rice plant assimilated much more soil nitrogen than fertilizer nitrogen. In all treatments receiving fertilizer nitrogen, only about one third of the nitrogen assimilated by the rice plant came from the fertilizers.

Table 2 The A-value and related values of four paddy
soils under different nitrogen sources applied

Fertilizer sources	Soil types	Total-N uptake mg-N/pot	Nitrogen uptake (mg N/pot)		S:F	A-value mgN/100g soil
			Soil-N (S)	Fert.-N (F)		
Crotalaria	A	560.5	444.2	108.8	4.083	34.03
	C	393.1	291.0	94.3	3.086	25.96
	D	290.7	193.0	90.2	2.140	17.83
	B	224.8	136.2	81.1	1.680	14.00
1/2 Crotalaria	A	567.9	446.0	114.3	3.901	31.86
+1/2$(NH_4)_2SO_4$	C	427.6	317.2	102.8	3.084	25.18
	D	352.4	223.7	121.1	1.847	15.08
	B	294.1	181.8	104.8	1.736	14.18
$(NH_4)_2SO_4$	A	597.6	457.0	135.9	3.363	26.90
	C	490.6	346.9	136.1	2.549	20.39
	D	383.5	230.1	145.8	1.578	12.62
	B	310.2	164.5	138.1	1.191	9.53

It is shown if Table 2 that the amount of fertilizer nitrogen taken up by rice plant is highest in the ammonium sulphate treatment, lowest in the crotalaria treatment and intermediate in the combined application. In addition, the ratio of soil nitrogen to fertilizer nitrogen taken up by rice appears to be closely related to soil types. In general, the more fertile the soil, the more soil nitrogen is taken up in spite of the organic or inorganic fertilizer nitrogen applied, the uptake of indigenous soil nitrogen by rice in fertile paddy soil equals two or three times those

----
[*] Non-labelled.

571

of low fertile ones. Hence, rice production essentially depends on the soil nitrogen supply.

2. Nitrogen Uptake Pattern of Rice Plant from Different Nitrogen Sources during Rice Growth Period

In the case of basal dressing, as shown in Fig. 1, at the first two weeks after transplanting, the uptake of fertilizer nitrogen by rice is more or less greater than that of soil nitrogen, while the rice plant absorbs more nitrogen from $(NH_4)_2SO_4$ than from crotalaria. Thereafter, the contribution of native soil nitrogen to rice plant is gradually increasing. As shown in Fig. 6, in combined application of crotalaria and $(NH_4)_2SO_4$, rice absorbs relatively more $(NH_4)_2SO_4$-N than crotalaria-N in the first 15 days. After 30 days, it absorbs approximately equal amount of nitrogen from $(NH_4)_2SO_4$ and crotalaria respectively.

As compared with $(NH_4)_2SO_4$ and crotalaria treatment shown in Fig. 2, the nitrogen supplying status of combined application reveals more evenly steady and long-lasting. This characteristic of combined application of organic and inorganic nitrogen sources is of great value in agriculture.

3. The Mineralization and Immobilization of Fertilizer Nitrogen

As shown in Fig. 3, it is obvious that the simultaneous mineralization of organic fertilizer nitrogen and the immobilization of inorganic fertilizer nitrogen take place as soon as the fertilizers are applied into soil. The percentage of crotalaria nitrogen mineralized increases rapidly within first two weeks after application. Then, it keeps a slow increasing tendency more than one month. On the other hand, about forty percent of the nitrogen of ammonium sulphate is immobilized within two weeks after flooding. Thereafter, more than half of the organic nitrogen newly formed from ammonium sulphate applied is gradually subjected to mineralization, and again, becomes available to the rice plant. Finally, about 16% of the immobilized nitrogen derived from $(NH_4)_2SO_4$ is still remained in the soil after harvest.

Combined application of ^{15}N labelled $(NH_4)_2SO_4$ and crotalaria shows that inorganic nitrogen promotes the mineralization of organic nitrogen. Also, the percentage of immobilization of the inorganic nitrogen is considerably increased by the existence of organic nitrogen. These mutual effects seem to be an essential characteristic of such a fertilization system(Fig. 6).

4. Correlations of A-value with the Nitrogen Uptake of Rice Plant

The A-value determined by the isotopic dilution method indicates the availability index of soil nitrogen. A good correlation is found between soil nitrogen uptake in rice plant and the A-value (Fig. 4), although the latter is affected by different nitrogen sources. Moreover, Broadbent has indicated that the correlation between A-value and total soil nitrogen uptake by rice plants is highly significant in spite of diverse climatic zones(2). Thus, the A-value is considered to be a good index of the nitrogen supplying capacity of soils.

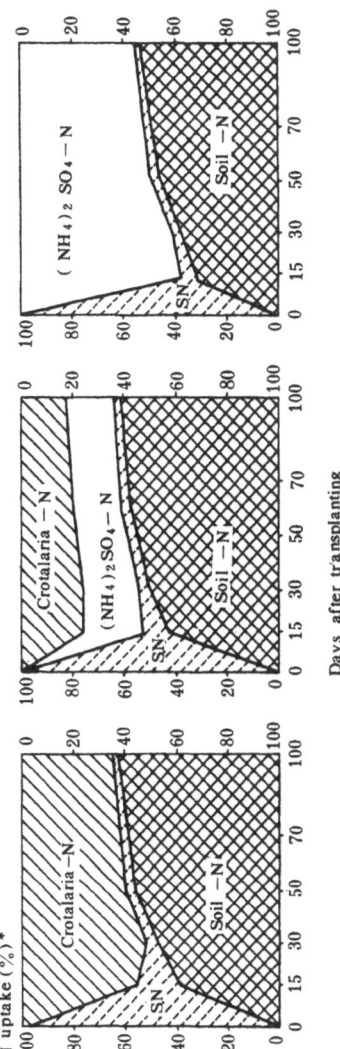

* Based on the total assimilated nitrogen
SN-Seedling N
Soil No. B = Acid clay loam derived from loess nearby Nanjing
Fig. 1 Variation of uptake of soil and fertilizer nitrogen during rice
growth period (Soil No. B)

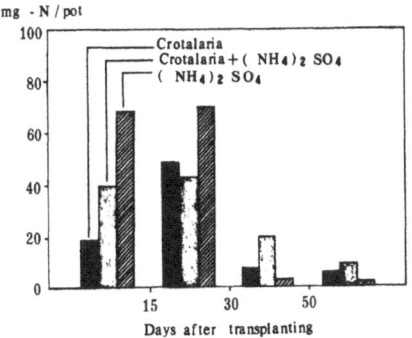

Soil No. B = Acid clay loam
derived from loess nearby
Nanjing
Fig. 2 Added amount of fertilizer
nitrogen uptake in rice in
at different growth dura-
tion (Soil No.8)

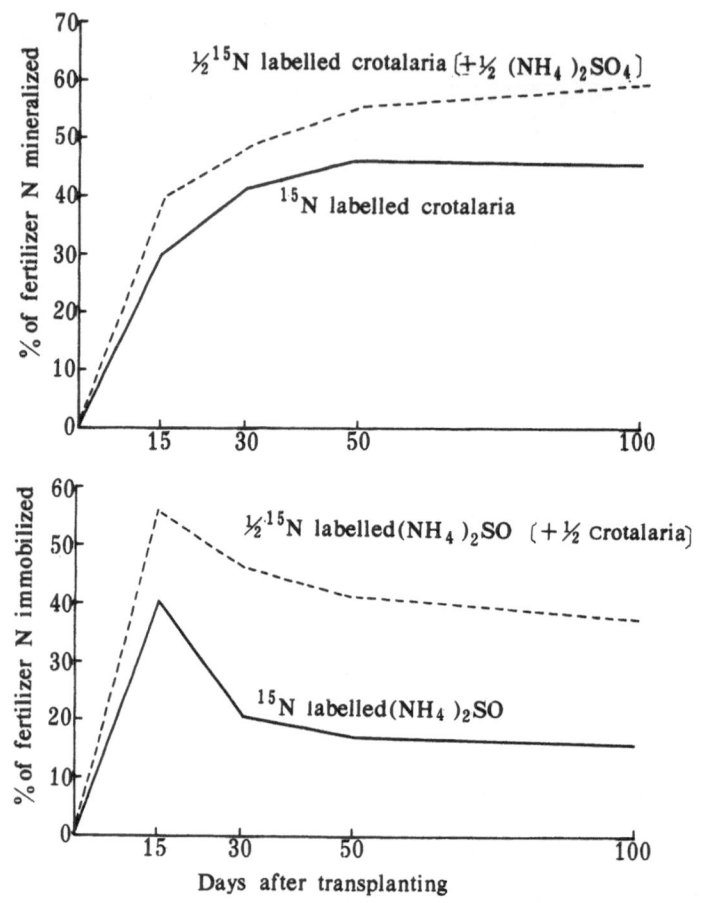

Fig. 3 The mineralization and immobilization of fertilizer nitrogen during rice growth period (soil No. B)

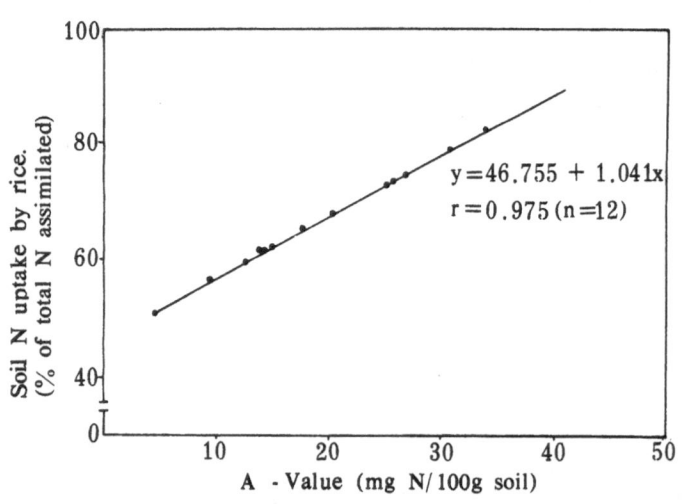

$$y = 46.755 + 1.041x$$
$$r = 0.975 (n = 12)$$

Fig. 4 Correlation of A-Value of different soil with uptake of soil nitrogen by rice plant

5. The Priming Effect in Relation to Nitrogen Sources

Comparative studies on the priming effect show that it is only significant under the application of chemical nitrogen fertilizer, but still questionable when organic manure is applied alone (Table 3).

Table 3   The priming effect on four paddy soils under different nitrogen sources applied

Soil	Treatment	Soil-N consumption (mg N/pot)	Rate of priming effect
A	Crotalaria	425.2	0.991
	1/2 Cro. + 1/2 A/S	495.5	1.150
	A/S	558.2	1.301
	Check	439.0	
B	Crotalaria	121.4	0.897
	1/2 Cro. + 1/2 A/S	168.6	1.222
	A/S	153.0	1.108
	Check	138.0	
C	Crotalaria	400.7	1.335
	1/2 Cro. + 1/2 A/S	468.0	1.418
	A/S	492.0	1.488
	Check	330.0	
D	Crotalaria	167.4	1.16
	1/2 Cro. + 1/2 A/S	209.8	1.46
	A/S	225.7	1.57
	Check	144.0	

A/S—Ammonium sulphate
Cro. —Crotalaria

6. Nitrogen Balance Sheet

Fig. 5 shows that in the case of basal application of crotalaria the amount of residual crotalaria nitrogen retained in soil is greater than the amount taken up by rice plant. The reverse is true for the treatment of ammonium sulphate. As to the treatment of combined application, the nitrogen balance sheet reveals an intermediate feature.

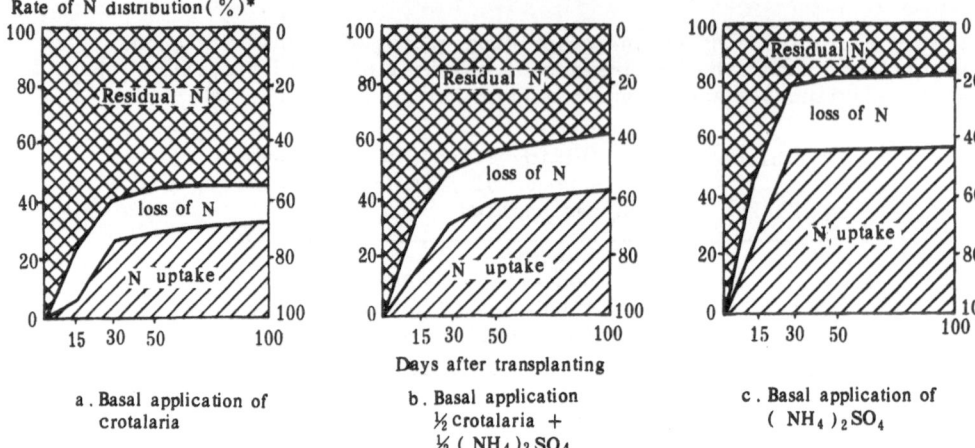

a. Basal application of crotalaria

b. Basal application ½ crotalaria + ½ (NH₄)₂SO₄

c. Basal application of (NH₄)₂SO₄

Soil No. B = Acid clay loam derived from loess nearby Nanjing
* Based on total amount of fertilizer nitrogen
Fig. 5 The balance of fertilizer nitrogen during rice growth period
(Soil No. B)

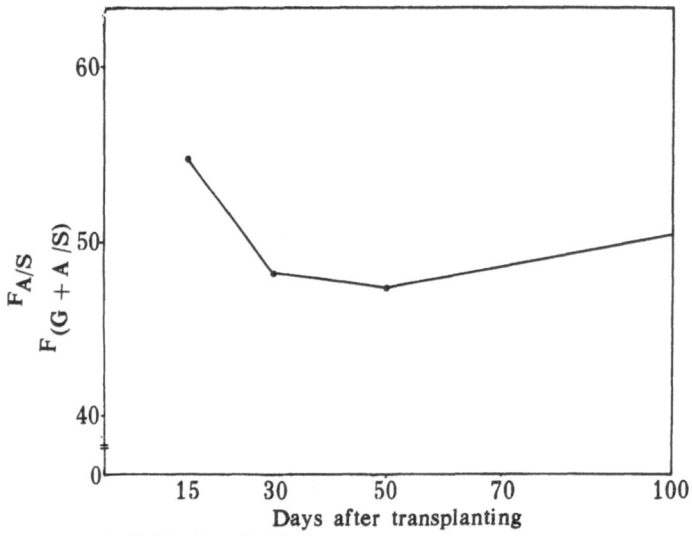

Soil No. B - Acid clay loam derived from loess nearby Nanjing

F_G - N uptake from crotalaria

$F_{A/S}$ - N uptake from $(NH_4)_2SO_4$

Fig. 6   The N uptake ratio from crotalaria and $(NH_4)_2SO_4$
during rice growth period (soil No. B)

The loss of applied nitrogen in the treatments of ammonium sulphate, combined application, and crotalaria averages 23.2%, 20.4%, and 16.2%, respectively. It indicates that the loss of chemical nitrogen could be reduced when applied with organic fertilizers.

REFERENCES

(1) Jansson, S.L., Experimental techniques with nitrogen-15. In "Report of the FAO/IAEA Technical Meeting", 415-422 (1966).
(2) Broadbent, F.E., Nitrogen transformation in flooded soils. In "Soils and Rice", IRRI, Philippines, 543-559 (1978).

TRANSFORMATION OF FERTILIZER, STRAW AND
SOIL NITROGEN IN PADDY SOIL

Hideaki Kai, Sadao Kawaguchi and Wittaya Masayna
(Faculty of Agriculture, Kyushu University, Fukuoka, Japan)

INTRODUCTION

The recycling of plant nutrients by returning crop residues to the soil can play an important part in the maintenance of soil fertility. The efficient use of crop residues can also decrease the need for chemical fertilizers. This is an important consideration from the viewpoint of conserving resources and also because of the possible adverse environmental effects of the widespread use of chemical fertilizers.

In various countries rice straw has been utilized in the form of compost, incorporated directly into the soil, or returned to the soil as ash after burning. In Japan, rice straw which is a main crop residue in paddy fields has been customarily returned to soil after being made into compost. The practice is considered to have an important effect upon the fertility of paddy soil and the yield of rice. However, in the past ten years or more, the return of rice straw as compost has been progressively decreasing due to the increasing application of chemical fertilizers, the change from draft animals to agricultural machinery and labor shortage on farms. Furthermore, in some parts of the country there has been an increasing tendency to burn away rice straw.

Thus, crop growers are now anxious about a possible deterioration of soil fertility. Application of rice straw directly into paddy soil is a very useful alternative to composting, because less labor is required.

It is therefore important to obtain a better understanding of the nitrogen transformations associated with the decomposition phase of rice straw in soil.

The present investigations were conducted (a) to make detailed investigations into the factors affecting immobilization and release of nitrogen in soil to which rice straw has been added, (b) to compare the availability of the immobilized nitrogen with that of the native soil nitrogen, (c) to elucidate the priming and residual effects of rice straw application on the nitrogen status of soil, (d) to obtain information on the chemical characteristics of the nitrogen fractions involved in these processes, and (e) to study the stimulative effect of rice straw application on the microbial nitrogen fixation in submerged rice field.

MATERIALS AND METHODS

Soil Samples

The soils used were three lowland fertilized soils sampled from their plow layers. Some properties of the soil samples are given in Table 1.

Incubation and Determination

Experiment 1. Moist Kasuya soil (pH 6.2) corresponding to 20 g dry soil was weighed into 50 ml Erlenmeyer flasks. The ground rice straw of 290 mg dry matter (35.28 % C and 0.42 % N, C/N ratio of 84) was added to each flask, and treated with different levels of nitrogen as $(NH_4)_2SO_4$ labeled with ^{15}N (17.70 atom % excess ^{15}N) to give different C/N ratios of 8, 16 and 32. All the flasks were incubated at 30 °C and 60 % of WHC (the maximum water holding capacity) for 20 weeks. Water was added at intervals to compensate for water loss during incubation by weighing the flasks. Flasks containing the mixture of C/N ratio 32 were also incubated at different temperatures (10,20,30 and 40 °C) for 20 weeks. Each treatment was run in triplicate and corresponding control were carried out.

In all experiments mineral nitrogen, total nitrogen and various forms of organic nitrogen were determined by Conway's micro-diffusion method (4) and Bremner's method (2), respectively. ^{15}N concentration in fractions was measured by massspectrometry.

Experiment 2. The three moist soils corresponding to 350 g dry soil were treated with 14 g of the ground rice straw tagged with ^{15}N. The straw contained 37.00 % C and 1.875 % N with 23.90 atom % excess ^{15}N. Each soil was contained in a polyethylene bag and the moisture was brought to 60 % of WHC. The bags were closed and incubated at 30 °C for 28 weeks. Moisture lost by evaporation was made up frequently by adding distilled water. Each treatment was run in duplicate. The term "check soil" refers to soil incubated without straw.

Inorganic nitrogen released from straw and native soil organic matter during decomposition was determined by the same way as in Experiment 1. ^{15}N concentration of the inorganic nitrogen was determined by emission spectrometry(3).

The effect of drying on mineralization of the tagged and native nitrogen in soil was measured in the following way: after 24 weeks incubation some of the samples were removed and dried at room temperature for 5 days, and then remoistened and reincubated for 4 weeks under the same condition as above. The corresponding controls were carried out. The accelerated mineralization of nitrogen due to drying of soil (the so-called drying effect) was computed from the difference between the nitrogen mineralized by the air-dried samples and controls.

Experiment 3. After 28 weeks incubation soils from Experiment 2 were air-dried at room temperature for 5 days and put into glass filters. After inoculation with suspension of each non-treated soil, the soil moisture was brought to 60 % of WHC. The filters were covered with polyethylene film and incubated at 30 °C for 3 weeks. The soils were then washed with distilled water to remove all nitrates, and the drying, remoistening, incubation and washing cycle were repeated 6 times. The flush of decomposition of the tagged and native nitrogen was measured.

Experiment 4. A field experiment was carried out at Kasuya soil with a NPK plot (75-56-38 kg / ha) and a rice straw plot (NPK plus 4 t rice straw / ha). Heterotrophic, phototrophic (10.000 Lux), aerobic and anaerobic acetylene reduction activities of the submerged plow layers at depth of 0-1 and 1-10 cm were measured (6) 17 days after the application of rice straw.

RESULTS AND DISCUSSION

A condition of dynamic equilibrium exists between inorganic and organic nitrogen in soil. The availability of nitrogen to plants is largely controlled by the relative magnitudes of the two opposing processes involved in this equilibrium, immobilization and mineralization. The conditions that control these processes are therefore of great practical importance.

1. Influence of C/N Ratio on Immobilization and Mineralization of Nitrogen in Soil

Fig. 1 shows the influence of the C/N ratio of the mixture of straw and mineral nitrogen added to soil on the immobilization of the inorganic nitrogen and on the length of time elapsed before the rate of remineralization exceeds immobilizaiton.

The principal results obtained are as follows; (a) Depending upon the C/N ratio of the addition, mineral nitrogen in soil is immediately immobilized and the release of immobilized nitrogen occurs after a time. The wider the C/N ratio, the larger the amount of nitrogen immobilized. (b) The rate of nitrogen release in soil was largest immediately after reaching maximum immobilization and thereafter decreased. (c) In the case of the widest C/N ratio of 84, where no mineral nitrogen was added, immobilization of soil nitrogen by the straw required 20 weeks to reach its maximum, thereafter nitrogen was released.

Immobilization of nitrogen in soil following addition of rice straw may result in nitrogen deficiency for crops. For this reason, it is recommended that some additional nitrogen fertilizers should be added to promote the decomposition of the straw and to prevent the nitrogen starvation, and also that the straw should be incorporated into soil as long as possible before growing the crops.

2. Influence of Soil Temperature on Immobilization and Mineralization of Nitrogen in Soil

When the soil temperature is lower, decomposition of rice straw in soil is slower. This may result in a slower and smaller immobilization and release of nitrogen in soil. The influence of temperature on these processes in the soil amended with straw is shown in Fig. 2. It shows the following tendencies: (a) The higher the temperature, the faster the rate of nitrogen immobilization in soil. Maximum nitrogen immobilization, however, occured at rather short incubation periods and its amounts were not greatly affected by temperatures ranging from 10 to 40 °C. (b) The amounts and rates of nitrogen release increased with increasing temperature. Therefore, nitrogen immobilized by the straw will become available to the plant at a later stage if the temperature is low. Even at 20 °C its release is considerably delayed compared with 30°C or 40°C.

These results suggest that incorporation of straw should be made much earlier in low temperature areas than in high temperature areas. For example, in Japan, rice straw can be incorporated without adverse effects, one month before transplanting of rice in the spring in southern areas, but it should be done in the previous autumn in northern areas. Further, in general, all of the rice straw produced can be incorporated without serious injuries to the plant in southern areas, while the maximum amount of straw application is said to be 4 t/ha in northern areas (1).

3. Measurement of Mineralization and Immobilization of Rice Straw nitrogen Incorporated into Soil and the Priming Effect

Mineralizaiton of rice straw-N and native soil-N in the course of 28 weeks incubation is shown in Fig. 3.

Three observations are pertinent: (a) Both native soil nitrogen and rice straw nitrogen were rapidly mineralized in the first 12 weeks of incubation and thereafter mineral nitrogen remained at stable levels in Kasuya and Nagano soils, whereas in Miyakonojo soil mineralizaiton was slow during the first 4 weeks of incubation, and only reached a stable level after about 20 weeks. (b) Eventual mineralization rates of rice straw-N in Kasuya, Nagano and Miyakonojo soils were 22,18 and 25 %, and those of native soil-N were 5.3,5.3 and 3.8 % ,

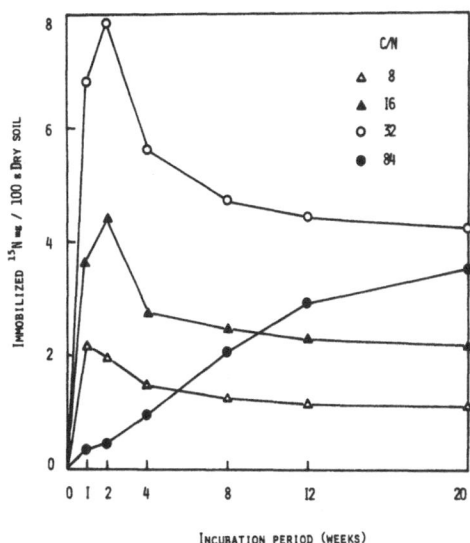

FIG. 1 IMMOBILIZED NITROGEN IN KASUYA SOIL AMENDED WITH RICE STRAW AND MINERAL NITROGEN GIVING DIFFERENT C/N RATIOS (AT 30 °C)

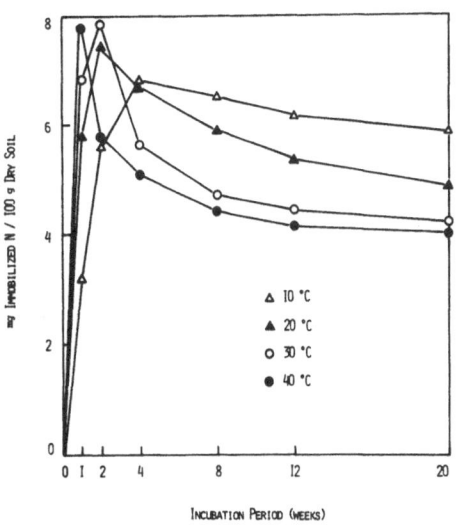

FIG. 2 INFLUENCE OF TEMPERATURE ON IMMOBILIZATION AND RELEASE OF NITROGEN IN KASUYA SOIL AMENDED WITH RICE STRAW AND MINERAL NITROGEN GIVING C/N RATIO OF 32

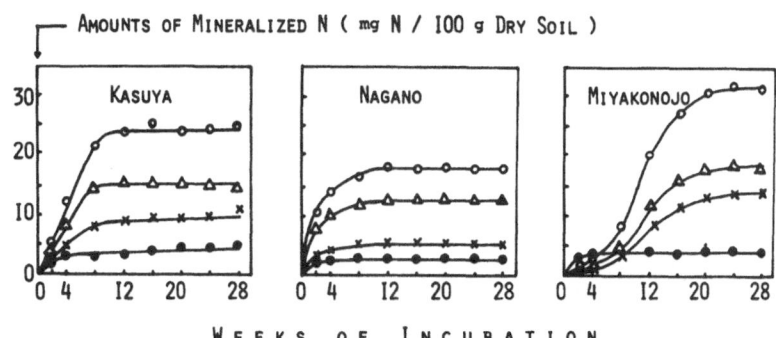

FIG. 3 AMOUNTS OF MINERALIZED NITROGEN AT VARIOUS TIME INTERVALS, FOLLOWING ADDITION OF RICE STRAW TO SOILS

respectively. Thus, eventual mineralization rates of both nitrogen fractions did not differ greatly among the soils. (c) Mineralization rates of soil-N in check soils were 2.2, 2.7 and 1.2 %, respectively. The differences in mineralization rates of native soil-N between the soils amended with straw and the check are the accelerated mineralization due to application of rice straw to soil.

This priming effect is given in Table 2. It was larger in Miyakonojo soil and smaller in Nagano soil. This order was the same as that of the total nitrogen contents of soils (see Table 1.). In addition, the magnitude of the priming effect correlated with the amounts of rice straw-N mineralized (Fig.4).

4. Effect of Rice Straw Application on the Accumulation of Decomposable Organic Nitrogen in Soil

Measurment of the effect of drying on the mineralization of nitrogen is often used for a routine determination of the nitrogen-supplying power of soil. Fig.5 shows the effect of drying on the mineralization of soil organic nitrogen derived from native soil-N and rice straw-N after 24 weeks incubation of three soils amended with ^{15}N-rice straw. The amounts of nitrogen mineralized due to the drying effect were considerably larger in the amended Kasuya and Nagano soils than in the corresponding check soils. Further, a larger proportion of mineralized nitrogen was derived from the nitrogen immobilized during the decomposition of rice straw than from native soil-N.

These results indicate that the application of straw is effective in accumulating decomposable organic nitrogen in soil and increasing the nitrogen-supplying power of soil. This was not true for Miyakonojo soil; the amended soil mineralized less nitrogen after drying and rewetting than the check soil. The reason is not known but it may be a result of the high rate of mineralization of both straw-N and native soil-N due to the priming effect following the addition of straw (see Table 2 and Fig. 4). The reserves of decomposable nitrogen may have been depleted during this period so that less nitrogen remained to be rendered decomposable by drying than in the check soil.

5. Residual Effect of Rice Straw Application on the Nitrogen Supply of Soil

To assess the residual nitrogen-supplying power of the amended soils after 28 weeks incubation, the soils were given 6 cycles of drying, remoistening and incubation for 3 weeks. The nitrogen mineralized during the treatments (a total of 18 weeks incubation) are shown in Fig. 6.

The mineralization of nitrogen originating from rice straw-N was greatest in the first and second cycles (the first 6 weeks of incubation), very little more was mineralized thereafter. This might suggest a rapid reduction in the residual effect of rice straw application on the nitrogen-supplying power of soil about 30 weeks after the application.

The mineralization of native soil-N was also greatest during the first two cycles but, in contrast to straw-derived N, mineralization continued at a fairly steady rate during the remaining 4 cycles and by the end of the experiment the amount of native soil-N mineralized equalled or exceeded the amount of straw-derived N mineralized.

These results would account for the necessity of successive long-term application of rice straw to improve soil fertility.

TABLE 1   DESCRIPTION OF SOILS USED

SOIL	pH H₂O	pH N-KC1	TOTAL N mg/100 g DRY SOIL	TOTAL C	C/N	C E C me/100 g DRY SOIL	TEXTURE	MAJOR CLAY MINERAL	MWHC*
KASUYA	5.6	4.8	239.7	2411	10.1	15.8	LiC	METAHALLOYSITE	60.5
NAGANO	6.1	5.1	138.8	1443	10.4	18.4	C L	MONTMORILLONITE	60.2
MIYAKONOJO	6.3	5.7	368.6	5389	14.8	24.9	C L	ALLOPHANE	116.3

* MAXIMUM WATER HOLDING CAPACITY,   g H₂O / 100 g DRY SOIL

TABLE 2   PRIMING EFFECT OF ^{15}N-LABELED RICE STRAW APPLICATION

ON MINERALIZATION OF NATIVE SOIL ORGANIC NITROGEN.

(mg MINERALIZED N / 100 g DRY SOIL)

INCUBATION PERIOD (WEEKS)	KASUYA	NAGANO	MIYAKONOJO
2	0 . 1	1 . 2	- 1 . 6
4	1 . 5	1 . 3	- 2 . 9
8	6 . 0	2 . 2	- 0 . 4
1 2	6 . 0	2 . 7	4 . 5
1 6	6 . 0	3 . 1	8 . 3
2 0	5 . 0	2 . 8	9 . 3
2 4	5 . 5	2 . 8	9 . 2
2 8	6 . 5	3 . 0	1 0 . 3

FIGURES IN PARENTHESES SHOW THE RATE OF PRIMING EFFECT EXPRESSED AS THE PERCENTAGE OF NATIVE SOIL ORGANIC NITROGEN

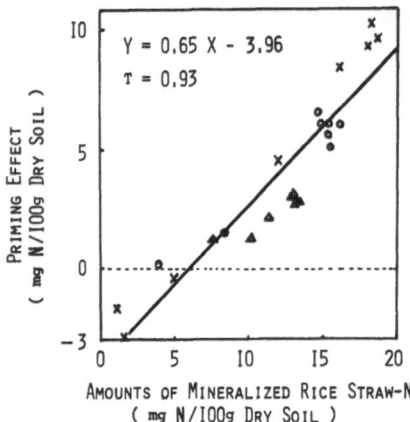

$$Y = 0.65 X - 3.96$$
$$r = 0.93$$

FIG. 4 RELATION BETWEEN PRIMING EFFECT AND AMOUNTS
OF MINERALIZED RICE STRAW-N ADDED TO SOILS

O KASUYA SOIL, △ NAGANO SOIL, ✕ MIYAKONOJO SOIL

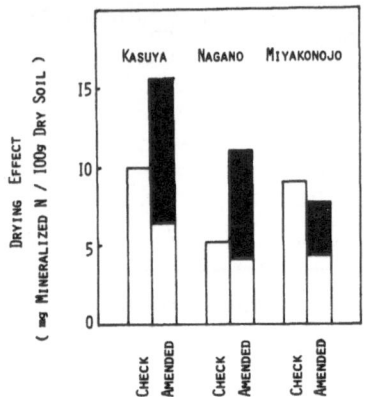

FIG. 5 DRYING EFFECT ON MINERALIZATION OF SOIL-N AND
RICE STRAW-N IN SOIL SAMPLES AFTER 24-WEEK INCUBATION

 SOIL - N ▮▮▮ RICE STRAW - N

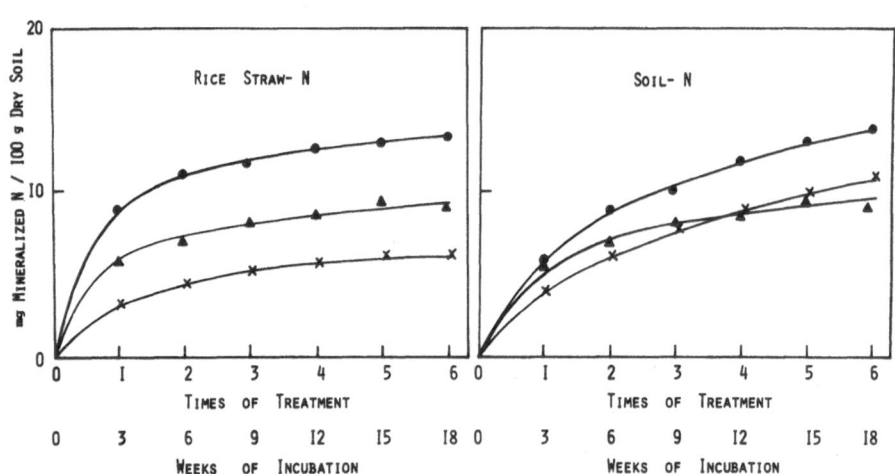

FIG. 6 NITROGEN MINERALIZATION OF THE AMENDED SOILS AFTER 28-WEEK INCUBATION, FOLLOWING
THE REPEATED TREATMENTS OF AIR-DRYING

● KASUYA SOIL ▲ NAGANO SOIL ✕ MIYAKONOJO SOIL

6. The Chemical Characteristics of Nitrogen in Immobilization and Mineralization Processes

Table 3 shows the results of a chemical fraction of the Kasuya soil amended with rice straw and tagged mineral N, as used in Experiment 1.

At the time of maximum immobilization (2 weeks incubation) the amino acid fraction was considerably enriched with applied N (^{15}N) whereas the non-hydrolyzable fraction contained very little. The distribution of applied-nitrogen between the other fractions was similar to that of the native-N. After 20 weeks incubation, when much mineralization had occurred, the amino acid fraction had decreased considerably and the proportion of added-N in this fraction had also decreased. It seems that the amino acid fraction is more susceptible to mineralization than the other fractions and makes a large contribution to this process.

The effects of air-drying of soil on the decomposition of organic nitrogen were examined in a separate experiment, and the chemical distribution of nitrogen in soil was also determined. This experiment showed that (a) the mineralization of soil organic nitrogen, both native and that derived from applied-N, was markedly accelerated by air-drying (Table 4), and (b) the susceptibilities of amino sugar-N and unidentified-N to mineralization were considerably enhanced by drying.

These results strongly support previous findings (5) which suggest that mucopeptides and structural proteins derived from microbial cell walls are the principal forms of residual nitrogen in soil following additions of straw and fertilizer, and that these are an important source of the nitrogen mineralized in soil.

TABLE 3 CHEMICAL DISTRIBUTION OF ORGANIC NITROGEN ORIGINATED FROM NATIVE-N AND APPLIED-N IN KASUYA SOIL

INCUBATION PERIOD		ORGANIC- N	NONHYDRO- LYZABLE N	HYDROLYZABLE N			
				AMMONIUM	AMINO SUGAR	AMINO ACID	UNIDEN- TIFIED
				(mg N PER 100 g DRY SOIL)			
2 WEEKS	NATIVE-N	232.5	50.3	26.4	20.4	92.3	43.1
		(97.0)	(21.0)	(11.0)	(8.5)	(38.5)	(18.0)
	APPLIED-N	7.8	0.4	0.9	0.6	4.6	1.3
		(78.0)	(4.0)	(9.0)	(6.0)	(46.0)	(13.0)
20 WEEKS	NATIVE-N	214.8	53.0	35.5	21.0	83.7	21.6
		(89.6)	(22.1)	(14.8)	(8.8)	(34.9)	(9.0)
	APPLIED-N	4.2	0.4	0.7	1.0	1.5	0.6
		(42.0)	(4.0)	(7.0)	(10.0)	(15.0)	(6.0)

FIGURES IN PARENTHESES SHOW THE PERCENTAGE OF TOTAL NATIVE-N OR APPLIED-N

TABLE 4 PERCENTAGE MINERALIZATION OF THE NATIVE - AND THE APPLIED - N OF THE NONTREATED, AND THE AIR-DRIED AND REMOISTENED SOIL SAMPLES AT 6 - WEEK INCUBATION

	NONTREATED	AIR-DRIED
NATIVE - N	6.3	11.0
APPLIED - N	32.1	46.2

7. Effect of Rice Straw Application on the Microbial Nitrogen Fixation in Soil

To investigate the stimulative effect of rice straw application on the microbial nitrogen fixation in submerged rice soil a field experiment was conducted in Kasuya soil with plots of NPK and NPK plus rice straw. Acetylene reduction activity (ARA) of the plow layers at depths of 0-1 and 1-10 cm was measured as an index of the nitrogen fixing activity of the soil.

As shown in Table 5, both phototrophic and heterotrophic ARA of the surface soil were significantly stimulated with rice straw application, whereas those of the lower plow layer were slightly stimulated. Especially, the phototrophic and anaerobic ARA of the surface layer which is mainly attributable to photosynthetic bacteria was greatly increased when rice straw was applied.

In the rice straw plot, the threshold of the stimulative effect on ARA was delayed in the lower plow layer, although it was observed in the surface layer soon after transplanting.

TABLE 5 EFFECT OF RICE STRAW APPLICATION ON ACETYLENE REDUCTION
ACTIVITY OF THE PLOW LAYER IN KASUYA PADDY SOIL

(NMOLE C_2H_4 / G DRY SOIL PER HR)

TREATMENT	LAYER*	LIGHT		DRAK	
		AEROBIC	ANAEROBIC	AEROBIC	ANAEROBIC
RICE STRAW (PLUS NPK)	SURFACE	0.77	1.49	0.95	0.85
	LOWER	0.60	0.71	0.95	0.77
N P K	SURFACE	0.61	0.65	0.59	0.65
	LOWER	0.59	0.78	0.59	0.59

* SURFACE LAYER : 0 — 1 CM, LOWER PLOW LAYER : 1 — 10 CM

REFERENCES

(1) Agriculture, Forestry & Fisheries Research Council, Ministry of Agriculture, Forestry & Fisheries of Japan. The methods and standards of rice straw application to paddy soils [in Japanese]. Agr. For. & Fish. Res. Council, Japan. Pages 38-42 (1968).
(2) Bremner,J.M. Organic forms of nitrogen. in Methods of soil analysis. Agronomy No.9, Part 2, American Society of Agronomy, Inc., Madison. pages 1238-1255 (1965).
(3) Feigenbaum,S. and A.Hadas. Method of sample preparation for ^{15}N determination in soil extracts by emission spectrometry. Soil Sci. 117:168-170 (1974).
(4) Kai,H. and T.Harada. Determination of nitrate by a modified Conway microdiffusion analysis using Devarda's alloy as a reducing reagent [in Japanese]. Sci. Bull. Fac. Agr., Kyushu Univ. 26:61-66 (1972).
(5) Kai,H.,Z.Ahmad and T.Harada. Factors affecting immobilization and release of nitrogen in soil and chemical characteristics of the nitrogen newly immobilized. III. Transformation of the nitrogen immobilized in soil and its chemical characteristics. Soil Sci. Plant Nutr. 19:275-286 (1973).
(6) Matsuguchi, T. Assay method of the biological nitrogen fixing activity [in Japanese]. Chem. and Biol. ("Kagaku to Seibutsu") 16:328-335 (1978).

NITROGEN EFFICIENCY STUDY UNDER FLOODED PADDY CONDITIONS:
A REVIEW OF INPUTS STUDY I

Yoshio Yamada*, Saleem Ahmed**, Adelaida Alcantara**, and Nurul Huda Khan*
(* Faculty of Agriculture, Kyushu University, Fukuoka, Japan 812.)
(** Resource Systems Institute, The East-West Center, Honolulu, Hawaii, USA.)

INTRODUCTION

Deep dressing of ammoniacal fertilizer was recommended by Shioiri for paddy basal dressing in Japan to minimize its movement to the oxidised layer and thus to prevent nitrification followed by denitrification (1, 2). This was also confirmed by Broadbent, et al.(3), DeDatta, et al.(4), and by international organizations such as FAO/IAEA (5, 6) and IRRI (7). The general recommendation to minimize nitrogen losses in paddy fields called for: broadcasting nitrogen fertilizer subsequent to first plowing, incorporating it into the furrow slice by crushing soil clods or plowing it under, and then irrigating as soon as possible to minimize its loss by nitrification followed by denitrification. This seemed rather difficult even in fields with good irrigation systems. Broadcasting or surface dressing of urea incorporated into the soil by harrowing or mechanical weeding operation after flooding gives the nearest approximation to deep placement. This can be achieved by most rice farmers. It is, however, still an incomplete method. Thus, efficient practical techniques to prevent denitrification are yet to be developed, in spite of a lapse of 40 years since denitrification in paddy fields was first reported.

In order to seek ways to minimize nitrogen losses and thereby optimize fertilizer use, the project called Increasing Productivity Under Tight Supplies (INPUTS) was initiated in 1974, and nitrogen efficiency for paddy was adopted as one of the studies under this project (8).

THE FIRST AND SECOND SERIES OF EXPERIMENTS (1975-1977)

The first series of nitrogen efficiency studies were carried out at twenty fields in nine countries in 1975. In every experiments randomized block design was followed and each treatment was replicated 4 times. Individual plot size was 10 m^2 and plant spacing was 25 x 20 cm. Each site used a rice variety commonly grown in its area. Soil samples from every experimental fields were analyzed. Climatic data of all experimental sites were also collected.

Results of studying various fertilizers and methods of placement showed that single application of mudball was the best and slow release nitrogen fertilizers such as SCU and IBDU generally increased yield over broadcast and split application of urea (9).

Although mudball treatment increased nitrogen absorption and grain yield, it would not be economical since too much labour would be involved in preparing adequate numbers of mudballs. This called for innovative methods for deep basal application of nitrogenous fertilizer. The International Fertilizer Development Center (IFDC), Muscle Shoals, Alabama, USA, thus developed urea supergranules (USG), which do not require being wrapped in mudballs or other types of coverings. Application of mudballs and deep placement of nitrogen fertilizers, while originally devised as methods to prevent nitrogen losses, could also meet the nitrogen requirement of the plant over a longer time. Deep placement of mudballs as single application, however, did not always fully meet the nitrogen reqire-ment of rice plants in the first experiment. Therefore, the aforementioned aspect of deep placement of nitrogen was examined in the second series of INPUTS trials, by split application of nitrogen at rates of 60 Kg N/ha and 90 Kg N/ha. The effect of several nitrogen sources as basal applications was also investi-gated under different methods of application, to identify locally effective fertilizers. Twenty five field experiments were conducted in nine countries during 1976 and 1977. The comparison of grain yield among treatments in 21 experiments which indicate high reliability (C.V. <12%) is shown in Fig. 1.

Results indicate that: i) Higher yield could be obtained with higher nitrogen applications, but the magnitude of the increase was reduced at high rates; ii) Yield from single deep placement was better than that from split applications at same nitrogen levels; iii) At the rate of 60 Kg N/ha, the larger quantity of nitrogen applied in deep layer as basal application, the higher yield could be obtained; iv) Yield obtained from broadcasting and incorporating SCU and IBDU was superior to that obtained from the same treatments of urea, but was inferior than that obtained from deep placement of urea. The uptake rates of nitrogen from the deep placement is superior than that from broadcast and incor-poration method (10).

Reddy and Patrick (11) showed that highest grain yield were obtained when all nitrogen was applied early in the growing season or when part of the nitrogen was applied in early season and part applied no later than midseason. Present study results are in agreement. Shiga et al. (12) found that, for early maturing varieties, basal application was best for increasing grain number and yield, but for medium maturing varieties, deep placement of the ball type ferti-lizer seemed favorable for the early growth. Comparison of split application and single deep application regarding length of growth period in the field were carried out this time. Rice, which had a growth period longer than 120 days, prefered split application, but crop whose growth periods were less than 120 days prefered single deep basal application. The present study's findings is in agreement with Shiga's. Therefore, for early to medium maturing varieties, a single deep application is adequate.

When CEC of soil is very low, supplied ammonium nitrogen is absorbed earlier by plant and some ammonium nitrogen is easily leached out. In that case, the effect of split application may be expected. CEC of soil were determined in 16 experimental sites this time. Comparison of split application and single deep basal application in connection with CEC were studied. It was found that split application was slightly better than single deep application when CEC is below 10, but it is not significant. So it is still not necessary to change previous conclusions (13).

THE THIRD SERIES OF EXPERIMENTS (1978, 1979)

Sixteen field experiments were carried out in eight countries during 1978-1979 to compare the efficiency of a single deep application of USG with that of broadcasted and incorporated urea fertilizer with two topdressings. It also compared the yield response curve for USG with conventional urea application and evaluated the geometrical pattern of USG application on growth of rice and its utilization. Deep placement of peat ball and other nitrogen sources were also compared to conventional urea fertilization. To provide the flexibility to suit local conditions, the experiment consisted of two types of treatments: i) main treatments and ii) optional treatments. Main treatments and 3 optional treat-ments are indicated in Table 1 along with timings and methods of nitrogen application. The response of paddy to nitrogen application rate from conventional urea and USG were established by using five nitrogen levels of the locally recommended nitrogen application rates for commonly grown rice varieties. Urea was applied in three equal doses, with one-third broadcasted and incorporated

589

basally, and the balance topdressed in two equal doses.

USG, provided by IFDC, was applied as single deep placement one week after transplanting. In order to evaluate the effect of various patterns of placement of the fertilizer granule on nitrogen utilization and yield of rice, three geometrical patterns— one USG placement/four rice hills (GPA), one placement/two hills (GPB), and one placement/one hill (GPC)— were compared. This was done at the same nitrogen rates using different sizes of USG as optional treatments.

N-15 technique was carried out at site No. 16 to evaluate crop recovery of applied nitrogen by rice in relation to the geometric pattern of USG placement. In this experiment, four treatments included tagged-N which was applied in small areas (1.25 - 1.8 m^2) of the 10 m^2 plot at 92 Kg N/ha level. Labelled nitrogen was supplied by FAO/IAEA, Vienna, Austria.

All yield data were statistically analyzed with computer, using Statistical Analysis System Package. This includes Analysis of Variance (Randomized Complete Block), Duncan's Multiple Range Test and Regression. Nitrogen analysis was carried out at site Nos. 14, 15, and 16. N-15 was determined by JASCO NIA-1, ^{15}N analyzer.

Table 1 Treatments and Methods of Application

Treatment No.	Carrier	Rate of N Application at 3 Growth Stages* Kg/ha			Method of Application#
		T_0	T_1	T_2	
A. Main Treatments					
1)	----	0	0	0	----
2)	Urea	8	7	7	T_0=BC & IC T_1 & T_2=BC
3)	Urea	15	14	15	T_0=BC & IC T_1 & T_2=BC
4)	Urea	22	22	22	T_0=BC & IC T_1 & T_2=BC
5)	Urea	30	29	29	T_0=BC & IC T_1 & T_2=BC
6)	Urea	44	44	44	T_0=BC & IC T_1 & T_2=BC
7)	Urea	59	58	59	T_0=BC & IC T_1 & T_2=BC
8)	USG	22	0	0	DP (GPA)
9)	USG	44	0	0	DP (GPA)
10)	USG	66	0	0	DP (GPA)
11)	USG	88	0	0	DP (GPA)
12)	USG	132	0	0	DP (GPA)
13)	USG	176	0	0	DP (GPA)
B. Optional Treatment					
14)	USG	44	0	0	DP (GPB)
15)	USG	88	0	0	DP (GPB)
16)	USG	88	0	0	DP (GPC)

* T_0 = 1 week after transplanting for DP treatments; BC & IC are carried out just before transplanting

T_1 = at beginning of rapid tillering stage
T_2 = 5-10 days before primordial initiation

\# DP = deep placement; Nitrogen sources were placed every 4 hills (GPA), every 2 hills (GPB), every 1 hill (GPC)

BC = broadcasting; IC = incorporating

Table 2 shows the regression equations of grain yield with conventional urea and USG. Out of a total of 16 locations, 10 sites gave a significant response to urea and 11 to USG. In most cases, USG treatments outyielded urea at the same N levels. It will be observed that, in the case of conventional urea, equations are linear, with one exception. On the other hand, with USG application, the equations are quadratic in nearly 50% of the cases. The "law of diminishing returns" appears to be applicable here. The absence of a quadratic relationship in the case of conventional urea treatments may be explained on the basis of decreased nitrogen efficiency due to denitrification. The differences in the regression equations between USG and conventional urea also indicate the higher utilization of nitrogen from USG application.

Fig. 2 shows the yield response curves to nitrogen from both nitrogen sources as obtained at 13 locations. Significant differences between grain yields due to conventional urea and USG are indicated by arrows. With the exception of site 8, the grain yield from USG is equal or superior to that from conventional urea. In most cases, superiority of USG at the medium nitrogen levels is indicated. Fig. 2 also shows the differences in the magnitude of response to N and the slope of curves at various sites. A number of possible reasons may be advanced to explain this. The high positive response at sites 6 and 13 might be due to their large CECs (cation exchange capacity) and high content of available cations, possibly providing for better nutrient conditions for the crops' entire growth stage. Good climatic conditions, especially high solar radiation at the ripening stage, may also have contributed positively at these sites; and poor climatic conditions may have contributed adversely at site 10. Plant height is another factor which varied significantly from site to site. Garima, PR 106, and IR 36 are dwarf types; PK 177, BG11-11, and Asominori are medium statured; BR 4, RD 7, and Tainan 5 have longer culms. Dwarf varieties were used at sites 5, 6, and 13. In the cases, grain yield increases with increasing rates of nitrogen application was striking. The effect of USG was also remarkable at these sites. At sites 7, 8, 12, and 14, where longer culms varieties were used, the pattern of grain yield increase with increasing N application is not so marked. It is possible that mutual shading due to the vigorous growth of leaves resulted in the negative or less response in these taller varieties (14, 15). Consequently, grain yield did not increase at the same rate as with the dwarf varieties.

We pointed out earlier that single deep application of urea is more favorable than its split application when CEC of soil is higher than 10 me/100 g and the growing duration is shorter than 120 days (10, 13). In our current study, the CEC of soils and growing duration mostly conformed with the above conditions, and the postulated results were actually obtained. At site 15, which was the only exception the CEC was 7.3 me/100 g. Here topdressing of USG after basal treatment with USG indicated grain yield increase, but this was not significant. Accordingly, the superiority of single deep application of USG over conventional urea application was reconfirmed under the above mentioned conditions.

Nitrogen analysis of harvested plant was conducted for sites 14, 15, and 16. Comparison of nitrogen recovery percentage by subtraction method is shown in Fig. 3. This indicates the obviously higher recovery of nitrogen from USG than from conventional urea at every sites and at all N levels. This probably indicates that deep application of USG minimizes nitrogen loss due to denitrification, thereby increasing its recovery.

The effect on grain yield of the positioning of USG in the rice field (that is, of its placement in the soil in relation to rice hills), was studied for two patterns-- GPA and GPB. Six experimental sites show the superiority of GPA, and seven experimental sites show the reverse. The differences are, however, not significant at the 5% level with exception of site 13.

The recovery rate of nitrogen from different pattern of USG were calculated by subtraction method. These are seen to coincide with the pattern of USG placement on grain yield. At site 16, a comparison is made of deep placement of USG

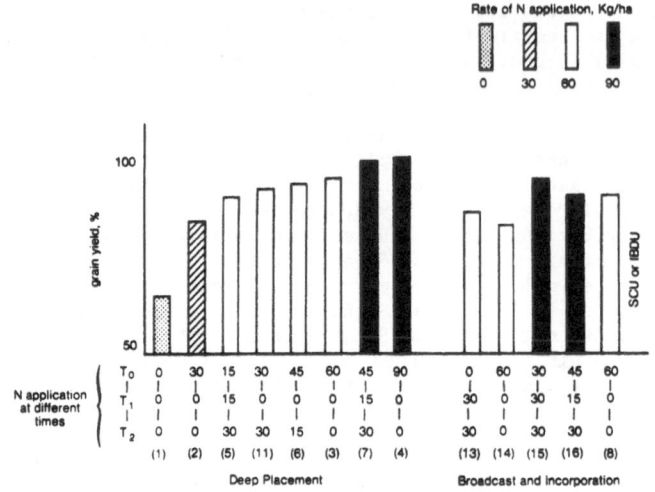

Fig. 1   Grain Yields of Main and 5 Optional Treatments
(average of 21 experiments)

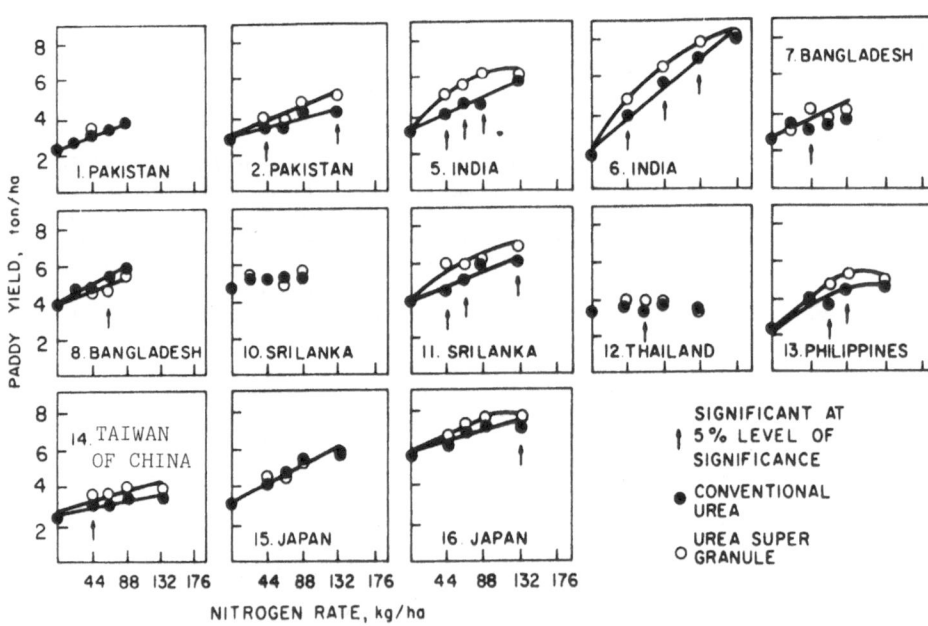

Fig. 2   Grain Yield response Curves to N from USG and Urea:Significant at 5%

Table 2 Regression of Grain Yield (Kg/ha) with Nitrogen Level (Kg/ha) of Two Nitrogen Sources.

No.	Site	Grain Yield with Urea	Grain Yield with USG
1.	Pakistan	$Y=2368 + 15N$	$Y=2343 + 18N$
2.	Pakistan	$Y=2815 + 12N$	$Y=3009 + 18N$
3.	Pakistan	NS	NS
4.	Pakistan	NS	NS
5.	India	$Y=3178 + 19N$	$Y=3059 + 50N - 0.199N^2$
6.	India	$Y=2182 + 35N$	$Y=1856 + 73N - 0.210N^2$
7.	Bangladesh	NS	$Y=2847 + 16N$
8.	Bangladesh	$Y=3936 + 22N$	$Y=4021 + 13N$
9.	Bangladesh	Different Treatment	Different Treatment
10.	Sri Lanka	NS	NS
11.	Sri Lanka	$Y=3941 + 18N$	$Y=4000 + 46N - 0.183N^2$
12.	Thailand	NS	NS
13.	Philippines	$Y=2524 + 33N - 0.132N^2$	$Y=2350 + 52N - 0.238N^2$
14.	Taiwan of China	$Y=2691 + 6.6N$	$Y=2904 + 11N$
15.	Japan	$Y=3310 + 19N$	$Y=3408 + 18N$
16.	Japan	$Y=5630 + 12N$	$Y=5435 + 35N - 0.140N^2$

NS = not significant at 5% level of significance

using the three geometric patterns and conventional urea application at 92 Kg N/ha level. The recovery of nitrogen was calculated by subtraction method and the N-15 technique. Grain yield, N-uptake, and N-recovery are shown in Fig. 4. USG was superior to conventional urea, and the GPB was found superior for all these three observations. It is seen that very low nitrogen recovery was found when conventional urea was applied in three split applications. Koyama, et al. (16), Reddy and Patrick (11), Watanabe and Padre (17), and IAEA (5, 6) also reported low recovery of fertilizer N from basally-applied conventional N fertilizer, indicating N loss due to denitrification. On the other hand, when USG was applied in deep placement with GPB pattern of application, recovery of nitrogen increased significantly 58.8%. Low recovery was observed with increased as well as with decreased number of placements (GPC and GPA, respectively); with GPC, it was 47.4%; with GPA, it was 46.8%. Grain yield from the three geometric patterns of fertilizer N placement was not significantly different, although more fertilizer N was absorbed in GPB. Further work should be undertaken along these lines. Studies on the concentration gradient of NH_4^+ in soil solution in relation to CEC of soil and nitrogen supplied is suggestive (18). However, we would like to point out again that we did not observe any significant difference in grain yield due to different geometric patterns of USG application. Consequently, we would like to recommend GPA, which requires less labour.

Table 3 compares the N rates needed to obtain the same grain yield by using urea and USG. These rates were calculated from the regression equations in Table 2. The results for only two nitrogen levels are being shown here for purposes of representation. With the exception of sites 8 and 15, it is seen that less N is required with USG treatment than with urea treatment to get the same grain yield. The average of ten locations show that, for obtaining one ton and two tons of increased grain yield, about 20 Kg N/ha and 35 Kg N/ha respectively can be saved by using USG instead of urea. This saving corresponds to about one-third of the nitrogen required if urea is used. USG, therefore, offers a potentially significant method for saving on fertilizer use.

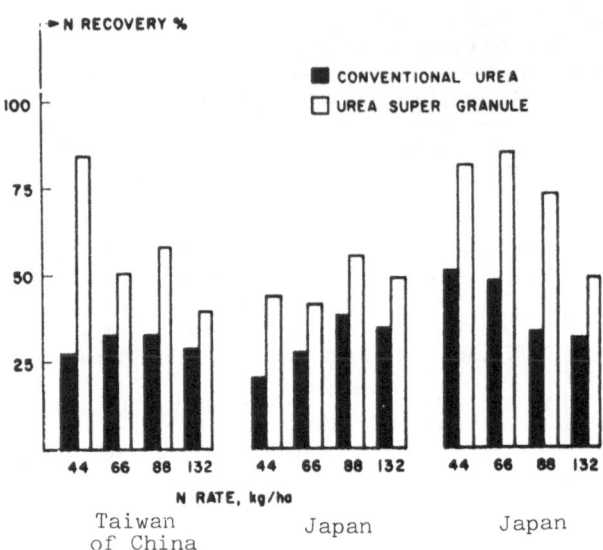

Fig. 3 Crop Recovery of Fertilizer-N by Rice as Affected
by N-sources

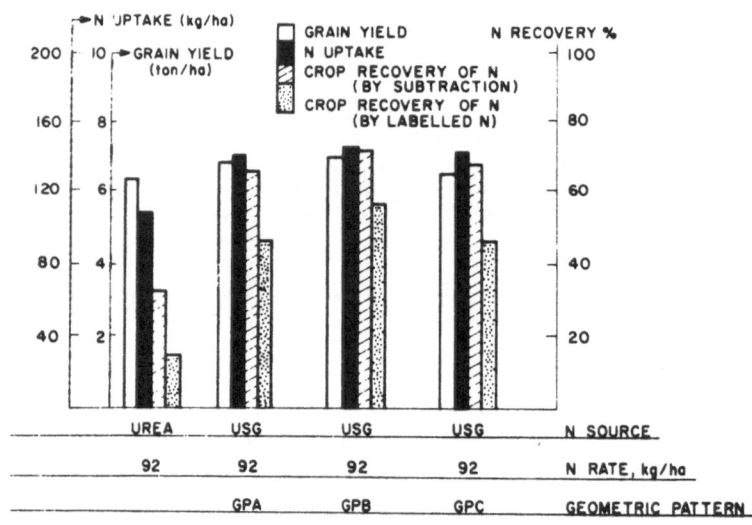

Fig. 4 Grain Yield and Crop Recovery of Fertilizer-N by Rice
as Affected by N-sources and Placement at Site No.16

Table 3 N Rates Needed to Get Same Grain Yield by Using Urea and USG

Expt. Site No.	N-Rate with Urea (Kg/ha)	N-Rate with USG (Kg/ha)	Calculated Grain Yield (Kg/ha)
1. Pakistan	60	51	3268
	120	101	4168
2. Pakistan	60	29	3535
	120	69	4255
5. India	60	28	4318
	120	65	5458
6. India	60	37	4282
	120	81	6382
8. Bangladesh	60	95	5256
	120	197	6576
11. Sri Lanka	60	25	5021
	120	60	6101
13. Philippines	60	39	4029
	120	59	4583
14. Taiwan of China	60	17	3087
	120	53	3483
15. Japan	60	58	4450
	120	121	5590
16. Japan	60	30	6350
	120	62	7070
Average	0	0	3248
	60	40.9	4360
	120	86.8	5367

ACKNOWLEDGEMENTS

This paper is presented on behalf of the members of INPUTS I team who belong to 11 countries 25 organizations. Countries are Bangladesh, Egypt, India, Iran, Japan, Korea, Pakistan, Philippines, Sri Lanka, China , and Thailand. The authors wish to thank the following organizations for supply of materials indicated: International Fertilizer Development Center (IFDC), Muscle Shoals, USA, USG and SCU; Nihon Hiryo Company, Tokyo, Japan, Peat Ball; Mitsubishi Chemical Industies Ltd., Japan, IBDU; FAO/IAEA, Vienna, Austria, Labelled N.

REFERENCES

(1) Shioiri, M., The Chemistry of Paddy Soil. (in Japanese) Dojo Hiryo Kowa, Asakura, Tokyo, pp. 181-236 (1943).

(2) Shioiri, M. and Tanada, T., The Chemistry of Paddy Soils in Japan, Ministry of Agriculture and Forestry, Dainippon Printing Co. Ltd., Tokyo (1954).

(3) Broadbent, P.E. and Mikkelsen, D.S., Influence of Placement on Uptake of Tagged Nitrogen by Rice Plant. Agron. J. 60, 674-677 (1968).

(4) DeDatta, S.K., Magnaye, C.P. and Moomaw, J.D., Efficiency of Fertilizer Nitrogen (^{15}N-labelled) for Flooded Rice. Int. Soil Sci. Trans. 4: 67-76 (1968).

(5) IAEA, Rice Fertilization, Technical Report Series No. 108. Joint FAO/IAEA Division of Atomic Energy in Food and Agriculture (1970).

(6) IAEA, Isotope Studies on Rice Fertilization, Technical Report Series No.181. Joint FAO/IAEA Division of Atomic Energy in Food and Agriculture (1978).

(7) IRRI, The IRRI Reporter, Annual Report for 1975. pp. 286-290 and 340-341 (1976).

(8) Ahmed, S., Proceedings, Planning and Organization Meeting, INPUTS Project, East-West Center, Honolulu, pp. 1-35 (1974).

(9) Yamada, Y., Nitrogen Efficiency Studies Under Flooded Paddy Conditions. Proceedings, First Review Meeting, INPUTS Project, East-West Center, Honolulu, pp. 105-130 (1976).

(10) Yamada, Y., Juang, T.C., Gunasena, H.P.M. and Ahmed, S., Nitrogen Efficiency Under Flooded Paddy Conditions. Proceedings, Second Review Meeting, INPUTS Project, East-West Center, Honolulu, pp. 183-215 (1978).

(11) Reddy, K.R. and Patrick, W.H., Yield and Nitrogen Utilization by Rice as Affected by Method and Time of Application of Labeled Nitrogen. Agronomy Jr. 68, 965-969 (1976).

(12) Shiga, H., Ventra, W.B. and Yoshida, T., Effect of Deep Placement of Ball Type Fertilizer at Different Growth Stages on Yield and Yield Components of Rice Plant in the Philippines. Plant and Soil. 47, 351-361 (1977).

(13) Yamada, Y., Khan, N.H. and Ahmed, S., Studies on Nitrogen Efficiency with Paddy Rice. Chemical Marketing and Economics Division, Amer. Chem. Soc., pp. 116-128 (1979).

(14) Tanaka, A., The IRRI Reporter, Annual Report for 1964, pp. 49-86 (1964).

(15) Ishizuka, Y., Physiology of the Rice Plant. Advances in Agronomy, 23, 242-315 (1971).

(16) Koyama, T., et al., Nitrogen Application Technology for Tropical Rice as Determined by Field Experiments Using [15]N Tracer Technique, Tech. Bull. TARC 3, 1-79 (1978).

(17) Watanabe, I. and Padre, B.C. Jr., Inorganic Nitrogen Movement in Tropical Soils - Nitrification and Nitrogen Loss in Paddy Drained Soils During Rainy Season. Soil Sci. Plant Nutr. 23, 217-224 (1977).

(18) Aomine, S., Movement of Ammonium in Paddy Soils in Taiwan. Soil Sci. Plant Nutr. 24, 571-580 (1978).

THE FATE OF NITROGEN FERTILIZER IN PADDY SOILS

Chen Rong-ye, Zhu Zhao-liang
(Institute of Soil Science, Academia Sinica, Nenjing)

The balance sheet of fertilizer nitrogen in the soil-plant system provides basic information for evaluating agricultural practices in general and for the rationalization of fertilization system in particular. Besides, it also has a referential value for the characterization of the fertilizer concerned and for the development of new fertilizers.

Early studies on the fate of fertilizer-N in soils were mainly carried out on upland fields. The recent review[1] made by Allison as late as 1966 also provided no information about the fate of fertilizer nitrogen in the paddy soil. In the past ten years, a series of papers dealing with this issue have been published which show that loss of fertilizer-N applied to the paddy soils was serious and the fate of fertilizer-N in rice fields was affected significantly by the type of soil, fertilizer, field management, etc[2-6].

In the present article, the fate of different chemical N-source in both calcareous and noncalcareous paddy soils was investigated using ^{15}N-tracer technique suggested by Carter et al.[7] with some modifications. At the same time, the effects of methods and time of nitrogen application, drainage and use of nitrification inhibitor on the fate of urea were studied.

MATERIALS AND METHODS

Microplot experiments were conducted in the calcareous as well as in noncalcareous paddy soils in the northern and southern Jiangsu Province respectively. They were arranged according to the randomized complete block design with four replications. The microplots, each with an area of about 0.07 m^2, were made by embedding into the soil the plastic cylinders with a diameter of 29 cm. Treatments were made to provide comparisons: 1) among three different N-sources at the transplanting time (i.e. urea, ammonium sulfate and ammonium bicarbonate), 2) among different methods of application (i.e. broadcasting on soil surface, incorporation into top-soil layer and dressing at a depth of 6 cm), 3) of urea application at two different stages of growth (i.e. at the transplanting time and at the middle stage of growth). The effects of nitrification inhibitor and draining the field at the middle stage of growth were also investigated for the calcareous and noncalcareous paddy soils respectively. To each treatment, superphosphate and potassium chloride along with the one of the three nitrogen sources were applied at the rates of 0.22g of P, 0.42g of K and 0.50g of N per microplot. ^{15}N-labelled fertilizers with around 10 % enrichment of ^{15}N atom were used only for the treatments in which the fate of fertilizer-N was to be investigated.

In the experiment laid on calcareous paddy soil, a split application of 0.50g of N was made: half was used as basal dressing at the rice transplanting and the other was used as top-dressing at the middle stage of rice growth.

Rice variety used was "Nong ken No. 57". Four hills of rice seedlings with 33 days old were transplanted in each microplot on June 22, 1978. Plant and soil samples were taken on September 1, 1978 for N-analysis.

In the experiment laid on noncalcareous paddy soil, 0.50g of N was applied for once at the time of transplantation except the treatment of top-dressing in which half of the fertilizer-N was applied as basal dressing and the remainder was top-dressed at the middle stage of rice growth. Rice variety used was" Guang lu ai No. 4". Four hills of rice seedlings with 38 days old were transplanted in each microplot on May 18, 1978. Plant and soil samples were taken on July 7,1978.

In both experiments, the sampling dates were so chosen as mentioned above that they corresponded to their full heading stage when the amounts of N recovered by rice plants were usually considered to have been at their maximum values. Since there was no grain in the plant samples thus taken, no extra-work for the separation of grain from the straw was needed during the preparation of samples. In all sampling work, all the rice plants including roots and the whole soil mass of 0-20 cm layer within each microplot were collected. The samples were dried, weighed, ground, sieved, mixed and subdivided into subsamples for analytical uses. The determination of total N and of the enrichement of ^{15}N atoms were carried out respectively by Kjeldahl and mass spectrography method as usual.

The properties of soils collected from the rice fields investigated were determined by conventional method and were shown in Table 1.

Table 1 Soil properties

Soil	pH (H_2O)	T.N. %	O. M. %	Mechanical composition <0.01 mm %	<0.001 mm%	CEC (m.e./100g)
Calcareous	8.2	0.139	2.10	34.6	15.6	14.3
Noncalcareous	6.6	0.155	2.53	49.6	20.4	18.6

RESULTS AND DISCUSSION

1. The Effect of Soil Type and Nitrogen Source

Table 2 gives the nitrogen balance sheets of ammonium sulfate, urea and ammonium bicarbonate surface-broadcast on calcareous and noncalcareous paddy soils. It shows that in the ammonium bicarbonate microplots the recovery of N by rice plant was the lowest while their N-loss was the greatest. For ammonium sulfate the situation of N-balance was quite different. In the noncalcareous paddy soil, the recovery of ammonium-N amounted to 50.1 %, which was much higher than that of urea: but in the calcareous paddy soil, it decreased to 22.5 %, which was close in value to that of urea. Obviously, when ammonium sulfate was surface broadcast on calcareous paddy soil, ammonia volatilization might play an important role in nitrogen loss other than denitrification. Generally speaking, the recovery of ammonium sulfate, urea and ammonium bicarbonate was higher in the noncalcareous paddy soil than in the calcareous paddy soil presumably because of the ammonia volatilization caused by the alkaline reaction in the calcareous paddy soil. The nitrogen loss through leaching during the growing season of rice seems to be negligible because it has been demonstrated that in the noncalcareous paddy soil the downward movement of ammonium sulfate surface-broadcast at the early stages of rice growth in no case exceeded 20-30 cm(8).

Tabel 2 Fate of ^{15}N-labelled fertilizer surface-broadcast
at transplanting time in rice field

Soil	N Source	Recovery by rice plant (%)	Retained in soil (%)	Deficit (%)
Noncalcareous	Ammonium sulfate	50.1	21.4	28.5
	Urea	27.5	18.6	53.9
	Ammonium bicarbonate	·24.0	18.6	57.1
	L.S.D. p = 0.05	4.7	5.3	7.4
	p = 0.01	6.3	7.2	10.1
Calcareous	Ammonium sulfate	22.5	26.3	51.2
	Urea	22.3	30.4	47,3
	Ammonium bicarbonate	17.1	12.7	70.2
	L.S.D p = 0.05	3.2	6.5	7.0
	p = 0.01	4.3	8.9	9.7

2. The Effect of Methods of Fertilizer Application

The common method for increasing the efficiency of N-fertilizer and decreasing nitrogen loss is deep dressing of N-fertilizer into the reduced horizon or to incorporate N-fertilizer with soil. Table 3 shows both of the above mentioned methods have increased the N-recovery and decreased the N loss in either calcareous or noncalcareous paddy soils. It should be mentioned that among all the methods of application investigated in the present paper, the deep-dressing supergranular urea, which was placed at a depth of 6 cm in the center of four hills of rice seedlings, had the highest recovery of nitrogen and the lowest N loss. In comparison with the deep incorporation of the ordinary urea in powered form, which is the most usual way of urea application prevailing in China now, deep dressing of the supergranular form increased the nitrogen recovery by one fold, while the N loss decreased by 1-2 folds. The same tendency was also reported in studies(9) using ammonium bicarbonate as N-sources. It is obvious that supergranular fertilizers can be more easily deep-dressed without the danger of being dispersed in the surface layer as compared with the powdered from. The greater extent of N-localization in the deep layer will mean that there would be less fertilizer-N loss through NH_3-volatilization and nitrification-denitrification reactions. As compared with the fertilizers in powder form, deep dressed supergranules will persist in the soil much longer because of its smaller interfacial contact with the soil. This will mean that N-release from deep-dressed supergranules would be slower and steadier. As a result, there will be less immobilization of fertilizer-N in the soil and more effective use of fertilizer-N by rice crops.

Table 3 Fate of ^{15}N-labelled urea in rice field applied as
basal dressing with different methods

Soil	Treatment	Recovery by rice plant (%)	Retained in soil (%)	Deficit (%)
Noncalcareous	Surface-broadcast	27.5	18.6	53.9
	Incorporated	37.2	22.7	40.1
	Dressing at a depth of 6 cm	37.6	18.9	43.5
	SGU, Deep dressing	74.5	12.4	13.1
	L.S.D p = 0.05	4.7	5.3	7.4
	p = 0.01	6.3	7.2	10.1
Calcareous	Surface-broadcast	22.3	30.4	47.3
	Incorporated	29.0	31.5	39.5
	Dressing at a depth of 6 cm	25.8	23.3	50.9
	SGU, Deep dressing	55.1	23.7	21.2
	L.S.D p = 0.05	3.2	6.5	7.0
	p = 0.01	4.3	8.9	9.7

3. The Effect of Time of Application

When ammonium sulfate and urea were top-dressed at the middle stage of
plant growth, N recovery values were raised up to 58-69 %. But at the same
time the amounts retained in .the soil were decreased considerably to low
values of only 5-16 %. Thus, the deficits in balance due to N loss still
remained to be considerable. They were about 21-30 % as shown in Table 4.

Table 4 Fate of ^{15}N-labelled fertilizer top-dressed
at middle stage of growth

Soil	N Source	Recovery by rice plant (%)	Retained in soil (%)	Deficit (%)
Noncalcareous	Urea	64.7	5.4	29.9
	Ammonium sulfate	68.9	10.3	20.8
Calcareous	Urea	61.8	12.4	25.8
	Ammonium sulfate	57.8	16.0	26.2

4. The Effect of Drainage

Draining of rice field at the middle stage of growth is generally employed to control the rate of vegetative growth of rice. It is commonly
recognized that drying and wetting of the soil by draining and reirrigation
of rice field will anhance the N loss. However, the present experiment indicated that draining of field at the middle stage of rice growth did not
appear to promote the loss of fertilizer-N applied at time of transplantation,
as shown in Table 5. Of course, the N loss might be severe, if the draining
and reirrigation cycles had been practised repeatedly or the top-dressing of
N-fertilizer, especially that of urea, had been practised not long before
draining.

Table 5   Effect of drainage on the fate of ^{15}N-labelled urea
applied as basal dressing(noncalcareous paddy soil)

Treatment	Recovery by rice plant (%)	Retained in soil (%)	Deficit (%)
CK	27.5	18.6	53.9
Draining at mid stage of growth	31.8	16.5	51.7

5. The Effect of Nitrification Inhibitor

It is generally believed that the N loss in rice field is mainly due
to nitrification-denitrification. Thus, various nitrification inhibitors
have been suggested to minimize such N losses. Table 6 shows that N-Serve
didnot exert any significant effect on the fate of N-fertilizer applied to
the paddy soil. It seems that N-Serve is either ineffective in improving the
efficiency of N-fertilizer in the rice field or its favorable conditions for
its effective use were not provided in the present experiment. Further
research for its valid use is advisable from athe practical as well as from
the theoretical point of view.

Table 6   Effect of N-Serve on the fate of ^{15}N-labelled urea
applied to rice field(calcareous paddy soil)

Treatment	Recovery by rice plant (%)	Retained in soil (%)	Deficit (%)
Basal, surface-broadcast	22.3	30.4	47.3
Ibid + N-Serve *	24.8	27.8	47.4
Basal, 6 cm deep	25.8	23.3	50.9
Ibid + N-Serve *	25.3	29.6	45.1
Top-dressed at midseason	61.8	12.4	25.8
Ibid + N-Serve *	61.2	12.4	26.4

* It was mixed with N-fertilizer at the rate of 3 % of the N
applied

REFERENCES

(1) Allison, F.E., The fate of nitrogen applied to soils. Adv. Agron.,
    18, 219-258 (1965).
(2) Zhu Zhao-liang, Chen Rong-ye, Xu Yong-fu, Xu Yin-hua, Zhang Shao-lin,
    The effect of forms and methods of placement of nitrogen fertilizer
    on the characteristics of the nitrogen supply in paddy soils. Acta
    Pedologica Sinica, 16,218-233 (1979). (in Chinese with English summary)
(3) Broadbent, F.E., Nitrogen transformation in flooded soils. In "Soils
    and Rice", IRRI, Philippines, 543-559 (1978).
(4) Broeshart, H., The fate of nitrogen fertilizer in flooded rice soils.
    In "Nitrogen-15 in Soil-Plant Studies", IAEA,Vienna, 47-54 (1971).
(5) Craswell, E.T., Vlek, P.L.G., Fate of fertilizer nitrogen applied to
    wetland rice. In "Nitrogen and Rice", IRRI, Philippines, 175-192 (1979).
(6) Patrick, W. H., Reddy, K. R., Fate of fertilizer nitrogen in a flooded
    rice soil. Soil Sci. Soc. Amer. J., 40, 678-681 (1976).
(7) Carter, J. N., Bennett, O. L., Pearson, R. W., Recovery of fertilizer
    nitrogen under field condition using nitrogen-15. Soil Sci. Soc. Amer.
    Proc., 31, 50-56 (1967).
(8) Chu Chao-liang, Tsai Kuei-hsin, Yu Chin-chou, A preliminary investigation
    on the fate of ^{15}N -labelled ammonium sulfate in rice field. Kexue Tongbao,
    22, 503 (1977). (in Chinese)
(9) Chen Rong-ye, Fan Qin-zhen, Effect of deep-application of prilled ammo-
    nium bicarbonate fertilizer on the nitrogen supplying status of noncal-
    careous paddy soils. Acta Pedologica Sinica, 15, 75-82 (1978). (in Chinese
    with English summary)

POTASSIUM STATUS OF PADDY SOILS IN SOME
COUNTRIES OF SOUTH AND EAST ASIA

G. Kemmler
(Büntehof Agricultural Research Station of
Kali + Salz AG, Bünteweg 8, 3000 Hannover 71
Federal Republic of Germany)

Soil potassium data from South and East Asian countries are
reviewed, partly in comparison with results of fertilizer trials
and demonstrations or fertilizer recommendations and estimates of
actual fertilizer use to rice.

A large percentage of soils high in exchangeable K is found
in the Philippines and in Java. Furthermore high K soils exist in
the Central Plains of Thailand, in Southern Burma and in parts of
Southwestern Bangladesh. On the other hand, most of the paddy soils
in Sri Lanka, Bangladesh, North Eastern Thailand, Kampuchea and in
large parts of West Malaysia are low in exchangeable K.

Comparison of soil analysis data with trial and demonstration
results in Bangladesh and in the Philippines confirms that crop
response to K application is high on soils testing low in K (Bangladesh) and low on soils with high K status (Philippines). However, there are exceptions from this general pattern. In various
cases potassium has become a yield limiting factor due to intensive
cropping on soils originally rich in exchangeable K, as can be seen
from the results of long-term fertilizer experiments carried out in
the Philippines.

INTRODUCTION

Much information on the fertility of rice soils in South and
East Asia is contained in recent publications, e.g. in the books
by Kawaguchi & Kyuma (1977), Moormann & van Breemen (1978) and in
the proceedings of the Soils and Rice Symposium at the International Rice Research Institute (1978). In these and other reports the
potassium status of the soils has been delt with in a rather global
way. More detailed studies are needed.

From the agronomic point of view it is advisable to consider
the soil K status in relation to the K requirements of the crops.
A soil sufficiently supplied with K to achieve paddy yields of
2 t/ha may be too low in available potassium to support a crop of
6 t/ha.
In the absence of generally accepted criteria for the assessment
of soil K availability with regard to the expected yield, soil test
data of paddy fields from South and East Asian countries are reviewed, partly in comparison with the results of recent fertilizer
experiments.

FIELD DATA ON SOIL TEST AND K RESPONSE

The country averages of the exchangeable K level in paddy soils of tropical Asia as reported by Kawaguchi & Kyuma (1977) are shown in Table 1.

Based on additional information from K. Kyuma, partly combined with local soil test results, maps have been prepared on the K status of paddy soils in a number of South and East Asian countries. For some of the countries data are available on fertilizer trials and demonstrations or fertilizer recommendations and estimates of actual fertilizer use to rice. They will be compared with the soil test results to obtain more insight into the suitability and the limitations of soil analysis as basis for fertilizer recommendations.

PHILIPPINES

Of the 8 countries listed in Table 1, Kawaguchi & Kyuma have found the highest average contents of exch. K in the soil samples from the Philippines. 42 % of the 54 paddy soils analysed tested high in exch. K with values above 0.45 me K/100g. Only 15 % were low (<0.15 me K/100g). However, the K saturation of the cation exchange capacity (CEC) was below 2 % in 56 % of all cases, including many of the soils with medium-low to medium status of exch. potassium. Figure 1 shows these soil test results by geographic location (except for the 10 sampling sites on the island of Mindanao). The symbols represent both, the exch. K and the K saturation of the CEC. Soil samples high in exch. K were found notably in the Provinces of Laguna and Iloilo as well as Cotabato Norte and Cotabato Sur/Mindanao.

134 sets of average results from 3225 fertilizer demonstrations and trials on rice carried out in the Philippines between 1976/77 and 1979/80 under the FAO Fertilizer Programme were available to evaluate the potash response. Each set represents the average results obtained per province and per rice growing season. The response to K application was calculated as yield difference between NPK and NP treatment and classified into response intervals of 200 kg paddy/ha. The result of this evaluation by provinces is presented in Table 2. It shows K responses below 200 kg paddy/ha in 62 % of all cases (15 % negative, 47 % between 0 and 200 kg/ha), indicating high soil K availability, notably in Iloilo, Albay, Camarines Norte, Isabela. High responses to potassium, on the other hand, were obtained in Mindoro Oriental, Quirino, Pangasinan. At the average potash rates of 30 kg K_2O/ha, used in most of the trials and demonstrations, yield increases of 200 - 400 kg paddy/ha represent very substantial responses (7 to 13 kg paddy/kg K_2O).

BANGLADESH

The compilation of K values for paddy soils of Bangladesh is based on data by Kawaguchi & Kyuma (1977) and Islam et al. (1974), in both cases completed by additional information obtained from the authors. Contrary to the situation in the Philippines, most of

Table 1   Exchangeable K in paddy soils of some Asian countries
          me K/100 g soil (Kawaguchi & Kyuma 1977)

Country	No. of samples	Mean	Mini- mum	Maxi- mum
Bangladesh	53	0.3	tr.	0.8
Burma	16	0.4	0.1	0.7
Kampuchea	16	0.2	tr.	0.8
Indonesia(Java)	44	0.4	0.1	1.2
W. Malaysia	41	0.4	0.1	1.8
Philippines	54	0.5	0.1	1.7
Sri Lanka	33	0.2	tr.	0.7
Thailand	80	0.3	tr.	2.6
Total/Mean	337	0.3	tr.	2.6

Table 2   Classification of K response in 3225 trials &
demonstrations on rice of the FAO Fertilizer Programme in the
Philippines, 1976-77 to 1979-80.

Province	No. of sets 1)	K response (NPK minus NP), kg/ha			
		> 400	200-400	< 200	Negative
Ilocos Norte	6	–	4	1	1
Ilocos Sur	6	–	–	5	1
La Union	13	–	2	9	2
Pangasinan	13	4	4	5	–
Isabela	12	–	3	6	3
Quirino	9	3	6	–	–
Nueva Vizcaya	10	–	1	7	2
Camarines Norte	10	–	2	5	3
Camarines Sur	10	–	4	5	1
Albay	12	–	1	8	3
Sorsogon	6	–	3	3	–
Mindoro Oriental	13	4	7	2	–
Iloilo	14	1	2	7	4
Total	134	12	39	63	20
	100%	9%	29%	47%	15%

1)One set = 24 trials or demonstrations on average

K status of rice soils in the Philippines
(Kawaguchi & Kyuma), 1977

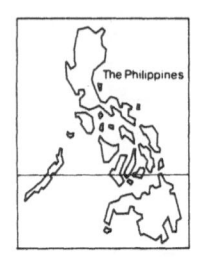

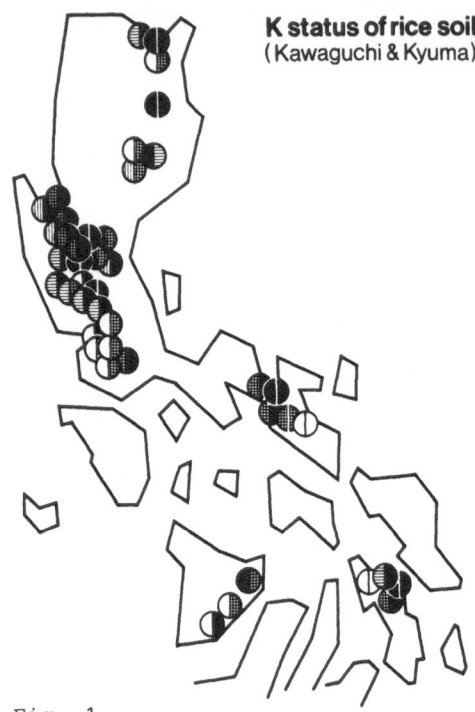

Rating	Exchangeable K		Symbol
	mg K/100g	me/100g	
Low	< 5,9	<0,15	
Med.-low	5,9-11,7	0,15-0,30	
Medium	11,8-17,6	0,31-0,45	
High	17,7-23,4	0,46-0,60	
Very high	>23,4	>0,60	

Rating	K saturation of CEC	Symbol
	% K sat.	
Low	< 2	
Med.-low	2 - 3	
Medium	3,1 - 4	
High	4,1 - 5	
Very high	> 5	

Fig. 1

K status of rice soils in Bangladesh (Kawaguchi & Kyuma and Islam et al.)

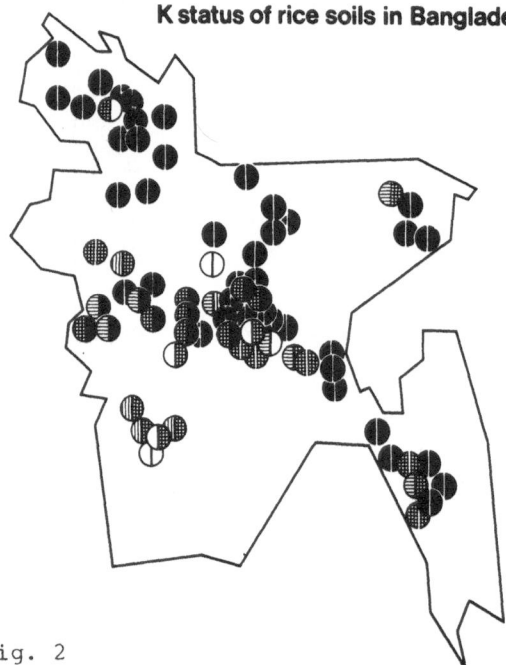

Rating	Exchangeable K		Symbol
	mg K/100g	me/100g	
Low	< 5.9	<0.15	
Med.-low	5.9- 11.7	0.15-0.30	
Medium	11.8- 17.6	0.31-0.45	
High	17.7 - 23.4	0.46-0.60	
Very high	>23.4	>0.60	

Rating	K saturation of CEC	Symbol
	% K sat.	
Low	<2	
Med.-low	2 - 3	
Medium	3.1- 4	
High	4.1- 5	
Very high	>5	

Fig. 2

the rice soils in Bangladesh are low in exch. potassium. 52 % of
the 73 samples analysed contained less than 0.15 me K/100g, and in
69 % of all cases the K saturation of the CEC was below 2 %. Only
13 % of the samples tested were higher than 0.45 me K/100g. Figure
2 shows the soil test results according to geographic location. The
few samples with high to very high K status ($>$ 0.45 me/100g) originate particularly from Khulna District in the South West of the
country where the soils are classified as subrecent Gangetic alluvium or recent brackish alluvium of high clay and silt contents with
clay minerals predominantly of the smectite and vermiculite type,
partly of the illite type (Kawaguchi & Kyuma). In most of the other
districts soil samples with low K status prevail.

Field data of the FAO Fertilizer Programme were evaluated for
the years 1975/76 to 1978/79, comprising 134 sets of results from
800 trials and demonstrations. Their K response pattern, worked out
in the same way as for the Philippines, is presented in Table 3
together with a compilation of the low and high soil K values for
the respective districts. Medium to high responses to potash application ($>$ 200 kg paddy/ha) have been obtained in 66 % of all cases.
More than half of them were above 400 kg paddy/ha, representing a
response of more than 8.5 kg paddy/kg K_2O at the average rate of
47 kg K_2O/ha used in the trials and demonstrations of the Bangladesh FAO Fertilizer Programme. K responses below 200 kg/ha were
recorded in 32 % of the 134 sets, negative responses in 2 cases
only, namely in the districts of Mymensingh and Sylhet in the
Northeast of Bangladesh.

SRI LANKA

Detailed information on the fertility characteristics of rice
soils in Sri Lanka is available from the basic report by Panabokke
& Nagarajah (1964). Its assessment of the status of exchangeable K
has been superimposed on the data of Kawaguchi & Kyuma in Figure 3.
Obviously, most of the paddy soils of Sri Lanka are low or medium-
low in exchangeable K. Accordingly, the Dept. of Agriculture
recommends the use of potash in all the districts of the country.

The recommendations range from 38 to 56.5 kg K_2O/ha, depending
on the agroclimatic zone and on the variety grown. The average is
about 44 kg K_2O/ha. Recommended rates of nitrogen and phosphorus
are approx. 68 kg N and 50 kg P_2O_5/ha which brings the total to
about 162 kg NPK/ha. The actual average fertilizer consumption for
rice in 1978 (based on figures published by the Ceylon Fertilizer
Corporation) was 47 kg N, 13 kg P_2O_5 and 12 kg K_2O/ha, together
72 kg NPK/ha at an average yield level of 2.37 t paddy/ha. While the
application of nitrogen is not much below the recommendation, actual
P and K levels are only about 1/4 of the recommended rates. Consequently, steps are required to improve the nutrient ratio in fertilizer use for rice in Sri Lanka. Increasing the level of K application would raise the efficiency of fertilizer N (and P), as can
be seen from results by Balasuriya et al. (1977), summarized in
Table 4, and from the long-term NPK experiment carried out in cooperation with IRRI.

Table 3 Classification of K response in 800 trials & demonstrations on rice in Bangladesh (FAO 1975-76 to 1978-79) and of low & high soil K values by districts (Kawaguchi & Kyuma and Islam et al.)

District	K response (NPK minus NP), kg/ha					Exch.K, me/100g		
	No. of sets 1)	>400	200-400	<200	Negative	No.of sites	<0.15	>0.45
Dinajpur	12	6	4	2	0	2	2	0
Rangpur	5	2	2	1	0	8	5	0
Bogra	9	1	5	3	0	3	3	0
Rajshahi	12	5	2	5	0	1	0	1
Mymensingh	15	2	5	7	1	5	4	1
Tangail	4	2	1	1	0			
Sylhet	8	1	2	4	1	4	3	0
Pabna	4	1	2	1	0	6	3	0
Kushtia	6	1	3	2	0	3	0	0
Dacca	14	7	1	6	0	21	10	1
Jessore	9	4	1	4	0	1	0	1
Faridpur	2	1	1	0	0	1	0	1
Comilla	8	3	2	3	0	6	4	0
Khulna	13	8	3	2	0	4	0	4
Noakhali	2	0	1	1	0			
Chittagong	11	6	4	1	0	8	4	0
Total	134	50	39	43	2	73	38	9
	100%	37%	29%	32%	2%	100%	52%	12%
		66%		34%				

1) One set = 6 trials or demonstrations on average

Table 4   Response of rice to K application in Amparai
Distr./Sri Lanka, East coast, dry zone (Balasuriya et al.1977)

| Treatment (kg/ha) | | | Gley soil[1], 23 trials | | | Alluv.soil,[2] 7 trials | | |
N	P_2O_5	K_2O	Yield t/ha	K response kg/ha	kg/kg K_2O	Yield t/ha	K response kg/ha	kg/kg K_2O
112	112	0	3.93	–	–	4.72	–	–
112	112	34	4.36	428	13	5.83	1108	33
112	112	67*	4.74	805	12	6.19	1464	22
112	112	100*	4.80	867	9	6.27	1542	15

*2 splits

1) Sandy gley soil, exch. K 0.12 me/100g
2) Loamy alluvial soil, exch. K 0.29 me/100g

Table 5   Frequency of low and high K status in samples
of paddy soils from 5 countries of South and East Asia

Country	No. of samples	<0.15 me K/100g	>0.45 me K/100g
Sri Lanka	33	58 %	9 %
Bangladesh*	73	52 %	13 %
Thailand	80	49 %	16 %
Indonesia (Java)	44	18 %	36 %
Philippines	54	15 %	42 %

Source: Kawaguchi & Kyuma

*incl. 20 soils analysed by Islam et al.

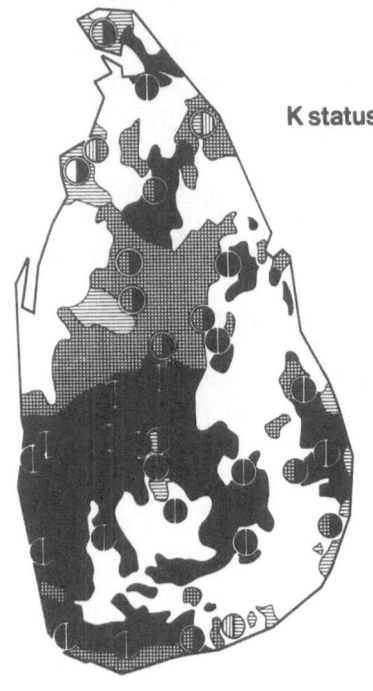

K status of rice soils in Sri Lanka (Kawaguchi & Kyuma = A, Panabokke & Nagarajah = B)

Rating	Exchangeable K		Symbol	
	mg K/100g	me/100g	A	B
Low	< 5.9	<0.15		
Med.–low	5.9 – 11.7	0.15 – 0.30		
Medium	11.8 – 17.6	0.31 – 0.45		
High	17.7 – 23.4	0.46 – 0.60		
Very high	>23.4	>0.60		

Rating	K saturation of CEC	
	% K sat.	Symbol
Low	<2	
Med.–low	2 – 3	
Medium	3.1 – 4	
High	4.1 – 5	
Very high	>5	

Fig. 3

K status of rice soils in Java (Kawaguchi & Kyuma)

Rating	Exchangeable K		K saturation of CEC	
	mg K/100g	Symbol	%	Symbol
Low	< 5.9		<2	
Med.–low	5.9 – 11.7		2 – 3	
Medium	11.8 – 17.6		3.1 – 4	
High	17.7 – 23.4		4.1 – 5	
Very high	>23.4		>5	

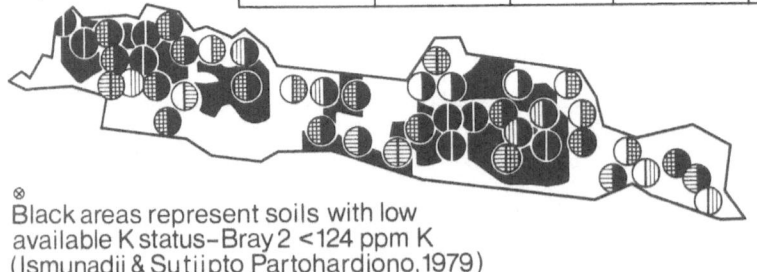

⊗ Black areas represent soils with low
available K status–Bray 2 <124 ppm K
(Ismunadji & Sutjipto Partohardjono, 1979)

Fig. 4

610

INDONESIA, THAILAND

A fairly large number of samples of paddy soils from these two countries were analysed by the Kawaguchi & Kyuma team, namely 80 samples from Thailand and 44 samples from the island of Java/Indonesia.

Although most of the soils of Java, being derived from volcanic ash, are considered to be rich in available potassium the data of Kawaguchi & Kyuma show that only 36 % of the 44 rice soil samples tested high in exchangeable K (> 0.45 me/100g). There was even a certain number of low-K soils (18 % below 0.15 me K/100g). This finding is confirmed by soil analyses reported by Ismunadji & Sutjipto Partohardjono (1979a). The data of both sources are incorporated in the map of Java, shown in Figure 4. Paddy soils testing low in exchangeable K are situated in West Java, classified as gray hydromorphic and red-yellow podsolic soils. and in certain parts of Central and East Java.

Potitive responses to potash application are to be expected on these soils. They have been obtained at several Stations of the Central Agricultural Research Institute, such as Singamerta and Serang (West Java), Chihea (South-East of Bogor), Jakenan (Central Java) and other places (Ismunadji & Partohardjono 1979b).

In Thailand 49 % of the 80 paddy soils, for which analysis data by Kawaguchi & Kyuma are available, tested low in exchangeable K. These soil cover extended areas in the North-East of the country. High-K soils are concentrated in the Bangkok plain. However, part of them have K saturation values below 2 % of the CEC.

COMPARISON OF VARIOUS COUNTRIES

Grouping the five countries considered above (p. 2-4) according to the K status of their paddy soils the order is as follows (Table 5). A large percentage of low-K soils is found in Sri Lanka, Bangladesh and Thailand, a large percentage of high-K soils in the Philippines and in Java. In the other countries covered by the Kawaguchi & Kyuma analysis (Table 1), paddy soils with low status of exchangeable K are widespread in Kampuchea and scattered in West Malaysia. High K soils occur in Southern Burma.

Correlation of K responses with soil test values could be attempted if data for both were available in the same areas. However in the Philippines as well as in Bangladesh the sites for fertilizer trials and demonstrations were not identical with the sampling sites. Nevertheless, the global comparison shows the expected relationship between soil test and crop response (Tables 6&7).

The figures for average yield at NP treated plots (no K) are substantially higher in the Philippines than in Bangladesh, indicating a generally higher level of soil fertility, including the K status, and hence lower average responses to K application. On the other hand, paddy soils in Bangladesh are generally poor in available K and respond to application of potash fertilizer in 66 % of all cases (Table 7).

Table 6 Average values for exch. soil potassium, paddy yields in FAO trials & demonstrations without K treatment and response to K application. Philippines and Bangladesh

Country	Soil K Status		FAO trials & demonstrations			
	No. of samples	Exch. K me/100g	No. of sites	Yield of NP-plots t/ha	Response to K	
					kg/ha	kg/kgK$_2$0
Philippines[1]	54	0.5	3225	4.61	139	4.6
Bangladesh [1][2]	73	0.25	800	3.55	356	7.6

1) Kawaguchi & Kyuma, 2) Islam et al.

Table 7 Comparison of soil K status and K response pattern in the Philippines and in Bangladesh

Country	Soil K Status		K response	
	Low <0.15 me/100g	High > 0.45 me/100g	Low or negative	Medium to high >200 kg/ha
Philippines	15 %	42 %	62 %	38 %
Bangladesh	52 %	12 %	34 %	66 %

Table 8 Initial soil K status and K response of the dry season rice crop at five long-term NPK experiments in the Philippines and in Sri Lanka

Experimental site	Initial soil status			Response to K (NPK minus NP)	
	CEC me/100g	Exch.K me/100g	K satu-ration	K rate kg K$_2$0/ha	Response t paddy/ha
IRRI, Los Banos,Laguna/ Philippines	45	1.5	3.4%	30	0 (14yr.av.)
Visayas Rice Exp.Station Iloilo/Philippines	59	0.9	1.5%	60*	0.4 (7 yr.av.)
Maligaya Rice Research & Training Center, Nueva Ecija/ Philippines	36	0.5	1.4%	60*	0.9 (11yr.av.)
Bicol Rice & Corn Exp. Station, Camarines Sur/ Philippines	29	0.2	0.7%	60*	1.9 (11yr.av.)
Mariawatte, Gampola Distr. Kandy/Sri Lanka	11	0.06	0.5%	40	0.6 (4 yr.av.)

* split application of K

There are, of course, exceptions from this general pattern. In Khulna/Bangladesh, for example, all the soil samples analysed tested high to very high in exchangeable K, yet in this district the largest number of trials and demonstrations with high K response (above 400 kg paddy/ha) was recorded. From the data available, however, it cannot be determined whether the soils analysed were also representative for the fertilized field plots. According to the FAO/UNESCO Soil Map of the World, soils in Khulna district are mainly Dystric Histosols (infertile peaty soils) or Calcaric Gleysols. In both cases K responses can be expected.

While there is no question that rice soils testing low in exchangeable K require application of potash (along with the other fertilizer nutrients) to avoid yield reductions due to potassium deficiency, the situation is more complicated for soils of medium and high K status. Generally they are supposed to contain sufficient available potassium to support high paddy yields. However, there are a number of reports about responses to potash application on soils testing high in potassium (Mahapatra & Prasad 1979, Kemmler 1974, Ramamoorthy et al. 1975, Ali et al. 1976, Goswami & Banerjee 1978). Therefore only long-term fertilizer experiments combined with soil analyses will indicate for how many years a soil may maintain its high K status and at what stage potash application will become necessary.

LONG-TERM EXPERIMENTS TO MONITOR SOIL-K STATUS

Data are available from four long-term NPK experiments in the Philippines, carried out on heavy clay soil (> 45 % clay) with initial K status from medium-low to very high. Their results have been periodically published in the IRRI annual reports and by several authors (De Datta & Gomez 1975, von Uexküll 1976, a,b, Kemmler 1978). The original level of exchangeable K at each site and the average response to potassium are shown in Table 8, together with the respective data of a Sri Lanka clay soil from the IRRI International Long Term Fertility Trial.

Three of the Philippine paddy soils had exchangeable K values above 0.45 me/100g, two even above 0.6 me/100g, usually rated as very well supplied with plant available K so that no K response was anticipated. At the IRRI site (exch. K 1.5 me/100g) there was indeed no response to K application while the yield level at 140 kg N and 30 kg P_2O_5/ha was 7.2 t paddy/ha. But at Visayas (Iloilo) an average increase by about 400 kg paddy/ha was recorded over the average yield of 5.6 t paddy/ha obtained at N_{140} P_{60} in the 7 dry season crops for which data are available. At 0.9 me/100g = 35 mg/100g exch. K, this result is unexpected. A possible explanation is the very high CEC coupled with the high clay content of 61 % which results in a rather low K saturation of the CEC, an indication of a too low K concentration in the soil solution, at least during the period of maximum vegetative growth. At Maligaya (exch.K 0.5 me/100g) the soil supplied sufficient potassium for dry season yields of 6 - 7 t paddy/ha during the first two years. Later on the response to 140 kg N/ha declined drastically on those plots which had not been treated with K, owing to a strong decrease in available soil potassium. As the Bicol soil was medium-low in exch. K (0.2 me/100g) and very low in K saturation, substantial

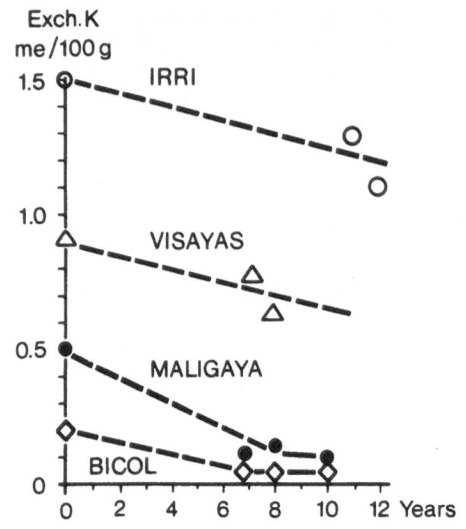

(soils of NP plots)

	Av.yield NP plots t/ha	Response to K
IRRI	7.2	0.0
VISAYAS	5.6	0.4
MALIGAYA	5.2	0.9
BICOL	5.0	1.9

Fig. 5   Change of soil K status in Philippine long-term experiments (partly based on unpublished data of Dr. S.K. De Datta from IRRI)

responses to K application were obtained right from the beginning, similarly at the Gampola site in Sri Lanka with an extremely K deficient soil. Average NP yield levels were 5.0 t/ha in Bicol and 4.6 t/ha in Gampola.

The gradual decrease of exch. K in the soils of the four Philippine sites is shown in Figure 5 for the plots which have received N and P but no K during the course of the experiment. At Bicol and Maligaya, within a few years, a level was reached which is generally regarded as K-deficient (< 0.15 me/100g). Under these conditions K application is recommended to correct the deficiency.

The Visayas soil still contained very high amounts of exch. K after 8 years. There was nevertheless a marked effect of potash application. Besides the exchangeable amount of K (very high in this case) other criteria, such as the K saturation of the CEC which has come down to 1.1 %, should therefore also be taken into account, when interpreting the soil test results.

At the IRRI long-term NPK experiment the exch. K has decreased by about 0.4 me/100g during 12 years. Yet the level is still so high that a response to K is unlikely. However in pot experiments it has been shown that the K reserves even of a soil like Maahas clay (IRRI) can eventually be exhausted (Busch 1980).

CONCLUSION

To eliminate potassium deficiency as a possible constraint to high yields it is advisable to assess the K status of paddy soils as part of the regular soil testing programme and to apply potash fertilizer on soils of low K status. On soils with medium and high levels of exchangeable K it is recommended to carry out long-term fertilizer experiments combined with soil analyses at regular intervals to monitor the changes in the potassium supply capability of the soils under intensive cropping, so that deficiencies can be avoided.

ACKNOWLEDGEMENTS

The author wishes to express his thanks to Dr. K.Kyuma, Kyoto University and to Dr. A. Islam, University of Dacca, for providing additional information on soil analysis results and to Dr. S.K. De Datta, Head, Department of Agronomy, IRRI, for the permission to use some of his unpublished data, furthermore he is grateful to FAO for supplying computerized data on trial and demonstration results.

REFERENCES

(1) Ali. M.H., S.P. Dhua and K.P. Dasgupta. 1976, Available potassium status of soils of West Bengal and response of high yielding rice to potassium. In Bull.No. 10, p. 164 - 169. Indian Soc. Soil Sci. J., New Delhi

(2) Balasuriya, I.K. Ratnasingham and M.W. Thenabadu. 1977. Response of rice (Bg 11-11) to potassium fertilizer on the non-calcic low humic gley soil complex and alluvial soils of Amparai,

Maha 1972/73. Tropical Agriculturist, Agr.J. of Sri Lanka, 133, No. 1, p. 51 - 70

(3) Busch, R. 1980. Der Einfluß der K^+-Konzentration der Bodenlösung und der K^+-Pufferung auf die K^+-Aufnahme und das Wachstum von Lolium multiflorum. Diss. Justus Liebig Univ., Giessen, Fed. Rep. of Germany

(4) De Datta, S.K. and K.A. Gomez, 1975. Changes in soil fertility under intensive rice cropping with improved varieties. Soil Sci. 120, p. 361 - 366

(5) Goswami, N.N. and N.K. Banerjee. 1978. Phosphorus, potassium and other macroelements. p. 561 - 580 in Soils & Rice. International Rice Res. Inst., Los Banos, Philippines

(6) Ismunadji, M. and Sutjipto Partohardjono. 1979. Recent research on potash application to lowland rice in Indonesia
a) Manuscript in Indonesian language,
b) Potash Review,
Subj. 91, 43rd suite, p. 1 - 9

(7) Islam, A., T.H. Khan, B. Ahmed and G.M. Panaullah. 1974. Phosphate and potassium status of some soils of Bangladesh. Proc. Planning and organization meeting fertilizer INPUTS project, Honolulu/Hawaii, p. 146 - 155

(14) Ramamoorthy B., M. Velayutham and V.K. Mahajan. 1975. Recent treads in making fertilizer recommendations based on soil tests under fertilizer resource constraints in India. Proc. FAI-FAO Seminar on optimising agric. prod. under limited availability of fertilizers. Fert. Assoc. of India, New Delhi, 1974, p. 335 - 346

(15) Tanaka, A. and S. Yoshida. 1970. Nutritional disorders of the rice plant in Asia. Int. Rice Res. Inst. Techn. Bull. 10

(16) von Uexküll, H.R. 1976,a. Response of HYV rice to potassium in the long-term fertilizer trials in the Philippines. In Bull.No. 10, p. 177 - 185. Indian Soc. Soil Sci. J., New Delhi

(17) von Uexküll, H.R. 1976,b. Fertilizing for high yield. Rice IPI-Bull.No. 3, International Potash Institute, Berne/ Switzerland

(18) von Uexküll, H.R. 1978. Agronomic and economic evaluation of crop responses to potash fertilizer. Rice: The Far Eastern & S.E. Asian Experience. p. 241 - 259 in Potash Res. Inst. of India. Proc. Symposium on potassium in soils & crops, New Delhi,India

(8) Kawaguchi, K. and K. Kyuma. 1977. Paddy soils in tropical Asia, their material nature and fertility. The Univ. Press of Hawaii, Honolulu.

(9) Kemmler, G. 1974. Kaliumverfügbarkeit verschiedener Böden und Kaliumdüngerwirkung beim Sumpfreis. Mitteilungen Deutsche Bodenkundliche Gesellschaft. 20, p. 201 - 210

(10) Kemmler, G. 1978. Economic evaluation of results of long-term NPK-experiments: Experience in some developing countries. p. 399 - 414 in Potash Res. Inst. of India. Proc. Symposium on potassium in soils & crops, New Delhi, India

(11) Mahapatra, I.C. and R. Prasad. 1970. Response of rice to potassium in relation to its transformation and availability under waterlogged condition. Fertiliser News 15 (2), p. 34- 41

(12) Moormann, F.R. and N. van Breemen. 1978. Rice: Soil, Water, land. Internat. Rice Res. Inst., Los Banos, Philippines

(13) Panabokke, C.R. and S. Nagarajah. 1964. The fertility characteristics of the rice-growing soils of Ceylon. Tropical Agriculturist, 120, p. 3 - 30

ON THE POTENTIAL OF K-NUTRITION AND THE REQUIREMENT
OF K-FERTILIZER IN IMPORTANT PADDY SOILS OF CHINA

Xie Jian-chang,   Ma Mao-tong,

Du Cheng-lin,   Chen Ji-xing
(Institute of Soil Science, Academia Sinica, Nanjing)

The appearing of potassium deficient symbols on rice plant and the good response of rice crops to potassium fertilizer have been recently revealed in important types of the paddy soil in China(1).

The supply of K-nutrition for a growing crop from the soil depends largely on its available potassium, but the potentiality of affording available potassium lies in the characteristics of K-bearing minerals in soil. Most of the paddy soils occur in humid and humid-tropic regions, where the clay minerals are highly weathered. An extensive survey on the potentiality of K-nutrition of important paddy soils may serve as a reference for the supply and distribution of K-fertilizer in this country.

To attain this end, we used the following materials and methods. Samples of important paddy soils were collected from different soil regions of China(2). They were analysed for available, slowly available and total K*(3). Pot culture experiments for 61 representative paddy soils were conducted. Field experiments for potassium fertilization on rice field were carried out in red earth region, the Taihu Lake region and cultivated meadow soil region(4).

The clay fraction of the soils was separated. Clay minerals were identified by X-ray diffraction. The quantity of hydrated micas in clay portion was estimated from their content of potassium(5).

Results from pot culture experiments on paddy soils of different K-nutrition potential revealed that successive rice cropping leads to exhaustion of K-nutrition repidly even in soils containing high levels of slowly available K (Table 1). The higher the slowly available K, the less increase by K-dressing. With the increase of cropping times, the effect of K-fertilizer is raised greatly. In some soils, there is no growth in control treatment.

Similar experiments were made on 26 soils of different K-nutrition level. The mean variations of available K and slowly available K before and after each cropping were determined. The K-uptake of the whole rice plant was determined at each harvest. Results obtained from 4 successive crops were analysed as presented in Table 2.

---

*Available K was determined as exchangeable K. Slowly available K by boiling in N HNO_3 for 10 min from which the content of exchangeable K is subtracted.

Table 1 Reponse of rice to K-fertilizer on soils
containing different amounts of slowly
available K
(pot experiment)

Slowly available K (K mg/100g)	No. of Soils tested	Increase by K-dressing (%)		
		1st crop	2nd crop	3rd crop
8	6	236.5	*	*
8-20	4	117.4	1870	*
20-35	6	53.6	392	*
35-50	5	15.8	98.0	631
50-75	5	4.3	72.8	594
75-120	6	2.8	57.2	187
> 120	3	0	10.6	39.6

* indicates no growth in control treatment

Table 2 K-uptake from soils by successive rice
(Pot experiments, mean value for
26 paddy soils)

Successive rice crop	Uptake of K (mg/100g)			Relative proportion (%)	
	From avail.K	From slowly avail. K	Total	From avail.K	From slowly avail.K
1st	2.56	3.37	5.93	43.2	56.8
2nd	1.05	1.40	2.45	42.9	57.8
3rd	0.43	1.04	1.47	29.3	70.7
4th	0.16	0.49	0.65	24.6	75.4
Sum	4.20	6.30	10.5	40.0	60.0

The results give such evidences that available K affords only a relatively small portion of K-nutrient to rice. A greater part of K-uptake by the rice comes from the slowly available K. The correlation coefficient of slowly available K to the total uptake of K is 0.983, while that to available K is 0.852. High correlations also have been found between the contents of illite, biotite and other hydrated micas in soil clay to the slowly available K. It is believed atthat K fixed on the cleavages or in the lattice of highly weathered hydrated micas released K-nutrient during continuous cultivation, the quantity of which, as illustrated by the present experiment, seems however far from being sufficient for good growth under the intensive cultivation and high application of nitrogen and phosphate fertilizers.

Extensive field trials on the important soil types in China also showed that K-deficiency occurs in soils of low slowly available K. For example, the Quaternary red clay in Zhejiang Province is deficient in potassium nutrition. As shown in Table 3, application of K-fertilizer to these soils have a significant effect.

Table 3 Response of crops to K-fertilizer on soils low in K
(Field experiment)

Crop	No. of experiments	Increase		kg grain/kg K
		kg/ha	%	
Early rice	21	608	16.9	8.4
Late rice	8	1140	39.7	19.3
Barley	4	623	459.5	9.2
Milk vetch*	9	6197	36.5	99.2

Note: Experiments on paddy soils derived from Quaternary red clay

  * Fresh wt.

The late rice gave better response to K fertilizer than early rice. Legume crop and barley also have a sensitive effect on K-fertilizer.

According to the content of slowly available K of regional soil types and the above-mentioned biotic tests, a generalized map for K-potentiality of China was compiled(1,2).

The K-potential in the map was divided into 7 grades. In different soils, there is different potential of K-nutrition. Potassium dificiency usually occurs in the acid soil regions of the south of China. From the south to the north, the potential goes from the lowest to the highest, and the amount of micas follows the same order, and the distribution of clay minerals is from kaolinite to hydromicas(5).

The potential of K-nutrition of the important paddy soils (as shown in Table 4) is in good consonance with regional soil group from which they originated. Five grades range from very low to high. The status of K-nutrition is quite different.

Agricultural statistic in 1979 reports that a large number of counties in the south of China, with an average of rice field around 30000-70000 hectares, gave crop yield at a rate of about 7.5 tons of rice grain per hectare. At such a rate of production, the removal of K from the soil by the tops of rice plant is about 155 kg K/ha. As given in our present investigation, most of important paddy soils in the south of China possess available K < 125 kg K/ha and slowly available K < 440 kg K/ha. It is estimated that about 30 % of the cultivated soils would have better crop yields if sufficient K-fertilizer should be provided. Statistical data also indicate that the consumption ratio of chemical fertilizers in 1978 is 1:0.28:0.001, in terms of $N:P_2O_5:K_2O$ and the appropriate proportion in the coming decades would be 1:0.7:0.2.

Table 4 Potential of K-nutrition of important paddy soils in China

Grades	Status of K-nutrition	Original soils and dominant clay mineral
Very low	Rice stunted in growth by serious K-deficiency	Lateritic red soils, red soils kaolinite (vermiculite)
Low	K-deficiency limiting the yield of rice	Red soils, yellow soils kaolinite (hydrated micas)
Medium	Potential K-nutrition not satisfied for high yield	Deposits (Zhujiang River, Changjiang River and Taihu Lake) Hydrated micas and kaolinite (montmorillonite)
Above Medium	K-fertilizer needed as rate of N, P application increasing	Yellow brown soils, purplish soils Hydrated micas and montmorillonite
High	K supply sufficient, except for a few high yield economic crops	Cultivated meadow soils, brown soils Hydrated micas (montmorillonite)

It may be concluded that as far as the present cropping system goes on, and the production and supply of nitrogen and phosphorous fertilizers are continuously increasing, the need of K fertilizers in agriculture would become more urgent in China.

REFERENCES

(1) Institute of Soil Science, Academia Sinica, Nanjing (ed.), Soils of China. Science Press, Beijing, 392-403 (1978).(in Chinese)
(2) Xie Jian-chang, The K-nutrition potential of the Chinese soils and the outlook on the need of potassium fertilizer in China(1979, ms.). (in Chinese)
(3) Ekpete, D.M., Comparison of methods of available potassium assessment for Eastern Nigeria soils. Soil Sci.,113, 213-221(1972).
(4) Zhang Xiao-pu, Du Cheng-lin, Ma Mao-tong, Chen Ji-xing, Jia Yi, Xie Jian-chang, The supply of soil potassium and the effect of potassium fertilizer on crop response in Jiangsu Province. Acta Pedologica Sinica, 15, 61-74 (1978). (in Chinese with English summary)
(5) Jiang Mei-ying, Luo Jia-xian, Study on the potassium-bearing minerals in soil, (1) The forms and release of potassium. Acta Pedologica Sinica, 16(4), 414-421(1979). (in Chinese with English summary)

BIOCHEMICAL AND NUTRITIONAL DIAGNOSIS OF K-DEFICIENCY
OF PADDY RICE PLANTS WITH REGARD TO SOIL FERTILITY

Sun Xi, Ma Guo-rui, Lin Rong-xin, Yin Xian-xiang
(Zhejiang Agricultural University, Hangzhou)

INTRODUCTION

Potassium is required for a number of biochemical and physiological processes of plant life. In rice plants potassium plays a major role in activating various enzymes which bring about growth and promote the resistance to environmental stress. Deficiency of potassium in rice plants usually shows symptom of reddish brown spots on lower leaves and a high percentage of blackened or rotten roots.

The brown spots on the leaves are assumed to be caused by the oxidation of polyphenol compounds into quinones by peroxidase and quinones thus formed combine with amino acids to produce brown substances(1-3). The rotten roots are assumed to be caused by the deficiency of potassium resulting in the suppression of "glycollic acid pathway" leading to depression of oxygen secretory power in rice roots(4). Based on the above assumptions, it may be recognized that the peroxidase activity in the leaves and the oxidative power of roots can be used for the diagnosis of potassium status of paddy soils.

MATERIALS AND METHODS

To investigate the effect of potassium status of paddy soils on the peroxidase activity in leaves and the oxidizing power of roots of different rice varieties, four field experiments were conducted over a 2-year period on the two types of paddy soils yellow-mottling silty clay and bluish-purple clay, which are widely distributed in Zhejiang Province. The available potassium of the experimental plots varies within the range of 33-42 ppm in surface soil.

The first experiment was carried out on the yellow-mottling silty clay at the station of Xiangfu Commune. Split plot design was used with 3 replications. Four late rice varieties were arranged on the subplots, while two treatments with and without potash were assigned to the main plots. Each plot was given a basal dressing of N45 kg/ha as ammonium bicarbonate, P 52.5 kg/ha as superphosphate and K 63 kg/ha as potassium sulphate. Another N 67.5 kg/ha as ammonium sulphate was used for top dressing. The plot size was 67 m^2.

The second experiment was carried out in the following year on yellow-mottling silty clay and bluish purple clay. Before transplanting, approximately 150 t fertile mud per ha containing 137 and 107 ppm available K was applied to the above two types of soils respectively to improve the potassium status. Randomized block design was used. There were three treatments: K_0, no potassium (check); K_1, 63 kg and K_2, 125 kg K per ha as potassium sulphate. The same amounts of nitrogen and phosphatic fertilizers were applied as that of the 1st experiment. Potassium fertilizer was used as basic dressing and top dressing at the ratio of 50:50.

The 3rd experiment was carried on the same soil as the first experiment containing 40 ppm of available potassium in the surface soil. Only one variety of late rice was used. The last experiment was conducted to investigate the effect of potassium on the oxidative power of rice roots and and the absorption of iron in paddy soil.

Plant samples were taken at different stages of growth. Total potassium in plants, activtiy of peroxidase in leaves, active iron (6 N HCl extracted) of terminal leaves and oxidation power of roots were determined. Total potassium was determined by atomic absorption spectrophotometer, active iron by O-phenanthroline, peroxidase activity by guaiacol and oxidation power of roots by α-naphthylamine method(5). All determinations were done in duplicate, while the determination of enzyme activity was in triplicate.

At the end of the experiment, statistical analysis was performed for the test of significance of the grain yields.

RESULTS AND DISCUSSION

1. Effect of Potassium on the Activity of Peroxidase in Leaves and Grain Yield of Rice Plants

In an attempt to clarify the effect of potassium on the peroxidase activity in the leaves of rice plants, it is necessary to find out which leaf is the most sensitive. Table 1 shows that the peroxidase activity in the

Table 1 The peroxidase activity of different leaves on the main stem of rice at tillering stage

Variety	Leaf	Peroxidase activity (g.u./gfw)	Brown spot on leaves
Yuanfeng zao	Terminal	24	0
	1st	41	0
	2nd	43	0
	3rd	58	++
Guang lu ai No. 4	Terminal	17	0
	1st	43	0
	2nd	52	+
	3rd	54	++

+ a few brown spots; ++ more brown spots; g.u. Guaiacol unit

622

leaves is correlated with the brown spots of plants and is increased progressively from the termal leaf to the lower leaves, while the 4th leaves began with necrosis at tillering stage in the experiment. From this fact, taking 3rd leaf of rice plants to determine the activity of peroxidase seems to be appropriate.

Four varieties of rice plants grown in potassium deficient paddy soil (yellow-mottling silty caly) showed very poor growth with disturbance in metabolism, such as the higher peroxidase activity in leaves and many reddish brown spots on the lower leaves. Due to the deficiency of potassium in the rice plants, the panicle weight per hill, number of perfect grains per panicle and weight of 1000 kernels all decreased, thus resulting in a decrease of the grain yield. With the application of potassium, the reddish brown spots and the peroxidase activity in the leaves both at vegetative and reproductive stage decreased and the grain yields of all varieties increased significantly. From the experiment, it was found that the average activity of peroxidase in the 3rd leaves among the four varieties decreased from 34.8 to 21.6 g.u. (guaiacol unit) per gram leaf at panicle initiation stage. (Table 2).

Table 2 Effect of K on the activity of peroxidase in
leaves and the grain yield of late rice
(Yellow-mottling silty clay)

| Variety | Treat-ment | Brown spot | Peroxidase activity in 3rd leaves (g.u./gfw) | | Grain yield | |
			Tillering stage	Panicle initiation stage	kg/ha	Increase (%)
Nonghu No. 6	K_0	+++	58.0	34.6	4245	
	K	+	21.0	17.3	5325	25.5**
Gongnong	K_0	+++	−	44.1	3330	
	K	+ +	−	34.0	3990	19.5**
Jiahu No.4	K_0	+ +	−	25.6	4545	
	K	+	−	18.5	5445	19.8**
Tongqing	K_0	+++	−	34.8	4245	
	K	+	−	16.5	5850	38.4 **

**Significant at the 1 % level; K_0, check; K, K 63 kg/ha

On paddy soils with a K-status favorable for rice growth, no brown spots were found on the lower leaves and the activity of the peroxidase did not decrease, sometimes it increased by the application of K-fertilizer. The reason for the increase of the peroxidase activtiy is not known and will be a subject for further investigation. Although the total potassium in the plants increased with the increase of the level of K-fertilizer applied, no significant yield difference among the early rice varieties was found. The experimental data of early rice cultivated at yellow-mottling silty clay and bluish-purple caly with an application of approximately 150 t fertile

623

mud per ha are shown in Table 3.

Table 3 Relationship between total K in plants, peroxidase activity in leaves and grain yields of rice

Rice	Soil	Treatment	Tillering stage		Panicle initiaton stage		Grain yield
			Total K (%)	Peroxidase activity (g.u./gfw)	Total K (%)	Peroxidase activity (g.u./gfw)	kg/ha
Early rice	Yellow-mottling silty clay	K_0	1.46	21	0.90	40	5670
		K_1	1.58	28	1.59	44	6180
		K_2	1.99	27	1.98	45	6120
	Bluish-purple clay	K_0	1.33	26	1.20	44	5145
		K_1	1.81	24	1.81	44	5340
	Yellow-	K_2	2.24	25	2.62	48	5370
Late rice	Yellow-mottling silty clay	K_0	–	59	–	60	5280
		K_2	–	45	–	53	5925*

* Significant at 5% level; K_0, Check; K_1, K63 kg/ha; K_2, K125 kg/ha

When the potassium status in the soil was not favorable for rice growth, the grain yield of rice plants without potassium fertilization was reduced even though no visible symptom of K-deficiency appeared on the lower leaves. Under such condition, disturbance in metabolism such as peroxidase activity similarly took place as soon as growth was reduced by potassium starvation. The application of potassium to the above potassium deficient soil might improve the normal metabolism in plants and decrease the activity of peroxidase in the 3rd leaves. Under such condition, the grain yield of the tested veriety increased (Table 3).

From the facts mentioned above, it may be suggested that variation in the activity of peroxidase in the 3rd leaves of rice plants with and without potassium fertilization can be used as a relative index for the diagnosis of potassium status of paddy soils.

2. The Effect of Potassium on the Oxidative Power of Rice Roots

It is generally accepted that there is a positive correlation between the "vitality of the rice roots" and the oxidative power of the roots. The oxidative power of the roots can be evidenced by measuring the oxidation of α-naphthamine (α-NA). The application of potassium to K-deficient soils usually increased the activity of α-NA oxidative power of rice roots. Our previous experiment showed that the grain yields of both early rice and late rice were significantly correlated with the activity of α-NA oxidation

by roots only at heading stage(6). The effect of potassium on the grain
yields of late rice and the activity of α-NN oxidation by roots at heading
stage is shown in Table 4.

Table 4 Effect of potassium on the activity of α-NA
oxidation by rice roots at heading stage

| Variety | Treatment | α-NA oxidation by roots | | Grain yield |
		μg/gfw/h	μg/hill/h	kg/ha
Nong hu No.6	K_0	146	495.2	4245
	K	213	536.9	5325**
Gong nong	K_0	151	545.4	3330
	K	153	666.4	3990**
Jia hu No.4	K_0	109	534.1	4545
	K	126	559.4	5445**
Tong qing	K_0	122	440.4	4245
	K	145	620.0	5850**

** Significant at 1 % level

   Table 4 shows that rice plant grown on the potassium deficient soil
results in the depression of α-NA oxidation by roots. By potassium
application, the normal metabolism of "glycolic acid pathway" in the rice
roots which generated the oxidative power of the roots was improved. The
average α-NA oxidation power of the roots among the four varieties increased from 132.2 to 159.3μg/g/h and 504 to 596 μg/hill/h at heading stage
by the application of potassium, therely the rhizosphere of the rice roots
under reduced condition might be oxidized to prevent the toxicity of hydrogen
sulfide, organic acids and excessive iron in paddy soils. Table 5 shows that
the iron absorbed in the leaves is reduced by potassium application. Due to an
increase of potassium in plants, the oxidizing power of rice roots increased
significantly.

Table 5 Effect of K on the active iron content
in the terminal leaves of rice

| Treatment Treatment | Active iron (ppm) | |
	Late tillering stage	Panicle initiation stage
K_0	82.0	62.4
K_1	75.0	44.5
K_2	67.0	45.0

Why does potassium usually increase the oxidative power of rice roots? One of the physiological functions of K is to activate various enzymes in plants. It promotes the biosynthesis of nucleic acid. Purine and pyridine base as well as ribose and phosphate are the chief constituents of nucleic acid. While pyrine biosynthesis in plants requries formate, glycine and CO_2, which are the products of "glycolic acid pathway" directly or indirectly in rice roots (Fig. 1). Application of K to K-deficient soils may improve both the nucleic acid synthesis and the "glycolic acid pathway" in rice roots. The latter produces H_2O_2, which may easy be decomposed in the roots by catalase into O_2 to give oxidizing power(7). Therefore the oxidizing power of rice roots usually increases by K-application.

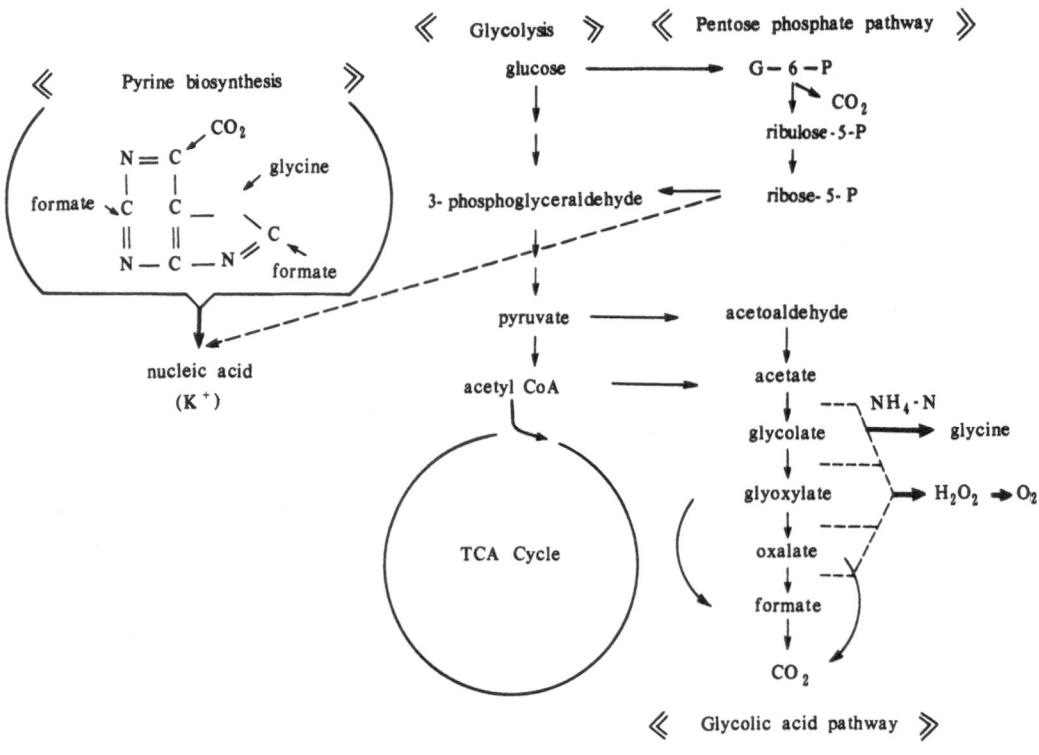

Fig. 1 K^+ in relation to nucleic acid synthesis and "glycolic acid pathway" in rice roots(7)

From these facts, it seems that determining the variation in the activities of α -NA oxidation power of the rice roots with and without potassium application can possibly be used as a complemental diagnosis of potassium status of paddy soils.

CONCLUSION

Field experiments and biochemical analysis were carried out to study the effect of potassium status of paddy soils on the peroxidase activity in the leaves and the oxidizing power of the rice roots. The experiments were conducted on two types of paddy soils: yellow-mottling silty clay and bluish-purple clay, widely distributed in Zhejiang Province. The available K of the experimental plots varied at a range of 33-42 ppm.

Four different varieties of late rice grown on yellow-mottling silty clay soil showed poor growth with accompanying disturbance in metabolism, such as higher peroxidase activity in the leaves and lower oxidizing power of the rice roots. Application of potassium caused an increase of K content in the rice plants, thereby improving their normal metabolism. The average activity of peroxidase in the 3rd leaves among the four varieties decreased from 34.8 to 21.6 g. u. per gram leaf at panicle initiation stage and the average α-NA oxidation power of rice roots increased from 504 μg/hill/h at heading stage. The grain yields of all tested varieties in the K-treated plots increased significantly.

On paddy soils with a K-status favorable for the rice growth, no significant yield difference among the early rice varieties was found by the application of K-fertilizer, irrespective of the amount applied. Under such conditions, the activity of peroxidase in the 3rd leaves did not decrease, sometimes it increased by K fertilization.

By comparing the peroxidase activity in the 3rd leaves and the α-NA oxidative power of the rice roots with and without K-fertilization, difference in activities can be used for the diagnosis of K-status of paddy soils.

REFERENCES

(1) International Potash Institute, Potassium in Biochemistry and Physiology, 8th Colloquium, IPI, Berne(1971).
(2) Bada, I., Harada, T., Physiological Diseases of Rice Plants in Japan. Ministry Agriculture & Forestry, Japan(1954).
(3) James, W.O., Plant Respiration. Clarendon Press, Oxford, 193-196(1953).
(4) Noguchi, Y., Sugawara, T., Studies on Some Indications of Potassium Deficiency in Rice Plants. Symposium on Potassium, Yokendo Press, Tokyo, 56-70(1957).
(5) Department of Agrochemistry, Zhejiang Agricultural University, Diagnosis of activity of reice roots and its application in agriculture. Fenxi Huaxue, 4, 252-257(1976). (in Chinese)
(6) Department of Agrochemistry, Zhejiang Agricultural University, Diagnosis of activity of root systems of rice and its relations to the yield of rice. Turanng, 6, 270-275(1977). (in Chinese)
(7) Mitsui, S., Kumazawa, K., Yazaki, J., Hirata, H., Ishizuka, K., Dynamic aspects of N.P.K. uptake and oxygen secretion in relation to the metabolic pathways within the plant roots. Soil Sci. Plant Nutr. 8(2), 25-30(1962).

SULPHUR CONTENT AND DISTRIBUTION IN PADDY
SOILS OF SOUTH CHINA

Liu Chong-qun, Chen Guo-an, Cao Shu-qing
(Institute of Soil Science, Academia Sinica, Nanjing)

In the hilly areas of South China, where the content of available
sulphur in paddy soils usually appears low, farmers used to dip the
roots of rice seedlings into a suspension of gypsum or sulphur before
transplantation. Old practice has been proved beneficial to the growth
and yield of rice plant. Over recent decades, most of the S-containing
fertilizers, such as $(NH_4)_2SO_4$ and superphosphate have been replaced by
NH_4HCO_3, urea and alkaline fused rock phosphate. The return of plant
ashes to agricultural field, which is actually a complete mineral
fertilizer, has been greatly reduced and the most of the K-fertilizer
is given in the form of KCl. Thus, the elimination of sulphur-containing
fertilizers induces the expanding of sulphur deficiency areas. On the
other hand, owing to the increasing yield of rice, the removal of soil
sulphure by the crop has also increased and consequently more sulphur
fertilizer is needed.

The present article reports the content and status of sulphur in
some paddy soil areas in South China, where the regional distribution of
sulphur usually appears low. The effect of sulphur fertilizers to the
growth and yield of rice in those areas is also dealt with.

It was reported by earlier investigators that the sulphur content
in world agricultural land varies from 0.01% to 0.05% (1). According to
the statistics from the IFDC (International Fertilizer Development
Center), the average content of total sulphur in the surface soil of
mollisols is 0.054%, 0.021% in the alfisols and 0.011% in the tropic soils
(2). Our present investigation covers 152 surface soils from the paddy
fields of various districts in nine southern provinces of China, where
the deficiency of sulphur in soils was usually regarded as suspectable.
Analytical results of total sulphur and available sulphur for these soils
are given in Table 1.

The foregoing investigation of sulphur on upland soils usually
reported organic sulphur by substracting available sulphur from the total
sulphur in soils and neglected the minute amount of mineral sulphur
compound. Under the low redox potential in the flooded paddy field, the
toxicity of sulphide does appear in highly gleyed profile. However the
quantity of sulphide was still found neglectible.

Available sulphur in soils has long been regarded as an index for the
nutritive status of soil sulphur. In the present paper, the paddy soils
under investigation are separated into the following three categories
according the status of soil sulphur.

Table 1 Mean values of the sulphur content in the
cultivated layer of paddy soils of South
China

Locality	Available S (ppm S)	Total S (S %)	Organic S (S %)	Category
Taihu Lake area	51 (12)	0.029 (12)	0.024	
Guizhou	38 (8)	0.058 (9)	0.054	High
Dongting Lake area	35 (17)	0.034 (16)	0.031	
Yunnan	27 (12)	0.021 (12)	0.018	
Guangdong	22 (10)	0.025 (10)	0.023	Medium
Northern Fujian	16 (36)	0.024 (36)	0.023	
Zhejiang	14 (15)	0.021 (12)	0.020	
Jiangxi	13 (33)	0.020 (34)	0.019	Low
Sichuan Basin	11 (9)	0.019 (9)	0.018	
Mean Value	22(152)	0.026(150)	0.026	

Note: Available S extracted with 0.03 M NaH_2PO_4-2N HOAc(5). Total S
was determined by combustion-iodimetric method (6). Numbers
in parentheses denote samples analysed

Category I: available soil sulphur ranging 30-50 ppm S, including
paddy soils developed on the lacustrine deposit of the Taihu Lake and
the Dongting Lake and the clayey paddy soils derived from limestone in
Guizhou Province. They contain high levels both of total and available
sulphur, S-nutrition for rice crop is sufficient.

Category II: available soil sulphur ranging 20-30 ppm S, total
sulphur below the average in the areas under investigation. Paddy soils
developed on the lateritic and the red soils of tropic region belong mostly
to this category. The high contents of kaolinitic clay and sesquioxides
in these paddy soils hold the SO_4^- from leaching. although there is no
evidence of S-deficiency in the present rice cultivation, owing to the
low potential of S-nutrition, S-fertilizer will be needed as cultivation
goes on and the yield of rice is on the increase.

Category III: available soil sulphur below 16 ppm S and total S below
the average, mostly found in paddy soils in the hilly regions of Southeast
China. The application of S-fertilizer to rice crop used to be an old
practice for these soils. A limited number of soil samples were collected
from the paddy field in the purplish soil region of Sichuan Basin. They
were found also low in S-nutrition but have not been verified by field
trial.

It is revealed through investigation that there are good correlations between the contents of soil organic matter and total soil sulphur.
Fig. 1 represents such correlations from 24 paddy soils derived from
different parent materials in Jiangxi Province. Oragnic matter can be
regarded as the storage of the S-nutrition in soils and further more,
plays an effective role of holding SO_4^{2-} against leaching.

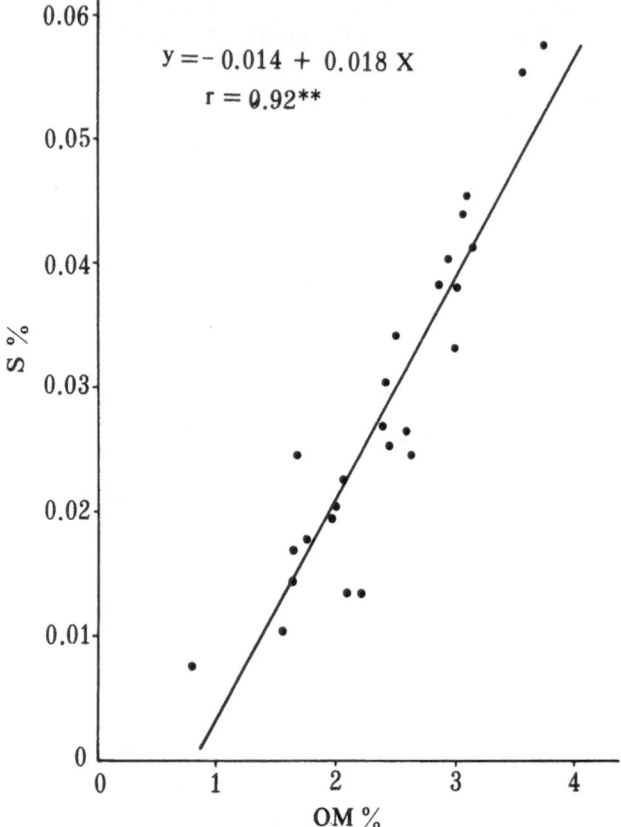

$$y = -0.014 + 0.018\,X$$
$$r = 0.92{**}$$

S %

OM %

Fig. 1 Correlation between the content of
total soil sulphur and soil organic
matter

In a given soil area, soil texture affects the content of total
soil sulphur. Fig. 2 shows the correlation between fine soil particles
(<0.01 mm) and the total sulphur in the paddy soil of the Taihu Lake
area.

Irrigation water is a source of S for crops. Wang (1976) suggested
that irrigation water with a sulfur content of 6 ppm would be enough to
supply the entire needs of a rice crop (2). A report from IRRI declared
that the mean value of sulphur content in world river water is about
3.7 ppm S(3). Within the soil areas under investigation, the mean value
of SO_4^{2-} in the streams generally appears low. Very low figures have
been reported from the streams in the hilly regions of Fujian and Yunnan
Provinces, where deficiency of S-nutrition in paddy soils is conspicuous.

As usual, pollution of river water near by the industrial districts
sometimes happens. Table 3 shows the very high average values of S-
content in the water of such rivers as the Zhujiang River, Guangzhou and
the Wusong River, Shanghai but variations in S-content are in wide
ranges.

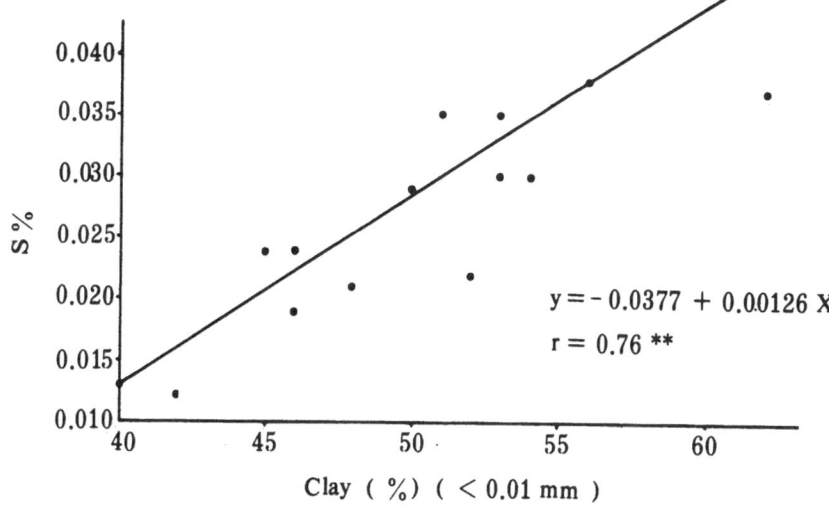

Fig. 2 Correlation between the content of
total soil sulphur and soil texture

Table 2 Content of sulphate in some river water
of South China

Rivers	Mean value[*] (ppm S)	Variations (ppm S)
A branch of Zhujiang River	2.72 (57)	0.03–11.04
From streams of Taihu Lake area	2.90 (12)	0.60–6.31
From Poyang Lake area, Ganjiang River	1.80 (159)	0.33–5.57
From Dongting Lake area, Xiangjiang River	1.50 (409)	0.13–4.66
Streams in northern Fujian	0.47 (6)	0.40–0.51
From streams of central Yunnan	1.37 (3)	1.25–1.50
Mean value	1.67 (646)	0.03–11.04

[*] Numbers in parentheses denote samples analysed

Table 3 Contents of sulphate in river of
industrial districts

Rivers	Mean value[*] (ppm S)	Variations (ppm S)
Zhujiang River (Guangzhou)	7.77 (52)	1.51–30.48
Wusong River (Shanghai)	22.72 (12)	1.94–36.99

[*] Numbers in parentheses denote samples analysed

A survey of the field trials on S-fertilization for the paddy soils
in the hilly regions of Zhejiang and Jiangxi Provinces showed that 54%
out of 93 field experiments had a significant response of rice crop to

Table 4 Response of rice crop to S-fertilizers by
different methods of application (4,7,8)

Soil type and locality	Parent material	S-fertilizer (S kg/ha)	Method of application*	Grain yield (t/ha)		Increase in yield (%)
				With sulphur	Without sulphur	
	Alluvial deposit	S 37.5	B	6.59	6.08	8**
Sandy soil, Jiangxi	Granite	Gypsum 75	B	4.19	3.89	8**
	Sandstone	S suspension 4.5	Di	8.25	7.27	13**
		Gypsum suspension 22.5	Di	8.09	7.27	11**
Cold spring paddy soil, Wuchang	Quaternary red clay	CaSO$_4$, MgSO$_4$ Na$_2$SO4 75-150	B	3.39	2.70	25*
Sandy soil, Yunnan	Sandstone	S 30	B	9.08	6.71	35*
		Gypsum 75	B	9.46	6.71	42*

* Method of Application: B=broadcasting Di=dipping seedling roots

the application of S-fertilizers (4). Sulphur and gypsum are the common fertilizers to be applied. Detailed descriptions for the method and rate of application are given in Table 4.

Typical hunger signs for sulphur nutrient on rice plants are shown in Fig. 3.

Fig. 3 Hunger signs for S-nutrient
on rice plants sandy soil
9.6 ppm available S.
left: + 3 gm Gypsum; right:
no S

Plants are grown in pot culture on a paddy soil of sandy texture, containing available sulphur 9.6 ppm S. Normal growth of rice, provided with 3 g of gypsum per pot, appears on the plant of the left pot. The plants grown on the right plot are stunted in growth, with a yellowish coloration of leaves. The maturity of the crop is delayed in comparison with the left.

Our investigation comes to the conclusion that in the hilly regions of Fujian, Yunnan and Jiangxi, the content of the available S in paddy soils is low. When S-fertilizers are applied, there's significant effect.

REFERENCES

(1) Beaton, J.D., Burns, G.R., Platou, J., Determination of sulphur in soils and plant material. Tech. Bull. Sulphur Inst., 14, 2-4 (1968)
(2) International Fertilizer Development Center, Sulphur in the Tropics. Sulphur Inst. & IFDC, 28-30 (1979).
(3) Wang, C.H., Liem, T.H., Mikkelsen, D.S., Development of Sulphur Deficiency as a Limiting Factor for Rice Production. IRI Tech. Bull., 47 (1976).

(4) Liu Chong-qun, Chen GuO-an, Liu Yuan-chang, A study on soil sulphur status in Zhejiang and Jiangxi provinces and application of S-fertilizer (1979, ms.). (in Chinese)

(5) Copper, M.A., Comparison of five methods for determining the sulphur status of New Zealand soils. Trans. 9th Int. Congr. Soil Sci., Adelaide, 2, 263-271 (1968).

(6) Institute of Soil Science, Academia Sinica, Nanjing, Soil Physical and Chemical Analysis. Shanghai Science & Technology Press, Shanghai, 277-281 (1978). (in Chinese)

(7) Chen Hua-kui, Zhuang Zheng-de, He Dian-yan, Lei Zi-qiang, Influence of sulphur on the rice yield of paddy soil with stunted seedlings. Acta ·Pedologica Sinica, 2, 34-36 (1952). (in Chinese)

(8) Deng Chun-zhang, Yao Tian-gin, Zhang Shi-yu, Li Dai-fang, Jiang Jin-yi, Li Sheng-de, Studies on sulphur deficiency of rice plant (1979, ms.). (in Chinese)

THE STATUS OF MICROELEMENTS IN RELATION TO CROP PRODUCTION IN PADDY SOILS OF CHINA: III. ZINC

Zhu Qi-qing, Liu Zheng *
(Institute of Soil Science, Academia Sinica, Nanjing)

Zinc deficiency is the most common problem in paddy soils. Since the disorder was reported, it has become recognized as an important and widespread problem(1). In China, zinc deficiency is also found and it distributes mainly on calcareous paddy soil(2,3). However, the reported zinc dificiency is restricted to small areas. Systematic survey remains to be done. As a reconnaissance work, the status of zinc in paddy soils in China was investigated. Its relation to growth of rice has been confirmed by field experiments.

STATUS OF ZINC IN PADDY SOILS

Zinc content in soils of China varies from 3 to 790 ppm with an average of 100 ppm(4). There are differences of zinc content among acid, neutral and calcareous paddy soils. The status of zinc is influenced either by the nature of parent materials or by the soil conditions, especially the soil reaction.

Zinc content in paddy soils of China is shown in Table 1. Total and available zinc contents of some profiles of paddy soils are listed in Table 2. Zinc status of acid paddy soils seems adequate. Total and available zinc

Table 1 Zinc content in paddy soils of China

Type of soil	Total zinc (ppm)		Available zinc (ppm)	
	Range	Average	Range	Average
Acid paddy soils	18 – 345 (144)***	113	trace – 19.90 (172)	3.49*
Neutral paddy soils	50 – 156 (61)	99	0.20 – 6.50 (62)	2.95*
Calcareous paddy soils	44 – 135 (23)	82	trace – 1.12 (32)	0.35**

    * Extractant: 0.1 N HCl
   ** Extractant: DTPA solution
  ***Figure in paratheses indicates number of samples investigated

* Tang Li-hua, Xu Jun-xiang and Yin Chu-liang also participated in this research.

Table 2 Zinc content of some paddy soil profiles

Type of soil	Parent material	Depth (cm)	Total zinc (ppm)	Available zinc (ppm)
Acid paddy soils	Granite	0–15	172	5.6*
		15–30	161	5.0
		67–82	128	4.1
	Sandstone	0–10	28	0.7*
		10–18	32	0.7
		37–80	28	0.6
Neutral paddy soils	Lacustrine deposit	0–17	136	5.9*
		17–26	120	3.5
		85–95	142	3.0
	Leached loess	0–15	120	3.4*
		15–27	106	3.1
		76–94	100	2.9
Calcareous paddy soils	Alluvium	0–20	99	0.84**
		20–29	110	0.28
		48–90	109	0.20
	Alluvium of Huanghe River	0–5	112	0.28**
		5–10	59	0.22
		10–20	55	0.22
		40–60	50	0.15

  * Extractant: 0.1 N HCl
 ** Extractant: DTPA solution

contents of acid paddy soils are higher than neutral and calcareous paddy
soils. Acid paddy soils derived from basalt, granite or limestone are richer
in zinc than those derived from sandstone or Quaternary red clay (Fig. 1).
The available zinc in most soils is generally higher than 1.5 ppm (the critical
value for an adequate supply) as extracted with 0.1 N HCl, with an exception
of those derived from sandstone and Quaternary red clay in which the available
zinc content is below the critical value.

Total and available zinc contents in neutral paddy soils are lower than
those in acid ones. The available zinc is higher than critical value and may
be considered sufficient. No deficient symptom of zinc is observed on acid and
neutral paddy soils in field condition.

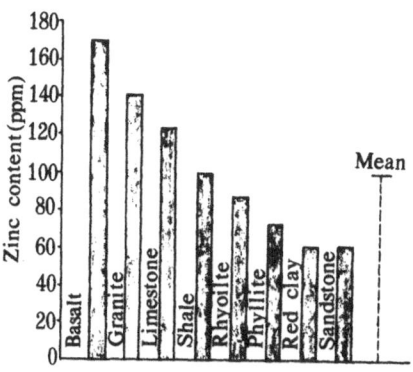

Fig. 1 Average content of zinc of
acid paddy soils in relation to parent
material

The average content of total zinc in calcareous paddy soils is lower
than that in the above-mentioned paddy soils, while the available zinc content of most samples is much lower than 0.5 ppm as extracted by DTPA solution. The fact of deficiency of zinc under alkaline soil condition might be
demonstrated by the analytical results of 150 samples of calcareous soils
including paddy soils. The content of available zinc of 150 samples and the
percentage of various levels of the contents are listed below:

trace	12 %
<0.5 ppm	47 %
0.5-1.0 ppm	27 %
>1.0 ppm	14 %

The data indicate that 59 % of the samples analysed is deficient in zinc,
27 % is of the marginal value, i.e. 0.5-1.0 ppm, and 86 % is below 1 ppm.
Field investigation has shown that the deficient symptom of zinc of rice is
widely spread on the zinc deficient soils mentioned above.

A map of content of available zinc in soils of China in scale of
1:10,000,000 is compiled. The available zinc is extracted by two extractants,
and that in acid and neutral soils is extracted by 0.1 N HCl, while that in
calcareous soils is extracted by DTPA solution. The contents of available
zinc in soils are divided in the following 5 grades.

1) Available zinc in acid and neutral soils extracted by 0.1 N HCl:

Very low	less than 1.0 ppm
Low	1.0-1.5 ppm
Medium	1.6-3.0 ppm
High	3.1-5.0 ppm
Very high	greater than 5.0 ppm

2) Available zinc in calcareous soils extracted by DTPA solution:

Very low	less than 0.50 ppm
Low	0.5–1.0 ppm
Medium	1.1–2.0 ppm
High	2.1–4.0 ppm
Very high	greater than 4.0 ppm

This map indicates that the zinc deficient soils are mainly distributed the calcareous soil region of North China.

An apparent response to zinc for rice in calcareous soils is found in the northern parts of Jiangsu and Anhui provinces, and northern and central parts of Hubei Province as well as in the coastal saline soil region as shown in Fig. 2 This map shows localities where crops are responsive to zinc fertilizer. All marks denote the localities only and give no idea about acreage.

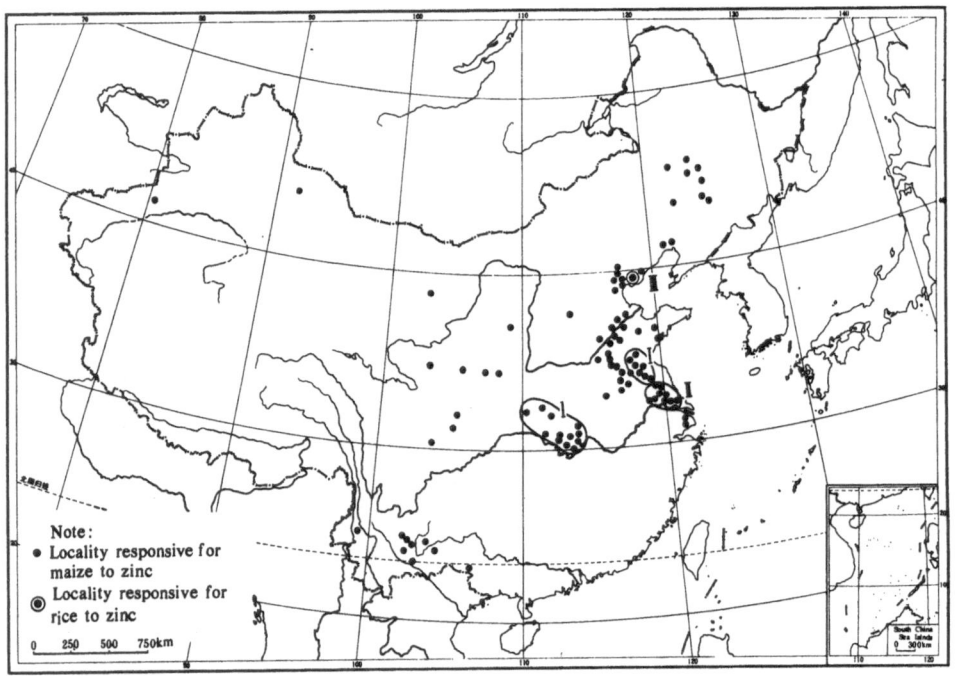

Fig. 2 Localities where response to zinc were confirmed by field experiments

Thus, zinc-deficient paddy soils so far reported may be classified into three categories, i. e.:

i) Zh-deficient calcareous paddy soils, e.g. paddy soils of northern and central parts of Hubei Province.

ii) Zn-deficient neutral paddy soils of swampy soil region, e.g. some paddy soils of Yangzhou Prefecture of Jiangsu Province.

iii) Zu-deficient paddy soils of coastal saline soil region, e.g. paddy soils of Baigezhuang at the Bohai Gulf of Hebei Province.

Although the soils at risk are commonly alkaline, paddy soils with impeded drainage or permanently flooded may be deficient in zinc also when soil reaction is less critical.

RESPONSE OF RICE TO ZINC FERTILIZER

Field trials have been conducted on calcareous paddy soils derived from alluvial deposit of the Huanghe River at Tongshan County in northern Jiangsu Province. The available zinc of the soil is 0.3 ppm as extracted with DTPA solution. No deficiency symptom of zinc can be observed on rice plant. It shows in the experiment that there is an increase of yield of rice from 6-23 % by application of zinc(5). Different methods of application of zinc fertilizer are adopted in the experiment; they include soaking seed with $ZnSO_4$ solution, foliar spray with $ZnSO_4$ solution, dipping seedling roots with ZnO suspension during transplanting, applying in seedling bed with ZnO during transplanting and application of $ZnSO_4$ in soil.

It seems that soaking seed or dipping the seedling roots are the most convenient and economic methods in the application of zinc for rice. Zn-fertilization may increase the numbers of tillers, grains per ear, and the grain weight, and decreases the percentage of empty grains (Table 3).

Table 3 The effect of zinc on the growth and yield of rice
(Plot experiment)

Treatment	Tillering	Grain/ear	Empty grain (%)	Weight of thousand grain(g)
C K	0.56	86.8	19.0	23.2
Dipping seedling with ZnO suspension	0.71	90.8	11.2	25.2

CONCLUSION

1) Zinc deficient paddy soils are mainly calcareous in reaction and are found in North China.

2) Zinc-deficient paddy soils may be classified into three types:
   i) Zn-deficient calcareous paddy soils.
   ii) Zn-deficient neutral paddy soils of swampy soil region.

iii) Zn-deficient paddy soils of saline soil region.

3) Zinc fertilization is beneficial to growth and production of rice. Soaking seeds or dipping seedling roots are the most convenient and economic ways to apply zinc fertilizer.

REFERENCES

(1) Tanaka, A., Yoshida, S., Nutritional disorders of the rice plant in Asia. Tech. Bull. IRRI, 10(1970).
(2) Rice Experiment Station, Linyi, Shandong Province, Symptoms of zinc deficiency of rice. Turang Feiliao, 6, 43-44(1976). (in Chinese)
(3) Institute of Soils and Fertilizers, Academy of Agricultural Sciences of Hubei Province, Report on zinc fertilizer experiment. Hubei Agricultural Sciences, 4, 4-5(1978). (in Chinese)
(4) Liu Zheng, Tang Li-hua, Zhu Qi-qing, Han Yu-qin, Ouyang Tao, Content and distribution of trace elements in soils of China. Acta Pedologica Sinica, 15, 138-150(1978). (in Chinese with English summary)
(5) Liu Zheng, Zhu Qi-qing, Tang Li-hua, Han Yu-qin, Xu Jun-xiang, Yiu Chu-liang, Qian Cheng-liang, Zinc in soils and zinc fertilization. In "Proceedings of /Symposium on Trace Elements" (ed. by Liu Zheng, Wu Zhao-ming), Science Press, Beijing, 154-161(1980). (in Chinese)

A REPORT ON THE SALING-SODIC SOILS OF BAIGEZHUANG
AND THE ZINC DEFICIENCY OF PADDY CROP

Wang Zhong-lian, Wang Wan-zhang
(Agricultural Reclamation Institute of
Hebei Province, Tangshan)

Qi Ming
(Chinese Academy of Agricultural Sciences, Beijing)

THE NATURAL ENVIRONMENTS AND THE CHARACTERISTICS OF THE
SOIL FORMATION OF BAIGEZHUANG DISTRICT

Beigezhuang is located on Luanhe   Delta of the Bohai Gulf, which belongs to the partly dry and partly wet monsoon belt. The parent materials of the soils are delta deposit and marine deposit, with a texture of mostly silt loam.

The types of soils developed in Baigezhuang are closely related to the elevation of the land and the time elapsed from the influence of the sea water. The belt about 10-20 km from the low tide line where the elevation of the land is round 2.5 m, used to be drowned before the dike was built. They were flooded periodically be the high tidal water. The salinity of these soils are very high with a salt content of about 2.5 %. No vegetation and germination will take place even with cultivation.

The belt further away from the sea with an elevation of above 3.0 m is effected by rain, desalinization occurs gradually, salt content decreases to 0.6-1.5 %, grows salt-tolerant grasses such as yellow beard fleabane (Suaeda salapull) and horsewhip weed (Aeluropus littoralis). The cation exchange capacity was generally low, below 20 m.e./100g, mostly composed of calcium and magnesium with more than 15 % of sodium ions (Table 1).

Table 1 Cation exchange capacity and exchangeable cations
of the soils before reclamation in general

Depth (cm)	CEC (m.e./100g)					pH
	Total	Ca^{2+}	Mg^{2+}	Na^+	K^+	
0-10	14.98	4.66	4.75	4.75	0.82	8.0
19-40	15.85	4.73	4.47	6.24	0.41	8.1

Available zinc of the soils of Baigezhuang is generally low, round 1.0-05 ppm, with many fields lower than 0.5 ppm, as shown in Table 2.

Table 2 Available nutrients of the soils

Sample No.	Soil type	Organic matter (%)	Total nitrogen (%)	Available phosphorus (ppm)	Available zinc (ppm)	Carbonate (%)
1	Heavy loam	0.94	0.036	20.2	0.42	6.70
2	Light loam	0.86	0.036	8.0	0.49	3.13
3	Heavy loam	1.11	0.071	17.0	0.55	6.84

At present, land with salt-tolerant grasses of this area mostly were reclaimed for paddy growing and obtain good yields, but some large area of lands will not grow good paddy with the ordinary reclamation measures. When the paddy seedling recovers from transplanting, the seedling grows slowly with fewer tillering, the plant becomes small and dwarf with brown leaves, root growth is reduced with fewer white roots, local people call that kind of symptom as "withdrawal (shrinkage) of paddy seedling". During the year f 1959, the State Farm of Baigezhuang planted more than 6,700 ha of paddy with one-third of the fields effected by "withdrawal" and caused a great loss to the harvest. Such symptom appears every year and effects further expansion of reclamation.

Recently, through the diagnosis of the paddy plant and analysis of the soils and plants with field experiment on the application of zinc sulphate, it has demonstrated than the "withdrawal of paddy seedling" of this area is caused by zinc deficiency. Degree of deficiency can be divided into four categories:

1) Growing normally: Available zinc in soil, as determined by DTPA + TEA method after W. L. Lindsay and W. A. Norvell, is greater than 1.0 ppm, and zinc content of the seedling is greater than 20 ppm.

2) Slight zinc deficiency: Young leaves are green, older leaves at the base are dotted with brown specks at the tillering period. In general, this condition will last for half month. Yield will be increased by application of zinc sulphate but not very much, available zinc in soil is between 1.0-0.5 ppm, zinc content of the leaves between 20-15 ppm.

3) Moderate zinc deficiency: Leaves are light green, one or two leaves near the base show large amount of brown specks, the midribs of the new leaves of some of the plants have chlorosis. The growth of the plant is stunted and this symptom will last for 20 days or more. Symptoms will disappear gradually with the monsoon rain. Available zinc of the soil is round 0.5-0.3 ppm zinc contents of the leaves are around 10 ppm. Effect of sulfate of zinc is significant.

4) Severe zinc deficiency: The midribs of most of the new leaves, have chlorosis. All of the older leaves are covered with brown specks, leaves on the base become brown with their points withered and brittle. The growth of the plant is checked and leaves small; some of the plant will not bloom or die; yield will be effected seriously. The available zinc of the soil is round 0.3 ppm, zinc content of the leaves will be below 10 ppm.

The increment of yield due to the application of zinc varies with the degree of zinc deficiency of the soil. Results of 39 field experiments show a general trend as follows: for slightly deficient soils, the paddy yield increases 7.1 % (438 kg/ha); for moderately deficient soils, the paddy yield increases 30.9 % (1.3 t/ha), and for severely deficient soils, paddy yield increases more than two-fold (2.6 t/ha).

Zinc deficiency of the paddy plant is closely correlated with yield. A generalization from many yield data and plant analysis of zinc content, as extracted with 1 N HCl, is shown in Fig. 1.

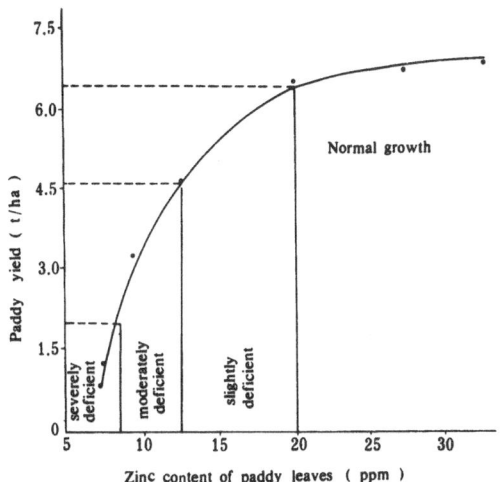

Fig. 1   Relationship between paddy yield
and zinc content of paddy leaves

THE EFFECT OF THE KIND OF ZINC FERTILIZER
AND THE METHOD OF APPLICATION

Experiments on moderate zinc deficient soils, an average available zinc of 0.4 ppm, foliar application, soil application or root dipping are all effective (Table 3). Soil application of zinc sulphate at 15 or 30 kg/ha the best, increase in yield of 18.4 %; soil application of zinc oxide at 15 kg/ha the least, increase in yield of 14.3 %, as shown in Table 3.

Table 3   The effect of kind of zinc fertilizer and
the method of application

Treatment (kg/ha)	Yield (t/ha)	Increase in production (t/ha)	Increase in production (%)
Check	6.69	---	--------
Zinc sulfate (15)	7.92	1.23	18.4
Zinc sulfate (30)	7.92	1.23	18.4
Zinc oxide	7.65	0.96	14.3
Foliar application 0.2 % ZnSO$_4$, twice	7.68	0.99	14.8
Root dipping 1 % ZnO suspension	7.77	1.08	16.1

LSD (0.05) 8.25 kg;   (0.01) 12.30 kg

643

There is some zinc in the paddy seed, but when the available zinc of the soil is very low, seedling in the seed bed will show zinc deficiency symptom of yellowing and poor germination with missing spots. The application of zinc fertilizer to a seed bed with available zinc of 0.36 ppm will improve the quality and the standing of the seed bed (Table 4).

Table 4 The effect of zinc fertilizer on seed bed

Treatment (kg/ha)	Germination (%)	Height of seedling (cm)	Number of leaves	Sick seedling (%)	Weight of 100 seedlings (g)	
					Fresh weight	Dry weight
Check	67.8	8.4	2.7	100	19.2	3.56
Zinc sulfate (30)	91.5	10.2	4.2	17.0	28.0	5.63

When we transplanted the seedlings from the seed bed which have been fertilized with zinc sulfate in comparison with the seedlings from the check plot, we found that the former grows nomally, while the latter recovers slowly and "withdrawal" developed. An inspection on the eighteenth day after transplanting, we found that the former grew 22 cm height with 94 % alive, and the latter was only as high as 12 cm with 67 % alive.

The residual effect of zinc fertilizer applied in quite significant. In the fields where zinc sulphate was applied once in a certain year, yet its effect can be seen in the next year. An experiment in 1978 with three replications were conducted for the application of zinc sulfate and zinc oxide before transplanting of paddy seedling. On the same plots paddy seedlings were transplanted again in 1979 without applying any zinc fertilizer. The residual effect of the zinc applied in 1978 was significant, as shown in Table 5.

Table 5 Residual effect of zinc fertilization on paddy

Treatment (kg/ha)	Available zinc in soil (ppm)	Zinc content of the plant (ppm)	Yield (t/ha)	Yield increase t/ha	Rate of increase (%)
Check	0.49	15.0	6.69	-----	-----
Zinc sulfate (30)	0.90	26.0	7.50	0.81	12.1
Zinc oxide (15)	1.02	26.0	7.45	0.76	11.4

FACTORS AFFECTING THE ZINC DEFICIENCY OF PADDY
CROP AND ITS REMEDY

1) It is quite evident that the newly leveled land is more apt to suffer from the "withdrawal of paddy seedling", especially when the soil is heavy, so we apply zinc to such spots and get uniform stand.

2) Antagonism between phosphorus and zinc has been well studied. Experiment at Baigezhuang shows that when the zinc status is marginal, the effect of phosphorus application is very significant. Without applying phosphorus, paddy grows normally; but when phosphorus is applied, zinc deficiency symptom becomes more numerous. The differences are: without application of phosphorus, plant height 30 cm, number of sick plant 3.7 %; with the application of phosphorus fertilizer, plant height increases to 35 cm, rate of growth increases, the number of sick plant raises to 28 %. Treatment with both phosphorus and zinc, the average plant height is 44 cm and no sick plant appears.

3) Experiment with zinc deficient soils shows that the hills with less seedlings suffer more from zinc deficiency than those with more seedlings. The causal effect has not been studied. Therefore, it is advisable to transplant more seedlings in each hill at the field where the zinc deficiency is apt to occur. The yield eventually increases. Experiment on a moderately deficient soil, the effect of the number of seedling in each hill shows that the yield of large hill planting, i.e.more seedlings in each hill,is 5 t/ha, while the yield of small hill planting, i.e. less seedlings in each hill, only 3.5 t/ha.

4) The salt contents of the newly reclaimed soils are relatively high, which will encourage the development of zinc deficiency of paddy plant, therefore, in the old practice we dig open ditches for drainage and keep shallow water irrigation for nromal growth.

From the observation, we recognized that the depth of water layer above the soil surface makes quite a difference in the degree of zinc deficiency. There are instances: the plot where water layer is kept above the soil surface between 2-3 cm in depth, no sick plant is found; while the depth of water layer above the soil surface raises to 10 cm, the number of sick plant raises to 54 %. The causal effect has not been studied.

5) From many years experiences at Baigezhuang we have learned that when the weather before full tillering was cold and dry, serious "withdrawal of paddy seedling" would probably appear, and naturally low yield be obtained. Therefore, our common practice was late transplanting and only robust seedlings were planted to decrease the symptom of "withdrawal of paddy seedling". But now we can transplant early and obtain bumper crop by soil analysis for available zinc and apply zinc sulfate accordingly.

ACKNOWLEDGEMENT

Appreciation is expressed to Professor Zhang Nai-feng, Vicedirector, Soils and Fertilizers Institute, The Chinese Academy of Agricultural Sciences, for hisinstruction in the experiments and translation of the report.

SOME METHODS TO MINIMIZE ZINC DEFICIENCY IN TRANSPLANTED WETLAND RICE

A first comparison of land ridging and zinc application; other possibilities

R.Y. Reyes, Robert Brinkman
(International Rice Research Institute, P.O. Box 933, Manila, Philippines)

INTRODUCTION

Zinc deficiency of wetland rice generally occurs in two main situations: in calcareous soils, which may or may not be very poorly drained, and in noncalcareous soils that are very poorly drained or remain reduced for long periods by perennial flood irrigation. The present paper deals with the second kind.

Zinc deficiency of wetland rice can be minimized or remedied by application of soluble zinc salts, chelates or oxide, or by improving the availability of soil zinc, for example through oxidation or acidification.

After the problem was initially recognized, large amounts of zinc salts were broadcast and incorporated in the field before transplanting, e.g., 100 kg/ha Zn as $ZnSO_4$[9]. Use of zinc oxide for broadcasting, coating the seeds for direct-seeded rice with zinc oxide powder or application of zinc oxide suspension to the roots of seedlings before transplanting (seedling dip), have made possible a drastic reduction of effective dosage, to amounts of the order of 4-8 kg Zn per ha per crop. Recent data by Bautista, Bunoan and Feuer[1] indicate economic yield increases from use of 1 kg ZnO per ha as a seedling dip or 5 kg $ZnSO_4$ broadcast per ha on large areas of slightly or marginally zinc-deficient soils.

Cost in itself does not seem to be major objection to zinc use at present. The present (1980) cost to the Philippine farmer of 4 kg Zn (as oxide) is about 20 US dollars, roughly 4 percent of the gross value of a 3.5 t/ha crop and a fraction of the cost of chemical or hand weeding, or of the major fertilizers required.

Still, there are both immediate and long-term objections to continual applications of zinc compounds. One is the very low efficiency of use. A rice crop with total dry matter above ground about 12 t/ha would contain about 0.3 kg Zn per ha, part of which is returned to the soil with the straw or its ash. Zinc removal would thus be of the order of 0.2 kg/ha per crop at a fertilizer efficiency of about 5 percent, or less if soil zinc is taken into account. In slightly zinc-deficient soils, the use efficiency of the low zinc applications (1-2 kg/ha) is of the same order of magnitude. As found by Kirschey (pers. comm. 1979) in earlier experiments at Natividad, the residual effect of Zn applications is a small fraction of the effect in the first season. It would therefore be less economic to apply a large storage dose than the usual regular amounts.

Regular zinc sulfate or oxide applications may result in a slow increase of total soil Zn, but also of cadmium and other heavy metals, which the commercial zinc compounds contain in small but variable proportions. Such additions to the soil might not remain within safe limits indefinitely.

Several countries do not have indigenous zinc supplies. An interruption of supply even for one or two seasons could drastically lower yield levels over large areas of rice land.

Successful use of the seedling dip method requires careful handling of the seedlings before transplanting. Much of the zinc oxide suspension washes off from the roots if the bundles of seedlings are placed in a wet field, even if the water is shallow; results of the method are often erratic.

Possible alternative methods to ameliorate zinc deficiency in wetland rice include drainage and aeration of the upper soil horizons before puddling; providing a degree of internal drainage (percolation) during crop growth; growing the crop in an aerated or partly aerated soil; acidification of the soil. Some of these alternatives are impractical or costly or cause other kinds of trouble.

Most of these methods would require a major change in soil hydrology. This would be difficult and expensive to achieve in the poorly or very poorly drained soils on level or concave landforms that constitute much of the zinc-deficient land. Where perennial irrigation causes continuously wet conditions or where both irrigation and drainage are possible, however, an alternation of wetland rice and dryland crops would eliminate zinc deficiency induced by continuous reduction.

Observations in different field experiments on zinc deficiency and in zinc-deficient farmers' fields had shown that often, plants located immediately adjacent to or on the lower slope of paddy bunds, or on high spots within the plot or the field, tended to grow better and mature earlier than the plants in level, properly inundated parts. No quantitative data on this effect were available, however. It seemed useful, therefore, to test the effect of ridging treatments and of drainage and aeration in comparison with conventional zinc oxide applications.

Two other approaches may minimize zinc deficiency or increase the efficiency of applied zinc compounds in transplanted wetland rice. Both involve charging the seedlings before transplanting with zinc in amounts adequate for further growth particularly in the crucial early stages. One possibility, without use of zinc applications, is to produce relatively old seedlings in seedbeds outside the zinc-deficient area. This may not be practicable in calcareous soils with a high pH in semiarid areas, where the zinc deficiency problem may be regional. In humid climates as in the Philippines, however, where zinc deficiency mainly occurs on neutral, slightly or noncalcareous soils in poorly or very poorly drained parts of the landscape, the approach may have promise.

It is also possible to charge seedlings with zinc by treating the seedbed with Zn compounds. Yoshida et al[9] applied zinc at 100 kg/ha to the nursery bed (equivalent to about 5 kg/ha field), in comparison with 10 and 100 kg/ha applied to the field, on a zinc-deficient, calcareous soil near Lahore, Pakistan (Kala Shah Kaku Rice Research Station). They found no significant differences between these treatments, each of which increased yields by roughly 2 t/ha (50%). Yield increases of about 1.5 t/ha were obtained by Mawardi et al[8] in Egypt with $ZnSO_4$ applications equivalent to 10 kg Zn per ha to the main field or 20 kg Zn per ha to the seedbed, and with Zn (EDTA) applications of 1 kg/ha to the seedbed, at Sakha Exp. Res. Station, presumably on a poorly drained, calcareous soil.

Castro[2] in a preliminary greenhouse trial corrected zinc deficiency in transplanted rice by application of zinc sulfate to the seedbed at a rate of 20 kg Zn per ha, equivalent to 1 kg Zn per ha main field, or by coating the germinated seeds with ZnO at about the same rate (1% of dry seed weight) before sowing in the seedbed.

These application rates should be compared with the Zn uptake by a rice crop, which is less than 0.5 kg/ha. Moreover, much of this is derived from the soil even with Zn fertilization as shown by isotope uptake experiments in Hubei Province, China (Shen Zhongquan, pers. comm. 1980). Very low amounts of zinc compounds, applied to the seedbed, may therefore be feasible and effective. A high efficiency of uptake might be achieved by somewhat older seedlings than usual, which would have more time to thoroughly explore a seedbed treated with low doses of zinc oxide. During the current wet season, an experiment is in progress to compare the effects of growing seedlings to different ages on a soil with adequate zinc availability and on a zinc-deficient soil amended with small amounts of zinc oxide.

THEORETICAL ASPECTS

Equilibrium zinc activities may be governed by the solubility of its sulfide, both in aerobic and in anaerobic soils[7]. Sulfide equilibria combined with some kinetic considerations can be used to explain the following qualitative observations about the occurrence and correction of zinc deficiency.

- the slow onset of zinc deficiency after the start of soil reduction, generally affecting the second or third and later crops rather than the first crop.

- the rapid improvement of a zinc-deficient crop by temporary drying of the soil.

- the generally lower application rates for Zn as oxide than as sulfate that are required to remedy deficiency and the failure of Zn sulfate even at fairly high rates to remedy deficiency in certain severe cases, especially in soils with high contents of decomposed organic matter.

- the small amounts of Zn oxide needed as seed coating or root dip compared with the rates needed to achieve the same results with broadcast and incorporated zinc oxide.

- the low residual effect of zinc applications in perennially wet soils.

In aerobic conditions, sulfide activities in soils are very low. In anaerobic soils, they are limited by the FeS solubility product and the generally moderate Fe^{2+} activities in the soil solution during the first weeks of inundation, and formation of ZnS proceeds very slowly if at all. After several months, or earlier in soils with much easily decomposed organic matter, continued sulfide production may bring about sufficient precipitation of FeS to lower the Fe^{2+} activity to much lower values. Then, the sulfide activity may rise with a consequent decrease in equilibrium Zn^{2+} activity to below a deficiency limit. Once air re-enters the soil, sulfide activities drop and the equilibrium Zn^{2+} activity is increased again to above the deficiency level.

When zinc sulfate is broadcast and incorporated in a zinc-deficient, puddled wetland soil, the Zn is evenly distributed throughout the soil in soluble and exchangeable form. Then, the rate at which the zinc is immobilized would depend on the rate of sulfide formation in the soil, after an initial very rapid precipitation by the sulfide already present. In extreme cases, this initial precipitation could remove most or all of the applied zinc.

If ZnO is broadcast and incorporated, it is distributed much less finely - for example, as grains 0.1 mm in size and at 1 cm intervals. Then, the proportion immediately precipitated would be lower because the movement of the sulfides in the soil solution is diffusion-controlled. The rate of subsequent immobilization could be lower than for soluble Zn, too, especially in conditions of rapid formation of sulfides by anaerobic microbial activity. This parallels the benefits from slow-release forms of other fertilizers.

Where diffusion of sulfide ions is indeed the rate-limiting process, the rate at which a given dose of Zn compound is immobilized decreases drastically with increasing distance between the particles. If in a given soil, ZnO broadcast at 2 kg/ha and distributed with a mean distance of 1 cm would be immobilized in three months, essentially all the ZnO would remain available for the whole growing season if it were placed at 20 cm mean distance, as in seed coating. This effect parallels the benefits of placement of other fertilizers.

The ZnO remaining from a previous crop, whether broadcast or placed, would be distributed more finely by ploughing and repeated puddling for the next crop, which would make it liable to more rapid immobilization.

MATERIALS AND METHODS

In a field experiment on a zinc-deficient Tropaquent, loamy, mixed, isohyperthermic, in the Philippines, the effects of two land management treatments were compared with low and high ZnO applications to the field and with root dipping of seedlings in a ZnO suspension.

A brief description of the soil follows, adapted from Haupenthal[4]. Location Barrio Canarem, Natividad, Pangasinan Province, Philippines, 16° 2.5' N, 120°46' E. Flat depressional area in nearly level alluvial plain, part perennially wet, part wet for most of the year, bunded, used for irrigated wetland rice. Dark greenish grey (5 G 4/1) silt loam with few fine faint dark reddish brown mottles to 60 cm, unmottled below that depth. Fine roots common to 22 cm, few to 40 cm. Some analytical data are listed in Table 1.

In two treatments, powdered ZnO was broadcast and incorporated before transplanting at rates of 4 and 8 kg Zn per ha ("low Zn" and "high Zn"); one treatment ("zinc dip") consisted of dipping the roots of the seedlings into a ZnO suspension, 4% w/v in water. This requires about 4 kg Zn per ha, the same as the "low Zn" treatment.

The experiment was done in the dry season of 1980, in three replications. Plot size was 3 x 5 m, plots were separated by bunds. Urea, triple superphosphate and KCl were applied at rates of 100 kg N, 60 P_2O_5, 60 K_2O per ha. Seedlings of IR44, 45 days old, were planted 2-3 per hill at a spacing of 30 x 13 cm (250,000 hills/ha); weed and insect control measures were as usual in the area.

A parallel experiment using surface soil from the same location in drums, 0.6 m diameter and 0.5 m high, three replications, was set up at the Inter-

Table 1 Analytical data for a soil profile[1]) in Bo. Canarem, Natividad adjacent to field experiment, and for surface soil material from field and drum experiments

Depth cm	pH (H₂O) 1:1 w/v	EC mS/cm	Exch. K	Na	1/2Ca	1/2Mg	CEC	Org. C	Total N	Clay	Olsen P mg/kg
					------ mmol/kg ------			------ % ------			
0-22	7.7	0.10	1.7	5	212	74	270	1.2	0.17	5.7	4
22-40	6.2	0.06	0.6	4	154	66	273	0.7	0.09	11.4	6
40-60	5.8	0.05	0.6	4	133	54	280	0.8	0.08	13.2	16
60-85	5.4	0.05	0.5	4	122	50	267	0.9	0.08	15.7	7
85-100	5.7	0.06	0.5	3	129	53	258	0.8	0.08	17.8	8
surface soil											
field experiment	6.9							2.4	0.36		5
drum experiment	6.9							1.8	0.27		12

Org. C = Walkley-Black; total N = Kjeldahl
K&P Zn 0.09 and 0.15 mg/kg in surface soil material from field and drum experiments, respectively
K&P Zn = 5 min. extraction by HCl 0.05 mol/1, soil/solution 1/2 w/v(6)

Cr total 100 mg/kg soil, extracted by NH₄ ac. 0.5-0.8 mg/kg, in soil solution from wet ploughed
layer less than 0.05 mg/1; Fe extracted by NH₄ ac. 180-470 mg/kg; pH 6.6-6.8; Eh 90-100 mV in
puddled ploughed layer during 1980 wet season

[1])Summarized from Haupenthal(4)

national Rice Research Institute, Los Baños, Philippines, with the expect-
ation that test conditions there could be more closely controlled, and in
order to include a treatment with drained soil material. Variety was IR34,
20-day-old seedlings.

Treatments were ZnO incorporation before transplanting, 0, 1, 2, 4 and
8 mg Zn per kg soil; "ridged high" and "ridged low" as in the field experiment, but with a narrower central furrow and with ridges adjoining the walls
of the drums; and "aerobic", with subsurface drainage, without standing water
except for short periods after heavy rain. Fertilizers were incorporated
before transplanting: 100 mg N, 50 mg P_2O_5 and 50 mg K_2O per kg as urea,
triple superphosphate and K_2SO_4. The soil material was kept wet from the
time of collection.

RESULTS, DISCUSSION AND OUTLOOK

During early growth, there was irrigation water shortage in the field
so that the surface soil in all treatments was aerated for a short period.
This must have increased zinc availability: in earlier seasons, crops
failed in this field without zinc applications. Both planting and harvest
in the field experiment were late compared with surrounding fields, and there
was severe rat damage in the last stage before harvest. Mean grain yield was
3.6 t/ha (s = 0.4) without significant treatment effects. Yields of the best
treatments in an earlier dry season were about 6 t/ha.

The drum experiment, too, suffered from hazards naively unexpected or
underestimated. Fallout from indiscriminate herbicide spraying beyond our
control damaged and weakened the plants more than once during the experiment,
with consequent insect and disease problems. Therefore, the results obtained
at 8 weeks after transplanting are used to estimate treatment effects.

Total dry matter and total Zn uptake (above-ground parts) in the field
experiment are listed in Table 2. Total dry matter increased by about 1.8
t/ha over control in the "ridged low" treatment as well as the low Zn (broadcast) and zinc dip treatments, and by 3.6 t/ha in the high Zn treatment.
Ridging and planting in the furrow adjacent to the ridges thus appears to
have a similar effect on total growth as the application of 4 kg zinc per
ha as ZnO, broadcast or as seedling dip. In contrast to this effect, the
total zinc uptake was not significantly increased in the ridged low treatment compared with the control, whereas all zinc treatments (both rates, 4
and 8 kg/ha) increased zinc uptake by about 140 g/ha (about 65 percent) over
control. Although the "ridged high" treatment produced more dry matter,
total zinc uptake was lower than in the control. Therefore, this treatment
does not appear to have promise compared with zinc application or "ridged
low" treatments.

Table 3 lists visual scores for zinc deficiency symptoms[5] and zinc
contents of plants, 8 weeks after transplanting, in the drum experiment.
The "ridged low" treatment scores better than the control but worse than all
zinc treatments. The same is the case with the zinc contents of the above-
ground parts. Thus, in the drum experiment the ridged low treatment seems
to be less effective than zinc applications. The "ridged high" and "aerobic"
treatments show no improvement over the control and do not appear to be
viable alternatives to zinc application or "ridged low" treatments in the
dry season.

The experiments are being repeated in the current wet season under more
closely controlled conditions. First indications are that in the wet season,

Table 2   Total dry matter and total zinc uptake at maturity
           in field experiment, dry season 1980

Treatment	Total dry matter above ground t/ha	Total zinc in grain and straw g/ha
Control	12.0	216
Low zinc	13.7	373
High zinc	15.6	360
Ridged low	13.6	259
Ridged high	12.9	183
Zinc dip	14.0	335
s.d.	0.4	27

Table 3   Deficiency scores and zinc content, 8 weeks after
           transplanting in drum experiment, dry season 1980

Treatment	Visual score of Zn deficiency[1]	Zn content mg/kg D.M. above ground
Control	4.5	20.5
Zn 1 mg/kg	2.3	33.3
2	2.3	35.8
4	2.0	37.0
8	2.0	46.2
Ridged low	3.3	23.3
Ridged high	4.5	16.6
Aerobic	6.0	16.8
s.d.		3.4

[1]Means of 3 replicates. Scores 1 = less than 1% of
leaves discolored or dead; 3 = 1 to 5%; 5 = 5 to 25%;
7 = 25 to 50%; 9 = 50 to 100%

the "ridged high" treatments shows better results than control but the "ridged low" treatment does not. Weed incidence is greater and more diffi-cult to control in the ridged treatments than in the level fields, in both dry and wet seasons.

Ridging a zinc-deficient paddy field and transplanting in the furrows, adjacent to the ridges, in the dry season or on the edges of the ridges in the wet season, seems to have a favorable effect on growth and zinc uptake of wetland rice. In the first comparisons reported here, the method was less effective than zinc oxide applications, whether broadcast and incorpo-rated or as zinc oxide seedling dips. This point, combined with the greater weed problems, seems to eliminate the ridging methods as practical solutions on zinc-deficient, very poorly drained soils.

First estimates of crop development suggest that zinc deficiency in the wet season is severe, both in the field and in the drum experiment. Plants in the control plots have virtually all died; those in the control drums are barely surviving. In the field, only treatments with high-strength (4% w/v) zinc oxide seedling dip and with 8 and 16 kg Zn as oxide per ha main field are expected to give high yields. In the drum experiment, the same applica-tion rates give similar results, while the low rates (2 and 4 kg/ha) are expected to yield considerably less.

Seedlings of different ages performed differently. The 20-day-old seed-lings did not survive in the main field, whether produced in a seedbed on soil with adequate available zinc or on a zinc-deficient soil with or with-out addition of 8 kg Zn per ha seedbed. Sixty-day-old seedlings both from a soil with adequate zinc and from a deficient soil treated with 8 kg Zn per ha seedbed are expected to give low yields, the latter with a considerable delay in maturity as well. Only the 40-day-old seedlings from a soil with adequate zinc are expected to produce a moderate yield.

One other kind of treatment would merit an experiment: spreading small amounts of zinc oxide or sulfate on the soil surface after transplanting, along the rows or in alternate inter-row spaces, and protecting it against incorporation. The zinc would then largely remain in the thin oxidized layer that covers most wetland rice soils except for those with strong upwelling of groundwater. Thus, it should escape immobilization in the reduced zone. The roots of the rice plants should reach the zinc soon after transplanting. Downward diffusion of Zn from the sulfate might be enough to guide plant roots to the soil surface at an early stage. Zinc oxide would allow much less downward movement of zinc into the reduced zone. Although this would prevent immobilization, it might also delay to an unacceptable degree effective use by the young rice plant. Surface application of Zn sulfate has been reported as more effective than incorporation[3], but no critical comparison between the use efficiencies of suface-applied and incorporated zinc oxide has been published to our knowledge.

Fig. 1 Wetland rice in a zinc-
deficient field adjacent to a
paddy bund. Lower half: mature
plants growing against the bund,
presumably with part of their
roots in the bund

Upper half: stunted plants,
still immature, growing further
from the bund

Fig. 2 Rice plants growing in the edges of furrows adjacent to ridges in
a "ridged, low" treatment. Field experiment Natividad, dry season 1980

REFERENCES

(1) Bautista, A.C., J.C. Bunoan Jr., R. Feuer. Development of the zinc
     extension component for irrigated rice in the Philippine Masagana 99
     production program. This Symposium.
(2) Castro, R.U. Zinc deficiency in rice: A review of research at the
     International Rice Research Institute. IRRI Res. Pap. Ser. 9,
     August 1977. 18 p. (1977).
(3) Giordano, P.M. Soil temperature and nitrogen effects on response of
     flooded and nonflooded rice to zinc. Plant Soil 52 (3), 365-372 (1979).
(4) Haupenthal, C. Studies on environmental factors influencing rice
     yields: relief-soil-drainage-rice yield interrelationship of reported
     problem soils in the Philippines. Preliminary report, Internat. Rice
     Res. Inst., Los Baños, Philippines. 101 p. (1979).
(5) International Rice Testing Program. Standard evaluation system for
     rice. 2nd ed. Internat. Rice Res. Inst., Manila. 64 p. (1976).
(6) Katyal, J.C., F.N. Ponnamperuma. Zinc deficiency: A widespread
     nutritional disorder of rice in Agusan del Norte. Phil. Agriculturist
     J. 58 (3 & 4), 79-89 (1974).
(7) Kittrick, J.A. Control of Zn^{2+} in the soil solution by sphalerite.
     Soil Sci. Soc. Amer. J. 40, 314-317 (1976).
(8) Mawardi, A., S. Ghaly, A. Serry. Zinc practices to nursery
     bed for efficient paddy crop production. Agric. Res. Review (Cairo)
     55, 157-163 (1977).
(9) Yoshida, S., G.W. McLean, M. Shafi, K.E. Mueller. Effects of different
     methods of zinc application on growth and yield of rice in a calcareous
     soil, West Pakistan. Soil Sci. Plant Nutr. 16 (4), 147-149 (1970).

DEVELOPMENT OF THE ZINC EXTENSION COMPONENT
FOR IRRIGATED RICE IN THE PHILIPPINE
MASAGANA 99 PRODUCTION PROGRAM

Aniceto C. Bautista
(Bureau of Agricultural Extension, Ministry
of Agriculture, Ouezon City, Philippines)

Juan C. Bunoan, Jr.
(Bureau of Soils, Ministry of Agriculture,
Manila, Philippines)

Reeshon Feuer
(International Rice Research Institute,
P.O. Box 933, Manila, Philippines)

Most Philippine soils used for irrigated rice-growing are moderately-
fine to fine-textured, high cation exchange capacity vertisols, and medium-
textured wet entisols, and inceptisols. Irrigation waters which come
primarily from stream diversions, contain considerable amounts of calcium,
magnesium and potassium. Because of increasing pH resulting from lime
residuals remaining from the transpiration and evaporation of irrigation
water, a zinc deficiency is intensifying although the dominant (60%)
currently grown variety, IR36, is tolerant of zinc deficiency (rates 2 by
the IRRI zinc screening method -- see Table 1). The second most widely
grown variety in the M99 program is IR42 (30%).

Prior to 1976 the junior author, who works closely with the technology
of the Masagana 99 program, was of the opinion that zinc deficiency in the
Philippines only was locally serious in small areas of high pH, except for
the Agusan del Norte Province where severe zinc deficiency occurred throughout the rice lands of that province. Soils used for rice in that province,
which is in the high rainfall climate Type II rain zone, are both high in
pH and high in organic matter (Katyal, 1972 [4]; Katyal and Ponnamperuma,
1974 [5]; Bunoan, 1974 [11]). And most of the IR rice varieties (IR20, IR28,
IR30 and IR34) in use in the early years of M99 had good tolerance to zinc
deficiency, except for IR26 which, however, was giving high yields on
farmers' fields [6, Table 9].

At the suggestion of Dr. F. N. Ponnamperuma, Head, Soil Chemistry
Department, IRRI, the junior author encouraged the Bureau of Soils in late
1975 to add an IR26 zinc treatment to one of their "verification" applied
research farm trials. The results showed a startling 13-cavan (44 kg)
average increase in yield per hectare (77 gin per mu*or 0.6 metric tons/ha)
from 1 kg of zinc oxide costing ₱15 (¥3 or US$2) for root-dipping the IR26
seedlings before transplanting from 17 farms, with soil pH 6.8 or above,

---

*One mu = 1/15 hectare; cavans (44kg) per hectare x 5.9 = gin
per mu (approximately) one gin = 0.5kg; gin per mu x 7.5 = kg per
hectare; kg per hectare ÷ 1,000 = metric tons per hectare.

in 11 provinces (Table 2). In addition several farms with pH below 6.8 also showed profitable zinc responses over the regular 90+30+30 dry season M99 fertilizer input.

During 1976 the junior author began checking with a field kit wet soil pH's of rice paddies showing zinc deficiency symptoms. He concluded that zinc deficiency was widespread on many Masagana 99 fields with pH 6.8 or above.

When the Bureau of Agricultural Extension-National Food and Agriculture Council-FAO Fertilizer Project reported lower yields in Mindoro Oriental Province irrigated farm trials from nitrogen plus phosphate plots compared with nitrogen-only plots, the junior author suggested field checking the sites to determine if wet soil pH's were 6.8 or above. If so, zinc deficiency was the probable cause of the yield differences because soil zinc combines very readily with phosphate, thus inducing zinc deficiency if available zinc was low. Field tests confirmed that most Mindoro Oriental wetland irrigated soils used for rice had pH's of 7.0 and above.

Dipping the roots of the rice seedlings before transplanting in a 2 percent zinc oxide slurry [1] or harrowing in 20 kg of zinc sulfate before transplanting corrected the problem. In addition yield increases of 10 to 20 cavans/ha from the addition of zinc were obtained*.

MASAGANA 99 APPROVES USE OF ZINC

Based on these findings the use of zinc oxide, 2 percent slurry (1 kg zinc oxide/ha at ₱15 /¥3, US&2/ was recommended and approved as an official component of the Masagana 99 "16 Step Masagana 99 Rice Culture" package of technology on irrigated rice soils with a wet soil pH of 6.8 and above.

To determine wet soil pH in the field, a brom thymol blue color-metric kit was developed, using a color chart prepared by the Department of Soils of the University of the Philippines at Los Baños, as well as preparing the standard brom thymol test solution with a toulene antibacterial agent, a locally cast spot plate, and a 60 cc plastic dropper bottle. Assembled in a plastic sack this simple pH kit was sold to the Bureaus of Agricultural Extension, Soils, and Plant Industry field staff by the hundreds for ₱10 (¥2.25, US$1.50) per kit, delivered. If a tested wet soil showed "blue" it was rated zinc-deficient. If the color showed "green" the soil was presumed not deficient.

Within a year it was determined that it was no longer necessary to test for pH. Most irrigated soils used for rice were showing increased yields of rice from the application of zinc by root dipping.

---

*Ho, C.T. 1977. Personal communication.

Table 1   Zinc tolerance of IR rices used in the Masagana 99
           rice production program in the Philippines.
           Approximate average ratings based on four
           seasons, 1978-79 IRRI tests

| Variety | Rating | |
	Numerical	Interpretative
IR20[1]/	3	MT[2]/
IR26[1]/	5	MS
IR32	3	MT
IR36	2	T
IR42	5	MS
IR44	5	MS
IR46	4	MT
IR48	5	MS
IR50	3	MT
IR52	5	MS
IR54	3	MT

[1]/Former M99 used varieties, shown only for reference
[2]/ T = Tolerant
    MT = Moderately tolerant
    MS = Moderately susceptible.

Table 2   Yield of IR26[1]/rice from root-dip of 2% zinc oxide,
           1976 dry season, Philippine Bureau of Soils
           "verification" farm trials, 11 provinces,
           17 farms.   Cavan = 44 kg palay

| Units* | Fertilizer/ha | | | |
	0	M99 (90+30+30)	M99 +Zn	Inc.
PHILS.: cav*/ha	76	111	124	+ 13
PROC.: gin*/mu*	448	655	732	+ 77
INT.: m.t.*/ha	3.3	4.8	5.4	+0.6

[1]/IR26 is moderately susceptible to zinc deficiency, rates 5
   by IRRI screening method (Table 1)

* 1 cavan = 44 kg x 23 = 1 metric ton
  1 gin = 0.5 kg, 1 mu = 1/15 ha;
          cavans (44 kg) per hectare x 5.9 = gin per mu (approx.)
     gin per mu x 7.5 = kg per ha, kg per ha ÷ 1,000 = metric
     tons per ha

BUREAU OF SOILS - BUREAU OF AGRICULTURAL EXTENSION
COOPERATIVE FARM TRIALS

During the 1977 wet season the Bureau of Soils and the Bureau of
Agricultural Extension carried out a series of cooperative zinc-phosphate
"verification" specialist trials on farms with the new IR36 very early
maturing (VEM) rice variety that was becoming popular. IR36 is tolerant
(rating of 2 - see table 1) of zinc deficiency and yielded very well
without zinc in earlier research trials. The results fully confirmed that
zinc deficiency was much more widespread on irrigated rice lands than
heretofore recognized (Table 3).

"ZINC INSURANCE", PHASE I FARM TRIALS

Concurrently, during wet season 1977, the Bureau of Agricultural
Extension sent out 10,000 packets of 10 g of zinc oxide, enough to dip the
roots of rice seedlings for 1/100 ha (1/7 of a mu). Extension rice
technologists were encouraged to ask farmers to try the zinc oxide seedling root dipping on a 1/100 ha plot, with a comparable adjacent 1/100 ha
plot without zinc. The farmer used his own choice of variety, inputs and
management. The instructions for the trials, named "Zinc Insurance",
Phase I, were printed in a four-page folder to spread knowledge of zinc
deficiency on irrigated rice as widely as possible[7]. Many extension
workers, however did not believe that 10 gm of zinc oxide was worth working
with; such a tiny amount of white powder couldn't possibly make any
difference in yield.

The results again startled us -- causing us to believe that most
irrigated rice in the Philippines was being grown on soils with some
degree of zinc deficiency, irrespective of variety or of pH status
(Table 4).

Two publications in 1977, one by R. U. Castro[8] and the other by
M. R. Orticio and F. N. Ponnamperuma[9], served to document the research on
zinc deficiency in rice, and, the extent of zinc deficiency of rice in the
Philippines. The chapter on "Zinc fertilization and behavior in flooded
soils," by D. S. Mikkelsen and Shiou Kuo in "The fertility of paddy
soils and fertilizer applications for rice, 1976,"[10] contains an
excellent discussion of the zinc problem in rice.

INDUSTRY - GOVERNMENT COOPERATE

With a formal request from the Minister of Agriculture to agribusiness, Bureau of Soils and Bureau of Agricultural Extension specialists
encouraged two agribusiness concerns to package high-grade zinc oxide in
1 kg packets for sale to rice farmers through their retail dealers of
fertilizer and/or insecticide outlets at ₱15 (¥3, US$2) per kilogram.
One kilogram is sufficient for treating the roots of rice seedlings for 1
ha for control of zinc deficiency, except where the deficiency is severe.
Two kg per hectare is used where the zinc deficiency is severe.

659

Table 3   Yield increase of IR36[1] rice from 2 percent zinc oxide
           root dip. Joint Bur. of Soils-Bur. of Agricultural
           Extension specialist "verification" farm trials,
           44 provinces, 98 farms, Philippines, 1977
           wet season

| Units | 69 + 30 + 0 | | | 69 + 0 + 0 | | |
	w/o Zn	w/Zn	Inc.	w/o Zn	w/Zn	Inc.
PHILS.: cav/ha	105	118	+ 13	101	114	+ 13
PROC.: gin/mu	620	696	+ 76	596	673	+ 77
INT.,: M.T./ha	4.6	5.2	+ 0.6	4.4	5.0	+ 0.6

---

[1] IR36 is tolerant of zinc deficiency, rates 2 by IRRI screening
method (see Table 1)

Table 4   Response of modern rice varieties to zinc[1]. "Zinc
           Insurance," Phase I, 1977 wet season, Philippines.
           Bureau of Agricultural Extension farmers'
           choice of varieties and inputs, 33
           provinces, 846 farms (irrigated)

Units	w/o ZnO	w/ZnO root dip	Inc.	Cost of ZnO	Value of Inc. (1980)
PHILS.: cav/ha	95	111	+ 16	₱ 15	₱ 960
PROC.: gin/mu	561	655	+ 94	¥ 3	¥ 192
INT.: M.T./ha	4.1	4.8	+ 0.7	US$2	$ 128

---

[1] 2% zinc oxide root-dip method; 1 kg ZnO/ha, plus 5 hours
labor to dip roots

Rice farmers began using zinc oxide but objected to the somewhat messy and time-consuming task of dipping the washed roots of rice seedlings before transplanting. Transplanting crews disliked handling the root-dipped seedlings. Consequently, we considered the use of zinc sulfate as an alternative method, because water-soluble zinc sulfate (heptahydrate, 22% zinc) can be broadcast, like basally applied fertilizer, before the last harrowing before transplanting. Although the estimated retail cost of zinc sulfate was only ₱5/kg, we were of the opinion that small-scale rice farmers would not be willing to spend ₱120 (¥24, US$16) per hectare for 20 kg of zinc sulfate, the minimum amount research scientists considered necessary to use to treat 1 ha of zinc-deficient wetland rice soil in the tropics (Bunoan, 1974[11]*).

BUREAU OF SOILS - INTERNATIONAL ATOMIC ENERGY COMMISSION "BASIC" RESEARCH

In late 1977, while the junior author was on an interagency consulting team working with the Pangasinan Provincial Masagana 99 staff in re-organizing to increase rice production in that province, Mr. Modesto Recal, Bureau of Soils provincial soil technologist, showed us a cooperative Bureau of Soils, Research Division-International Atomic Energy Commission replicated field experiment designed to study the residual effects of irradiated zinc applications to zinc-deficient wetland puddled irrigated soils growing transplanted rice.

We noted that there were no apparent differences in field appearance of the nearly mature IR26 rice under four of the many treatments: 2 percent zinc oxide root-dipping; 20 kg, 10 kg and 5 kg zinc sulfate per hectare harrowed in. The rice in the no-zinc plot and the adjacent area was very poor, confirming that the zinc deficiency of the site was severe.

We checked the pH of the wet soil with the brom thymol blue pH kit. It was 6.8 plus, again indicating zinc deficiency by our imperical pH method. We estimated that the zinc-treated plots would average to yield 80 cavans/ha (472 gin/mu; 3.5 M.T./ha), a figure later confirmed in a paper presented by Ms. C.M. Rosales, et al., of the Bureau of Soils, Research Division[12], presented at an International Atomic Energy Commission conference in Indonesia in September 1978 and in popular form in October 1978 in the Philippines (unpublished, see Table 5).

Immediately, on the spot, we designed a simple "verification" applied research trial of four treatments: check, 2 percent zinc oxide root-dip; 5 kg; and 10 kg zinc sulfate; and named the trial "Zinc Methods". We chose to use IR36, the newly popular zinc deficiency tolerant rice variety, as the test variety. By December 15 we delivered 50 units of the trial, complete with all inputs, to Mr. Recal. His results confirmed that a 5 kg zinc sulfate soil treatment before transplanting was a viable alternative to the 2 percent zinc oxide root-dipping method (Table 6). Another 50 of

---

*Ho, C.T., 1976; India. Personal communications.

these trials were sent to other regions. The results further confirmed that a 5 kg zinc sulfate soil treatment was adequate in most casts, but not in all, to correct the zinc deficiency for a crop of irrigated rice.

EXTENSION MAXIMUM EFFICIENCY FARM TRIALS

Again, concurrently with the Bureau of Soils "Zinc Methods" 1978 dry season applied research trials, the Bureau of Agricultural Extension provincial subject matter specialists implemented a series of 250 "maximum efficiency" evaluation trials on irrigated rice farms of farmer leaders. Five kilograms of zinc sulfate per hectare, harrowed in before transplanting, was the evaluation practice studied using the new high-potential IR42 or IR44 rice varieties, both of which are moderately susceptible to zinc deficiency (see Table 1). The results from 34 provinces, 64 farms, are shown in Table 7.

ZINC SULFATE RECOMMENDED

Based on these results the Inter-agency Masagana 99 Fertilizer Technical Advisory Committee recommended that 5 kg of fine granular zinc sulfate (hyptahydrate) be used per crop on all irrigated rice in the Masagana 99 rice production program. The Masagana 99 national Management Committee approved the recommendation on June 20, 1978.

Bureau of Agricultural Extension and Bureau of Soils specialists worked with the major agribusiness supply companies in designing packaging and writing instructions for the retail sale of fine granular zinc sulfate in 5 kg plastic sacks at ₱30 (¥6, US$4) per sack.

By late August 1978 limited supplies of zinc sulfate were available at retail dealer outlets.

As a result of mimeograph [13], radio[14] and news[14] releases on the use of zinc sulfate to increase rice yields, the demand for zinc sulfate soon exceeded supplies. Because local manufacture of zinc sulfate was insufficient to meet this new need, supplies of fine granular zinc sulfate (hyptahydrate) are being imported from at least three countries.

During wet season 1978 the Bureau of Agricultural Extension implemented two series of extension teaching zinc trials on irrigated rice farms. One thousand "Zinc Methods" trials, complete with all inputs and instructions, for a 1/100 ha area, with four treatment plots, were distributed through provincial rice program officers to rice extension technologists, at the rate of one trial per six extension technologists, or in terms of irrigated rice farms, one trial per 500 farms, or, in terms of hectares, one trial per 1,000 hectares (15,000 mu), for implementation on farms of farmer-leaders.

The results again confirmed the widespread existence of zinc deficiency on Philippine irrigated rice farms, and the highly profitable results that

Table 5   Yield response of IR26 rice to zinc, International
           Atomic Energy Commission - Bureau of Soils,
           Philippines, 1977 wet season, one farm,
           irrigated, Pangasinan Province,
           severe zinc deficiency[12]

Units	No Zn	2% ZnO root dip	Methods and amounts with 60 + 30 + 30 Soil-incorporated ZnSO$_4$ per hectare		
			5 kg	10 kg	20 kg
PHILS.: cav/ha	52	71	84	83	78
PROC.: gin/mu	307	419	496	490	460
INT.: M.T./ha	2.2	3.1	3.7	3.7	3.4

Table 6   Increase in yield of IR36 rice from "Zinc Methods"
           verification farm trial, Bureau of Soils,
           Pangasinan Province, 23 farms, 1978
           dry season (late) Philippines

Units	No Zn	2% ZnO root-dip	ZnSO$_4$	
			5 kg	10 kg
PHILS.: cav/ha	94	101	101	98
PROC.: gin/mu	555	596	596	578
INT.: M.T./ha	4.1	4.4	4.4	4.3

Table 7   Yield increase of IR42[1] and IR44[1] rices in "maximum
           efficiency" extension specialist evaluation farm
           trials from zinc, 34 provinces, 64 farms,
           5 kg ZnSO$_4$ soil-incorporated, 1978
           dry season, Philippines

Units	w/o ZnSO$_4$	w/ ZnSO$_4$	Inc.	Cost	Value of inc. (1980)
PHILS.: cav/ha	130	149	+ 19	₱30	₱ 1,140
PROC.: gin/mu	767	879	+ 112	¥6	¥ 228
INT.: M.T./ha	5.6	6.5	+ 0.9	US$4	US$ 152

---

[1] IR42 and IR44 are moderately susceptible to zinc deficiency,
    rate 5 by IRRI screening method (Table 1)

can be obtained from a ₱30 (¥6, US$4) investment for 5 kg of zinc sulfate per hectare (Table 8).

When Bureau of Agricultural Extension rice specialist Antonio H. Cruz further analyzed the results of the 10 kg zinc sulfate treatment for the moderately susceptible to zinc deficiency IR42 variety, he found that 45 percent of the 119 farms that used the IR42 variety, that is, 54 farms from 18 provinces, had twice the average yield increase per hectare, 30 cavans compared with 16 where 10 kg of zinc sulfate yielded more than 5 kg (Table 9).

The results shown in Tables 8 and 9 indicate that 1) zinc deficiency is very widespread in irrigated soils used for rice in the Philippines, 2) there is a likely need to compare rice varieties differing in tolerance to zinc deficiency, and 3) to further identify and delineate those areas of soils used for irrigated rice where a 10 kg zinc sulfate application rate is more profitable than the currently recommended 5 kg zinc sulfate per hectare rate [15].

To date, 1980, no additional "verification" applied research or extension specialist "evaluation" farm trials have been established to determine either items (2) or (3). It is hoped that with the regionalization of the Ministry of Agriculture into 12 regions, and influenced by the increasing price of fertilizers, further work on these two aspects of rice production efficiency through the use of zinc sulfate will be undertaken.

Upon review of the IR42 zinc results shown in Table 9, the second author suggested a simple 3-plot "Zinc Alert" extension farm trial that could be used locally to identify and delineate severe zinc deficiency areas. Each of the three plots are 1/100 ha in size with one end plot using the 5 kg zinc sulfate per hectare rate and the other end plot the 10 kg rate. The center plot would be the "0 zinc" check. With cooperating farmers furnishing inputs and management, 30 of these trials from one 5-kg sack of ZnSO₄ could be used in a local area at a cost of ₱2 each, ₱1 for the ZnSO₄ and ₱1 for two plastic sacks and instructions. This type of trial was suggested to provincial rice program officers in early 1979 (Cruz, A.H., et al., 1979[16]). To encourage use of the concept, 1,000 of these "Zinc Alert" farm trials are being distributed nationally, two per extension district, during the 1981 dry season by the Bureau of Agricultural Extension.

Informationally, however, the existence of these still undefined areas of severe zinc deficiency in 18 provinces of the Philippines (25% of all provinces) has been called to the attention of provincial rice program officers through an interagency four-page mimeograph entitled "Zinc Alert" (Cruz, A.H., et al., March 24, 1979). The "Zinc Alert" text is being reissued as part of the November 15, 1980 issue of the "Rice Specialist," a semimonthly mimeographed subject matter letter being sent airmail to the 250 provincial and district rice specialists by the Bureau of Agricultural Extension[17].

The April 15, 1979 issue of "The Rice Specialist,"[13] a four-to six-page "transfer-of-technology" extension subject matter letter prepared

Table 8   Yield increases from the use of zinc on irrigated rice farms
          in the Philippines. "Zinc Methods" Extension teaching
             farm trials, 75 percent IR42, 25 percent BPI-Ri4
               rice varieties, averages from 43 provinces,
                 186 farms. 1978 wet season, Bur. of
                   Agricultural Extension. Cavans
                        (44 kg) per hectare

Treatment	Yield, cav/ha	Inc.	Value (1980) ₱	Cost, ₱	Benefit/ cost ratio
No zinc	107	-	-	-	-
2% ZnO dip	117	+ 10	600	25[1]/	24/1
5 kg ZnSO$_4$	123	+ 15	900	30	30/1
10 kg ZnSO$_4$	127[2]/	+ 20[2]/	1,200	60	20/1

[1]/ Includes ₱10 for 5 hours labor to dip roots.

[2]/ Further analysis of these results by A.H. Cruz, Bur.of Agricultural
Extension, showed that available zinc in soils of 18 provinces is
so low that 10 kg of zinc sulfate per hectare should be used for
growing the moderately susceptible IR42 variety (see Table 9).

Table 9   Average yield increase of irrigated IR42 rice from zinc.
          Extension teaching "Zinc Methods" farm trials, Bur.
             of Agricultural Extension, Philippines, 18
               Provinces, 54 farms, where 10 kg ZnSO$_4$
                 outyielded 5 kg ZnSO4

Unit	No Zn	2% ZnO root-dip	ZnSO$_4$ 5 kg	10 kg
PHILS.: cav/ha	110	116	126	140
PROC.: gin/mu	649	684	743	826
INT.: M.T./ha	4.8	5.0	5.4	6.1
119 farms, IR42, 22 provinces, Philippines	115	124	134	135

by Bureau of Agricultural Extension national subject-matter specialists and sent airmail to field extension specialists twice a month, was entirely devoted to the use of zinc on irrigated ricelands in the Philippines.

To further popularize the use of zinc sulfate on irrigated rice farms the second series of 1978 wet season, extension teaching farm trials were called "Zinc Insurance, Phase II". These were patterned after the successful "Zinc Insurance, Phase I" 1977 wet season trials (Table 4), but consisted of 0.5 kg zinc sulfate, with instructions, for a 1/10 ha area, one or two paddies (1.5 mu), and an adjacent 1/10 ha area without zinc. All inputs and management were supplied by the cooperating farmer. One trial per municipality, a total of 1,500, were sent out so that there would be an observation-extension teaching zinc sulfate site in each rice-producing municipality (approximately 10 to 15 barangays, villages, with 100 rice farmers per barangay). The results are shown in Table 10.

Irrespective of whether it is extension teaching zinc farm trials complete with all inputs using the "16 Steps Masagana 99 Rice Culture" package of technology or the farmer's choice of inputs and management with modern rice varieties, the increase in yield from either the 2 percent zinc oxide root-dipping, 1 kg ZnO/ha, or the 5 kg $ZnSO_4$ soil incorporation method of correcting zinc deficiency on soils used for irrigated rice production in the Philippines, the results are the same, a 15 cavan/ha (89 gin/mu or 0.65 M.T.) increase in palay (paddy or rough rice) yield at a negligible investment for zinc[18, 15, 19, 20].

USE OF ZINC FOR RICE IN THE PHILIPPINES

In 1978, approximately 50,000 ha of irrigated rice lands (25,000 rice farms, 2.8% of irrigated rice lands) in the Philippines were treated with zinc, mostly 2 percent ZnO seedling root-dip method (Bautista, A.C., et al., February 1979[18]). Based on a 15 cavan (44 kg) increase in yield per hectare from the addition of zinc, the 1978 Philippine national production of palay was increased by 0.75 million cavans worth ₱36 million for a "farm gate" cost of zinc of ₱1.5 million.

In 1979, approximately 200,000 ha of irrigated rice lands, 11 percent of the 1.8 million ha of irrigated rice lands in the Philippines, were treated with zinc, mostly with $ZnSO_4$ at the 5 kg rate per hectare (Bautista, A.C. et al., January 1980[21]). At the expected 15 cav/ha increase in yield from the use of zinc, Philippine 1979 national palay production was increased by 3 million cavans (130,000 M.T.) worth approximately ₱170 million (¥35 millions, US$23 million) for a "farm gate" cost of ₱6 million (¥1.2 million, US$0.8 million).

The 3 million cavan increase from the use of zinc comprised one-half of the 6 million cavan increase in national production for the crop year ending June 1980 of 169 million cavans.

The potential of using zinc sulfate for increasing palay production in the Philippines is tremendous! The following projections in Table 11 illustrates this (Bautista, A.C. et al., April 1979[15]).

Table 10   Response of modern rice varieties to zinc.[1] "Zinc Insurance,
            Phase II," 1978 wet season, Philippines, Bur. of Agricultural
            Extension, farmer's choice varieties and inputs, 5 kg zinc
            sulfate soil incorporated with basal fertilizer per hectare
            rate, irrigated farms, two 1/10 ha plots, 27 provinces,
            195 farms

Units	w/o Zn	w/ZnSO$_4$ soil-incorporated	Inc.	Cost ZnSO$_4$	Value of inc. (1980)
PHILS.: cav/ha	97	113	+ 16	₱30	₱ 960
PROC.: gin/mu	572	667	+ 95	¥ 6	¥ 192
INT.: M.T./ha	4.2	4.9	+ 0.7	US$4	US$128

[1] 5 kg of zinc sulfate (22% Zn) per hectare rate (0.65 gin/mu rate)
harrowed in mixed with basal fertilizer

Table 11   Estimated increase in production of palay from the use of zinc
            on irrigated rice farms in the Philippines. Based on average
            increase of 15 cavans (44 kg) palay per hectare from using 5 kg
            ZnSO$_4$/ha on an estimated 1.2 million ha irrigated during wet
            seasons and 0.8 million ha irrigated during dry seasons

Percent adoption	Millions of cavans (44 kg) of palay			Millions of pesos annually	
	Wet season	Dry season	Increase	Value at ₱60	"Farm-gate" cost of ZnSO$_4$
10 (1980)	1.85	1.20	3.05	183	6
20 (1981)	3.70	2.40	6.10	366	12
50 (?)	9.25	6.00	15.25	915	30

[1] Approximate exchange rates: ₱7.5 = ¥1.5 = US$1 (1980)

If by the end of 1981, 20 percent of the irrigated rice lands in the Philippines were treated with zinc sulfate at the rate of 5 kg/ha, the expected potential annual increase in production is estimated at 6.1 million cavans (44 kg) worth more than ₱350 million (¥65 million, US$41 million) at July 1980 prices in additional gross income to Philippine rice farmers for a cost of only ₱12 million (¥2.4 million, US$1.6 million). This is a return of 30 times the cost of the zinc sulfate.

If by the mid 1980's half of the 1.8 million hectares of irrigated rice lands in the Philippines were treated with 5 kg of zinc sulfate per hectare, the expected annual increase in palay production would exceed 15 million cavans worth nearly one billion pesos at a "farm-gate" cost of approximately 30 million pesos.

Cited References

(1) _____, undated (1977?) R_x4 tablespoons of zinc. Reprinted from 1/73 issue of the IRRI Reporter, 2 pp. includes section on "Hidden Hunger of Rice". Ministry of Agriculture, Manila.

(2) Ho, C. T. 1976. Personal communication. Food and Agriculture Organization Special Fertilizer Program, Ministry of Agriculture, Manila.

(3) Bureau of Agricultural Extension. 1978 and later revisions. 16 Steps Masagana 99 rice culture. Ministry of Agriculture, Manila. 20 pp.

(4) Katyal, J.C. 1972. A study of zinc equilibria in flooded soils and amelioration of zinc-deficient soils of Agusan del Norte. International Rice Research Institute. 115 pp, 59 tables, 108 references. (Unpublished).

(5) _____ and F. N. Ponnamperuma. 1974. Zinc deficiency. A widespread nutritional disorder of rice in Agusan del Norte. Journal of Philippine Agriculture. 58(3,4):79-89.

(6) Orticio, M.R. 1979. Zinc deficiency: A widespread nutritional disorder of rice in the Philippines. Saturday Seminar. International Rice Research Institute, Philippines, 8 pp., 9 tables, figure (map), 17 references. (Unpublished mimeo.).

(7) Bureau of Agricultural Extension. 1977. "Zinc Insurance", Phase I, Zinc Oxide Extension Rice Farm Trial, 4 pp., Ministry of Agriculture, Philippines.

(8) Castro, R.U. 1977. Zinc deficiency in rice: A review of research at the International Rice Research Institute. IRRI Research Paper Series No. 9. 18 pp.

(9) Orticio, M.R. and F. N. Ponnamperuma. 1977. Zinc deficiency: A widespread nutritional disorder of rice in the Philippines. In Proceedings Eighth Annual Meeting of the Crop Science Society of the Philippines, May 1977, Benguet, Philippines.

(10) Mikkelsen, D.S. and Shiou Ku. 1976. Zinc fertilization and behavior in flooded soils. In The Fertility of Paddy Soils and Fertilizer Applications for Rice. Food and Fertilizer Technology Center for the ASEAN and Pacific Region, Taiwan. pp. 170-196.

(11) Bunoan, J.C., Jr., F.M. Melchor and E.G. Sabornido. 1974. Zinc effect on the yield of C4-137 grown in Agusan problem soils at varying levels of NPK fertilizers. Presented at the 5th Annual Meeting of the Crop Science Society of the Philippines, Naga City. 11 pp., tables.

(12) Rosales, C.M. 1979. The efficiency of fertilizer-zinc by rice grown under wetland condition. Technical Bulletin No. 2. Bureau of Soils, Ministry of Agriculture, Manila. 19 pp. 16 tables. 12 figures. Appendix.

(13) _____. 1979. Results of zinc farm trials. The Rice Specialist. Masagana 99. Ministry of Agriculture, Manila. Vol. 1 No. 2, Apr. 15. 4 pp.

(14) Bautista, A.C., G.V. Bautista, J.C. Bunoan Jr., R. Feuer. 1978. 7-12-13-14-16 = Zinc! Ministry of Agriculture, Manila. 2pp. (Unpublished mimeo.).

(15) _____, _____, M.E. Protacio, J.C. Bunoan Jr. and R. Feuer. 1979. Developing a zinc extension program for irrigated rice in the Philippines. Presented at the Crop Science Society of the Philippines 10th Annual Meeting, April 1979, Los Baños, Philippines. 12 pp.

(16) Cruz, A.H., A.C. Bautista, J.C. Bunoan Jr., M.E. Protacio and R. Feuer. 1979. Zinc Alert! Bureau of Agricultural Extension, Manila. 4 pp.

(17) _____. 1980. Efficiency: Zinc will give it. "The Rice Specialist". Masagana 99. Ministry of Agriculture, Manila, November 1, 1980. Vol. 2 No. 21. 6 pp. (In press).

(18) Bautista, A.C., A.H. Cruz and R. Feuer. 1979. Zinc boosts palay production in 1978. News Bulletin, February 1979. Fertilizer and Pesticide Authority, Manila, Philippines. Vol. 1 No. 2.

(19) Bunoan, J.C., Jr., 1979. Fertilization practices of the Masagana 99 rice production program of the Philippines. Soils Technical Bulletin No. 1. Bureau of Soils, Ministry of Agriculture, Philippines. pp. 18-50.

(20) University of the Philippines College of Agriculture. 1977. Zinc starvation of rice. "Crops and Soils" special issue, August 1977, Los Baños, Philippines. 4 pp.

(21) Bautista, A.C., A.H. Cruz, J.C. Bunoan, Jr., and R. Feuer. 1980 Breakthrough in use of zinc for rice production in the Philippines. Ministry of Agriculture, Philippines. 2 pp. (Mimeo.).

Other References

(1) Yoshida, S. 1968. Occurrence, causes, and cure of zinc deficiency of
the rice plant in calcareous soils. International Rice Commission
11th Session Working Party on Rice Soils, Water and Fertilizer
Practices, Kandy, Ceylon, September 2-5, 1968.

(2) Tanaka, A. and S. Yoshida. 1970. Nutritional Disorders of the Rice
Plant in Asia. pp. 22, 23, 39, 40 zinc deficiency, Philippines.
International Rice Research Institute, Philippines. 51 pp. 38 tables,
11 figs. 150 citations. Second printing 1975.

(3) _____. 1974. 500,000 Hectares Zinc Deficient in the
Philippines. IRRI Research Highlights for 1974. p. 45. International
Rice Research Institute, Philippines.

(4) Castro, R.U., M. R. Orticio, F.N. Ponnamperuma, F.N. Bunoan, Jr.
and R. Feuer. 1977. Use zinc for higher rice yields. 4pp. Ministry
of Agriculture, Manila. (Unplished mimeo.).

(5) _____. 1977. Zinc and rice. Zinc Deficiency Script A,
Agricultural Information Division, Department of Agriculture,
Philippines. 3pp. (Unpublished mimeo.).

(6) Moorman, F.R. and N. van Breemen. 1978. Zinc deficiency. In "Rice:
Soil, Water, Land". International Rice Research Institute,
Philippines. pp. 137-140.

(7) Babiera, V.V. 1980. Response of rice to zinc and phosphorus in
flooded soils. M.S. thesis, University of the Philippines at Los
Baños. 74 pp. illus. tables. (Unpublished).

ON ZINC DEFICIENCY OF PADDY SOILS

Xie Zhen-chi, Deng Kai-yu, Yang Hai-qing,
Gong Yu-xi, Wang Zhen-wen, Wang Qin-sheng

(Hubei Academy of Agricultural Sciences, Wuhan)

Zinc deficiency has been reported in tropical and temperate soils for many crops[1,2]. In paddy soils, zinc deficiency has been reported in India[3], Philippines[4] and in other parts of Asia[3].

The present paper summarizes field observations, field experiments, greenhouse experiments and laboratory experiments on zinc deficiency of rice plant grown in calcareous soils of Hubei Province. Field experiments on application of zinc sulphate were carried out in 1976-1979 in 5 locations of Jianghan Plain of central Hubei Province.

THE ZINC DEFICIENT PADDY SOILS

A number of workers have found that the zinc deficiency is associated with soils of high pH value and high levels of phosphorus and organic matter[1,5]. The bog type paddy soil of Jianghan Plain derived from lake sediment is calcareous, has a high ground-water table and is low in rodox potential. $CaCO_3$ content is 1-7 %, bicarbonate content of irrigation water is 45-77 m.e./1 (Table 1).

Table 1 Chemical Properties of a typical zinc deficient soil

Depth (cm)	pH	$CaCO_3$ (%)	O. M. (%)	N (%)	Available nutrients(ppm)	
					$P^{1)}$	$Zn^{2)}$
0-20	7.9	3.9	2.37	0.165	9.3	0.51
20-40	8.1	3.9	2.05	0.173	5.5	0.47

1) 0.5N $NaHCO_3$ extractant   2) DTPA extractant

The calcareous bog type soil is widespread in Jianghan Plain. It covers an area of 300,000 hectares. In 1978, zinc fertilizer was applied to 15,000 hectares in this region and resulted in a marked increase in grain yield of rice.

ZINC DEFICIENCY SYMPTOMS OF THE RICE PLANT

Zinc deficiency symptoms are distinctive and useful for visual diagnosis. Two to three weeks after the transplanting of the rice seedlings, when the plants possess 5-6 leaves, new leaves become decolourized. Decolourization starts from the middle vein and spreads to the whole leaf. The youngest leaves are light yellow in colour, particularly at the base. Brown spots appear on old leaves at first, then coalesce to brown streaks. The affected leaves finally wither. The leaf veins of the lower part of plant become brittle. The leaves hang down and arrange in disorder. The plant is dwarfed. Tillering time delays. The root system is sparse and short. The percentage of sterile flowers increases. Growth and ripening of rice are delayed. Both the dry matter accumulation and the zinc content are lower than normal plant, as shown in Table 2.

Table 2 Effect of zinc deficiency on dry weight and zinc content (Calcareous soil, early rice)

Treatment	Tillering stage				Ripen stage	
	DM(g/pot)		Zn level(ppm)		DM(g/pot)	
	Tops	Roots	Tops	Roots	Tops	Roots
Normal plant	18.18	4.35	15	31	21.00	5.94
Zn deficient plant	12.81	2.21	8	11	19.62	5.16

EFFECT OF ZINC FERTILIZER(6)

Table 3 shows that basal dressing, spraying with solution and dipping of seedling with suspension of zinc fertilizer are all effective measures for curing zinc deficiency. Basal dressing of $ZnSO_4$ is most practical. In 24 experiments, application of $ZnSO_4$ at a rate of 15 kg/ha gave an increase of grain yield by 630-840 kg/ha over the control plots (Table 4). Meantime, the number of rice panicles per ear increased by 10 %, the number of grains per ear by 30.8 %, 1000-grain weight by 8 % in $ZnSO_4$ treated plots.

Table 3 Comparison of method of zinc application on the yield of rice

Treatment[1]	Average yield		Zn in
	(kg/ha)	difference	roots (g/ha)
Control	4910		4.2
15kg $ZnSO_4$ ha basal dressing	5730	820**	48.0
Rice seedling dipping in Zn suspension	5740	830**	31.5
$ZnSO_4$ foliar sprays	5240	330*	7.2

* 1) All plots received adequate amount of N and P, with three replications

Table 4 Effect of zinc application on the yield of rice

Year	Number of experiments	Average yield(kg/ha) Control	Average yield(kg/ha) ZnSO$_4$	Difference
1976	2	4410	5170	760**
1977	7	4690	5530	840**
1978	12	4800	5430	630*
1979	3	5370	6010	640*

AVAILABLE ZINC IN PADDY SOILS

The lime content and pH value of the soil are important factors affecting soil zinc availability(5). Table 5 shows the effect of liming on the availability of soil zinc. The available zinc decreases with the increase of soil alkalinity.

Table 5 Effect of liming on available Zn (acid paddy soil)

Treatment slaked lime (kg/ha)	Soil pH	Available nutrients (ppm) Zn	Available nutrients (ppm) Mn	Available nutrients (ppm) P
0	5.4	0.65	9.6	5.4
2500	7.1	0.50	4.1	9.1
5000	8.0	0.35	2.6	10.5

Table 5 also shows that as the rate of liming increases, the available manganese decreases, while the available phosphorus increases. The available zinc content of soil shows seasonal variation (Fig. 1). A series of available zinc content determinations were made at a given field. The soil pH is 8.1 and the organic matter content is 2.78 %. The available zinc content in early spring was 0.38-0.43 ppm and 0.58-0.76 ppm in summer and autumn. High values (0.61-1.0 ppm) appeared in June, July and August, above the critical value of zinc deficiency (0.5 ppm). Low value 0.3-0.4 ppm below the critical value of zinc deficiency appeared in winter and early spring. The beneficial effect of zinc on the early and medium-maturing varieties of rice was greater than late-maturing ones. The zinc deficiency symptom appeared during seedling stage on the early maturing rice but gradually disappeared in medium and late stage of growth. Zinc deficiency in Jianghan Plain is generally pronounced during cool and wet spring season. Root system of rice plant is expected to expand repidly just prior to mid-season, this would afford a greater soil volume from which

the plant can absorb available zinc. The another possible reason is that a
larger fraction of the available zinc in the soil comes from the decomposition of organic residues, while at low soil temperature the reduced
microbiological activity does not release sufficient available zinc for
proper plant growth.

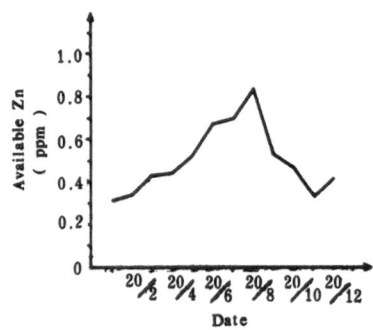

Fig. 1 Seasonal variation of the
content of available Zn in soil

EFFECT OF PHOSPHORUS ON Zn DEFICIENCY

Earlier literatures(1) reveal that precipitation of available zinc by
application of phosphate fertilizer is the main cause of zinc deficiency(7).
Later, the effect of phosphate has been regarded as the interfering with the
uptake, translocation and utilization of zinc by plant(8-11).

Field experiment in calcareous alluvial soil of Jianghan Plain approved
that when the rate of superphosphate dressing was raised to 600 kg/ha,
visual symptoms of zinc deficiency appeared and the rice plant grew slowly
and the grain yield was reduced but available zinc in the soil did not decrease
(Table 6). However, when superphosphate was applied together with zinc sulphate, both the uptake of phosphorus and zinc by rice were increased. Experiments with ^{32}P and ^{65}Zn showed that in treatment of superphosphate plus zinc
sulphate, the P uptake increased from 12.8% to 18.7 % and Zn uptake from 2.16
to 3.53 % than they were applied alone.

High level of available soil phosphorus has been implicated as interfering with the uptake of zinc by rice plant. During the rice growing stage,
the ratio of available P to available zinc in the soil may be an useful
index to fertilizer application. It seems that R value 8-56 is beneficial to

the growth of rice crop. When R value is higher than 74, rice yield increase can be obtained by application of zinc fertilizer. But when soil available zinc content is below 0.5 ppm, it is becoming the limiting factor to the growth of rice crop, and application of P to raise the R value to 56 is unfavorable to the growth of rice crop leading to the reduction of yield.

Table 6 Effect of phosphatic fertilizer on Zn availability
of a calcareous paddy soil

Treatment[1]	1977		1978		1977
	Yield (kg/ha)	Avail. Zn in soil (ppm)	Yield (kg/ha)	Avail. Zn in soil (ppm)	Avail. P ─────── Avail.Zn (R)
N	3480	0.52	3660	0.33	15.5
N-Zn	3680*	1.10	3970*	0.74	3.4
N-P$_1$	3400	0.50	3990	0.33	36.2
N-P$_1$-Zn	4540**	0.96	4543**	0.65	8.0
N-P$_2$	2800	0.65	3940	0.33	56.8
N-P$_2$-Zn	4660**	1.04	4520**	0.65	12.4
NK-P$_2$	3910	0.48	4310	–	56.0
NK-P$_2$-Zn	5250**	–	4820*	–	12.0

1) P$_1$: 300 kg/ha

2) P$_2$:600 kg/ha

ZINC UPTAKE IN RICE

Water culture experiments with labelled $ZnSO_4$ showed that the radioactivity of Zn was detected in one hour after its application. After 72 hours, 10.9 % of the amount of Zn applid was absorbed by rice(12).

Plant analysis showed the zinc contents in different parts of the rice plant were roots > stems and leaves > grains, being 73.3. 35.3 and 19.5 ppm respectively on average. After the application of zinc sulphate to the culture, zinc absorbed was distributed more in the root, being 12.3-24.4 ppm; next in the stems and leaves, being 4.1-16.4 ppm; and least in the grains, being 1.6-9.7 ppm. Zinc deficiency in rice plant led to a significant reduction on dry matter accumulation. The reduction was more pronounced in roots than in stems and leaves, being 10.6 % in stems and leaves and 96.3 % in roots at tillering stage, and 7.1 % and 15.1 % respectively at ripen stage. Field tests also showed that grain yield of rice was positively correlated with zinc reserve in roots at the panicle formation stage.

The results obtained in field experiment, 15 kg/ha of zinc sulphate applied to the early rice, only 1.2-3.5 % was absorbed by the early rice plant, and 0.6-1.5 % was absorbed by the late rice plant. Large amount of Zn fertilizer was remained in the soil available to the succeeding crops. Field experiment showed that rice yield of the third season after application of zinc fertilizer was higher by 17.4 % than that of the control polt (Table 7).

Table 7 Residual effect of Zn fertilizer on rice

| Applied Zn Sulfate (kg/ha) | | | | Grain yield (kg/ha) | | | | Sum of |
| 1978 | | 1979 | | | | 1978 | 1979 | 1978-1979 |
Early rice	Late rice	Early rice	Late rice	Early rice	Late rice	Early rice	Late rice	
0	0	0	0	4080	1930	4560	3170	13740
15	15	15	15	4800**	2020	5660**	3220	15700
15	15	0	0	4740**	1980	5620**	3170	15510
15	0	0	0	4840*	1970	5450**	3270	15530

REFERENCES

(1) Thorne, W., Zinc deficiency and its control. Adv. Agron., 9, 31-65 (1957).
(2) Lindsay, W. L., Zinc in soils and plant nutrition. Adv. Agron., 24, 147-181 (1972).
(3) Ou, S. H., Rice Disease. Eastern Press Ltd., London, 354 (1972).
(4) Zinc deficiency. In "The International Rice Research Institute Annual Report for 1970", 102-107 (1971).
(5) Yoshida, S., Tanaka, A., Zinc deficiency of the rice plant in calcareous soils. Soil Sci. Plant Nurt., 15, 75-80 (1969).
(6) Department of Soils & Fertilizers & Department of Agrophysics & Agrochemistry, Former Hubei Institute of Agricultural Sciences, Report on zinc fertilizer experiment of Hubei Province. Hubei Agriculatural Sciences 4, 20-23 (1978). (in Chinese)
(7) Olsen, S. R., Micronutrient interaction. In "Micronutrients in Agriculture" (ed. by Mortvedt, J. J., Giordano, P. M., Lindsay, W. L.), Madison USA, 243-264 (1972).
(8) Brown, A. C., Krautz, B. A., Eddings, J. L., Zinc-phosphorus interaction as measured by plant response and soil analysis. Soil Sci., 110, 415-420(1970).
(9) Warnock, K. E., Micronutrient uptake and mobility within corn plants (Zea mays L.) in relation to phosphorus-induced zinc deficiency. Soil Sci. Soc. Amer. Proc., 34, 765-769 (1970).

(10) Safaya, N. M., Phosphorus-zinc interaction in relation to absorption rates of phosphorus, zinc, copper, manganese and iron in cron. Soil Sci. Soc. Amer. Proc., 40, 719-722 (1976).

(11) Giordano, P. M., Efficiency' of zinc fertilization for flooded rice. Plant Soil 48, 673-684 (1977).

(12) Gong Yu-xi, Wang Qin-sheng, Research work in zinc and phosphorus uptake by rice with radioactive isotopes ^{65}Zn and ^{32}P. Hubei Agricultural Sciences, 6, 24-26 (1979). (in Chinese)

RECOVERY OF FERTILIZER-NITROGEN BY RICE GROWN IN A GREENHOUSE UNDER VARYING SOIL- AND CLIMATIC CONDITIONS

A.C.B.M. van der Kruijs, J.C.P.M. Jacobs, P.D.J. van der Vorm and A. van Diest
(Department of Soil Science and Plant Nutrition
Agricultural University, Wageningen, Netherlands)

INTRODUCTION

Under field conditions, the efficiency with which urea is utilized by paddy rice is commonly low. Recovery values below 50% are often reported[1] Volatilization of NH_3 formed in alkaline media, after hydrolysis of the urea to $(NH_4)_2CO_3$, and denitrification of NO_3, after formation from NH_4, are usually looked upon as the main causes of low efficiency of urea-N.

In greenhouse experiments carried out in the Netherlands to test the efficiency of utilization of various N-fertilisers applied to rice, the efficiency values of urea-N are usually found to be much higher than 50%. One factor which is lacking in these greenhouses and which under field conditions may exert an influence on N-losses, is wind. Ordinarily, summer air temperatures in the Netherlands are low enough to allow tropical crops to be grown in the greenhouse without the use of mechanical ventilation. Lack of air movement could be responsible for a reduction in NH_3- volatilization losses, thus causing unrealistically high values for recovery of urea-N by paddy rice.

The factors that in general may contribute to low efficiency of fertilizer nitrogen in wetland rice culture are listed by de Datta et al. [1] as: 1. NH_3 volatilization, 2. denitrification following nitrification, 3. biological immobilization, 4. fixation of NH_4^+ by clays, 5. leaching, 6. runoff, and 7. seepage. The aim of the experiment to be discussed in this paper was to eliminate or to minimize the influences of all these factors, except NH_3 volatilization, and to impose variations in environmental conditions which may exert influences on volatilization losses of NH_3 arising from N-fertilizer application to rice.

In greenhouse pot experiments, the factors 5, 6, and 7 can be disregarded. When use is made of soils that do not fix NH_4^+, also factor 4 can be left out of consideration. In such a case, of the factors potentially responsible for low efficiency of fertilizer-N only the nos. 1, 2, and 3 may remain. Since denitrification is dependent on the availability of energy material, this factor may be expected to be minimized when use is made of soils that are extremely low in organic matter. Furthermore, if denitrification is important, its importance is not likely to be affected much by variation in type of NH_4 fertilizer used. For the present experiment this means that losses due to denitrification may be expected to be the same for urea and $(NH_4)_2SO_4$, as long as these fertilizers are applied in the same quantity, time, and manner. To some extent, the same holds for the factor "immobilization" except that the influences that urea and $(NH_4)_2SO_4$ exert on the pH of the floodwater and, through pH, on the intensity of algal growth in the floodwater, may differ. Variations in algal growth may cause variations in N immobilized in algae tissue.

It is known that the rate of mineralization of soil organic N can be affected by the addition of fertilizer-N to a soil. When the efficiency of a N-fertilizer is estimated from differences in N contained in plants from fertilized pots and from unfertilized pots, this so-called priming effect can lead to overestimations of the efficiency values[2]. Such a complication can be avoided when use is made of ^{15}N-enriched fertilizer. However, in that case the phenomenon of isotopic exchange may interfere with a proper evaluation of the efficiency of the N-fertilizer used.

In the present experiment, these complications were avoided through the use of a soil having a very low organic-matter content and, consequently, a very low level of residual N availability.

NH_3 volatilization is likely to be enhanced by high pH levels in the floodwater, when during the daytime algae consume CO_2 in the water for their photosynthesis. Hence, NH_3-volatilization losses are a function of the presence of floodwater and, consequently, these losses might be expected to be lower when rice is grown under dryland conditions. Therefore, in the present experiment, for comparative purposes, rice plants were grown under both wetland and dryland conditions.

Large daily fluctuations in pH of irrigation water can be expected as long as a still open canopy of rice plants allows enough light to penetrate to the water surface to accomodate the growth of algae. Once the canopy is closed, the algae population disappears and the daily fluctuations in pH of the irrigation water become less pronounced. Hence, during daytime, conditions for NH_3 volatilization are not as favorable any more as they were during the early part of the growth period, and on-surface application of ammonium fertilizers can be expected to become more efficient than when the ammonium is applied on the surface as a basal dressing.

To investigate the validity of the abovementioned hypotheses, in a greenhouse a factorial experiment was conducted which comprised
- 2 moisture regimes,
- 2 N-fertilizer forms,
- 2 levels of air movement,
- 2 ways of placing basal dressings of N-fertilizer,
- 3 ways of distributing fertilizer-N over basal and topdressings.

MATERIALS AND METHODS

The soil used in the experiment was a loam taken from a sandpit in which the loam was found buried under a few meters of sand. The use of this subsoil which hardly contains any nitrogen, practically ruled out the risk of a priming action exerted by added fertilizer-N on the mineralization of residual soil organic N.

The lack of organic matter in the soil excluded the possibility of a pH increase after inundation. To account for this anomaly, lime was added to the soil to raise its pH level to 6.

The pots used were each filled with 6.5 kg soil. The nutrients P, K, and Mg were added in quantities considered adequate to meet the needs of the rice plants grown. Nitrogen was added in a quantity of 1 g N per pot, either as $(NH_4)_2SO_4$ or as urea. The quantities of N used as basal dressings were 1 g, 0.5 g or 0.25 g. These basal dressings were either uniformly mixed throughout the soil, or broadcast onto the surface. Whenever visual symptoms indicated that the plants had exhausted the basal supply, the supplemental quantities of 0.5 g or 0.75 g N were applied superficially as topdressings.

To study the direct and indirect influences of inundation, for one-half of the pots the rice plants were grown on soil that each day was moistened to 60% of its water-holding capacity. The soil in the other half of the pots was kept permanently flooded. Samples of an algae culture were inoculated into the water. To stimulate algae growth, 3 mg P as KH_2PO_4 were added to the water. It was observed that two types of algae developed, one being a green alga and the other a blue-green alga. The blue-green one was found to be of the non-N fixing type.

Variation in air movement was obtained by installing overhead ventilators over one-half of the pots in one section of the greenhouse. The ventilators were turned on at low speed for 22 hours per day, and at high speed for 2 hours per day (12.00-14.00 p.m.). The wind velocities measured at the edges of the pots (approximately 5 cm above the soil - or water surface) were 0.28 ± 0.14 m/sec and 0.43 ± 0.17 m/sec, respectively. The pots were moved regularly to other positions on the table to account for variations in wind velocity per table. In the nonventilated section of the greenhouse, the average wind velocity was 0.03 m/sec. Eight weeks after transplanting when, except for the -N pots, the canopy had closed, the ventilators were turned off to prevent desiccation damage to the plants. Air temperatures in the two greenhouse sections were kept equal as much as possible by regulating the aperture of the overhead windows.

The rice variety used was IR-8. At the time of transplanting (3 plants per pot), the seedlings were 2 weeks old. The quantity of N applied was purposely kept too low to carry the plants to maturity. Harvesting of the above-ground portions of the plants was carried out whenever the plants in a treatment clearly showed the symptoms of N-deficiency.

All treatments were present in triplicate pots, which led to a total of 156 pots. The soil in the control pots received no N-fertilizer, but was otherwise treated as the soil in the other pots.

RESULTS

In the inundated soil, the pH levels (measured in situ at 4-cm depth) in the urea-treated pots rose about one-half pH unit above the initial level of pH 6, whereas in the $(NH_4)_2SO_4$-treated pots, the levels fell by about 0.7 pH unit. In the pots with aerobic soil, as expected, urea initially caused a pH rise followed by a decline to the starting level of pH 6. $(NH_4)_2SO_4$ applications caused the soil pH to drop gradually to an approximate level of pH 5.2.

As an example of diurnal fluctuations of pH in the inundation water, Fig. 1 supplies data collected 12 days after transplanting, for those pots which had received basal N-dressings mixed with the soil. The topdressings had not yet been applied. The weather on the two days of the measurements was sunny. It can be observed that the pH-levels are considerably higher with urea- than with $(NH_4)_2SO_4$ applications, and that in the water of the urea-treated pots and the control pots, the diurnal fluctuations were as expected, whereas on the $(NH_4)_2SO_4$-treated pots, the fluctuations were practically absent, indicating that at low pH levels in the inundation water, photosynthetic activities of algae hardly affect the CO_2 concentrations in the water.

Fig. 2 shows the pH values of the inundation water measured periodically during the daytime over a period of 8 weeks. Here it can be seen that after 30 days the pH values of the water on urea-treated soil did no longer rise above the value of pH 7.

In Table 1, the values of above-ground dry-matter yields are given. A statistical analysis of these data indicates that all main effects were highly significant, which means that yields were higher
A. without than with wind,
B. under dryland- than under wetland conditions,
C. when basal dressings were mixed with the soil than when they were applied on the surface,
D. with $(NH_4)_2SO_4$ than with urea,
E. when fertilizer applications were split than when all N was applied as a basal dressing.
The highly significant first-order interactions AB and BC disclose that the absence of wind was more beneficial under dryland than under wetland condi- tions, and that the advantage of mixing basal N-dressings with the soil was more pronounced under wetland than under dryland conditions.

Statistical analysis of the data on fertilizer-N recovery in the rice tops (Table 2) shows that all main effects were highly significant, except the effect of wind which proved to be nonsignificant. Of the first-order inter- actions, all involving the factor "form of fertilizer", except one, were found to be significant. The same can be said for the second-order interactions, involving this factor. Also the highest-order interaction (ABCDE) was found to be significant.

For an evaluation of the efficiency of utilization of applied fertilizer-N, it would be preferable to have insight in the N-recovery by the roots as well. However, the procedure for harvesting roots from potted soil is very time- consuming and the validity of the results obtained is questionable. For a limited number of pots, the procedure was carried out and the results indicated that, on the average, 17% of the N in the rice plants was contained in the

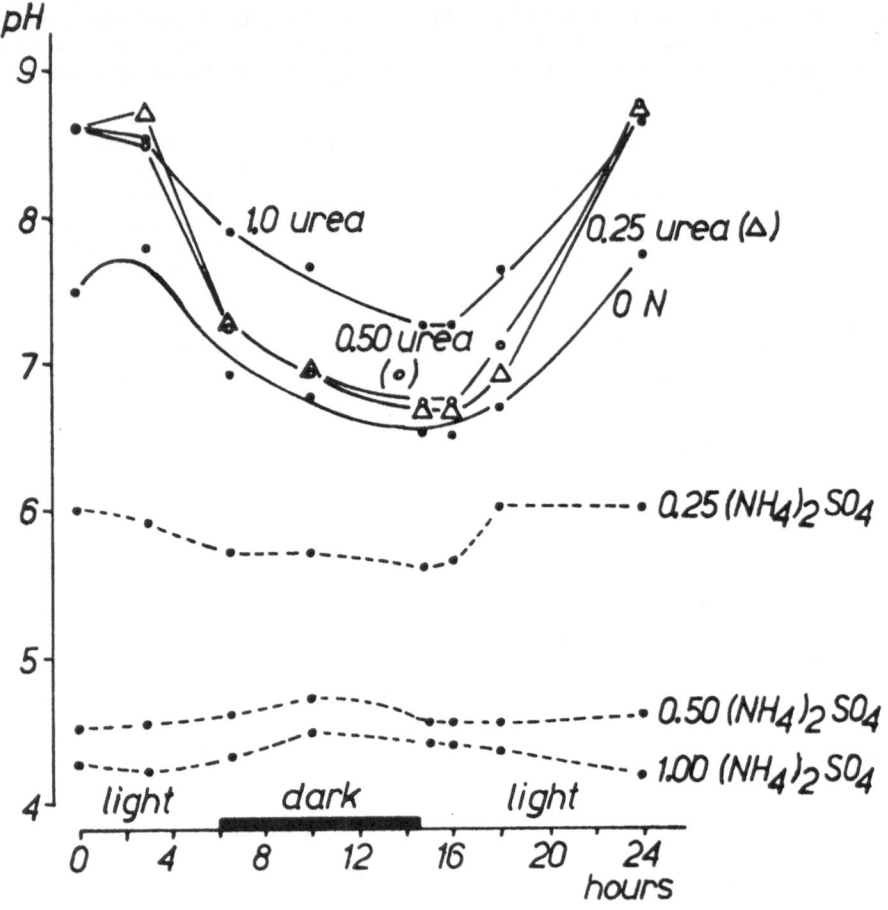

Fig. 1 Diurnal course of the pH in irrigation water standing on soil treated with N fertilizer in two forms and three quantities, expressed as fractions applied as basal dressings and mixed with the soil. The pots were exposed to wind

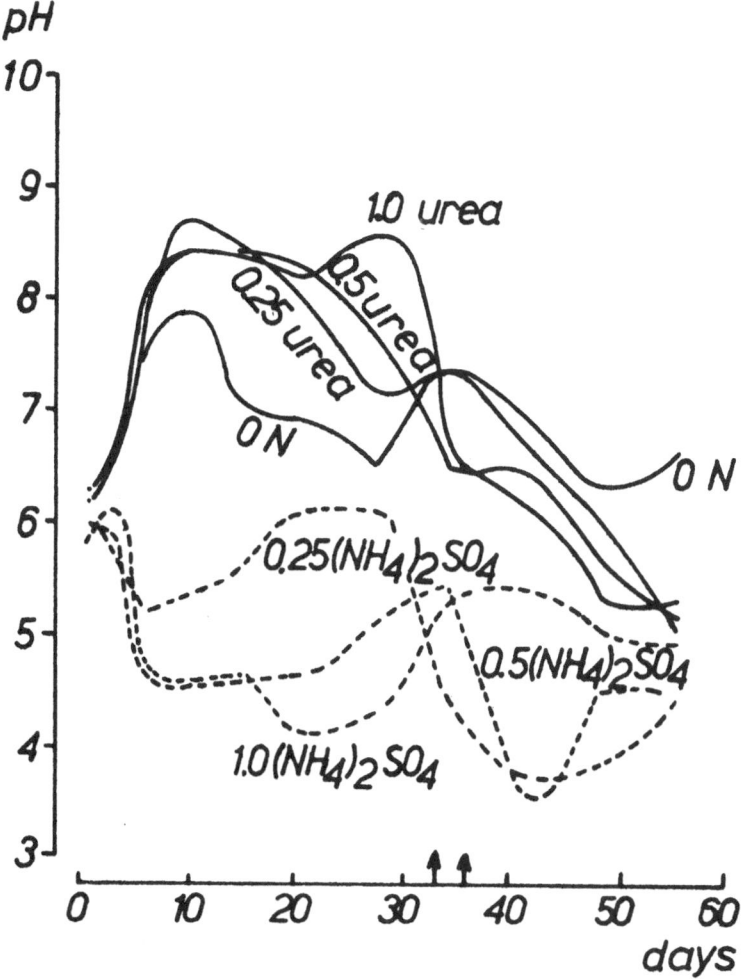

Fig. 2 Course of pH in irrigation water during
a 55-day period in which rice plants developed
on soil treated as mentioned under Fig. 1. All
pH-measurements were made during daytime. The
arrows indicate the days topdressings were applied

Table 1 Dry-matter yields of rice grown under varying water-, wind-, and N-fertilizer regimes in a greenhouse pot experiment

| Fertilizer application schedule | Form of fertilizer | Above-ground dry matter, g/pot | | | |
| | | Wetland | | Dryland | |
		wind	no wind	wind	no wind
control	–	1.7	1.7	1.9	1.5
all N applied on surface as basal dressing	urea	27.7	37.1	64.1	56.7
	$(NH_4)_2SO_4$	54.4	49.3	64.9	74.5
all N worked in as basal dressing	urea	50.9	50.8	62.1	62.2
	$(NH_4)_2SO_4$	58.7	49.8	62.5	75.7
50% of N applied on surface as basal dressing, 50% as topdressing	urea	44.2	43.9	65.7	69.1
	$(NH_4)_2SO_4$	51.2	54.4	69.1	79.1
50% of N worked in as basal dressing, 50% as topdressing	urea	52.7	55.8	61.6	72.1
	$(NH_4)_2SO_4$	56.7	55.7	74.3	77.6
25% of N applied on surface as basal dressing, 75% as topdressing	urea	43.9	46.7	63.9	72.2
	$(NH_4)_2SO_4$	49.4	49.8	73.6	73.8
25% of N worked in as basal dressing, 75% as topdressing	urea	49.4	55.9	64.4	74.6
	$(NH_4)_2SO_4$	55.4	55.2	70.3	79.2

Table 2 Recovery values of two forms of fertilizer N applied in various combinations of timing and placement to rice grown under wetland and dryland conditions, with and without wind, in a greenhouse pot experiment

Fertilizer application schedule	Form of fertilizer	Recovery values, %			
		Wetland		Dryland	
		wind	no wind	wind	no wind
all N applied on surface as basal dressing	urea	25	43	64	56
	$(NH_4)_2SO_4$	64	69	68	68
all N worked in as basal dressing	urea	54	48	67	64
	$(NH_4)_2SO_4$	63	66	71	67
50% of N applied on surface as basal dressing, 50% as topdressing	urea	47	49	69	66
	$(NH_4)_2SO_4$	62	62	73	74
50% of N worked in as basal dressing, 50% as topdressing	urea	59	63	68	66
	$(NH_4)_2SO_4$	64	63	68	67
25% of N applied on surface as basal dressing, 75% as topdressing	urea	47	55	65	66
	$(NH_4)_2SO_4$	62	62	70	70
25% of N worked in as basal dressing 75% as topdressing	urea	56	64	67	67
	$(NH_4)_2SO_4$	69	68	73	74

roots. This would mean that for wetland and dryland conditions, the highest total recovery values will lie in the vicinity of 83% and 89%, respectively, both for $(NH_4)_2SO_4$, and that the highest values for urea would be 77% and 83%, respectively. These values show that losses of N other than through volatilization are likely to have been of limited importance.

DISCUSSION

Even when mixed with the soil, the basal dressings of urea appear to be able to affect the pH of the inundation water to such an extent that during daytime in the early part of the growth period of the rice, conditions are favorable for NH_3 volatilization (Fig. 1). In itself, NH_3 volatilization has a pH-decreasing effect. The observation that with the highest N-dose applied, also during nighttime the pH remains above 7 suggests that under this condition considerable diffusion of NH_4^+ from the soil to the water can be expected to take place. The data of Fig. 2 reveal that after 30 days of rice growth the conditions have changed enough to rule out the possibility of substantial NH_3 loss.

Since the experiment was designed in such a way that N was to become the growth-limiting factor a certain parallelism could be expected in the ways in which plant growth and N-recovery were affected by the various treatments. In this respect, it was rather surprising to notice that wind negatively affected rice growth, but had no effect on N-recovery. A highly significant -interaction between water regime and wind velocity, however, pointed out that wind did have an effect on N-recovery when the soil was flooded. The signifi-cant second-order interaction ABD furthermore showed that this effect was more pronounced for urea than for $(NH_4)_2SO_4$. The negative effects of wind and inundation could be partially overcome when the basal urea dressing was mixed with the soil, but the extent of the positive effect of mixing the basal dressing with the soil was dependent on the size of the fraction of the N that was applied as a basal dressing. This, in essence, is the meaning of the significant fourth-order interaction ABCDE.

As can be read from Table 2, the lowest N-recovery value (25%) was obtained when all urea was applied as a basal dressing on the surface of a flooded soil exposed to wind. In the following, it will be shown through which measures this value could be improved.

An important improvement to 64% could be obtained when rice was grown under dryland conditions. An improvement to 43% was possible when rice was grown under wetland conditions in the absence of wind. Both improvements are interesting mainly from a theoretical standpoint, as rice farmers in general will show little inclination to abandon growing rice under wetland conditions just to raise the efficiency of the fertilizer-N applied. Nevertheless, it is worth remembering that in a 20-year period of rice-growing contests in Japan, in 17 years the highest yields were obtained under dryland conditions (moisture level permanently below water-holding capacity) [3]. Furthermore, wind is usually something rice farmers have to live with.

In the realm of fertilizer-application management, a considerable improvement in N-recovery to 47% could be achieved, when only 50% of the urea was applied on the surface as a basal dressing. A further reduction in the on-surface basal dressing to 25% of total N applied yielded no further improvement in urea-N efficiency. An increase in recovery of urea-N to 59% was obtained when only 50% of the urea was mixed with the soil as a basal dressing. Larger (100%) and smaller (25%) fractions worked in as basal dressings resulted in somewhat lower recovery values (54% and 56%, respectively).

This value of 59% of urea-N recovered in the rice tops under wetland conditions in the presence of wind was the highest obtained for urea. Assuming 17% of the absorbed N to be present in the root system, the total N-recovery value would lie around 71%. The corresponding value for recovery of $(NH_4)_2SO_4$-N in the rice tops was 64%, a value not appreciably higher than that obtained with urea-N. However, for $(NH_4)_2SO_4$ a still higher recovery value (69%) was obtained when under wetland conditions in the presence of wind only 25% of the total N was mixed with the soil as a basal dressing.

Another important practical finding was that, irrespective of the way and the timing of application, $(NH_4)_2SO_4$ applied to rice exposed to wind under wetland condition, always yielded higher recovery values that did urea. The values pertaining to $(NH_4)_2SO_4$ were relatively little affected by variations in fertilizer-application management. They ranged from 62% to 69%. The presence of wind usually had no effect on the level of recovery, but all values, except one, were lower than those obtained when rice was grown under dryland conditions.

However, for urea the differences in recovery values for wetland- and dryland conditions were much larger. This finding emphasizes once more the role that irrigation water plays in reducing the efficiency of urea-N for wetland rice. As pointed out above, this role of the irrigation water is only an indirect one, in that the water serves as growth medium for algae which in the case of urea-use are responsible for daytime-pH levels in the water which are high enough to facilitate a rapid conversion of NH_4^+ to NH_3 gas which will volatilize, thus imposing a sharp concentration gradient between NH_4^+ in the soil and NH_4^+ in the water, which in turn will be responsible for diffusion of NH_4^+ from the soil to the irrigation water.

That such NH_3 losses are enhanced by wind can be seen from the data of Table 2, which reveal that when urea is applied under dryland conditions, the recovery of N is generally not reduced by the presence of wind. This differential response to wind was responsible for the finding that in the statistical analysis the main effect of wind proved to be nonsignificant.

The practical conclusions that can be drawn from the results of this experiment are that
1. under comparable fertilizer management, the efficiency of N applied as $(NH_4)_2SO_4$ was always higher than that of urea-N,
2. the recovery of urea-N was particularly low, when all urea was applied as a basal dressing on the surface of a flooded soil,

3. a considerable improvement in the recovery of urea-N could be obtained when only one-half of the urea was applied as a basal dressing, and when this basal dressing was mixed with the soil,
4. the low recovery values of urea were associated with the presence of irrigation water, which is conducive to the development of conditions under which NH_3 volatilization is enhanced,
5. the presence of wind reduced the recovery of urea-N, especially when wet-land conditions prevailed,
6. the negative influences of algae in the irrigation water and wind on urea-N efficiency were gradually suppressed by the growing rice plants themselves, so that circumstances favoring NH_3 volatilization had largely disappeared by the time the rice canopy was closed,
7. the use of $(NH_4)_2SO_4$ prevented the pH of the irrigation water from reaching values at which wind could exert a negative effect on N-recovery.

Any recommendation for a switch from urea to $(NH_4)_2SO_4$ runs counter to present-day developments in tropical and subtropical countries where for wetland rice culture $(NH_4)_2SO_4$ is gradually replaced by urea. In view of an increase in occurrence of S-deficiency cases, and considering the low recovery values of urea-N, particularly so when the urea is applied on the surface as a basal dressing, it seems justified to recommend $(NH_4)_2SO_4$ for use as a basal dressing and urea for use as topdressings during growth stages in which a closed canopy has suppressed on the one hand the growth of algae which can be responsible for high daytime pH values of the irrigation water, and on the other hand the velocity of wind at the surface-water level. The possible benefit arising from the use of $(NH_4)_2SO_4$ as a basal dressing and of urea as a topdressing will be examined in a next experiment.

REFERENCES

(1) De Datta, S.K., Stangel, P.J. and Craswell, E.T. Evaluation of nitrogen fertility and increasing fertilizer efficiency in wetland rice soils. Proc. Symp. Rice Soils, Nanjing, People's Republic of China, - (1981).
(2) Broadbent, F.E. Nitrogen transformation in flooded rice. Soil and Rice, Int. Rice Research Inst., Los Banjos, Philippines, 543-559 (1978).
(3) Ishizuka, Y. The rice yield competition in Japan, Potash Review, subject 9, no. 6, Int. Potash Inst., Berne, Switzerland (1979).

EFFECT OF WATER CONDITIONS AND ORGANIC MATTER
ON MODULUS OF RUPTURE OF PADDY SOILS

Yuan Jian-fang, Zhou Yue-hua
(Institute of Soil Science, Academia Sinica, Nanjing)

In recent years the tilth of some paddy soils in Jiangsu Province becomes inferior owing to the change in cropping system. The increase in modulus of rupture of the soil is one of the important manifestations in this respect. Some scientific workers (1,2) have devoted themselves to the study of relationship between modulus of rupture and its affecting factors such as clay content, organic matter content, porosity, free iron oxides and amorphous silica. The purpose of the present work is to study the effect of water condition and organic matter on modulus of rupture under laboratory conditions.

METHOD

A permeable paddy soil and a silty paddy soil were used for the study. A red earth was also used for comparison. After adding different amount of rice straw or milk vetch, the samples were kept for a period of 4 months under different water conditions to model after the different cropping systems:

(A) Submerging;
(B) Intermittent submerging and drying;
(C) Intermittent moistening to 60% of capillary water capacity and then drying.

The modulus of rupture of soils were measured with dried soil cores about 2.7 cm in height and diameter respectively. The cores were ruptured by applying pressure vertically, and the modulus of rupture were then calculated according to the following formula (3).

$$M = \frac{F}{\pi r\, l}$$

where M——modulus of rupture, kg/cm^2;
F——crushing force, kg;
r——radius of the soil core, cm;
l——length of soil core, cm.

RESULTS AND DISCUSSION

1. Effect of Water Condition

Fig. 1 shows the effect of water condition on modulus of rupture for three soils. It is seen that the modulus of rupture for a long-term submerging treatment is 1-3 times higher than that for other treatments.

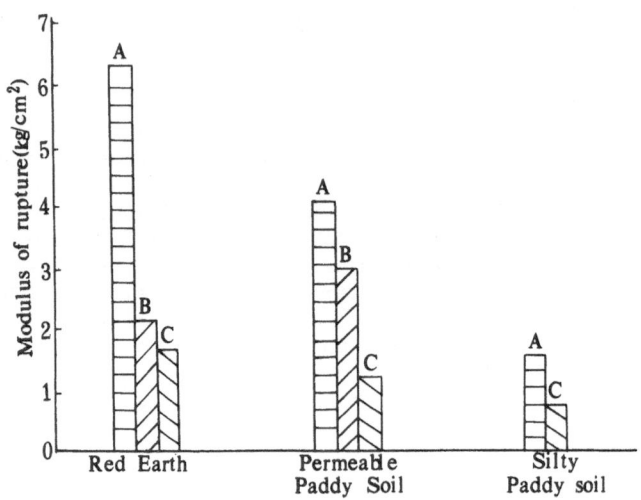

Fig. 1 Effect of water condition on modulus of
rupture of three soils
A. Submerging;
B. Intermittent submerging and drying;
C. Intermittent moistening and drying

This probably related to the destruction of aggregates caused by submerging
as can be seen from Table 1(4,5). In addition, the long-term submerging
treatment possibly causes more parallel orientation or greater close packing
of small particles, especially clay particles (6). This is also responsible to the increase of modulus of rupture. Another possible cause for
the difference of modulus of rupture is that in the intermittent submerging
and drying treatment there is a possibility of forming of crackings due to
uneven swelling and shrinking, as evidenced by the decrease in volume
weitht shown in Table 2. The volume weight of soil cores decreases with the
increase in rounds of wetting-drying circle. Take for example an experiment

Table 1 Effect of water condition and organic matter on percentage
of clay particles ($<$0.001 mm) of permeable paddy soil

Treatment	Content of clay particles ($<$0.001 mm) %
Submerging	14.0
Submerging, applying rice straw and milk vetch	12.3
Intermittent submerging and drying	4.3
Intermittent moistening and drying	3.4

Table 2 Effect of water condition on volume weight of
soil cores of permeable paddy soil

Treatment	Volume weight (g/cm^3)
Submerging, with o.m. applied	1.39
Intermittent submerging and drying, with o.o. applied	1.20
Intermittent moistening and drying, with o.m. applied	1.21
Submerging, with no o.m. applied,	1.67
Intermittent submerging and drying, with no o.m. applied	1.64
Intermittent moistening and drying, with no o.m. applied	1.42

with silty paddy soil. The modulus of rupture were 0.53, 020 and 0.17
kg/cm^2 after 1,3 and 7 rounds of wetting-drying circle respectively.

2. Effect of Organic Matter

Fig. 2 shows that the modulus of rupture of the soil decreases with
the increase in organic matter content.   It is possibly due to the fact
that the organic matter promotes the aggregation(7), hence the decrease
of the amount of surface contacts among soil particles, as can be seen
from Table 1.

The magnitude of effect of organic matter on modulus of rupture is
correlated to its kind. The effect of rice straw is stronger than that
of milk vetch, as can be seen from Fig. 2 lines 5 and 6, and the rice
straw is more favourable to decrease the modulus of rupture than milk
vetch. This is probably related with the difference in their ability of
aggregation as they were decomposed. The undecomposed straw is also more
effective in decreasing the modulus of rupture than undecomposed milk
vetch, and this is probably related with the difference in magnitude of
their influences on coherence between the soil aggregates.

The magnitude of effect of organic matter on modulus of rupture
differs among soils. Fig. 2 shows that for the original Red earth, the
modulus of rupture is high and the effect of organic matter is strong. On
the contrary, for the silty paddy soil, the modulus of rupture is low and
the effect is also weak.

The effect of organic matter is influenced by the water condition of
the soil. It is seen from Fig. 2 that the modulus of rupture for submerging treatment is higher than that for other treatments with comparable
organic matter content.   And. what is more important, the effect of water
condition is sometimes so strong that the effect of organic matter can be
overlaid.

From the above it can be seen that the modulus of rupture of soil is
influenced by the water condition and the content and kind of organic
matter through their influences on the soil aggregation and orientation

of small particles and the coherence between soil aggregates. It is of
great importance to shorten the period of submerging of soil by changing
the cropping system and increasing the amounts of organic manure in order
to decrease the modulus of rupture and improve the tilth of some paddy
soils in Jiangsu Province.

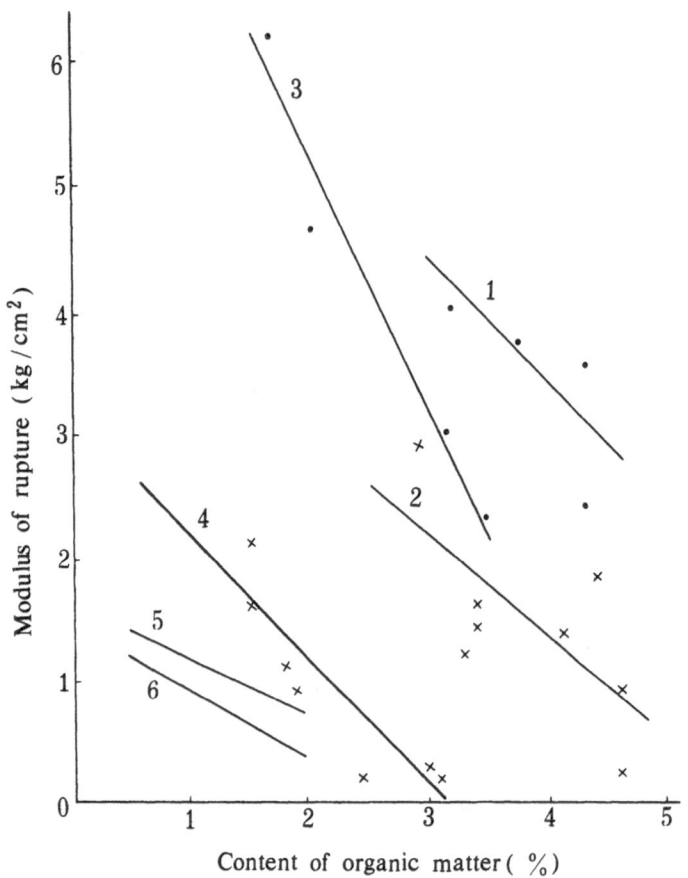

Fig. 2 Relationship between modulus of rupture and
       organic matter content
       1. Permeable paddy soil, submerging;
       2. Permeable paddy soil, intermittent submerging
          and drying, intermittent moistening and drying;
       3. Red Earth, submerging;
       4. Red Earth, intermittent submerging and drying.
          intermittent moistening and drying;
       5. Silty paddy soil, submerging, milk vetch;
       6. Silty paddy soil, submerging, rice straw

REFERENCES

(1) Baver, L.D., Gardner, W.H., Gardner, W.R., Soil Physics.
4th edition, John Wiley & Sons, Inc., New York, 83-87 (1972).

(2) Yao Hsian-liang, Chao Wei-sheng, Yu Teh-fen, Hsu Hsiu-yun,
Preliminary investigation of structural characteristics of fertile
paddy soils. Acta Pedologica Sinica, 15(1), 1-12 (1978). (in
Chinese)

(3) Don Kirkham, De Boodt, M.F., De Leenheer, L., Modulus of rupture
determination on undisturbed soil core samples. Soil Sci., 87 (3),
141-145 (1959).

(4) Lygo, H.M., Weed, S.B., Sopher, C.D., Hilton, H.G., The effect of
clay mineralogy on soil strength and cotton stand. Soil Sci., 120
(2) 117-125 (1975).

(5) Richardson, S.J., Effect of artificial weathering cycles on the
structural stability of a dispersed silt soil. J. Soil Sci., 27
(3), 287-294 (1976).

(6) Ferry, D.M., Olsen, R.A., Orientation of clay particles as it relates
to crusting of soil. Soil Sci., 120 (5), 367-375 (1975).

(7) Tiarks, A.E., Mazurak, A., Leon Chesnin, Physical and chemical
properties of soil associated with heavy applications of manure from
cattle feedlots. Soil Sci. Soc. Amer. Proc., 38 (5), 826-830
(1974)

THE EFFECT OF WATER-STABLE AGGREGATES IN VARIOUS
SIZES ON AIR AND WATER REGIME OF SOILS

Xu Fu-an

(Institute of Soil Science, Academia Sinica, Nanjing)

Studies in recent years indicate that some of the paddy soils in the Taihu Lake region have become degenerated in structure owing to the management practice unadaptable for the current triple cropping system with two crops of rice (1,2). Both the macro- and micro-aggregates in soils have been destroyed, with a consequent increase in fine fraction. The amount of macropore space decreased. The soil became compact and hard. In order to elucidate the effect of the size of soil aggregates on air and water regime, simulated aggregates of different sizes were prepared in the laboratory, and a preliminary study of the effect of these various aggregates on soil physical properties was carried out.

METHODS

The soil samples were collected from the subsoil of a yellow-brown earth and the bleached horizon of a whitish soil respectively. The former contains 30.9% of clay while the latter 18.3%. The samples were air-dried, ground and passed thoroughly through a sieve with 1 mm meshes. In order to obtain highly stable artificial aggregates without any change in aggregate size during the experiment, hydrolyzed polyacrylonitrile in a dose corresponding to 0.1% by weight of soil was spread over the samples, and then a force of 50 g/cm^2 was applied to compress them. The sample blocks were air-dried, ground and passed through a series of sieves to obtain three groups of aggregates, 1-2, 0.25-1 and <0.25 mm in size respectively.

Soil porosity and water retention curve were determined by the pressure reduction method. Each sample was placed in a glass tube, 2.25 cm in internal diameter and 5 cm in height and sealed with a filter paper at the bottom of the tube. After the aggregates were put into the glass tube, the tube was gently tapped 15 times to get an uniform packing. The height of the soil column in the tube after tapping was 2cm. Each test was replicated three times. While measuring the change in water content under different suction, the height and the diameter of the soil column were recorded, so that the change in volume weight can be calculated. Using these data the total pore space, capillary pore space and air pore space under different suctions were calculated. The value of the specific water capacity under different suction range was obtained by graphics of the water characteristic curve.

RESULTS AND DISCUSSION

The results in Fig. 1 indicated that the volume weight of the soils composed of aggregates smaller than 0.25 mm increased rapidly as the soil suction increased, while that of the soils with aggregate sizes of 1-2 and 0.25-1 mm did not increase much. It implies that when dehydrated the soil would readily become compact as the fine particles were increased in it.

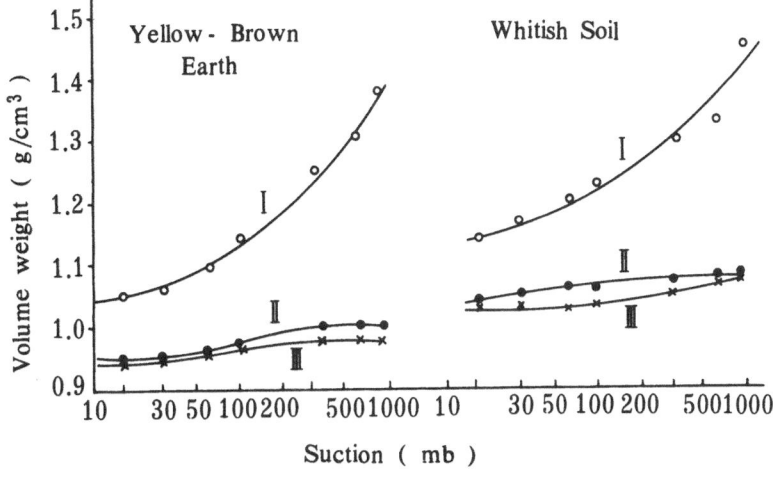

I: < 0.25 mm ; II: 0.25−1 mm ; III: 1−2 mm

Fig. 1 Variation in volume weight under different
       suctions

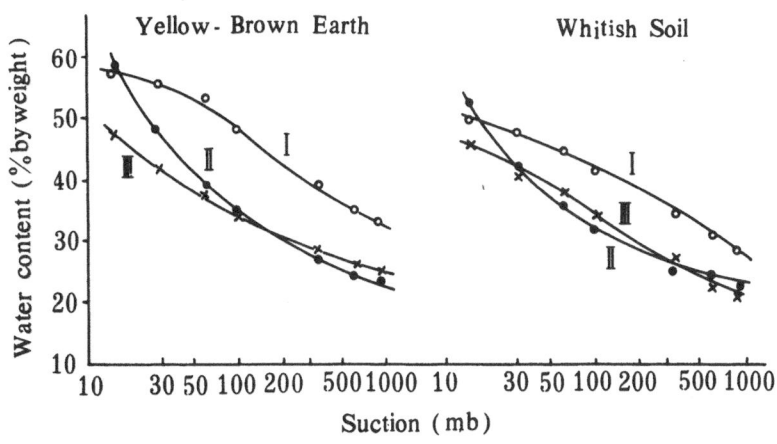

I : < 0.25 mm ; II : 0.25−1mm ; III: 1−2 mm

Fig. 2 Water retention curves of soils

The soil water characteristic curves (Fig. 2) showed that the finer aggregates (<0.25 mm) retained more water than the coarser ones (1-2 and 0.25-1 mm). The difference in water content was even more evident in the case of yellow-brown earth with heavier texture. This may be the principal cause by which the water stagnancy in some of clayey paddy soils in the Taihu Lake region developed further when the aggregates were broken down.

It must be pointed out that although the finer aggregates (<0.25 mm) have the highest water retention, its specific water capacity (the volume of water yielded under unit suction per gram soil) is the lowest within the suction range of 5-100 mb (Fig. 3). It is known that the soil under a suction less than 100 mb is beneficial to the growth of dry farming crops(3). As compared with the three fractions of aggregates under the same suction, it will be seen that the finer aggregates (<0.25 mm) are less suitable for the growth of plants owing to its lower specific water capacity.

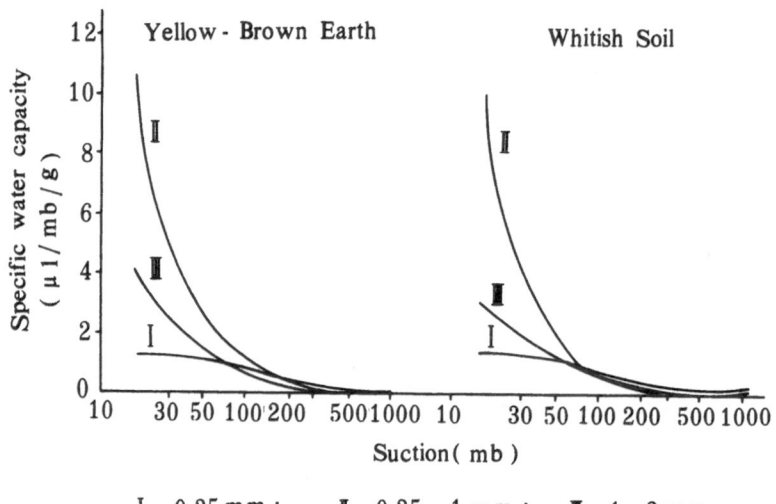

I: 0.25 mm ;   II: 0.25−1 mm ;   III: 1−2 mm

Fig. 3 Variation in water capacity under different suctions

The results in Fig. 4 indicated that under different suctions the air pore space in soil composed of the aggregates smaller than 0.25 mm was smallest. When the suction increased from 0 to 600 mb, the air pore space of yellow-brown earth and whitish soil was 6.8% and 8.6% respectively. In the case of aggregates of 0.25-1 and 1-2 mm in size the air pore space significantly increased as the suction reached to 30 mb. It means that even under low suction and high water content of the soil, the coarser aggregates may improve soil aeration yet. This favorable effect on soil' aeration appeared clearly even in the case of 0.25-1 mm aggregates. On the contrary, the proportion of capillary pore space to total pore space in soil composed of the aggregates less than 0.25 mm was the highest all along as the suction gradually increased. Even under a suction of 600 mb the capillary pore space of yellow-brown earth and whitish soil with aggregates smaller than 0.25 mm was 86.9% and 82.8% respectively. This is due to the fact that in the case of aggregates smaller than 0.25 mm the dehydration of soil was mainly accompanied by the shrinkages of the solum as the suction was on the increase. The loss of water upon dehydration did not leave much air pore space correspondently. In the case of coarser aggregates the loss of water upon dehydration was mainly brought about by the replacement of water by air from the structural pore space. Thus, as the water decreased from the coarser aggregates, the air pore space rapidly increased.

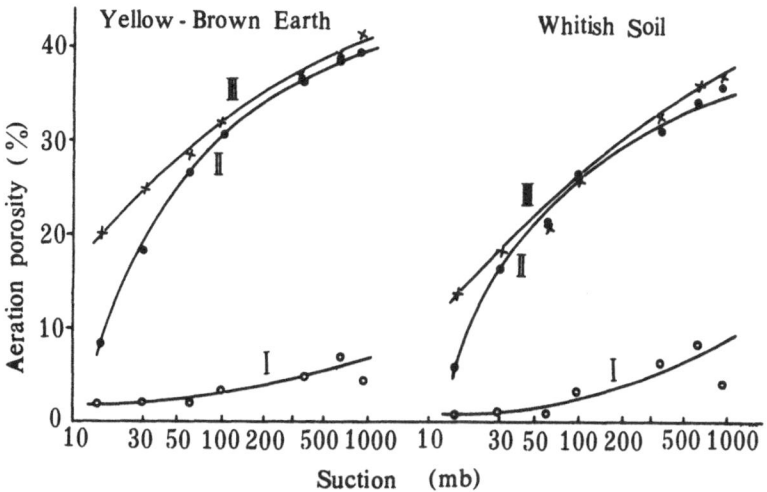

Fig. 4 The air pore space of soil under different suctions

CONCLUSTION

Although this experiment is merely a simulation under laboratory conditions, it is revealed that following the carrying out of the tripe cropping system with two crops of rice, the destruction of soil aggregates has a significant bearing on the unfavorable physical conditions of paddy soils, such as compactness, poor aeration, increased water retention, poor permeability, etc. It is of great importance to cultivate the soil in such a way that a definite quantity of coarser aggregates should be present to ensure favorable physical conditions of the paddy soil.

REFERENCES

(1) Yao Hsian-liang, Chao Wei-sheng, Yu Teh-fen, Hsu Hsiu-yun, Preliminary investigation of structural characteristics of fertile paddy soils. Acta Pedologica Sinica, 15(1), 1-12 (1978). (in Chinese with English summary)

(2) Hseung Yi, Xu Qi, Yao Xian-liang, Zhu Zhao-liang, Effect of cropping system on the fertility of paddy soils. Acta Pedologica Sinica, 17 (2), 101-119 (1980). (in Chinese with English summary)
(3) Yuan Jian-fang, et al., Effect of soil suction on the uptake of plant nutrient (1979, ms.). (in Chinese)

INFLUENCE OF TRANSFORMATION OF IRON OXIDES
ON SOIL STRUCTURE

He   Qun, Xu Zu-yi
(Institute of Soil Science, Academia Sinica,Nanjing)

Although iron oxides in soils are the secondary component of soil clay, they have attracted the attention of many soil chemists. This is because iron oxides are highly active, are easily changed with a change in soil condition, and can exert influences on chemical, physical and biological properties of the soil. The purpose of the present work is to study the transformation of iron oxides during the course of submergence or dehydration of the soil and its influence on soil structure under laboratory conditions.

METHODS

A yellowish brown earth, a red earth and a laterite were added with various amounts of dry powder of milk vetch, and were incubated under submerged conditions with or without permeation. After 8 months the incubated samples were treated as (A) dehydrated at pF 2; (B) air-dried; (C) overdried at $105^{\circ}C$; (D) check.

Physical properties of the soil were measured by routine methods. Complexed iron was extracted with pyrophosphate. Amorphous iron oxides and free iron oxides were extracted with Tamm's solution and dithionite-citrate solution respectively. Total iron was digested with mixed acids of $HClO_4$ and HF. The iron was determined by colorimetric method.

RESULTS AND DISCUSSIONS

The results for treatments with permeation are similar to those without permeation, and therefore are omitted in the following discussions.

It is seen from Table 1 that the contents of complexed iron of three soils increase markedly under the combined effect of the addition of organic matter and submerging. This is in   agreement with the previous results that there is a positive correlation between the complexed iron and the organic matter content of soils(1). The amorphous iron oxides content also increases as a result of the addition of organic matter and submerging. According to the detection of Q value on the basis of analysis of variance, the influence of dehydration strength on the complexed iron content is not obvious.

The total iron content and the free iron oxides content of three soils did not show remarkable change after incubation. However, if we take the ratio between amorphous iron oxides and free iron oxides as an index of activity of iron oxides(2), it is seen from Table 2 that it is markedly influenced by conditions of incubation and dehydration strength.

Table 1 Amounts of complexed iron and amorphous iron oxides

Soil	Added milk vetch (%)	Complexed iron(Fe,mg/100g)				Amorphous iron oxides (Fe_2O_3%)			
		Sub-merged*	pF 2	Air-dried	Oven-dried	Sub-merged	pF 2	Air-dried	Oven-dried
Yellow-ish brown earth	0	9.5	10.8	11.1	11.6	0.85	0.33	0.49	0.44
	2	50.9	20.6	25.9	22.0	1.32	0.90	0.86	0.49
	5	69.8	35.4	40.2	30.0	1.26	0.88	0.85	0.52
Red earth	0	6.9	7.4	8.2	9.2	0.62	0.27	0.31	0.30
	5	52.2	18.4	24.1	16.2	1.82	0.52	0.63	0.46
Laterite	0	10.1	6.8	5.6	8.0	0.64	0.22	0.51	0.46
	5	75.4	15.4	23.5	18.5	1.68	0.59	0.85	0.72

* Water soluble and exchangeable iron were removed with N NH_4OAc before the determination

Table 2 Activity of the free iron oxides

Soil	Added milk vetch (%)	Activity of free iron oxides ($Fe_o:Fe_d$)				Free Fe_2O_3 (%)
		Sub-merged	pF 2	Air-dried	Oven-dried	
Yellow-ish brown earth	0	0.35	0.14	0.20	0.18	2.41
	2	0.59	0.41	0.39	0.22	2.22
	5	0.56	0.40	0.38	0.23	2.23
Red earth	0	0.13	0.06	0.06	0.02	4.77
	5	0.38	0.11	0.13	0.09	4.77
Laterite	0	0.05	0.02	0.04	0.04	13.0
	5	0.12	0.04	0.06	0.06	13.1

The activity increases under the influence of organic matter, but decreases after dehydration. The change is largest in the yellowish brown earth and least in the laterite. This seems to be in accordance with the difference in crystallinity (expressed as $1-Fe_o/Fe_d$) of yellowish brown earth (0.65), red earth (0.87) and laterite (0.95).

The organic matter plays an important role in the activation of iron

oxides. For, a part of iron oxides may be reduced during the course of the
decomposition of organic matter and then transformed into iron hydroxide
by hydrolysis or oxidation, or the organic matter can inhibit the crystallization of iron hydroxide(3). Besides, the amounts of OH^- ions desorbed
by treatment with NaF solution are in the order laterite > red earth >
yellowish brown earth, showing highly significant positive correlation
(r = 0.94, n = 24, p<0.001) with the crystalline iron oxides content. It
appears that the essence of the activation of iron oxides may include protona-
tion on the surface of iron oxides.

As can be seen from Fig. 1, the degree of dispersion (<0.01mm) of the
soil decreases markedly after dehydration. This may perhaps be related to
the transformation of iron oxides acting as cementing agent. Among the three
soils the order of degree of dispersion is yellowish brown earth > red earth >
laterite.

The shrinkage of soil blocks as calculated by volume weight ranges from
0 to 43 % (by volume). In addition to the effect of organic matter, it is
also influenced by amorphous iron. There is positive correlation (r =0.455*)
between shrinkage and amorphous iron oxides content for 28 incubated
samples.

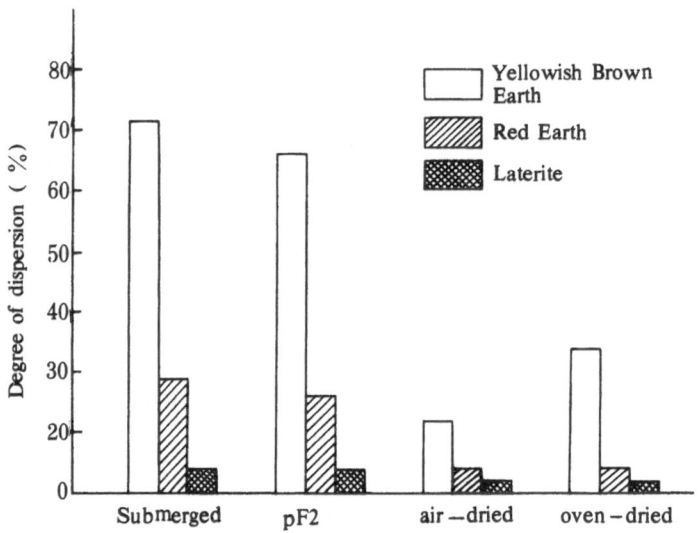

Fig. 1 Effect of dehydration on degree of
dispersion, < 0.01 mm of three soils

There is positive correlation between the shrinkage of soil blocks and
the apparent destruction percentage of aggregate with diameter larger than
0.25 mm for the yellowish brown earth and red earth, with correlation coefficients of 0.717 and 0.681 (n = 10) for air-dried samples and oven-dried

samples respectively. Laterite is an exception, where the destruction percentage may be negative in value. It appears that shrinkage caused by excessive amorphous iron oxides is not favorable to the formation of water-stable structure. This supposition seems to be supported by the phenomenon that there is significant positive correlation between the activity of free iron oxides (represented by X) and the apparent destruction percentage of aggregate with diameter larger than 0.25 mm (represented by Y) ($Y= 13.5\ X^{0.44}$ $r= 0.680$, freedom is 33) (Fig. 2).

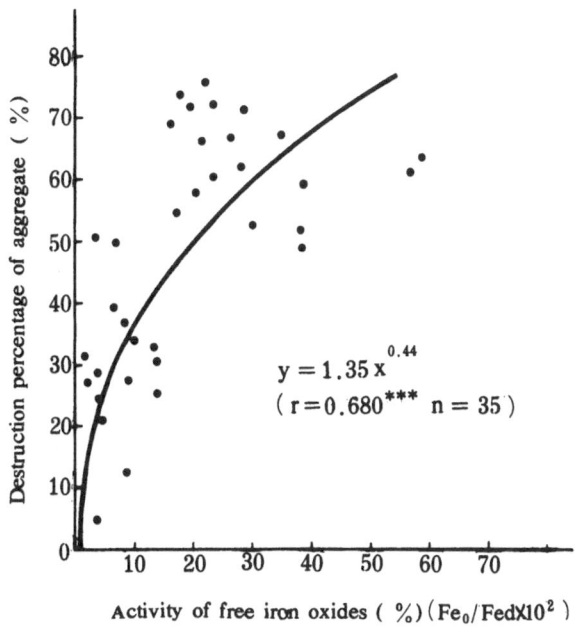

Fig. 2 Relationship between destruction percentage of aggregate ($>$0.25 mm) and activity of free iron oxides of the soil

It is worthy to be noticed that although the degree of dispersion is positively correlated with the activity of free iron oxides, the slopes of the straight lines are different under various dehydration conditions. For the submerged control, dehydrated at pF 2, air-dried and oven-dried samples the b values and correlation coefficients are 0.71 (0.554*), 1.54 (0.752**), 0.25(0.788***)and 1.00 (0.981***) respectively (freedom = 12) (Fig. 3).

CONCLUSION

In conclusion, it is considered that for paddy soils with a high degree of dispersion agricultural pratices leading to the weakening of conditions for the activation of iron oxides (such as avoiding prolonged submerging) and the strenthening of conditions for the ageing of iron oxides (such as drainage) are beneficial to the formation and maintenance of good structure of soils.

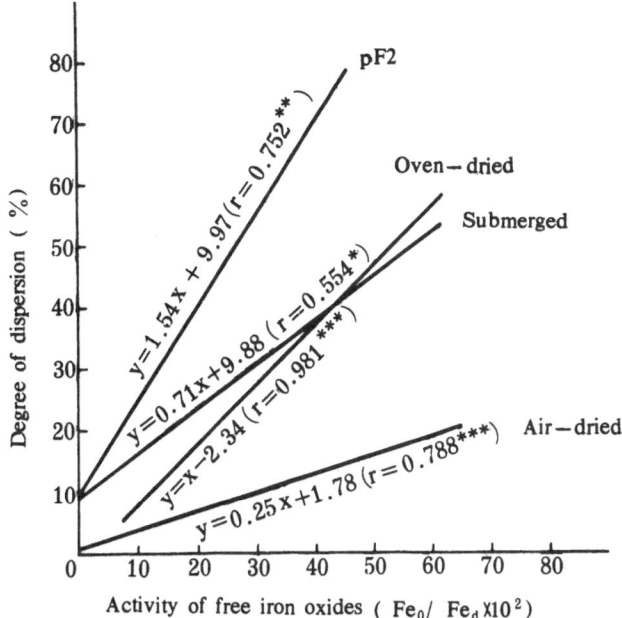

Fig. 3 Relationship between degree of
dispersion (<0.01 mm) and activity of free
iron oxides of the soil

REFERENCES

(1) Chen Jia-fang, He Qun, Content of colloid in neutral paddy soils and
    its influence on the physical properties of soils. Turang, 2, 45-50(1979).
    (in Chinese)
(2) Schwertmann, U., The differentiation of iron oxide in soil by extraction
    with ammonium oxalate solution. Z. Pflanzenernahr. Dung. Bodenkd.,
    105, 194-202 (1964). (in German)
(3) Schwertmann, U., Inhibitory effect of soil organic matter on the
    crystallization of amorphous ferric hydroxide. Nature, London, 212,
    645-646(1966).

DETERMINATION OF STABILITY CONSTANT
OF Mn (11)-COMPLEX BY VOLTAMMETRIC METHOD

Bao Xue-ming, Ding Chang-pu
(Institute of Soil Science, Academia, Sinica, Nanjing)

In our previous work (1) it was shown that in submerged paddy soils the ferrous iron complexed by water-soluble organic substances accounted for 10-40% of the total water-soluble iron, and the stability constants as determined potentiometrically ranged from 2.5 to 4.0. Since the chemical properties of Mn^{2+} are similar to that of Fe^{2+}, it seems reasonable to suspect that Mn^{2+} can also be complexed by water-soluble organic substances of paddy soils.

For the determination of stability constant, the commonly used methods in soil science are ion exchange and potentiometric titration. In recent years the ion-selective electrodes have also been used. In view of the fact that Mn^{2+} ions can produce a well-defined anodic wave at a carbon electrode, in this work an attempt was made to use the voltammetric (polarographic) method to determine the stability constant of Mn^{2+}-complex in soils.

PRINCIPLE

After the complex formation of Mn^{2+} with ligands, the shift in peak potential, ΔE_p, is measured with an oscillograph, and the stability constant, log K, and mean coordination number, $\bar{n}$, are calculated from the following equation: $\Delta E_p + 0.059 \log \frac{i_{Mn}}{i_{MnL}} = 0.059\, \bar{n} \log(L) + 0.059 \log K$,

where

ΔE_p —— shift of peak potential in presence of ligands,
i_{Mn} —— peak current of Mn^{2+} alone,
i_{MnL} —— peak current of Mn^{2+} in presence of ligands,
$\bar{n}$ —— mean coordination number,
L —— concentration of free ligands,
K —— apparent stability constant.

If plot is made for the logarithm of the concentration of complexing agents (L) against the $\Delta E_p + 0.059 \log \frac{i_{Mn}}{i_{MnL}}$, the log K can be found from the intersect of a straight line, and the $\bar{n}$ is calculated from the slope(2),

DETEMINATION OF THE CONCENTRATION OF COMPLEXING AGENTS

In the above equation, it is necessary to know the concentration of complexing agents. In the present work three methods have been used: (A) Potentiometric titration; (B) Method of complexing capacity; (C) Titration with $KMnO_4$.

The method (B) is briefly as follows.

A certain amount of Fe^{2+} is added to a certain amount of complexing agent. After the completion of complex formation α-α-dipyridyl is added, and the color intensity is measured at different time intervals. The concentration of Fe^{2+} measured is plotted against time(t). The concentration of ionic iron is obtained by extrapolating to time o, and the concentration of complexed iron is taken as the difference between the total iron and the ionic iron (Fig. 1). In a series of determinations a limit is reached where the amount of complexed iron does not change with the increase in added Fe^{2+}. We take this value as the complexing capacity. This method is a development(3) of a method for determining the complexed iron in paddy soils in a previous work (1).

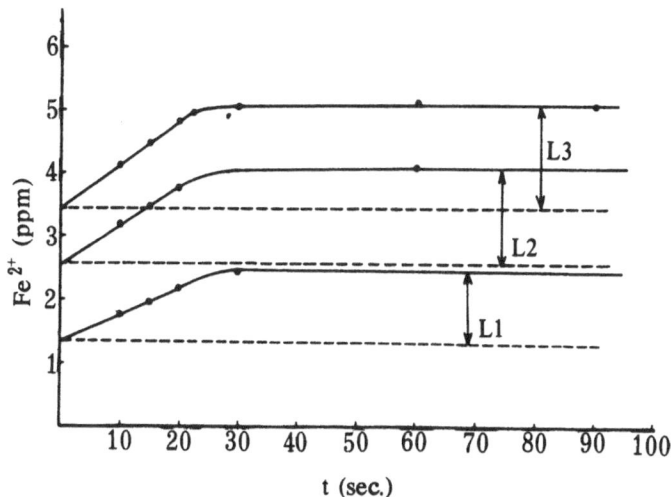

Fig. 1 Determination of complexing capacity
of organic substances (color intensity
after the addition of dipyridyl)

PROCEDURE AND THE CALCULATION OF LOG K

The water solution from the incubated soil or green manure is treated with beds of cation-exchange resin packed in a cellophane bag to remove the metallic ions, and then let it react with Mn^{2+} ions in a 0.2 M ammonium acetate solution of pH 7.4. The E_p is measured with a glassy carbon electrode, and then log K and ñ are calculated. Since the composition of the complexing agents is rather complicated and the mechanism of complex formation with Mn^{2+} ions is not known we call the obtained iog K as "apparent mean stability constant".

RESULTS AND DISCUSSIONS

1. Comparison of Results Obtained by Different Methods of Determining the Concentration of Complexing Agents

The potentiometric titration and the method of complexing capacity are based on different principles, and as a consequence the results obtained by these two methods differ. Generally the concentration as determined by the former method is higher. But, owing to the logarithmic relationship in calculation, the computed log K values by the two methods do not show large difference (Table 1). Although the concentration as determined by titration with $KMnO_4$ does not have direct relationship with the ligands, it was observed that there was a roughly proportional relationship between the two. Therefore, sometimes we also use this method to get results of relative significance.

Table 1 Computed log K based on different methods for determining the concentration of complexing agents

Stage of decomposition	Method	Log K			
		Astragalus	Vetch	Grass	Crotalaria
Initial	K_p^*	3.6	3.7	3.7	4.0
	K_m^{**}	3.6	3.6	3.8	3.9
Middle	K_p	3.7	3.9	3.9	3.7
	K_m	3.8	4.1	3.9	4.0
Final	K_p	2.2	2.2	3.0	3.0
	K_m	2.3	2.2	3.0	2.9

* Potentiometric titration

** Complexing capacity

2. Effect of Kind of Plant Material and Decomposition stage

5% suspensions of different plant materials in water were incubated, and the log K were determined at different stages of decomposition. The results are shown in Table 2.

It is seen from the Table 2 that: (A) There are differences in stability constant among different organic substances, amounting to 3 orders of magnitude. The complexing power of decomposition products from vicia sativa is strongest, and that from rice straw weakest. (B) The log K changes with the stage of decomposition of plant materials, with a maximum at the stage of intensive decomposition, and a minimum at the final stage of decomposition.

Table 2 Log K of Mn-complex with decomposition
products of different plant materials

Plant material	Log K_R^*			
	2-5 days	5-6 days	12-14 days	20days
Vicis sativa	5.0	6.0	3.0	
Vetch	5.0	5.4	3.0	3.0
Astragalus	4.7	4.8	3.0	2.9
Rice straw	4.0	4.0	2.7	2.9

*Titration with $KMnO_4$

3. Relationship Between Stability Constant and Electric Charge of Complexing Agents

We have attempted to distinguish the electric charge of the complexing agents by passing through columns of ion-exchange resin, and then determine the stability constants separately. The following are indicated in Table 3.

Table 3 Relationship between stability constant
and electric charge of complexing agents*

Plant material	Log K_m		
	Original solution	Negatively charged fraction	Positively charged fraction
Alfalfa	4.7	2.2	3.1
Hairy vetch	4.6	4.0	2.9
Crotalaria	4.o	4.1	3.9
Vetch	3.8	3.9	2.9
Grass	3.7	3.7	3.1
Astragalus	3.9	4.0	3.1
Rice straw	2.9	2.9	—

*No electrical neutral fraction isolated

1) With the exception of alfalfa the comlexing power of the ligands carrying negative charge is about the same as that of the original solution.

2) With the exception of alfalfa, the log K values for complexing agents carrying positive charge are smaller than those carrying negative charge.

3) No complexing agent carrying positive charge was isolated from the decomposition product of rice straw. The log K for the negatively charged fraction is equal to that of the original solution. Infrared spectrogram show that there is only one intensive absorption peak by carboxyl group for this fraction.

4. Mean Coordination Number

It is seen from Table 4 that the mean coordination number($\bar{n}$) is 1 or so. Thus it may be supposed that the complex with Mn^{2+} is primarily 1:1 type.

Table 4 Mean coordination number in Mn^{2+}-complex

Plant material	Initial stage		Middle stage	
	$\bar{n}_p{}^*$	$\bar{n}_m{}^{**}$	$\bar{n}_p$	$\overset{6}{n}_m$
Astragalus	0.9	0.9	0.8	0.8
Vetch	0.9	0.9	0.9	1.0
Grass	1.0	1 .0	1.0	1.0
Crotalaria	0.9	0.9	1.0	1.0

* Potentiometric titration

** Complexing capacity

REFERENCES

(1) Bao Xue-ming, Yu Tian-ren, Studies on oxidation-reduction processes in paddy soils. VIII. Characterization of the water-soluble ferrous iron. Acta Pedologica Sinica, 15 13-21 (1978). (in Chinese with English summary)
(2) Inczédy, J., Analytical Applications of Complex Equilibria. Ellis Horwood, Chichester, 147-165 (1976).
(3) Zunino, H. P. Determination of maximum complexing ability of water soluble complexants. Soil Sci., 114, 414-416 (1972).

DETERMINATION OF pH OF PADDY SOILS IN SITU

Cang Dong-qing, Yu Tian-ren

(Institute of Soil Science, Academia Sinica, Nanjing)

In 1937 McGeorge (1) first determined the pH of soils in situ with a spear-type glass electrode. In the early forties several papers(2-5) were devoted to factors affecting the measured value. The commonly accepted view was that the results were not very reliable if the water content of the soil was below a certain limit, say, the field capacity. In the last thirty years research work in this field was not active. In view of the progress in theory and technique in the determination of pH in the last decade, we think that it is feasible to reconsider this problem, starting primarily from the obvious advantages of determining pH directly in the field.

FACTORS AFFECTING THE RESPONSE
TIME OF GLASS ELECTRODE

There should be no special difficulty for the determination of pH of paddy soils if the soil is water-saturated. However, under unsaturated conditions the discontinuity of water films around soil particles and also on the glass electrode may raise some problems. Inasmuch as the essential requirement for the glass electrode to indicate a reliable potential is the establishment of chemical equilibrium between its whole surface and the soil solution with respect to hydrogen activity, it may be supposed from the diagram shown in Fig. 1 that it is the factors affecting the speed and distance of diffusion of H^+ ions from soil solution to the water film originally existing on the glass surface that is important for the rapid attainment of equilibrium potential of glass electrode.

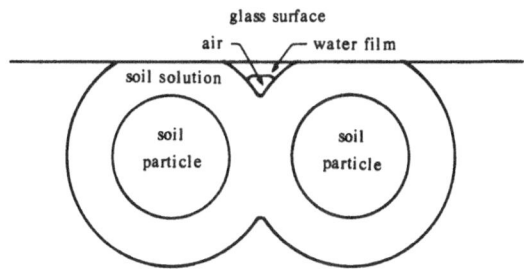

Fig. 1 Diagram of the contact between
glass electrode and unsaturated soil

1. Effect of Water Content

Figs. 2,3 show that the higher the water content of the soil, the more rapid the attainment of equilibrium potential. It is worthy to be noted that it is possible to get an equilibrium reading even under conditions where the water content is conditions where the water content is as low as 30% of the water-holding capacity, if sufficient time is allowed to elapse.

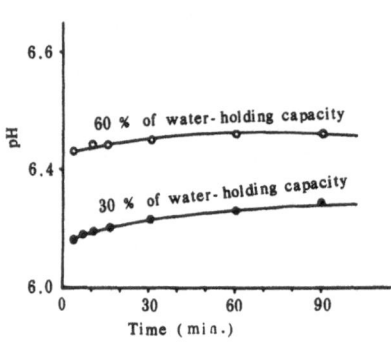

Fig. 2 Effect of water content of a paddy soil derived from red earth on response time of glass electrode

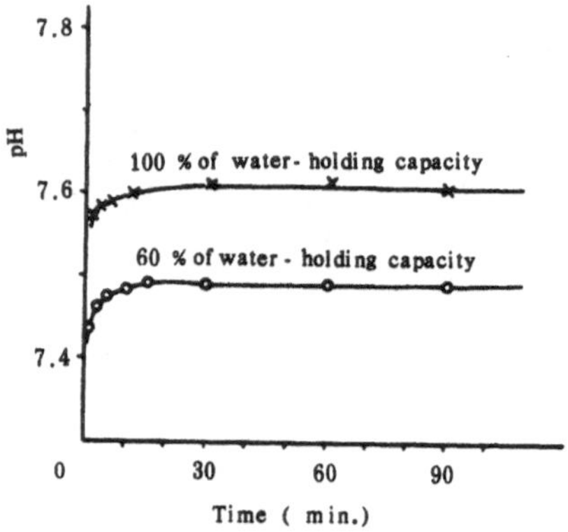

Fig. 3 Effect of water content of a paddy soil derived from calcareous alluvium on response time of glass electrode

2. Effect of Aggregate Size

It is seen from Fig. 4 that the smaller the aggregate size, the more rapid the attainment of equilibrium potential. This is obviously due to the relatively short distance for the diffusion of H^+ions.

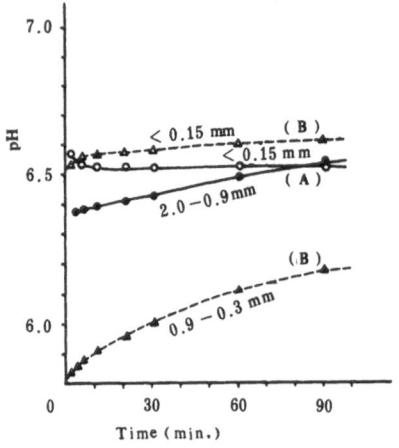

Fig. 4 Effect of aggregate size of the soil on response time of glass electrode soil
A derived from loess soil
B derived from red earth

3. Effect of Buffering Capacity

Fig. 5 shows that for the original soil the equilibrium potential attains more rapidly than when the soil is mixed with three times its weight of sand.

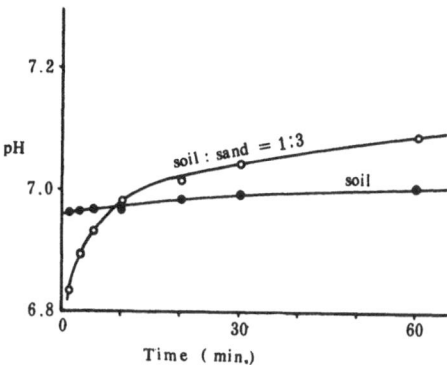

Fig. 5 Effect of buffering capacity
of a neutral paddy soil on
response time of glass
electrode (water content=50%
of water-holding capacity)

In summarizing, it may be said that although in some field soils it needs hours for the attainment of true equilibrium potential if the water content of the soil is very low, experiences show that the pH reading three minutes after the insertion of glass electrode into the soil generally differs by not more than 0.1 pH unit from the steady reading, and thus the results are sufficiently accurate for general diagnostic purposes.

LIQUID-JUNCTION POTENTIAL

Although the liquid-junction potential caused by charged soil particles— the so-called suspension effect—is a troublesome problem common to all pH determinations in soils, for field soils with low water content there may sometimes arise an additional trouble, i.e. the time needed for a steady potential is longer than when the soil is wet, as is shown in Table 1.

The water content of the soil may also affect the magnitude of the liquid-junction potential. It is seen from Table 2 that for an acid paddy soil derived from yellow earth the liquid-junction potential caused by soil particles is slightly larger when water cotent is 10% than when it is 25% or when the soil is wetted with N KCl solution.

Table 1 Effect of water content of paddy soil on establishment of liquid-junction potential

Water content (%)*	Time for steady reading (min.)			$\Delta E(mV)$**		
	Acid sandy soil	Acid clayey soil	Neutral soil	Acid sandy soil	Acid clayey soil	Neutral soil
30	7–10	7–10	–	4	3	–
50	7–10	–	7–10	2	–	3
75	–	5–7	3–5	–	2	2
100	2–5	3–5	2–3	–	2	1

* As % of water-holding capacity
** Difference between reading at 1 min. and steady reading.
Mean of 5 measurements

Table 2 Effect of water content on liquid-junction pontential caused by an acid paddy soil

Water content (%)	$Ej \ (mV)$*				
	Beginning	1 min.	3 min.	5 min.	10 min.
10	−1.3	1.5	2.9	3.5	4.2
15	−6.9	−3.3	0.4	1.3	0.9
25	−4.2	−1.7	−0.8	−0.5	−0.8
25(N KCl)	−3.6	−0.3	−0.1	−0.5	–

* Potential difference between two identical calomel
electrodes when one is in direct contact with the
soil and another one is through 2 layers of wet
filter paper

In order to know the actual magnitude of liquid-junction potential in situ, some measurements were made in spring. It is seen from Table 3 that if the water content of the soil is sufficiently high the liquid-junction potential caused by the soil generally lies below 6 mV after 3–5 minutes of contact between the salt bridge and the soil, corresponding to about 0.1 pH unit.

From above it may be concluded that the problem of liquid-junction in the determination of soil pH in situ is not especially serious for ordinary paddy soils. And actually, experiences show that if care is taken the error due to this cause generally does not exceed 0.1 pH unit. Of course, the liquid-junction potential caused by charged soil particles per se is an unsolved problem theoretically and technically in the determination of soil pH.

Table 3 Liquid-junction potential caused by some paddy
soils in situ

Soil type	Water content (%)	Ej(mV)			
		Beginning	1 min.	3 min.	5 min.
Acid red	32.4	-4.7	-6.1	-6.1	-5.9
Neutral alluvial	31.2	-5.5	-6.2	-5.1	-3.3
Neutral lacustrine	23.9	-8.4	-3.2	-2.1	-1.7
Calcareous alluvial	22.9	-	-	-7.0	-

SOME MEASUREMENTS IN SITU

In recent years some measurements were made in some places of southern China. The followings are some of the results:

1) Table 4 shows that for two acid paddy soils the pH as determined in situ is about 1 unit higher than when dried and then determined in the laboratory, while for another two acid paddy soils where the superficial water has been drained the pH is about the same as those determined in the laboratory.

Table 4 Soil pH during the growing period of rice

Parent material	Location	pH	
		In situ	Laboratory[**]
Sandstone	Jiangxi	5.9	4.9
Yellow earth	Zhejiang	6.4	5.4
Red earth[*]	Guangdong	5.3	5.4
Yellow earth[*]	Guangdong	5.5	5.6

[*] Drained wet soil
[**] Soil: water = 1:1

2) Table 5 shows that when paddy soils are planted with wheat or green manure the pH as determined in situ is slightly lower than that determined in the laboratory for acid and neutral soils, whereas the Fig. 5 in situ may be 0.3-0.8 pH unit lower for calcareous soils.

3) Fig. 6 shows that the pH as determined in the laboratory differs significantly from those determined in situ, the magnitude of difference being dependent on the water to soil ratio, especially for calcareous soils.

4) Data in Table 6 show that the soil in the field is rather heterogeneous with respect to pH, although reproducible results can be obtained if the soil is mixed.

5) In Fig. 7 is shown pH of three paddy profiles. It is to be noted that for a young calcareous paddy soil derived from the Changjiang River alluvium, the pH as determined in situ when planted with wheat is about 7.5, quite differrent from the general impression that calcareous soils have a pH of 8-8.5 as determined in the laboratory.

713

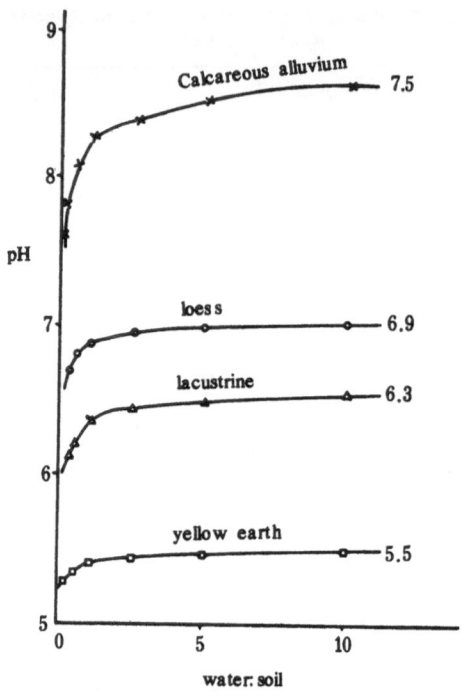

Fig. 6 Effect of water to soil ratio
on pH of paddy soils derived from
different materials (figures in
parenthesis are pH in situ)

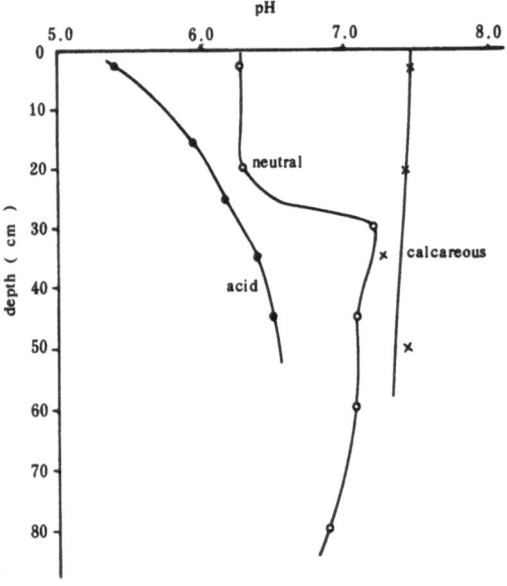

Fig. 7 pH of three paddy soil profiles
as determined in situ

714

Table 5 pH of some paddy soils when planted
with wheat or green manure (Jiangsu)

Parent material	Location	pH	
		In situ	Laboratory
Red earth	Yixing	5.4	5.5
Allubium	Yixing	6.3	6.4
Lacustrine	Wujin	6.3	6.4
Loess	Nanjing	6.9	7.0
Limestone	Yixing	7.6	7.9
Calcareous alluvium	Nanjing	7.5	8.3

Table 6 Heterogeneity of soil pH in situ

Soil	Method	pH*	Standard deviation	Coefficient of variation (%)	Range
Calcareous paddy soil derived	In situ	7.51	0.04	0.53	0.13
from alluvium	Mixed	7.40	0.02	0.31	0.08
Neutral paddy soil derived from la-	In situ	6.30	0.19	3.02	0.68
custrine	Mixed	6.10	0.03	0.49	0.10

* Mean of 10 measurements

CONCLUSION

If care is taken, the determination of soil pH can be made in situ
within 3-5 minutes after the insertion of two electrodes into the soil,
with an error of 0.1 to 0.2 pH unit for paddy soil with water contents
commonly encountered when dry-farming crops are planted.

REFERENCES

(1) McGeorge, W. T., The determination of soil reaction under field
    conditions by means of a spear-type glass electrode. J. Amer. Soc.
    Agron., 29, 841-844 (1937).
(2) Chapman, H.D., Axley, J.H., Curtis, D.S., The determination of pH
    at soil moisture contents approximating field conditions. Soil Sci. soc.
    Amer. Proc., 5, 191-200 (1940).
(3) Huberty, M. R., Haas, A.R.C., The pH of soil as affected by soil moisture
    and other factors. Soil Sci., 49, 455-478 (1940).
(4) Haas, A.R. C., The pH of soils at low moisture content. Soil Sci.,
    51, 17-39 (1941).
(5) Davis, L. E., Measurements of pH with the glass electrode as affected
    by soil moisture. Soil Sci., 56, 405-422 (1943).

MERCURY POLLUTION OF SOME PADDY SOILS

Yang Guo-zhi, Rong Jie

(Institute of Soil Science, Academia Sinica, Nanjing)

This paper reports the results of an investigation on the mercury pollution of calcareous paddy soils in a commune with special reference to its effect on Hg content in grains of rice and dry-farming crops.

MERCURY CONTENT OF SOILS

Table 1 shows the mercury content of some soils. It is seen that the mercury content of soils from four production teams under investigation is higher than the background level (30 ppb) of Team No.5 by several to over ten times. The pollution of mercury in Production Team No.1 is especially serious.

Table 1 Mercury content of soils (cultivated horizon)

| Team No. | No. of Sample | Hg (ppb) | |
		Range	Mean
1	8	191–834	455
2	9	61–328	199
3	4	97–149	115
4	5	52–107	76
5(background)	3	24–36	30

When extracted with 0.1 N HCl, no Hg could be detected for all the soils, although the pH of the equilibrium solutions were in the range of 6-7. When extracted with 0.5 N HCl the amount of soluble Hg was below 10 ppb for most of the samples, accounting for only a few percent of the total Hg in soil (Table 2). Since most of the carbonates in soils have been decomposed in an equilibrium solution of pH about 1, it seems reasonable to assume that in the form of compounds the mercury in these soils is more insoluble than mercury carbonate. A supplementary experiment (Table 3) [*] shows that the fixation of Hg by the soil happened rapidly (within 5 minutes) and the relationship between the fixed amount, m (μg/g soil), and the dose of added Hg, C (μg/g soil) may thus be characterized by the equation:

$$m = 1.10 \ C^{0.79}$$

---

[*] Yang Guo-zhi, Xia Jia-qi, Rong Jie, A preliminary study on the fixation and release of mercury in soils. Acta Pedologica Sinica, 16 (1), 38-43 (1979). (in Chinese with English summary)

Table 2 Content of acid-soluble Hg in some paddy soils
(cultivated horizon, 0.5 N HCl extracted)

		Hg	
Sample No.	Total (ppb)	Acid-soluble (ppb)	Acid-soluble % ──────── Total
101	191	6	3.14
102	583	8	1.37
103	239	5	2.09
104	545	7	1.28
105	414	7	1.69
106	834	22	2.64
107	380	15	3.93
108	458	9	1.96
201	61	tr.	—
202	137	2	1.46
203	183	4	2.18
205	246	5	2.03
206	203	7	3.45
207	162	5	3.08
208	347	18	5.18
209	328	10	3.05
background	30	—	—

Table 3 Fixation of Hg by a
calcareous paddy soil
(shaken for 30 min)

Hg added (μ g/g soil)	Fixation(%)
1	99.9
2	98.1
5	93.2
10	57.5

MERCURY CONTENT OF GRAINS

Analyses showed that the Hg content in grains of some dry-farming
crops such as wheat, corn and sorghum grown on the soils was very low.
However, when the soil was submerged, the Hg content of brown rice frequently exceeded the critical level (20 ppb) as established by sanitary
authorities. Fig. 1 shows the relationship between the Hg content of
brown rice and of soil for 21 pairs of samples. The regression equation
may be expressed as:

$$y = 0.155 \ x + 2.65 \ (r = 0.665)$$

Where y and x are Hg content of brown rice and soil in ppb respectively.

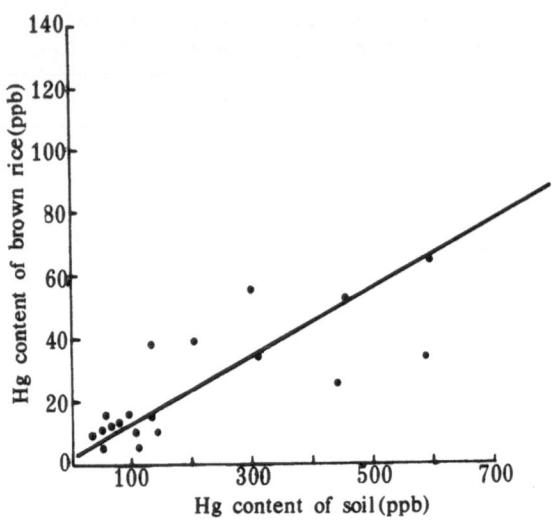

Fig. 1 Relationship between Hg contents of brown
and soil

Thus, submerging can lead to the liberation of fixed Hg, and result in a higher Hg content of brown rice.

GRADATION OF SOIL POLLUTION

In the region under study the soils in Production Team No. 5 did not suffer from Hg pollution caused by irrigation with Hg-containing sewage water. The mean Hg content of the cultivated horizon was 30 ppb. With a standard deviation of 6 ppb. If we take the background level plus three-fold standard deviation—in this cass 30+(3 x6) = 48 ppb as the critical value, it is seen from Table 1 that the severity in Hg pollution was not the same for the four production teams. Statistical data show that if the Hg content of the soil exceeds 150 ppb there is the liability for the Hg content of brown rice to exceed the critical level of 20 ppb. Therefore, we distinguish the paddy soils of this region into three grades with respect to Hg pollution:

48 ppb Hg————————————non-polluted
48-150 Ppb Hg————————slightly polluted
150 ppb Hg————————————moderately polluted

In this commune there was no such severe case in pollution as to affect the growth of rice by excessive Hg. A supplementary experiment showed that manuring or the application of phosphate fertilizers were effective in decreasing the Hg content of brown rice.

EFFECT OF NITROGEN SOURCES ON SOME PHYSIOLOGICAL CHARACTERISTICS OF AZOLLA

You Chong-biao, Li Jing-wei, Song Wei
(Institute of Application of Atomic Energy in Agriculture,
Chinese Academy of Agricultural Sciences)

Wei Wen-xiong
(Soil and Fertilizer Research Institute, Fujian Academy
of Agricultural Sciences)

The exterior nitrogen sources seem to have a notable effect on azolla[1,2]. At a high concentration, they will inhibit the growth of azolla and nitrogenase activity[2,3]. Of these exterior nitrogen sources, $CO(NH_2)_2$ produces the most serious effect, $NH_4^+ - N$ somewhat less, and NO_3^--N the least or no effect al all[4]. However, during the cultivation of azolla it is imperative to apply adequate amount of nitrogen fertilizer in order to bring about early budding[1], and as often as not, the ammoniacal nitrogen predominates in the case of paddy soil. This paper aims to look into the effects of various nitrogen sources on growth and nitrogen fixation of azolla, and find out the relationship between Azolla imbricata grown in paddy soil and nitrogen nutritional conditions and effect of various nitrogen sources on growth and nitrogen fixation of azolla, with special emphasis on the function of nitrate nitrogen.

MATERIALS AND METHODS

1. Materials

In the experiment, a local strain of Azolla imbricata (ROXB) NaKai, "Azolla Putian" and the algae-free one of the same azolla were used. The latter was obtained through stem-tip culture by the Soil and Fertilizer Research Institute of Fujian Academy of Agricultural Sciences, with no activity of acetylene reduction whatever.

2. Gulture Conditions

1) In the experiments of various nitrogen sources, "Zhejiang Agr. 6302" nitrogen-free culture solution (hereafter referred as 6302(5) was used as the basic culture solution, into which was added the nitrogen in different concentrations as required in the experiment, and the balance of other kinds of ions in which was properly adjusted pH value of the solution maintained at 5.8-6.2. The cultivations went on under natural conditions with nocturnal temperature of 12-17ºC and diurnal temperature of 22-28ºC, with culture solution being changed every four days.

2) In the experiments concerning the effect of NO_3^--N on nitratase activity of azolla, the same basic culture solution was employed. Before testing, algae-free azolla was removed into 6302 culture solution containing 0.23 mM $(NH_4)_2SO_4$

and 2% sucross and cultivated there for a week. Before experiment started, (N (NH$_4$)$_2$SO$_4$ was replaced by KNO$_3$ in different concentrations and the concentration   of sucross was lowered down to 0.2 %. The experiment was conducted exclusively under controlled conditions: light, intensity 20,000 Lux, temperature 28°C in day time and 19°C at night, with a change of the culture solution every day.

3. Methods of Assay

1) Assay on nitrogen fixation activity: 1-5 grams of azolla was put into a flask filled with 200 ml of culture solution and cultivated under the cinditions of light intensity of 20,000 Lux and temperature between 22-25°C for five hours after acetylene being added. Then the acetylene amount generated was measured with a model SP-02 gas chromatograph(2).

2) Assay on ^{15}N: Total nitrogen was determined by Kjedahl method and the abundance of ^{15}N with a mass spectrograph(6).

3) Measurement of free amino acid: Free amino acid was extracted from azolla with hot 70 % alcohol and measured with a 835-50 type amino acid analyzer.

4) Assay on nitratase activity: Cut up the azolla into pieces of about 2 mm in size, put it under anaerobic conditions, add 0.04 M KNO$_3$ as substrate and chloromycetin to inhibit reproduction of nitratase, keep in darkness under required temperature and oscillate. Then, measure by Nason method(7).

RESULTS AND DISCUSSION

1. Effects of Various Nitrogen Nutritional Condition
on Physiological Characteristics of Azolla

As indicated in Fig. 1 unlike the case of <u>Azolla</u> <u>caraliniana</u>(8) , all kinds of nitrogen sources will inhibit the growth and nitrogen fixation of <u>Azolla</u> <u>imbricata</u> while reaching a certain concentration. The effect of NO$_3^-$ -N is the slightest of all. Having been cultivated at a concentration of 44.6 mM for ten days, azolla still has a certain biomass increment. The nitrate concentration plotted against nitrogen fixation of azolla varies in a curve of dual parabola (Fig. 1 A).

The harmfulness of ammonium sulfate appears to be much serious. At a concentration of 22.7 mM, it will cause a large portion of azolla plant to die off after a period of ten days. 0.076 mM seems to be the optimum concentration but at this point nitrogenase activity notably drops down, signifying the significant inhibitory function of NH$_4$. Effect of NH$_4$ concentration on nitrogen fixation of azolla varies in a curve of parabola (Fig. 1 C).

The same effect of ammonium nitrate lies between that of NO$_3^-$ and NH$_4^+$. Half of the azolla plant still survive after a period or ten days at 37.5 mM, but its nitrogenase activity disappears totally at 12.5 mM. The optimum concentration is 0.063 mM. The nitrogenase activity will drop down sharply in linearity with the enhancement of NH$_4$NO$_3$ concentration I (Fig. 1 D). It appears that the co-existence of NH$_4^+$ and NO$_3^-$ will give an additive result to the inhibitory function on nitrogenase activity, and produce an antagonistic effect to the toxicity to the vegetative.

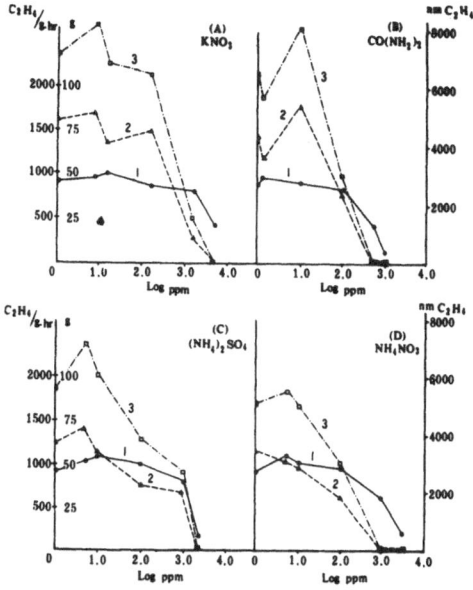

1. Yield (g); 2. Specific activity (nm C_2H_4/g
fr. wt. hr); 3. Total nitrogenase activity

Fig. 1 Effect of various N sources on the growth
and nitrogenase activity of azolla "Putian" (incubation
time 8 days)

The toxicity of $CO(NH_2)_2$ to azolla is most serious. Almost all of the
azolla cultivated at 16.7 mM will die off after 10 days. The optimum concentration for culture is ·0.017 mM. Nitrogenase activity will drop to nil
in 10 days at 8.33 mM therewithal. The effect of $CO(NH_2)_2$ concentrations
on nitrogenase activity alos varies in a parabola curve (Fig. 1 B).

Various nitrogen sources will to some extent affect the composition
and the amount of free amino acid in vivo (Table 1). The total amount of
free amino acid in azolla will notably increase when nitrogen is applied.
So much better if urea or ammonium nitrate is applied. Obvious impacts
are detected on aspartic acid and threonine. It seems to indicate that the
change of exterior nitrogen sources and their concentrations will affect the
nitrogen metabolism way of azolla, but its mechanism still calls for further
study.

2. Effect of NO_3^- on Nitrogenase and Nitratase Activity of Azolla

Having been carried out for a period, the researches on effect of NO_3^-
on azolla have not yet so far reached a view which most scientists would
share in common (3,8). According to our research work mentioned above on
this particular aspect, at a nitrate concentration of 1.49 mM or below, a
unique up-trend peak emerges in nitrogenase activity. In order to define

Table 1 Effect of nitrogen resource on free amino acid
of "Azolla Putian"

Kind of nitrogen compound	Concentration (mM)	Amino Acid (nm/mg fresh weight)		
		Aspartic acid	Threonine	Total
Check	-----	0.70	-----	1.17
Potassium nitrate	0.1	0.68	0.01	1.13
	0.8	0.20	0.85	1.42
	6.34	-----	1.33	1.98
Urea	0.833	-----	12.72	13.41
Ammonium sulfate	0.758	-----	2.87	5.43
Ammonium nitrate	0.75	-----	4.00	4.84

Note: Incubation period is one week

such a phenomenon, more experiments in a wider range of concentration of
KNO_3 and under more strict control have been carried out. The period of
experiment covered 144 hours, with the result shown in Fig. 2. It illustrates that being affected by nitrate of different concentrations, the
symbiotic algae azolla shows certainly a double parabola curve of nitrogenase
activity, irrespective of either the length of period being 24 hours, 72
hours or 144 hours, or the calculation based on fresh weight, dry weight or
chlorophyll. It is noteworthy that at the end of 72 hours, when the concentration of NO_3^- remains 0.8 mM, its nitrogenase activity will far surpass
that of control. In the case of 144 hours experiment, the nitrogenase
activity is generally higher than those of 24 hours or 72 hours, unless the
concentration of NO_3^- is above 25.6 mM. This typifies that NO_3^- exerts a
slighter inhibition to the nitrogenase activity of azolla than other nitrogen
compounds.

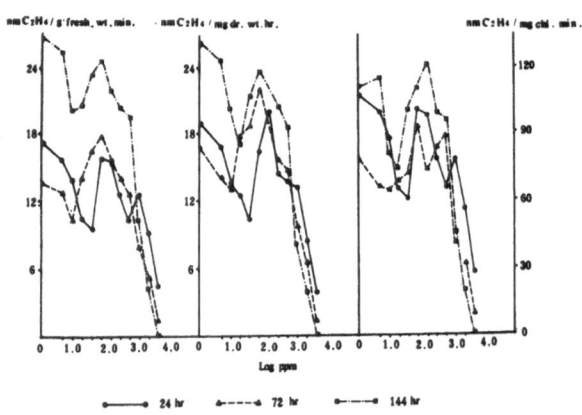

Fig. 2 Effect of concentration of KNO_3 on
nitrogenase activity of azolla "Putian"

In the 144 hours' experiment the biomass output of azolla registers a notable increment of over 20 % when concentration of NO_3^- remains 1.6 mM; but when it reaches 51.2 mM, the nitrogenase activity disappears. At this point, the output will decrease by 20 % as compared with the control.

With a view of revealing the mechanism of its function, experiments on determining nitratase activity have also been conducted. Results indicated that its variation of peak and valley is complemental with the variation curve of nitrogenase activity (Fig. 3). When the concentration of NO_3^- remains 0.8–1.6 mM, the nitrogenase reaches a high peak, whereas the nitratase activity a low valley. In other words, the accumulation of NO_2^- greatly decreases. To clarify such relationship, we use algae-free azolla of the same strain "Putian" to induce the nitratase activity by adding different concentrations of NO_3^- under sterilized conditions (Fig. 4). The result indicated that nitratase activity of algae-free azolla is higher than that of symbiotic algae azolla. A peak emerges in the variation of nitratase activity as the concentration of nitrate in the media changes, that is to say, at such a concentration, within the vegetative of azolla much more NO_2^- are accumulated during the absence of nitrogenase. Then, what is the reason that in the meantime the variation curve of NO_2^- within algae azolla records a deep valley? It seems that nitrate directly absorbed by azolla through its root system is capable of being induced to form nitratase. In algae-free azolla such nitratase will rapidly reduce NO_3^- into NO_2^-, and maximize its accumulation at medium concentration. In the symbiosis of algae and azolla, thsi process is fairly complicated. Although the physiological relationship between these two is yet to be fully interpreted, one can observe its close relationship with nitrogenase activity of the algae. In most likelihood a part of NO_2^- is transferred from azolla to the algae and consumed by the nitrogenase as electronic donor during the process of nitrogen fixation, and subsequently diminished the accumulation of NO_2^- in the azolla plant and promotes the nitrogenase activity to upgrade and raise the biomass of azolla (Figs. 2 and 3).

In order to verify the above-mentioned conclusions, an experiment was conducted on inducing the nitratase activity of algae and algae-free azolla with original set of concentration by ^{15}N-labelled K [15]NO_3 (8.7 % abundance). The following result are indicated in Table 2.

1) For these two kinds of azolla under the conditions of same concentration the actual amount of [15]NO_3^- remains equal.

2) Following the upgrade of concentration of [15]NO_3^- in the media, the amount of reduction and absorption of integrated nitrogen also proportionately increases, showing a totally positive correlation. Again it testifies to the fact that the above-mentioned variation curve holds ture.

As to some of the phenomena occuring under high concentration, e.g., the accumulation of ^{15}N rapidly increases in vivo, but nitratase activity remains unlifted, nitrogenase activity drastically drops down, output of biomass acutely diminishes, azolla plant gradually breaks up and dies off, etc , all of these can be looked upon as a certain kind of intoxication of pathology or physiology.

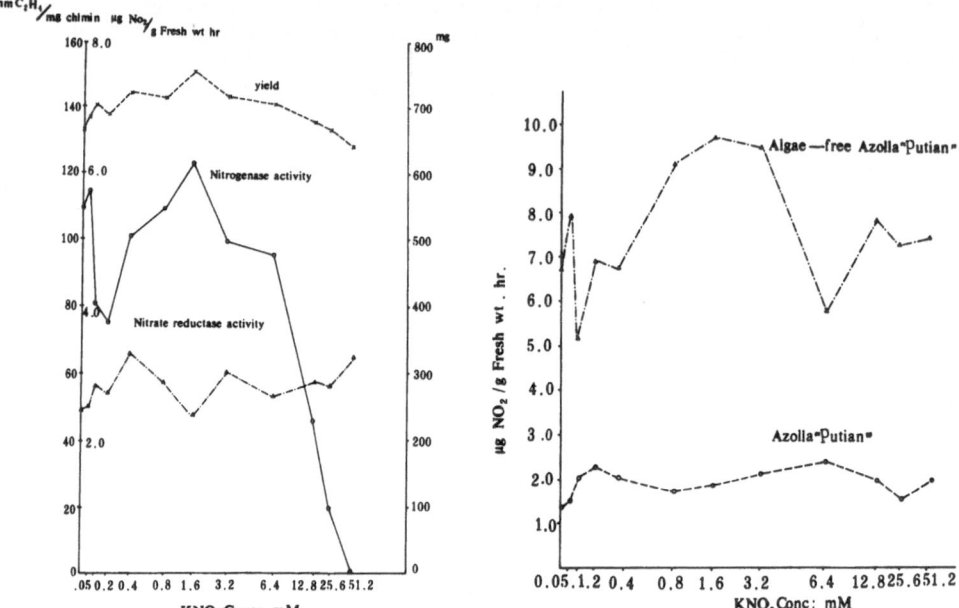

Fig. 3 Effect of KNO3 concentration on nitrogenase activity yield and nitrate reductase activity of Azolla "Putian"

Fig. 4 Effect of KNO3 concentration on nitrate redutase activity (72 hrs)

Table 2 Effect of nitrate concentration on the amount of $^{15}NO_3^-$ reduced and nitrogen content of "Azolla Putian" in vivo

$K^{15}NO_3$ concentration (mM)	Total N (mg N/g fresh wt.)	^{15}N abundance (atom %)	$^{15}NO_3$-reduced (μ mole/g/day)
0.00	1.92	0.379	-----
0.05	2.24	0.480	0.6
0.10	2.19	0.564	1.03
0.20	2.26	0.624	1.39
0.40	2.21	0.673	1.62
0.80	2.29	0.804	2.41
1.60	2.51	0.876	3.08
3.20	2.49	1.097	4.38
6.40	2.58	1.178	5.05
12.80	3.04	1.366	7.33
25.60	2.95	2.183	13.10
51.20	3.43	2.651	18.94
Algae-free Azolla	0.88	0.646	0.59

Note: 1) Incubation period is 72 hours
2) Algae-free azolla was cultivated in culture solution containing 0.05 mM KNO3

REFERENCES

(1) Liu C. C. Use of Azolla in rice production in China. In "Nitrogen & Rice", IRRI, Philippines, 375-394(1978).

(2) Liu Zhong-zhu et al., Preliminary study on some physiological aspects of Azolla. Scientia Agricultura Sinica, 2, 63-70(1979). (in Chinese)

(3) Peters, G.A., The Azolla Anabaena azollae symbiosis. In "Genetic Engineering for Nitrogen Fixation" (ed. by Hollaender, A.), Plenum Press, New York, 231-258(1977).

(4) Peters, G.A. et al., The Azolla, Anabaena azollae relationship II. Plant Physiol., 53, 820-824(1974).

(5) Institute of Soils and Fertilizers, Agricultural Academy of Zhejiang, Cultivation and Utilization of Azolla. Agriculture Press, Beijing(1974). (in Chinese)

(6) You Chong-biao, Li Yu-gui, Liu Dong-lai, Hu Tie-min, Ling Ming-de, Analysis of ^{15}N by mass spectrograph in the biological samples. Yuanzineng, 6, 535-540(1965). (in Chinese)

(7) Garrett, R.H., Nason, A., Further purification and properties of Neurospora nitrate reductase. J. Biol. Chem., 244, 2870-2882(1969).

(8) Peters, G.A., Mayne, B.C., Ray, T.B., Toia, R.E. Jr., Physiology and biochemistry of Azolla- anabaena symbiosis. In "Nitrogen & Rice", IRRI, Philippines, 325-344(1978).

STUDIES ON NITROGEN FIXATION BY ASSOCIATION OF BACTERIA WITH RICE ROOT[*]

Lin Cang, Huang Shi-zhen, Tang Long-fei, Liu Zhong-zhu
(Soil and Fertilizer Research Institute, Fujian Academy of
Agricultural Sciences, Fuzhou)

Li Jing-wei
(Institute of Application of Atomic Energy, Chinese Academy
of Agricultural Sciences, Beijing)

It is known that nitrogen fixation in the paddy fields is more effective than that in the upland soils, and is therefore of great significance to an intensive study on the associated function of rice and bacteria on the field of biological fixation (1-3). Since 1975 we have been engaged in the study of the bacteria associated with rice root system. 627 strains of bacteria were isolated from the roots of 45 rice varieties. According to the measurement by acetylene reduction method, 31 strains out of 435 exhibit the nitrogenase activity, suggesting that N-fixing bacteria do exist in the rice root system. Two of them, m-sm-1612 and st-sm-9021, were most active. They are both rod in shape (Pls. 1 and 2), 0.55 x 1.73-3.3 μ in size, Gram negative, with peritrichous flagella (Pls. 3 and 4), of non-diffusing pigment, no capsule: on broth-agar media their colonies were round, slightly convex and yellow in colour. The physiological properties of the both are alike (Table 1).

In the last few years we have carried out an investigation further on the factors affecting nitrogen fixing activity of the above-mentioned m-sm-1612 and st-sm-9021 strains and their associated role on nitrogen fixation with rice roots in comparison with the strain of N-9, i.e. Beijerinckia sp..

MATERIALS AND METHODS

1. Strains of Bacteria under Test

Strains m-sm-1612 and st-sm-9021, both were cultured in N-free media.

Strains N-9, i.e. Beijerinckia sp., isolated in our Institute, was cultured in Starkey's N-free medium.

Strain HI-3, isolated from rice roots by I. Watanabe in IRRI, was cultured in IR medium.

---

[*] The authors are very grateful to Dr. Watanabe for the present of the nitrogen fixer HI$_3$.

Table 1 Some physiological properties of N-fixing bacteria in rice roots

Species	Litmus milk	Protein with H_2S	Decomp. with NH_3	Voges-proskauer reaction	Gelatin lique-fication	Hydrolysis of starch
M-sm-1612	Acid production & precipitation	+	-	-	+	-
St-sm-9021	ibid	+	-	-	+	-

Nitrate reduction	Indole production	Saccharide fermentation
-	-	-
-	-	-

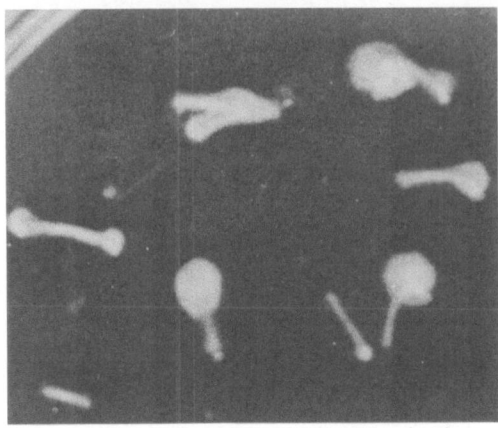

Plate 1 Bacteria grown from both ends of rice-root section

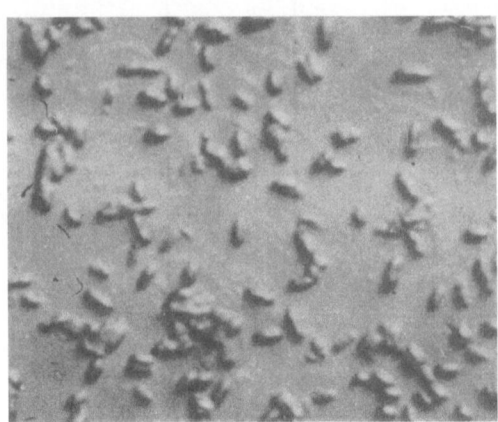

Plate 2 Microscopic photo of bacteria strain 9021

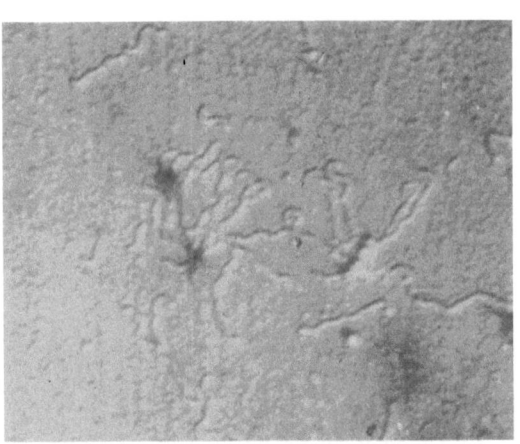

Plate 3 The microscopic photo of the peritrichous flagella of bacteria strain 1612

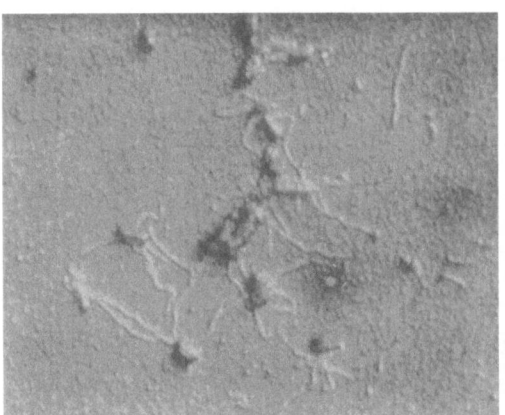

Plate 4 The microscopic photo of the peritrichous flagella of bacteria strain 9021

2. Rice Variety under Test

Rice variety Agriculture No. 1 was used.

3. Measurement of Nitrogenase Activity

1) Measure by acetylene reduction method (4). By using acetylene of 10% in concentration, the age of bacteria isolated was measured as about 60 hours.

2) Measure by tracer technique. In order to carry on the test, $^{15}N_2$ was obtained first through the oxidation of ammonium salts with LiBrO(5). After purification, it was mixed with O_2 in the ratio of 8:2. Putting the bacteria into a 100 ml oval bottle (or a test tube) together with 30 ml culture solution and sealing it with a rubber stopper, we pumped out the air from the bottle (or test tube) and filled it with mixed gas mentioned

above. The bacteria were grown at 30°C and under shaking condition for 48 hours, then the amount of nitrogen fixed was measured by Kjeldahl method; and the abundance of ^{15}N, by mass spectrometer.

4. Experiment Inoculating Isolated Bacteria to Rice Seedlings

After being sterilized, the germinant rice seeds were put into the bottle containing Starkey's agar medium, and inoculated with m-sm-1612 after two days. When the seedlings were on the 17th day, we broke the bottle, took the rice roots out and isolated the bacteria from the root under sterilized conditions, then measured the nitrogenase activity by acetylene reduction method.

5. Measurement of Associated Activity of Nitrogenase of Bacteria-Rice System

We grew the rice seedlings inoculated with bacteria tested in the bottle at 30°C and 20,000 Lux light intensity for 48 hours, took them out of the bottle, rinsed with sterilized water three times, and then measured the activity of nitrogenase.

Pot experiment was carried out for the same purpose simultaneously. Baterial manure consisting of bacteria and absorbent was made. The absorbent is a mixture of soil, zeolite and sucrose (8:2:0.05)sterilized by auto-claving, approximates to pH 7.0. After the bacteria were absorbed, the bacterial manure was kept under 28°C for 5 days, then was applied to rice during booting and flowering stages. 5-30 hours after application the activity of nitrogenase was measured in situ.

RESULTS AND DISCUSSION

1. Relation Between Nitrogenase Activity and Time of Culture, Temperature and pO_2

We made experiment on strains 1612, 9021 and N-9 to study the influence of culture time, temperature and pO_2 on nitrogenase activity.

Fig. 1 shows that the nitrogenase activity of N-9 is highest at the 18th hour after inoculation, then it drops sharply, and entirely vanishes after 56 hours. But the nitrogenase activity of both 1612 and 9021 do not appear within 24 hours. Like <u>Azospirillum</u> (6) their highest nitrogenase activity appears at the 72nd and 80th hour respectively after inoculation. The activity continues till the 96th hour, thus bringing favourite condition to the application of N-fixing bacteria(7).

Fig. 2 shows that 30°C is their optimum temperature.

Fig. 3 shows the relationship between nitrogenase activity and pO_2. The highest nitrogenase activity of 1612 and 9021 appears at $0.04pO_2$. From then on the more the pO_2, the less the nitrogenase activity, so both 1612 and 9021 are microaerophitic N-fixing bacteria. As for N-9, the nitrogenase activity increases as the pO_2 was enhanced, therefore it is a typical aerobic bacterium.

729

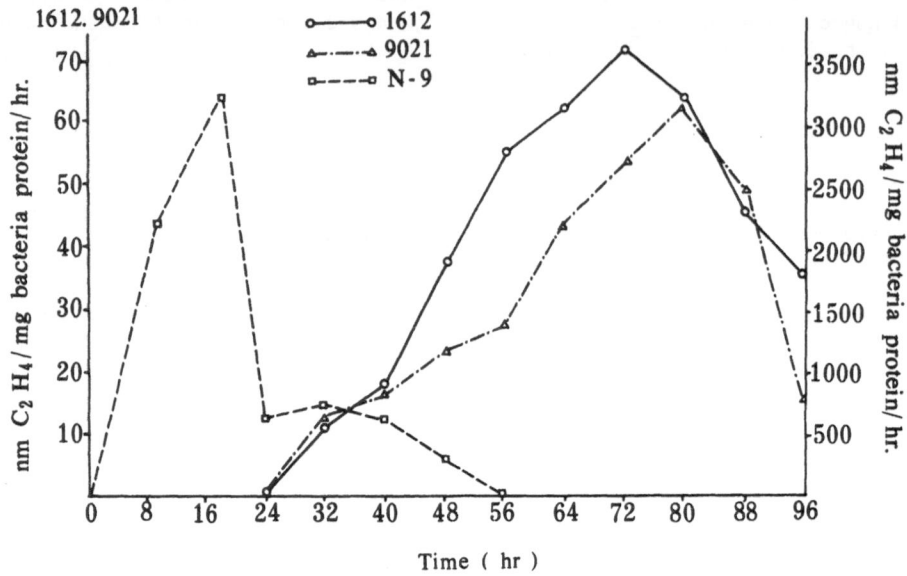

Fig. 1 Relationship between nitrogenase activity and time course of
       bacterial cell propagation

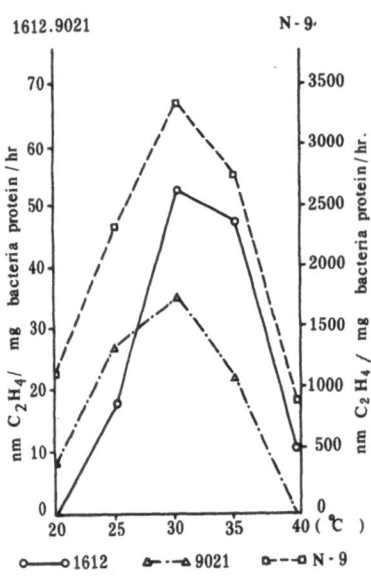

Fig. 2 Effect of temperature
       on the nitrogenase
       activity

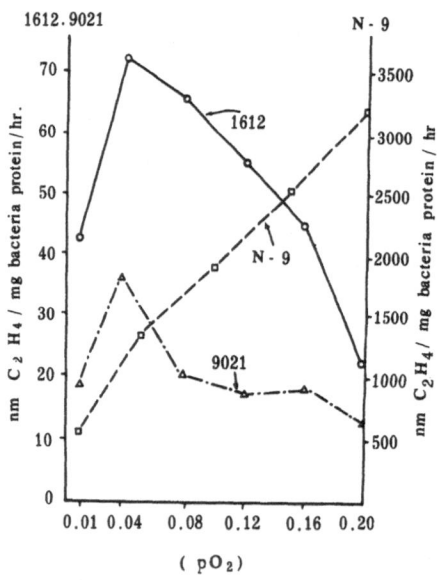

Fig. 3 Effect of pO_2 on the
       nitrogenase activity

2. Change of N-fixing Activity of Isolated Bacteria after Associating with Rice

The m-sm-1612 was inoculated to the rice variety tested growing under sterilized conditions. After 17 days we isolated the bacteria from the rice roots and measured their nitrogenase activity. The result shows that five strains isolated from the rice roots have nitrogenase activity, while three strains of the check have not; the former resists the streptomycine while the latter does not.

The nitrogenase activity of the associated system is quite different from the pure culture (Table 2). The results show that under pure culture conditions the nitrogenase activities of 1612 and 9021 are both far lower than that of N-9 (being only 1/15-1/40 of the latter) but somewhat higher than that of HI_3: whereas after associating with rice, the nitrogenase activities of 1612 and 9021 as well as HI_3 are all higher than that of N-9 (1612-rice, 8.5 times that of the latter; 9021-rice, 2 times).

Table 2 Acetylene reduction of bacteria under various conditions

Bact. strain	Pure culture (nM C_2H_4/mg bact. protein/hr)	Rice root-bact. system (nM C_2H_4/g dryed root/hr)
1612	13.1	3111
9021	8.7	769
HI_3	4.0	1877
N-9	173	368

The above-mentioned results have been further proved by ^{15}N method. The results (Table 3) show that although the bacteria show low nitrogenase activity under pure culture, yet they show high activity when associated with rice. Special attention should be paid to the fact that mixed inoculation of 1612 and 9021 to the rice seedlings greatly promotes the activity of the nitrogenase, probably because these two strains may compensate each other in physiological functions. The results are similar to those reported by Dommergues[7].

Table 3 Nitrogenase activity of bacteria under pure culture conditions

Bact strain	^{15}N fixed (μ g)	Total protein (μg)	Specific activity (nM^{15}NH$_4^+$/mg bact. protein/hr)
1612	16.6	0.615	31.6
9021	8.3	0.715	14.0
HI_3	31.5	1.26	29.1
N-9	325	5.04	74.5

Table 3 (Continued) Nitrogenase activity of
rice-bacteria system[*]

Table 3 (Continued) Nitrogenase activity of
rice-bacteria system[*]

| | Fixed nitrogen being absorbed (μg) | | | Protein | Specific activity |
	Root	Leaf	Total	(μg)	($nM^{15}NH_4^+$/mg bact. protein/hr)
1612	15.47	18.97	34.44	19.81	1.43
9021	13.54	18.11	31.64	19.46	1.34
N-9	8.62	7.59	16.21	18.24	0.97
HI_3	16.98	11.32	28.30	17.73	1.75
1612-9021	26.68	30.75	57.43	23.72	1.99
1612-HI_3	8.83	15.51	24.34	19.27	1.38
9021-HI_3	4.52	3.56	8.08	16.98	0.52
CK	1.63	0.78	2.41	16.13	0.16

[*] Incubation period of 48 hours

We applied 1612 bacterial manure to rice at booting and flowering
stages and by using acetylene reduction method measured the nitrogenase
activity of rice-soil system (Table 4). The results show no matter when
we apply the bacterial manure the activity of nitrogenase in situ is twice
that of N-9; and the activity increases with the increase of time within
30 hours as C_2H_2 injecting in.

Table 4 Effect of applying bacterial manure on nitrogenase
activity of rice-bacteria association

| Time of application of bact. manure | Time of reaction with C_2H_2(hr) | Nitrogenase activity (nM C_2H_2/g wet soil) | |
		Addition of 1612	Control
Booting stage of rice	24	230	124
	31	324	179
Flowering stage of rice	5	18.6	—
	16	49.5	—
	22	124	49
	26	186	92.9
	30	217	112

CONCLUSION

The results from our experiment indicated that bacteria of 1612 and
9021 strains isolated from rice roots exhibit close relation with rice
roots. Although their nitrogenase activity is much lower than that of
azotobacter under pure culture, yet it will raise dramatically when
associated with rice, being higher than that of azotobacter. It seems that
this specific type of bacteria plays an important role in biological nitrogen
fixation of rhizosphere of rice roots. To make further studies on N-fixing

capacity and field application of these bacteria will open up a new way or explore new sources of nitrogen fertilizer to rice.

REFERENCES

(1) Rinaudo, G., Fares-Hamad, I., Dommergues, Y., Nitrogen fixation in the rice rhizosphere: methods of measurement and practices suggested to enhance the process. In "Biological Nitrogen Fixation in Farming System of the Tropics" (ed. by Ayanaba, A., Dart, P.J.), John Wiley & Sons, Chichester, 313-311 (1977).

(2) Watanabe, I., Barraquio, W.L., Low levels of fixed nitrogen required for isolation of free-living N_2-fixing organisms from rice roots. Nature (London), 277, 565-566 (1979).

(3) Balandreau, J., Rinaudo, G., Fares-Hamad, I., Dommergues, Y., Nitrogen fixation in the rhizosphere of rice. In "Nitrogen Fixation by Free-living Micro-organisms" (ed. by Stewart, W.D.P.), Cambridge University Press, London, 57-70 (1975).

(4) Liu Zhong-zhu, Lin Cang, Huang Shi-zhen, Investigation on nitrogen-fixing bacteria in rice roots. Fujian Agricultural Sciences & Technology 4, 21-25 (1979). (in Chinese)

(5) You Chong-biao, Li Yu-gui, Liu Dong-lai, Hu Tie-min, Ling Ming-de, Analysis of ^{15}N by mass spectrograph in the biological samples. Yuanzineng, 6, 535-540 (1965). (in Chinese)

(6) Dobereiner, J., Isotopes in Biological Dinitrogen Fixation. IAEA, Vienna, 51-70 (1978).

(7) Dommergues, Y., Balandreau, J., Rinaudo, G., Weinhard, P., Non-symbiotic nitrogen fixation in the rhizospheres of rice, maize and different tropical grasses. Soil Biol. Biochem., 5, 83-90 (1973).

GEOGRAPHICAL DISTRIBUTION OF PADDY SOILS
IN CHINA

Chen Hong-zhao
(Institute of Soil Science, Academia Sinica, Nanjing)

Paddy soils are developed under anthropogenic actions. Their geographical distribution comes under influence by local conditions and is to a
great extent affected by human activities. Here is a discussion about the
general characteristics of this distribution.

GENERAL CONSIDERATION

In China paddy soil is one of the most significant of the cultivated
soil, making up about one quarter of the total area of arable land. It is
found practically in every province of the country, from the Heilong River
in the far north to Hainan Island in the south and from Taiwan in the east
to the western extremities in Xinjiang and Xizang (Tibet), with the upper
limit of elevation for paddy rice being 2,200-2,400 meters, occasionally
up to 2,600 meters. There is a tendency for its occurrence to decrease
gradually from the tropical and subtropical zones to the higher latitudes
in the temperate zones. The same is true from the coastal plains in the
southeast towards the high plateaus in Yunnan and Guizhou. But it is in
the area to the south of the Qinling Mountains, the Huaihe River and the
Bailong River that over 90% of the paddy soil are found, with the highest
concentration on the alluvial plains of the middle and lower reaches of
the Changjiang River, Chengdu Plain and Zhujiang Delta. More recently,
as more areas are extensively put under irrigation, come into being more
paddy soil areas on the North China Plain, in Nei Monggol Inner Mongolia,
Autonomous Region, Xinjiang and the provinces in Northeast China.

SOIL CHARACTERISTICS IN DIFFERENT REGIONS

The paddy soils in China may be roughly divided into five regions
(Fig. 1).

(I) South China Region. This region lies in the tropics and subtropics
and geomorphologically, is characterized by medium and low mountains, rolling
hills along with their terraces, basins, valley plains, and coastal plains.
It is possible to plant three crops a year; and paddy rice, i.e. currently
two crops of rice in succession, takes up some 40-60% of the cropping area.
The patches of paddy soil are mostly in big continuous areas, as seen in
the western part of Taiwan, the lower reaches of the Hanjiang River and
Zhujiang Delta (Fig. 2a). They may also branch out along the river banks
as seen in the regions of rolling hills and low mountains in Fujian,
Guangdong and Guangxi, or they may be just like spots dotting here and
there in the karst areas of Yunnan, Guangxi and Guangdong, forming a
series of chains as in certain valleys in southern Yunnan.

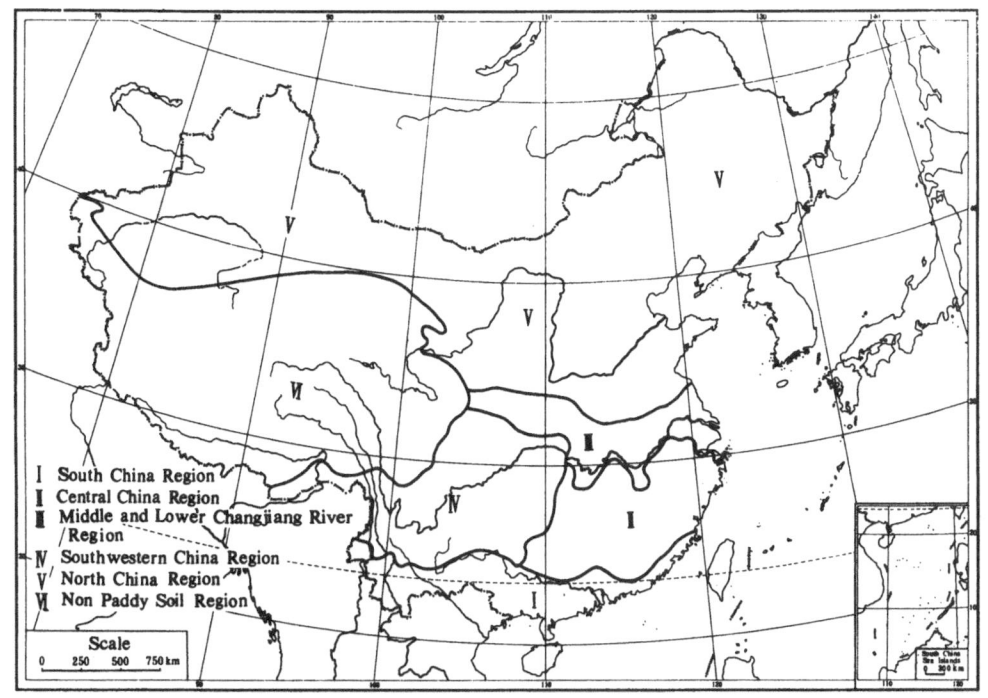

Fig. 1 Distribution of paddy soils in China

(II) Central China Region. On mountains, rolling hills along with their basins, valley plains and coastal plains in this subtropical region, rice takes up some 40-60% of the cropping area. Instead of two rice crops on other lands, hilly land is devoted to only one rice crop for each year. The paddy soils are mainly branch-shaped, as is the case in the valleys among the hills in Hunan, Jiangxi, Fujian and Zhejiang (Fig. 2b).

(III) Middle and lower Changjiang River Region. Paddy soils in this subtropical region occur chiefly as broad plains along the rivers, the sea coast and the lakes, and to a less extent among hills and in basins. Rice is the main crop, occupying 20-40% or more of the cropping area. Under a cropping system of wheat-rice or rice-rice, i.e. two main crops a year, the paddy soil are mostly in large and continuous patches, as seen in Taihu Plain, Poyanghu Plain and the plain around the junction of the Changjiang and the Hanshui Rivers (Fig. 2c). In the hilly areas also occurs branching paddy land.

735

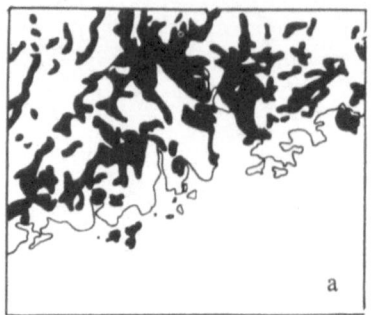

a. a largely and continuously patched pattern in the coastal plains, South China rigion

b. a branch-shaped pattern, in the valleys of the hills, Human and Jiangxi

c. a largely and continuously patched pattern, in the estuary delta of the Changjiang

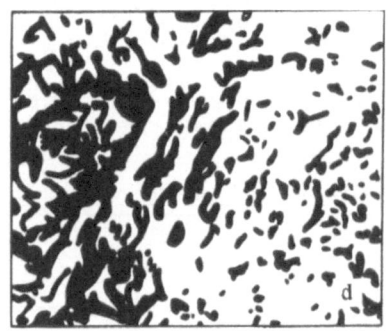

d. a band-shaped pattern, on the parallel hill, eastern Sichuan

e. spot-shaped pattern, on the plateau basins of Yunnan and Guizhou

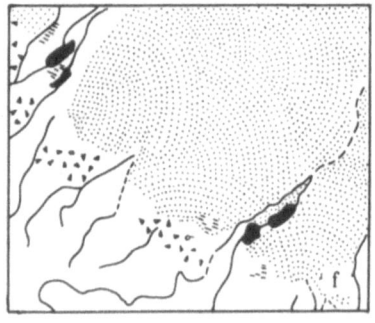

f. a spotted and sparse pattern, on the interior basin fan edge in Xinjiang

Fig. 2 Characteristics of different regions of paddy soils in China

(IV) Southwest China Region. This region is mainly on plateaus, their elevation being mostly above 1,500-2,000 meters, intermingling with hilly land, rolling hills and plateau basins. This region is, therefore, the highest among rice-producing areas in China. The climatic factors are subtropical, with clearly defined dry and wet seasons. Rice generally occupies around 20% of the cropping area, but in certain localities it may be as high as 40%. The cropping system is usually just one crop of rice, occasionally wheat-rice or rice-rice. Paddy fields are either in continuous patches on the plains of western Sichuan or in bands on the parrallel hills in eastern Sichuan (Fig. 2d). But in other localities most of them are in the form spots (Fig. 2e).

(V) North China Region. Topography in this region is mainly plains and plateaus. It lies in the temperate zone, with a short frost-free season and a long period of frozen soil. Rice takes up only less than 20% of the cropping area, with only one crop each year. Rice fields are sparsely spotted (Fig. 2f).

SOIL ASSOCIATION

Paddy soils have long been under the effect of modification by human efforts for making it adaptable to rice culture, such as land-levelling and terrace-building; and in so doing, soil has been excavated, transported and piled up. Thus, four distinctive types of soil association in the micro-relief may be identified.

(a) In the mountainous and hilly areas measures must be taken to prevent erosion. The result is the formation of bench terraces, basin-listing, etc. The soils are mainly oxidizing paddy soils, occasionally associated with reducing paddy soils. On Zhujiang Delta, from the hill-top to the valley bottom appear successively sloping fields, bench terraces and excavated beds, forming the association of cultivated laterites, sandy-loam, loamy and clayey lateritic paddy soils (Fig. 3b). In the lower Changjiang hilly regions and among the valleys of mountain ranges, the flood terraces belong to the same form of association (Fig. 3a, 3c).

(b) On the plains, fields of all descriptions often have a marked difference in their soil association, though their soil is mainly paddy soils of the oxidizing reducing type. On Zhujiang Delta, for a long distance, successive stages of enclosing more and more beaches along the coast by building dykes around them give rise to paddy soils of strongly reducing, lightly reducing, and oxidizing-reducing types (Fig. 4).

(c) In the fields where soil has been piled up after enclosure by embankments, there come into being local depressions. The paddy soils in such depression are chiefly the reducing type. But the lacustrine soils, as it get farther and farther away from the water, are successively turned into paddy soils of the medium (strong) hummus-swamp type, the muck-swamp type, the swamp glei type respectively (Fig. 5).

(d) On coastal plains, the effect of tide often quite prominent, including paddy soils reclaimed from saline waste land and a very limited area from acid sulphate soils. In South China, the soils newly formed by coastal deposits show the effect of sea water and mangrove vegetation and have an association of strongly acidic saline paddy soils, salty soils and coastal dunes (Fig. 6).

737

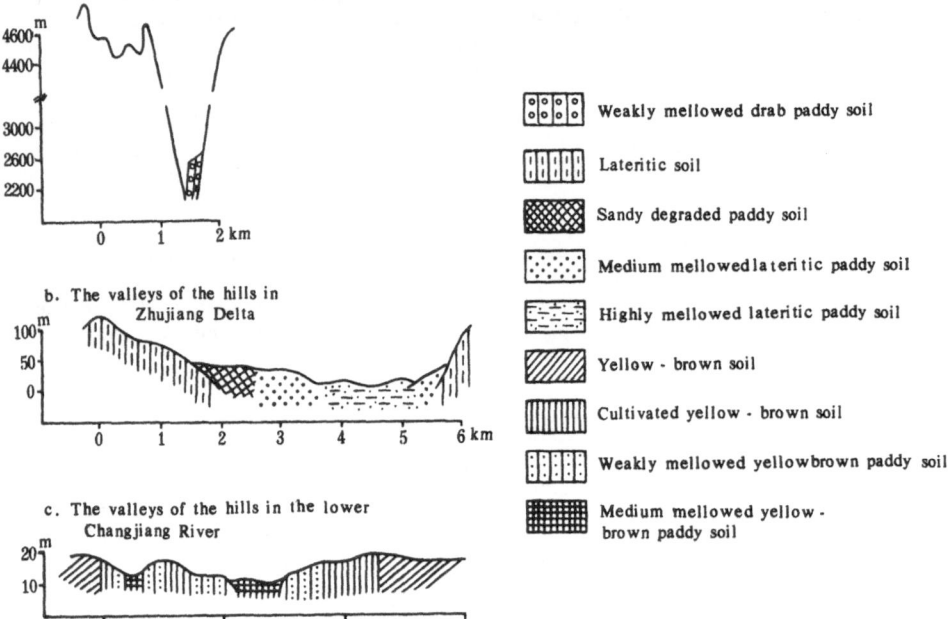

a. The fan platform in mouth of river tributary of the Lancang River

b. The valleys of the hills in Zhujiang Delta

c. The valleys of the hills in the lower Changjiang River

Weakly mellowed drab paddy soil	
Lateritic soil	
Sandy degraded paddy soil	
Medium mellowed lateritic paddy soil	
Highly mellowed lateritic paddy soil	
Yellow - brown soil	
Cultivated yellow - brown soil	
Weakly mellowed yellow brown paddy soil	
Medium mellowed yellow - brown paddy soil	

Fig. 3 Paddy soil association in mountainous and hilly area

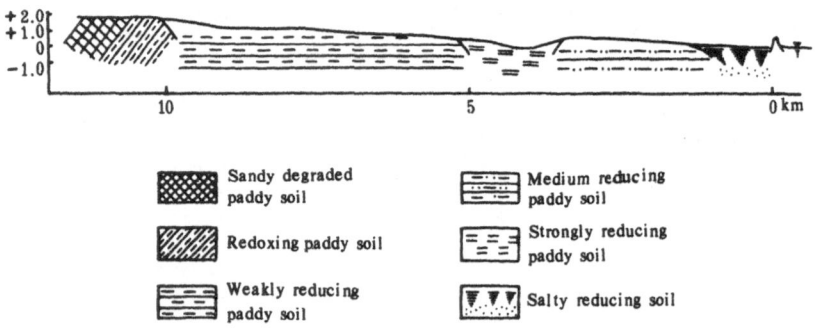

Sandy degraded paddy soil	Medium reducing paddy soil
Redoxing paddy soil	Strongly reducing paddy soil
Weakly reducing paddy soil	Salty reducing soil

Fig. 4 Paddy soil association in plain area

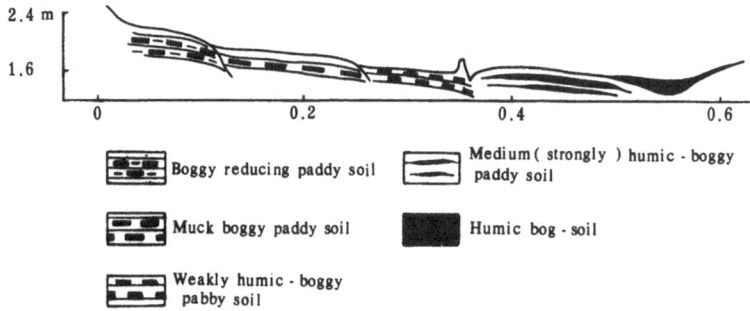

Fig. 5 Paddy soil association in local depression area

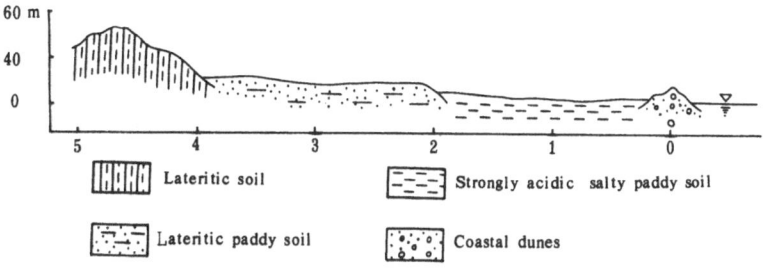

Fig. 6 Paddy soil assciation in coastal area

Since rice paddy soils are so varied and so widespread geographically, it is quite necessary to find out their characteristic properties and the rules governing them in different localities so as to give adequate scientific grounds to the measures for their utilization and amelioration.

REFERENCES

(1) Institute of Soil Science, Academia Sinica, The Environmental Conditions of Soils for High Yield of Rice. Science Press, Beijing, 1-18 (1961). (in Chinese)
(2) Ding Ying (ed.), Cultivation of Rice in China. Agriculture Press, Beijing, 147-178 (1961). (in Chinese)
(3) Institute of Soil Science, Academia Sinica, Nanjing (ed.), Soils of China. Science Press, Beijing, 465-491 (1978). (in Chinese)
(4) Kawaguchi, K., Kyuma, K., Paddy Soils in Tropical Asia, Their Material Nature and Fertility. The University Press of Hawaii Honolulu, 25-34 (1977).
(5) Fu Su-xing, Compilation of rice field map of South China. Geography, 3, 114-117 (1962). (in Chinese)

(6) Institute of Geology, Academia Sinica, Map of Land Utilization in China (1:6,000,000). Atlas Press, Beijing (1980). (in Chinese)
(7) Gong Zi-tong, Chen Zhi-cheng, The soil of Zhujiang Delta. Turang Zhuanbao, 36, 69-129 (1964). (in Chinese)
(8) Lei Wen-jin, Zhu Hong-guan, The soils of Lixiahe region and their amelioration and utilization in Jiangsu Province. Turang Zhuanbao, 36, 130-178 (1964). (in Chinese)

CHARACTERISTICS OF SOME TYPES OF PADDY SOIL
IN NORTHEAST CHINA

Lan   Shi-zhen
(Academy of Agricultural Sciences of Jilin
Province, Gongzhuling)

Northeast China situated in temperate zone is charaterized by the
continent and monsoon climate. It is hot and humid in summer, cold and
dry in winter; the annual rainfall is about 300-1000 mm; and the minimum
temperature ranges from -30 to -10°C; Its frost-free season is about
100-160 days. Therefore, the climatic conditions of this region is suitable
for rice growth. Paddy soils are scattered in this region as Fig. 1, and
among the districts of rice cultivation, Korean Autonomous Prefecture of
Yanbian is the most important one of rice cultivation in a long period, and
the paddy soils in the Prefecture are well-developed and widely distributed.
The present article deals with the genesis, development and ulitization of the
paddy soils in the Prefecture.

THE CHARACTERISTICS OF THE FORMATION OF PADDY SOILS

In Yanbian district, the paddy soil is usually about 150 days under
submergence, 150-160 days under frozen condition, and only 30-40 days under
dry condition without freezing. Therefore, the predominate effect on soil
is freezing and thawing rather than wetting and drying. The frozen layer of
the soil in winter and early spring is usually about two meter thick. Owing
to the alternation of freezing and thawing, the soil structure is stratified
or in plate form, hence it is more friable.

Under the condition mentioned above, the eluviation process in soil is
so weak that the translocation of clay is negligible. It is indicated in
Table 1 that the translocation of iron and manganese in soil is not very
intensive either(1).

The paddy soils are derived from various original soils in Yanbian
region(2): those drived from alluvial meadow soil make up 50-60 % of total
area of paddy soil, those from brown forest soil, 20 %, those from peaty
soil, 5 %, and those from white bleached paddy soil, 15 %. The morphological
features of the paddy soils are gradually changed with the continuous activities of cultivation, irrigation and fertilization. Generally, only A-C or
A-G horizon are formed after 3-5 years rice cultivation; A-P-C(G) horizons
after 20-30 years; and A-P-B-C(G) horizons after 20-30 years of rice cultivation.

As to the chemical properties, the soils in this area contain such kinds
of minerals as hydromica and montmorillonite with the former in a predominate
position; and in soil organic matter, the value of humic acid/fulvic acid is
more than 1. All the characteristics of the paddy soils in this area are
given in Table 2.

Fig. 1 **A SKETCHMAP OF DISTRIBUTION**

PADDY SOILS IN

NORTHEAST CHINA

1 ∶ 10000000

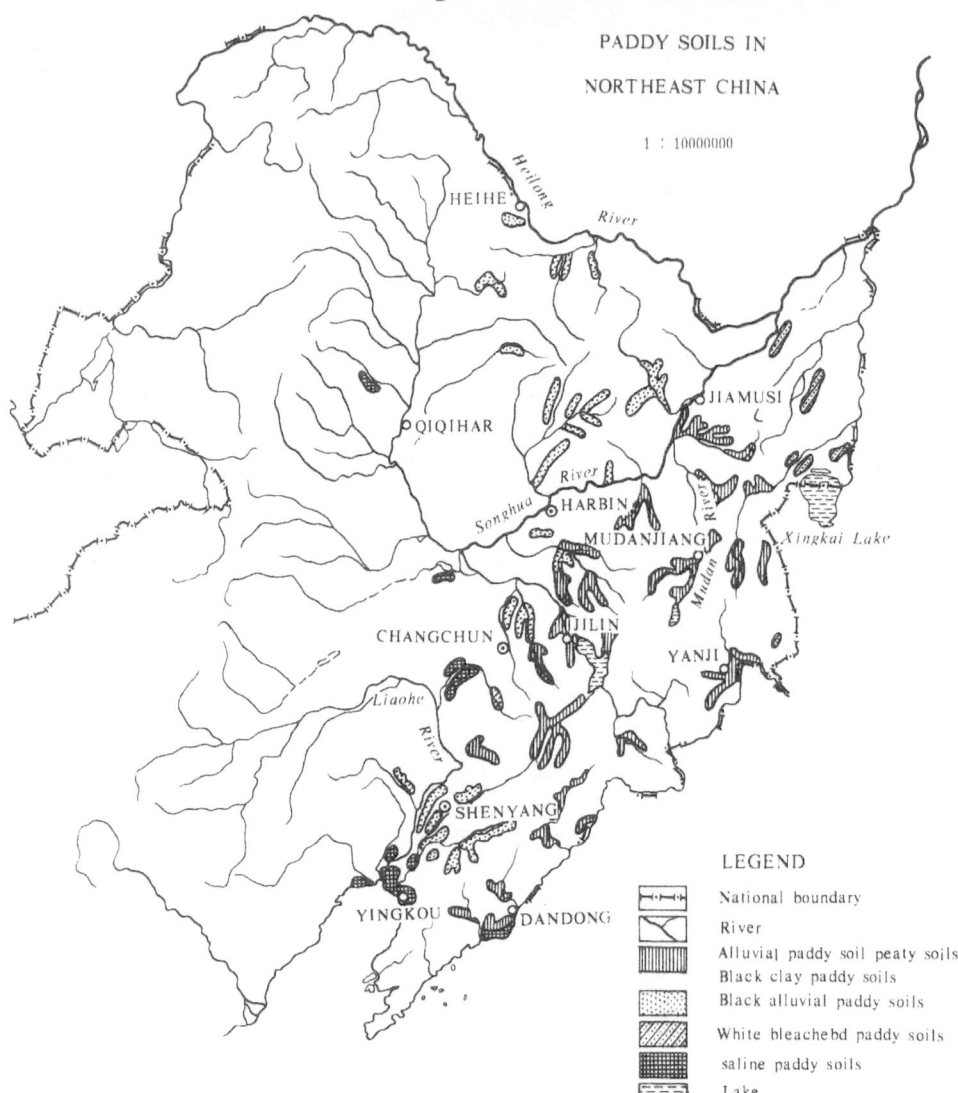

LEGEND

	National boundary
	River
	Alluvial paddy soil peaty soils
	Black clay paddy soils
	Black alluvial paddy soils
	White bleachebd paddy soils
	saline paddy soils
	Lake

Table 1   Translocation of Fe-Mn in paddy soils(1) derived
from alluvial meadow soil

Depth	Total Fe (Fe_2O_3)	Total Mn (MnO)	Active Fe*		Active Mn*	
(cm)	%	%	%	$\dfrac{active}{total}$ x100	%	$\dfrac{Active}{total}$ x100
0-13	6.20	0.097	1.14	18.4	0.044	45.4
13-20	5.52	0.082	1.09	19.7	0.050	57.5
20-27	4·66	0.095	0.89	19.1	0.042	44.2
50-80	2.87	0.061	0.56	19.5	0.019	31.1
80	4.78	0.099	0.87	18.2	0.025	25.3

* Extracted by 0.05 N H_2SO_4

Table 2   General characteristics of paddy soils in
various regions(1)

Region	Characteristics of soil formation	Composition of humus		Minerals
		Humic acid fulvic acid	Composition of humic acid	
Brown-black earth	Submergence acc-omponied with fre-ezing and thawing process	>1	-	Hydromica predominate
Yellow-brown earth	Remarkable alter-nation of submer-gence and drying	<1	Combined with Ca	Hydromica predominate, montmorillonite
Red earth	Submergence with recover of base saturation	<1	Combined with Fe, Al	Hydromica pre-dominate, Vermi-culite, mont-morillonite

MORPHOLOGICAL FEATURES OF THE PADDY SOILS

1) Alluvial paddy soils are derived from alluvial meadow soil, widely
distributed over the plains and valleys in this area. The soils include
Funi alluvial paddy soil and black alluvial paddy soil.

The Funi alluvial paddy soil stretches on smooth terrains with
sufficient water supply and thick soil layer; its texture is sandy loam,
and belongs to well-drained permeable paddy soil with a A-P-B-C profile. Its
physical and chemical properties are listed in Table 3. The soil contains
abundant mineral nutrients and organic matter of more than 2 %. The soil
is friable and with good aeration and permeability. It is the best soil
suitable for rice cultivation always with a promise of high yield.

743

Table 3 Physical and chemical properties of alluvial
paddy soil

Location	Depth (cm)	pH	Organic matter (%)	T. N. (%)	T. P. (%)	Hydrolyzable N (mg/100g)	Physical clay (%)	Physical sand (%)
Xinfeng, Yanbian	0-13	5.7	2.59	0.234	0.25	8.7	31.00	69.00
	13-37	6.4	2.25	0.186	0.07	6.3	34.56	65.45
	27-37	6.7	1.83	0.171	-	3.8	32.64	67.36
	40-50	6.8	0.78	0.082	-	4.0	18.15	81.85

This soil is widely spreaded in the valleys of the Mudan River and the Songhua River and the plains of Jilin, Tonghua and Dandong districts including Yanbian district.

The black alluvial paddy soil is distributed over the allvial plains along the Songhua River, the Liaohe River and the Heilong River. Its parent material is black earth or black meadow soil. This soil has a thicker plowed horizon brown-grey color, 3-5 % organic matter and a higher level of fertility.

2) Black clayey paddy soils are distributed over the area farther from the river, and afjacent to the mountain with a higher topography and very good irrigation condition. This paddy soil is highly fertile, and has a good retention ability of water and nutrients with A-P-C profile of the pattern of surface water paddy soils. It is characterized by a brown-grey plowed horizon with clod structures and rusty mottlings, by a hard and dark brown plowpan with prismatic structures and few rusty mottlings, and by a dark brown subsoil horizon with a friable structures of plate form or in granular form while broken. Its physical and chemical properties is showed in Table 4. The surface soil is acid in reaction while the sub-soil is near neutral. It is spreaded in the area of black earth or black meadow soil of Heilongjiang and Jilin provinces including Yanbian.

Table 4 Physical and chemical properties of black
clay paddy soil

Location	Depth (cm)	pH	Organic matter (%)	T. N. (%)	T. P. (%)	Physical clay(%)	Physical sand (%)
Xinfeng, Yanbian	0-12	5.7	2.55	0.205	0.23	66.92	33.08
	12-16	6.25	2.32	0.163	0.23	70.30	29.70
	16-25	6.72	2.11	0.169	-	77.50	22.50
	25-35	6.84	1.33	0.140	-	70.80	29.20

3) White bleached paddy soils are developed on white bleached soil. This soil is characterized by a thin grey brown plowed horizon with plenty of rusty streaks and spots which are clay-loamy in texture; a grey-whitish bleached horizon under the surface horizon with plate-like structure; a brown, compact subsoil layer with medium rusty structure, coated by

mottlings, rusty spots and Fe-Mn concretions; and a clayey of bottom soil
with grey-blue "silty granules" and more Fe-Mn spots. The physical and
chemical properties(3) of the soil are showed in Table 5. The soil is very
infertile, especially deficient in phosphorus. Rice can not grow well on
this kind of soil because of soil its poor permeability and low fertility.
Besides a small area of this soil in Yanbian, most parts of the soil are
distributed over Sanjiang Plain in Heilongjiang and some areas along the
Mudan River.

Table 5   Physical and chemical properties of white bleached
paddy soil(3)

Location	Depth (cm)	pH	Organic matter (%)	T.N. (%)	T.P (%)	T.K. (%)	Base exch.(m.e./100g) Ca	Mg
Hunchun,	0-17	6.5	2.03	0.106	0.064	2.3	9.34	5.7
Yanbian	17-29	6.6	2.05	-	0.060	2.3	7.74	4.1
	29-57	6.8	0.85	0.060	0.040	1.5	17.75	10.4

MANAGEMENT AND IMPROVEMENT OF PADDY SOILS

The people of Yanbian are experienced in growing rice and also very
good at cultivating and improving paddy soil. For example, in Xinfeng
Brigade in Yanji County, rice yield has been retaining the high and stable
record up to 6 - 6.6 t/ha for more than 30 years. An even more yield has
been acquired in recent years. Their main experiences as follows: To
counter the low temperature, freeze   injury, cold water, clayey and
depression, they have all along set great store by rice cultivation and the
management and improvement of paddy soil and approprirate measures have been
taken accordingly.1) It is important to combine the land use with the fertility maintenances, to combine the application of chemical fertilizers with
that of organic manure, to combine basal application with top dressing of
fertilizer, and to manage the soil in line with local conditions. 2) The
infertile paddy soil is improved with various methods such as ditching trenches for drainage in depression, applying lime and coal ash for imporoving
the paddy soil, applying compost and organic manure for improving black
clayey soil, etc. All these measures have enabled the production brigade
to obtain high and stable yields for years in succession.

REFERENCES

(1) Institute of Soil Science, Academia Sinica, Nanjing(ed.), Soils of
    China. Science Press, Beijing, 23-35(1978). (in Chinese)
(2) Li Zong-tie, Main types and productivities of rice field soils in
    Yanbian region, Jilin(1979, ms.). (in Chinese)
(3) Wang Ru-yong, Agricultural properties and improvement of paddy soils
    of Jilin Province(1959, ms.).(in Chinese)
(3) Wang Ru-yong, Agricultural properties and improvement of paddy soils of
    Jilin Province(1959, ms.). (in Chinese)

THE FORMATION AND CHARACTERISTICS OF COLD
SPRING PADDY SOIL

Lin Zeng-quan, Chen Jia-ju
(Fujian Academy of Agricultural Science, Fuzhou)

Being long immergad in cold water, the cold spring paddy soil is a
kind of paddy soil with a low yield. Such soil is scattered widely over
the mountainous areas, hilly land and depressions of alluvial plain in the
southern provinces of China. It was long before that the cold spring paddy
soil yielded only one crop with 1.5-2.25 tons per hectare. Over the past
one or two decades, since various kinds of improving measures have been taken
and single-crop has been replaced by double-crop, the yield has been raised
to 4.5-5.25 tons. The status of low-yield has initially turned for the
better.

Based upon the moisture conditions, physiological and chemical characteristics, the cold spring paddy soil falls into three categories, namely,
slush field, cold water field and rusty-water field. Among them, the slush
field covers a largest area, from which the other two categories are developed, some of them being complex in form.

Most of the slush fields are formed by the colluvial deposits and erosion
sediment from the nearby red soils. Through land levelling and cultivation,
the paddy soil has smaller acreages with high deviations in altitude between
two adjacent fields. The characteristics of slush field are predominated
by many factors such as topography, plant cover, catchment area, ground-
water level and drainage. Those located in the mountainous areas, on account
of less arable land and bigger catchment area, have higher ground-water
levels. According to some investigations, the average catchment area is
9.8 hectares for each hectare of slush field. Shunchang County is located
in mountainous area, covering a total acreage of 0.16 million hectares,
out of which there is only 13.8 thousand hectares of arable land. The ratio
between the mountain and arable land is 11.5 to 1. The catchment area is ten
times bigger than that of the field.

Owing to the fact that both surface water and ground-water tend to flow
towards depressions and retain there, the ground-water level moves up
gradually and eventually close to the surface of soil. Therefore, the soil
is constantly immersed in water all the year round without the possibility
of being dried out. The strong gley process inevitably induces a dispersed
clayey subsoil, appearing structureless and faint stratification. Bog type
paddy soil normally has a slush layer of about 33 cm, but in some cases, it
could be as thick as 166-200 cm. It is reported under investigation that
the yield of rice is usually declining with the increase of the thickness of
slush layer. For example, slush soil with a slush layer of 50 cm thick
gives a yield of 5.26 tons of grain per hectare; while that of 70 cm in
thickness 3.90 tons and 85 cm in thickness 3.22 tons respectively (Table 1).

The formation of cold water field is mainly affected by cold springs.
In some cases and in some areas, the ground water of the slush field, under
the pressure of the springs nearby the hills, crops up out of the soil

surface. Usually, the more the springhead, the lower the water temperature, and the higher the degree of slush of the soil. Spring water is the ground-water under a given geological formation, and hardly affected by the atmospheric temperature. According to climatic records, the soil temperature of ordinary paddy field in summer (July) is up to 28-30°C, while that of cold water field is only 21°C. In winter (January) the ordinary paddy field is frozen, but the cold water field is not, because the temperature of the springhead remains at about 19°C. As is known from the data available, the temperature of the cold water fields becomes ever lower when the distance is getting near to the springhead. When the temperature at the springhead is 19.6°C, that of the water around it is about 20.9°C, while the water 2 or 3 metres away is at about 25.9°C. Some of the cold water fields are irrigated by the cold spring water originated from the dense forest of the mountains. Although the temperature of the water is low, it does not develop any gleization because of the better condition of oxidation. It is quite a different type of cold water field.

Table 1 Effect of the thickness of slush
layer on the yield of rice

Depth of slush layer (cm)	Yield of rice (grain) (t/ha)
50	5.26
70	3.90
85	3.22

Rusty water fields are mostly located at the hillsides, the foot of slope or the verge of waterways of the red earth hills. Owing to the higher contents of ferrous components in strongly acid red soil, the soluble ferrous contents often permeate into the fields. After contacting with air, it becomes rusty sediments. Having flowed down to the paddy field, they condense into rusty flocculence, and convert into ferrous ions under the anaerobic conditions. Field investigation reveals that rice may grow normally in rusty water field of light degree, containing ferrous iron of 44 ppm Fe; while in rusty water field of medium degree, containing 126 ppm Fe, rice could hardly grow well and will partly turn yellow; in the case of rusty water field containing ferrous ion up to 259 ppm Fe, rice grows badly, roots become black, some of them decay with bad odour and die (Table 2). It seems feasible to set ferrous iron of 100 ppm Fe as a critical level in soil. As to the rusty film floating on the water, it is a ferrate compound formed by oxidation. It obstructs the healthy growth of rice as it tends to handicap the interflow and renewal of the atmosphere air of the standing water, and to aggravate the anaerobic environment of the paddy soil.

To summarize, the water in all kinds of slush field is cold but uncertainly rusty, although a rusty water field always gets hold of cold water and slush.

The bad conditions of water and temperature characterise the poor agricultural properties of cold spring paddy soil. The soil is flooded all the year round and the temperatures of water and soil are both low.

During the growth of rice the atmospheric temperature goes up rapidly in
early spring but the temperature in cold spring paddy soil follows up slowly.
Especially for such fields located in the valley, on account of shorter
period of sunshine, the temperature of atmosphere, water or soil is usually
lower than that of ordinary field. By data available, in 1960-1961, the
annual average air temperature for ordinary field is $18.7^{\circ}C$, water temperature $19.3^{\circ}C$, while for cold spring paddy soil, they are $17.9^{\circ}C$ and $18.1^{\circ}C$
respectively. The low temperature of water and soil, together with the
strongly reducing potential of such kind of field, not only obstructs the
growth of rice, but also inhibits the activities of soil microbes. It has
been observed that in the cold spring paddy soil, the quantities of
bacteria, actinomyces and fungi are obviously less than those in ordinary
field, especially aerobes. Azotobacter and cellulose-decomposing bacterium
activate weakly. Intensity of ammonification is very low. Consequently, the
accumulation of soil organic matter is over the effect of mineralization
(Table 3).

Table 2 Effect of Fe^{++} in rusty water field on the
growth of rice plant

Content of Fe^{++} in soil (ppm Fe)	Growth of rice plant
44	Normal growth
126	Stunted in growth, yellowish coloration on leaves
259	Roots turning black, sometimes rotten, growing very poor

The content of organic matter in the cold spring paddy soil generally
ranges from 3 to 5%, the quality of which is however rather poor, usually
with a C: N between 12 and 14, sometimes up to 20. The contents of nitrogen,
phosphate and potassium are relatively high, but the available nutrients are
lower. Furthermore, owing to the low excahangeable capacity, it is rather
difficult to hold the available nutrients (Table 4).

Under strong reduction conditions, the cold spring paddy soil has a
low redox potential. Therefore, it facilitates the accumulation of harmful
reductive matter. The redox potential in most of the cold spring paddy
fields is about 100 mV. When the redox potential of the soil is at 145 mV,
the ferrous content is about 9.9 m.e./100g. It will go up to 35.4 m.e./100g.
at 85 mV. Manganous compounds, hydrogen sulphide, organic acids and marsh
gas have been found along with ferrous substances in the cold spring paddy
soil. All these reducing substances, when reaching to a certain concentration, will do much harm to the healthy growth of rice.

In summarizing the above assertions, low temperature of water and
soil, lack of available nutrients, and excess of reducing substances are
the major characteristics of the low yielding cold spring paddy soils.
These characteristics are resulted from long period of over-high level of
ground water. Only through measures like digging ditches to drain,
installing drainage tiles and pipings, land levelling, planting green

manure, liming and phosphate fertilization can satisfactory results be achieved.

Table 3 The population of micro-organisms in cold spring paddy field

Soil type		Bacteria 10^6/g	Actinomyces 10^3/g	Fungi 10^3/g	Azotobacter 10^3/g	cellulase-decomposing bacterium 10^3/g	Intensity of ammonification Mg/g/day
Cold spring paddy soil	Slush field	0.41	0.51	5.72	4.4	1.1	28.6
	Cold water field	0.61	4.20	3.21	0.9	1.0	22.7
	Rusty water field	1.21	3.48	47.96	7.1	2.8	24.7
Fertile paddy soil		2.00	13.66	51.75	12.8	4.1	35.2

Table 4 Chemical properties of cold spring paddy soil

soil type		Total nutrients (%)			Available nutrients (ppm)			Organic matter (%)	C E C (m.e./100g)	pH
		N	P	K	NH_4-N	P	K			
Cold spring paddy soil	Slush field	0.192	0.088	2.151	13.3	7.6	141	3.41	13.2	6.4
	Cold water field	0.229	0.046	3.540	21.9	9.2	119	4.43	10.8	5.5
	Rusty water field	0.220	0.036	4.180	18.0	0.4	67	3.85	7.5	5.9
Fertile paddy soil		0.168	0.088	2.420	32.4	16.3	162	3.15	15.6	6.5

GENETIC AND ANTHROPOGENIC CHARACTERISTICS OF
PADDY SOILS DERIVED FROM SWAMPY LAND IN   LI-
XIAHE DISTRICT, JIANGSU PROVINCE

Lei   Wen-jin
(Institute of Soil Science, Academia Sinica, Nanjing)

Lixiahe district is located in northern marginal region of Changjiang
Delta between latitudes 32° 20' 02"-33° 40'08" N and longitudes 119°4'40"
-120°52'27" E (Fig. 1). As early as about six thousand years ago, this
district was still a gulf. With the expansion of the elta, it was turned
gradually into a lagoon area and after it was enclosed an inland swampy
depression came into being.

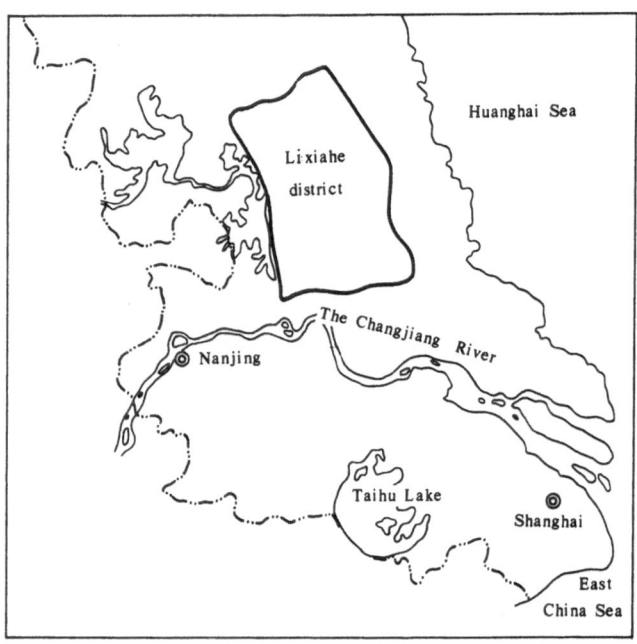

Fig. 1 Location of Lixiahe district

Since it was formed as land, this area has undergone three stages. i.e.
inundation by inverted tidal current, divagation of rivers and inland
flooding. Simultaneously, there has been corresponding stages in soil forming process, i.e. salinized swamping, desalinized swamping, and alternation

of swamping and deswamping. Only after the natural calamities of inverted
tiding and river flooding were stopped could the polder fields be made for
rice cultivation extensively in this district.

The polder fields were built on swampy land generally with a size of
about 7-20 hectares around the fields, drainage canals were dug and the
subspil from the canales was used to build dikes on the four sides around
the field in order to prevent the field from river water-flowing. After the
polder fields were built, the land surface have been raised gradually with
long term application of river mud. The river mud applied to the polder field
is estimated at 900-1125 t/h per annum,i.e the land surface of polder field
have been raised by 5 mm in height on the basis of the river mud applied to
2/3 of the total area annually. Therefore, the land surface could be rised
by 1 meter in 300 years. Being nearby the river, it is more convenient for
the margin area of the polder field to collect and apply river mud and raise
its land surface, As a result, there is a micro-relief·differentiation of
the polder fields with a higher surroundings and a lower centre. Because of
the difference in micro-relief, different patterns of soils are developed
respectively. According to the location of the soils, they are divided into
four patterns, i.e. high level polder soils, intermediate level polder soils,
low level polder soils and central polder soils (Fig. 2). At the same time,
the groundwater table becomes higher gradually from the side area to the central part of the polder fields. In the central part, the groundwater is generally connected with the surface water, and the profile is entirely saturated
by water (soil Nos. 4 and 5 on Fig. 3). The soil located in this part
belongs to swampy paddy soil and can be used only for single cropping rice.
As the ground water table in the soils of side area of the polder fields is
lower, the lowest water content of surface soil occurs in the dry season,
while the highest water content in surface soil and the lowest water content
in subsoil are found in the rainy season soil Nos. 1 and 2 on Fig. 3),
indicating that the soils are being separated from the process of swamping.
Rotation of rice and wheat can be steadily carried out on this type of soils,
known as permeable paddy soils. The soils located between the two above
mentioned patterns of soils are in the initial stage of separation from the
swamping process. The water content of the surface horizon is between that
of permeable paddy soils and swampy paddy soils both in dry and rainy
seasons; but the water content of subsoil is markedly affected by the
seasonal variation of ground water (soil No. 3 on Fig. 3) Unstable rotation
of rice and wheat can only be carried out on these soils known as gleyed
paddy soils.

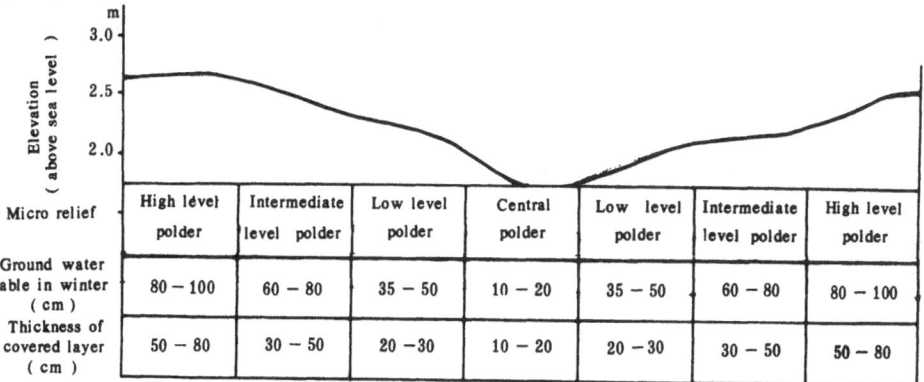

Fig. 2 Typical micro relief of Lixiahe district

751

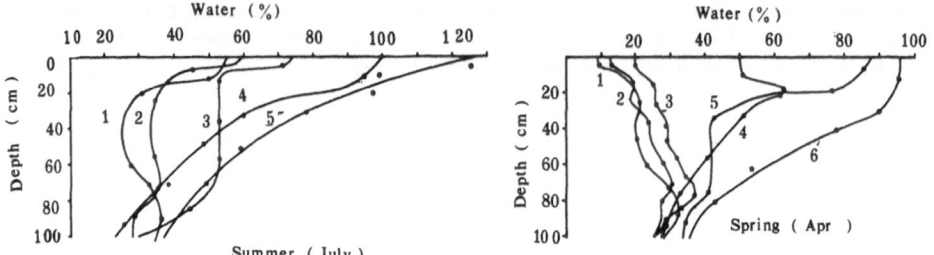

Fig. 3 Water regime of paddy soils. 1. Fertile permeable paddy soil
2. Ordinary permeable paddy soil, 3. Gleyed paddy soil.
4. Old swampy paddy soil. 5. New swampy paddy soil. 6 -
Virgin swamp soil

Under the influence by human activity of cultivation, the paddy soils
originated from swamp land are characterized by following unique genetic
and anthropogenic features.

CHARACTERISTICS OF OXIDATION-REDUCTION PROCESS

Because of the high ground-water table and the soil solum under
submergence all the year around, the whole profile of swampy paddy soils is
in a strongly reduced state. While the redox potential is below 100 mV, the
total reduced substances range from 12-16 m.e./100g soil, ferrous iron--4-8
m.e./100g soil (Table 2).The silica-iron ration and aluminium-iron ratio
of the clay fraction amount to more than 12 and 4 respectively (Table 1),
showing the remarkable characteristics of the deironization under a reducing
condition. In permeable paddy soils, two reducing horizons are formed in
the surface soil layer and bottom soil layer in the season of rice cultivation and irrigation. The ferrous and manganous oxides formed in the upper
and bottom reducing layers move either downward or upward to the oxidizing
subsoil horizon and are oxidized and illuviated there. In the season of
upland crops, the middle and upper parts of the profile except the bottom
layer are all in an oxidizing state (see potential curves of profile Nos.
1 and 2 on Fig. 4). In short, reducing deironization is the main characteristics of the swampy paddy soils with a profile pattern of AG-G. The
permeable paddy soils are characterized by the alternation of reducing
eluviation and oxidizing illuviation and mechanical leaching, and have the
profile pattern of A-P-B-G.

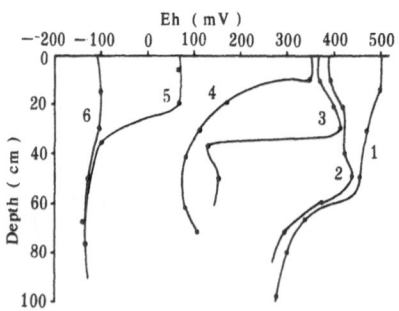

Fig. 4 Redox potential of paddy soils
1. Fertile permeable paddy soil.
2. Ordinary permeable paddy soil.
3. Gleyed paddy soil. 4. Old
swampy paddy soil. 5. New swampy
paddy soil. 6. Virgin swamp soil

CHARACTERISTICS OF MECHANICAL LEACHING

The marked mechanical leaching of paddy soils originated from swampy land only occurs in the stage when the permeable paddy soil has been developed. According to data of total chemical composition and mechanical composition, no variation of aluminium-iron ratio and translocation of clays less than 0.001 mm were found in the profile except the original soil buried below the depth more than 80 cm, which indicates that mechanical leaching can not occur under submergic conditions. As to the gleyed paddy soils developed in the early stage of converting the waterlogged field into dry farming, its characteristics are still similar to those of swampy paddy soil except the marked variation in water regime and redox potential. However, in the process of alternation of submergence and drainage, soil crack that may induce the leakage of water and nutrients often appears. In permeable paddy soils, no significant variation in aluminium-iron ratio has been found in the profile except G horizon, this indicates that destruction and translocation of clay in the soil are insignificant under the condition of alternation of oxidation and reduction. However, the data of the contents of clay less than 0.001 mm have shown that there are remarkable mechanical leaching and illuviation in the soil.

ANTHROPOGENIC CHARACTERISTICS

At the first stage in the process of reclamation and utilization of the soil for rice cultivation, the original surface soil is buried by a thick soil layer due to large amount of river mud applied year by year. The covered soil layer thickened gradually many alleviate or weaken the further development of gleization of the soil. At the second stage, the profile is differentiated from the pattern of AG-G to the patterns of A-BG-G and A-B-BG-G, but the plowpan has not developed yet. At the third stage, owing to the continuous raising of soil surface and lowering of the ground water level, the soil tends to the process of degleization. Rotation of rice and upland crop can be carried out steadily on the soil. The alternation of oxidation and reduction in the soil is evident, and so is the differentation of the profile, which has been developed from the pattern of A-BG-G to that of A-P-B-BG. Under long term intensive cultivation, the fertility of the soils is promoted constantly, and the highly matured paddy soils are formed.

On the basis of recognition of the genetic characteristics of the soils and adoptation of adequate improving measures and cropping system, it is expected that a high and stable yield could be obtained on the soils in this region.

GEOCHEMICAL CHARACTERISTICS OF SOME TRANSITIONAL AND RARE EARTH ELEMENTS IN PADDY SOILS OF RED EARTH REGION, GUANGDONG PROVINCE

Yang Xue-yi

(Institute of Soil Science, Academia Sinica, Nanjing)

The investigation of the contents and distribution patterns of elements in soils constitutes an indispensable part of environmental science and soil geochemistry. This paper deals with the contents of transitional and rare earth elements in paddy soils and their relations to the geochemical characteristics.

SELECTION OF THE SITES FOR COLLECTING THE REPRESENTATIVE SOIL SAMPLES AND METHODS OF DETERMINATION

Fourteen soil profiles of paddy soils were used for study. The parent materials included recent seashore sediment, delta deposit, river alluvium and ancient marine sediment. The total contents of 15 elements—lithium, beryllium, chromium, manganese, iron, cobalt, nickel, copper, zinc, arsenic, strontium, lead, cadmium and mercury were determined. Some of the profiles were analyzed for the rare earth elements and natural radicactive elements. Soil samples were treated with nitric and hydrofluoric acid for determining Li, Cr, Cu and Mn. The others were digested with aquaregia and perchloric acid for the analysis of Ni, Co and Pb. The elements were determined by atomic absorption spectrophotometer. Soil samples were also digested with 1:1 HCl to extract Cd and Be which were determined by atomic absorption spectrophotometer with graphite furnace. Zr, Sr, Zn and Fe were determined by X-ray fluorometric method. As and Hg were determined by diethylamine DDC-Ag colorimetric method and 590 type Hg analyzer respectively. The rare earth and radioactive elements were determined by neutron activation method.

CONTENTS OF THE ELEMENTS IN PADDY SOILS

The contents of the elements are listed in Table 1. It is evident that the amount of each element varied greatly. For example, the content of Co varied hundredfold; Be 50-fold; Mn and As 25-to 30-fold and the rest, 5- to 10-fold . This reflects the variations in compositions of parent materials. The absolute amounts of various elements decreased according to the following order: Fe > Zr > Mn > Zn > Cr > Li > Sr > Pb > Ni > Cu > As > Co > Be > Hg > Cd.

Table 1   Contents of some elements in paddy soils of
the red earth region

Table 1   Contents of some elements in paddy soils of
the red earth region

Element	No. of samples	Content (ppm)		Standard deviation	Coefficient of variation
		Range	Mean		
Li	14	11.0–75	38.4	19.4	0.51
Be	10	0.05–2.5	0.77	0.67	0.87
Cr	14	12–90	46.8	25.7	0.55
Mn	14	25.5–57.5	178	151	0.85
Fe*	14	0.55–5.67	2.73	1.47	0.54
Co	14	0.2–20.1	7.0	6.0	0.85
Ni	14	7.5–60.5	26.2	16.4	0.63
Cu	13	6.2–55.4	18.5	15.0	0.81
Zn	14	15.0–113.5	60.7	33.9	0.56
As	14	2.4–68.0	16.1	1.7	0.105
Sr	14	2.5–84	37.3	28.2	0.76
Zr	14	200–1188	464	290	0.62
Pb	14	10.0–57	31.7	15.2	0.48
Cd	10	0.01–0.08	0.04	0.02	0.50
Hg	14	0.005–0.2	0.075	0.065	0.87

*Fe in %

FACTORS AFFECTING THE ABUNDANCE OF THE ELEMENTS IN PADDY SOILS

1. Effect of Parent Material on the Abundance of Elements

Parent material has a direct effect on the abundance of the elements
in paddy soils. The mineral composition and the amounts of the elements
in paddy soils are similar to those in their preceding soils. The kinds
and contents of the elements of soils developed from river alluvium bear a
resemblance to the kinds of rocks in that river realm.

The results in Fig. 1 indicate the abundance of the elements in paddy
soils developed from different parent materials of this district. It can
be seen that: 1) the coastal paddy soils often flooded with sea water have
higher content in alkali metals K and Na of 1 A group than those derived
from the river alluvium. 2) The amount of Cr, Mn, Fe. Co, Ni, Cu and Zu
of the fourth group in periodical table is highest in paddy soils developed
from the delta sediment, next in the coastal paddy soil and lowest in the
paddy soil developed from river alluvium. 3) The river alluvium in granitic
region is relatively deficient in transitional elements, but the contents
of Zr, Cs. Ta, Lu, Th, U, etc. rich in granite are relatively high in
the paddy soils. The paddy soils developed from delta sediment are relatively
low in Th and U. Owing to the differences in soil composition and properties,
the Th/U ratio in the paddy soils developed from river alluvium is 3.58, while
that in the paddy soils developed from delta sediment is 5.34. 4)   The contents of La, Ce, Sm, Eu, etc. are more or less the same.

The Zhujiang Delta is composed of deposits of the Xijiang River, Beijiang
River and Dongjiang River. The Xijiang River and Beijiang River basins are
rich in basic and metamorphic rocks while the Dongjiang River basin is rich
in granite, so the soils developed from deposits of the Xijiang River and

Beijiang River contain much higher amounts of transitional elements than those developed from deposits of the Dongjiang River.

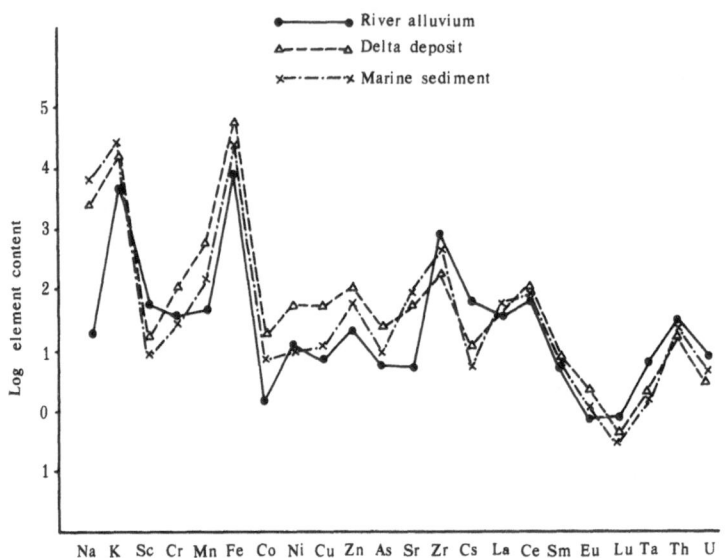

Fig. 1 Comparison of the abundance of elements in
paddy soils derived from different parent materials

The amounts of the elements in paddy soils developed from the same parent material also varied with the texture, the age of materials deposited as well as the physico-chemical properties of the soil itself. It can be seen from Table 2 that although the well drained clayey paddy soil and the sandy paddy soil are developed from the same ricer alluvium, the former is heavier in texture containing higher amount of organic matter, while the latter is lighter in texture containing low amount of organic matter. The absolute amount of Fe, Mn, Cr, Ni, Zn, As,Pb, Li and Sr in the former is 2 times larger than that in the latter, while the amount of Zr in the former is only half of that in the latter. The acid sulfate soil and the tidal sandy soil are developed from the same coastal sediment(1). The former contains higher amounts of S and sulfate due to the biological enrichment of S by mangrove, and the amounts of Fe, Mn, Cr, Ni, Li and sulfur-loving elements are also higher than those contained in the latter.

2. Effect of Submerging on the Translocation of Elements

Under the periodically water-logged conditions, iron and manganese suffered leaching by reduction and depositing by oxidation. Also owing to the decrease in Eh value and the followed reduction of sulfate and increase in solubility of phosphate, some minor elements went into soil solution and translocated downward(2,3). The results in Table 3 show that the content of Fe and Mn in PB horizon of the loamy alluvial paddy soil was increased by 4300 and 140 ppm respectively in comparison with A horizon. The amounts of Co, Zn, Ni and Li in PB horizon compared with A horizon were increased by 2.6, 5.5, 2.4 and 5.3 ppm respectively. This tendency of downward movement of elements is even more evident in coastal paddy soils of Hainan Island.

Table 2 Contents of some elements in different types of paddy soils

Soil tyoe	Depth (cm)	Fe (%)	Mn	Cr	Co	Ni	Cu	Zn	As	Pb	Li	Sr	Zr
Well-drained clayey	0-12	0.91	41	31.4	0.2	12.5	7.2	27.0	5.6	32.0	27.4	5.0	8.45
Sandy	0-15	0.55	26	12.0	0.9	7.5	7.2	15.0	3.9	15.0	13.4	2.5	11.87
Acid sulfate	0-20	4.27	203	65.0	11.2	25.3	14.8	88.0	16.2	36.0	75.0	200	46
Tidal sandy	0-40	2.05	115	28.1	10.0	20.0	8.0	50.0	8.2	36.0	38.8	215	35

Table 3 Distribution of some elements in paddy soil profiles

Soil type	Horizon	Depth (cm)	Fe (%)	Mn	Cr	Co	Zn	Ni	Li
Loamy alluvial paddy soil	A	0-15	3.0	118	75.3	6.6	63.0	30.1	48.8
	PB	25-50	3.43	258	74.8	9.2	68.5	32.5	54.1
Coastal paddy soil	A	0-14	1.41	245	27.9	6.9	31.0	16.3	14.9
	PB	25-50	2.18	257	41.0	9.5	44.5	27.3	21.3
Sandy paddy soil	A	0-15	0.55	26	12.0	0.9	15.0	3.5	13.4
	B	23-50	1.33	61	36.9	1.7	16.5	10.0	24.0

RELATIONSHIP AMONG DIFFERENT ELEMENTS IN PADDY SOILS

During the processes of translocation and deposition the minerals in parent materials underwent mechanical separation. The physico-chemical properties of the soil also experienced changes under particular hydrological conditions. Therefore, the amount, state of existence and interrelationship of these elements are rather complex. Correlation analyses showed that in the paddy soils of this region elements correlated significantly ($P < 0.001$) with Fe and Mn are iron-loving elements such as Co, Ni, copper-loving elements such as Cu, Zn, As, Pb and lithophile elements such as Li, Cr. There was no correlation between Hg and other elements. A negative correlation between Zr and nearly all the transitional elements was found. The test for the correlation coefficient among different elements and the establishment of linear regression equation may not only aid in the estimation of approximate amount of the coordinated elements, but also be significant for studying the translocation pattern of the elements in

757

soils (Table 4).

Table 4　Relationship between Zn and some other elements in
lateritic paddy soils (* P　0.001)

Regression equation	Correlation coefficient (γ)	
$[Fe] = 403.7 + 2720\ [Zn]$	0.93	*
$[Mn] = 3.6\,[Zn] - 40.1$	0.81	*
$[Pb] = 9.9 + 0.4\,[Zn]$	0.80	*
$[As] = 0.11 + 0.2\,[Zn]$	0.79	*
$[Li] = 9.4 + 0.5\ [Zn]$	0.84	*

REFERENCES

(1) Gong Zi-tong, Zhou Rui-rong, On the genesis of strongly acid salty
    paddy soils of southern Guangdong.　Acta Pedologica Sinica,12(2),
    183-190(1964). (in Chinese with English summary)
(2) Ponnamperuma, F. N., The Chemistry of Submerged Soil.　Adv. Agron.,
    24,29-96(1972).
(3) Randhawa, N. S., Sinha, M. K., Takkar, P. N., Micronutrients.　In
    "Soils and Rice", IRRI, Philippines, 581-598(1978).

PRELIMINARY STUDIES ON PRIMARY MINERALS OF SOME
PADDY SOILS IN TAIHU LAKE REGION

Luo Jia-xian
(Institute of Soil Science, Academia Sinica, Nanjing)

Making an investigation of the primary minerals can provide useful
information for the study of soil genesis. When studying the weathering
degree of soil, emphasis is always laid on heavy minerals. For this purpose,
the Pettijhon System is commonly employed(1), i.e. to arrange 25 kinds of
heavy minerals in order according to their weathering degrees. In the present investigation, we take the Pettijhon System as a reference, referring
the minerals above the garnet to stable mineral, and those below unstable
one. The sphene being, however, classified as stable mineral and chlorite
as unstable one. For the separation of heavy minerals, bromoform with SG
2.9 is used(2). The heavy minerals have been identified and calculated under
polarizing microscope(3), and the mineral composition in the profiles of 7
paddy soils of the Taihu Lake region determined. There is also a preliminary
discussion on the mineral composition, weathering degree and the genesis of
bleached horizon.

Among the tested samples, their pH values, texture, and parent materials
are given in Table 1.

Table 1 The pH texture and parent materials of 7 paddy soils

No	Soil	Location	pH (surface soil)	Texture	Parent materials
1	Silty paddy soil	Jiangning	4.7	Clay loam	Highly leached loess
2	Surface bleached paddy soil	Liyang	5.7	Clay loam	Highly leached loess
3	Bleached paddy soil	Wuxian	6.0	Clay loam	Lacustrine deposit of the Taihu Lake
4	Fertile permeable paddy soil	Wuxi	7.1	Clay loam	Lacustrine deposit of the Taihu Lake
5	Waterlogged paddy soil	Kunshan	6.9	Clay loam	Lacustrine deposit of the yang cheng Lake
6	Percolating paddy soil	Shanghai	8.2	Clay loam	Alluvial deposit of the Changjiang River
7	Permeable paddy soil	Jiaxing	7.5	Clay loam	Alluvial deposit of the Qiantang River

COMPOSITION OF PRIMARY MINERAL IN SOILS

The distribution of soil primary minerals is essentially concentrated in the fraction of 20-50 micron. In the > 50 micron fraction, the kinds of minerals are somewhat incomplete. Thus, the present work is all based on the samples of 20-50 micron fraction.

The tested samples contain nearly 20 kinds of primary minerals, a majority of which are light minerals, such as quartz and feldspar, but poor in heavy minerals. If expressed in % by weight, the amount of heavy mineral in the surface bleached paddy soil of Li yang County is less than 2.2 %, that of bleached paddy soil of Wu xian County and that of fertile permeable paddy soil of Wu xi County are 3.3-4.5 %; and the other soils are more than 5 % respectively. Among the heavy minerals, if their amounts are expressed in percentage of grain, the content of epidote is the greatest, amounting to 35-50 %; clinozoisite and chlorite 3-10 %; tourmaline 2-6 %; grammatite 1-5 %; zircon 1-3 %; hornblende, sphene, rutile and garnet less than 2 % for all of samples; mica varying greatly from trace to about 10 %; ferrous and opaque minerals having a considerable proportion, amounting to about 20-35 %.

It could be seen from Table 2 that among the unstable minerals in the 7

Table 2 The composition of heavy minerals * (and variation coefficient) (20-50 μ fraction)

Soil	No. of horizons	Unstable mineral		Stable mineral	
		X + S %	C.V(S:X)	X + S %	C.V.(S:X)
Silty paddy soil	5	64.1±2.36	0.04	5.6±0.82	0.15
Surface bleached paddy soil	5	49.9±5.67	0.11	10.0±1.67	0.17
Bleached paddy soil	6	58.8±3.90	0.07	9.3±0.57	0.06
Fertile permeable paddy soil	5	59.4±2.97	0.05	10.8±1.04	0.10
Waterlogged paddy soil	5	65.9±1.71	0.03	7.5±0.44	0.06
Percolating paddy soil	5	63.7±5.03	0.08	6.3±0.75	0.12
Permeable paddy soil	5	59.7±2.05	0.03	6.5±1.25	0.19

* The opaque minerals and mica not included

soil profiles, the surface bleached paddy soil of Li yang County is the
least, with about 50 % in each horizon, while the waterlogged paddy soil
of Kunshan County is the most abundant, containing about 66 %. More stable
minerals have been found in the surface bleached paddy soil of Li yang County,
in the bleached paddy soil of Wu xian County, and in the fertile permeable
paddy soil of Wu xi County (about 10 %), with the least in the silty paddy
soil of Jiang ning County (about 5.5 %). The content of stable mineral
within a profile varies moderately but the content of zircon varies to some
extent between the horizons in the surface bleached paddy soil, being higher
in bleached horizon than both upper and lower horizons in the profiles of
bleached paddy soil and fertile permeable paddy soil.

RATE OF SOIL WEATHERING

The sum of zircon and tourmaline divided by the total amount of unstable
minerals is regarded as the weathering rate of heavy minerla, while the ratio
between quartz and feldspar represents the weathering rate of light minerals
(Table 3). As for the 4 profiles of non-bleached horizon, the weathering
rate of heavy minerals is rather low, not exceeding 0.10, with little variation in various horizon, while that of light minerals in these 4 profiles
varied greatly, highest in silty paddy soil, followed by the percolating paddy
soil and waterlogged paddy soil, and lowest in the permeable paddy soil, very-
ing also to a certain extent within the same profile.

Table 3   Weathering rate of light and heavy minerals in 4 kinds of soil *

Horizon	Silty paddy soil		Waterlogged paddy soil		Percolating paddy soil		Permeable paddy soil	
	Heavy mineral	Light mineral	Heavy mineral	Light mineral	Heavy mineral	Light mineral	Heavy mineral	Light mineral
Surface (A)	0.06	5.97	0.10	2.97	0.08	2.87	0.06	1.70
Plowpan (P)	0.06	5.90	0.07	3.11	0.08	3.15	0.07	1.40
Percogenic (W)	0.05	2.34	0.08	2.90	0.05	5.00	0.07	2.14
Illuvial (Bg)	0.06	3.54	0.07	0.38	0.05	5.00	0.10	1.67
C (C)	0.08	2.74	0.07	1.79	0.04	0.33	0.06	1.64

* Weathering rate of heavy minerals= $\dfrac{\text{Zircon + tourmaline}}{\text{Unstable minerals}}$

Weathering rate of light minerals= $\dfrac{\text{Quartz}}{\text{Feldspar}}$

761

As for the three soil profiles of bleach-containing horizon, the weathering rates of both heavy and light minerals are all greater than the above-mentioned profiles (Table 4). Generally speaking, the weathering rate of heavy minerals is 0.12-0.22 and that of light minerals is 5-14, among whith the highest is in the surface bleached paddy soil of Li yang County. The weathering rate of light mineral of bleached paddy soil (Wu xian County) is higher than that of fertile permeable paddy soil (Wu xi County), but that of heavy minerals is just similar to each other. The weathering rate of heavy and light minerals in the upper horizon of these 3 profiles is increased markedly as compared with the lower horizons. This seems to indicate that during the course of paddy soil formation from the parent material, the primary minerals passed through different degrees of weathering.

Table 4 Weathering rate of light and heavy minerals in
3 kinds of bleached soil

Horizon	Surface bleache paddy soil		Bleached paddy soil		Fertile permeable paddy soil	
	Heavy mineral	Light mineral	Heavy mineral	Light mineral	Heavy mineral	Light mineral
Surface (A)	0.14	13.50	0.14	10.5	0.15	5.27
Plowpan (P)	0.20	13.50	0.13	9.71	0.13	5.18
Bleached (WL)	0.22	13.90	0.15	9.88	0.14	9.00
Percogenic (W)			0.12	2.00	0.12	5.50
Illuvial(Bg)	0.20	11.34				
C (C)	0.12	5.30	0.1-0.13	1.9-2.3	0.13	3.50

MINERALS IN BLEACHED HORIZON

In the soil profiles collected from Li yang, Wu xi and Wu xian Counties, there is a bleached horizon with a thickness of 10-25 cm. Compared with other horizons, the light minerals in the bleached horizon are relatively high, where the quartz and feldspar amount to more than 96 %, but the heavy minerals are relatively low with generally about 3 % in weight. Comparing with the content of the heavy minerals below and above, the unstable minerals in the bleached horizon decrease significantly, but the variation of stable minerals is little (Table 5). As for the weathering rates of heavy and light minerals, only the light minerals in the bleached horizon of fertile permeable paddy soil (Wu xi County) are apparently at a high rate, but not for the other two (Table 4).

It can be seen from Table 6 that the varieties of heavy minerals in the C horizon of these soils are all similar and so are the amount of various minerals. The types of heavy minerals in these soils are similar to those of the loess in Nothwest China(4). The composition of the heavy minerals in the bleached horizon shows that, except for the variation in the content of unstable minerals, the total amount of stable mineral remains consistent to each other and, moreover, it is very close to its own C horizon. It comes to the conclusion that parent materials of bleached layer and C horizon in these three paddy profiles are all related to loess-like material. However, sedimentation conditions above and below the bleached horizon might be different. The unstable minerals in the bleached horizon of surface bleached paddy soil of Li yang County are lower than the bleached horizons of the other two. It is suggested that the weathering rate of surface bleached paddy soil be higher than those of the other two.

Table 5　The content of heavy minerals in the bleached horizon(WL) as compared with those above and below it (%)

Soil	Location	Horizon	Unstable mineral	Stable mineral
Surface bleached paddy soil	Liyang	P	52.9	12.7
		Wl	43.7	10.4
		W	45.0	10.4
Bleached paddy soil	Wuxian	P	59.2	8.8
		WL	51.0	8.7
		W	57.3	9.1
Fertile permeable paddy soil	Wuxi	P	63.9	12.4
		WL	56.1	10.1
		W	58.5	9.6

Table 6　The content of heavy minerals in three soils with bleached horizon

Mineral		Surface bleached paddy soil		Bleached paddy soil		Fertile permeable paddy soil	
		C*	WL*	C	WL	C	WL
Unstable mineral	Epidote	48.7	32.4	44.7	37.5	46.0	36.8
	Chinozoisite	4.3	6.9	9.5	8.8	5.6	10.5
	Chlorite	2.4	0.6	3.9	1.8	2.0	4.8
	Hornblende	0.7	1.9	0.8	0.4	0.7	1.2
	Grammatite	2.9	1.9	3.5	2.5	2.6	2.8
Stable mineral	Garnet	0.1	0.3	1.9	0.2	1.9	1.4
	Sphene	1.0	0.5	0.6	0.4	0.5	0.5
	Rutile	0.1	0.1	1.0	0.2	0.6	0.3
	Zircon	2.1	3.5	1.8	2.9	1.5	3.0
	Tourmaline	4.9	6.0	5.6	5.0	5.8	4.9

* C: parent material;　WL: bleached horizon

REFERENCES

(1) Mitchell, W.A., Heavy minerals. In"Soil Components" (ed. by John E. Gieseking, Springer-Verlag, Berlin, 455-457 (1975).
(2) Robert, E. Carver., Heavy mineral separation in Procedures. In "Sedimentary Patrology" (ed. by Robert E. Carver), Wiley-Interscience, 427-452(1971).
(3) Carroll, D., A statistical study of heavy minerals of the south river, Augusta county, Virginia. J. Sediment Petrol., 27, 387(1957).
(4) Liu Dong-sheng, Components and fabric of loessal materials. Science Press, Beijing, 4, 46-55(1966). (in Chinese)

DOUBLE CROPPING OF RICE IN TRIPLE CROPPING SYSTEM AND SOIL FERTILITY

Yang Wen-yuan, Liang Dun-fu, Xie Chun-qing, Wan Zong-yi
(Soil and Fertilizer Institute, Sichuan Academy of
Agricultural Sciences Chengdu)

Since the 1970s, the continual extension of double cropping of rice in triple cropping system has promoted the production to a certain extent but has brought about some adverse effects on soils in the western plain of Sichuan Province. For the solution of this problem, a long-term experiment has been projected tp assess the above-mentioned influence on soil fertility, and to provide a guiding principle for a rational cultivation system in this region. Because the content and composition of soil humus are important criteria in differentiating soil fertility, and the development of gley horizon is an important factor in assessing the deterioration of soil properties, this article presents a preliminary investigation on the influence of double cropping of rice in the triple cropping system on soil organic matter, on the development of gley horizon and on crop yield as well.

METHODS OF EXPERIMENT

Experiment was mainly done in the field in combination with simulating test and observation made in typical fields. The soil under test was paddy soil developed on the redeposited platform of yellow soil (zheltozem). Experiment was designed as in Table 1.

Table 1 Experimental design

Series	Abbreviation	Rotation	N applied (kg/ha/yr)	
			Org. manure	Chem. fert.
1	RORR	Milk vetch, rape,or wheat rotated yearly-early rice-late rice	94	278
	MRR	Milk vetch-early rice-late rice	63	172
	RaRR	Rape-early rice-late rice	110	330
	WRR	Wheat-early rice-late rice	110	330
2	WCR	Wheat-corn-late rice	110	315
	MR	Milk vetch, rape, or wheat-midseason rice	75	225
	DCR	Milk vetch, rape, or wheat-early rice-late rice	94	278

765

RESULTS AND DISCUSSION

1. Effects on Content of Soil Organic Matter and Constituents of Humus

Table 2 showed that: 1) five years after various types of double cropping of rice in the system were carried out, the content of soil organic matter tended to increase and the effects of various winter crops were insignificant; 2) after four years, the ratio of humic: fulvic in the soil humus extracted with 0.1 N NaOH differed little from that of the original one. This might be due to the short duration in conducting the experiment.

Table 2   Influence of double cropping rice on the content of soil organic matter in the cultivated horizon

Treatment	Organic matter (%)		Humic acid/fulvic acid*	
	Apr., 1974	Dec., 1979	Jul., 1974	Nov., 1979
RoRR	2.12	2.50	0.36	0.38
RaRR	2.24	2.53	0.35	0.37
MRR	2.26	2.48	0.32	0.33
WRR	2.19	2.45	0.35	0.36

*Extracted with 0.1 N NaOH

2. Effect of Double Cropping of Rice on Development of Gley Horizon

During the middle of November to the middle of December in 1979, the observation of soil profile showed that no gley horizon was found in the WCR plot and the MR plot; but, in the plots with double cropping rice, an apparent gley horizon occurred at the depth of 13-16 cm. In the experiment with double cropping rice in the triple cropping system, different types of winter crop had various influences on the development of gley horizon. According to the observation at the middle of November 1979, apparent gleying appeared at a depth of 16-21 cm in MRR plot; in plots RaRR and RoRR, only a small number of gley spot appeared; and the WRR plot lay intermediate. The determination of ferrous ion also showed a similar tendency.

Fig. 1 showed that among the different types of rotation with double cropping of rice in the triple cropping system, the content of ferrous ion was highest in MRR, lowest in the RaRR and RoRR.

In the depth of 16-26 cm, ferrous content of the former was 20 ppm; of the latter, only about 3 ppm. This was closely related to the high water content in the soil of MRR, because its tested value in November 1979 was 39.7-39.9 %, which exceeded those of the other treatment by 2.8-8.0 %. Mechanical pressing was the important factor for the occurrence of gley horizon (Table 3). The simulating experiment showed that soil rammed three times was most severe in the formation of gley horizon, and the ferrous content was also the highest; soil rammed twice was second to the above; those rammed once ranked third; and the non-rammed soil was the least in the formation

766

of gley horizon.

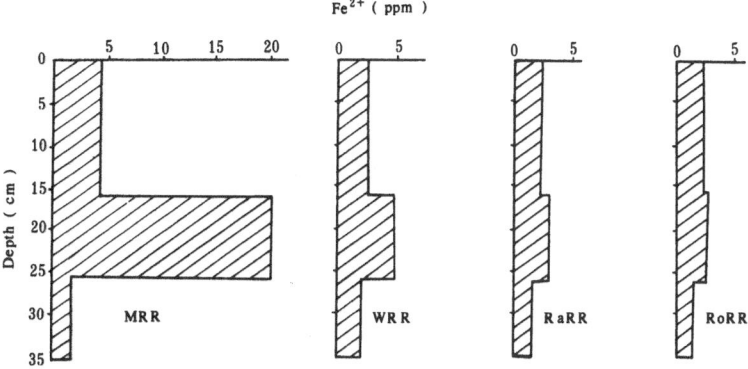

Fig. 1 The effect of various types of double cropping of rice in the triple cropping system on ferrous iron content of the soil
(Extracted with 0.1 M $Al_2(SO_4)$ solution)

Table 3 Simulating experiment on mechanical pressing

Treatment	Volume weight (g/cm^3)	Ferrous ion* (ppm)
Non-rammed	1.28	8-47
Rammed once	1.32	11-56
Rammed twice	1.36	27-85
Rammed three times	1.39	80-98

*Extracted with 0.1 M $Al_2(SO_4)_3$ solution

Even after long duration of winter and spring draught, the cultivated horizon which had been rammed two or three times still left odds and ends of gley spots and the ferrous content within 0-16 cm was 8-12 ppm.

It can be seen then that under the conditions of double cropping of rice in the triple cropping system, improper mechanized cultivation will cause the soil to become compact, enhance the development of gley horizon and to a certain degree affect crop yield (e.g., the yield of wheat was decreased by about 10 %).

3. Yield of Rice and Spring Crops under Different Types of
Double Cropping of Rice in Triple Cropping System

The yields of double crop rice in the past 5 years were as follows:
The average annual yield of RaRR was the highest (9945 kg/ha), 7 % higher

than that of the control RoRR (9338 kg/ha). The MRR ranked second, whose yield was 9443 kg/ha, 1 % higher than that of the control. The yield of WRR was 9150 kg/ha which was equivalent to 98.0 % that of the control.

As for the spring crops, according to the data obtained in the past 5 years, the annual yield of rapeseed in the RaRR was 2228 kg/ha which was 99.3 % of that in the control, its coefficient of variation for 5 years, 12.2 %; the yield of green manure in the MRR was 37358 kg/ha (fresh matter) which was 91.6 % of the annual output of that in the rotation plot RoRR, its coefficient of variation, 12.0 %; the average yield of wheat in the WRR plot was 4268 kg/ha which was 92 % of that in the control, its coefficient of variation, 27.6 %.

Hence, it should be noted that when rape was used as successive crop to the double crops of rice from year to year, the yield of rapeseed was high and stable; but when wheat was used instead, its yield was neither high nor stable. So when WRR was adopted, it would be unfavorable either to the rice or to the wheat for a high and stable yield even though the fertilizing levels in both cases were similar (e.g., annual application of fertilizer was 440 kg N/ha in which organic nitrogenous fertilizer was 110 kg).

CONCLUSIONS

1) Through a rotation of triple cropping system, with double rice cropping as the base for 5 years, the average content of soil organic matter increased, but its composition showed no significant difference.

2) Field observation revealed that no gleization process took place in soils under crop rotation systems of single rice and upland crops. However, gleyed subsurface layers appeared in all plots with double cropping rice in triple cropping systems at different degrees. Strong development of gleyed soil clods to a depth down to 21 cm below soil surface appeared in double cropping rice followed by milk vetch.

3) The highest annual yield of rice grain in general was obtained in rice-rice-rape rotation. The wheat or rice yield in wheat-rice-rice rotation was neither high nor stable.

CHARACTERISTICS OF HIGH-YIELD PADDY SOILS IN SUBURBS OF SHANGHAI

Fu Ming-hua*

(Soil and Fertilizer Institute, Shanghai Academy of
Agricultural Sciences)

Shanghai is located at the mouth of the Changjiang River, close to the East China Sea. It is one of the earliest farming areas in the history of China. The yield of grain has already reached up to 12 t/ha per year. The parent material is alluvial deposit of the Changjiang River, thick in depth and even in texture. Most of the soils are medium to heavy loam; pH 6.8-7.5. According to various conditions during the process of soil formation, the soil distribution in the suburbs of Shanghai may be divided into two parts: in the eastern part, including the ancient upland and the wide expanse of eastern region, there is mainly a kind of temporary submerged type of paddy soils originated from meadow soils, while in the western part (Dianmor lowland) there are gley paddy soils originated from bog soils (Fig. 1). As is known, the high-yield paddy soils are anthropogenically developed in parallel with the continuous increase of soil fertility through proper soil management. All paddy soils of various origin can develop into high-yield paddy soil through soil improvement.

THE CHARACTERISTICS OF HIGH-YIELD PADDY SOILS

1. Soil with Good Profile Construction

The differentiation of soil horizons of paddy soils is due to the influence of the interaction of irrigated water, percolating water and ground water, as well as the cultivation and fertilization. Generally a high-yield paddy soil comprises a cultivated horizon, a plowpan, a mottling horizon, an illuvial horizon and a gley horizon(1). Within one mater's depth of soil body, there is not any barrier layer. The characteristics of cultivated horizon are: 15-20 cm in thickness fine in structure, rich in nutrients, loose, soft and fertile. The thickness of plowpan is usually 10 cm or so which is maintainable and pervious to water. The mottling horizon is generally more than 50 cm in thickness with a marked perpendicular joint and good in permeability and aeration, known as a typical horizon of high-yield paddy soil. According to the investigation, the developed degree and thickness of mottling horizon of paddy soils originated from bog soils will reflect correctly the level of fertility of the soil. As shown in Fig. 1 the blue-mud soil originated from the same bog soil is a low-yield paddy soil because of its high ground water table, so the gley horizon is near the soil surface with no mottling horizon in the profile. The greyish-blue lowland paddy soil is not a high-yield paddy soil either. Although its gley horizon sinks to

---

* Wang 'Fang-tao, Yang Jin-lau, La Zheng-yi, Wang Chao-jun, Gu Zhong-lan, Yao Nai-hua, Liao Zhao-xiong, Cheng You-song also took part in this work.

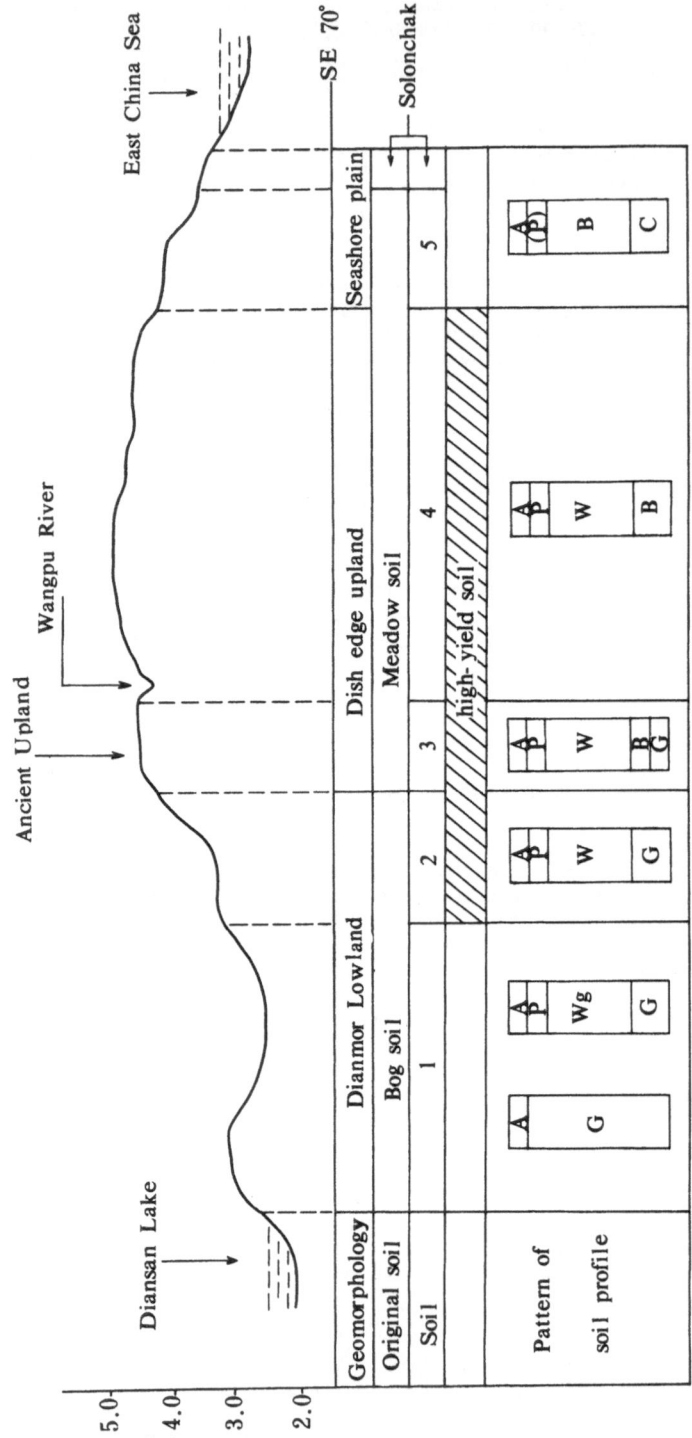

Fig. 1 The profile of geomorphology and soil distribution in the suburbs of shanghai
1. Greyish-blue lowland paddy soil; 2. Greyish-yellow lowland paddy soil;
3. Greyish-brown upland paddy soil; 4. Brownish-yellow upland paddy soil;
5. Sandy loam paddy soil

a lower level from the surface thus forming mottling horizon, it is not
well developed and rather thin, whereas the greyish-yellow lowland paddy soil
is a kind of high-yield paddy soil, having a thicker mottling horizon, well-
developed structure, and a distinct leaching and illuvial phenomenon. The
illuvial horizon, where Fe and Mn accumulated, has marked perpendicular
and horizontal joints and water holding capacity. A desirable occurence
depth is 80 cm below the soil surface. The gley horizon exists where ground
water often fluctuates and it is more desirable to be found below more than
120 cm.

2. Soil with Good Water Permeability

Most of high-yield paddy fields are water-permeable. People often say:
"Where there is permeability, there will be a high-yield." "Permeability"
means that there is a proper amount of leakage in the soil. It is known
through investigation that during the flooding period, the seeping rate
of high-yield paddy soil is 3-5 mm/day, and after a short man-made drought
period of soil surface becomes 9-15 mm/day[1,2]. The permeability of paddy
field is closely related to the bulk structure of soils and the status of
porosity. As a rule, the soil which has a well-developed and thick mottling
horizon must have good permeability. According to the test, the rate of non-
capillary pores in cultivated horizon of the permeable fields is usually
7-8 %, and 10-15 % during dry cropping season, whereas the rate of capillary
pores is 50 % and 35-40 % respectively. Non-capillary pores in the plowpan
is rather stable (2-3 %). The rate of non-capillary in mottling horizon is
4-8 %. But for the waterlogged lowland paddy soil, the rate of non-capillary
pores in each horizon is apparently low.

Results of Eh values determination show that the circulation of air
and supply of oxygen in the soil of permeable fields are fairly well.
During the waterlogged period, the Eh value falls slowly and always remains
above 120 mV. During the middle and later period after irrigation, Eh
value is increased to 200 mV or so when the soil is wet. But in the waterlogged field after flooding, Eh value drops drastically even reaching a
negative value[1] (Fig. 2).

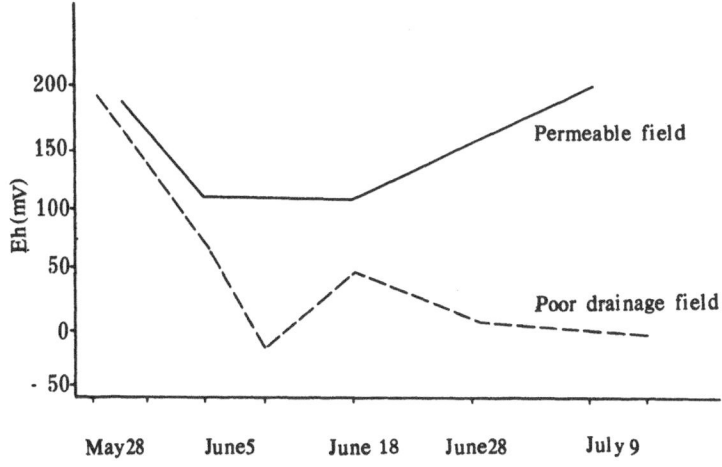

Fig. 2 Changes of Eh in two soils during the growing
period of early rice (at depth of 5-13 cm)

3. Soil with Rich Potential of Nutrient Supply

High-yield paddy soils in the suburbs of Shanghai usually contain 0.5-3.5 % of organic matter, 0.15-0.18 % of total N, 0.065-0.079 % of total P and 2.0-2.5 % of total K, so they have not only rich nutrients for plant to take in, but high ability of supplying available nutrients as well. Take the organic matter for example: the oxidizable organic matter in the soil is relatively high (Table 1). The content of available phosphorus is also higher than 11 ppm (extracted with 0.5 M $NaHCO_3$). Table 2 shows the relativity between the available phosphorus in the soil and the yield of paddy rice(3). High-yield paddy soils generally contain more than 100 ppm of alkaline soluble nitrogen and more than 80 ppm of available K (extracted with 1 N NH_4Ac). The contents of available B, Mn, Cu, Mo are all over the critical values.

Table 1 Comparison of relative oxidizability of organic matter between two soils

Soil	Location	Fertility	O.M.* (A %)	Oxidizable** O. M. (B%)	$\frac{B}{A}$ (%)
Brownish-yellow	Jinwei Commune,	Low	2.47	1.40	56.7
upland paddy soil	Jinshan County	High	2.23	1.39	62.4
Greyish-blue low-	Chengdong Commune,	Low	4.22	2.04	48.3
land paddy soil	Qingpu County	High	4.44	2.49	56.0

 * Boiled for 5 minutes at 170–180°C with 0.4 N $K_2Cr_2O_7$-1:1 H_2SO_4 mixed solution
 ** Boiled for 5 minutes at 170–180°C with 0.2 N $K_2Cr_2O_7$-1:3 H_2SO_4 mixed solution(5)

Table 2 Relationship between available phosphorus and the yield of early rice

Site	Available P (ppm)	Yield (t/ha)	Correlation coefficient	Site	Available P (ppm)	Yield (t/ha)	Correlation coefficient
Zhujia-	3.0	3.9		Jinwei	9.1	7.9	
jiao	4.0	4.4		Commune,	11.4	7.4	
Commune,	5.1	5.2		Jinshan	18.7	7.6	
Qinpu	6.0	5.4	r=0.91**		18.1	7.7	Insig-
County	6.6	5.5		County	15.9	7.4	nificant
	8.0	5.7			12.9	7.6	
	9.3	5.6					
	11.4	6.2					

PROPER MANAGEMENT FOR HIGH-YIELD PADDY SOILS

The suburbs of Shanghai is well known as one of the highest rice yielding areas in China, but with the development of triple cropping system (wheat–rice–rice, or green manure–rice–rice), some soil problems emerged. The long-term waterlogging of the soil induces poor physical characteristics and poor ability in supplying nutrients and draining of water. Below the cultivated horizon an undesirable blue-mud horizon has been formed in consequence of long-term waterlogging in the cultivated horizon. This blue-mud horizon was 6–8 cm in thickness, the soil mass is strong in reduction and dark-grey in colour. During the season of upland crops, the Eh value of blue-mud horizon is 100–150 mV lesser than that of cultivated horizon, the lowest Eh value even reaching 100–120 mV. It contains more reducing substance (Table 3). The existence of such horizon has an ill influence on the growth of plants. So the most important measure to be taken in culturing high-yield paddy soils in the suburbs of Shanghai is to intensify the drainage of water and improve the circulation of water and air in the soil in order to enhance the fertility of soil.

Table 3 Contents of reducing substances in various horizons

Site	Horizon	Reducing substance			
		Total (me/100g)	Active (me/100g)	Fe^{2+} (ppm)	Mn^{2+}
Chengdong Commune, Qingpu County	Cultivated	0.42	0.29	75	157
	Secondary gley	1.73	0.72	593	296
	Plowpan		0.56	189	247
	Cultivated	0.41	0.40	85	125
	Secondary	3.11	2.77	860	150
	Plowpan	0.42	0.38	104	143

Date: Dec., 1979

In the high yielding areas, the following measures of regulating water are to be taken: establishing a complete system of drainage and irrigation, bringing rivers and trenches under control at the water level one meter below the ground all the year round, and digging three kinds of trenches in various depths in the field for draining. Besides external drainage, such internal drainage as tubes and holes is being used for draining water. The whole system plays a major role in draining off the surface water, hanging water in cultivated horizon, and depressing the level of ground water. Furthermore, the quality of cultivation of dry ploughing and harrowing has been taken into consideration because ploughing and harrowing under submerged conditions will make the soil become more compact and harder after drying and the non-capillary pore space decrease, the granular structure destroy and the ability of internal drainage decrease (Table 4)(4). The experience of high-yielding unit shows that crop rotation, especially rice alternating with upland crop (such as cotton) once a year, is good for the improvement of soil fertility. The alternative drying-wetting in soil will induce quite a good structure and a circulation of water and air in soil. Our experiment shows that after planting cotton instead of rice for a year, the yield of rice both early and late increased by 12.9 % and 14.3 % respectively more than that of the same kind of soil which was still planted with rice. Another experience of high-yielding areas in the course of culturing high-yield paddy soils is to apply organic fertilizer. It has been proved beneficial to imporove soil, enhance and balance the supply of nutrients.

Table 4 Effect of plowing and harrowing under different moisture conditions on soil structure

Date of determination	Moisture condition when plowed and harrowed	Capillary porosity (%)	Non-capillary porosity (%)	Aggregate (%)					
				<0.5 cm	0.5-1 cm	1-2 cm	2-4 cm	4-6 cm	>6 cm
One month after plowing	Desirable	38.7	15.1	4.9	5.5	14.1	43.4	32.1	0
	Saturated	42.6	5.1	2.1	2.8	6.3	16.1	21.7	51.0
	Sith water layer	43.2	3.6	0.5	0.3	1.3	1.9	5.7	90.1
After freezing in winter	Desirable	37.6	10.8	22.7	19.3	32.0	19.4	6.7	0
	Saturated	41.0	7.1	15.6	14.0	23.8	25.2	22.7	0
	With water layer	43.0	5.6	12.6	10.4	23.4	35.1	18.6	0

REFERENCES

(1) Soil Section of the Institute of Soil, Fertilizer and Plant Protection, Shanghai Academy of Agricultural Sciences, Characteristics of the soil fertility and its improvement of the high-yielding paddy soil in the Shang-hai suburbs. Scientia Agricultura Sinica, 2, 66-72(1978). (in Chinese)
(2) Soil Section of the Institute of Soil, Fertilizer and Plant Protection, Shanghai Academy of Agricultural Sciences, Studies on the soil fertility of high and stable-yielding paddy soil of Ba Er brigade, Shanghai. Turang, 4, 156-162(1975). (in Chinese)
(3) Fu Ming-hua, Chen You-song, Studies on the phosphorus status of the soil in Shanghai. Acta Pedologica Sinica, 16, 372-379 (1979). (in Chinese with English summary)
(4) Soil Section of the Institute of Soil, Fertilizer and Plant Protection, Shanghai Academy of Agricultural Sciences, Soil structure under the influence Óf different soil moisture status. Turang, 2, 52(1978). (in Chinese)
(5) Yuen Ku-nun, Studies on the organo-mineral complex in soil. I.The oxidation stability of humus from different organo-mineral complexes in soil. Acta Pedologica Sinica, 11, 286-292(1963). (in Chinese with English summary)

ON THE FORMING OF WATERLOGGED CONDITION IN THE CULTIVATED HORIZON OF PADDY SOILS IN SHANGHAI DURING DRY FARMING PERIOD AND THE APPROACH TO ITS ELIMINATION

Yang Jin-lou, Zhu Ji-cheng, Jiang Su-zhen, Shi Nan-chang,
Zhu Lian-long
(Soil and Fertilizer Institute of the Shanghai Academy of
Agricultural Science, Shanghai)

The rainfall in Shanghai is abundant, amounting to 500 mm during the growing season of wheat, barley and naked barley. The waterlogged condition of the cultivated horizon of paddy soils is a grave obstacle to a high and stable yield of wheat, barley and naked barley.

THE FORMING OF WATERLOGGED CONDITION IN CULTIVATED HORIZON OF PADDY SOILS

In consequence of the development of triple cropping system (rice-rice-upland crop), waterlogged period is prolonged. Wet ploughing and suppressing makes the plowpan thickened by 6-8 cm and non-capillary porosity decreased (generally only 0.5-2.0%, the saturated conductivity also decreased (generally below 1×10^{-6} m/sec). The permeated rainwater tends to be retained at the interface of cultivated horizon and plowpan, forming a gleyed horizon with distinct reduction feature. When it happen to rain for sucessive days and the amount of transpiration is checked, the plowpan will stop the water from seeping down and harmfulness caused by logged-water will then result in the cultivated horizon. Drainage in the field is poor because the water level of canals and trenches is so high that it will lead to the elevation of underground water. Thus, the clayed soil becomes more sticky, and capillary water increases. All these factors lead to waterlogging in the cultivated horizon.

EVIDENCES OF WATERLOGGING IN THE CULTIVATED HORIZON OF PADDY SOILLS

After raining more water retains within 30 cm of the upper soil than in the capillary. This indicates the existence of gravitation water in the cultivated horizon (Fig. 1).

In the field, a 20 cm shallow well and a 1.5 m deep well were dug simultaneously. After raining the water in both wells was found to be from two isolated water sources. Obviously, the seeping rate of the rain water into cultivated horizon is greater than that of translocation of water to the plowpan. Thus, water accumulated at the interface of the cultivated horizon and the plowpan can afford water supply to the shallow well (Fig. 2).

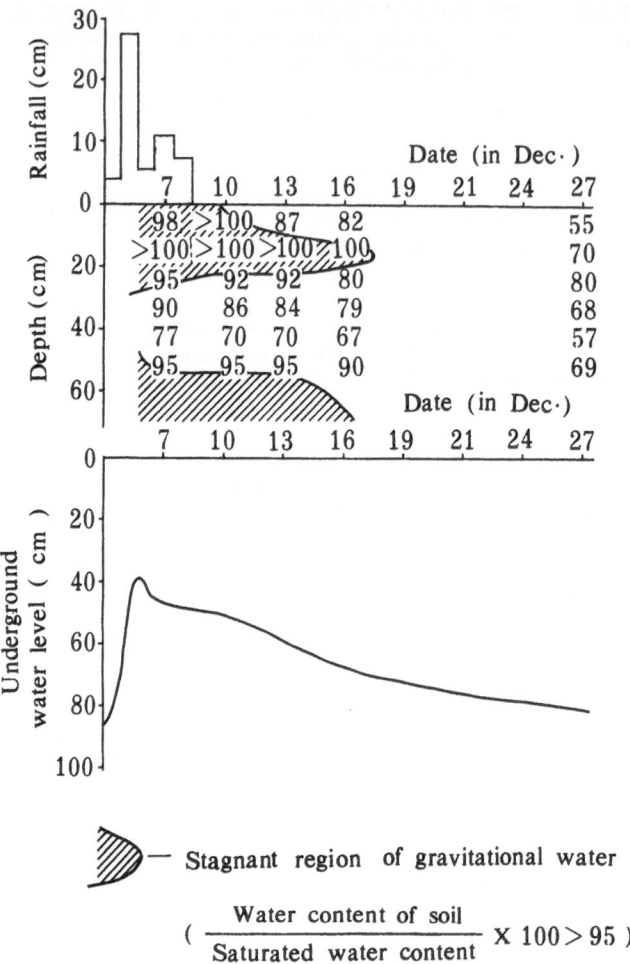

Fig. 1 Stagnant water region in cultivated horizon and
underground water level after raining in plot of
string-ditching (gleyed paddy soil)

HARMFULNESS OF WATERLOGGING IN THE CULTIVATED HORIZON

About 70% of the absorbing root systems of wheat, barley and naked
barley concentrate at the upper 30 cm of this horizon. The non-capillary
pores of this horizon are generally less than 10%, and the logged water
after raining occupies parts of these pore spaces; thus, an anaerobic
condition in the cultivated horizon results with the accumulation of
ferrous iron up to 300–800 ppm. Censequently, it causes the weakening of
root activities and abilities in absorbing potassium nutrient, and the
ratio N/K in plant being too high (>1.3), results in physiological
hindrance (Table 1).

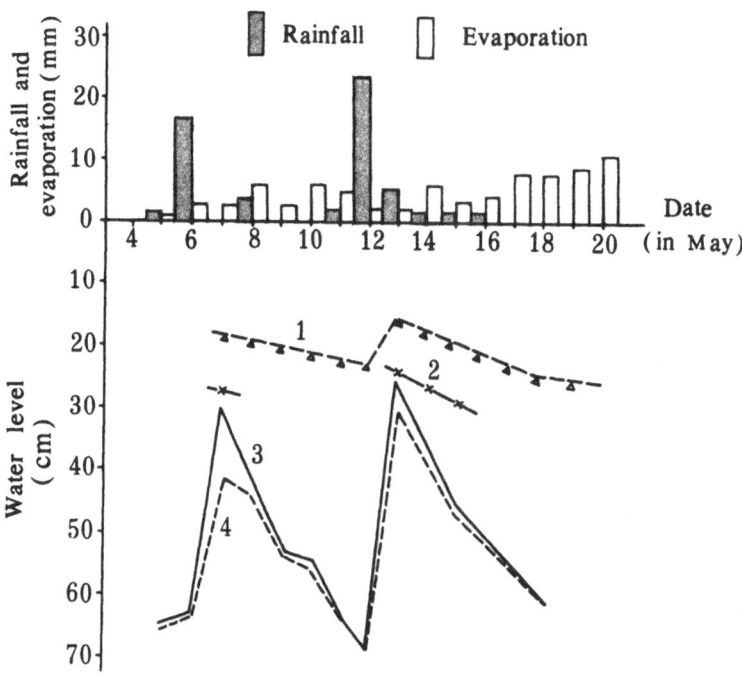

Fig. 2 The formation of two water levels in soil after
raining (gleyed paddy soil)
1.Stagnant water level in control plot;
2.Stagnant water level in deep-ditching plot;
3.Underground water level in deep-ditching plot;
4.Underground water level in control plot

Table 1 Absorption of N, K by wheat and naked barley under
different water conditions of the soil (in April)

Crop	Water condition	Dry weight (g/plant)	Nutrient content of plant			N and K absorbed by single plant	
			N%	K%	N:K	N(mg)	K(mg)
Naked barley	Serious water logging	0.21	1.95	1.48	1.32	4.10	3.11
	Median water logging	0.47	1.53	1.60	0.96	7.19	7.52
	Non-water logging	1.04	1.68	1.83	0.92	17.47	19.05
Wheat	Serious water logging	0.21	2.24	1.61	1.39	4.70	3.38
	Median water logging	0.58	2.49	1.67	1.49	14.44	9.68
	Non-water logging	0.84	1.54	1.96	0.79	12.93	16.46

APPROACH TO GETTING RID OF LOGGED WATER IN THE
CULTIVATED HORIZON OF PADDY SOILS

Drainage through string-form trenches in plowpan. In 1976-79, drainage trenches of deep cutting string-form were developed to get rid of the water in the cultivated horizon. It was done by using a 50 H.P. (Pl. 1) tractor to draw a string-form trench plough, with the bullet-head plough share removed and with the blades alone (Pl. 2). In front of the plough are the round plates for cutting soil. The plough-depth is 0.3-0.4 m and the distance between them is 0.3-0.8 m (Pls. 3,4). In this way, the plowpan that holds water is destroyed. Results obtained from 188 field experiments indicated that the water content of the soils at the depthe of 0-20 and 20-30 cm provided with such drainage trenches was decreased by 5.6% and 9.8% respectively as compared with that of the control field. The drainage rate of free water in the cultivated horizon is 40% earlier (Table 2). Results from 44 field experiments showed a yield increment of 406.5 kg/ha, a figure which was 16.5% higher than that of the control plots (Table 3).

Table 2 The effect of deep string-ditching on the moisture content of soil in the growing season of upland crop (n = 188)

Depth (cm)	Decrease (%) in moisture content of soil $\bar{X} \pm s_{\bar{X}}$	T test	Item	Different proportion of decrease of the relative value (%)					
				0-5	5-10	10-20	20-30	>30	Σ
0-20	5.6±7.0	**	Amount	97	50	38	3	0	188
			(%)	51.6	26.6	20.2	1.6	0	100
20-30	9.8±10.0	**	Amount	87	24	41	25	11	188
			(%)	46.3	12.8	21.8	13.3	5.8	100

Table 3 The effect of deep string-ditching on the yield of wheat and naked barley (n = 44)

Increase in yield	Mean $\bar{X}$	Standard deviation $s_{\bar{X}}$	T test	Item	Extension of increased yield and decreased yield (%)					
					Decreased yield	-5-0	0-5	5-10	10-20	>20	Σ
kg/ha	407	510	**	Amount	3	1	8	6	15	11	44
	16.5	18.7	**	(%)	6.8		2.3	18.2	13.6	34.1	25.0	100

Plate 1 Haddy whell
with strake

Plate 2 String-form
trenches
plough

Plate 3 Working in
the field

Plate 4 The string-
form trenches

Besides, increasing the upland crop in rotation is beneficial. The non-capillary pore in the plowpan for planting cotton and maize is 55% greater than that for rice-rice-upland crop, and in the sub-horizon, 30% greater. As a result, seeping of water was strengthened unlikely to become waterlogged.

779

A COMPARATIVE STUDY ON METHODS OF TILLAGE OF
PADDY SOILS IN TAIHU LAKE REGION

Zhao Cheng-zhai, Zhou Zheng-du, Dong Bo-shu
(Institute of Soil Science, Academia Sinica, Nanjing)

The principal rotation system in Taihu Lake Region is rice-wheat rotation. The traditional method of tillage is such that in the cultivated layer
there is an upper sub-layer with muddy paste and a lower sub-layer with small
clods before the transplanting of rice. It is generally considered that the
muddy paste is beneficial to the liberation of nutrients and so is the clod
to aeration. For wheat crop the requirement is that the whole cultivated
layer be soft and friable. In order to fulfil these requirements large amount
of labour and energy has to be spent every year. Since the seventies the
tilth of some soils becomes worse due to the development of another rotation: rice-rice-wheat, and as a consequence it becomes even more difficult
to crush the clods(1).

For the purpose of trying to find out more rational method of tillage,
we conducted in 1978-1979 some comperative experiments on the effect of
method of tillage on some properties of the soil on a permeable paddy soil
in Wuxi County.

TILLAGE IN SPRING

The soil contains 3.2 % of organic matter. The texture is clay loam.
The preceding crop before plowing was barley. The field was manured with 75
t/ha of compost as basal dressing, and fertilized with nitrogen equivalent to
772.5 kg/ha of ammonium sulfate.

The treatments included:
(A) Rotary tillage after flooding.
(B) Zero tillage after stubble cleaning.
(C) Loosening without turnover of furrow slice.
(D) Turnover, drying, and smoothing manually after flooding.
(E) Conventional tillage, including turnover, drying and rotary tillage
    after flooding.

Observations on the growth of rice showed that before midsummer drying
of field rice grown in plots (B) and (C) was darker in color with more tillering, while in plot (E) was shown some symptom of browning at leaf-tips.
After midsummer drying the difference in growth became indistinct. The
yields of rice for plots (B) and (C) were slightly higher than those with other
other treatments, but not significant statistically. Other experiments showed
similar tendency (Table 1).

Analyses of the soil showed that there was marked decrease in hardness
after submerging (Fig. 1). It is worthy of note that the hardness in the
lower part of the cultivated layer is only slightly higher for the zero
tillage treatment than for the tilled treatment. At the harvest time the

soil in the zero tillage plot became yellowish brown in color, with a Eh value 150-250 mV higher than the tilled plot, and was not distinct in the secondary gleyization. And, there appeared air space rapidly after drainage of the superficial water. The order of puddling among the treatments were: B<C<D<E<A, in conformity with the degree of disturbing caused by tilling. The soil suffering rotary tillage was especially soft, and changed into very hard clods after drying (Pl. 1). All these seem to imply that zero tillage or cultivating practices leading to the disturbance of the soil as little as possible are beneficial to the maintenance of soil structure with-out adverse effect on the growth or yield of rice.

Table 1 Effect of method of tillage on yield of rice

Soil	Exp. No	Crop	Treatment	Yield (kg/ha)
Permeable Paddy soil	1	Early rice	Conventional	5198
			zero tillage	5250
	2	Early rice	Conventional	5138
			zero tillage	5363
	3	Vesting rice	Conventional	7983
			zero tillage	9908
	4	Early rice	Conventional	7223
			zero tillage	7275
	5	Later rice	Conventional	4695
			zero tillage	5033
Bleached paddy soil	6	Early rice	Conventional	4845
			zero tillage	4905

A supplementary experiment showed that the upper part of the cultivated layer was quite different from the lower part in some respects. The structure improved more remarkably than the lower part after the planting of dry-farming crops. The content of organic matter in this part is also slightly, high. Fig. 2 shows that the dry matter of rice grown in soil samples collected from the upper part (0-4 cm). Lower part (4-13 cm) and plowpan is 103,32 and 21 g/pot respectively. Thus, if the field is tilled in the conventional manner, there is the liability that the fertile soil of the upper part of the cultivated layer may be turned downward. thus leading to an unfavorable effect on rice.

A pot experiment showed that the effect of drying the soil on rice growth was very remarkable (Fig. 3). While the influence of mechanical disturbance was insignificant statistically.

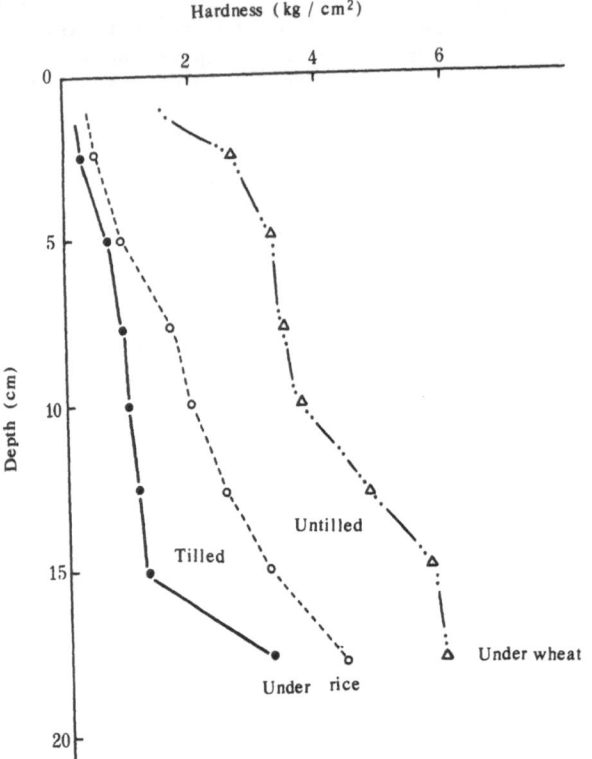

Fig. 1 Hardness of the tilled and untilled soil after submerging, and under wheat

rotary tillage   zero tillage

Plate 1 Soil structure in rotary tillage and zero tillage

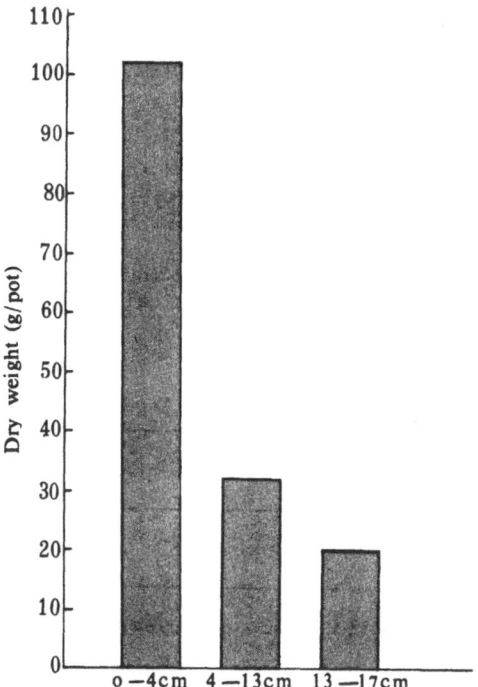

Fig. 2 Dry weight of rice plant grown
on soils from different depth (pot experiment)

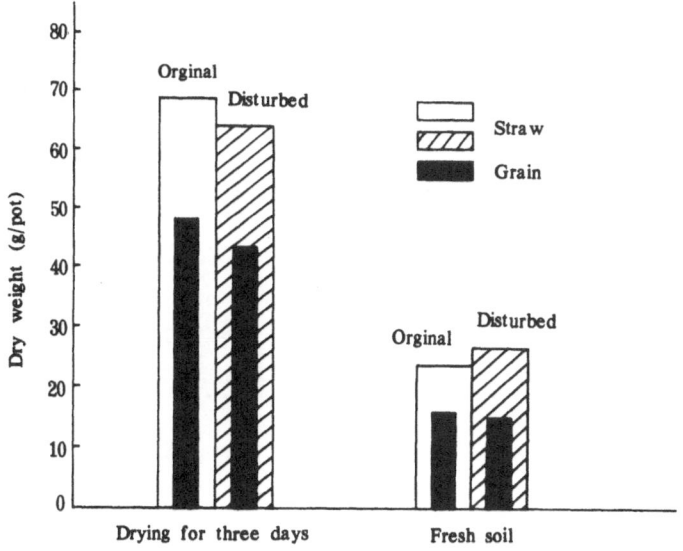

Fig. 3 Effect of drying and disturbance of the soil
on rice growth

TILLAGE IN AUTUMN

The experiment was conducted on the same type of soil including the following treatments:

(A) Zero tillage after stubble cleaning.
(B) Conventional: plowing for 14 cm, and then the upper part pulverized.
(C) As (B), with the whole cultivated layer pulverized.
(D) Plowing for 20 cm, and the upper part pulverized.

The results of experiment are indicated in Table 2. The difference in alphabets after the mean value means statistical significance (P = 0.05). The yields of deep plowing treatment were higher.

Table 2 Effect of method of tillage on yield of wheat

Treatments	Straw/grain	Yield(kg/ha) mean	
Zero-tillage after stubble cleaning	0.64	4770	a
Conventional tillage (14 cm)	0.64	4785	a
Whole layer crushed (14 cm)	0.61	4800	a
Deep plowing (20 cm)	0.56	5115	b

Observations revealed that the high water-holding capacity of the cultivated layer was beneficial to the emerging and tillering of wheat in plots of shallow plowing and zero tillage. However, the comparatively low suction and low hardness in the lower part of deep-plowing plot provided better nutrient supply at the latter stage of wheat growth. The structure of the plowed soil especially its upper part, improved considerably (Pl. 2) in contrast to soils after the planting of rice (Pl. 3).

plate 2 Development of soil
structure after wheat

Plate 3 Puddling of paddy soil
caused by conventional tillage

DISCUSSION

The alternation of drying and wetting of the soil can promote the transformation of nutrients(2-4) and the improvement of structure. This is the principal cause for the development of surface soil fertility after the planting of wheat. Therefore, plowing in autumn is helpful both to the growth of wheat and to the improvement of soil structure. Nevertheless, excessive pulverization of the soil not only has no beneficial effect on the growth of wheat but is also liable to the pudding of the soil if plowed under high moisture conditions(5).

The plowing and harrowing before the transplanting of rice can turn the fertile surface soil downward, thus unfavorable to the early tillering and leading to the destruction of soil structure. Obviously the conventional method of tillage has some disadvantages. We suggest that the soil should be plowed and then pulverized for the upper surface layer before the seeding of wheat in autumn, and transplanted with rice by zero tillage aftar stubble-cleaning in spring. In this way the soil structure formed during the growing period of wheat may be destroyed to the least extent by the planting of rice. In cases where the soil moisture is excessive due to the heavy raining in sutumn, it is also feasible to plant wheat with zero tillage after stubble-cleaning as an alternative under unfavorable climatic conditions.

REFERENCES

(1) Chen Jia-fang, Wu Mei-ling, He Qun, Liu Bin, Studies on the compacting of paddy soils in Suzhou, Jiangsu. Turang,6, 286-290(1975). (in Chinese)
(2) Brich, H. F., The effect of soil drying on humus decomposition and nitrogen availability. Plant Soil, 10,9-31 (1958).
(3) Laura, R. D., Short communication on the stimulating effect of drying a plant material. Plant Soil, 44, 464-465 (1976).
(4) Baver, L. D., Gardner, W.H., Gardner W.R., Soil Physics, 4th edition, John Wiley & Sons, Inc., New York, 146-149 (1972).
(5) Koeniges, F. E., The puddling of clay soils. Netherlands J. Agric. Sci., 11, 145-155 (1963).

EFFECT OF RICE PLANTING ON IMPROVEMENT AND
UTILIZATION OF SALINE SOILS IN TARIM BASIN

Li Li-qun, Dong Han-zhang, Wang Zun-qin
(Institute of Soil Science, Academia Sinica, Nanjing)

It has been proved in a long period of practice that the planting of
rice is effective on the improvement of saline soils. This can be applied
to the extremely arid areas in southern Xinjiang(1). Flooding in the rice
growing season makes it possible for the irrigation water to dissolve salt
and thence remove it in drainage. In the meantime, the salt content of
underground water in paddy fields is gradually on the decrease. Even in
places where drainage conditions are unfavorable, such practices as irrigation, drainage, fertilization, and rotation of upland crop and paddy rice are
still beneficial to the control of salt content in the soil and an increase
of crop yield is expected.

In Tarim Basin, one year's rice cultivation on soils with high salinity
brought about a decrease in salt content to the extent of 20 to 80 % (Table
1). In the soil profile, desalinization is most remarkable in the plowed
layer.

Table 1 Salt content of the soil before and
after the planting of rice

| Depth (cm) | Soluble | | Salt (%) | |
| | 1973 | | 1974 | 1975 |
	Before planting rice	After harves- ting rice	Under upland crop	
0–25	4.19	0.89	2.01	4.15
25–50	1.90	1.40	1.63	2.01
50–75	2.26	1.58	1.34	3.01
75–100	1.59	1.62	1.19	1.77
100–125	1.10	1.20	0.87	0.86
125–150	1.19	1.24	0.96	1.14
0–150	2.04	1.32	1.33	2.16

Under equal conditions, the longer the period of rice cultivation is,
the thicker the layer of desalinization is formed. Great changes took
place in the composition of soil salts after a rice cultivation. In a farm
of Aksu Prefecture, for example (Fig. 1), the main component of salt in soil
was mainly chlorides, which changed into sulphates ranging from 60 to 90 % of
total anions after four year's rice cultivation. At the same time, the sodium
was replaced by calcium as the main component of cations up to 50-80 % of
the total cations. Such desalinization and changes in the components of salt

after rice cultivation provided a favorable condition for the succeeding crop.

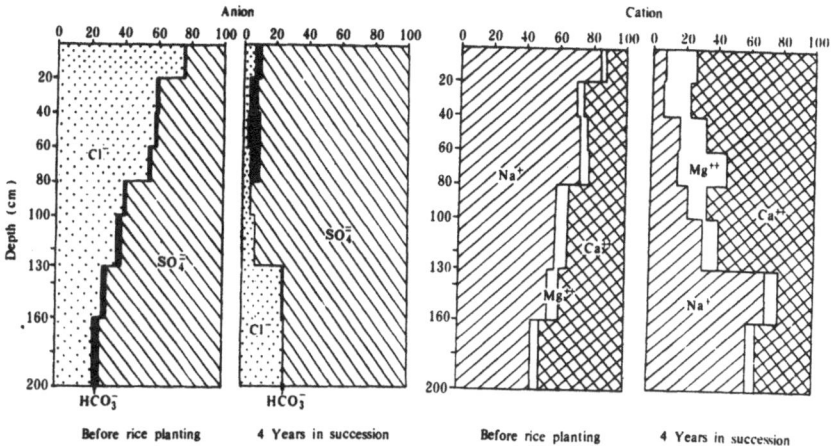

Fig. 1 Changes in composition of soluble salts in soil profile after rice planting

The downward water movement in paddy fields not only desalinizes the soil to a certain depth, but also decrease the salinity of ground water(2). These changes depend chiefly on local hydrogeological and drainage conditions.

In the area along the Tarim River (Table 2), the original salinity of ground water is lower and easy to form the layer of desalinized ground water through rice cultivation and irrigation. But in the joint areas between the flooding plain and diluvial fan where the original salinity of ground water is generally higher up to 20-60 g/l, the desalinization of the soil and ground water in this region is slower and more difficult, even under favorable drainage condition.

Table 2 Soluble salt content of underground water in paddy fields

Depth (cm)	Soluble salt (g/1)	
	Along the banks of Tarim River	Pluvial-alluvial plain
100	3.8	30.6
150	3.4	42.3
200	12.4	56.0
250	11.8	52.8
300	—	62.1

Taking for instance a field in Mongolian Autonomous Prefecture of Bayingolin, the original water table was 2.6 meter below the surface and the salt concentration was above 40 g/l. After one year's paddy planting, the water table was raised to 1.2 meter and the salt concentration was decreased to only 14-30 g/l. However, after the field was kept dry and planted with cotton for one year, the surface evaporation caused the underground water table to drop and the comparatively dilute soil water to be used up, and when the water table restored its original depth of 2.6 meter, the salt concentration in soil water restored its original concentration to above 40 g/l as well. Therefore, in the area of poor drainage, rice cultivation may only bring about temporary benefits. In this case, a portion of salts was removed from the land through drainage outlets, but under a dry condition, the salts may rise again along with the upwared capillary water and accumulate in the surface soil (Table 1). By keeping on dry farming for one to three years, the beneficial effects of rice cultivation will be of no avail completely. Then it is necessary to plant rice once more, and a rotation system of paddy rice and upland crops is thus developed, in order to control the salt regime in the plowed horizon of the soil. The period of such a rotation system depends largely upon the rate of desalinization of the soil and ground water.

It is obvious, therefore, that the drainage is of most importance to the improvement of saline soil. As there is little difference in the elevation of land surface, it is impossible to drain the soil with drainage ditches in adequate depth. For speeding up the desalinization of soil and ground water, it is preferable to adopt such reasonable drainage systems suitable to local conditions as combining gravity with pump, shaft with ditch and shallow with deep drainage.

It seems certain through practice that in order to obtain full benefit from utilizing paddy rice to improve saline soil, the following should be considered.

1) There should be water supply sufficient to insure the rice crop all water it needs.

2) A drainage system should be carefully designed and duly installed.

3) There should be supplementary farming operations such as land levelling, liberal application of organic fertilizers or growing sufficient green manure crops to be turned under.

4) Paddy rice and upland crops should be properly arranged to avoid the possibilities of producing ill effects due to admixture of paddy and upland fields. In this regard, paddy field should be better put in to a rather concentrated area.

REFERENCES

(1) Wen Zhen-wang (ed.), Soil Geography of Xinjiang. Science Press, Beijing, 445-446(1965). (in Chinese)
(2) You Wen-rui, Hong Qing-wen, Bi Si-ming, Zhao Hong-jun, The water-salt regime and its regulation of coastal saline soils in the process of melioration by rice planting in Hopei Province, Acta Pedologica Sinica, 12(2), 107-119(1964). (in Chinese with English summary)

ON RICE NUTRIENT DISORDER AND ITS DIAGNOSIS IN CHINA

Luo Zhi-chao, Tang Yong-liang, Cui Rong-hao
(Institute of Soil Science, Academia Sinica, Nanjing)

Rice nutrient disorder has been widely studied in Asiam countries(1). In recent years, with the rapid expansion of cultivated area of rice in our country, this problem involved in both new and old rice regions alike calls for investigation and solution. The present article gives a brief account of the situation of rice nutrient disorder, the main symptoms, and its diagnosis in our country.

GENERAL SITUATION ON RICE NUTRIENT DISORDER IN CHINA

The issue of nutrient disorder found in rice production may be summerized as follows: 1) A deficiency in nitrogen brings about an undesirable growth, and in some cases, excessive application of nitrogenous fertilizer leads to superfluous uptake. Besides, a long-term application of nitrogenous fertilizer leads to tipping of balance in phosphorus and potassium (2). 2) With the increase in yield and cropping index, the area lacking in phosphorus is greatly expanded, not only in the strongly acid and calcareous soil, but also in certain area of the fertile meutral paddy soil region. At the same time, the problem of zinc deficiency caused by the over use of phosphorus emerges. 3) Due to the lack of potassium fertilizer resources, problems become even more critical in the paddy soil of South China where the content of available K is rather low and where the latent K deficiency under high yielding conditions is rather serious because of unbalance in N:K. This leads to the occurrence of yellowish stiff-seedling, Helminthosporium leaf spots, bronzing strip disease and withered disease. 4) The application of sulfur fertilizer to the cold submerged field of the hilly region in South China has an apparent effect, and in certain region yellowing and autumn sprout seedling have been found. 5) In some calcareous paddy soil and coastal salt-affected soil, deficiency in zine gives rise to stiff-seedling, red-seedling and rice dwarf. 6) The content of available boron in some paddy soil is low, and a good response will be obtained upon applying it.

THE NUTRIENT DISORDER FOR SEVERAL MAJOR ELEMENTS

1. Nitrogen

Though it is easy to diagnose nitrogen deficiency, it remains a problem to be fully solved to control adequate amount of nitrogen supply and enable a balamced growth, lacking an appropriate measure, especially a quick diagnosing method, to better reflect mitrogemous nutrient status in rice (3).

We have made some investigation on the diagnostic method for rice

nitrogenous nutrient and found that colorimetric method of ninhydrin could
be used for determining the total amount of amino acid nitrogen in the
diagnosis of rice nitrogenous status in leaf-sheath and this method is
closely related with the yield of rice (4). When amino acid nitrogen is
higher than 250 ppm, nitrogenous fertilizer is unnecessary, otherwise plant-
lodging would result due to excessive vegetative growth; when it is below
100 ppm, it is indicated that the plant is lacking in nitrogen. The
referential criteria in diagnosis are shown in Table 1.

Table 1 Diagnostic criteria for N-nutrient status
in rice plant (panicle differentiation stage)

Content of amino acid-N (ppm)	100	150-200	250
N-nutrient status	Low	Medium	High

2. Phosphorus

In the strongly acid and calcareous paddy soils of China, stiffen
symptom in rice seedlings is a common occurrence due to phosphorus deficiency,
indicating that after transplanting seedling, growth is extraordinarily
slow (5), especially in years when atmospheric temperature is low. Its main
characteristics are:the leaf becoming dark green, plant upright, non-
tillering, the cluster of the plant not dishevelled. Here are some more
complicated phenomena. For example, when nitrogen supply is low, the
phosphorus-lacking leaf doesn't show a dark-green color but light green
instead. In acid soil, P-deficient rice seedlings show distinct brown
spots as well as burnt leaf tips showing that there is toxic effect of
some ions due to low pH in soil. Furthermore, in some P-deficient soils,
when rice seedlings become stiff as a result of P-deficiency, and when
nitrogenous fertilizers are further applied, perishing seedlings would
happen.

Experiments show that Mo-blue test for the sap from rice leaf sheath
can be used for diagnosis of P status in rice plants. The referential
criteria are shown in Table 2.

Table 2 Diagnostic criteria for P status in rice(tillering stage)

P status	Total P in the whole plant(%)	P in sap from leaf sheath(ppm)	Available P in soil (ppm)[*]
Stiffen seedling due to P deficiency		< 40	< 3
Latent P deficiency	< 0.165	40-80	3-7
Normal growth	> 0.165	> 80	> 7

[*] by 0.5N $NaHCO_3$ extractant

3. Potassium

In case of K deficiency, besides typical symptom, there are different
phenomena under different situation. for example, the K deficiency in
late rice appears to be more pronounced than that in early rice. More

brown spots are seen on leaves and burnt tip is more severe. Being different from N and P deficiency, K deficiency has more detrimental effects on root system which tends to become slender, short and dark brown in color, and somewhat transparent, and around which the soil is comparatively low in redox potential, becoming greyish black in colour.

Both the paper test with dipicrylamine and the turbidimetric method with cobalt sodium nitrite are generally applied to the diagnosis of K status for rice plant. The criteria used for soil available K are exchangeable K. In some cases, the content of slowly available K is also taken into account. The criteria for the diagnosis of plant and soil K status are summarized in Table 3.

Table 3 Criteria for the diagnosis of K nutrient
status(panicle differentiation stage)

Nutrient status	Total K in plant (%)[*]	K in basal leave sap (ppm)	Available K in soil (ppm)
K-deficiency(brown spots & burnt tip)	<1.0	1000	<50
Latent K-deficiency	1.0–1.5	1000–2000	50–100
Normal growth	>1.5	2000	>100

[*] Dry weight basis

4. Zinc

In recent years, Zn deficient symptom has been reported in the plain to the north of the Huaihe River, the calcareous alluvial soils in the Jianghan Plain and the saline soils along the coastal regions (6,7). Its typical symptoms are: the mid-rib of young rice leaf turns to white color and becomes more and more distinct towards the base. There are irregular brown spots on both sides of the mid-rib, and the distance between the leaf cushion of the top three leaves is shortened and parallelly arranged. Moreover, the base of the rice plant is brown in color, and dying seedlings are seen in some places; the growth of seedlings uneven and the developing stage delayed. As a result, the yields of rice are ill affected.

Reports from other regions on the critical criteria of Zn deficiency were rather similar. These could be summarized as Table 4.

Table 4 Criteria for the diagnosis of Zn nutrient in rice

Nutrient status	Zn content in leaves[*] (ppm)	Available Zn in soil[**] (ppm)
Deficiency	<10	<0.3
Latent deficiency	10–15	0.3–0.5
Normal growth	>15	>0.5

[*] Extracted with 1 N HCl

[**] Extracted with DTPA

Experiments also show that the application of Si-fertilizer helps promote healthy growth of rice plant in Si-deficient soil. The induction of trace element deficiency caused by excessive application of lime in acid soil region has been received more and more attention.

In summary, the diagnosis of nutrient disorder is a complicated and tough issue. The criteria for its diagnosis have been dealt with by researchere concerned with a view to better reflecting the nutrient status in plant. Based on the viewpoints of nutrient balance, the ratios of N:K, N:P and N:S have been used for this purpose in rice plantation. For example, if N:K is greater than 0.95, then the application of K is beneficial to the yield of rice. Also, studies on the ratio of certain elements in young and old tissues have been made to detect the nutrient status.

REFERENCES

(1) Tanaka, A., Yoshida, S., Nutritional disorder of the rice plant in Asia. Tech. Bull. IRRI, 10(1970).
(2) Chu Chao-liang, Liao Hsian-ling, Tsai Kuei-shihn, Yu Chin-chou, Soil nutrition status under "rice-rice-wheat" rotation and the response of rice to fertilizers in Suchow district. Acta Pedologica Sinica, 15(2), 126-137 (1978). (in Chinese with English summary)
(3) Mervyn, W. Thenabadu, Evaluation of the nitrogen nutrition status of rice by plant analysis. Plant Soil, 37(1), 41-48 (1972).
(4) Nutrition Diagnosis Research Group, Institute of Soil Science, Academia Sinica, Total amino nitrogen content in rice plant as a diagnostic index of manuring. Acta Botanica Sinica, 19(3), 200-208 (1977). (in Chinese) with English summary)
(5) Faculty of Soil, Zhejiang Agricultural University, Chemical diagnosis of stunted seedling of early rice resulted by phosphorus deficiency. Turang, 1, 9-17 (1974). (in Chinese)
(6) Coordinated Research Group of Rice Zinc Deficiency, Anhui Agricultural College, The symptoms of stunted rice seedling resulted by zinc deficiency and effect of zinc fertilizer. Turang Feiliao, 1, 29-34 (1978). (in Chinese)
(7) Wang Zhong-lian, Wang Wan-zhang, Studies on the symptoms of rice zinc deficiency and effect of zinc fertilizer. Turang Feiliao, 1, 36-40(1979). (in Chinese)

CHARACTERISTICS OF NITROGEN MINERALIZATION OF PADDY SOILS AND THEIR EFFECT ON THE EFFICIENCY OF NITROGEN FERTILIZER

Cai Gui-xin, Zhang Shao-lin, Zhu Zhao-liang
(Institute of Soil Science, Academia Sinica, Nanjing)

As revealed by the papers published in recent years (1-4), the minera-lization pattern of soil nitrogen during a rice growing season in combina-tion with the total amount of soil nitrogen supply is an important factor that should be taken into consideration in improving the efficiency of nitrogen fertilizers in rice production. So it is worthwhile to investigate: 1) the characteristics of mineralization of soil nitrogen in relation to soil type and fertility and 2) the efficiency of nitrogen fertilizer in relation to soil properties and characteristics of mineralization of soil nitrogen. Here are the results obtained from pot experiment.

MATERIALS AND METHODS

A total of soil samples of the plowed-layer were collected from the paddy field in the southern part of Jiangsu Province early in May, 1979, when the fields were covered with wheat or barley. The soil properties are shown in Table 1. The fresh samples were separated by hand into clods of 3-4 cm in diameter, and meantime the plant residues in the samples were picked out. An adequate amount of the prepared samples, equivalent to 2.3 kg of oven-dried soil, was weighed and put into each pot. These pots covered with polyethylene film were kept in room beforr flooding. On May 26 they were placed outdoor and flooded with tap-water. Rice seedlings were transplanted on May 30. On the basis of P and K fertilization, two treatments on each soil were made: the one with no N, and the other, applying 197 mg N per pot in the form of ^{15}N-labelled urea, with 5.25% ^{15}N. All the fertilizers were applied as solution and were mixed with the top soil of 5 cm. The wilting leaves were collected from each pot throughout the experiment and were combined with the plant samples taken from the same pot for determination. Soil and plant samples were taken at intervals of 14, 28, 49, 98, and 145 days after transplanting. Three pots were harvested for each treatment at each interval, and only the tops of the plant were analysed, except that at maturity (145 days after transplanting), where the plant roots and the tops were analysed separately. The total N was determined by Kjeldahl method and the ^{15}N enrichment by mass spectro-graphy. The maximum and minimum temperature in the pot at a depth of 5 cm were recorded every day throughout the experiment.

RESULTS AND DISCUSSION

1. Characteristies of Nitrogen Mineralization of Paddy Soils without Additional N-fertilizer

As shown in Table 2, at the maturity of rice plant, the N mineralized* from the soils in control pots and the N mineralization coefficient** ranged from 4.08 - 6.38 mg N per 100 g of soil and 2.16 - 3.86 % respectively. Bleached paddy soils were characterized by a higher N mineralization coefficient than the permeable paddy soils, possibly owing to the coarser texture of the former. On the other hand, the N mineralization coefficients of the waterlogged paddy soils were the lowest among the soils investigated. It may be due to the nature of the organic matter formed under the poor-drainage conditions.

Table 1 Properties of the soils

Soil No.	Location	Soil type	Ferti-lity	PH	O.M. (%)	N(%)	Texture	CEC (m.e./100g)
1	Wujin	Bleached	High	6.3	2.75	0.158	Silty soil	21.0
2		paddy soil	Low	6.6	2.01	0.120	Silty soil	17.7
3	Wuxian	Permeable	High	6.5	3.16	0.195	Clay loam	20.5
4		paddy soil	Low	6.2	2.52	0.159	Clay loam	18.4
5	Danyang		High	6.2	3.40	0.194	Clay	28.3
6		Waterlogged	High	7.5	3.47	0.201	Silty soil	20.7
7	Changshu	paddy soil	Low	8.1	3.22	0.190	Silty soil	21.6

The mineralization pattern of the 7 paddy soils showed a general increase in N mineralization rate from July 18 through August 8 (Fig. 1), which was in consonance with the peak in the curve of the soil temperature. Furthermore, the mineralization pattern of the soils fits well to the equation of $y = kx^n$ (r^2 higher than 0.98), where y denotes the N mineralized, X the summation of effective soil temperature, K and n the parameters(5).

Table 2 and Fig 1 show that the mineralization coefficient of the paddy soil of high fertility in each of the three soil types was higher than that of the low fertility though the n value in the mineralization equation, as an index of the characteristic of N mineralization pattern, did not show any definite difference among those soils differed in fertility.

2. Balance of Urea-N Applied in Basal Dressing

As shown in Table 3, when urea was applied on bleached paddy soils as basal dressing, the N recovery by rice plant and the N remaining in soil was higher, and the N loss was lower than that on the other two soil types. In the case of the soilty waterlogged paddy soils, the N recovery from urea

---

*N mineralized from soil = soil N supply - Initial soil NH_4-N;
Soil N supply = N assimilated by rice plant + Soil NH_4 - N in the same pot;
N assimilated by rice plant = N accumulated in the top of rice plant - N in rice seedlings;
N accumulated in the top of rice plant = Dry wt. of the top x N%.

**N mineralization coefficient (%) = $\dfrac{\text{N mineralized}}{\text{Total content of soil N}}$ x 100.

by rice plant was lowest, and the N loss was highest among the soils investigated. The higher pH value of the silty waterlogged paddy soils may induce a higher N loss through ammonia volatilization.

Table 2 Mineralization of soil nitrogen in relation to
grain yield of rice (at maturity)

Soil No.	Uptake of N by rice plant from soil (mgN/100g)	Soil -N supply (mgN/100g)	N minera- lized[**] (mgN/100g)	Minera- lization coefficient (%)	Grain wt (g/pot)	Grain wt[*] / Soil-N supply
1	6.02	6.17	6.10	3.86	9.64	67.9
2	4.00	4.14	4.08	3.40	7.29	76.6
3	6.05	6.48	6.38	3.27	9.97	66.9
4	4.80	4.97	4.84	3.04	8.22	72.1
5	5.49	5.76	5.60	2.89	9.89	74.9
6	5.20	5.40	5.31	2.64	8.46	68.2
7	4.02	4.19	4.10	2.16	7.23	75.1
L.S.D. 5%			0.66	0.38		

[*] $\dfrac{\text{grain wt (g/pot)}}{\text{soil-N supply (mgN/100g)} \times \dfrac{23}{1000}}$

soil-N supply = N assimilated by rice plant + NH_4-N remained in the soil

[**] N mineralized = soil N supply - initial NH_4-N in the soil before flooding

Within the same soil type the N recovery by rice plant was slightly higher on the soil of low fertility than that of high fertility and the reverse was true for the N loss. However, statistical test of significance did not verify such differences.

3. The Priming Effect of Urea Applied as Basal Dressing

The increment[*] of N mineralized from soil through the application of urea was 2.13 - 3.03 mg per 100 g of soil (Table 4). Most of the increment of N mineralized from soil through priming effect was compensated for by the urea-N remaining in the soil, mainly as a result of biological interchange. Consequantly, the nitrogen nutrition level for the growing of rice would not be raised considerably by the priming effect.

4. Efficiency of Urea Applied as Basal Dressing

In the control pots, the grain weight depends primarily on the soil N supply (r^2 = 0.927), but the efficiency of the soil N supply (the weight of grain produced per unit weight of soil N supply) tends to be decreased with the increase of the soil N supply (Table 5).

---

[*] Increment of N mineralized from soil through the application of urea = Mineralized N in N-pot - Mineralized N in control pot.

795

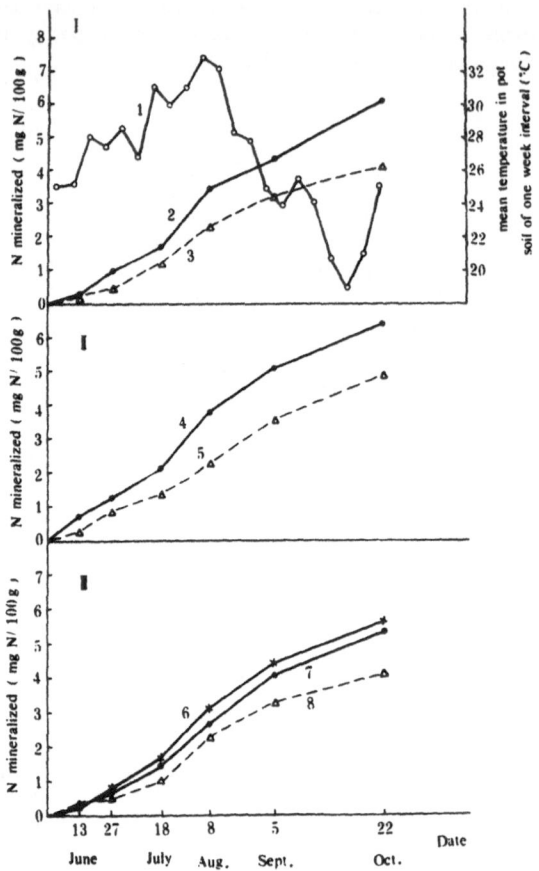

Fig. 1 Mineralization patterns of soil organic N
    I. Bleached paddy soil:
      1. Mean temperature;
      2. High fertility,
        $Y = 0.000231 \; X^{1.37}$ $(r^2 = 0.986)$;
      3. Low fertility,
        $Y = 0.000101 \; X^{1.44}$ $(r^2 = 0.996)$.
   II. Permeable paddy soil:
      4. High fertility,
        $Y = 0.00294 \; X^{1.03}$ $(r^2 = 0.996)$;
      5. Low fertility,
        $Y = 0.000387 \; X^{1.27}$ $(r^2 = 0.986)$.
  III. Waterlogged paddy soil:
      6. High fertility (clay),
        $Y = 0.000114 \; X^{1.47}$ $(r^2 = 0.994)$;
      7. High fertility (silty soil)
        $Y = 0.000171 \; X^{1.39}$ $(r^2 = 1.00)$;
      8. Low fertility (silty soil)
        $Y = 0.000367 \; X^{1.25}$ $(r^2 = 0.980)$

Table 3 Balance sheets of ^{15}N-labelled
urea (at maturity)

Soil No.	Plant recovery (including tops and roots) (%)	Remaining in soil (%)	Deficit (%)
1	47.4	33.1	19.5
2	51.9	34.0	14.1
3	41.9	27.7	30.4
4	46.2	27.5	26.3
5	43.3	27.3	29.4
6	37.3	27.5	35.2
7	39.6	29.8	30.6
L.S.D. 5%	5.0	4.2	7.6

Table 4 The priming effect of ^{15}N-labelled urea (at maturity)

Soil No.	Increment* of N mineralized from soil via priming effect (mg N/100g)	^{15}N-labelled urea remained in soil (mg N/100g)	Net result of priming effect (mg N/100g)
1	2.96	2.83	0.13
2	2.97	2.90	0.07
3	2.89	2.37	0.52
4	2.77	2.35	0.42
5	2.13	2.34	−0.21
6	2.40	2.36	0.04
7	3.03	2.54	0.49

* Soil-N mineralized in N-pot-Soil-N mineralized in the control.
In this case the N accumulated in the tops and roots both were
taken into account.

Table 5 Correlations between soil N supply and
grain production in the control

Item	Soil N supply (at maturity)	n value in the mineralization equation
Grain wt	0.963**	−0.190
Grain wt / Soil N supply	−0.796*	0.490

The efficiency of the urea N in grain production, was expressed in the increment of grain weight per unit weight of urea-N applied (Δ Grain wt/N applied) which varied from 24.2-34.2 g/g (Table 6) and was positively correlated with the apparent recovery of urea-N by rice plant (Table 7). However, the efficiency of the assimilated urea-N in grain production (Δ Grain wt/ ΔN uptake in the top) was positively correlated to the n value in the mineralization equation (r = 0.76[*]), that is to say, under the same apparent recovery of fertilizer-N, the less the amount of N mineralized from soil during the early stage of growth, the higher the efficiency of the fertilizer-N applied as basal dressing. Therefore, increasing the recovery of fertilizer-N by rice plant is generally of primary importance. In the meantime consistent coordination between the nitrogen mineralization pattern of the paddy soil and the N demand of the rice plant has also to be taken into consideration in the nitrogen fertilizer management practices, especially in the case of high-yielding cultivation, as given in the previous report (6). In this respect, the n value in the mineralization equation might be taken as a criterion.

Table 6 Efficiency of urea applied as basal dressing

Soil No.	Grain wt in N-treatment (g/pot)	$\dfrac{\Delta \text{Grain wt}}{\text{N applied}}$ (g/g)	Apparent recovery (%)[*]	$\dfrac{\Delta \text{Grain wt}}{\Delta \text{N uptake}}$ (g/g)[**]
1	15.41	29.3	58.2	50.3
2	14.02	34.2	65.9	51.9
3	14.73	24.2	53.5	45.2
4	13.84	28.5	59.9	47.6
5	15.41	28.0	50.8	55.1
6	13.23	24.2	43.4	55.8
7	12.55	27.0	50.3	53.7

[*] $\dfrac{\text{Increment of N accumulation in the tops after N application}}{\text{N applied}} \times 100$

[**] $\dfrac{\text{Increment of grain wt. after N application}}{\text{Increment of N accumulation in the tops after N application}}$

Table 7 Correlations between the efficiency of urea-N and the factors concerned

Item	$\dfrac{\Delta \text{Grain wt}}{\text{N applied}}$	n value in the mineralization equation
Apparent recovery of urea-N by rice plant	0.846*	0.080
$\dfrac{\Delta \text{Grain wt}}{\Delta \text{N uptake}}$	0.0532	0.760*

CONCLUSIONS

1) The N mineralization pattern of the paddy soils is readily in consonance with the effective temperature equation, $y=kx^n$.

2) Most of the increment of N mineralized from soil organic nitrogen
through priming effect was compensated for by the urea-^{15}N remaining in
the soil, mainly as a result of biological interchange. Consequently, the
nitrogen nutrition level for the growing rice would not be affected considerably by the priming effect.

3) At medium level of productivity, the efficiency of applied fertilizer-
N in grain production is determined mainly by the apparent recovery of
fertilizer-N. However, there is a significant positive correlation between
the n value in the mineralization equation and the efficiency of the
assimilated fertilizer-N in grain production, thus the n value might be taken
as a criterion in timing the nitrogen fertilizer dressing.

4) Among the three types of paddy soils investigated, bleached paddy
soils were characterized by the highest mineralization coefficient, highest
percentages of fertilizer-N recovered by rice plant and remained in soil,
and the least in N loss. The reverse was true for the alkaline silty
waterlogged paddy soils, and the permeable paddy soils were in the intermediate.

REFERENCES

(1) Hseung Yi, Xu Qi, Yao Xian-liang, Zhu Zhao-liang, Effect of cropping
    system on the fertility of paddy soils. Acta Pedologica Sinica, 17,
    101-119 (1980). (in Chinese with English summary)
(2) Koyama, T., Practice of determining potential nitrogen supplying
    capacities of paddy soils and rice yield. J. Sci. Soil Manure
    Japan, 46, 260-269 (1975). (in Japanese)
(3) Onikura, Y., Yoshino, T., Maeda, K., Mineralization patterns of soil
    nitrogen during the growth period of rice plant. J. Sci. Soil Manure
    Japan, 46, 255-259 (1975). (in Japanese)
(4) Zhu Zhao-liang, Chen Rong-ye, Xu Yong-fu, Xu Yin-hua, Zhang Shao-lin,
    The effect of forms and methods of placement of nitrogen fertilizer on
    the characteristics of the nitrogen supply in paddy soils. Acta
    Pedologica Sinica, 16, 218-233 (1979). (in Chinese with English
    summary)
(5) Yoshino, T. Dei, Y., Prediction of nitrogen release in paddy soils by
    means of the concept of effective temperature. J. Cent. Agric. Exp. Sta.
    Konosu Japan, 25, 1-62 (1977). (in Japanese with English summary)
(6) Dongting Experiment Unit, Institute of Soil Science, Academia Sinica,
    Nanjing, Problems on the promotion of the nitrogen fertilizer effect
    on the yield of rice in the triple-crop system with double cropping
    of rice in Suzhou district. Turang, 3, 127-135 (1977). (in Chinese)

UPTAKE OF NITROGEN BY RICE PLANT FROM STRAW MANURE, UREA AND SOIL

Mo Shu-xun, Qian Ju-fang
(Institute of Soil Science, Academia Sinica, Nanjing)

Nitrogen uptake by rice plant in its growing stage mainly comes from manure, inorganic fertilizer and native soil nitrogen. Most investigations on the availability of nitrogen from these sources have been made separately. Yoneyama[1] studied nitrogen uptake by rice plant in pot experiment dressed by the mixtures of rice straw incorporated with $(NH_4)_2SO_4$. The fertilizer mixtures were prepared by incorporating labelled rice straw with $(NH_4)_2SO_4$, and vice versa by rice straw and $(^{15}NH_4)_2SO_4$. Present investigation on the uptake of nitrogen by rice plant from different sources was made on paddy soils derived from red earth, using microplot dressed with labelled and unlabelled nitrogen of both straw and urea as nitrogen source. Experiment was carried out in the red earth region of Jinhua County, southwestern Zhejiang Province.

MATERIALS AND METHODS

Plastic cylinders (26 cm in diameter, 40 cm in height and bottomless) were inserted into the paddy field to a depth of 28 cm leaving the upper 12 cm above the ground. In order to reduce the heterogeneity of the paddy field, the surface soil (10 cm) in all cylindric plots was taken out, air-dried, put in one pile and mixed thoroughly and then returned to each plot in an equal quantity. After the soil in each plot had been flooded for one day, fertilizers were dressed into the surface soil 6 cm in depth according to the following treatments with each treatment in triplicate.

1) Twenty-five grams of powdered rice straw, labelled with ^{15}N containing 235 mg N, were thoroughly mixed with 316 mg of fertilizer nitrogen in the form of urea.

2) A mixture of the same amount of rice straw and ^{15}N-labelled urea.

3) 316 mg ^{15}N in the form of labelled urea.

Rice straws, both ^{15}N labelled and unlabelled, were prepared as follows: Cut off the ear at the full earing stage of rice plant, then harvested when the leaves began to turn yellow. To the $(^{15}NH_4)_2SO_4$ dressed rice field, the straw contained 0.94 per cent of nitrogen with an abundance of 27.1 ^{15}N, The labelled urea was 10.3 abundance of ^{15}N. All plots received 2 grams of alkaline fused rock phosphate and 1 gram of K_2SO_4. The experiment was carried out on two paddy soils derived from red earth in Jinhua County, Zhejiang Province. The characteristics of these two paddy soils are:

Soil type	Location	pH	O.M.%	T.N.%	C:N	<0.001 mm clay %
Yellowish argillic paddy soil	upland	5.02	1.55	0.093	9.7	29.7
Submergic paddy soil	lowland	5.22	3.15	0.175	10.4	20.8

Two holes about 1 cm in diameter were drilled on each side of the microplot just above the soil surface and were stopped with rubber taps. The taps were removed 15 days after fertilization to provide a free circulation of soil water. On the eighth day after fertilization, twelve rice seedlings (20 days old) were transplanted in each microplot. The whole plant was harvested 90 days after transplanting, dried at 85°C, weighed and analyzed for nitrogen, the content of ^{15}N in the plant and remained in 10 cm surface soil was analyzed by mass-spectrometer.

RESULTS AND DISCUSSION

1. Fate of Urea-N and Straw Manure-N

Results on the fate of fertilizer nitrogen from straw manure incorporated with urea on a yellowish argillic paddy soil and submergic paddy soil are shown in Fig. 1.

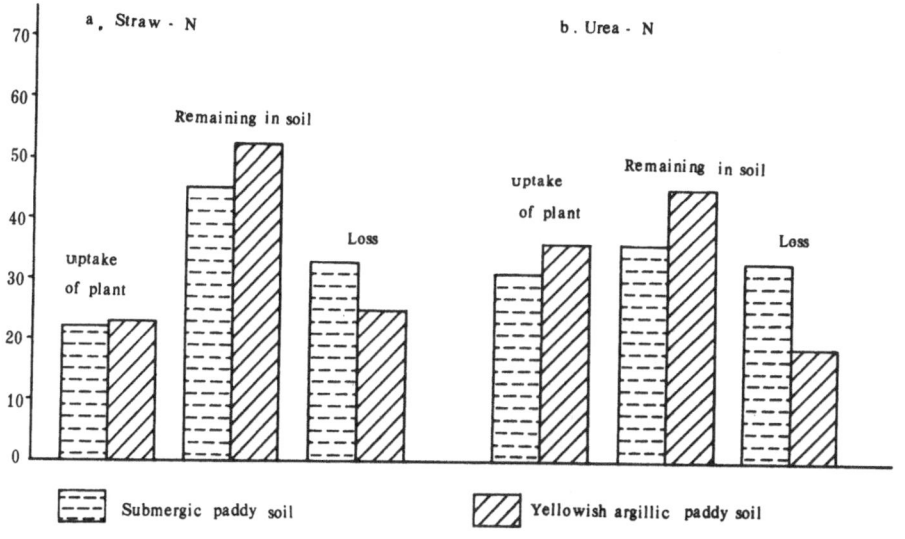

Fig. 1 Fate of fertilizer nitrogen from urea and straw in two paddy soils

Under the acid condition of these two paddy soils, the rice plant uptook more N from urea than from straw, i.e. 31% and 36% from urea and 22% and 23% from straw respectively. As shown in Fig. 1, more urea nitrogen was remained in the yellowish argillic paddy soil (45%) than in the submergic paddy soil (36%). It is believed that the heavier texture of the former soil might hold more ammonical-N than that of the latter.

The loss of nitrogen from straw manure had up to 33% in submergic paddy soil and 25% in yellowish argillic paddy soil. Thus, the rate of decomposition of rice straw, as expressed by the sum of the plant uptake and loss, had up to around 50% in these two paddy soils within a short period of rice growing season. Although the submergic soil condition favoured decomposition of raw rice straw and which surely promoted the loss of nitrogen (2). However, judging by the rather low-N content in straw manure, 0.94% in dry matter, present results for such a high rate of the liberation of the nitrogen from rice straw remain to further investigation.

A series of separate trial was laid for studying the fate of urea nitrogen in the submergic paddy soil, where 316 mg ^{15}N from urea alone was dressed to each microplot. The uptake rate of urea nitrogen by rice plant was 35%, but the urea nitrogen remained in the soil dropped to 19%. This could be due to the absence of straw manure, which would surely hold a part of ammonical nitrogen.

2. UPTAKE OF NITROGEN BY RICE PLANT

The uptake of nitrogen by rice plant from straw manure, urea and native soil nitrogen is given in Table 1. In these two paddy soils, when

Table 1 Uptake of nitrogen by rice plant

Soil type	Total uptake (mg)	Fertilizer-N applied (mg)			N uptake (mg)		
		Urea-N	Straw-N	Total-N	Native[*] soil	Urea	Straw
Yellowish argillic paddy soil	770±17	316	235	551	601	115±7.1	54±1.0
Submergic paddy soil	923±25	316	235	551	771	99±3.8	53±2.3
Submergic paddy soil	950±39	316	—	316	839	111±5.0	—

[*] Calculated from total uptake substrated by uptake from fertilizers. It included nitrogen brought in by the rice seedlings

they were dressed by straw manure incorporated with urea, around 6 or 7 percent of plant nitrogen was uptaken from straw, while the uptake from urea-N in yellowish argillic paddy soil was 15% and that in submergic paddy soil was 11%. In both soils around 80% of the plant nitrogen was originated from the native soil. A comparatively low dress of urea in such microplots , equalivalent to anout 55kg nitrogen per hectare might result a higher contribution of native soil nitrogen to the total N uptake of rice plant. This figure was found up to 88% in the case of submergic soil when urea was dressed alone. However, these figures included nitrogen brought in by the rice seedlings.

Further examination on the relationship between the content of nitrogen in soil organic matter and its amount taken up by rice plant is shown in Table 2. On the assumption that the supply of soil nitrogen was originated from the 15 cm surface soil.

Table 2 Uptake of nitrogen by rice plant from native soil nitrogen

Soil type	Treat-ment	Grain whole plant		native soil N in surface soil (15cm)			Uptake of native N (mg)	Rate of utiliza-tion (%)
		g/plot*		N(%)	mg	C:N		
Yellowish argilic paddy soil	Straw and urea	40	76	0.093-0.074	7400	9.7	600	8.1
Submergic paddy soil	Straw and urea	52	107	0.175-0.183	11800	10.4	770	6.5
	Urea	49	103	0.175-0.183	11800	10.4	839	7.1

*Roots plus tops

Table 2. reveals that under the local climatic conditions, the two paddy soils had a fairly uniform C:N ratio(about 10) in their organic matter, and the rice plant had a similar rate of utilization of native soil nitrogen (about 7-8). Although the content of native nitrogen as shown in Table 2 had a wide difference in the two paddy soils, the rate of utilization of native soil nitrogen in submergic paddy soil with straw manure added to it was 6.5%. This figure increased to 7.1% in the microplot without additional organic material.

CONCLUSION

Experiment in micro-plot was carried out on two acid paddy soils in the subtropic region of Zhejiang Province, with straw manure incorporatied with urea as a mixed fertilizer. The fate of nitrogen from the urea, and straw manure was studied with labelled nitrogen technique. The recovery of urea-N in yellowish argillic paddy soil was 36% and 31% in submergic paddy soil, with 23% and 22% for straw manure respectively.

The rate of nitrogen liberation from straw manure under present condi- tion, as expressed by the sum of plant uptake and loss, has been found around 50% under experimental condition. Judging by the rather low nitrogen content (0.94%) in straw manure and the short period of rice growing season, the present results should be suggested for further investigation.

The uptake of native soil nitrogen by rice plant in these two paddy soils amounted to 80% in the presence of urea and organic manure, and 88% with urea alone. The rice plant shows a similar rate of utilization of native soil nitrogen, i.e. 8.1% in yellowish argillic paddy soil and

6.5% in submergic paddy soil, in spite of the fact that the content of organic matter in the former soil is much less (1.55%) than the latter (3.15%).

REFERENCES

(1) Yoneyama, T., Yoshida, T., Decomposition of rice residue in tropical soil. I. Nitrogen uptake by rice plants from straw incorporated, fertilizer $(NH_4)_2SO_4$ and soil. Soil Sci. Plant Nutr., 23 (1), 33-40 (1977).
(2) Broadbent, F. E., Tusneen, M. E., Loss of nitrogen from some flooded soils in tracer experiment. Soil Sci. Soc. Amer. Proc., 35 (6), 922-926 (1971).

PHOSPHORUS BALANCE ON PADDY SOILS RECLAIMED FROM THE SALINIZED WASTE LAND, NORTHERN JIANGSU

Li Qing-kui, Qin Sheng-wu
(Institute of Soil Science, Academia Sinica, Nanjing)

RECLAMATION OF SALINE SOILS

The present experiments on the reclamation of saline soil by rice plantation were carried out in 1969-1977, in the coastal areas of northern Jiangsu Province. The soils were derived from the loessial deposit of the old Huanghe River, and underlaid by marine mud. Solonchak soils of the experimental areas may be divided into two groups according to their salt content:

(1) Heavily salinized soils containing soluble salts 2.5-6 %. Large areas of such soils remain waste, while the depressions of which are covered by salt tolerant plants such as reed, cogongrass, etc. (2) Cultivated saline soils. They are provided with drainage ditches for lowering the water table, and used for barley, wheat or cotton. However, during the drought spring season, plant seedlings usually appear burned by high concentration of soluble salts in surface soil. Cultivated saline soils contain about 0.1-0.2 % of soluble salts. The present experiments were laid on the heavily salinized soils, covering an acreage about 25 hectares.

After the establishment of the irrigation and drainage system of the coastal areas, northern Jiangsu, at the fifth decade of this century, the reclamation of saline soils through rice plantation became feasible. Strips of of paddy fields, provided by appropriate system of irrigation and drainage ditches, have been laid in sizes of 50-100 m wide and 300-500 m long. Rice seedlings stand healthy under water flooded condition.

PROCESS OF SOIL RECLAMATION

The heavily salinized soils are strongly calcareous loam, containing organic matter around 1 %. They are very low in available phosphorus (Table 1).

Table 1 Soil properties of the experimental field

Texture	Loam*
$CaCO_3$	13.0 ± 1.9 %
pH	8.9 ± 0.2
Organic matter	1.06 ± 0.2 %
Total P	0.07 ± 0.004 %
Available P	1.8 - 2.2 ppm (very low)**

 * With small areas of light loam
 ** By 0.5M $NaHCO_3$ extractant

After the establishment of irrigation and drainage system, amelioration of saline soils can be accomplished by proper soil management. During 1969-77, the land was used for rice - milk vetch rotation annually.
The top portions of the rice plant were removed at harvest and the milk
vetch was turned over as green manure about two weeks before the transplantation of succeeding rice seedlings.

Adequate amount of nitrogen fertilizer, in the form of NH_4HCO_3, were
given in split applications for rice plant. Superphosphate at a rate of
33 kg/ha(P) was dressed annually as basal fertilizer, about two third of
which was given on milk vetch and one third on rice seedlings. Changes
on the status of soil phosphorus, and the response of crops to phosphatic
fertilizer are given in Table 2.

Table 2 shows that after about 3 years of the application for phosphatic
fertilizer, the rate of response of rice to the fertilizer appears declined
and becomes insignificant to soils having received 5-6 years of phosphatic
fertilizer. However, as for the leguminous crops, beneficial effect seems
still significant even after 6 years of phosphatic dressing. Changes of
available phosphorus in soils give good positive correlation with the plant
response.

Table 2 Variations of the crop response to phosphatic
fertilizer in years 1970-1977

History of fertilization	Available soil P	Fertilizer P added	Yield (grain)	Crop response to P fertilizer
Rice	ppm	t/ha	t/ha	%
1970, first year rice, field cultivated in 1969	1.7-2.2	0	3.72	75 **
		21	6.52	
1973, having received P-fertilizer for 5-7 crops	3.1-3.9	0	6.52	15 **
		21	7.53	
1977, having received P-fertilizer for 11-13 crops	4.4-5.2	0	6.68	
		7.9	6.97	N.S.
		24	7.16	N.S.
Vetch				
P-fertilizer for 3-5 crops	3.1-3.9	0	11.8 *	208 **
		21	36.5 *	
Milk vetch				
P-fertilizer for 11-13 crops	4.4-5.2	0	32.6 *	20 **
		21	39.1 *	

  * Green wt
 ** Significant at 0.01, N.S. = No significant effect

PHOSPHORUS BALANCE UNDER PRESENT SYSTEM OF SOIL MANAGEMENT

Generally speaking, the balance of phosphorus is rather simple in comparison with that of nitrogen. It can be estimated, as shown in Table 4, by the added phosphatic fertilizer, the phosphorus removed by rice straw and grain, and the change of phosphorus in surface soil. The balance of soil phosphorus in the years 1969–1977 are given in Table 3.

Table 3 Change of the status of soil phosphorus
during 1969–1977

History of fertilization	Total P	Available P	Organic P	Organic matter
	%	ppm	%	%
1969, waste land, no fertilizer	0.060	1.7	0.010	1.05
1970, after dressing P-fertilizer for 2 crops	0.060	2.3	0.012	1.01
1973, after dressing P-fertilizer for 7 crops, green manure turned over	0.065	4.0	0.010	1.22
1977, after dressing P-fertilizer for 11 crops, green manure turned over	0.066	5.4	0.014	1.31

Both total phosphorus and available phosphorus in soils were increasing as the cultivation went on. Slight accumulation of soil organic matter appeared under present soil management. However, only 1/5 of total soil phosphorus is in the   form of organic phosphorus.

Table 4 Balance sheet of phosphorus under present method
of soil management (1969–1977)

(A) P added in years 1970–1977 (in form of superphosphate)	229 (kg/ha)
(B) P removed by the tops of rice plant in years 1970–1977	164 (kg/ha)
(C) P remained in soil, calculated as (A)-(B)	65 (kg/ha)
P remained in soil, calculated from the increase of total P in 10 cm surface soil	69 (kg/ha) *

* $(0.066 - 0.060)\% \times (113 \times 10^4 \text{kg}) = 69$ (kg/ha)

807

Table 4 gives a balance sheet of phosphorus. Seven harvests of rice plant have removed 164 kg/ha of P, and the increment of P in 10 cm surface soil from 1969 to 1977 is 69 kg/ha. These two figures add up to 233 kg/ha while the P added in the form of superphosphate in the 7 years (13 dressings) is 229 kg/ha.

Our investigation comes to the conclusion that accumulation of phosphorus nutrition in the soil occurs under present method of fertilization. Rice plant needs no more phosphatic fertilizer after about 5 years of fertilization. Legume crops, as milk vetch and vetch, still give 20 % increase in yield after 6 years of phosphorus fertilization.

Present experiments alos show the good correlation between the level of soil available phosphorus and the response to crop yield. Rice gives no significant increase in yield at available P above 5.2 ppm (P), while significant response of milk vetch and vetch still can be found in soils with available P up to 8.7 ppm.

A STUDY ON THE RESPONSES TO FERTILIZER POTASSIUM ON PADDY SOILS IN GUANGDONG PROVINCE

Zhu Wei-he, Wen Ying-chang
(Soil and Fertilizer Institute, Guangdong Academy of
Agricultural Sciences, Guangzhou)
Shen Dao-ying
(Guangdong Institute of Soil Science, Guangzhou)

Rice is cultivated widely all over Guangdong Province. As a general practice, the farmers used to apply farm manure, nitrogen and phosphorous fertilizer to increase the yields of rice, with little attention to potassium fertilizer. In order to have a better understanding of the responses to K fertilizer of various paddy soils and the efficient use of potash, a series of works have been done by the authors of the present paper on this problem since 1958-1979 (1-4). The results obtained of the responses of main crops to K fertilizer, K status of various paddy soils and practices for efficient use of K fertilizer are summarized in this paper.

RESPONSES OF MAIN CROPS TO K FERTILIZER

Prior to 1976, 830 field experiments were carried out throughout the province on the responses of various crops and soils to potassium fertilizer, the main results being given in Table 1. On the average, the yields are increased by 15.7%, 27.8%, 20%, 13.2%, 10.9%, 13.5%, 31.7% and 32.5-47.9% for the rice, wheat, peanut, sugar cane, sweet potato, jute, soybean and green manure crops respectively. The average increment per 1 kg K applied is 8.5 and 13.5 kg grains of rice and wheat respectively, 10.6 kg of peanut, 141.1 kg of sugar cane, 52.9 kg of sweet potato, 4.9 kg of jute, 47.3 kg of soybean and 104.6-167.5 kg of green manure.

Table 1 Responses of main crops to K ferilizer on paddy soils

Crop	No.of trials	Aver.rate of K kg/ha	Aver. yield(kg/ha) Crotrol	K	Yield increase %	kg/kg K
Rice	328	43.0	3.56	4.12	15.7	13.0
Wheat	41	51.7	1.61	2.05	27.8	8.5
Peanut	164	38.6	2.04	2.45	20.0	10.6
Sugar cane	144	77.2	66.72	75.53	13.2	114.1
Sweet potato	12	52.9	16.58	19.88	19.9	62.2
Jute	4	107.7	3.76	4.27	13.6	4.9
Soybean	3	47.3	0.62	0.81	31.7	4.1
Astragalus	18	34.9	12.24	18.10	47.9	167.5
Sesbania	1	31.1	12.00	16.50	37.4	144.0
Vetch	4	52.9	17.07	22.63	32.5	104.6

Table 2 The status of K of paddy soil different
parent material in Guangdong Province

Soils (Parent material)	Available K ppm		1 N HNO_3 soluble K ppm		Total K %	
	Range	Ave.	Range	Ave.	Range	Ave.
Sandy soil (Residues of sandstone, shale and granite)	8.3–54 (88)*	50	23–158 (33)	71	0.25–2.46 (75)	1.11
Silt soil (Residues of sandstone and shale)	21–50 (11)	35	31–66 (6)	41	0.38–0.90 (6)	0.69
Black loam (Shallow-sea deposit)	19–108 (22)	50	42–120 (8)	167	0.23–0.86 (13)	0.37
Lateritic clay (Residues of basalt)	17–120 (21)	68	37–133 (6)	97	0.30–0.51 (14)	0.35
Sandy loam (Residues of sand-stone, shale and granite)	19–199 (169)	79	15–521 (46)	159	0.37–3.15 (101)	1.41
Yellow loam (Same as above)	31–149 (48)	79	42–216 (17)	76	0.30–2.62 (31)	1.31
Clay soil (Residues of sandstone, shale, granite and delta deposit)	61–87 (113)	75	21–415 (58)	239	0.50–2.81 (74)	2.04
Loam (Residues of granite, sand-stone, shale, lime-stone, delta and river deposit)	56–115 (86)*	84	29–797 (36)	198	1.55–2.64 (47)	1.91
Highly gleyed clay loam (Delta deposit)	42–116 (14)	77	118–420 (14)	297	1.88–2.08 (14)	1.99
Gleyed clay loam (Delta deposit)	45–303 (50)	131	123–1010 (47)	330	1.17–2.76 (51)	1.93
Saline soil (Coastal deposit)	62–415 (19)	133	148–1010 (15)	398	1.19–2.31 (18)	1.96
Acid saline soil (Coastal deposit)	54–299 (9)	168	196–430 (9)	334	1.52–2.18 (9)	1.91
Purple soil (Residues of purple shale)	83–498 (7)	212	174–506 (4)	360	2.02–2.87 (7)	2.51
Beach soil (Coastal deposit)	166–374 (8)	251	255–396 (8)	246	1.50–1.93 (8)	1.68
Mean	8.3–498 (665)	83	15–1010 (307)	213	0.25–3.15 (468)	1.56

* Figures in parentheses indicate the number of samples under investigation

THE STATUS OF POTASSIUM IN SOILS OF DIFFERENT PARENT MATERIALS

The status of potassium in over 900 soil samples of different parent materials selected from different parts of this province has been investigated (Table 2). Table 2 shows the wide variation in K content of various soils. The ranges of the contents of the available K (extracted with 1 N neutral NH_4OAc) are 8.3-498 ppm, slowly available K (extracted by 10-minutes' boiling with 1 N HNO_3) 15-1010 ppm and the total K 0.25-3.15%.

Form Table 2 we can see that the sandy soil, silt soil, beack loam, lateritic clay, sandy loam and yellow loam are relatively low in K content, and may be classified as the K-deficient soils or seriously K-deficient soils; clay soil, loam soil and highly gleyed clay loam are the moderate K level soils; the gleyed clay loam, saline soil, acid saline soil, purple soil and beach soil are the K-sufficient soils. If the above-mentioned soil potassium status is compared with that of the other countries in Asia, it may be found that the available K is rather lower in Guangdong (5,6) and the increasing need of K fertilizer in this province would be expected.

RESPONSES OF RICE TO POTASSIUM IN VARIOUS AGRICULTURAL SOILS

The responses of rice to K in various agricultural soils are shown in Table 3.

Table 3 Responses of rice to K on main agricultural

soils in Guangdong Province

Soil class	Average grain yields (t/ha)		Yield increased by applied k(%)
	Control	K*	
Sandy soil	3.51(11)**	4.30	22.5
Silt soil	3.44(5)	4.28	24.4
Sandy loam	4.30(10)	4.85	12.8
Black loam	3.68(11)	4.37	18.8
Lateritic clay	2.96(4)	3.32	12.2
High gleyed clay loam	5.11(10)	5.48	7.2
Clay soil	5.31(14)	5.63	6.0
Loam	5.32(10)	5.73	7.7
Purple soil	5.24(4)	5.31	1.3
Saline soil	3.19(6)	3.20	0.3

 * Rate of application: 46.7 K kg/ha.
** Figures in parentheses indicate the number of trials.

According to the status of K in different paddy soils (Table 2) and the responses of rice (Table 3), the paddy soils in Guangdong Province may be classified into the following three groups:

1) Soils with good responses to K: The contents of available, slowly available and total K of such soils are usually below 66 ppm, 166 ppm and 1.25% respectively. The yields of rice increased by K dressing are usually 10-30%.

2) Soils with moderate responses to K: The contents of the above-stated three K values are 66-125 ppm, 166-332 ppm, and 1.25-2.08% respectively. The efficiencies of K fertilizer vary with the crops and conditions of farming. Generally, there is an increase of 5-10% of rice yieldsby using K fertilizer.

3) Soils with no gignificant responses to K: The contents of available, slowly available and total K are more than 125 ppm, 332 ppm and 1.66-2.08% respectively.

RATES OF POTASSIUM APPLICATION

A number of experiments on the rates of potash application were carried out on potassium-deficient paddy soils. Results of three trials on the sandy soil and sandy loams are given in Table 4. From the results it seems that the rates of K ranging from 30-90 kg/ha appear adequate to the yield level of 3-6 t paddy/ha in k-deficient soils. These rates are similar to those used in some countries of Southeast Asia (5,7,8).

Table 4 Effect of rates of K fertilizer on the yields of rice

Location	Soil	K applied (kg/ha)	Grain yield (t/ha)	Yield increased (t/ha)
Chikan, Kaiping County	Sandy soil	0	2.933	
		31	3.638	0.705
		62	3.383	0.450
		93	3.128	0.195
Changsha, Kaiping County	Sandy loam	0	5.138	
		49	5.678	0.540
		90	5.805	0.667
		180	5.805	0.667
		215	5.565	0.427
Taishan County	Sandy loam	0	3.765	
		22.5	4.275	0.510
		45	4.320	0.555
		90	4.590	0.825
		135	4.500	0.735

EFFECT OF TIME OF APPLICATION ON POTASH RESPONSES

A series of field experiments were conducted on the optimum timing of K application. The results of 4 trials shown in Table 5 indicated that the efficient time of K application is affected by the soil K status. On the seriously K- deficient paddy soils (such as black soil and sandy soil), 1/2 potassium applying at transplanting and 1/2 at tillering stage is more effective. On the moderate K content soils (such as clay soil and sandy loam), the application of K before (or after) panicle differentiation stage or split application (1/3 basal, 1/3 at panicle differentiation and 1/3 at booting stage) is usually favorable.

Table 5 The effects of different timing of K application on some paddy soils

Treatment 46.7 kg/ha(K)	Black soil Rice yield (t/ha)	%	Sandy soil Rice yield (t/ha)	%	Clay soil Rice yield (t/ha)	%	Sandy loam Rice yield (t/ha)	%
Control	3.97	100	4.88	100	4.21	100	5.70	100
Basal	4.61	116	5.18	106	4.64	110	6.29	110**
1/2 Basal, 1/2 Tillering stage	4.94	124	5.40	111	4.79	114	6.29	110**
Tillering stage	4.95	125	5.03	103	4.81	114	6.12	107*
Panicle differentiation stage	4.51	114	5.03	103	5.03	119	6.39	112**
Booting stage	4.30	108	4.95	101	4.66	111	6.00	105
1/3 Basal, 1/3 Panicle differentiation and 1/3 Booting stage	4.73	119	5.18	106	5.03	119	6.36	112**

* Significant at P = 0.05 (0.35 t/ha)

** Significant at P = 0.01 (0.48 t/ha)

POTASSIUM RESPONSES AND CHANGES IN AVAILABLE SOIL
POTASSIUM IN THE ROTATION SYSTEM

Generally, the rotation of wheat-rice-rice three crops a year is commonly practised in Guangdong Province. The responses to K fertilizer of the late rice are usually higher than those of the early rice (Table 6), and that of wheat are higher than the rice.

Table 6 Effect of K on the yields of double-cropping rice

| Location | Soil | Yield increased by K applied | | | |
| | | Early rice | | Late rice | |
		t/ha	%	t/ha	%
Shima, Xingning County	Yellow loam	0.34	5.6	0.74	24.1
Longtian, Xingning County	Clay soil	0.20	3.2	0.65	16.0
Tangjiao, Huiyang County	Clay soil	0.33	6.3	0.58	15.6

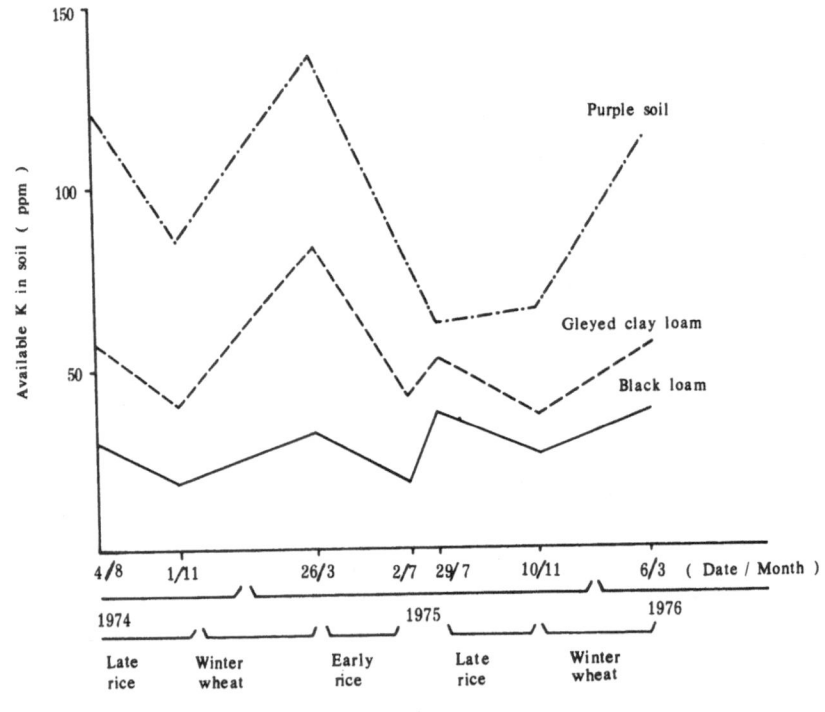

Fig. 1 Variation of available soil potassium in
rice-rice-wheat rotation

As shown on Fig. 1, the content available soil potassium in the transplanting stage of early rice is much higher than that in the same stage of late rice, and that of late rice usually higher than in the seeding time of winter wheat. This phenomenon occurs on the purple soil, high gleyed clay loam, black loam and also on some other soils. From these phenomena it is supposed that the different responses to K of the afore-mentioned crops may be more or less due to the periodical variation of available soil K in the rotation system.

REFERENCES

(1) Institute of Soils & Fertilizers, Guangdong Academy of Agricultural Sciences, A summary report of potassium trials in Guangdong in 1958-1972. Guangdong Agricultural Sciences, 4, 10-19 (1973). (in Chinese)
(2) Institute of Soils & Fertilizers, Chinese Academy of Agricultural Sciences, Effect of potassium fertilizer. Turang Feiliao, 1, 21-32 (1974). (in Chinese)
(3) Agricultural Bureau of Kaiping County, Efficiency and practical use of potassium fertilizer. Guangdong Agricultural Sciences, 2, 30-32 (1980). (in Chinese)
(4) Zhu Wei-he, Wen Ying-chang, Shen Dao-ying, Potassium status of main soils in Guangdong Province and the efficiency of K fertilizer Guangdong Agricultural Sciences, 3, 27-34 (1980). (in Chinese)
(5) Uexkull von H.R., Agronomic and economic evaluation of crops responses to potash fertilizer. Rice, The Far Eastern and SE Asian Experience. In "Potassium in Soils and Crops" (ed. by Sekhon, G.S.) Potash Research Institute of India, 241-259 (1978).

(6) Kemmler G., Potassium deficiency as a yield-limiting factor. Potash Review, 5 (1980).
(7) Uexkull von H. R., Potassium Nutrition of Tropical Crops. In "The Role of Potassium in Agriculture" (ed. by Kilmar, V.J., Young, S.E., Brady, N.C.), Amer. Soc. Agron., etc. 385-398 (1968).
(8) Uexkull von H.R., Potash and rice production in Asia. Potash Review, 8 (1978).

IRON AND CHROMIUM UPTAKE BY CROPS ON WELL DRAINED SOILS
AND ON POORLY DRAINED WETLAND SOILS

Robert Brinkman, R.Y. Reyes; H.W. Scharpenseel, E. Eichwald
(International Rice Research Institute, P.O. Box 933, Manila, Philippines;
Hamburg University, Von-Melle-Park 10, 2000 Hamburg 13, W. Germany)

INTRODUCTION

In two upland crops, corn (maize) and sugarbeets, a drastic increase
in iron content of the above-ground parts was found below a limiting zinc
content[18, 11, 24]. Above that limit, there was no relation between these
two. Only in corn roots, higher iron contents were recorded as zinc contents dropped to a deficiency limit. At still lower Zn contents, iron
contents in the roots dropped again, in the same treatments in which a
drastic Fe increase occurred in the above-ground parts.

High Fe contents and bronzing symptoms attributed to "iron toxicity"
are found in wetland rice grown on certain acid soils with low fertility
levels. These may be caused by very high Fe (II) concentrations in the
soil solution (100-500 mg/1), but are also associated with moderate Fe (II)
concentrations (10-40 mg/1) together with very low levels of K or Ca[4].
The high Fe contents and the toxicity symptoms of the rice plant may thus
be induced by K or Ca deficiencies or caused by high Fe concentrations in
the root environment.

The rice plants on the soils used in the present study had a wide range
of Fe contents, in some cases exceeding 2000 mg/kg dry matter (2×10^{-3} w/w
D.M.) without any toxicity symptoms. These high Fe contents occurred in
plants with low as well as with adequate Zn contents, and on soils with
adequate or high levels of available K and Ca.

Chromium contents in plants generally range from 1 to 5×10^{-5} w/w
(D.M.) in roots, 5×10^{-7} to 10^{-5} in vegetative parts above ground, and
from less than 10^{-8} to 2×10^{-7} in grains and seeds. This variation occurs
almost regardless of the total Cr (III) contents in soils, which range
from less than 10^{-5} to more than 10^{-3} w/w. The present study reports Cr
contents well above these ranges in rice straw and grain on poorly
or very poorly drained soils.

MATERIALS AND METHODS

Wetland rice roots, straw and grain were collected together with a
soil sample of the ploughed layer from a range of poorly and very poorly
drained, noncalcareous soils, some of which have upwelling ground water.
In the wet condition, their pH is near to neutral. The soils have a wide
range of chromium contents. All are moderately to severely zinc-deficient;
samples were taken from plots with and without added zinc. Samples were
ground and analysed for total Zn, Fe and Cr by X-ray fluorescence spectrometry. When Cr values in grain appeared incredibly high, a set of
samples was analysed again, this time by atomic absorption spectrometry,
with similar results. A double check showed Cr (and Ni) contamination of
the grain from a hard metal mill. Only Cr values for the few grain samples
ground in an agate mill are reported here.

Plant samples of different growth stages were collected from a drum experiment on zinc deficiency of wetland rice, as well as samples of rice straw at maturity from a parallel field experiment on the same soil (description and analytical data in 17). These samples were wet ashed and analysed for Zn, Fe and Cr by atomic absorption spectrometry.

IRON CONTENTS OF DRYLAND PLANTS AND WETLAND RICE

The aerial parts of many dryland crop plants have iron contents between about 10 and 100 mg/kg dry matter. Only some iron accumulators such as spinach, turnip or beet normally contain up to several hundred mg/kg [5]. Iron contents up to about 1000 mg/kg have been recorded in zinc-deficient plants containing less than 10-20 mg Zn per kg dry matter. At higher Zn contents there appears to be no relation between Fe and Zn contents within a given experiment [18, 11, 24]. Between experiments on the same crop (corn), however, there may be a factor 2-3 difference in Fe contents.

Iron contents of wetland rice shoots or straw commonly range between about 50 and 500 mg/kg. In rice plants grown on acid wetland soils, iron toxicity symptoms generally occur at Fe contents above 300-1000 mg/kg, depending on variety and nutrient status. In rice grown on neutral, very poorly drained soils, however, iron contents between 500 and 5000 mg/kg were measured, in straw collected from several locations in the Philippines as well as in shoots from a drum experiment on a similar soil 4 and 8 weeks after transplanting. No symptoms of iron toxicity were observed in any of these plants. Only in mature rice straw from the field and drum experiments in the dry season of 1980, iron contents were about 60-100 mg/kg, within the normal range for rice and close to common values for dryland crops. Rice roots were found to have still higher iron contents, about 2000 to 9000 mg/kg, similar to corn roots, with perhaps a slight decrease at high zinc contents. These data on wetland rice are summarized in Fig. 1, together with iron and zinc contents in sugarbeet leaves and in corn shoots and roots [18, 24].

It is clear that at zinc contents higher than about 10-20 mg/kg, the great differences in iron contents of rice plants should be ascribed to another cause than zinc deficiency. The major differences in Fe contents of wetland and dryland crops lead to the known suggestion that the Fe (II) level in the soil solution influences Fe contents in the aerial parts. Since the roots of rice and maize appear to contain similar proportions of Fe, this suggestion does not seem to apply to uptake per se, but rather to the mobility of iron within the plant.

Rice plants grown in conditions of marginal zinc deficiency on drained sites (Fig. 1) were found to have iron contents similar to those in zinc-deficient dryland crops and much lower than rice plants with the same or higher Zn contents but grown in very wet sites. This finding, and the still lower Fe contents of wetland rice commonly reported, indicate that the largest differences in iron transport to the aerial parts are not between dryland and wetland crops but between wetland rice grown on usual sites, with some percolation, and wetland rice grown on extremely wet sites, with or without upwelling water.

Fe content, log w/w (D.M.)

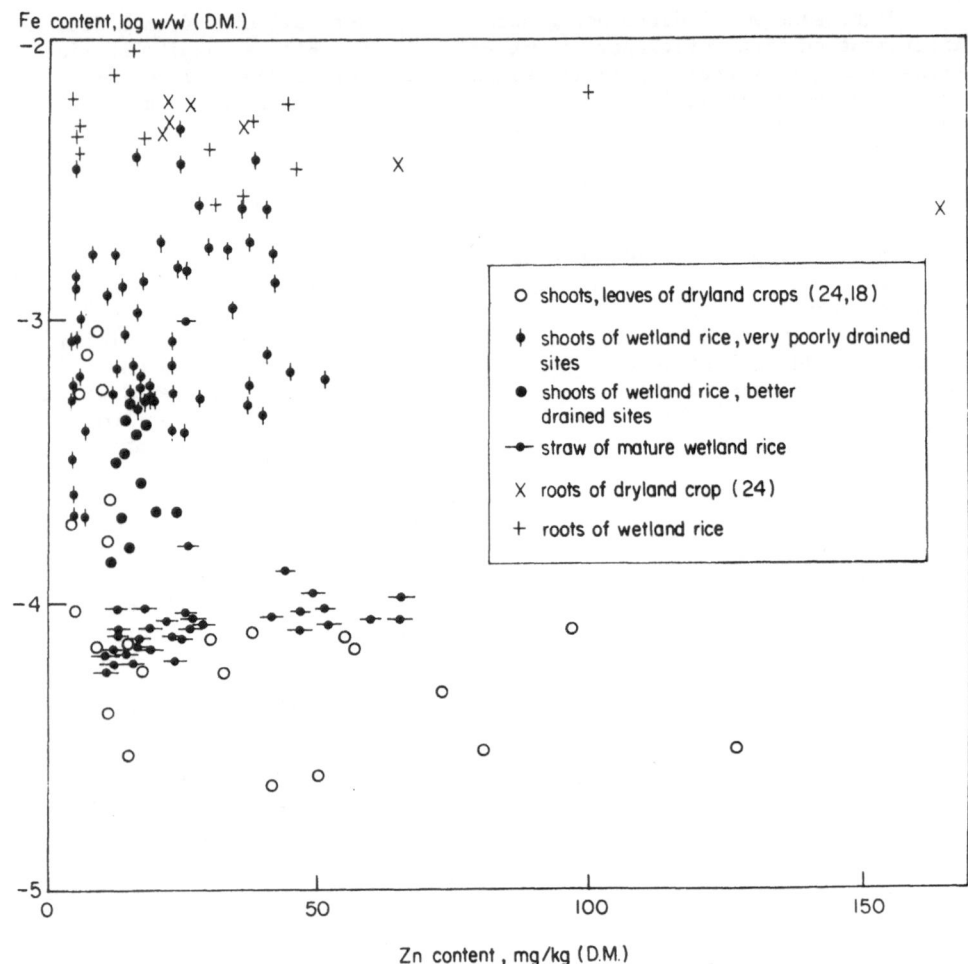

Fig. 1   Relation of Fe and Zn contents in shoots and roots of wetland rice and corn and in sugarbeet leaves

CHROMIUM IN SOILS: SPECIES, MOBILITY, CONCENTRATION IN SOLUTION

In non-polluted soils, chromium appears to exist mainly as hydrated Cr(III) oxides mixed with, or occluded in Fe oxides[5]. Cr (III) added to neutral soils also appears to be precipitated as hydroxide[8]. This agrees with the finding[20] that the rate of Cr(III) absorption by rice plants went down with increasing pH to zero at pH 6 to 7.

Although Cr(VI) is mobile in neutral soils[8], this is only found in soils recently polluted by, e.g., smelter or tannery waste. Within weeks to a few months, it is reduced to Cr(III) in a wet soil environment (e.g., 2) and is not reoxidized even after air drying[12].

The Cr concentration in the soil solution of a soil from ultrabasic rock, where plants showed signs of Ni toxicity, was 0.01 mg/l in most of the area and 0.02 mg/l in the most severely affected part[1]. In several very poorly drained soils where rice showed Zn deficiency symptoms, Cr concentrations in solution were less than the 0.05 mg/l routine detection limit; NH_4 acetate-extractable Cr was less than 0.8 mg/kg in the wet, puddled plough layer of the Natividad soil which was studied in some more detail[17]. It seems reasonable to assume that in soils with a wide range of total Cr contents, Cr concentrations in the soil solution range from less than 0.01 to, at the most, 0.1 mg/l.

CHROMIUM CONTENTS OF DRYLAND PLANTS AND WETLAND RICE

Roots and aerial parts of different dryland crops and of rice contain Cr in roughly linear proportion to the Cr(III) concentration in the nutrient solution. The mean Cr content in roots (mg/kg dry matter) is 3 times the Cr concentration in solution (mg/l); the total range of the available data points lies between one tenth and ten times that factor (Fig. 2). Cr contents in shoots, leaves and stems cover the same range but at levels a hundred times lower. Contents in grain (2 data points only) are a hundred times lower again (Fig. 2). All nutrient solutions were presumably aerated or at least not strongly reducing. At Cr concentrations in solution of about 0.05 mg/l, Cr contents in roots, shoots and grain are 200-2, 2-0.02 and about 0.002 mg/kg D.M. respectively.

In contrast to the near-linear relationship between Cr concentrations in plant parts and solution, there is a small influence of total Cr contents in soil on Cr contents in parts of crop (Fig.2). These increase about 4 times (in roots and grain) to 30 times (in shoots) with a 1000-fold increase in total soil Cr content. The ratios of Cr contents in roots, shoots and grain are narrower than in the solution experiments. At total soil Cr contents of about 100 mg/kg, mean Cr contents of roots, shoots and grain are about 20, 2 and 0.04 mg/kg D.M., respectively. On low-Cr soils, mean values are 8, 0.2 and 0.01 mg/kg D.M. The data for the dryland crops summarized in Fig. 2 are most likely from aerobic soils. For the wetland rice, presumably most data refer to "normal" rice soils that were not very poorly drained or perennially wet. Cr contents of rice roots from very poorly drained sites estimated by X-ray fluorescence were within the range for roots grown in nutrient solutions with less than 0.1 mg Cr per liter even though an error in pretreatment may have caused an overestimate.

A few exceptions have been reported to the relationships in Cr contents between plants and soils summarized above. Two rice grain samples out of

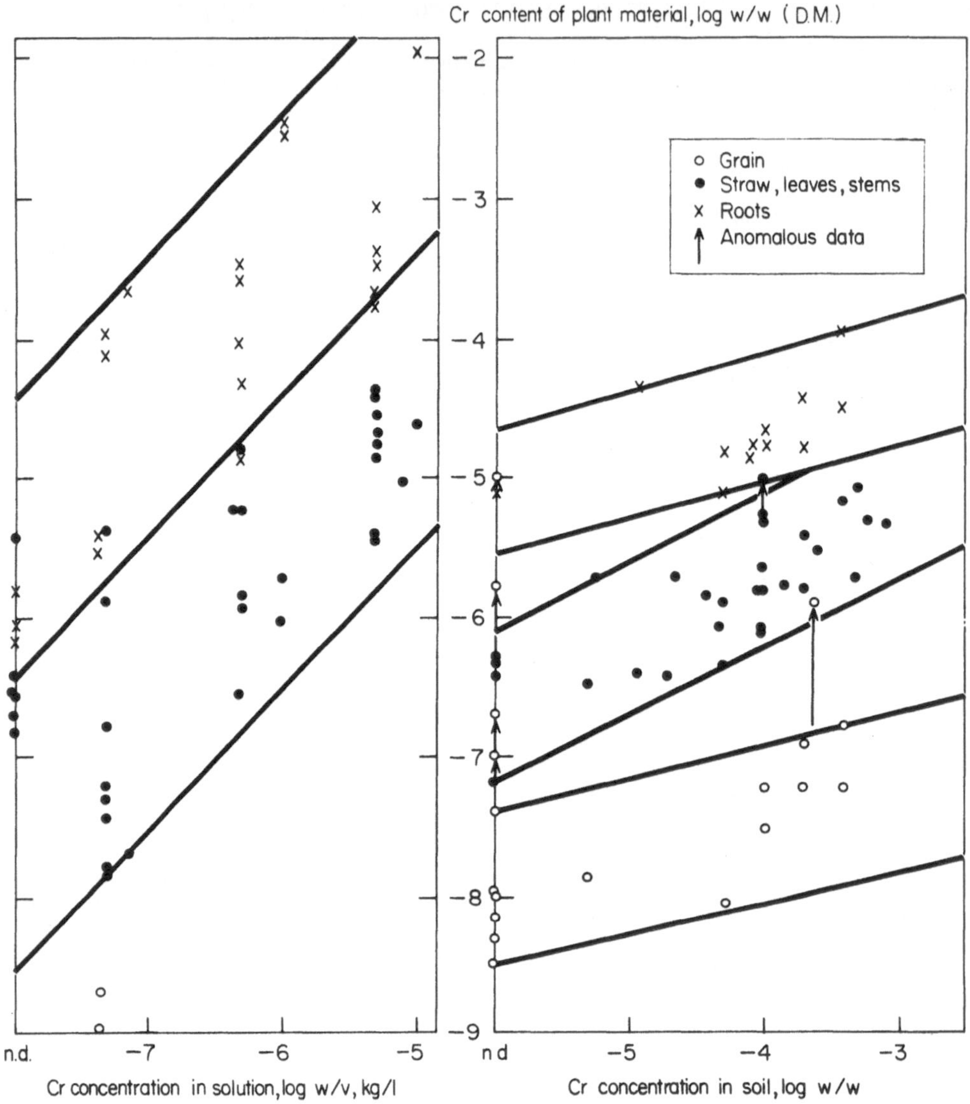

Fig. 2   Relations of Cr contents in roots, shoots and grain of wetland rice
and other crops to Cr(III) concentrations in nutrient solution and to total
Cr contents in soil

Data sources (nutrient solution). Rice: recalculation from (20); (14, 22);
wheat (10); barley (6); beans (10, 23, 21); cotton (16); cabbage (9)

Data sources (soil). Rice: Depts. Soil Science of Hamburg University and
International Rice Research Institute; (13, 3); wheat (25, 5, 26); barley
(19); herbage plants (7); beans (23)

45 listed in [13] contained 10 and 20 times the normal amount of Cr. They contained less Zn than any other sample. This suggests that they were grown on a zinc-deficient soil that was possibly very poorly drained. As reported in [3] "The values of Cr found in rice were high with respect to the data published in the literature. Therefore, without repeating the experiments we cannot say anything about the accumulation of Cr in the rice-plant and the resulting distribution in its different parts". From Table 2 in [3], a Cr content of 1.2 mg/kg in the grain can be calculated, 100 times the normal amount. Although the infiltration rate was several cm/day, about 5 m excess irrigation water flowed out of the field during the crop season. The field may thus have been completely water-saturated throughout. A rice sample from IRRI and a commerical rice sample with hulls, analyzed by X-ray fluorescence, contained 1.75 and 10.4 mg Cr per kg respectively.

In our drum experiment, Cr contents of the shoots were about 5 mg/kg 4 weeks after transplanting and 6 mg/kg at 8 weeks, in the upper part of the normal range. At maturity, the straw from the drained and ridged high treatments contained about 2 mg/kg, while the straw from the permanently wet treatments contained 11 mg/kg. High Cr contents did not appear to be related with especially low Zn contents. These exceptions to the normal range of data are marked in Fig. 2.

Only several metal accumulators of the many wild plants analyzed from serpentine soils [15] contained high amounts of Cr: for example *Euphorbia nicaeensis, Leptospermum scoparium, Diccoma niccolifera* (= nickel-bearing), *Pearsonia metallifera*.

DISCUSSION

Although the data are still few, especially for grain, there is sufficient reason to consider the parallels that seem to exist between uptake and transport of Fe and Cr in crop plants, on aerobic sites or "normal" wetland sites as well as in very poorly drained conditions.

In brief, both elements occur mainly as M (III) hydrous oxides or oxides in aerobic soils. Both are found in relatively large quantities in roots of dryland plants and "normal" wetland rice: Fe 10^{-2} to 10^{-3} w/w and Cr 10^{-4} to 10^{-5} w/w. In vegetative parts above-ground, Fe contents are about a hundred times lower and Cr contents, about 10 to 100 times lower than in the roots.

In very poorly drained soils, Fe (II) is mobile and, at least at a pH about neutral, Fe (III) is not. Cr(III) is also immobile at a pH near neutral; any Cr(VI) that might be present is quickly reduced in a reduced soil. Still, there is considerable plant uptake. Standard chemical and thermodynamic data indicate that Cr(II) is either unstable with respect to Cr(III) or stable only in a small region about pH 7 and at very low redox potentials.

Rice roots in very poorly drained soils contain about the same amounts of Fe (and, by rough estimate, of Cr) as roots in better drained soils. However, the Fe and Cr concentrations in aerial parts are higher on very poorly drained soils than on better drained soils. On very poorly drained soils both elements thus seem to be transported from the roots to the shoots more efficiently than in better drained conditions. Iron and chromium toxicity symptoms have been observed in plants growing on nutrient solutions (and, for iron, on acid soils). Neither seems to be clearly toxic in rice plants on neutral, very poorly drained soils.

On dryland soils and most wetland rice soils, Fe and Cr may be mainly trapped by the roots in the trivalent form. The much higher mobility of Fe in rice plants on neutral, very poorly drained soils is probably due to the predominance of Fe (II) in the soil and the plant. The increased mobility of Cr in these conditions could be explained in the same way, if the Cr(II) species would be stable at the redox potentials prevailing in the very poorly drained soils. Complexation by an organic anion might increase the range of stability of divalent chromium. Masking by a complexing anion might also explain the apparent lack of toxicity of the high Fe concentrations observed in the rice shoots.

Apparently, the range of conditions under which anomalously high Fe and Cr concentrations occur in rice shoots or straw and seed is very limited - otherwise, they would have been reported in the literature before now. Similar but incidental observations may sometimes have been discarded as discrepancies before publication.

High Cr contents, especially in seeds, may have implications for human nutrition. If the present report is substantiated by further data this aspect would deserve attention as well.

ACKNOWLEDGEMENTS

We thank K.G. Kirschey for the collection of the soil and plant samples analysed in Hamburg, B. van de Lustgraaf for a computer-aided literature search and H.-U. Neue for stimulating discussions.

REFERENCES

(1) Anderson, A.J., D.R. Meyer, F.K. Mayer. Heavy metal toxicities: levels of nickel, cobalt and chromium in the soil and plants associated with visual symptoms and variation in growth of an oat crop. Aust. J. Agric. Res. 24, 557-571 (1973).
(2) Bartlett, R.J., J.M. Kimble. Behavior of chromium in soils: I. Trivalent forms, II. Hexavalent forms. J. Environ. Qual. 5 (4), 379-386 (1976).
(3) Bigliocca, C., F. Girardi, E. Orthmann, P. Reiniger. The balance of Cd, Pb, Cr and Cu in an experimental rice-field for one growing season. Il Riso 28 (1), 23-35 (1979).
(4) Breemen, N. van. Landscape, hydrology and chemical aspects of some problem soils in the Philippines and in Sri Lanka. Terminal Report, Internat. Rice Res. Inst., Los Baños. 295 p. (1978).
(5) Cary, E.E., W.H. Allaway, O.E. Olson. Control of chromium concentrations in food plants. 1. Absorption and translocation of chromium by plants, 2. Chemistry of chromium in soils and its availability to plants. J. Agric. Food Chem. 25(2), 300-309 (1977).
(6) Davis, R.D., P.H.T. Beckett, E. Wollan. Critical levels of twenty potentially toxic elements in young spring barley. Plant Soil 49 (2), 395-408 (1978).
(7) Dijkshoorn, W., L.W. van Broekhoven, J.E.M. Lampe. Phytotoxicity of zinc, nickel, cadmium, lead, copper and chromium in three pasture

plant species supplied with graduated amounts from the soil. Neth.
J. Agric. Sci. 27 (3), 241-253 (1979).

(8) Griffin, R.A., A.K. Au, R.R. Frost. Effect of pH on adsorption of
chromium from landfill-leachate by clay minerals. J. Environ. Sci.
Health, A 12 (8), 431-449 (1977).

(9) Hara, T., Y. Sonoda. Comparison of the toxicity of heavy metals to
cabbage growth. Plant Soil 51 (1), 127-133 (1979).

(10) Huffman, W.D., W.H. Allaway. Chromium in plants: distribution in
tissues, organelles, and extracts and availability of bean leaf
chromium to animals. J. Agric. Food Chem. 21 (6), 982-986 (1973).

(11) Jackson, T.L., J. Hay, D.P. Moore. The effect of Zn on yield and
chemical composition of sweet corn in the Willamette Valley. J.
Amer. Soc. Hort. Sci. 91, 462-471 (1968).

(12) Kamada, K., K. Doki. The differences and factors of chromium reduction in two types of soil. 1. Movement of chromium in submerged
soil and its effects on the growth of rice (Japanese, translation
used). J. Sci. Soil Manure 46 (11), 478-482 (1975).

(13) Masironi, R., S.R. Koirtyohann, J.O.Pierce. Zinc, copper, cadmium
and chromium in polished and unpolished rice. Sci. Total Environ.
7 (1), 27-44 (1977).

(14) Myttenaere, C., J.M. Mousny. The distribution of chromium-51 in
lowland rice in relation to the chemical form and to the amount of
stable chromium in the nutrient solution. Plant Soil 41 (1), 65-72
(1974).

(15) Proctor, J., S.R.J. Woodell. The ecology of serpentine soils.
Advances in Ecol. Res. 9, 255-366 (1975).

(16) Rehab, F.I., A. Wallace. Excess trace metal effects on cotton: 3.
Chromium and lithium in solution culture. Commun. Soil Sci. Plant
Anal. 9 (7), 637-644 (1978).

(17) Reyes, R.Y., Robert Brinkman. Some methods to minimize zinc deficiency in transplanted wetland rice. A first comparison of land
ridging and zinc application; other possibilities. This Symposium.

(18) Rosell, R.A., A. Ulrich. Critical zinc concentrations and leaf
minerals of sugarbeet plants. Soil Sci. 97, 152-167 (1964).

(19) Singh, B.R., E. Steinnes. Uptake of trace elements by barley in
zinc-polluted soils. 2. Lead, cadmium, mercury, selenium, arsenic,
chromium, and vanadium in barley. Soil Sci. 121 (1), 38-43 (1976).

(20) Verfaillie, G.R.M. Kinetics of chromium absorption by intact rice
plants. pp. 315-331 *in* Beck, E.R.A. (ed.). Comparative studies
of food and environmental contamination. Proc. Symposium Otaniemi,
27-31 Aug. 1973, FAO, IAEA, WHO. IAEA, Vienna, STI/PUB/348. 623 p.
(1974).

(21) Wallace, A., G.V. Alexander, F.M. Chaudhry. Phytoxicity of cobalt,
vanadium, titanium, silver and chromium (Phaseolus vulgaris).
Commun. Soil Sci. Plant Nutr. 8 (9), 751-756 (1977).

(22) Wallace, A., J.W. Cha, F.M. Chaudhry, J. Kinnear and E.M. Romney.
Tolerance of rice plants to trace metals. Commun. Soil Sci. Plant
Anal. 8 (9), 809-817 (1977).

(23) Wallace, A., S.M. Soufi, J.W. Cha, E.M. Romney. Some effects of
chromium toxicity on bush bean plants grown in soil. Plant Soil
44 (2), 471-473 (1976).

(24) Warnock, R.E. Micronutrient uptake and mobility within corn plants
(zea mays L.) in relation to phosphorus-induced zinc deficiency.
Soil Sci. Soc. Amer. Proc. 34, 765-769 (1970).

(25) Welch, R.M., E.E. Cary. Concentration of chromium, nickel and
vanadium in plant materials. J. Agric. Food Chem. 23 (3), 479-484
(1975).

(26) Yamaguchi, T., S. Aso. Studies on chromium from the standpoint of plant nutrition. 1. Effect of chromium concentration on germination and growth of several kinds of plant. (Rice, wheat, Japanese radishes). (Japanese, translation used). Nippon Dojohiryogaku Zasshi 48 (11/12), 466-470 (1977).

THE STATUS OF MICROELEMENTS IN RELATION TO CROP PRODUCTION IN PADDY SOILS OF CHINA: I. BORON

Liu Zheng, Zhu Qi-qing*

(Institute of Soil Science, Academia Sinica, Nanjing)

The paddy soils are a kind special cultivated soils, Their chemical characteristics are radically different from those of normal upland soils. Some nutrients are moblized and translocated, while others remain un-affected, including both the macroelements and microelements. Deficiency of microelements is a common problem in certain type of paddy soils.

Microelements in soil influence crop growth in a variety of ways, of which the following two may be highlighted:

(a) The content of microelements in soil may not be sufficient to meet the requirement of crops;

(b) The microelements may not be in a form readily available to the crops.

The former is determined by the type of soils and parent materials and the latter is influenced by soil conditions, especially the soil reaction.

This series of publications is based on the point of view mentioned-above to discuss the status of microelements in paddy soils and their relations to the growth of crops.

The paddy soils of China may be divided into three types according to soil reaction, i. e. acid, neutral and calcareous paddy soils. The acid paddy soils distribute in acid soil region of southern China and the calcareous ones distribute in the northern China. The neutral paddy soils form an intermediate zone located between them. Distinct differences of status of microelements may be found in paddy soils of different reaction. The influence of parent materials is discussed also. There are certain differences of content in paddy soils derived from various parent materials.

STATUS OF BORON OF PADDY SOILS

The total content of boron in soils of China varies from trace to 500 ppm with an average of 64 ppm(1). However, in paddy soils, this range is narrower. Table 1 shows the total and available boron contents of paddy soils. They are grouped into three types according to soil reaction. Dis-tinct differences of available boron, as shown in Table 1, can be found in soil types of different reaction.

---

*Tang Li-hua, Xu Jun-xiang and Yin Chu-liang also participated in this reseach.

Table 1 Boron content in paddy soils of China

Type of soil	Total boron (ppm)		Water-soluble boron (ppm)	
	Range	Average	Range	Average
Acid paddy soils	5-351 (170)*	68	Trace-0.60 (310)	0.18
Neutral paddy soils	11-116 (64)	69	0.04-0.76 (67)	0.30
Calcareous paddy soils	12-61 (21)	45	0.10-1.79 (32)	0.72

*Figure in paratheses indicates number of samples investigated

The status of boron in soil is determined by parent materials and conditions of soil. Unfavourable soil conditions lower the availability of boron. The total content of boron in acid paddy soils is usually low, much lower than the average value in the soils of China. Effect of parent materials is profound. The average content of total boron in acid paddy soils derived from different parent materials is shown in Fig. 1. Acid paddy soils derived from limestones,quaternary red clay and shale are rich in boron, while those derived from granite and rhyolite are poor in boron. The difference between the two extremes is about fivefold. In southern China, granite and other acid igneous rocks distribute widely. Thus, there exist vast areas of paddy soils low in total boron, including Guangdong, Fujian, southern part of Jiangxi and central and western parts of Zhejiang Province.

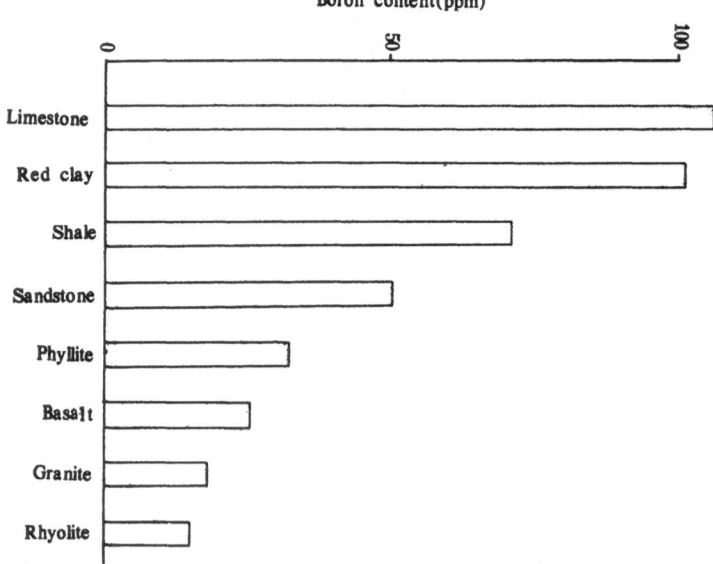

Fig. 1 Average content of total boron in acid paddy soils

Total boron content is not a reliable guide of boron nutrition for crop growth. Water-soluble boron is a more appropriate index. In acid paddy soils the content of water-soluble boron is very low. In 325 samples of acid soils analyzed including both upland soils and paddy soils, 90 % of them are lower than 0.5 ppm, which is usually regarded as the critical value for boron deficiency; while 82 % of the samples are lower than 0.3 ppm, i.e.severely deficient in boron (Table 2). Table 3 shows the distribution of total boron and water-soluble boron in soil profiles of three types of paddy soils.

Table 2 Content of water-soluble boron in acid soils
(Number of samples: 325)

Range (ppm B)	%
> 0.50	1
< 0.50	99
< 0.30	82

Table 3 Boron content in profiles of paddy soil

Soil type (parent material)	Depth (cm)	Total boron (ppm)	Water-soluble boron (ppm)
Acid paddy soils (red clay)	0–13	135	0.06
	13–36	102	0.14
	60–90	115	0.14
Acid paddy soils (granite)	0–15	9	0.09
	15–30	9	0.07
	47–67	9	0.06
Neutral paddy soil (lacustrine deposit)	0–17	79	0.54
	17–26	70	0.50
	85–95	93	0.28
Calcareous paddy soils (alluvium)	0–20	48	1.45
	20–29	50	1.19
	48–90	54	1.09
Calcareous paddy soils (Quaternary deposit)	0–26	40	1.14
	26–42	44	1.53
	82–120	60	1.79

In neutral paddy soils both total and water-soluble boron are richer than in acid paddy soils. The content of water-soluble boron in neutral paddy soils is usually lower than 0.5 ppm, but water-soluble boron content in the paddy soils derived from lacustrine deposit in southern Jiangsu Province is above the critical value and may be considered sufficient.

Again, the ratio of water-soluble boron to total boron is higher in neutral paddy soils as compared with acid ones.

Calcareous paddy soils usually contain less total boron than neutral paddy soils. However, the water-soluble boron content in calcareous paddy soils,is high. Ratio of water-soluble boron to total boron is greater than 1 % as shown in Table 3.

In conclusion, the content of water-soluble boron is low in acid paddy soils,medium in neutral paddy soils and high in calcareous paddy soils. A map of water-soluble boron content in soils of China in scale of 1:10,000,000 is provided. According to the legend of the attached map, the water-soluble boron in soils of China is given in five grades:

< 0.25	ppm	very low
0.25-0.50	ppm	low
0.51-1.00	ppm	medium
1.01-2.00	ppm	high
> 2.00	ppm	very high

Present map reveals that acid paddy soils in the acid and lateritic soil regions, are highly deficient in boron. Most soils in eastern China are comparatively rich in boron and the soils in western China, mostly calcareous in reaction, are high in boron.

RESPONSE OF CROPS TO BORON ON PADDY SOILS

The boron requirement of crops varies greatly. In general, cereal crops have the least requirement. However, in certain conditions, especially in case of overliming, rice shows response to boron on acid paddy soils. Winter rape and leguminous green manure crops grown on paddy soils are more susceptible to boron deficiency. Symptoms of boron deficiency were observed on rape in large areas of the acid soil region.

In acid paddy soils, water-soluble boron content is very low. Application of lime also lowers the availability of boron. Boron fertilizer increases the tillering, length of ear, grain weight and also strengthens the root developement of rice plant (Table 4). An increase of grain yield of 7-15 % may be expected(2). Plate 1 and plate 2 show the response of rice to boron under greenhouse and field conditions. Soaking rice seeds in a dilute solution of boron fertilizer is the simplest and cheapest way to apply boron.

Plate 1 Response of
         rice plant to
         boron fertilizer
         (plot experiment)

Rape grown on paddy soil has good response to boron fertilizer. Typical deficiency symptom of rape is failure of blossoming. Besides, terminal growth shows rosette and dieback(3). Symptomatology in connection with soil analysis usually gives correct diagnosis for boron deficiency in crops. Foliar spray with solution of boron fertilizer has been proved a successful way for rape. An increase of yield from 48 % up to three-fold may be expected as shown in plate 3 and Table 4(4). Boron deficiency in rape has been reported from ten provinces in southern China.

Leguminous green manure crops such as milk vetch show response to boron fertilizer on acid paddy soil. Fresh weight, weight of seeds as well as weight of root nudules increase after boron fertilization.

Plate 2 Response of rice plant to boron fertilizer (plot experiment)

Plate 3 Response of rape to boron fertilizer on acid paddy soil (plot experiment)

Table 4 Response of paddy rice to boron fertilizer
on paddy soil

Variety	Treatment	Tillering	Weight of thousand grains(g)	Percentage of empty grains (%)	Weight of roots(g)	
					Fresh weight	Dry weight
Nante	Soaking seeds	1.45	27.1	10	24.0	6.8
	C K	1.26	26.5	17	9.0	2.3
Nongken 58	Soaking seeds	2.34	28.9	12	---	5.6
	C K	1.48	26.8	21	---	4.6

Table 5 Response of rape to boron fertilizer
on acid paddy soil*

Locality	Parent material	Treatment	Yield(kg/ha)	
Jiangxi	Sandstone	Boron fertilizer, ** 375 kg/ha	814	+333 %
Yujiang	and red	Boric acid, foliar spray	701	+273 %
	clay	C K	188	---
Jiangxi	Phyllite	Boron fertilizer, ** 750 kg/ha	439	+318 %
Jinxian		Boric acid, foliar spray	221	+110 %
		C K	105	---
Jiangxi	Granite	Boron fertilizer,** 375 kg/ha	945	+49.6 %
Zixi		C K	635	---

Note: * Boron-containing industrial waste, applied to soil
     ** Plot experiment, in coorperation with local institutes
        and state farms

CONCLUSION

1) The content of water-soluble boron is ususally low in acid paddy
soils,medium in neutral paddy soils and high in calcareous soils.

2) In boron-deficient acid paddy soils, boron fertilizer is beneficial
for growth and yield of rice and rape. Typical boron deficiency symptoms
of rape is widespread in southern China. An increase of yield may be expected after boron fertilization. Legumineous green manure crops usually
show response to boron fertilizer also.

3) Soaking seeds and foliar spray have been proved the successful ways to apply boron fertilizer with the least danger of overdose and toxicity.

REFERENCES

(1) Liu Zheng, Tang Li-hua, Zhu Qi-qing, Han Yu-qin, Ouyang Tao, Content and distribution of trace elements in soils of China. Acta Pedologica Sinica, 15, 138-150(1978).(in Chinese with English summary)
(2) Liu Zheng, Ouyang Tao, Zhu Qi-qin, Sun Xiu-ting, Xu Jun-xiang, Xing Guang-xi, Status of microelements and its relation to plant growth in red earth of central China. Turang, 2, 76-85(1975).(in Chinese)
(3) Ren Hu-seng, The sterility of winter rape and its cure method. In "Proceedings of Symposium on Trace Elements" (ed. by Liu Zheng, Wu Zhao-ming), Science Press, Beijing, 87-98(1980).(in Chinese)
(4) Liu Zheng, Zhu Qi-qing, Ouyang Tao, Qian Cheng-liang, Tang Li-hua, Han Yu-qin, Yin Chu-liang, Xu Jun-xiang, Boron in soils and boron fertilization. In "Proceedings of Symposium on Trace Elements" (ed by Liu Zheng Wu Zhao-ming), Science Press, Beijing, 78-86(1980).(in Chinese)

THE STATUS OF MICROELEMENTS IN RELATION TO CROP PRODUCTION IN PADDY SOILS OF CHINA: II. MOLYBDENUM

Tang Li-hua*

(Institute of Soil Science, Academia Sinica, Nanjing)

The status of molybdenum of soil is important for growth and yield of certain crops. It has long been known to be required for fixation of nitr en in legumes. Leguminous green manure crops are important sources of nutrients for paddy rice under the rotation system in southern China. Although paddy rice shows no response to molybdenum, leguminous green manure crops are usually responsive to molybdenum fertilizer. The status of molubienum in soil and the effect of Mo-nutrient on the growth and yield of crops planted in paddy soils in China are summarized in this paper.

THE STATUS OF MOLYBDENUM IN PADDY SOILS

The total content of molybdenum of soils in China varies from 0.1 to 6 ppm (Mo) with an average of 1.7 ppm(1). In paddy soils, its content and availability vary with soil reaction, parent materials and soil condition.

The total and available molybdenum contents of paddy soils in China are listed in Table 1.

Table 1   Molybdenum content in paddy soils in China

Soil type	Total Mo (ppm)		Available Mo (ppm)*	
	Range	Average	Range	Average
Acid paddy soils	0.25-3.26 (142)**	1.32	0.02-0.60 (211)	0.14
Neutral paddy soils	0.27-1.38 (61)	0.69	trace-0.34 (72)	0.09
Calcareous paddy soils	0.26-1·21 (24)	0.57	trace-0.44 (31)	0.12

* Extractant: Tamm solution
** Figure in paratheses indicates number of soil samples investigated

In acid paddy soils derived from granite, for example, molybdenum is enriched. However, due to the effect of acidity, the availability of molybdenum in acid paddy soils is rather low and molybdenum deficiency may

---

*Zhu Qi-qing, Xu Jun-xiang, Yin Chu-liang and Liu Zheng also participated in this research.

be expected.

Fig. 1 shows the relation of molybdenum content of acid soils to their parent materials. Analytical results indicate that acid paddy soils derived from granite are richer in total molybdenum than those derived from sandstone. However, molybdenum may be easily fixed into unavailable form under acid condition, and its availability is irrelevant to the high content of total molybdenum. The results of 200 samples of red soil and acid paddy soils analysed for available molybdenum have revealed that 85 % of the samples are deficient in molybdenum, i.e. less than the critical value (0.15 ppm Mo). The results are listed below:

Trace	11 %
0.01-0.05 ppm	17 %
0.06-0.10 ppm	40 %
0.11-0.15 ppm	17 %
>0.15 ppm	15 %

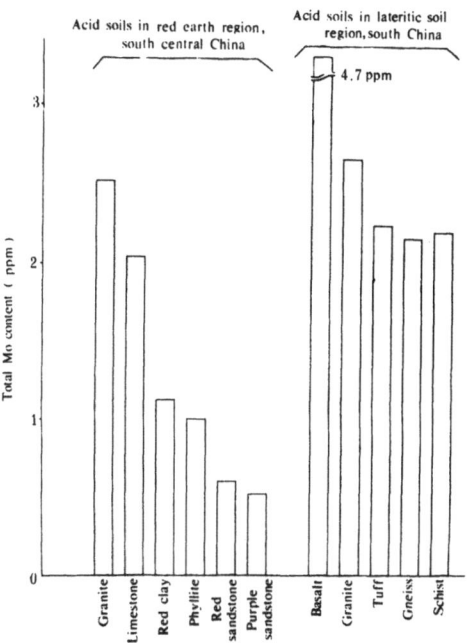

Fig. 1 Relationship between soil molybdenum content and parent materials

The total molybdenum in the neutral paddy soils is lower than the acid ones. Their available molybdenum is usually below the critical value for an adequate supply. It is resulted from the low content of molybdenum in parent materials, e.g. lacustrine and loessal deposits showed in Table 2.

833

Table 2 Molybdenum content of some paddy profiles

Type of soil	Parent material	Depth (cm)	Total Mo(ppm)	Available Mo(ppm)
Acid paddy soils	Sandstone	0–13	0.36	0.04
		13–36	0.40	0.04
		60–90	0.72	0.04
	Granite	0–15	2–19	0.15
		15–30	3.04	0.52
		67–82	3.02	0.30
Neutral paddy soils	Lacustrine deposit	0–17	0.62	0.12
		17–26	0.56	0.04
		85–95	0.70	0.04
	Leached loess	0–15	0.09	0.06
		15–27	0.76	0.05
		76–94	0.67	0.03
Calcareous paddy soils	Alluvium	0–20	0.73	0.21
		20–27	1.06	0.44
		27---	1.20	0.27
	Alluvium of Huanghe River	0–10	0.31	0.04
		10–20	0.40	0.04
		20–40	0.30	0.08

In calcareous soils, the status of molybdenum is rather complicated. In accordance with the relationship between molybdenum and soil acidity, it is more available under alkaline reaction. The soils derived from parent materials rich in molybdenum are usually adequate in molybdenum supply, conversely, the soils derived from parent materials low in molybdenum are always deficient in molybdenum. Analytical data reveal that the total molybdenum content of loess is quite low, ranging from 0.4–0.8 ppm. For example, both the total and available molybdenum contents of paddy soil derived from leached loess are low as listed in Table 2.

In conclusion, the supply of molybdenum is usually insufficient in acid and neutral paddy soils, while in the calcareous paddy soil, it is mainly determined by the nature of parent materials. Molybdenum deficient soils are distributed both in northern and southern China, but their causes of molybdenum deficiency are quite different from each other. Molybdenum content of soils in China is shown in a map; paddy soils are also included. On the map, according to the available molybdenum level, the contents of molybdenum in soils are divided into five grades (Table 3).

It is indicated in the map that there is a large area of molybdenum-deficient soils in China, and this is confirmed by field experiment and agricultural practice. Molybdenum fertilizer is one of the microelement fertilizers extensively used in China.

Table 3 Interpretative guides for available molybdenum

Interpretative guides	Range (ppm)
very low	< 0.10
low	0.11-0.15
medium	0.16-0.20
high	0.21-0.30
very high	> 0.30

MOLYBDENUM AND THE GROWTH OF CROPS

Generally, rice plant shows no response to molybdenum fertilizer. In the meantime, leguminous green manure crops grown on paddy soil have good response to molybdenum fertilizer.

The increase of fresh weight and seed yield of milk vetch by application of Mo is shown in plate 1 and plate 2(2) and similar response is found for the other leguminous crops(3). The deficiency of phosphorus may seriously restrict the effect of molybdenum on the crops. The experiment carried out on the basis of phosphorus fertilizer shows that little or no response to molybdenum could be found unless adequate phosphorus is also applied (Pl. 2).

Plate 1 Effect of molybdenum fertilizer on growth of milk vetch
(Soil: whitish soil of Liyang county, Jiangsu Province)

Plate 2 Effect of molybdenum fertilizer on
growth of milk vetch and relation to phosphorus.
Notes: left (dark-coloured) Mo+P;
       right (light-coloured) P
(Soil: whitish soil of Liyang county,
Jiangsu Province)

CONCLUSION

1) The supply of molybdenum is usually insufficient in acid and
neutral paddy soils, while in the calcareous paddy soils, it is mainly
determined by the nature of parent materials.

2) Leguminous green manure crops, such as milk vetch, growing on Mo-
deficient paddy soils usually show good response to molybdenum fertilizer.

REFERENCES

(1) Liu Zheng, Tang Li-hua, Zhu Qi-qing, Han Yu-qin, Ouyang Tao, Content
    and distribution of trace elements in soils of China. Acta Pedologica
    Sinica, 15, 138-150(1978). (in Chinese with English summary)
(2) Liu Zheng, Xu Jun-xiang, Xing Guang-xi, Sun Xiu-ting, Zhu
    Qi-qing, Status of microelements in soils and its relation to growth
    of crops in southern Jiangsu Province. Turang, 1, 28-36 (1974).
    (in Chinese)
(3) Liu Zheng, Zhu Qi-qing, Ouyang Tao, Han Yu-qin, Tang Li-hua, Xu Jun-
    xiang Yin Chu-liang, Qian Cheng-liang, Molybdenum in soils and molybdenum
    fertilization. In "Proceedings of Symposium on Trace Elements" (ed. by
    Liu Zheng,Wu Zhao-ming), Science Press, Beijing, 114-123(1980). (in
    Chinese)

EFFECT OF NITRAPYRIN ON THE INHIBITION OF NITRIFICATION IN SOME PADDY SOILS OF CHINA

Li Liang-mo,   Zang Shuang,
Zhou Xiu-ru,   Pan Ying-hua
(Institute of Soil Science, Academia Sinica, Nanjing)

Nitrification-denitrification is an important mechanism of N loss in paddy soils. How to prevent such loss is a matter of primary importance (1-4). The present article deals with the effect of Nitrapyrin (2-chloro-6-(trichloromethyl)-pyridine) on nitrification, nitrogen loss and percentage of fertilizer nitrogen recovery by crops, and on crop yield.

MATERIALS AND METHODS

Six kinds of soils including two strongly calcareous loams, two clay loams of about neutral reaction, one acid clay loam and one acid loamy clay were collected from Jiangsu and Zhejiang provinces for the present investigation. Nitrapyrin, abbreviated as CP in the following paragraphs, was synthesized by the Institute of Light Chemical Industry in Lüda City. The rate of application of "CP", as nitrification inhibitor, amounts to 3 percent of fertilizer nitrogen for experiments in field end pot culture.

Experiments for the effect of CP on soil nitrification, loss of soil nitrogen, and on the yield and growth of rice plant were carried out in three ways: 1) 49 field trials with small plots of $20m^2$ or $135m^2$ were made in 1973-1979 in the paddy field of Jiangsu. A series of supplementary micro-plot for labelled nitrogen fertilizer was occasionally provided to trace the effect of CP on the loss of fertilizer nitrogen. 2) 6kg pots were used for pot culture experiment. 3) culture test was made in 100 ml beakers, using 100 g of air dried soil with the addition of 60 mg N in the form of $(NH_4)_2SO_4$. For details of experimental methods, see the foot note under the corresponding table or figure.

RESULTS AND DISCUSSION

1. Effect of CP on the Nitrification in Soils

Effects of CP on the nitrification in strongly calcareous loam and slightly acid clay loam were studied in culture test under flooded conditions. Experiments showed that CP is an effective nitrification inhibitor in both soils. The rate of nitrification in cultures without CP is 3 to 33 times, sometimes 100 times, larger than that of CP-treated ones by applying equal quantity of nitrogen. In both treatments, the rate of nitrification is much higher in calcareous soil than in slightly acid soil (Fig. 1). Labelled $(^{15}NH_4)_2SO_4$ was used to trace the recovery of fertilizer-N in NH_4-form at different intervals of culture. Even at the very beginning, for 10 days, the $^{15}NH_4$-N found in the NaCl extract was only about 40% of the added $(^{15}NH_4)_2SO_4$, while at that time the rate of nitrification is far less than 10%. A great depletion has been found between the addition of NH_4-N and NO_3-N in comparison with the total N of added $(NH_4)_2SO_4$. It is believed that, without the effect on the uptake of nitrogen by plant, a large portion of the NH_4 ions from the applied $(NH_4)_2SO_4$ was fixed by the soil complex. However, the effect of CP on the stabilization of NH_4-N can be still observed in both soils (5-8). More $^{15}NH_4$-N has been found in NaCl extract from the CP-treated cultures than that from the controls. The rate of inhibition of nitrification ranges from 4-10% ($P \leqslant 0.01$, Fig. 2).

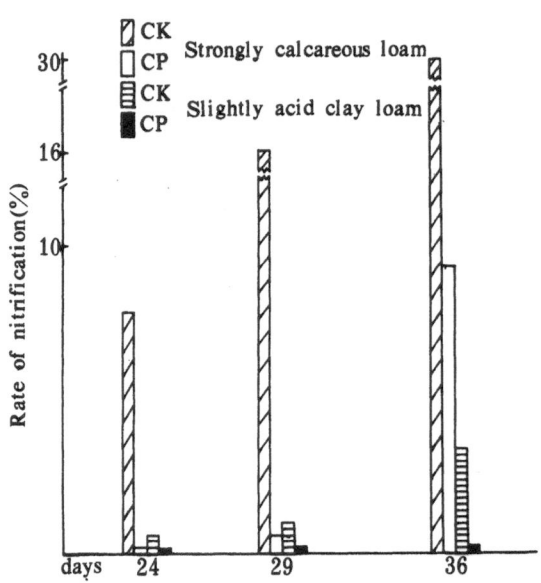

Fig. 1 Effect of CP on the inhibition of
    nitrification
    CK: $(NH_4)_2SO_4$ added at 60 mg
        N/100g soil
    CP: $(^{15}NH_4)_2SO_4$ added at 60 mg
        N/100g soil CP added at a
        rate of 5 ppm in soil

The effect of CP on the inhibition of nitrification was also observed in the greenhouse and in the field for both rice and wheat plants. The content of NH_4-N, NO_3-N and amount of nitrite bacteria in soils were analysed. Results in Figs. 3 and 4 show the significant effect of CP on the inhibition of the activity and development of nitrite bacteria. Figs. 3 and 4 give the results from greenhouse trial for wheat, and from field plot for rice respectively.

The above stated figures show that changes of the number of nitrite bacteria and the rate of nitrification are not always consistent. In Fig. 3, numbers of nitrite bacteria in the control pots dropped down during January 7th to 17th, while the rate of nitrification somewhat increased. During February 5th to 20th, when the numbers of nitrite bacteria somewhat increased, the rate of nitrification also declined. However, significant effect of CP has been found on the amount of nitrite bacteria and on the rate of nitrification. Both of them are at a low level. Results from field experiment for rice plant (Fig. 4) show that the amount of nitrite bacteria decreases while the rate of nitrification increases in the interval of 10 to 20 days after the application of nitrogen fertilizer both under the treatment of CP and without it.

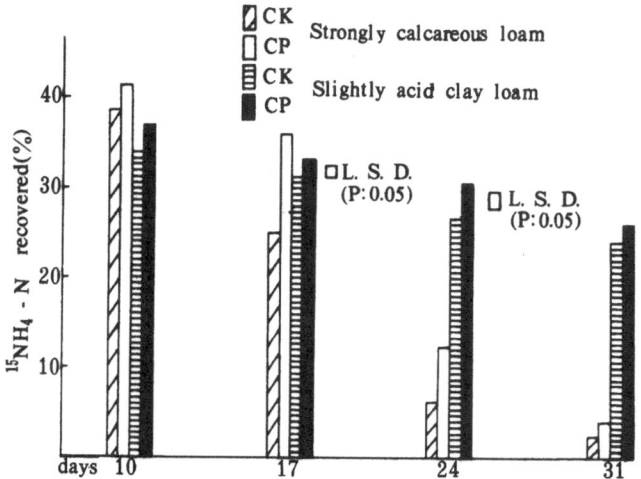

Fig. 2 Effect of CP on the recovery of $^{15}NH_4-H$
CK: $(^{15}NH_4)_2SO_4$ added at 60 mg N/100g soil
CP: $(^{15}NH_4)_2SO_4$ added at 60 mg N/100g soil
CP added at a rate of 5 ppm in soil

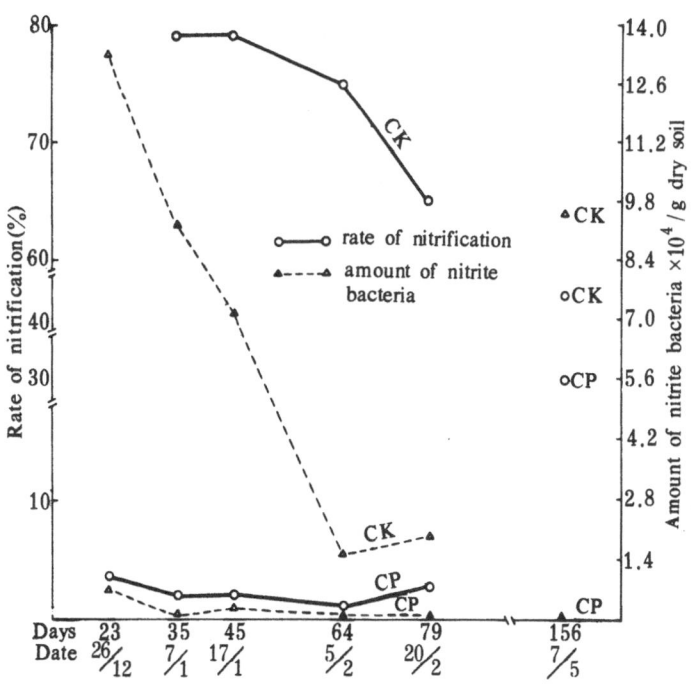

Fig. 3 Effect of CP on the rate of nitrification and the amount of nitrite bacteria in pot culture with wheat

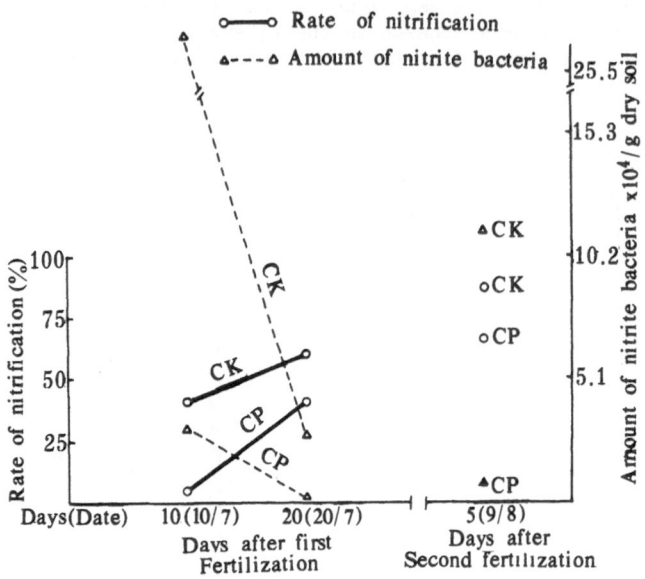

Fig. 4 Effect of CP on the rate of nitrification and the amount of nitrite bacteria
in rice field

It is also observed in Figs. 3 and 4 that the effect of CP in dryland
lasted longer than that under waterlogged soil conditions due probably to
the faster degradation of CP in submerged soil.

The effect of CP on the loss of fertilizer nitrogen was under survey
in culture test under submerged condition for five soil types. Investigation was made on the determination of total-^{15}N retained in soils after one
month's culture. Obviously, the problem is much complicated in calcareous
soils, where volatiligation of NH_3 inevitably occurs at alkaline reaction (9,10). As shown in Table 1, the effect of CP in two strongly
calcareous loams is both insignificant. On the other hand, the effect of
CP on the decrease of N loss in two soils of slightly acid clay loam and
acid loamy clay are significant. Increments of residual ^{15}N have been
found by 7-8% in CP-treated cultures in comparison with the controls in 1
month. However, the reason why CP was ineffective on the acid clay loam
soil, pH 5.7, remains to be investigated.

2. Effect of CP on the Uptake of Fertilizer Nitrogen
and the Growth of Rice Plant

It is demonstrated that the inhibition of nitrification by CP in paddy
soils may retain more fertilizer nitrogen in the form of ammonium. The
persistant presence of NH_4-N in paddy soil would be helpful to the assimilation of soil nitrogen by rice. Thus, field microplot trail was investigated
with rice plant.

Table 1 Effect of CP on N loss ($^{15}NH_4)_2SO_4$ added at 40 mg N/100g soil, CP added at a rate of 5ppm in soil, incubated 1 month under submerged condition)

Soil	Treatment	Residue of ^{15}N in soil (%)	Decrease of loss (%)
Strongly calcareous loam (pH8.4, O.M. 0.93%)	^{15}N ^{15}N+CP	45.5 47.3	1.8
Strongly calcareous loam (pH7.9,O.M. 2.30%	^{15}N ^{15}N+CP	50.1 50.8	0.7
Slightly acid clay loam (pH6.1, O.M. 3.14%)	^{15}N ^{15}N+CP	83.1 91.1	8.0**
Whitish soil, acid clay loam (pH5.7, O.M. 1.60%)	^{15}N ^{15}N+CP	81.4 79.8	-1.6
Acid loamy clay (pH5.5, O.M. 2.88%	^{15}N ^{15}N+CP	69.2 76.3	7.0**

Note: The abundance of ^{15}N in ($^{15}NH_4)_2SO_4$ applied was 10.78%

841

Microplots for studying the effect of CP on the assimilation of fertilizer nitrogen, including labelled $(NH_2)_2CO$, NH_4HCO_3 and $(NH_4)_2SO_4$, by
rice plant at full earing were laid on the strongly calcareous loam in
northern Jiangsu Province. Experiments were designed for 7 blocks, with
4 microplots in each block. As shown in Table 2, the mean increment of
assimilated nitrogen by rice plant for 7 blocks was 1.72 % ($p < 0.05$). The
very slight increase of the uptake of fertilizer-N of rice plant by the
application of CP is not reflected by the response of crop yield. Therefore,
as a whole, the effect of CP on the uptake of ferilizer nitrogen at such a
dressing rate, 75 kg/ha, by rice plant on strongly calcareous soil is
considered to be insignificant.

Table 2 Effect of CP on the increase of nitrogen recvery (Field
microplot trial with ^{15}N labelled fertilizer, strongly
calcareous loam)

Treatment	^{15}N assimilated by rice (%)	Net increase (%)
Prilled urea, top-dressing	22.3	2.5
Ibid + CP	24.8	
Prilled urea, incorporated	29.0	2.7
Ibid + CP	31.7	
Prilled urea, deep-dressing	25.8	-0.5
Ibid + CP	25.3	
SGU, deep-dressing	55.1	1.6
Ibid + CP	56.7	
Prilled urea, top-dressing at panicle formation stage	61.9	
Ibid + CP	64.6	3.2
Powdered ammonium bicarbonate, top-dressing	17.1	
Ibid + CP	18.2	1.1
Powdered ammonium sulphate, top-dressing	22.5	
Ibid + CP	23.8	1.5

Note: X = 1.72 ($P < 0.05$)
Microplot was formed by a plastic cylinder without bottom pressed
into soil. Its area was about 0.062 m^2

N-fertilizer added at 75 kg N/ha, of which 37.5 kg N/ha was applied
as basal dressing, 37.5 kg/ha as top-dressing

Forty-nine field experiments were carried out in 1973-1979 to study
the effect of CP on the yield of rice on the calcareous paddy soils in
Jiangsu. Among 40 experiments out of 49 at the yield level of above 5250
kg/ha, only 7.5% of them had a remarkable effect of CP, with an increase of
yield up to 11.0-20.5%, while 9 of them were below 5250 kg/ha and 6 out of
9 had a significant effect of CP, with an increase of yield by 8.5-16.9%.
From these results, one can see that the effect of CP on rice yield is
unstable. As shown in Table 3, on paddy field with grain yield of over

842

Table 3 Effect of CP on rice yield in soils with different fertility

Soil	Soil fertility	Treatment (urea, kg N/ha)	Mean yield (kg/ha)	Increase kg/ha	%
Strongly calcareous loam	Low fertility	75.9	5750	730^{**}	12.6
		75.9 + CP	6480		
	High fertility	75.9	6370	$-20^{n.s.}$	-0.2
		75.9 + CP	6350		
Sandy soil	Low fertility	69.0	4580	770^{**}	16.9
		69.0 + CP	5350		
	High fertility	86.5	7890	$380^{n.s.}$	4.9
		86.5 + CP	8272		
Slightly calcareous clay loam	Low fertility	86.5	4910	560^{**}	11.5
		86.5 + CP	5480		
	High fertility	52.0	6170	$120^{n.s.}$	2.3
		52.0 - CP	6290		

6000 kg/ha, CP has no effect, while on those with yield of less than 5250 kg/ha, CP shows an impressive effect. To heavily manured field, as in the case of a slightly calcareous clay loam, the application of CP proves also ineffective.

CONCLUSION

From the data mentioned above, CP is significantly effective on the inhibition of nitrification. However, its effect on the decrease of N loss is governed by soil properties. Since calcareous soils, as usual, enhance the volatilization of NH_3, the beneficial role for retaining nitrogen in ammonium form in acid paddy soil may not be true as in calcareous soil by the application of CP. The slight increase of the uptake of fertilizer nitrogen of rice plant by the application of CP, as showed in the present investigation, is not reflected by the response of crop yield. Field experiments show that in soils of low fertility, with low dressing of nitrogen fertilizer or manures, CP gives major effect on the response of rice yield even in calcareous soils.

REFERENCES

(1) Li Ching-kwei, Agrochemical problems of nitrogen fertilizer. Turang, 3, 130-133 (1975). (in Chinese)
(2) Zhu Zhao-liang, Recent study on transformation and removal of nitrogen in soils. Turangxue Jinzhan, 2, 1-16 (1979). (in Chinese)
(3) Alexander, M. (ed.), Advances in Microbial Ecology. Plenum Press, New York, 2, 82-90 (1978).
(4) Walker, N. (ed.), Soil Microbiology (A Critical Review). Butterworths, London, 11-12, 133-144 (1975).
(5) Rajendra Prasad, Rajale, G.B., Lakhdive, B.A., Nitrification retarders and slow-release nitrogen fertilizers. Adv. Agron., 23, 337-358 (1971).
(6) Cleve, A.I. Goring, Control of nitrification of ammonium fertilizers and urea by 2-chloro-6-(trichloromethyl)-pyridine. Soil Sci., 93, 431-439 (1962).
(7) Goring, C.A.I., Hamaker, J.W. (ed.), Organic Chemicals in the Environment. Marcel Dekker, Inc., New York, 2, 653-666 (1972).
(8) Bundy, L.G., Bremner, J.M., Inhibition of nitrification in soils. Soil Sci. Soc. Amer. Proc., 37, 396 (1973).
(9) Bundy, L.G., Bremner, J.M., Effects of nitrification inhibitors on transformation of urea nitrogen in soil. Soil Biol. Biochem., 6 (6), 369-376 (1974).
(10) Yu Suo-fu, Zhao Mei-zhi, Primary study of nitrogen loss in calcareous soil. Turang, 1, 31-33 (1979). (in Chinese)

EFFECT OF SOIL FERTILITY ON THE GROWTH AND YIELD OF RICE

Kaoru Seino

(Tohoku National Agricultural Experiment Station)

It is well know that rice yield is affected by the nitrogen fertility of paddy soils (1-5), but there are many problems to be solved about its mechanism.

In this paper, I tried to make clear the growth pattern of rice plant in in some paddy soils with different nitrogen fertility, and discussed the significance of soil nitrogen fertility on the basis of experimental results of the soils with and without application of nitrogen fertilizer. Furthermore, I attempted to describe the influence of soil fertility on the growth pattern of rice plant.

EFFECT OF SOIL TEXTURE AND CLAY MINERALS ON THE
GROWTH PATTERN OF RICE PLANT

The experiments have been carried out in 1967 at Kyushu National Agricultural Experiment Station in cooperation with seven prefectural experiment stations in Kyushu districts (6,7). In these experiments, the same rice variety as well as the same treatments in nitrogen fertilizer application were adopted. The soils of the paddy fields used were classified into four groups on the basis of soil texture and crystalline structure of clay minerals. There were characteristic growth patterns shown in each group. The mechanical compositions and chemical properties of four groups of soils with the locality names are shown in Table 1. In order to make clear the response of nitrogen application on the rice growth pattern, the method used in this study is as follows:

rop Growth Rate (CGR: $g/m^2/day$)
$\Delta N/\Delta W$: ΔN (amount of nitrogen absorbed within a given period of time),
ΔW (amount of dry matter produced within a given period of time).
Productive efficiency of absorbed nitrogen: efficiency of absorbed nitrogen to grain production (weight of brown rice per unit of absorbed nitrogen at harvest).

The pattern of rice growth in the vegetative phase are shown in Table 2. The CGR and $\Delta N/\Delta W$ in the vegetative phase of rice plant in the group A with fine texture and 2:1 lattice clay composition are small compared to those of other groups. In this group, the increase in brown rice due to the application of nitrogen fertilizer is large and decrease in productive efficiency of absorbed nitrogen by the nitrogen application is small. However, the $\Delta N/\Delta W$ in the vegetative phase of rice plant in the group C with coarse texture and

1:1 lattice clay composition is larger than that in the group B with medium texture and 1:1 lattice clay composition. It becomes clear from the above that the clay contents in soils and their composition have direct effects upon the rice growth pattern.

Table 1   Relation between the classification of paddy fields in Kyushu district and the response of nitrogen application for rice plants

Group	Crystalline structure of clay minerals	Soil texture by mechanical composition	Clay content (%)	Main minerals in clay	C.E.C. (meq/ 100g)	Locality	Increase amount of brown rice by nitrogen application (ton/ha)
A	2:1 lattice	Fine	31.8	Montmorillonite	21.7	Saga	2.6
B	1:1 lattice	Medium	28.1	Kaolin,14A	20.2	Chikugo	2.3
C	1:1 lattice	Coarse	11.8	Kaolin,14A	8.7	Fukuoka	1.9
D	Amorphous	Medium	23.9	Allophane	24.7	Kumamoto	1.7

	Productive efficiency of absorbed nitrogen	
	With no nitrogen application	With* nitrogen application
A	57	49
B	62	41
C	66	40
D	51	36

* Basal application 80 kg/ha
  Top dressing at the ear formation period
  (25 days before heading) 30 kg/ha
  (10 days before heading) 30 kg/ha

Table 2   The pattern of rice growth process in soils

Group	Nitrogen application	CGR(g/m^2/day)			$\Delta N/\Delta W(\%)$		
		I	II	III	I	II	III
A	With no	4	8	12	1.6	0.7	0.5
	With	8	14	23	2.2	0.3	1.2
B	With no	5	7	16	1.9	0.0	0.7
	With	9	14	26	2.0	0.6	1.1
C	With no	3	16	13	2.0	1.0	0.6
	With	7	27	21	2.8	1.1	1.0
D	With no	5	16	18	2.1	0.6	0.6
	With	8	19	35	2.9	0.2	1.2

I: Transplanting - Maximum tillering stage
II: Maximum tillering stage - Young ear formation stage
III: Young ear formation stage - Heading stage

RELATION BETWEEN THE PATTERN OF NITROGEN SUPPLY TO RICE PLANT
FROM PADDY SOILS AND THE RICE GROWTH PATTERN

This experiment has carried out from 1970 to 1973 in Chikugo field
(Kyushu Agricultural Experiment Station) which belongs to the group B(8).
The purpose of this experiment is to make clear the relation between the
nitrogen fertilizer application and the soil fertility to get a high yield
of rice grain. Many evidence have been obtained about this subject through
the yield of rice cultured with no fertilizer nitrogen(1-5). As the greater
part of nitrogen absorbed by the roots of rice plant without nitrogen fertilizer is considered to be derived from the soil nitrogen, the pattern of
nitrogen uptake in the plants are closely related to the mineralization of
soil nitrogen. This suggests that the pattern shows indirectly the nitrogen
fertility of soils.

Generally speaking, in the soils which shows high yield of rice with no
nitrogen fertilizer, there is a similar tendency found also in the case with
nitrogen fertilizer application. But, sometimes this is not true. Consequently, it is necessary to clarify the relation between the pattern of nitrogen uptake in the rice plant showing a high grain yield and that in the rice
plant with no nitrogen fertilizer.

Table 3 shows the results of the experiment of nitrogen application to
rice plant in Chikugo(8). As the difference of the amounts of nitrogen
removed by the plant between years is small, thc high yield with no additional
nitrogen indicates the high productive efficiency of absorbed nitrogen. When
the yield of rice grain in a plot with no additional nitrogen is high, both
the grain yield and the productive efficiency of absorbed nitrogen in the
nitrogen application plot are high. CGR and $\Delta N/\Delta W$ in each growth period of
rice cultured without nitrogen application show the difference yearly and a
similar tendency is shown also with nitrogen application (Table 4).

In the plot with no nitrogen application, there is an increase in the
grain yield when $\Delta N/\Delta W$ (I) and CGR (III) are high, but $\Delta N/\Delta W$ (III) is low.
This tendency can also be seen in the plot with nitrogen application.

In this field, according to the results of the experiment of nitrogen
application carried out from 1969 to 1975., positive correlations were observed
between the grain yield of rice and $\Delta N/\Delta W$ (I,II) and CRG (III,IV) during the
rice growth.

Therefore, the results shown in Table 3 and 4 also agree very well with
the fundamental properties to obtain the high grain yield in rice. Namely, the
high soil fertility indicates that the process of the mineralization of the soil
nitrogen agrees with the desirable patterns of nitrogen uptake for the high
yield of rice.

When we use a recently developing method with transplanting machine and
young rice seedlings, however, the relation may differ somewhat from the above
(Table 5). With this change, it is necessary to reconsider relation between
the soil fertility and the pattern of rice growth.

Table 3 Relation between rice grain yield in the no additional
nitrogen and effect of nitrogen application

Plots	Years	Yield of brown rice (ton/ha)	Amount of nitrogen uptake (kg/ha)	Productive efficiency of absorbed nitrogen
With no nitrogen application	1970	3.78	74	51
	1971	3.96	68	58
	1972	4.28	66	65
	1973	4.97	72	69
With[*] nitrogen application	1970	6.74	160	42
	1971	7.03	169	42
	1972	7.58	170	45
	1973	7.81	174	45

* Nitrogen level is not the same in each years. The highest grain yield in
each years are given　　　　　　　Rice variety: Reihou

Table 4　Growth process of rice plants

Growth[*] process	Years	CGR $(g/m^2/day)$ Nitrogen application		$\Delta N/\Delta W$ (%) Nitrogen application	
		With no	With	With no	With
I	1970	1.8	3.0	2.3	3.0
	1971	3.2	5.4	2.6	3.5
	1972	1.6	3.5	3.0	3.9
	1973	2.4	4.9	3.1	3.5
II	1970	5.9	15	1.6	1.4
	1971	8.7	16	1.2	1.9
	1972	7.4	19	1.7	2.6
	1973	9.4	17	1.3	1.9
III	1970	11	19	0.82	1.9
	1971	9.4	16	0.65	0.52
	1972	13	23	0.65	0.52
	1973	19	24	0.64	0.32
IV	1970	16	21	0.62	0.93
	1971	14	22	0.37	0.99
	1972	15	21	0.46	0.81
	1973	17	21	0.49	1.2

* I:　　Transplanting - Active tillering stage
　II:　　Active tillering stage - Maximum tillering stage
　III:　Maximum tillering stage - Young ear formation stage
　IV:　Young ear formation stage - Heading stage

Table 5   Correlation between the process of rice growth and
the yield components

Process of growth		Coefficient of correlation between weight of brown rice		Coefficient of correlation between percentage of perfect rice grain	
		Manual transplanting of mature seedlings	Use of transplanter with young seedlings	Manual transplanting of mature seedlings	Use of transplanter with young seedlings
CGR	I	0.081	0.027	0.147	-0.154*
	II	0.104**	-0.628*	-0.264	0.648*
	III	0.558**	0.314	-0.022	-0.417
	IV	0.472**	0.201	-0.403*	-0.383
ΔN/ΔW	I	0.481**	-0.412	-0.308	0.636*
	II	0.453*	0.459	-0.325	-0.540*
	III	0.336	0.515	-0.127	-0.561*
	IV	-0.280	0.295	0.137	-0.468

I,II,III,IV : See Table 4.

EFFECT OF IMPROVEMENT OF SOIL FERTILITY ON RICE GROWTH

The relationship between the percentage of yield increase by the application of improving materials and the rice yield components are shown in Table 6.

Table 6   Effect of successive application of soil improving
materials (Silicate slag and minor elements)

Years	Manual transplanting of mature rice seedlings				Years	Use of transplanter with young seedlings			
	Percentage of increased yield with improvement	Ratio of improvement to control (100)				Percentage of increased yield with improvement	Ratio of improvement to control (100)		
		R.	E.	G.			R.	E.	G.
1969	13	93	117	127	1976	17	112	94	99
	16	92	119	131	1977	-5	97	100	97
1970	19	98	101	123		0	96	92	101
	18	97	104	122		0	106	93	94
1971	3	97	91	105	1978	4	108	95	93
1974	3	101	99	101		6	106	90	100
	8	103	99	103		6	115	95	91
1975	-3	95	95	91		5	102	87	101
	7	96	107	117	1979	7	105	85	101
						8	102	88	106
						6	106	89	99
						6	102	93	104

R. : Percentage of perfect rice grain.
E. : Number of ears
G. : Number of grains

In the plot of improvement material application. Silicate slag (4 t/ha) and fritted trace elements, FTE (30 kg/ha) were applied each year since 1969. Beneficial effects of their application with manual transplanting of mature rice seedlings were found on increases in number of grains ($r = 0.892**$, significant at 1 % level), whereas in the case of using transplanter with young seedlings, there was an improvement of ripened grain ($r = 0.624*$, significant at 5 % level).

Table 7 shows the growth pattern of these rice plants. If we make calculation using the data of Table 7, in the case of the manual transplanting of mature rice seedlings, a possitive correlation ($r = 0.714*$) was observed between the ratio of increased yield with the improvement to the control and the ratio in $\Delta N/\Delta W$ (I) of the improvement to the control. In the case of transplanter use with the young seedlings, a positive correlation ($r = 0.684*$) was observed between the ratio of the former and the ratio in the $\Delta N/\Delta W$ (II).

Table 7   Ratio of increased yield with improvement and pattern of vegetative phase of rice plants

Years	Ratio of increased yield with improvement	Ratio of improvement to control(100)				Years	Ratio of increased yield with improvement	Ratio of improvement to control(100)			
		$\Delta N/\Delta W$ I	II	CGR I	II			$\Delta N/\Delta W$ I	II	CGR I	II
1969	13	109	157	78	131	1976	17	84	100	120	93
	16	106	115	106	105	1977	-5	102	73	129	68
1970	19	108	89	88	93		0	100	95	100	105
	18	106	186	90	94		0	97	80	139	100
1971	3	93	12	134	75	1978	4	105	80	126	96
1974	3	92	105	76	107		6	105	93	100	109
	8	92	123	104	106		6	105	85	100	108
1975	-3	100	100	74	100		5	105	86	100	100
	7	93	107	72	100	1979	7	104	96	120	120
							8	91	103	100	123
							6	91	103	100	115
							6	91	94	100	119

I,II : See Table 4

EFFECT OF ORGANIC MATTER APPLICATION

As the practice for the improvement of soil fertility, application of compost and farmyard manure are common in this district.

Table 8 shows the results of the successive application of compost in Chikugo fields, compost was applied at the rate of 2.5 tons per hectare (corresponding to the weight of dry matter) with the chemical fertilizer each year(9). When compost application brought about the beneficial effects, the rice growth showed a tendency that $\Delta N/\Delta W$ (I,II) was small and $\Delta N/\Delta W$ (IV) was large compared with the other years. Compost application generally increases the CGR in the period of I and II, but when there occurs the high $\Delta N/\Delta W$ simultaneously, the beneficial effects of compost application does not appear.

Table 8　Effect of compost application ȯn rice plants

Years	Treatments	Yield of brown rice (ton/ha)	CGR (g/m^2/day) I	II	III	IV	ΔN/ΔW (%) I	II	III	IV
1976	Only chem. fer.	5.25	2.3	12	16	17	3.7	1.5	0.81	0.98
	With compost	6.03	3.0	16	16	19	3.4	1.5	0.69	1.2
1977	Only chem. fer.	6.05	2.5	15	10	23	3.4	1.6	0.57	0.49
	With compost	6.18	3.0	16	16	22	3.6	1.7	0.21	0.68
1978	Only chem. fer.	6.47	2.4	13	18	22	3.9	1.5	0.86	1.2
	With compost	6.94	2.6	14	20	18	3.8	1.2	0.74	1.5

chem. fer. : chemical fertilizer　　　　I,II,III,IV : See Table 4

An effect of wheat straw application on the rice growth and yield are shown in Table 9 (10). The growth of rice plant in the active-vegetative phase was controlled by the application of wheat straw. But when a large quantity of superphosphate (corresponding amount to 10 % of absorption coefficient of soil for P_2O_5) was added at the transplanting, the growth of rice plant in the early stage was promoted greatly in spite of the presence of wheat straw.

Table 9　Effect of wheat straw application on the growth of rice plants

Treatments	Amount of wheat straw applied(t/ha)	CGR (g/m^2/day) I	II	ΔN/ΔW (%) I	II
Heavy dressing	0	2.4	9.9	2.8	1.8
of	3.75	4.6	19	2.8	1.6
superphosphate	7.50	3.3	15	2.8	1.7
Control	0	1.8	12	2.9	2.1
	3.75	1.5	14	2.9	2.1
	7.50	1.3	11	2.9	2.2

Remarkes : Straw was applied 2 weeks before transplanting with calcium cyanamide, fused magnesium phosphate, potassium chloride (50 kg N, P_2O_5, K_2O per hectare) and plowed in the land at once. In addition to the above, top dressing applied 4 times. Rice variety : Asominori I, II(See Table 4)

Table 10 shows the results of successive application of farmyard manure obtained at Kagoshima(11). This experiment started from 1975. The highest yields are given in the plots of 5 ton application in the farmyard manure and 10 ton application in the rice straw, respectively. With the increases in the organic matter application, the CGR (I,II) increases in the farmyard manure, but in the rice straw the ΔN/ΔW in the period of I decreases and the rice plant shows nitrogen starvation.

Treatments	Amount of materials (t/ha)	Yield of brown rice (t/ha)	CGR(g/m^2/day)		$\Delta N/\Delta W(\%)$	
			I	II	I	II
Control	0	5.0	2.7	11	2.8	0.87
Farmyard manure	5	7.2	3.1	15	3.4	1.8
	10	6.2	3.2	19	3.4	1.6
	20	5.6	4.4	15	3.5	1.5
Rice straw	5	5.9	3.0	15	3.3	1.0
	10	6.4	2.4	14	3.2	1.3
	20	6.1	2.7	8.9	2.6	1.1
	40	5.4	3.1	13	2.6	1.5

I, II : See Table 4

DISCUSSION

Some studies have been made about the relation between the soil fertility consisting especially of soil texture or clay minerals and the yield of rice grain(12-14).

From this report it becomes clear that the native fundamental property of some paddy soils affects the growth of rice plant regardless of the nitrogen fertilizer application. This fact has a significance in that for the increase of rice yield, it is necessary to make use of the characteristics of the soil effectively.

For example, the yield of rice in Saga soil does not increase by the practice of nitrogen application concentrating on the basal dressing as shown in this paper. This is based on the fact that the $\Delta N/\Delta W$ in the vegetative lag phase is low, and it is closely related to the soil texture and the pro-, perty of clay minerals. Therefore, rice cultivation in Saga district maintains its high yield by improving the practice of nitrogen application in reducing the amount of basal dressing and increasing the top dressings. This is just the practical use of the soil fertility.

On the other hand, in a certain soil like Chikugo soil, the pattern of nitrogen uptake of rice plant with nitrogen application was similar to that with no nitrogen. And in the year gaining the high grain yield of rice plant with no nitrogen application, the yield was also high in the case of applied nitrogen fertilizer. Moreover, in such years, the productive efficiency of absorbed nitrogen was high both with and without nitrogen application.

These facts mean that there is a regular pattern in the process of nitrogen uptake in gaining a high yield of rice plant, and the productive efficiency of absorbed nitrogen is high in accordance with this pattern.

Generally, the application of nitrogen fertilizer increases the yield of rice, but decreases the productive efficiency of absorbed nitrogen. In the cultivation techniques of the winners in the contest of high yield of rice grain, the productive efficiency of absorbed nitrogen however is also high in spite of the high yield. There are many cases for these facts, and the heavy dressing of farmyard manure is a factor which cannot be overlooked(15). They supply the greater part of absorbed nitrogen with such a organic material(16). An example of Table 11 proves that can be considered in the above results.

Table 11   Productive efficiency of absorbed nitrogen in the high yield rice plants

Fields		Yield of brown rice (t/ha)	Productive efficiency of absorbed nitrogen	Rate of nitrogen caused by farmyard manure to applied quantity
Farmer	Mr. Kato	10.84	69	67 (%)
	Hatakeyama	10.81	46	
	Kudo	10.60	60	55
	Jyoraku	10.22	49	25
	Kitahara	10.24	53	53
	Tan	9.24	75	47
	Koike	9.75	50	39
National Agricultural Experiment Station	Kantotosan	8.07	34	65
	Tohoku	7.95	48	56
	Kyushu	7.81	45	0
	Chugoku	7.37	46	0
	Hokuriku	7.32	38	0

In many experiments which were made in agricultural experiment stations, the nitrogen fertilization for increasing rice yield often depended upon the chemical fertilizers. As a result, even if the grain yield increased, the productive efficiency of absorbed nitrogen clearly became low.

When the compost was applied, rice plant produce the dry matter without too high nitrogen content in the early stages of rice growth(17). This is an important techniques to prevent from lowering of the productive efficiency of absorbed nitrogen. These facts also form a part of the reason why the productive efficiency of absorbed nitrogen of rice plant with no nitrogen application is high.

But the application of crude organic materials like rice straw or wheat straw to paddy fields gives a different effect on rice plant relatively to compost; as it is called the nitrogen starvation and the functional disorder of roots by soil reduction in the early stages of rice growth. In the application of these materials, some other techniques are necessary for an increase of rice growth in the early stages and for the retardation of damage caused by the decomposition of these materials.

As the above, rice cultivation in warm district of Japan realized the high yield of rice grain in harmony with the soil fertility and the techniques of nitrogen application of the chemical fertilizer.

In recent years, the developing methods with transplanting machine and young rice seedlings brought about the changes of rice growth pattern, and hence an increase in the yield of rice came to a halt.

These factors affecting it could be as follows:

(1) A fall in the productive efficiency of absorbed nitrogen.

(2) A decrease in the percentage of ripened grains.

These facts may come from that the pattern of nitrogen uptake is unfitable to the growth pattern of rice plant. As one of the countermeasures, the application of compost or silicate slag are suitable techniques. Recently, in paddy fields of warm district in Japan, the application of compost become less frequently and the direct application of rice straw increases. In autum application of rice straw it is not necessary to consider, but in the case of the application just before transplanting, rice development is checked in the early growth stage of rice plant. In the case of wheat straw, it will react more vigorously than in the case of rice straw. But, by the increase of nitrogen fertilizer and other techniques, rice growth recover to the normal in the later growth stage and the yield of rice grain is almost the same as the control. The successive application of crude organic materials often find some difficulties in controlling the rice growth than the first year. This phenomenon can be considered different from soil to soil, but in the case of the longer successive application, it is more and more difficult to control the growth of rice plant by the organic matter accumulated in soil and also to obtain the increase of rice grain yield.

Therefore in paddy fields, there is a limit in the amount of application of organic materials.

REFERENCES

(1) Katsumi F., Soil type and nitrogen response in paddy field. Bull. Fukui Agr. Exp. Stn., 3, 1-10 (1966)
(2) Kono M., Morita K., Some roles of soil fertility in grain production of rice plants. Bull. Hokuriku Natl. Agr. Exp. Stn., 22, 1-33 (1979).
(3) Oyama N., Sakai H., Kobayashi H., Kawasaki I., Nonoyama Y., Komoto Y., Characteristics of climates, soils and rice plants in high yielding paddy fields in Chugoku District. Bull. Chugoku Natl. Agr. Exp. Stn., Series E, 7, 49-94 (1972)
(4) Shiga H., Kakimoto A., Doi Y., Awasaki H., Miyake M., Sekiya S., Kataoka T., Environment and characteristics of rice plants in high-yielding paddy fields in Hokkaido. Bull. Hokkaido Natl. Agr. Exp. Stn., 99, 30-40 (1971).
(5) Yanagisawa M., Takahashi J., Studies on the factors related to the productivity of paddy soils in Japan with special reference to the nutrition of rice plants. Bull. Natl. Inst. Agr. Sci., Ser. B. No. 14, 41-171 (1964).
(6) Seino K., Yamashita K., Motomatsu T., Morooka M., The growth pattern of rice plants in several main paddy soils in the southern district of Japan. Bull. Kyushu Agr. Exp. Stn., 18, 133-156 (1976).

(7) Yamashita K., Fertilization techniques of rice plants in a warm region. Dojyo Hiryo Shin Gijyutsu, Hakuhoudo, Tokyo, 84-101.
(8) Seino K., Relation between the pattern of rice growth process and the nitrogen fertility of paddy soils in a warm region, Japan. J. Sci. Soil Manure, Japan, 46, 303-307 (1975).
(9) Akiyama Y., Ida A., Seino K., Effect of the application of compost for rice plants in a warm region. Annual Report on soils and fertilizers, Kyushu Natl. Agri. Exp. Stn. (Chikugo), 34-48 (1979).
(10) Wakimoto K., Seino K., Effect of the application of organic materials for rice yield and yield components. 1. Application of wheat straw. Abstr. Annu. Meet. Soc. Sci. Soil Manure, Japan, (25) 332 (1979).
(11) Kamimura Y., Utagawa Y., Increase of productivity in volcanic soil (Shirasu) by the application of organic materials. Annual Report on soils and ferti- lizers, Kagoshima Agr. Exp. Stn., 17-33 (1978).
(12) Harada T., Hashimoto H., Yumoto T., Studies on the clay minerals of paddy soils. Bull. Nat. Inst. Agr. Sci., Series B. No. 10, 81-113 (1960).
(13) Hashimoto H., Harada T., Yumoto T., Hara M., Studies on the organo-mineral colloidal complexes of paddy soils. Bull. Nat. Inst. Agr. Sci., Series B. No. 9, 201-256 (1959).
(14) Onikura Y., Some soil factors relating to regional differences in rice productivity in Kyushu, Japan. Bull. Kyushu Agr. Exp. Stn., 12, 225-233 (1967).
(15) Murayama N., Productive efficiency of applied nutrients in Japanese rice cultivation. Jour. of Agricultural Science, 29, 356-361 (1974).
(16) Yamasaki T., Cultivation techniques in the winner of high yield contest of rice grain in Japan. Potassium-Symposium, 98-111(1963).
(17) Sekiya S., Honya K., Nitrogen behavior in paddy soil, with reference to applied and soil organic matter. Bull. Touhoku Natl. Agr. Exp. Stn., 36, 1-25 (1968).

CHINA AND THE INTERNATIONAL COMMUNITY OF SOIL SCIENCE

Wim G. Sombroek
Secretary-General, International Society of Soil Science
(P.O. Box 353, 6700 AJ Wageningen, the Netherlands)

Mr. Chairman, esteemed officials of the Government and the Academia Sinica, Ladies and Gentlemen,

When you, Mr. Chairman, invited me to say a few words this afternoon, you gave me a free choice of the subject.

I gladly take this opportunity to say something on the International Society of Soil Science and on the participation of Chinese soil scientists in the international community of soil science.

Though this Symposium is not an official activity of the International Society, we gladly gave it some publicity in our Bulletin, and there will also be a report in the forthcomming issue. In the same line, I should like to convey to you the best wishes of our President, Dr. J.S. Kanwar of India, who unfortunately could not be here. My greetings are also on behalf of all other officers of the Society. In fact two of them, the past presidents Prof. Hallsworth and Prof. Bentley, are right now with us.

The International Society has already existed for nearly sixty years and has now about 7000 members, from practically all countries of the world. Recently, a group of 35 Chinese soil scientists joined, through the intermedium of the Soil Science Society of China, with its seat here in Nanking. We hope that many more will follow. Certainly they will be very welcome.

I am sure that I speak on behalf of all foreign participants, when I say that we are very much impressed by the good organisation of this Symposium, and also by the high quality of most of the papers submitted by Chinese soil scientists, either read or presented at the poster session. We have had too little opportunity to know of your achievements in the recent past. It is now obvious that you made great progress in the study of the genesis, the classification, the improvement, and the management of paddy soils, which are so wide spread in your country.

We already knew of your work in the subject of green manuring and organic recycling.

You also published a new soil map of China, at the scale of 1:4.000.000, which is a very impressive work. We do hope that the book, accompanying this map, will soon be translated into English, for the benefit of the advancement of soil geography and soil classification in the world at large.

It may be, Mr. Chairman, that you want to learn from foreign soil scientists. But certainly we can learn very much from your studies and experiences. We therefore hope that your scientists will actively participate in international meetings and programmes in soil science. The meetings of the Society are known to you. I may mention some of the other programmes for your special attention; some of them are also described in the latest issue of the Society's Bulletin:

- the international programme on the <u>combat against soil degradation and desertification</u>, initiated by UNEP, with the necessary mapping being executed by FAO.
  We know that you are very active in the combat of soil erosion, especially that in the loess area, so your experience is of great value elsewhere.

- the elaboration of a <u>World Soils Policy</u>, again organised by UNEP, with the support of FAO, Unesco and the ISSS. This will hopefully lead to a "World Soil Charter", that would alert the general public and governments alike to the fact that the soil resources of the world are limited and that they should be safeguarded for future generations of mankind.

- the envisaged <u>updating of the existing 1:5.000.000 soil map of the world</u> of FAO/Unesco, in which also the International Soil Museum will be involved. The present map is already ten years old, and since it was published many more relevant soil-geographical data have become available. These include your own 1:4.000.000 soil map of China.

- the proposed establishment of an <u>International Network of Benchmark Sites</u> These are sites where the complex relationships between soil, plantgrowth and climate will be studied and monitored in great detail, involving major crops under different management systems. This will hopefully allow the effective transfer of agro-technology to comparable soil- and agro-climatic units. It would also allow the identification of objective criteria to establish suitability classes in land evaluation schemes.

- last, but not least, the establishment of an <u>International Reference Base for Soil Classification.</u>
  At a recent informal meeting in Sofia of FAO, Unesco and the ISSS, with representatives of some major soil classification systems (American, Russian and French) an agreement-in-principle was reached that such reference base should be prepared. It is not the intention that national soil classification systems be replaced, but rather that a reference system be established through which the different existing systems could be correlated and harmonised, starting with the highest categories of classification. It was felt at this meeting that the existing legend of the FAO/Unesco soil map of the world should be used as much as possible as a basis for such an international soil classification system.
  At this major undertaking, an effective contribution will be required from soil classification specialists of all countries of the world. Because of the great size of your country, Mr. Chairman, and because of the many millions of people having to live from your soil resources, we count also on a substantial cooperation from your side. Your accumulated knowledge on a useful classification of these soil resources is indispensable at this effort.

In conclusion, I express the hope that this Symposium will mark the beginning of a new area of participation of Chinese soil scientists in the international community of soil science, for the benefit of mankind.

Thank you, Che-chi.

SOILS AND PEOPLE

C. F. Bentley
(Professor Emeritus, University of Alberta, Edmonton, Canada)

Mr. Chairman, officials of the People's Republic of China and Academia
Sinica, Members of the Organizing Committee, Members of the Nanjing Institute
of soil Science, our Chinese colleagues and wonderful hosts, friends:
I am highly honored to have been invited to participate in these closing cere-
monies.

This has been a most interesting symposium and one with very important
implications for the future. I am confident that I speak for all visiting
participants when I express genuing thanks and appreciation to our hosts for
their gracious hospitality as well as for the excellent program and arrangements. Thank you all very much indeed.

I am sure the foreign participants in the symposium came here confidently
expecting to learn a great deal about paddy soil research in China. They
certainly have not been disappointed: indeed, as the week progressed the
excitement and satisfaction grew as the extent, nature and quality of Chinese
research on paddy soils unfolded. It has been pleasing, too, that there has
been such broad representation at the symposium of the agencies and geographical
regions in China. We express warm congratulations and genuine thanks. Best
wishes for continuing indeed expanding success.

I have been delighted that some very senior officials of the People's
Republic of China have been present for parts of this symposium. Their
attendance is important because their support will be essential in the years
ahead to solving many of the problems we have been discussing. On a global
basis too few government ministers, policy-makers and senior decision-making
administrators have adequate understanding and appreciation of the importance
of land and soil science to all people. The life of every person is dependent
on food, almost all of which comes from land based plants and animals. That
dependence will not change while any of us here today is still living.

During 10,000 years of farming on earth, people have learned a great deal
about agricultrue and we now produce tremendous amounts of food. To do that
we have changed much of the land on earth. But today human numbers are
increasing at a   frighteningly rapid rate. Fortunately there
are encouraging signs that most nations and people are realizing the dangers
of excessive population and the pace of increase is beginning to slow. However,
unless there are terrible calamities, world population will double in the next
two or three decades. The food requirements for such a population increases;
plus the food needed to reduce hunger and malnutrition, which are still unac-
ceptably common in much of the world, constitute the greatest and most imperative challenge for the human family. Successful solution of that problem will
tax the agricultural productivities of the lands of the earth, and the skills
of farmers. . . as well as the scientific, managerial and governmental capabilities of people worldwide.

On a global basis most of the increase in food production will have to come from land that is already being farmed. FAO forecasts that in the next two decades less than 30 percent of the increase will come from new lands brought into agricultural use. Therefore, most of the additional food production will be achieved by expansions and improvements in irrigation, by additional crops per year on lands now in use, and especially by higher yields.

Many technical disciplines will be needed to achieve higher levels of food production as well as to improve the reliability and stability of such production. Plant scientists can breed improved crop cultivars, entomologists can develop better methods of insect control, animal scientists may create more feed efficient livestock, agricultural engineers can improve machines or the technologies for irrigation, and other scientists are also needed to make their various contributions. But ultimately actual increases in food production will depend on how the lands and soils of the earth are managed and maintained. Soils must provide plants with the nutrients they require. Soil and land use management must reduce erosion and the excessive runoff which causes floods while decreasing fertility and soil moisture for growth of plants. Thus, application of knowledge of soils and soil science has a fundamental role in enabling the potentials of the other agricultural disciplines to be transformed into actual increases in food production.

During the last three or four decades China and some other countries have succeeded in increasing food production dramatically. Land reclamation, terracing, planting of forests, building of reservoirs and dikes have in combination reduced erosion, conserved water for future use and reduced flooding. Fertilizer providing N, P and K and development of new high yielding varieties have also contributed to dramatic increases in production. There have been other new developments too. But such advances will not be sustainable without increasing contributions from soil scientists and agronomists. High levels of crop production from farmed and other lands increase the demands for all plant nutrients and so the frequency of defi-ciencies of S, Zn, Cu and other elements is becoming more common year by year. Nitrogen fertilizers speed the acidification of many soils, resulting in new toxicities and other problems. As we have heard this week, biological nitrogen fixation and improved methods of utilizing crop residues, composts and manures also require research and practical experimentation by soil related agricultural scientists.

Those of us at the symposium, as well as our fellow scientists elsewhere, have an obligation to explain these concepts to ministers, to those who form policies, and other decision-makers. If we do not transmit our message, if we not do it well, the support for mission oriented research and experi-mentation by soil scientists and agronomists is not likely to receive the funding and other resources essential for us to meet our responsibilities. Without adequate amounts and quality of soil and agronomic research, the needed quantities of food will not be produced. In that case many hundreds of millions of people will be the victims of the collective neglect and failure.

That comment prompts me to discuss briefly another responsibility. On a world basis, agricultural production of all kinds involves about as much work by people as all other human activities combined. And a great deal of agricultural production now entails arduous life-shortening drudgery and toil.

For too many millions of people in Latin America, parts of the Caribbean, Africa, india and other parts of Asia, life consists mostly of the four bends;

● bending to till the land and to prepare seedbeds;

● bending to plant crops;

● bending to control weeds or other pests; and

● bending to harvest crops.

I congratulate the People's Republic of China for its plans to seek ways to reduce those types of toil. Soil scientists and agronomists have knowledge of all four of those types of drudgery. I think they have responsibilities to seek much more vigorously than they have done during the past three decades to reduce such toil. There is great need to develop simple, effective, low-cost machines and procedures that will reduce the human labor of land preparation, of crop planting, of weed and pest control, and of harvesting. The work now in progress in China and at IRRI to develop rice transplanters illustrates the sort of things I have in mind. Perhaps most of the machines and tools needed will be muscle powered during the next decade or two. But well designed appropriate tools and machines, even if muscle powered, can make exceedingly important contributions to increasing food production and speeding the work while simultaneously reducing human toil of types which should no longer be necessary. Moreover, it has been clearly shown that appropriate mechanization of the types being advocated increases both income and employment. Surely if mankind can put people on the moon we can devise affordable ways to reduce food producing labor here on earth.

Perhaps nothing that people could do for each other has such great potential to improve the general human condition—especially for the more disadvantaged people of the world.

I hope this symposium will stimulate the participants and others to continue their endeavors in the needed activities. To earn and retain the funding and other resources required for such research and experimentation, practical results must be forthcoming.

New knowledge, whether of improved methods of paddy production or of producing other agricultural products, will result from research and experimentation on a worldwide basis. By sharing knowledge and discoveries, by discussing problems together as we have been doing at this symposium, we can speed discoveries as well as the development of improved practices and materials. Such achievements were the hope of the organizers of this symposium.

The benefits of international cooperation in paddy soil and other research were nicely illustrated in a conversation heard this week. The networks mentioned in the reports of Drs. Kyuma and De Datta were being discussed when it was asked: "Which benefit is the more important in such international networks — the interactions of the scientists involved, or the standardization and comparability of methods which enable valid comparisons?" The instant reply was: "They are equally important and desirable." When Dr. Brady concluded his address at the opening of the symposium he, too, emphasized the need for international exchange of knowledge —to speed the work, to make optimum use of resources and to benefit people.

It is fitting that this has been a symposium on paddy soils. Paddy soils produce the single most important human food —— rice, and many other crops too! If soil scientists and agronomists improve the productivity of paddy soils, and if they contribute to reduction of the associated human work requirement, they will improve the quality of life for a great many people . That would be satisfying to the organizers of the symposium.

SOME REMARKS ON THE DEVELOPMENT OF SOIL SCIENCE
IN CHINA

James   Thorp
(Professor Emeritus of Geology and Science, Earlham College, Richmond, Indiana, USA)

Interest of the Chinese people in the nature and use of soils goes
back beyond written historical records. The late and distinguished Dr.
Wong Wen-hao once told me that, according to Chinese legends, soils were
first classified by Engineer Yu of the Yao Dynasty more than four thousand
years ago. According to those legends, Yu classed Chinese soils according
to their color and structure. Whether or not Emperor Yao or Engineer Yu
actually existed, Chinese farmers have accumulated a vast fund of practical
knowledge from generation to generation. Modern soil science in China, as
in Western countries, draws on this practical knowledge when setting up
scientifically controlled experiments. Our hosts at this symposium have
sought and used both the practical knowledge of the farmers and the results
from scientific investigations.

Modern soil science (Pedology), based on observation of physical, chemical, and biological properties of soils, and on controlled experiments and
interpretation, had its early beginnings in Europe, Great Britain and North
America in the latter half of the 19th Century. During the last seventy-
five years, the growth of soil science has been exponential in these countries,
and interest in and investigation of soils are universal throughout the world
today. Generally speaking, the early field studies of soils in Russia and
other European countries were made on a regional basis, while more or less
detailed surveys were the rule in the United States. Geographical generalization followed detailed surveys in the United States.

Western soil science came first to China in 1930 when Professor J. Los-
sing Buck, Agricultural Economist at Nanking University, invited Professor
Charles F. Shaw of the University of California to make a regional study of
the soils of northeastern China, as a contribution to the University of
Nanking's research on land utilization in China. Professor Shaw, assisted
by Chinese colleagues spent three or four months in the field and made a
generalized soil map of the region from the lower Yangtze River valley to
the border of Northeast China. The map, with an accompanying report, was
published in 1931 as soil Bulletin Number 1 of the Geological Survey of China.

The first regional soil survey under the aegis of the Geological Survey
was made by C.Y. Hsieh and L.C. Chang in the Wei Valley of Shensi around the
ancient city of Sian, and the findings were published as Soils Bulletin No. 2
Soils Bulletin No. 3, by Dr. Dscheng Wang, dealt with the pH status of the
loessial soils of northern China; published in Chinese and German.

In 1931-32, Dr. Wong Wen-hao, Deputy Director of the Geological Survey,
invited Professor Robert L. Pendleton of the University of the Philippines to
organize regional surveys of soils and to train a staff of Chinese University

graduates in soil-survey and laboratory methods. Dr. Pendleton and Chinese
associates made brief visits to Suiyuan, Shansi and Kwangtung Provinces
and to the region around Harbin, Northeast China, and published bilingual
illustrated reports on their field studies (Soil Bulletins 4,5,6, and 11).

Dr. Pendleton returned to his teaching post at the University of the
Philippines in 1932 and spent his later years teaching in Thailand.

In 1933, Dr. Wong and Dr. V.K. Ting, who was director of the National
Geological Survey of China, asked Dr. C.F. Marbut, Chief of the U.S. Soil
Survey, to send one of his survey men to work with Chinese soil scientists
and help familiarize them with our survey methods. After consultation with
senior members of his staff, including especially Dr. Mark Baldwin, Dr.
Marbut recommended to Dr. Ting that I be invited to take this responsibility.

I arrived in Peking (Beijing) in early October, 1933, where I received
a warm welcome from Director Wong and members of the staff of the Soils Division, who were waiting eagerly to go ahead with their field and laboratory
studies.

During the nearly three years of my work in China, the Chinese staff of
the Soils Division comprised fourteen members of various ranks, including
eight soil scientists (called technologists). Of these eight soil scientists, six men are still living and hold important positions. Dr. Hseung Yi
and Dr. Li Ching-kwei are, respectively, Director and Deputy Director of
the Institute of Soil Science; Mr. K.C. Hou is Professor of Soil Science at
the Southwestern Agricultural College; Dr. Li Lien-chieh is Professor of
Pedology at Beijing Agricultural University; Dr. E.F. Chen is Director of
Shenyang Agricultural College; and Mr. L. T. Chu is Associat Chief of the
Bureau of Land Utilization. 5 of these six men are attending this symposium .

The late Mr. Y.T. Ma was Director of the Nanjing Institute of Soil
Science for ten years before his death in 1976. Dr. T.Y. Tschau died recently
in Taiwan.

It was my duty and also my great privilege to travel to all the agricultural provinces of China, accompanied always by one or more of the soil scientists, making careful studies of the soil profiles and collecting samples for
chemical and physical analysis. Each of the scientists had his individual
area of interest or one in cooperation with another member of the staff.
C.K. Li and Yi Hseung were concerned chiefly with chemical and physical
analyses and turned out a large amount of valuable work.

Results from field and laboratory studies appeared in a series of Soil
Bulletins of the Geological Survey of China-up to Number 20 for the period of
my stay in China. In addition, we put out several "Special Soil Publications.
"Many field and laboratory studies appeared in later Soil Bulletins, They may
be seen in the library of the Institute of Soil Science.

I summarized the work of the Soils Division for 1933 to 1936 in the book
"Geography of the Soils of China, "552 pages, illustrated. The book was
accompanied by a generalized map of the soils of China which was compiled
mainly from the work of numbers of the Soils Division and geologists of the
Geological Survey, with additional material from various scientists and
explorers. A number of Soil Bulleting and many other publications about
Chinese soils have appeared during the often- difficult years since 1936.

In recent years, since 1953, the old Soils Division has become the Institute of Soil Science, a subsidiary of the Academia Sinica, with headquarters at Nanjing. In addition, much work in soil research and in teaching about soils has been and is being carried out in Agricultural Colleges and Universities in many parts of China. Knowledge about soils is advancing geometrically as more and more men and women, including both those with formal training and those with practical experience, join the ranks of the researchers.

The rapid increase in knowledge about Chinese soils has led inevitably to the recognition of a need for a better understanding of soils used for centuries to produce rice, the "Staff of life" for many hundreds of millions of Oriental people and for an important but smaller number of Occidentals. The Symposium on Paddy Soils, so well organized by our hosts of the People's Republic of China, has provided a splendid opportunity for the international exchange of ideas in this field.

Recent knowledge of Chinese soils, supplemented by work handed down from earlier years, has been summarized in the splendid book "Soils of China" (762 Pages), illustrated, compiled and published by the Institute of Soil Science. I hope very much that this book will be translated into English or one of the western European languages so that it will be available to a much larger audience than it can reach in its present form. We of the West need to draw on the rapidly accumulating knowledge about the soils of China, India, Japan and other Oriental countries, and we are happy to share the findings of our own soil research with our Oriental friends.

Increasingly, people all over the world are becoming aware of the tremendous pressure of expanding populations on the supplies of food that can be produced to support them. During the 1930s it seemed as though the 500 million people of China were more than the land could support adequately. Now, we are told that the population is about 960 million. Our hosts tell us, they believe that, under peaceful conditions, the population will rise to about 1,170 million by the year 2040 or 2050, and they hope the numbers will be reduced somewhat after that. We hear of similar trends in both other Oriental and in Western countries, and we hope that control of family size may eventually bring a desirable balance between people and the land not only in the People's Republic of China, but in all countries. In the meantime, agronomic sciences must find ways to feed the hungry and clothe the naked. This symposium represents one of the ways we can find out how to go about the job.

Let me say a short word of a personal nature. My Chinese colleagues and friends have experienced difficulties and trials almost beyond imagination during the last more than 40 years; but they have "carried on" and surmounted their problems, one by one. They have trained a new generation of men to replace themselves, as all of us should do; and they pass along to their successors the knowledge that they and their predecessors have accumulated. To my Chinese colleagues and friends and to all the delegates, "xiexie nimen."— Thank you!

James Thorp